Trace Metals in the Environment 4

Trace Elements – Their Distribution and Effects in the Environment

Trace Metals in the Environment 4

Series Editor: Jerome O. Nriagu
Department of Environmental and Industrial Health
School of Public Health
University of Michigan
Ann Arbor, Michigan 48109-2029
USA

Other volumes in this series:

Volume 1: Heavy Metals in the Environment, edited by J.-P. Vernet
Volume 2: Impact of Heavy Metals on the Environment, edited by J.-P. Vernet
Volume 3: Photocatalytic Purification and Treatment of Water and Air, edited by D.F. Ollis and H. Al-Ekabi

Cover illustration:
Comparison of chronic and acute effect doses of toxic substances on living systems (modified after Markert, B., 1996. Instrumental Element and Multielement Analysis of Plant Samples – Methods and Applications, 2nd revised ed. Wiley, New York).

Trace Metals in the Environment 4

Trace Elements – Their Distribution and Effects in the Environment

Edited by

B. Markert
International Graduate School
Chair of Environmental High Technology
Zittau, Germany

and

K. Friese*
UFZ Centre for Environmental Research
Department Inland Water Research
Magdeburg, Germany

**on leave at:*
Universidade Federal de Ouro Preto
Departamento de Geologia
Ouro Preto, Brazil

2000

Elsevier
Amsterdam – Lausanne – New York – Oxford – Shannon – Singapore – Tokyo

ELSEVIER SCIENCE Ltd
The Boulevard, Langford Lane
Kidlington, Oxford OX5 1GB, UK

First edition 2000

Library of Congress Cataloging in Publication Data
A catalog record from the Library of Congress has been applied for.

British Library of Congress Cataloging in Publication Data
A catalog record from the British Library of Congress has been applied for.

ISBN: 0 444 50532 6

∞ The paper used in this publication meets the requirements of ANSI/ISO Z39.48-1992 (Permanence of Paper).
Printed in The Netherlands

Series Preface

Industrial activity inevitably has an impact on the environment and the link between industrial discharge of toxins, including metals, and human health has been established many times throughout the past millennium. According to the suggestion by Mieck (1990), the following sequence of changes in nature and cause of metal pollution can be identified: *pollution microbienne* or *pollution bactérielle* caused by microbial methylation processes which result in bioaccumulation and biomagnification of naturally occurring metals in the human food chain; *pollution artisanale* or cottage industries such as pottery making, gold mining, battery repair and tanning which contaminate the household environment; *pollution foncière* in which vast quantities of toxic metals and their compounds (such lead arsenate) are deliberately applied to land as fertilizers and biocides; *pollution industrielle* involving large-scale and pervasive pollution over major centers of industrial activity; and *pollution accidentelle* in which accidental spill of toxic metals is neither foreseen nor calculable (as in the Chernobyl accident). The escalating trajectory of changes in the forms and quantities of toxic metals being released into the environment has led to metal pollution being a worldwide phenomenon.

Toxic metal pollution has achieved high visibility and is garnering public concern globally. Most communities are less and less tolerant of any risks to their health, safety and quality of life due to discharges from local factories or even distant sources. The changing public perceptions have contributed to the drive for higher environmental standards and reduction in metal releases. Industry is responding to the challenge partly out of a desire to be a good neighbor and enhance goodwill and partly because of growing regulatory pressure. Because of their persistence, the traditional technological focus on end-of-the-pipe removal or destruction of contaminants does not always provide acceptable solutions. Furthermore, the remediation or clean-up of metal contaminated ecosystems is a complex and expensive proposition. Against this background, emphasis is shifting to eliminating emissions by either not creating pollutants in the first place or by recovering and reusing metals that would otherwise be discarded – a management strategy variously called pollution prevention, waste minimization or conservation-based environmental compliance. Problems of environmental metal pollution are clearly multifaceted, dynamic and generally reflective of changes in technology, law and public perception.

A response to the threat and challenge of fouling the human life support system with toxic metals has been a cascade of studies that have given rise to a huge database and very extensive research findings on many dimensions of environmental metal pollution. The information is available in disparate journals, magazines and reports and oftentimes in hard to locate sources. Volumes in the *Trace Metals in the Environment* book series are designed to provide timely, critical and integrative summaries of the global literature on various aspects of the sources, fate, effects, control and management of toxic metals in our environment. They cater primarily to the needs of multidisciplinary readership.

Jerome O. Nriagu
Series Editor
University of Michigan

May 2000

Mieck, L., 1990. Reflections of a typology of historical pollution: complementary conceptions. In: Brimblecombe, P., Pfister, C. (Eds.), The Silent Countdown. Springer Verlag, New York, pp. 73–80.

Preface

Our knowledge of the occurrence, distribution and effects of individual elements in the environment is increasing exponentially. The Seventh International Congress of Ecology, held from July 25th to July 29th 1998 organised by INTECOL (The International Association for Ecology), and well attended by around 2000 scientists, offered some specialists in the field of ecological trace element research an opportunity to meet and discuss "hot topics" in this growing scientific field. During this meeting it became evident that a great deal of work remains to be done before we achieve a general understanding of the effects of individual elements and their species. Although the development of efficient methods has led to a breakthrough in the analytical sector, major questions concerning the essential and/or toxicological effect of individual trace elements remain to be answered. Trace elements in terrestrial and aquatic environments are characterized by being spatially variable in their concentrations. In general the concentration of trace elements varies in different compartments of the environment according to their input quantities and dilution behavior, including transport and accumulation. The inputs of many trace elements to soil, animal or plant is greater than the losses leading to a net accumulation. On the other hand, the opportunity for dilution in the water phase is enormous, especially in the open oceans, where trace metal deficiency is widespread, although they receive huge quantities of trace elements in absolute terms by rivers and atmospheric fall out. Nevertheless, accumulation of toxic elements or trace elements up to toxic levels occurs in the ocean life world. These effects are not restricted to the estuaries where we can observe an enhanced input of trace elements of anthropogenic origin. Rivers and wetlands have lower outputs than inputs, so they can suffer from accumulation of potentially toxic elements. This situation is likely to continue in the foreseeable future, and environmental pollution with trace metals is a truly long-term problem.

In contrast to their great importance in nature we know relatively less about the interactions between trace elements in the biosphere. Solely continuous studies of the distribution and fate of trace elements in the environment, their biogeochemical behavior on molecular, microscopic, macroscopic, local, regional and global scales could help us to close this gap in knowledge. Case studies, field work and laboratory experiments are useful to enhance our knowledge about processes related to the biological response of trace metal stress under realistic environmental conditions. More work about the ecotoxicological response such as detoxification strategies or biotransformation of trace elements may lead to new insights into the behavior of nature under stress. Five important points arose from the papers presented by the participants at the Symposium:

- There is no convincing reason why current trace element research should be largely confined to just one-third of the chemical elements in the periodic table. In trace analyses, as much attention should be devoted to elements such as lanthanum, platinum and antimony as to cadmium, mercury and arsenic. In most cases the necessary techniques are already available.

- The total element concentration is an initial indication of the occurrence and distribution of individual trace elements in the environment, but in many cases it does not permit an assessment of their effects. To assess such effects it is necessary to determine the speciation, i.e. the chemical form of the element concerned in the environmental sample to be analyzed and deduce possible effects on organisms and communities of organisms from this.
- The quality of the analytical data obtained through investigations of this kind has to be ensured by implementing strict quality control; this involves the use and further development of reference materials, inter-laboratory tests and other checking methods.
- The enormous volume of research data on trace elements collected over the past 30 years has not been subjected to any kind of general analysis. It is therefore suggested that the data from the past 20 or 30 years be compiled in the form of a database – as far as such data are suitable from the analytical point of view – in order to identify possible connections and draw conclusions useful for further studies.
- Greater use should be made of bioindication and biomonitoring, as they permit time-resolved investigations of individual effects of specific trace elements over wide areas. The conclusions to be drawn from such methods in respect of possible remedial or preventive measures emerged from opinions expressed at several events throughout the INTECOL Congress as well as at this symposium.

Ecological trace element analysis is interdisciplinary by nature and includes a variety of experimental and theoretical methods from related sciences such as biology, chemistry, toxicology, geology and physics. From this list it is clear that a comprehensive treatment cannot cover all aspects of potential relevance. The present text discusses the following major areas that are of primary concern for the current understanding and future development of this field.

Part I presents the sources and fate of trace elements needed for an understanding of the complexity associated with ecological trace element research. For a deeper understanding of transformation processes, physiological effects, availability studies and remediation procedures, a strict and artificial deviation into inorganic and organic chemistry can no longer be accepted. The necessary analytical techniques for trace elements in the environment are given in Part II. The combination of inorganic (AAS, ICP/OES, ICP/MS) and organic (GC, HPLC, etc.) instrumental techniques is mainly focused on speciation analysis and multi-element investigation needed in the future.

Parts III, IV and V deal with the distribution of trace elements in biota and reservoirs (soils and sediments). In this context, it should be noted that organisms, populations, biocenoses and finally the entire ecosystem are influenced by a number of different biotic and abiotic stressors under natural conditions. These are for example climatic changes, variations in radiation regimes or availability of food resources, predator–prey relationships, parasites, diseases, intraspecific and interspecific competition. Stress is an existential prerequisite for all biological levels of organization, as it is an important characteristic of all living systems to respond to stress. Conversely, without these natural stress sources, no evolutionary development of species and therefore of the entire biological system is possible. Stress can be described as the "engine of evolution". In evolutionary periods of time, the variance of stress is more or less constant so that biological species have the opportunity to adapt to the existing stress levels. In the last centuries, these alterations of stress have reached a new qualitative and quantitative dimension. An increasing number of new substances were produced by man, which have not been exposed to species in the environment before (e.g. xenobiotics, radionuclides) or potentially harmful substances were deliberately introduced into the environment in previously unknown amounts (e.g. heavy metals, naturally occurring radionuclides). Furthermore, these new stress sources exhibit multiple effects, i.e. they cause an additional stress to the already existing natural stress levels mentioned above and they do occur normally in combination so that the tolerance range of the evolutionary adaptability of the organisms is exceeded. Many of the so-called "biomarkers of exposure" (e.g. metallothioneins in animals, phytochelatins in plants, EROD activities) give an estimation of the existing level of contamination in an ecosystem. But it is controversial whether or not biomarkers are also applicable for the assessment of biological effects of contaminants on higher levels of biological integration.

There is a strong need for more coordination of future research in this field. On the national and international level, better harmonization and quantification of the data is still in progress. This needs to be extended to standardization, mathematization and identification of suitable organisms that can be used for example as bioindicators or phytoremediators. Here, especially international organizations come into play such as the International Union of Biological Sciences (IUBS) in Paris, The International Atomic Energy Agency (IAEA) in Vienna, and many more. The main objective is to establish nationwide international projects in different continents of the world with the following goals in mind:

- gathering and supplying relevant scientific material
- proposing national specialists for participation in different phases of ecological trace element research
- supporting doctoral and post-doctoral fellows to be associated with trace element research
- appointing functional national groups to collaborate by providing case studies
- providing host facilities for workshops and training camps and
- supporting and coordinating reports and publications.

This book would not have been possible without the agreement of the authors to finish their contributions in a relatively short time; from the first inquiry to write a manuscript to the receipt of the final version, only one year passed. We would like to express our sincere thanks to the authors for their kind cooperation, stimulating input and valuable contributions. No attempt to provide a uniform terminology has been made by the editors and authors (e.g. for the terms bioindication and biomonitoring), and therefore it varies from chapter to chapter due to the many different views that coexist at present. The book rather gives a survey of the state of the art of this expanding field based on the authors' work. Both contradictory and complementary aspects of ecological trace element research are pointed out to the attentive reader.

There has also been excellent office support by our secretarial staff at IHI Zittau, Katrin Zimmermann, and Kathleen Kirschner from the UFZ Research Centre in Leipzig/Halle. Moreover, we would like to express our special gratitude to Mary Malin and Sandra Jenkins, guiding us from the first beginnings, through the ups and downs of preparing the book to the final product. Finally, Jerome Nriagu gave us his long experience in all questions arising during the overall composition of the book. Thank you! The editors and authors hope that a wide distribution of the book will stimulate discussion on perspectives of ecological trace element research and of ecology in general.

Any suggestions or critical comments are willingly accepted by the authors and the editors.

Univ.-Prof. Dr. Bernd Markert
Chair of Environmental
High Technology
International Graduate School
Zittau

Dr. Kurt Friese
UFZ – Centre for Environmental Research
Department of Inland Water Research
Magdeburg

January 2000

Contents

I
General aspects

Trace elements – Their distribution and effects in the environment
B. Markert and K. Friese, editors

Chapter 1

Distribution and effects of trace substances in soils, plants and animals

Bernd Markert, Gernot Kayser, Siegfried Korhammer, Jörg Oehlmann

Abstract

In this contribution, inorganic and organic chemicals are considered together as for both group of pollutants many aspects of biogeochemical distribution, environmental fate, toxicodynamic and toxicokinetic aspects as well as remedial possibilities are comparable. An overview of the occurrence and distribution of chemical substances in various environmental compartments is given. Special emphasis is placed on naturally-occurring substances and the influence of anthropogenic activities on their biogeochemical distribution. The significance of chemical substances in living organisms is considered in greater detail, and aspects of essentiality, the biological function of chemicals in plants, animals and man, the uptake form of various chemicals by living organisms and the interrelationship are discussed. Typical data are given for emissions of specific chemical substances into the environment by man. Special attention is given to the cycles and availability times of various chemical substances in the environment. Examples are given of corresponding effects in the animal and plant kingdoms. Finally, possible remedial and revitalizing measures for soils contaminated with trace substances, which as a field is expected to acquire much greater significance in the future, are outlined.

Keywords: Trace elements; Heavy metals; Contamination; Toxic effects; Remediation

1. Introduction

Inputs of chemical substances constitute the greatest pollutant burden on natural ecosystems. When considering their impact it is necessary to distinguish between poisonous substances that have a toxic effect on organisms even at low concentrations, and burdening substances (nutrients) that bring about undesirable changes in ecosystems at higher concentrations or loads. Table 1 gives examples of pollutant and burdening substances in different sectors of the environment. In this contribution, inorganic and organic chemicals are considered together as for both group of pollutants many aspects of biogeochemical distribution, environmental fate, toxicodynamic and toxicokinetic aspects as well as remedial possibilities are comparable.

The term "environmental chemicals" is used to describe substances that enter the environment as a result of human activity and sometimes occur in quantities that endanger living organisms and abiotic sectors of ecosystems, and especially man. The number of known substances has risen rapidly since the development of modern

Table 1. Pollutant and burdening substances in different sectors of the environment (modified after Förstner (1995))

Environmental sector	Burdening substances	Examples	Places of origin
Water	Degradable organic compounds	Faeces, tensides, solvents, pesticides, industrial process materials, fats, oils, soluble animal and vegetable residues, basic chemicals, intermediate and end products	Towns and villages, households, agriculture, textile industry, metal working, paintshops, food industry chemical industry, paper industry, landfills
	Persistent organic compounds	Tensides, solvents, pesticides, industrial process materials, basic chemicals, intermediate and end products	Agriculture, textile industry, metal working, paintshops, chemical industry, paper industry, landfills
	Inorganic compounds	Heavy metals, salts, cyanide, chromate, fertilizers	Metal working, mining, leather production, towns and villages, agriculture, landfills
Soil	Degradable organic compounds	Faeces, pesticides, animal and vegetable residues, basic chemicals, intermediate and end products, sewage sludge compost	Agriculture, landfills for domestic waste and waste requiring special monitoring, disposal facilities
	Persistent organic compounds	Tensides, solvents, pesticides, industrial process materials, basic chemicals, intermediate and end products	Landfills for waste requiring monitoring and for industrial waste
	Inorganic compounds	Heavy-metal compounds, salts, ash, slag	Landfills, incinerators, metal production
Air	Organic gases	Solvents, hydrocarbons, volatile pesticides, volatile industrial chemicals	Paintshops etc., refineries, tank farms, agriculture, industry
	Inorganic gases	Carbon monoxide, hydrochloric and sulphuric acid, nitrogen oxides (ozone), metal vapours, carbon dioxide, ammonia	Firing systems, incinerators, engines, industry
	Dust and smoke	Metal oxides, PAH, soot	Metal production, waste incineration, firing systems in general

chemistry. Between 1970 and 1993 the number of substances registered with the Chemical Abstract Service of the American Chemical Society increased 6-fold from 2 million to 12 million. Assuming a linear course, this corresponds to a synthesis of nearly 1200 new substances every day (Koch, 1995). But only a fraction of these compounds are produced and used in such quantities that measurable effects on the environment are to be expected. Of the 73 000 or so chemicals that were on the market in Germany in 1985, 312 were produced in volumes of over 10 000 t/a; 2200 compounds were produced in quantities of 10–100 t/a. About 90% of the total annual production of the chemical industry world-wide is made up of about 3000 substances: in 1980, for example, some 140 million tons of sulphuric acid were produced, and the amount of ethene and methanol produced annually is estimated at 37 million tons (Koch, 1995). The most rapid growth in the chemical industry is in special products such as pesticides, plasticizers, catalysts, detergents and cleaners. The main problem is that many of the products contain exotic chemicals that have not been shown to be harmless to man and the environment; in some cases, in fact, no attempt has been made to determine their possible toxicity (Förstner, 1995).

In recent years the widespread use of chlorine as a constituent of products and intermediate products for chemical synthesis ("chlorine chemistry") has become the focus of criticism by experts and the public. The most frequent target for attack is the plastic polyvinyl chloride (whose production accounts for about 30% of the volume of chlorine obtained annually) because of inadequate recycling methods and the release of hydrogen chloride during its disposal by incineration. Further problems are created by chlorine production, since it results in residues in the order of magnitude of several hundreds of thousands of tons per year that have to be disposed of as hazardous waste. A great disadvantage is that undesirable side reactions during the production or incineration of (intermediate) products containing chlorine may result in the formation of highly toxic substances such as polychlorinated dibenzodioxins and dibenzofurans. This makes the further use or disposal of the products contaminated in this way extremely difficult.

2. Dimensions of environmental pollution

Depending on their physical, chemical and biochemical properties and the specific conditions under which they are released, environmental chemicals may be more or less widely dispersed or accumulate in a particular sector of the environment. Both the global spread of substances and regional pollution have to be taken into account.

2.1. Localized pollution

2.1.1. Indoor rooms

Contamination of indoor rooms at the workplace and also in the personal environment makes a major contribution to the burden on human beings from environmental chemicals. The substances emitted may be gases or particulate matter. They are released by combustion processes or during painting and decorating, cleaning or cooking, and may be contained in a wide variety of household articles (adhesives, cleaners, construction materials, furniture etc.). In the past, two substances – formaldehyde and pentachlorophenol (PCP) – have been especially important because of their widespread use. Formaldehyde is used as a binding agent in wooden particle board and fibreboard, in textile auxiliaries and as a disinfectant. It also serves as a parent compound in the production of synthetic resins. Since formaldehyde is classified as carcinogenic, regulations have been issued to minimize emissions of the substance from working materials. Pentachlorophenol was used as a preservative and for protecting wood. In some cases the products treated in this way were a very considerable health hazard because of contamination with polychlorinated dibenzodioxins and dibenzofurans from the production process. The use of PCP was prohibited in Germany with effect from the end of 1989 (Hempfling and Stubenrauch, 1994; Förstner, 1995).

2.1.2. Historical burdens

The term "historical burdens" is used to describe anthropogenic contamination of soils and groundwater caused by manufacturing, transport or military activities since the beginning of industrialization but not by current processes. The number of detected or cases of such historical pollution is immense; in Germany it is about 250 000. The funds required for cleaning up such sites and rendering them safe are considerable; the present estimate is 200 billion DM (about 100 billion ECU).

The contamination may take many different forms, but to constitute a historical burden the substances at the polluted site have to be sufficiently persistent (non-degradable). Contamination is often in the form of oils, solvents and other liquid chemicals spilt as a result of accidents during transportation or careless handling. Dumps for industrial residues were usually set up on company sites without an impermeable bottom layer and filled with acids, lyes, sludge etc. Similarly, old gasworks sites are often found to contain products from coking such as benzole, toluene, phenol, polycyclic aromatic hydrocarbons, tar and ammonia.

2.2. Global pollution

Persistent or non-degradable environmental chemicals – even those with low volatility – may be dispersed around the globe. Polychlorinated biphenyls (PCB), polychlorinated dibenzodioxins and dibenzofurans (PCDD, PCDF), 1,1,1-trichlor-2,2-bis(4-chlorphenyl)-ethan DDT, hexachlorocyclohexane (HCH) and hexachlorobenzene (HCB) have been found in samples from the Arctic and Antarctic and in the tissues of marine animals. Although the open use of many of these chemicals has been prohibited in most industrialized countries, considerable residues still remain in the environment. It is estimated that about 20% of the global production of PCBs (230 000 t/a) persists in the upper layers of the oceans and that a further 790 t are present in the atmosphere (Fiedler and Lau, 1998). Such contamination may accumulate in organisms and have negative effects that go as far as the death of whole populations in ecosystems (Jørgensen, 1998).

Table 2. Climate- and ozone-relevant trace gases (Förstner, 1995)

	Concentration	Annual increase (%)	Residence time	ORP[a]	GWP[b]
CO_2	247 ml/m^3	0.4–0.5	≈100 a	–	1
CH_4	1.7–1.8 ml/m^3	0.8–1.0	10 a	–	2–3
CO	50–200 μl/m^3	?	0.1–0.5 a	–	–
NO*x*	0–100 μl/m^3	0.2–0.3	Few days	0.25	–
N_2O	310 μl/m^3	0.2–0.3	170 a	0.05	240
CH_3CCl_3	140 nl/m^3	3–4	6–7 a	0.15	–
CHF_2Cl	70 nl/m^3	12	17–20 a	0.0	2000
$CFCl_3$	225 nl/m^3	4	55–70 a	1.0	8600
$CFCl_2$	385 nl/m^3	5	100–150 a	1.0	18000
$C_2F_3Cl_3$	40 nl/m^3	10	90–100 a	1.0	22000
CCl_4	140 nl/m^3	2	50–70 a	1.2	≈4000 (?)
CF_2ClBr	2.2 nl/m^3	≈10	25 a	2–3	–
$CBrF_3$	2.1 nl/m^3	≈15	110 a	5–8	–

[a] ORP, ozone-reducing potential, relative per unit of mass to the destruction of ozone by $CFCl_3$.
[b] GWP, global warming potential, relative per emitted quantity to CO_2, weighted according to residence time.

In addition to these effects that are restricted to organisms or limited ecosystems, changes to the global system "Earth" have been found in recent years: an increase in the carbon dioxide content of the atmosphere that is causing the mean temperature of the planet to rise, and the partial destruction of the ozone layer in the stratosphere by trace gases.

In scientific circles there is no longer any doubt that the atmospheric CO_2 concentration has risen since the beginning of industrialization. But it is not certain what impact this increase has on the global climate or the climate of individual continents or countries, and projections differ according to the climatic model used. However, the mechanisms leading to the destruction of stratospheric ozone are largely explained.

Extensive measuring programmes supply more and more reliable data on concentrations of trace gases in the atmosphere and changes in these over the course of time. Table 2 shows concentrations of trace gases relevant to climate and ozone, their (estimated) annual increase and residence time and their ozone-reducing and/or warming potential. The ozone-reducing potential (ORP) is given per unit of mass relative to $CFCl_3$; the global warming potential (GWP) is given in relation to CO_2 and weighted according to residence time. It becomes evident from Table 2 that because of the long residence times a fall in trace gas concentrations in the stratosphere cannot be expected for several decades even if the emission of ozone-relevant gases were to be stopped immediately throughout the world.

3. Action of pollutants

To have any noticeable damaging effects an environmental chemical has to meet certain conditions. It has to persist in the environment long enough, that is: it must be difficult to break down and mobile enough to be taken up by organisms. The mobility of a substance also determines the speed with which it is disseminated in the environment. In toxicology a distinction is made between two main disciplines – dynamics and kinetics – by which the action of foreign substances or other substances with a toxic effect can be characterized. Toxicodynamics investigates the effects of toxic substances on the organism, i.e. why a substance has a particular effect on the site of action at certain concentrations. Toxicokinetics is the study of the movements of foreign substances in the organism and seeks to discover the laws governing uptake, distribution, metabolism, and finally excretion. It is only the combined effect of toxicodynamics and toxicokinetics that determines the extent of the toxic effect of xenobiotics in the organism (Oehlmann and Markert, 1997).

3.1. Toxicodynamics

The study area of toxicodynamics includes analysis of the effects of toxic substances on the organism in respect of mechanisms, time and space. In other words, toxicodynamics asks what harmful effects a substance provokes in the organism and by what mechanism and at what concentrations or doses the damage occurs. In this a very important role is played by the interaction of the xenobiotics with the body's own structures, for example membrane-bound or cytosolic receptors. The search for antagonists at the receptors opens up possibilities of developing antitoxicants (antidotes) and thus of treating poisoning. The relationship between structure and effect is thought to enable a prediction of the toxic effects of new compounds, and the quantitative effect of a substance on the organism can be determined by means of dose-response curves.

3.1.1. Characterization of the poisoning

A longer or shorter time may elapse between the action of a xenobiotic on the organism and the occurrence of its effects. If this lag phase (latency period) is short, i.e. if it is in the order of seconds, minutes, hours or a few days, the effect is described as acute. If weeks, months or years pass the effect is said to be chronic. These two terms are often used in another connection also. Acute intoxication (poisoning) occurs if a toxicant or xenobiotic enters the organism once or over a very short period of time, usually in large doses. Chronic poisoning usually involves longer periods of exposure to the pollutant, sometimes even years: in this case relatively small concentrations or doses of the substance are sufficient to cause what may be massive effects.

There are pollutants that can cause both acute and chronic damage. They include the bispyridinium compounds paraquat and diquat (Fig. 1) that were used as herbicides in agriculture and above all in horticulture from the late 1950s. Since their acute toxicity is very high in comparison with other herbicides (the lethal dose for man is less than 6 g of paraquat or 15 g of diquat) use of the compounds is no longer permitted. During an initial phase after intake spontaneous vomiting occurs, and within hours pathological cell changes take place in the liver and kidneys that may result in death. If the subjects survive these acute harmful effects, the damage to the liver and kidneys recedes within 5–10 days. But after three weeks, at the latest, paraquat causes cell

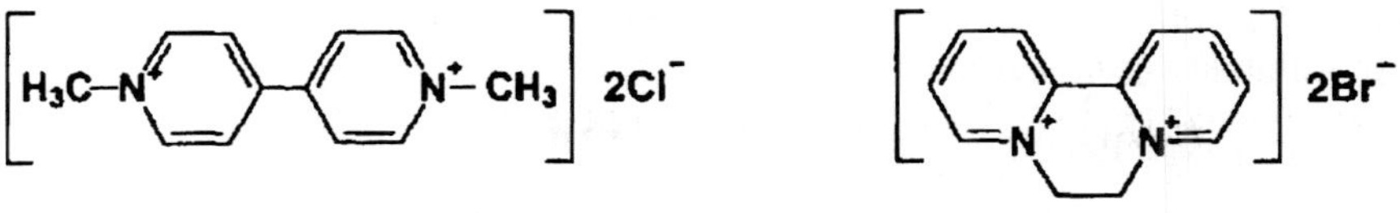

Figure 1. Structural formulae of paraquat (left) and diquat (right).

proliferation in the lungs that then results in death by suffocation (pulmonary hypoxaemia) (WHO, 1984).

Further examples of acute damage are pulmonary oedema provoked by toluene, cadmium oxide dust or 4-ipomeanol, a toxicant synthesized biogenically from fungi (mycotoxin). Examples of chronic damage are pulmonary fibrosis and finally lung cancer caused by asbestos and quartz dust. A distinction is made between reversible and irreversible effects, depending on the mechanism by which damage occurs. In the case of reversible damage the organism returns to its normal condition once the harmful agent has been removed, i.e. as soon as the concentration of the substance at the site of action falls below the critical threshold value there is no further detectable effect (Fig. 2). In the case of irreversible effects the damage to the organism remains even after the xenobiotic has been totally eliminated from the system (e.g. all carcinogens).

3.1.2. Sites and mechanisms of action

The term "mechanism of action" describes the biochemical or biophysical processes by which a xenobiotic acts; as a rule these are (intra)cellular, but in certain cases they

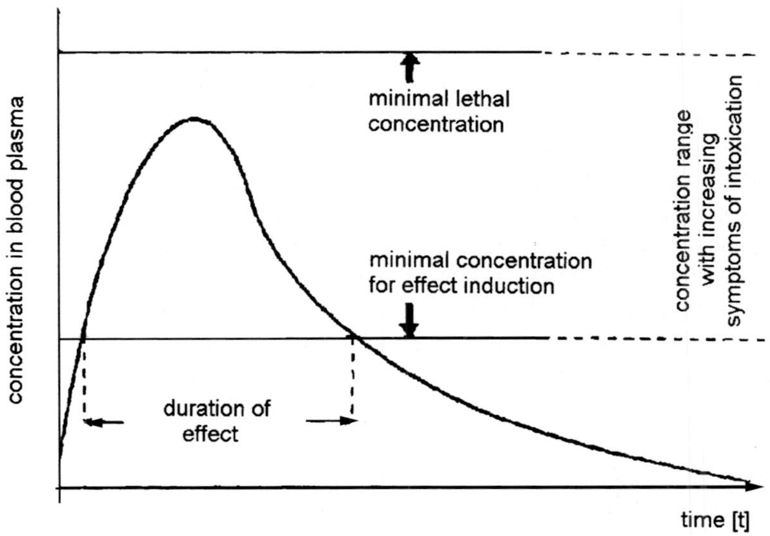

Figure 2. Degree of action and concentration of a chemical substance with a reversible effect in the blood after a single oral dose, shown in relation to time.

may be extra-cellular. It is possible to differentiate between five basic mechanisms by which pollutant substances act:

- reaction with defined receptors;
- inhibition of enzymes;
- interference with specific transportation processes;
- accumulation in membranes and alteration of their properties;
- covalent binding to essential substances of cell metabolism.

The above mechanisms of action are described in detail by Oehlmann and Markert (1997) with suitable examples.

3.1.3. Structure-activity relationships

Structure-activity relationships are based to a large extent on the concepts of the receptor theory. The theoretical principle can be imagined as a lock-and-key system. With a large number of substances it has in fact been possible to show what properties the chemical structure has to have in order to achieve certain biological or toxic effects. A good example is the toxicity of inorganic tin in comparison with the organic tin compounds of the butyl group (Fig. 3).

The acute toxicity of the compounds increases with the extent of organic substitution at the tin atom and reaches a maximum with tributyltin (TBT). The reason for this is the strong increase in their lipophilic properties that causes greater accumulation in membranes, especially in the mitochondria, and results in a breakdown of the organism's energy supply through the formation of ion channels. But quantitative structure-activity relationships (QSARs) are not deduced solely from linear or logarithmic

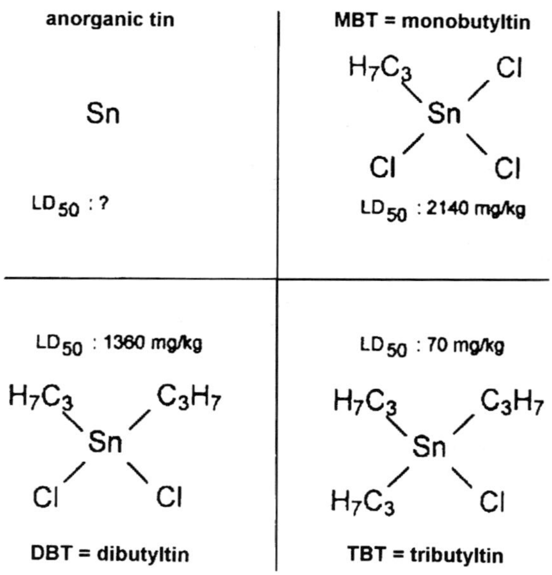

Figure 3. Comparison of the acute toxicity of inorganic tin and organic tin compounds of the butyl series in acute toxicity tests on rats.

relationships between the properties of the molecules and the toxicity of the substances. For example, the acute toxic effect of tetrabutyltin, which is even more lipophilic than TBT, is noticeably reduced; in the test with an LD_{50} of 1389 mg/kg body weight in rats it shows a much lower value similar to that of dibutyltin. In such cases toxicity takes a parabolic course as the lipophilic properties of the substances tested increase, a phenomenon that has already been detected in alkylmetal compounds with different chain lengths in the organic groups (Schüürmann and Marsmann, 1991; Schüürmann and Segner, 1994).

With simple test objects, especially, QSAR relationships are fairly clear – for example with isolated enzymes, organelles or organs. But if substances are used in intact organisms such QSAR relationships are overlaid to a greater or lesser extent by toxicokinetic processes concerned primarily with absorption, distribution and breakdown.

QSAR analyses can be carried out very successfully with classes of substances in which a non-specific effect predominates and thus determines their toxicity, as in the case of non-polar anaesthesia by hydrocarbons. Highly significant QSAR deductions can be made even if the toxic effect of a class of substances is dominated by a particular mechanism, for example by the inhibition of acetylcholinesterase in phosphoric acid esters (Schüürmann, 1992). It is at least much more difficult to deduce QSAR relationships if there is additional interaction between the chemical substances and the organism, for example intracellular bonding with enzymes, proteins, receptors or other endogenous molecules, or if specific protective intracellular mechanisms exist. In such cases much higher or much lower toxicity can be observed than that predicted by the QSAR analysis. In future QSAR relationships will be used increasingly for predicting what metabolites are likely to occur in the organism when organic pollutants are broken down. But as shown by the example of the halogenated hydrocarbons, that are oxidized primarily by the cytochrome-P-450 enzyme of the CYP2E1 family, very different metabolites may occur in the organism even with initial substances that are very similar in structure.

3.1.4. Dose-effect relationships

A positive relationship exists between the dose of a pollutant and the harmful effect it has on the organism, i.e. the more of substance X is taken in, the greater is effect Y. As the concentration of the harmful substance – for example an environmental chemical – increases, the system becomes saturated, i.e. a maximum effect is achieved by the quantity of the harmful substance taken in. The effect is no longer increased if the dose is raised further. The receptor theory assumes that a pollutant that acts as an agonist or antagonist is present in a concentration sufficient to occupy all the receptors of an organism, organ or tissue. Instead of the dose – i.e. the quantity of the substance taken in per kg of body weight – the concentration of a pollutant resulting from exposure through water in the case of aquatic organisms or through the ambient air may be used as a reference value on the abscissa (x-axis). Correspondingly, this results in a concentration-effect curve instead of a dose-effect curve. The effect of a pollutant may occur either on an all-or-nothing basis, for example as the death of a laboratory animal, or as a gradual response such as the inhibition of an enzyme system, restriction of mobility,

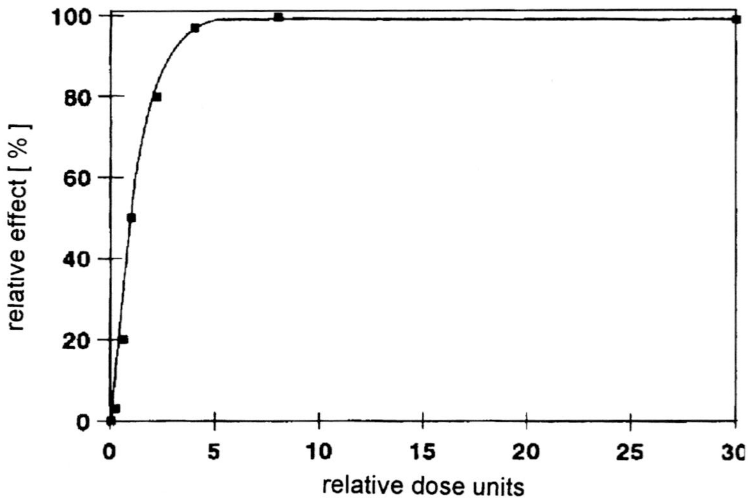

Figure 4. Typical dose-effect curve of a receptor-mediated effect as a linear representation with relative dose units.

a reduced rate of conduction in nerve fibres etc. In the case of all-or-nothing effects it is necessary to test a larger group of individuals in which the frequency with which the harmful effect occurs – its incidence – is determined. In the case of gradual responses it is theoretically possible to conduct tests on individual subjects, but here too a collective test is usually carried out to eliminate differences between individuals. This excludes the possibility that a highly sensitive or insensitive subject has been chosen merely by chance. Otherwise the result of the test would probably not be representative of the total population.

A typical asymptotic dose-effect curve is plainly visible in Fig. 4. Curves of this kind may be described as saturation kinetics after the formula $y = (\mathrm{a} \times x) \div (b + x)$. But this example is an ideal case, as the scatter of the results is normally much greater due to the heterogeneity within the groups tested (Fig. 5).

3.1.5. Environmental standards

Investigations in the field of ecotoxicology and human toxicology are carried out in order to assess and evaluate the risk constituted by a particular exposure to substances or radiation. Bodies of experts and public authorities then publish environmental standards based on the information thus acquired. In Germany, environmental standards are classified as "values for discussion or orientation", "recommended values" and "threshold values" (or limits). Threshold values are approved by public authorities and are legally binding. Values for discussion and orientation and recommended values are suggested by bodies of experts, commissions, associations or public authorities and are usually discussed in detail before being prescribed as limits.

Threshold values for contamination with inorganic and in some cases organic

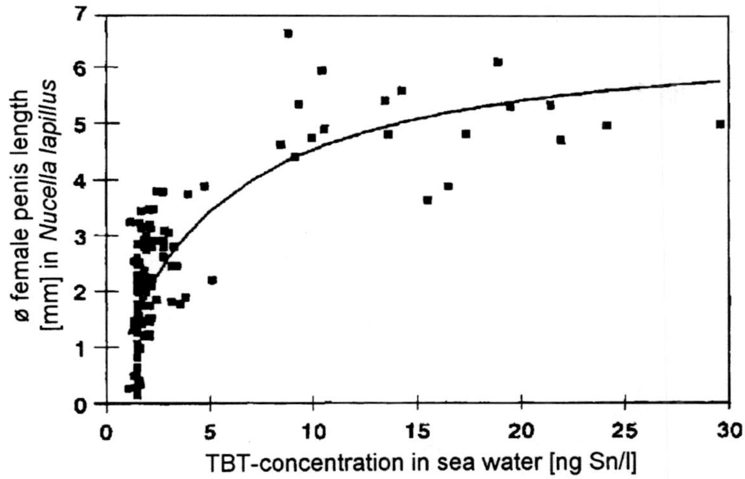

Figure 5. Relationship between the tributyltin (TBT) concentration of seawater and average penis length in the females of the Nordic dog whelk (*Nucella lapillus*) as an example of a concentration-effect curve. The figure shows the values determined (■; $n = 131$) and the calculated regression: $y = (6.63 \times x) \div (4.67 + x)$; $r = 0.831$; $P < 0.005$. Adapted from Oehlmann (1994).

substances have already been prescribed for numerous environmental media and substances connected with food:

- Water
  - Water Resources Act
  - Effluents Ordinance etc.
- Soil
  - Soil Protection Act
  - Guidelines of the Public Study Group on Waste Management (LAGA)
  - Sewage Sludge Ordinance etc.
- Air
  - Federal Emission Control Act
  - Administrative Regulation on Air Conservation etc.
- Food
  - Food and Utility Articles Act
  - Drinking Water Ordinance
  - Fertilizers Ordinance
  - Animal Feeding Stuffs Ordinance

In the case of harmful substances taken in by human beings through food or contact with "articles of daily use", a value is determined for the quantity of the substance that is considered safe or tolerable when taken in daily. This accepted daily intake (ADI) is derived from animal experiments. The highest dose found to be without effects in the experiments (no observed effect level (NOEL)) is divided by a safety factor that depends on the extent of the data available and the nature of the effect:

Table 3. Pollutants relevant to the soil and their potential hazard to human beings when taken in with the food (adapted from Förstner, 1995)

Pollutant	(Average) weekly intake with the food	Provisional tolerable weekly intake	Effects/special features
Lead	0.91 mg	3.5 mg	Increased absorption in children
Cadmium	0.284 mg	0.525 mg	In conjunction with Fe, vitamin D and Ca deficiency
Copper	Up to 700 mg	15 mg	Lethal dose of $CuSO_4$: >8 g
Nickel	2–4 mg	Not known	Oral inorganic intoxication not known
Mercury	0.063 mg	0.35 mg	PTWI for methyl-Hg: 0.23 mg
Thallium	Not known	0.1 mg	Lethal dose 8 mg/kg body weight
Arsenic	0.2–0.3 mg	1 mg	Lethal dose 100–300 mg; carcinogenic
PCB	Approx. 0.04 mg	0.5 mg	Accumulation in fatty tissue, milk, liver
TCDD equiv.	140–1400 pg	500 pg	Hand-to-mouth contact in children
Paraquat/diquat	No data	0.5/4 mg	
γ-HCH	2.6 μg	4.2 mg	Severe poisoning from 10 to 20 mg/kg
HCB	0.2 μg	0.25 mg	Intake with breast milk

- Factor < 100: for substances for which there is sufficient experience on toxicity to human beings;
- Factor $= 100$: normal factor when suitable long-term experiments have been carried out on animals;
- Factor > 100: when the toxicological data available is inadequate or the substance has carcinogenic effects.

Table 3 gives an overview of pollutants relevant to the soil and the potential risk to human beings constituted by them when they are taken in with the food. The stated values for "provisional tolerable weekly intake" correspond to the ADI but are based on a week instead of 1 day.

4. Inorganic pollutants

The effect of metals, metalloids and their compounds differs fundamentally from that of other toxic substances to which living organisms are exposed. Unlike most other pollutants that act on organisms they cannot be broken down: they can only be deposited in certain compartments of the body, accumulated, or excreted from the organism. The formation of organometallic compounds, that sometimes even takes place biogenically, usually results in substances that are considerably more toxic than the inorganic metal salts. Their often high level of toxicity is put to specific industrial use: they occur, for example, as biocides in various formulations. Such widespread use is leading to a contamination of the environment with metals and their compounds that is found to be increasing when viewed on a global scale.

The chemical form (compound) in which the metals occur is of great toxicological importance. It can be changed in the organism by metabolic processes (oxidation, reduction, methylation etc.) and show correspondingly varying toxic qualities. The toxicologically relevant form of a metal in a tissue or target organ is not necessarily identical to the form in which it is taken in or excreted. The analytical determination of the concentrations of different chemical forms of the elements in the organism – known as speciation – will acquire increasing significance over conventional element determination in chemical analysis and thus for the environmental specimen bank that permits retrospective monitoring through long-term storage of environmental samples (Hamilton, 1979; Ernst and Joosse van Damme, 1983; Rossbach et al., 1992; Wilken, 1992).

The toxicological significance of the chemical form does not only derive from the fact that the great majority of the elements in the periodic table are metals and metalloids: many metals also make it possible to demonstrate the dose-effect relationship very clearly. Copper, manganese, selenium, cobalt and zinc, for example, are essential trace elements whose absence inevitably results in deficiency symptoms or loss of proper functioning. On the other hand, large doses are cytotoxic or cause cancer.

4.1. Occurrence and distribution

The earth's crust may be regarded as a natural reservoir for all the chemical elements of the biosphere. Over 99% of the total mass of the earth's crust is made up of only eight of the 88 elements that occur naturally. It consists of 46.4% oxygen, 28.15% silicon, 8.23% aluminium, 5.63% iron, 4.15% calcium, 2.36% sodium, 2.33% magnesium and 2.09% potassium (Fig. 6). Of the 8 most common elements of the earth's crust, oxygen is the only non-metal. The remaining 80 elements of the periodic system contribute less than 1% to the composition of the earth's crust (Markert, 1996). By contrast, the greater part of organic life consists of non-metals. Since they occur in large quantities "in biomass" the elements C, H, O, N, P, S, Si, Ca and Mg are also termed macro-elements (Fig. 6). In addition to these there are so-called micro-elements that occur in low concentrations in organisms and are essential for the life of most organisms (Table 4).

Both macro- and micro-elements are nutrients that are necessary for the growth and normal development of organisms and whose function cannot be taken on by any other element. They are therefore termed "essential". For this reason, macro- and micro-elements are also called macro- or micro-nutrients. A division into essential and non-essential elements is not always appropriate, since there are numerous exceptions that become apparent when, for example, higher and lower plants are compared. We now know that calcium, boron and chlorine are not essential to some bacteria and fungi, and that sodium and silicon are not essential to higher plants (Marschner, 1983).

Like the division into essential and non-essential elements, the above classic division into micro- and macro-elements that is based solely on the physical mass of an element within an organism has undergone considerable modification in modern plant, animal and human physiology. Additions have had to be made to the list of macro-elements for certain groups of organisms. For example, the element silicon must be regarded as a macro-element for horsetails and diatoms. Moreover, conditions at particular sites often lead to element-specific or organism-specific accumulation

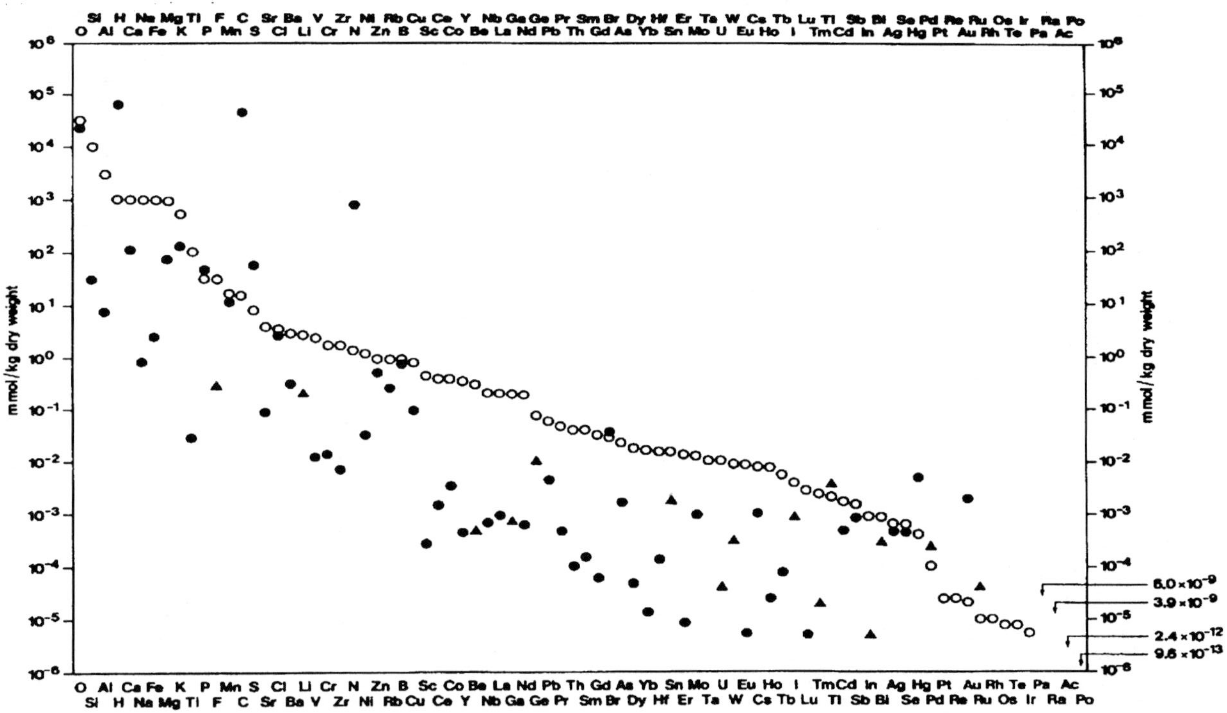

Figure 6. Concentrations of 82 naturally occurring elements (excluding the noble gases) in plants and in the earth's crust (from Markert (1992)). The elements are plotted in decreasing order of their occurrence in the earth's crust. ○, concentration in the earth's crust from Bowen (1979); ●, concentration in plants according to Markert (1986); ▲, concentration in plant material according to Bowen (1979).

processes: sodium, bromine and chlorine are accumulated by many halophytes; copper, nickel, zinc, lead, cadmium and other heavy metals are taken up to an increased extent by metallophytes.

4.2. Action and toxicity of metals

It is unlikely that accumulation can be equated with an increased physiological benefit of the element to the organism: many cases are doubtless no more than signs of adjustment to a particular site. Nevertheless, the above examples help to show some of the difficulties that result from a systematic physiological classification of the chemical elements according to their physical mass. It is not unusual for a macro- and a micro-nutrient to show much greater similarity in their physiological behaviour than two micro- or two macro-nutrients (Mengel, 1984). Since the importance of an element for the organism is not a function of its quantity in relation to the total substance of the organism it would seem more realistic to classify the elements according to physical and biochemical characteristics as attempted in Table 4 after Sansoni and Iyengar (1978). The term "structural elements" is used to describe those elements that are part of the structure of functional molecular components of cell metabolism (proteins, lipids, carbohydrates, nucleic acids etc.) or which have a direct supporting and reinforcing effect (calcium, silicon). Nitrogen and sulphur are inte-

Table 4. Classification of the chemical elements according to their physiological characteristics[a]

Structural elements	C, H, O, N, P, S, Si, Ca		
Electrolytic elements	K, Na, Ca, Cl, Mg		
Enzymatic elements	V[b], Cr, Mo, Mn, Fe, Co, Ni[b], Cu, Zn, B[c], Sn[b], Se[b], F[b], I[b], Mg		
As yet without affirmed significance			
1st main group	Li, Rb, Cs, (Fr)	2nd main group	Be, Sr, Ba, Ra
3rd main group	Al, Ca, In, Tl	4th main group	Ge, Pb
5th main group	As, Sb, Bi	6th main group	Te, Po
7th main group	Br, (At)		
8th main group	He, Ne, Ar, Kr, Xe, Rn		
1st sub-group	Sc, Y	2nd sub-group	Ti, Zr, Hf
3rd sub-group	Nb, Ta	4th sub-group	W
5th sub-group	(Tc), Re	6th sub-group	Ru, Os
7th sub-group	Rh, Ir	8th sub-group	Pd, Pt
9th sub-group	Ag, Au	10th sub-group	Cd, Hg
Lanthanoids	La, Ce, Pr, Nd, (Pm), Sm, Eu, Gd, Tb, Dy, Ho, Er, Tm, Yb, Lu		
Actinoids	Ac, Th, Pa, U, (Np), (Pu), (Am), (Cm), (Bk), (Cf), (Es), (Fm), (Md), (No), (Lr)		

[a] Elements in parentheses do not occur naturally (Hoffmann and Rüdorff (1973) after Sansoni and Iyengar (1978)).
[b] Essential significance affirmed for animal organisms only.
[c] Essential significance affirmed for plant organisms only.

grated in the carbon chain biochemically, i.e. after reduction of their generally high level of oxidation (nitrate or sulphate) they are bonded firmly to the organic substance. On the other hand phosphorus, boron and silicon are present on their highest level of oxidation and are not reduced; instead, they tend to form esters with OH groups of all kinds of molecules, especially sugars (Mengel, 1984). The "electrolytic elements" are required for building up specific physiological potentials and are important for maintaining defined osmotic relationships in cell metabolism. This means that the element calcium can occur both as a structural element and as an electrolytic element. A number of chemical elements, especially metallic ions, exercise a catalytic function in cell metabolism in the form of metal complex compounds. In Table 4 these elements are termed "enzymatic elements". The physiology of individual elements in metabolism is described in detail in the textbooks on plant, animal and human biology.

In many cases the relationship between the nutrients available in the substrate (soil, nutrient solution, atmosphere) and activity in the organism is not linear; it describes an optimum curve that is either symmetrical (Fig. 7a,c,d) or asymmetrical (Fig. 7b,e) and has either narrow (Fig. 7b,d,e) or wide (Fig. 7a,c) tolerances. The form taken by such dose-effect relationships depends of the organism concerned, the element species and conditions at the particular site. If an essential nutrient element is totally absent, activity can generally be said to be zero; the organism is incapable of life.

An inadequate supply of an element from the nutrient medium often results in deficiency symptoms. These can be prevented by adding a small amount (Fig. 7a,b) or a larger amount (Fig. 7c,d,e) of the nutrient substance. If the supply is adequate the activity of the organism reaches its maximum.

An increased supply of nutrients does not initially increase activity. This remains constant over a wider (Fig. 7a) or narrower (Fig. 7b,c,d,e) concentration range in the substrate (saturation with nutrients). A further supply from the substrate initially leads

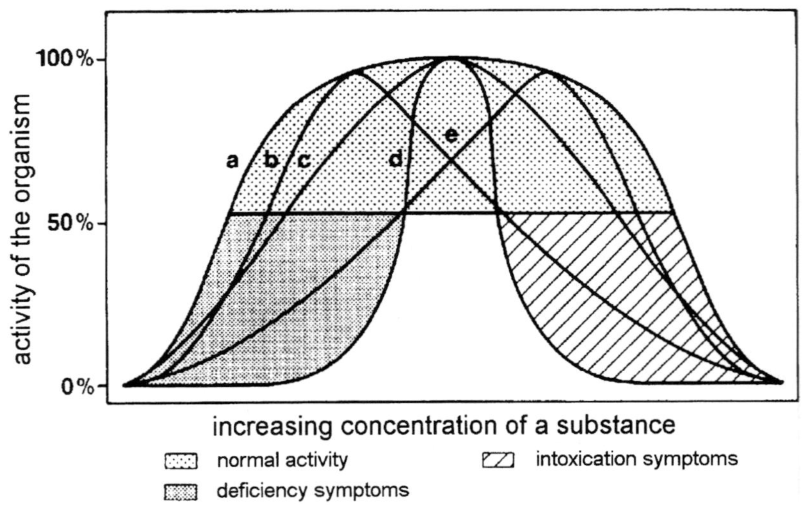

Figure 7. Idealized dosis effects of substrate concentration and activity of the organism. Reproduced from Markert (1996).

to symptoms of poisoning, then to the death of the organism. This reminds us of what Paracelsus said of all chemical substances, whether toxic, essential or indifferent, some 500 years ago:

> What is there that is not a poison? All things are poisons / it is the dose alone that makes a thing not a poison.

A reliable description of the effects of particular elements, some of which are given as examples in Table 5, is made more difficult by the presence of other substances that often have a similar ion radius and the same charge, since it is not usually the concentration of a single element or individual element species that is responsible for the healthy growth and normal development of an organism; the important factor is a proper balance between the individual elements and element species. Interaction between elements in the form of competitive inhibition or promotion may have a decisive influence on the physiology of certain organisms.

The intensity and extent of intake influence the actual concentration of an element in the organism (Baker, 1981). It is possible to distinguish roughly between three types of intake depending on the organisms concerned, the element species and the conditions at a specific location (Fig. 8). Ideally, intake by the organism is directly proportional to the availability of nutrients. In this case the specific element concentration in the organism reflects the concentrations in the environment. The chemical composition therefore has an indicative character. This connection, that has been observed both in experiments and in the field with a number of organisms and all manner of elements, is increasingly attracting attention for practical purposes such as ore prospecting and in the use of (mostly lower) organisms for environmental monitoring (biomonitoring). Because of unfavourable conditions at certain sites some organisms have developed the ability to accumulate large concentrations of individual elements, often irrespective of whether these elements are of physiological benefit to them or not. Some Ericaceae, for example, show a high level of manganese independently of the concentration in the soil. In recent years this accumulative behaviour, which is more likely to have pre-determined genetic causes than to be due to peculiarities of the location, has opened up the possibility of carrying out chemical fingerprinting on all manner of plant species. In future this may well lead to chemical characterization and thus a systematization of individual plant species providing information on evolutionary connections on the phytosociological level. Rejection or exclusion – i.e. reduced uptake of certain elements – may also occur and has been detected in numerous organisms, although it is less common than element accumulation (Baker, 1981). The reduction of the concentration of a element in the organism may take place by complete or partial exclusion. According to Ernst and Joosse van Damme (1983), bacteria, algae and higher plants have heavy-metal resistant populations that can considerably reduce their uptake of heavy metals by excreting mucin or by modifying their cell walls.

Detailed information on the essentiality, incidence, toxicity and mode of uptake of individual elements occurring naturally in the environment is given by Bowen (1979), Merian (1991), Markert (1996) and others. Examples of technical/industrial applications and the relevant estimated production figures for the year 2000 are also given.

Table 5. Selected metals and metalloids and their toxic effects on human beings (after Oehlmann and Markert, 1997)

Metals and metalloids	Mode of intake	Depot compartments	Mode of excretion	Target organs of toxicity
Lead	Mainly by inhalation: inorganic compound, scarcely percutaneous and enteral; organic compound, also percutaneous and enteral	Bones, erythrocytes; liver and muscles in children	Renal, biliary	CNS, peripheral nerves, red bone marrow, smooth muscles, kidneys
Mercury	Mainly by inhalation: inorganic compound, scarcely percutaneous and enteral; organic compound, also very ready percutaneous and enteral intake	CNS, liver, blood plasma proteins, pancreas, kidneys	Metabolism of organic Hg compounds in the liver, renal, biliary, saliva, sweat	CNS, kidneys, mucous membranes
Cadmium	Inhalation: scarcely percutaneous and enteral	Liver, kidneys, pancreas, thyroid gland	Renal, biliary.	Kidneys, lungs, mucous membranes, bones, gonads.
Chromium	Mainly by inhalation: not percutaneous and enteral	Lungs, liver, kidneys, spleen, cardiac muscle	Renal	Respiratory system, kidneys, mucous membranes, skin
Thallium	By inhalation: percutaneous and enteral	Kidneys, bones, muscles, hair and nails	Biliary, renal, saliva, sweat	CNS, peripheral nerves, cardiac muscle, skin, mucous membranes, smooth muscles, kidneys and liver

Nickel	By inhalation: not percutaneous and scarcely enteral; organic compounds by all paths	Kidneys, liver	Renal, biliary	CNS, lungs, mucous membranes, upper air passages
Copper	By inhalation and also enteral: scarcely percutaneous	Liver, kidneys	Biliary, renal	Liver, kidneys, mucous membranes, smooth muscles, cardiac muscle, CNS
Beryllium	Mainly by inhalation: scarcely percutaneous and enteral	Bones, liver, kidneys, lungs	Renal, biliary	Skin, lungs, respiratory system, mucous membranes, cardiac muscle
Aluminium	Mainly by inhalation: not percutaneous and enteral	Bones, muscles	Renal	CNS, muscles, bones, lungs, respiratory system
Arsenic	By inhalation, enteral and percutaneous	Hair, nails, erythrocytes, epididymis, thyroid gland, liver, kidneys	Renal, biliary, pulmonary, sweat	Cardiac muscle, lungs, mucous membranes, skin, liver, kidneys, smooth muscles
Tin	Mainly by inhalation: inorganic compounds, scarcely enteral; organic compounds, percutaneous	Bones, liver, kidneys	Organic Sn compounds: metabolized in the liver; biliary, renal	

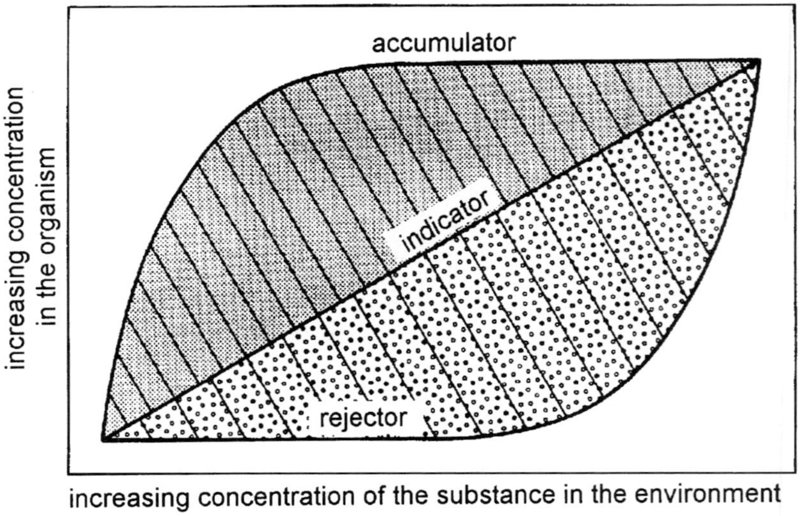

Figure 8. Different uptake behaviour of living organisms in relation to concentrations in the substrate. Reproduced from Markert (1996).

5. Organic pollutants

Organic compounds with an immense number of different structures are produced, used, and ultimately discharged into the environment. They can be classified according to their chemical structure (e.g. hydrocarbons, polyvinyl chloride) or according to their purpose (solvents, insecticides etc.). Substances with greatly differing chemical structures may often be used for the same purpose. An important criterion for the potential hazard to the environment from chemicals is the quantity produced and used. The production figures for the most important synthetic compounds are between 1000 and 3 500 000 tons/year (Koch, 1995).

5.1. Harmful effects

Organic compounds have different toxic effects according to their chemical structure. The effects may occur locally at the site of action on the organism or lead to systemic reactions after intake, transportation and possible distribution. They may be restricted to individual organs or organ systems or affect the organism as a whole (Oehlmann and Markert, 1997). Table 6 shows important classes of organic pollutants and their toxic effects on human beings. It should, however, be taken into account that not all the representatives of a class of substances have all the listed effects, nor are they used for all the purposes mentioned.

5.2. Behaviour in the environment

The behaviour of synthetic compounds in the environment is largely determined by their mobility and (milieu-dependent) degradability. A substance that is potentially

Table 6. Important classes of organic pollutants and their toxic effects on human beings

Pollutant substances	Toxic effects	Remarks	Applications
Aliphatic hydrocarbons	Anaesthetizing; irritant to mucous membranes; pulmonary oedema, death; a few compounds are carcinogenic	Metabolized to form fatty acids	Solvents; petrol
Aromatic hydrocarbons	Anaesthetizing; irritant to the skin; haemotoxic; teratogenic; circulatory failure; death; some are carcinogenic	Some metabolites are more toxic than the initial compounds; accumulation in fatty tissue and bone marrow	Cleaners and solvents; petrol
Polycyclic aromatic hydrocarbons	Eczema; clouding of the cornea and lens of the eye; anaemia; liver damage; death; some compounds are carcinogenic	Some metabolites are more toxic than the initial compounds; accumulation in fatty tissue, adrenal glands, liver and ovaries	Undesirable by-products of incomplete combustion processes
Aromatic amines	Haemotoxic effects; nausea; headache; disorders of balance; difficulty in breathing; loss of consciousness; death; some compounds are carcinogenic	Good absorption; excretion via the kidneys	Chemical industry; aniline dyes; tobacco smoke
Nitroaromatic compounds	Haemotoxic effects; liver damage; nausea; headache; disorders of balance; difficulty in breathing; loss of consciousness; convulsions; death	Excretion partly via the lungs	Chemical industry

Table 6 (continued)

Pollutant substances	Toxic effects	Remarks	Applications
Aliphatic halogenated hydrocarbons	Anaesthetizing; irritant to mucous membranes; liver damage; kidney damage; impaired vision; loss of consciousness; sometimes convulsions, oedema, death; some compounds are carcinogenic	Effects depend on the nature of the halogen atoms; excretion almost exclusively after metabolism	Chemical industry; solvents; refrigeration plant
Aromatic halogenated hydrocarbons	Difficulty in breathing; dizziness; vomiting; cardiac failure; death; damage to the skin and liver; bronchitis; irritation of mucous membranes; chlorine acne; fatigue; some compounds are carcinogenic	Accumulation in fatty tissue, liver, kidneys, central nervous system	Biocides; preservatives
Alcohols, ether	Anaesthetizing; irritant to mucous membranes; loss of consciousness; liver damage; death; teratogenic; eye damage; convulsions; damage to the heart and kidneys	Ethers are metabolized to form the corresponding alcohols	Alcoholic drinks; solvents; chemical industry; petrol substitutes; medicine
Phosphoric acid esters and carbamates	Nausea; convulsions; elevated blood pressure; rapid heartbeat; impaired vision; diarrhoea; pulmonary oedema; asthma attacks; loss of consciousness; respiratory paralysis; teratogenic	Very good absorption through the lungs, digestive system and skin	Insecticides; warfare agents

highly toxic cannot cause lasting damage if it is quickly degraded and its (intermediate) degradation products are less toxic.

5.2.1. Mobility

The uncontrolled spread (dispersion) of organic chemicals results from their mobility (determined by their chemical structure), the nature of their applications and the physicochemical conditions at the place where they are used. Increasing concentrations of halogenated hydrocarbons in the atmosphere would be impossible without the high volatility (vapour pressure) of these substances. Similarly, bioaccumulation factors of up to 100 000 (PCBs in oysters) (Förstner, 1995) can only occur because the chemicals concerned are highly lipophilic and the equilibrium of their distribution between water and fat is very much on the side of the solution in fat.

Although the connections between the structure of a substance and its chemicophysical properties are not yet fully understood, it is possible to identify definite trends:

- Low-molecular, non-polar substances are relatively volatile. They therefore occur as air-borne pollutants, can be dispersed diffusively and with movements of the air and may undergo oxidative, photolytic degradation.
- Halogenated, low-molecular substances have very high vapour pressures and are therefore extremely volatile.
- High-molecular, non-polar substances have low solubility in water and show a pronounced tendency to accumulate in fatty tissue etc. and a strong tendency to adsorb to surfaces.
- Polar substances are readily soluble in water and are easily conveyed in water; their sorption tendency is slight.

The mechanisms of the "sorption" of lipophilic substances to surfaces from an aqueous solution have more to do with displacement from the body of water and solution in a lipid-type surface layer than with the classic sorptive interactions (Förstner, 1995). On the basis of their chemical structure and possible surface charges it is possible to distinguish between the following categories of organic environmental chemicals that are relevant to water and soil (Förstner, 1995):

- Cationic or basic substances (e.g. the herbicides paraquat and diquat) that interact strongly with negatively charged surfaces and are therefore adsorbed almost irreversibly.
- Acid compounds (e.g. the herbicides 2,4-D and 2,4,5-T) that are repelled by negative surface charges on minerals or organic matter and therefore scarcely sorbed.
- Non-polar, readily volatile substances (e.g. toluene) that show slight, hydrophobic interaction with the surfaces of particles.
- Non-polar substances with low volatility (e.g. hexachlorocyclohexane, hexachlorobenzene, PCBs) that show strong hydrophobic interaction with organic solids.

5.2.2. Degradability and persistence

The possible applications of a substance are greatly influenced by its shelf-life, i.e. its stability under environmental conditions. On the other hand, a substance that is no longer used should be eliminated as quickly as possible. Heavy metals cannot really disappear from the environment since they are elements. These substances can only be shifted into compartments that are less at risk and better controlled, so that they are removed from the biosphere. But organic compounds can be destroyed completely and thus truly removed from the environment. Natural mechanisms for this exist in the form of photo-oxidation of volatile substances and in biodegradation.

When speaking of the degradation of chemicals we have to distinguish between the following terms:

- Primary degradation: a substance has undergone primary degradation if a single bond in the molecule has been broken and the original structure has thus been destroyed. This (very incomplete) degradation is detected by analytical procedures such as those prescribed for degradability tests on detergent tensides (DIN 38412 L24, 1981; L26, 1994).
- Partial degradation: if an organic substance has not been completely transformed into C_1 bodies and possibly salts or oxides (by heteroatoms) it has been partially degraded. Primary degradation is the first step in partial degradation.
- Total degradation (mineralization): if an organic compound has been completely split up into C_1 components (CO_2 through oxidation, CH_4 through reduction), water and possibly salts or oxides we speak of total degradation or mineralization of the compound.
- Persistence: a substance is persistent under the conditions prevailing in its surroundings if there is no attack on its original structure, i.e. if no primary degradation takes place. According to another definition a compound is also persistent if its degradation stops at an intermediate stage so that no mineralization takes place.

The degradability of a substance is determined primarily by its structure. Unbranched carbon chains are more readily degradable than branched ones; alkanes are less persistent than aromatic compounds. In the case of polynuclear aromatic compounds persistence increases with the number of rings; a substitution increases the persistence of the compound. Halogen substituents reduce degradability to a greater extent than carboxyl, alkyl or phenol substituents. A further factor directly determined by the structure of a compound is its solubility in water, which is also of central importance for degradation in solids and waters. The speed of degradation also depends on the following factors:

- pH
- redox potential
- temperature
- biomass density (in biodegradation)
- biomass activity (in biodegradation)
- availability of nutrients (N, P, trace elements; in biodegradation)
- pore volume and distribution (in the degradation of solids)
- particle size and distribution (in the degradation of solids)
- nature and quantity of clay minerals (in the degradation of solids)
- quantity of organic matrix (in the degradation of solids)
- insolation (in degradation in the atmosphere)

Hydrolysis, oxidation and reduction reactions play a major role in biodegradation. The substances may serve the microorganisms as a source of energy or carbon, or they may be degraded "cometabolically" without the organisms' putting this degradation to any noticeable use (for acquiring energy or building up biomass).

The length of time for which a substance persists in the environment is often stated in the form of a half-life of degradation. This is the time required for an initial concentration to be reduced by half. Strictly speaking, it is only possible to define of a half-life if degradation follows first order kinetics. This is the case if the speed of degradation is proportional to the current concentration of the substance. Since at least one reaction partner is necessary for the (bio)chemical conversion of a substance, this condition can only be fulfilled if an excess of such a reaction partner is present. This is usually the case with water and oxygen in suitable environmental media, which means that the reaction does in fact follow first order kinetics.

6. Radioactivity

High-energy radiation has various harmful effects on man, ranging from sunburn (from slight overdoses of UV-B radiation) to cancer or death from cell damage. Damage from electromagnetic radiation is to be expected below a wavelength of 280–315 nm (UV-B). The wavelengths of X-rays lie between 0.1 and 10 nm, and gamma radiation has wavelengths of 0.001–0.01 nm (Fritz-Niggli, 1975).

6.1. Use of ionizing radiation

Ionizing radiation is used chiefly in medicine, in nuclear engineering and for special analytical procedures. The burden on man and the environment resulting from such applications (and from natural sources of radiation) is measured by radiometric methods. It is the medical applications (in particular radiological diagnosis) that constitute the chief man-made source of the radiation burden on the population.

The measure for the burden on human beings is the equivalent dose, stated in Sieverts (1 Sv = 1 J/kg, standardized to electromagnetic radiation). It takes into account the varying degrees of potential hazard from the different kinds of radiation. Table 7 shows the applications of radiometry classified according to fields.

6.2. Harmful effects

Acute damage by radiation is only observed after exposure to large or very large doses, i.e. from about 0.1 to $>$10 Gy (1 Gy = 1 J/kg, irrespective of the nature of the radiation). The symptoms are fatigue, nausea and vomiting, headache, clouding of consciousness and changes in body temperature. At doses above approximately 6 Gy the chances of survival are slight (Bahadir et al., 1995). Delayed effects include the development of cancer and genetic damage to the DNA material with corresponding mutations in future generations (Bahadir et al., 1995).

Table 7. Applications of radiometry classified according to fields (adapted from Förstner (1995))

Geology	Uranium prospecting and exploration; geological and archeometric age determination
Medicine and health care	Diagnosis by radiological methods (e.g. scintiscanning); treatment by radiological methods (e.g. cancer therapy); irradiation of equipment (and foods)
Analysis	Materials testing (X-ray examination); indicator and tracer methods; radiochemical and biochemical labelling and trace analysis; organic trace analysis (ECD); activation analysis
Industrial measuring techniques	Control engineering (e.g. non-contact thickness gauging, level measuring); smoke alarm systems
Monitoring and observation	Monitoring of contamination; observation of nuclear plant; observation of nuclear weapons, military disaster procedures; radioactivity in the environment (gamma dose rate, nuclide-specific activity measurement)

6.3. Natural sources of radiation

When discussing the exposure of human beings to natural radiation we have to distinguish between cosmic and terrestrial radiation. Cosmic radiation results from the collision of high-energy particles with the atmosphere, during which neutrons, protons and alpha particles with a high ionization rate are released. At average populated altitudes cosmic radiation is about 0.5 μSv; at the altitudes at which jet aircraft fly it is 50–500 μSv.

The main source of terrestrial radiation is long-living isotopes of the uranium-radium series (^{238}U), the thorium series (^{232}Th) and the actinium series (^{235}U). When these isotopes decay they form numerous active daughter nuclides with different half-lives. Nuclear reactions of the neutrons from cosmic radiation produce tritium (^{3}H) and ^{14}C. These enter the biosphere by way of air currents and precipitation and are therefore classified as terrestrial radiation. In part, the decay of some extremely long-living isotopes (^{40}K, ^{87}Rb) that have existed since the earth was formed is also responsible for terrestrial radiation. These isotopes form inactive disintegration products. Rocks may also contain radionuclides, but to greatly differing extents: the average is 0.5 mSv/a.

The radionuclides in rocks result in exposure to radiation from building materials. Walls and ceilings act as a shield against external radiation, but the building materials themselves may constitute an additional burden. In the case of wooden and prefabricated buildings the two effects usually cancel each other out, or the shielding effect is greater. But with stone walls or materials such as concrete and bricks the natural radioactivity results in much greater radiation doses. Table 8 shows the influence of various construction materials on exposure to radiation in residential buildings.

The body's own radiation results in an average exposure of 0.2–0.3 mSv/a for the

Table 8. Influence of various construction materials on exposure to radiation in residential buildings (Förstner, 1995)

Construction material	Additional exposure to radiation (mSv/a)
Wood	−0.2–0[a]
Sand-lime bricks, sandstone	0.0–0.1
Bricks, concrete	0.1–0.2
Natural stone, industrially produced plaster	0.2–0.4
Slag blocks, granite	0.4–2.0

[a] By shielding against ambient radiation.

lungs, gonads and bone marrow. It comes from radionuclides that are taken in with the food and through air and water and incorporated in the body. It is a radioactive isotope of the noble gas radon (^{222}Rn) and its short-lived succeeding products that are largely responsible for this internal exposure (approximately 68%). It originates from the decay series of uranium and thorium. The most significant source of radon seems to be the soil, since the dose measured in buildings decreases from the basement to the attics. Like other inorganic pollutants, radioactive isotopes can be accumulated through the food chain. Algae that serve as food for fish accumulate radioactive iodine, terrestrial plants accumulate strontium and terrestrial animals may store radionuclides in certain organs (Förstner, 1995). The total natural radiation exposure of the population of Germany is estimated at 1.5–4 mSv/a, three quarters of this being from internal sources.

6.4. Man-made sources of radiation

Most of the radiation to which the population is artificially exposed is due to radiological diagnosis. There are considerable individual differences, depending on the medical requirements, the design of the equipment, the number of images taken and other factors. Table 9 shows the mean radiation dose used in X-ray examinations.

The contribution of radiotherapy and the use of radioactive substances for diagnostic purposes to overall exposure to radiation may be very great in individual cases, but in relation to the population as a whole it is comparatively small. It is estimated at only a few percent in the form of X-rays. Surface explosions of nuclear weapons released mainly dust-bound fission and activation products: about half of these were conveyed into the stratosphere, the other half caused local fallout. About three quarters of the total amount released have already decayed, the remaining external dose is estimated at approximately 0.27 μSv in mid-latitudes. It is due primarily to the decay of long-living ^{137}Cs.

Internal exposure following surface explosions of nuclear weapons is caused by incorporated ^{137}Cs; the resulting whole-body radiation dose of about 30 μSv in 1964 had fallen to about 1 μSv by 1979 (Förstner, 1995).

The total exposure of the population to radiation resulting from the operation of nuclear plant is said to be about 10 μSv. Special precautions are taken in respect of

Table 9. Mean radiation dose used in X-ray examinations (Förstner, 1995)

Applications	Skin surface dose (mSv)	Bone marrow dose (mSv)	Gonadal dose	
			Female	Male
Cardiac catheterization	410	90	36	17
Renal angiography	300	10	30	12
Gastro-intestinal tract	160	7	4	1.4
Gall bladder	45	1	5	0.4
Lumbar spine	35	0.6	3	1.3
Pelvis	20	1	4	2
Lungs	1	0.2	0.03	0.01

certain highly critical accumulation paths. For example the emission of ^{131}I, that is accumulated readily through milk, taken into the body and stored in the thyroid gland, is restricted to a very low level. For the planned fuel reprocessing plant in Wackersdorf – that was never built – the equivalent dose from emitted radionuclides was calculated at 10–100 μSv; this was wholly due to the activity of ^{129}I.

For ultimate disposal sites for radioactive waste a maximum permissible value of 0.3 mSv has been set: this is based on the value for natural ambient radiation. It is possible to conclude from models that the burden on human beings and the environment would be very slight even if unusual accidents were to occur (Förstner, 1995).

6.5. Total exposure

The mean total radiation to which an inhabitant of Germany is exposed is estimated at 3.2 mSv/a. This burden is due in approximately equal parts to natural sources, civilization-related exposure (for example through construction materials) and the use of diagnostic radiology. This dose is one of the factors responsible for the occurrence of cancer and leukaemia; the genetically significant dose that causes damage to offspring is considerably lower. Flying increases the dose by about 20 μSv. Luminous figures on the face of a watch, containing tritium, contribute 0.3 μSv/a to the total dose. Occupational exposure to radiation is (still) mainly a problem in mining, where there is greatly increased exposure to and intake of radon and its succeeding products. In uranium mines in France, Canada and the USA the mean equivalent dose for the workers is between 6 and 34 mSv/a; for underground workers in ore mines it is between 3 and 20 mSv/a (Förstner, 1995).

References

Bahadir, M., Parlar, H., Spiteller, M. (Eds.), 1995. Springer-Umweltlexikon. Springer, Berlin.

Baker, A.J.M., 1981. Accumulators and excluders – strategies in the response of plants to heavy metals. Plant Nutr. 3, 643–654.

Bowen, H.J.M., 1979. Environmental Chemistry of the Elements. Academic Press, London.

DIN 38412 L24, 1981. Determination of Biodegradability with the Use of Specialized Analytical Techniques. Wiley-VCH, Weinheim; Beuth, Berlin.

DIN 38412 L 26, 1994. Degradation and Elimination Test Detergants for the Simulation of Domestic Sewage Treatment Plants. Wiley-VCH, Weinheim; Beuth, Berlin.

Ernst, W.H.O., Joosse van Damme, E.N.G., 1983. Umweltbelastung durch Mineralstoffe. Gustav Fischer Verlag, Stuttgart.

Fiedler, H., Lau, C., 1998. Transformation of chlorinated xenobiotics in the environment. In: Schüürmann, G., Markert, B. (Eds.), Ecotoxicology. Wiley, New York, pp. 283–316.

Förstner, U., 1995. Umweltschutztechnik, 5th ed. Springer, Berlin.

Fritz-Niggli, H., 1975. Strahlengefährdung – Strahlenschutz. Huber-Verlag, Bern.

Hamilton, E.I., 1979. The Chemical Elements and Man. Wiley, New York.

Hempfling, R., Stubenrauch, S., 1994. Schadstoffe in Gebäuden. Erkennen, Bewerten, Sanieren. Eberhard Blottner Verlag, Taunusstein.

Hoffmann, U., Rüdorff, W. (Eds.), 1973. Anorganische Chemie. Vieweg Verlag, Brunswick.

Jørgensen, S.E., 1998. Ecotoxicological research – historical development and perspectives. In: Schüürmann, G., Markert, B. (Eds.), Ecotoxicology. Wiley, New York, pp. 3–15.

Koch, R., 1995. Umweltchemikalien, 3rd ed. VCH Verlagsgesellschaft, Weinheim.

Markert, B., 1986. Aufstellung von Elementkonzentrationskatastern in unterschiedlichen Pflanzenarten und Bodentypen in Deutschland, Österreich und Schweden. In: Stoeppler, M., Dürbeck, H.W. (Eds.), Beiträge zur Umweltprobenbank, Jül. Spez., 360.

Markert, B., 1992. Presence and significance of naturally occurring chemical elements of the periodic system in the plant organism and consequences for future investigations on inorganic environmental chemistry in ecosystems. Vegetatio 103, 1–30.

Markert, B., 1996. Instrumental Element and Multielement Analysis of Plant Samples – Methods and Applications. Wiley, Chichester.

Marschner, H., 1983. Mineral Nutrition of Higher Plants. Academic Press, London.

Mengel, K., 1984. Ernährung und Stoffwechsel der Pflanze. Fischer, Stuttgart.

Merian, E. (Ed.), 1991. Metals and their Compounds in the Environment. VCH, Weinheim.

Oehlmann, J., 1994. Imposex bei Muriciden (Gastropoda, Prosobranchia), eine ökotoxikologische Untersuchung zu TBT-Effekten. Cuvillier-Verlag, Göttingen.

Oehlmann, J., Markert, B., 1997. Humantoxikologie. Wissenschaftliche Verlagsgesellschaft, Stuttgart.

Rossbach, M., Schladot, J.D., Ostapczuk, P. (Eds.), 1992. Specimen Banking, Environmental Monitoring and Modern Analytical Approaches. Springer, Berlin.

Sansoni, B., Iyengar, V., 1978. Sampling and Sample Preparation Methods for the Analysis of Trace Elements in Biological Materials. Jülich Research Centre. Jül. Spez., 13.

Schüürmann, G., 1992. Ecotoxicology and structure-activity studies of organophosphorus compounds. In: Draber, W., Fujita, T. (Eds.), Rational Approaches to Structure, Activity and Ecotoxicology of Agrochemicals. CRC Press, Boca Raton, FL, pp. 485–541.

Schüürmann, G., Marsmann, M., 1991. QSAR-Modelle – Interpretation und Prognose der Biokonzentration und aquatischen Toxizität. UWSF. Umweltchem Ökotox 3, 42–47.

Schüürmann, G., Segner, H., 1994. Struktur-Wirkungs-Analyse von Trialkylzinnverbindungen. Ecoinforma 7, 439–453.

WHO, 1984. Paraquat and Diquat (Environmental Health Criteria 39). World Health Organisation, Geneva.

Wilken, R.D., 1992. Mercury analysis – a special example of species analysis. Fresenius J. Anal. Chem. 342, 795–801.

Trace elements – Their distribution and effects in the environment
B. Markert and K. Friese, editors

Chapter 2

Natural trace element input to the soil–sediment–water–plant system: examples of background and contaminated situations in Switzerland, Eastern France and Northern Italy

H.-R. Pfeifer, M.-H. Derron, D. Rey, C. Schlegel, O. Atteia, R. Dalla Piazza, J.-P. Dubois, Y. Mandia

Abstract

Studies on environmental pollution in a given region often suffer from the lack of reliable information on the natural level of trace elements. The natural contribution to the trace element contents and fluxes of local and regional ecosystems strongly depends on the local composition of the lithosphere, i.e. the dominating rock type. The dispersion of trace elements in the different environmental compartments is mainly governed by weathering and mass transport. Based on examples from Central Europe, this paper gives an insight into the typical natural input of six major, geochemically defined rock types ("natural background"). It also presents two cases of regional "natural contamination" related to ore deposits. Each rock type contributes in a characteristic manner to the major and trace element budget of soils, sediments, plants, surface and ground waters. Whereas well soluble elements such as Ca, Sr, Mg, Na and K are enriched in the water phase, less soluble elements such as Al, Fe and in part Si and most trace elements are enriched in the soil compartment. In the latter, these elements are either part of, or are adsorbed on organic material, oxy-hydroxides or clay type minerals. From soils, trace elements are transported either to plants or to ground waters. The extent and manner to which chemical elements of the underlying rocks are dispersed in their surroundings depends first of all on the chemical composition of the parent material, varying greatly from one rock type to the other. Secondly, the typical pH created in the soil and water compartments strongly influences the sorption behavior and thus retention or release of specific elements. Soils formed on silicate rocks usually exhibit acid conditions (pH 4–6.5), related soil and ground waters typically are near neutral (pH 6.5–7.5). However, soils and waters related to carbonate rock environments are usually neutral to alkaline (pH 7–8.5). Soils on granitic silicate rocks have been found to be slightly enriched in Zr, Ti, Pb and Rb and often host ore deposits rich in U or As, which influence the water and sediment composition of whole catchment areas. Basaltic silicate rocks enrich the soils typically in Sr, Ti, Fe, V, Cr, Ni, Cu and Zn in concentrations up to three times those of granites. Mg-rich silicate rocks, such as peridotites and serpentinites, typically enrich soils, plants and waters with very high amounts of Mg, Fe, Cr, Co, Ni and V, but other trace elements remain very low. Al-rich silicate rocks, such as shales and micaschists, are often rich in sulfur and metal traces such as Zn, Pb, As and Cd. In contrast, carbonate dominated rocks, such as limestones or sulfate bearing dolomites, typically contribute to enrich their environment with Mn, F, S, Cl, Ba, Sr, As, Cd and radon. In conclusion, insufficient knowledge of this natural contribution can lead to misinterpretations of contaminated sites, especially for elements such as Ni, Cr, Zn, V, Cd and As.

Keywords: Trace elements; Natural input; Rocks; Soils; Water; Plants; Weathering; Switzerland; Mountainous area

1. Introduction

1.1. Scope of this paper

During the last 20 years, elevated concentrations of a steadily increasing number of trace elements have been analyzed in the different compartments of the environment, i.e. air, soils, biota, sediments and water. In those cases, where concentrations were much higher than the average lithosphere and no direct relation to an anthropogenic source could be established, such as industrial activities, mine dumps, waste repositories etc., their origin often remained doubtful. Was it a diffuse anthropogenic input or a case of elevated natural concentrations in soils and waters, well known from mining exploration studies? In fact, natural background values for trace elements contained in soils, waters and biota, especially for the industrially important elements, are still badly known in many areas. This is due to the fact, that for a long time, earth scientists were only interested in trace elements important for modeling rock-forming processes which take place deep in the earth's crust and the mantle (lithosphere). However, these elements are only rarely relevant for environmental considerations.

Earth scientists involved in the search of metal ore deposits and a few others, interested in geochemical mapping in general, had known for years, that rock compositions of a given area strongly influence the chemical composition of the soils, plants and waters of this area (Rose et al., 1979; Govet, 1983; later volumes). In cases where a rock type, rich in particular elements, dominates a whole catchment area of a river system, local or regional natural contaminations are common and could be mistaken for industrial pollution.

This work deals on one hand with the normal characteristic trace element input from common rocks to the environment (natural background), but also shows cases of regional natural contamination. In the first section, an overview on the typical chemical variation of the different chemical rock types and general chemical trends of soils, waters, plants and sediments related to them will be given. In the second section, case studies representing typical local background situations of the most important chemical rock types of the lithosphere are discussed. The third section deals with a selected number of naturally contaminated situations at local and regional scale.

For both categories, the author team, composed of geochemists, hydrogeologists, chemists and soil scientists, looked at many examples and confirmed that the influence of a given rock type on its environment is considerable and changes rapidly with the underlying rock. In order to be as interdisciplinary as possible, specialized, branch specific terms have been avoided, hoping that the specialists of a given branch might forgive us. All examples stem from mountainous areas in Switzerland, Northern Italy and Eastern France (Fig. 1) and are therefore characteristic for temperate-humid climate zones with typical precipitation rates of 100–200 cm/a (Spreafico et al., 1992). Under different climatic conditions, another behavior can be expected.

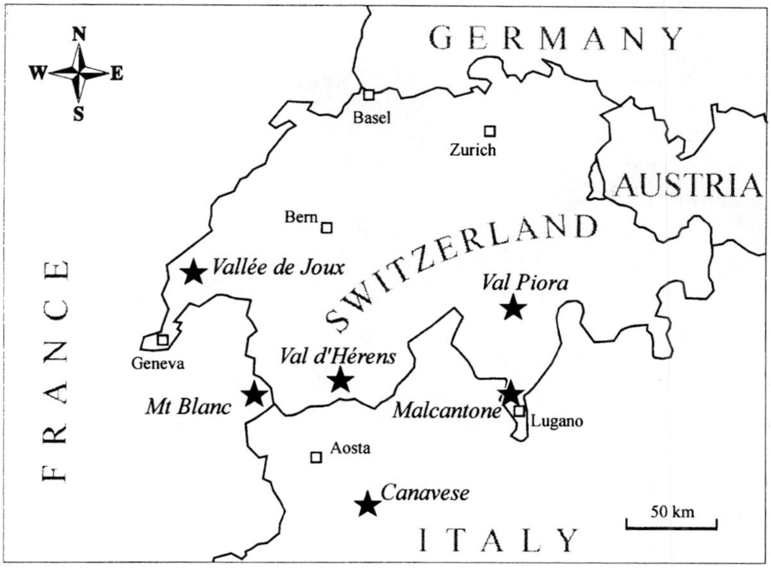

Figure 1. Location of case studies discussed in this paper (shown by asterisks). The exact location with coordinates is given in Appendix A.9.

1.2. Analytical methods

The analytical methods used to produce the results presented in this paper were adapted to the analyzed material. (1) Rocks and soils were analyzed with X-ray fluorescence (XRF) methods (Pfeifer et al., 1991). The obtained results correspond to the overall elemental content of the solid phases, independent of whether an element is part of a mineral, an amorphous phase or whether it is adsorbed on the surface of a solid. They differ in most cases from so-called "total concentrations" determined by acid extraction methods (HCl, HNO_3) and reported in some of the soil science literature. The latter correspond to 15–85% of the concentrations determined by XRF or aqua regia (Meyer, 1991). Results are traditionally reported as weight percent (wt.%) or as weight parts per million (ppm, equivalent to g/ton or mg/kg). (2) Plants were completely mineralized with a mixture of HNO_3 and H_2O_2 in a microwave oven and analyzed with inductively coupled plasma methods (ICP, optical atomic emission spectroscopy (AES) and mass detection (MS) depending on the element to analyze). Results are given in mg/kg dry plant. (3) Soil percolation and ground waters were analyzed by ionic chromatography and ICP-AES (major ions) and ICP-MS for trace elements. Some of the arsenic concentrations were determined with atomic absorption equipped with a hydride generator. Results are reported in μg /l or mg/liter.

Details about the preparation and analytical methods applied can be found in the original work cited with each example. Typical uncertainties (2 sigma) are between 2 and 5 rel.% for major elements and between 5 and 10 rel.% for trace elements. Detection limits are given in the appendix directly with the analyses.

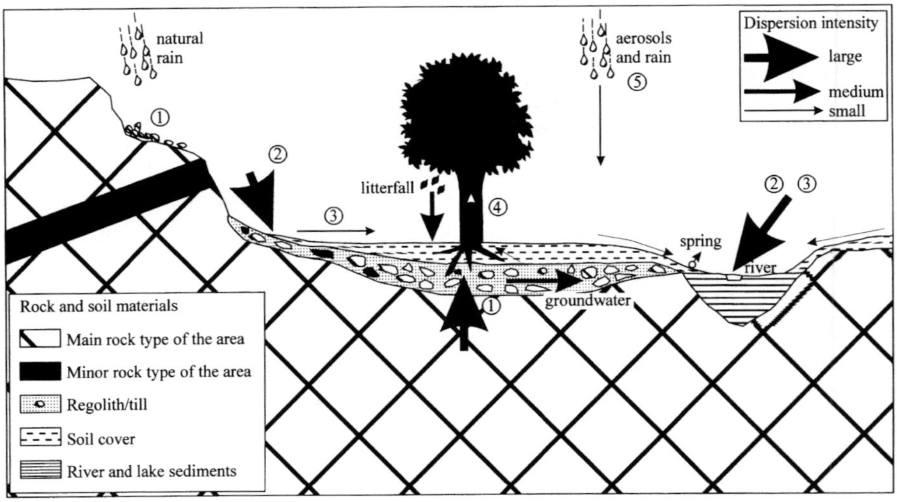

Figure 2. General scheme of the pathways to the different environmental compartments of chemical elements contained in rocks (not to scale). (1) Weathering, (2) mechanical transport (creep and erosion), (3) transport in the water phase, (4) accumulation in plants and (5) diffuse atmospheric input (aerosols, rain).

1.3. Natural dispersion of chemical elements originating from the lithosphere

1.3.1. General processes

Fig. 2 shows the different pathways in which chemical elements contained in rocks are released to the different environmental compartments. Five main processes are responsible for their dispersion into the different ecosystems: (1) Weathering, either directly by rain water on rock outcrops, by soil percolation water or by root exsudates, which interact with rock fragments, contained in the soil cover; (2) Down hill mechanical transport of weathered rock particles, such as creep and erosion and subsequent sedimentation as till material or alluvial river and lake sediments; (3) Transport in dissolved or low size colloidal form by surface and groundwater; (4) Terrestrial and aquatic plants growing in undisturbed natural situations will take up whatever chemical elements they need and which are available in the surface and shallow groundwater. Trace elements taken up from the soil will accumulate in the leaves and will possibly enrich the soil by litterfall; (5) Diffuse atmospheric input by aerosols and rain: rock particles from volcanic eruptions, desertic areas (Chester et al., 1996), seaspray and their reaction with rain water. A considerable part of this can be anthropogenic.

When the rock disintegrating process occurs on flat or only slightly dipping terrains, the result of this dispersion, i.e. the soils, sediments and waters, exhibiting the chemical signature of the altered rock, is only seen locally. However, in hilly and mountainous areas with steeper slopes, the resulting compositional halo can be displaced for tens to hundreds of meters or several kilometers – a detailed discussion on the various dispersion situations can be found in Rose et al. (1979). In all these cases, water, whose composition usually mirrors the chemistry of the adjacent rocks (as debris or in place), plays a major role in the dispersion process.

The chemical interactions between water, rocks and biota typically comprise the following steps: dissolution, complexation, precipitation, sorption, diffusion and/or advective transport – for a more detailed review see Bourg and Loch (1995); Bourg (1995). For all these reaction steps, the initial chemical environment (i.e. the lithosphere), temperature, pH and redox conditions (pe or Eh) are the master variables influencing the release and transport of chemical elements to the aqueous phase and their availability to biota. The following secondary solid phases control the fixation of metals and metalloids by adsorption: clay minerals, organic matter, oxides and hydroxides of Fe, Mn and Al. Others can fix trace elements by precipitation: sulfides, carbonates and phosphate minerals. This paper, on purpose, focuses on the observable facts, mainly as a function of changing natural chemical conditions and only occasionally enters into a detailed discussion on the speciation and its controlling parameters.

1.3.2. Characteristic chemical tendencies of common rock types

Earth scientists usually use a genetic rock classification and distinguish magmatic rocks formed from molten rocks at various depths in the lithosphere, sedimentary rocks formed as the result of weathering and transport at the surface of the earth and metamorphic rocks as a result of deformation under elevated temperatures and pressures of the first two categories in descending lithospheric plates. Rocks of each of these categories are composed of a series of characteristic minerals with variable proportions (Hamblin, 1992). Names given to the different rocks are based on tradition and are not very systematic.

For environmental purposes a chemical classification is more useful. Based on the dominant minerals, two main chemical categories can be distinguished: (A) Silicate rocks and (B) carbonate rocks (Table 1). Among the rocks of group A, the Si-content and the second major elements such as K, Na, Ca, Mg/Fe and Al can be used to classify them in more details (Fig. 3): (1) Si–K/Na-rich rocks (granites), (2) Si-reduced, Ca–Mg/Fe-rich rocks (basalts), (3) Si-poor, Mg/Fe-rich rocks (peridotites), (4) Al-rich rocks (shales). For group B only two types are distinguished: (5) limestones dominated by $CaCO_3$ and (6) gypseous dolomite rocks, dominated by carbonates of Ca–Mg and sulfates. In Table 1, another frequently encountered rock type, i.e. sandstone is distinguished. Since it is chemically close to granitic rocks, we do not treat it here separately. More rare rocks and materials rich in organic carbon, such as coal, bitumen or bog are not treated here. The reader interested in the trace element of these rocks is referred to Tuchschmid (1995). The values given in Table 1 correspond to typical average values and in reality there is quite some chemical variation within a given rock group. But, as the discussion of specific cases will show below, Table 1 is quite representative. Fig. 4 shows typical trace elements contained in the various rocks distinguished. Peridotites are characterized by high Cr- and Ni-concentrations and very low values for other traces. Basalts contain many metal traces in considerable amounts: Co, Cu, Zn, Ni, V, Cr, Sr and Ti. Granitic rocks, on the contrary, exhibit a

Table 1. Chemical characteristics of the major rock types of the lithosphere in ppm (mg/kg or g/t)[a]

Element (ppm)	A. Silicate rocks				B. Carbonaie rocks		
	1a Granite	1b Sandstone	2 Basalt	3 Peridotite	4 Shale	5 Limestone	6 Dolostone[b]
Li	24	15	16	0.5	66	5	–
B	9	35	5	2	100	20	–
F	520	270	385	100	740	330	–
Na (wt.%)	2.84	0.33	1.87	0.49	0.96	0.04	0.29
Mg (wt.%)	0.94	0.7	4.55	23.2	1.5	4.7	4.7
Al (wt.%)	8.2	2.5	8.28	1.2	8	0.42	0.24
Si (wt.%)	31.4	36.8	23.5	19.8	7.3	2.4	0.56
P	920	170	130	195	700	400	349
S	300	240	300	200	2400	1200	9
Cl	130	10	55	45	180	150	–
K (wt.%)	2.52	1.07	0.83	0.017	2.66	0.27	0.15
Ca (wt.%)	2.53	3.91	7.2	1.6	2.21	30.23	32.1
Ti	3400	1500	11400	300	4600	400	119
V	88	20	225	40	130	20	–
Cr	22	35	185	1800	90	11	–
Mn	540	0	1750	1560	850	1100	288
Fe (wt.%)	2.96	0.98	8.6	9.64	4.72	0.33	0.26
Co	7	0.3	47	175	19	0.1	33
Ni	15	2	145	2000	68	20	10

Cu	30	0	94	15	45	4	5
Zn	60	16	118	40	95	20	48
As	1.9	1	2.2	0.8	13	1	–
Rb	110	60	38	1.1	140	3	6
Sr	440	20	452	5.5	300	610	199
Y	35	40	21	0	26	30	18
Zr	140	220	120	38	160	19	19
Nb	20	0	20	9	11	0.3	4
Mo	1	0.2	1.5	0.3	2.6	0.4	–
Ag	0.051	0.01	0.11	0.05	0.07	0.01	–
Cd	0.13	0	0.21	0.05	0.3	0.035	–
Ba	420	10	315	0.7	580	10	14
Hg	0.08	0.03	0.09	0.01	0.4	0.04	–
Pb	15	7	7	0.5	20	9	2
Th	8.5	1.7	3.5	0.0045	12	1.7	1
U	3	0.45	0.75	0.002	3.7	2.2	–

[a] Analyses no. 1–5 from Faure (1992), no. 6 from Mandia (1991).

[b] Rock dominated by the minerals dolomite and calcite and often rich in gypsum, therefore called gypseous dolomites in this paper. The analysis shown is not completely respresentative, Mg can reach up to 35 wt.%.

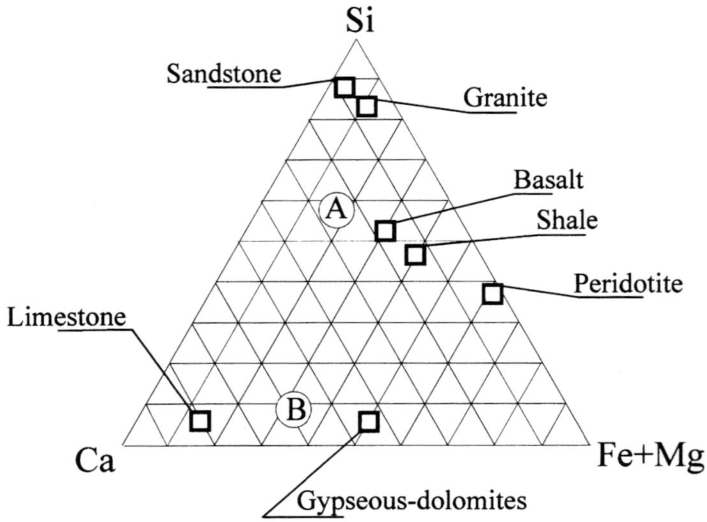

Figure 3. Typical proportions of the major elements Ca, Si and Fe + Mg (wt.%) of the seven chemical rock types distinguished in Table 1. A, silicate rocks; B, carbonate rocks.

low concentration trace spectrum, but with a few exceptions: U, Zr, Pb and F. Shales are again rich in trace elements, especially in rarer ones, such as As, U, V, F and Rb. Sandstones are especially rich in Zr. Carbonate rocks (represented in Fig. 4 only by limestones) have, apart from U, Sr and F generally low trace element concentrations.

1.3.3. General trends of soil and water compositions

During weathering a typical "differentiation" occurs in most rock-soil-water systems. The easily soluble cations, such as Ca, Mg, K, Na, Sr and Rb are fractionated in the soil percolation and ground water (local springs), Fig. 5. The less soluble elements such as Fe, Mn, Al, Si, and most metal traces are concentrated in and on the solid soil phases such as crystalline or amorphous oxy-hydroxides, clay type minerals and organic material. In the triangle $CaO + Na_2O - Al_2O_3 - K_2O$ after Nesbitt and Young (1982), during progressive weathering the soils lying on top of the different rocks distinguished above, all show a compositional trend towards the Al_2O_3 corner, while the corresponding water compositions move towards the Ca–Na-corner.

Ground and surface waters related to the different rock types introduced above show well discriminated major ion patterns. Fig. 6 shows about 700 typical spring water analyses, mainly from the Central Alps, plotted in a Ca–10∗Si–5∗Mg triangle in molar proportions (Derron, 1999). This diagram allows in many cases to identify the dominant rock type of a given watershed influencing the water composition.

Interestingly enough, the same is much less apparent for trace elements. Different rocks often yield almost similar trace element concentrations in waters (Derron and Pfeifer, 1998; Derron, 1999). Only Ni, plotted against conductivity and the Ba/Rb ratio vs. Sr allows to separate waters according to their parent rock type (Fig. 7). Interpret-

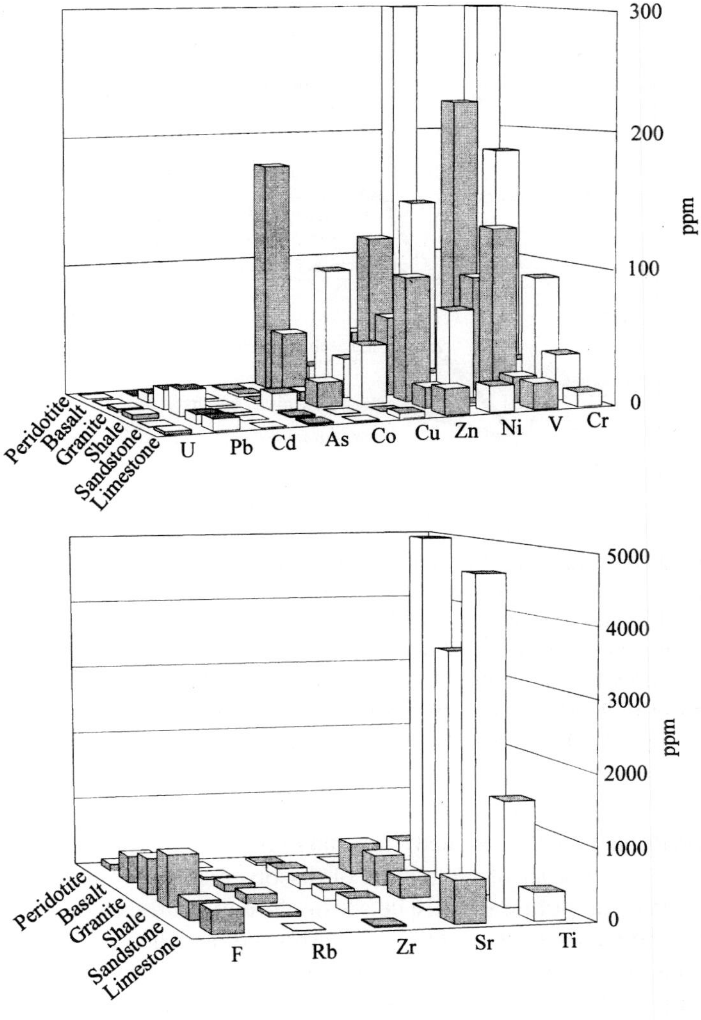

Figure 4. Typical absolute values of important trace elements of the various rock types distinguished (ppm). Gypseous dolomitic rocks are not shown because representative trace data are lacking.

ing this phenomenon is not straightforward, especially because part of the compiled analyses are based on non-filtered water samples (Parriaux et al., 1990). A more detailed discussion is contained in Section 4.3.

The two main compositional rock families discussed above, i.e. silicate and carbonate rocks, give also rise to two different water characteristics. The kinetics of silicate weathering reactions observed in the field are slow, compared to carbonate or sulfate rock weathering (Swoboda-Colberg and Drever, 1993; Drever and Clow, 1995). The solubility of most secondary minerals formed during silicate weathering (such as clay minerals and oxy-hydroxides) is small (Sposito, 1989). Therefore, in temperate regions of Europe, such as the Alps, dissolution of silicate rocks produces weakly

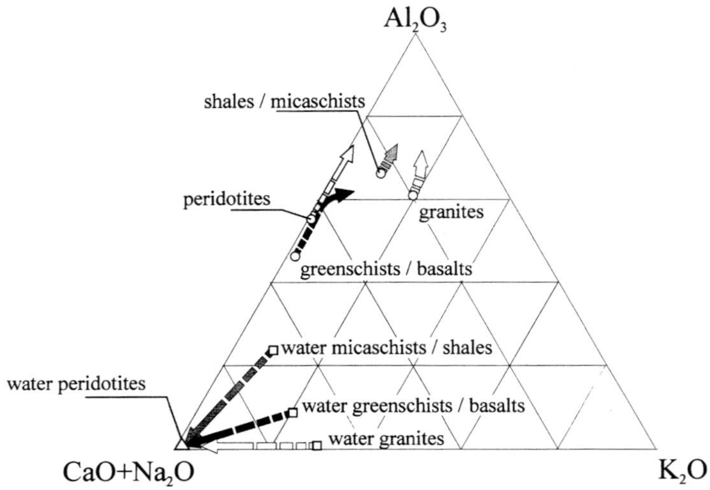

Figure 5. Typical differentiation of chemical elements between waters and soils occurring during weathering of silicate rocks (after Nesbitt and Young (1982), data from this paper): the open circles correspond to the initial rock compositions, the arrows pointing towards Al_2O_3 show the chemical evolution of the soil horizons lying on top of the parent rock (molecular proportions of oxides). They become more and more enriched in Al_2O_3. For each rock type, the typical evolution of water is shown in the left corner of the triangle: the squares correspond to the rain water and the tip of the arrows correspond to the spring waters (molecular proportions of cations), points in between correspond to the soil percolation water. They evolve towards proportionally Al-K-poor, but Ca-Na-rich compositions.

mineralized waters. Typical conductivities range from 50 to 150 μS/cm (corresponding to about the same order of magnitude of dissolved salts in mg/l, cf. Fig. 6A). pH values in the waters are often slightly below or close to neutral (Scheder et al., 1999).

However, conductivities of waters from carbonate terrains typically vary between 300 (limestones) and 2500 μS/cm (gypseous dolomitic rocks, in the Alps often rich in solid NaCl) and have water pH values between 7.5 and 8.5.

1.3.4. Rock-soil-plant relations

The chemical composition of the bedrock influences the above lying soil and plant cover in the same manner as in the rock–water system. In Central Europe, soils formed on silicate rocks show a typical progressive evolution from regosols through brown soils to podzols with soil pH values below 6.5. Such a typical series from the Alps was described by Huber (1997).

Soils formed on carbonate rocks should ideally contain only insoluble clay minerals of the rocks, some Fe–Mn-oxy-hydroxides and quite an important organic layer. All the other original constituents are removed by the water phase. Soil pH values typically lie between 7 and 8. As we will see below, in Switzerland, the soils in carbonate terrains are more complex (Section 2.6).

When dealing with the terms acid, neutral and alkaline/basic in the interdisciplinary

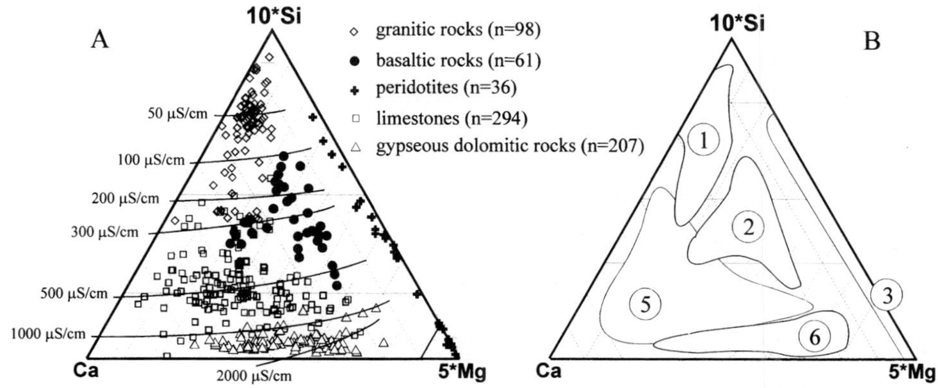

Figure 6. (A) Chemical discrimination diagram for major elements in spring waters formed in connection with common rock types (after Derron, 1999). The resemblance with the diagram for the rock composition of Fig. 3 is obvious. In contrast to the latter figure, sandstones and shale rocks (no. 1b and 4 of Table 1) are not distinguished from granitic rocks. The simplified version (B) of this diagram will be used below to show how rain water changes to soil percolation water and then to ground (spring) water. (1) Granites, (2) basaltic rocks, (3) peridotites, (5) limestones, (6) gypseous dolomitic rocks.

field of biogeochemistry, often some confusion comes up, especially because for rock compositions, these terms are used in a different manner. In soil science, the pH limits to define acid and neutral soils do not always correspond to the limits used in water chemistry. Table 2 presents current definitions for the four major compartments discussed in this paper.

Many plants grow on any kind of chemical soil type and mainly depend on relative humidity and amount of sunlight – xerophilous-hydrophilous (Baize and Jabiol, 1995). But some of them are more selective and prefer Ca-rich alkaline soils (calciphilous and calcicole plants, Table 2), other do not tolerate Ca-rich soils, soil percolation waters and/or prefer acid soils (calcifuge or acidophilous plants). For typical higher plants and mosses of Central Europe these preferences ("indicator values") are well known and often applied in ecological research (Landolt, 1977, 1992; Ellenberg et al., 1991; Delarze et al., 1998). Some plants have adapted themselves to special chemical conditions. Typical examples are the serpentine-type plant associations which have been adapted to high Mg-, Cr- and Ni-values with a high Ca deficiency (Brooks, 1987) and metal accumulator plants growing on weathered sulfide rich ore deposits and mine tailings (Ernst, 1988; Markert, 1993). In this paper we will present some selected plant analyses from situations rich in trace elements, but we will not deal with the complex physiological processes which are at the base of these phenomena.

1.3.5. Rock-alluvial sediment-aquatic moss relations

River and lake sediments and related trace elements have been the main target of mining exploration (Rose et al., 1979; Govet, 1983) and are among the earliest environmental compartments studied in interdisciplinary research (e.g. Salomons and

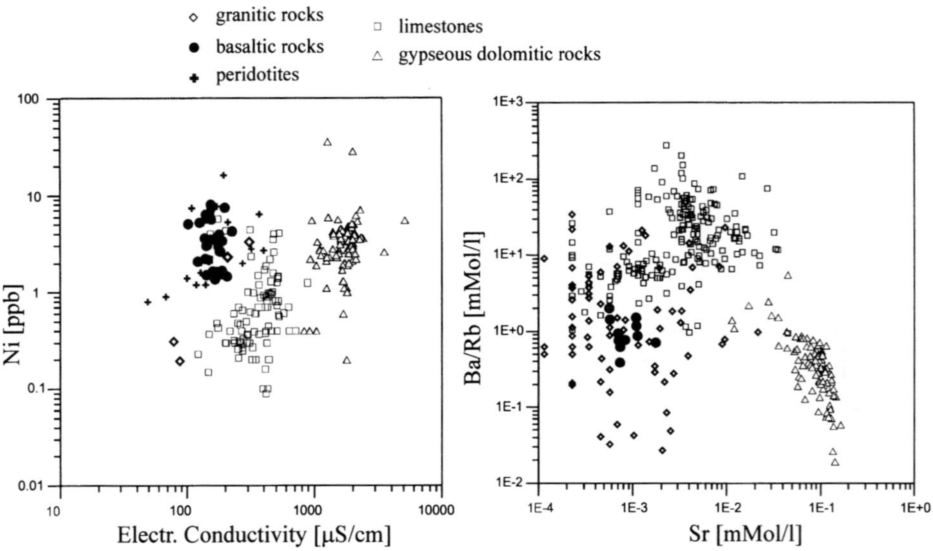

Figure 7. Variation of some trace elements in spring waters from common rock types (Derron, 1999). No simple discrimination diagrams exists for traces. The two diagrams shown here, involving Ni, Rb, Ba and Sr are more or less the only ones which discriminate the rock categories distinguished in Fig. 6.

Förstner, 1984; Förstner, 1988). Usually, the chemical compositions of these sediments are close to those of the regoliths and C-horizons in soil profiles of a given area. As in soils, trace elements can be contained in the mineral phases or adsorbed on them (Bourg, 1988). In this paper, we discuss river sediments in connection with naturally contaminated catchments (Section 3). In this field, a good knowledge of typical natural trace element ranges is particularly important in order to assess a possible anthropogenic contribution.

Aquatic mosses in rivers are often used as additional indicator of a possible pollution. In the same manner than the sediments, they represent a record of the average trace element content of the water (Mouvet, 1986). We present data from a naturally metal rich area and a limestone dominated catchment for comparison (Section 3).

2. Typical natural background situations

2.1. General aspects

Here we present, based on the chemical rock classification introduced in Section 1.3.2, typical cases of how major and trace elements contained in normal rocks are dispersed in the surrounding ecosystem. The different compartments studied are: (A) the solid parts of the soil column, (B) the meteoric water that percolates through the soil and is later collected as ground water, emerging in springs, and (C) plants growing in the soil, in cases where data are available (cf. Fig. 2). In the last chapter on contaminated

Table 2. Comparison of the terms acid and alkaline/basic used to describe different environmental compartments

Compartment	Extremely acid	Strongly acid	Moderately + weakly acid	Neutral	Moderately + weakly alkaline	Strongly alkaline	Extremely alkaline
Water[a]	pH < 3	pH 3–4.2	pH 4.3–6.7	pH 6.8–7.3	pH 7.4–8.3	pH 8.4–9.5	pH > 9.5
Soils[b]	pH < 4.5	pH 4.5–5.5	pH5.6–6.5	pH 6.6–7.3	pH 7.4–8.4	pH 8.5–9.0	–
Rocks[c]	$SiO_2 > 80$ wt.%	65–80 wt.%	Intermediate: 53–64 wt.%	Basic: 45–52 wt.%	Ultrabasic: <45 wt.%		
Plants[d]	Acidophilous + calcifuge (does not tolerate Ca)			Neutrophilous	Calciphilous (tolerate high Ca), calcicolous (Ca necessary)		Serpentinicolous

[a] Proposition in this paper (explanation: < 3: Al is the major element in mine waters, 4.3: inflection point when titrating the $H2CO_3$–HCO_3- system, 8.3: idem for the HCO_3–CO_3 system; for rain water, acid means pH < 5.6 (Appelo and Postma, 1993).

[b] Lozet and Mathieu (1990).

[c] Relative acidity definition based on the SiO_2 concentration in weight percentage (wt.%, g/l00g) (Carmichael et al., 1974).

[d] Brooks (1972).

situations, we complete the picture with information on the chemical dispersion in (D) glacial and alluvial sediments and (E) bryophytes (aquatic mosses). A good introduction to the way in which we look at this complex interactive system is given by De Caritat and Saether (1997) and in Chapter 2 of Appelo and Postma (1993). It can be summarized such as "How rain water becomes successively soil and ground water and the role of related rocks, solid soil phases and plants".

The chosen examples correspond to the different rock types presented in Table 1 and Figs. 3 and 4. The general geographic situation of the discussed examples is shown in Fig. 1, the exact location is given in Appendix A.9. The complete chemical information about the examples presented is also contained in the appendix and the synthetic sketches presented for each case study only contain a limited number of chemical elements. The typical major element tendency for the water and soil compositions for each example is illustrated on the triangular plot Ca–10$*$Si–5$*$Mg introduced previously in Fig. 6. Iron and the most common trace elements Cu, Zn and Pb appear in most figures and are completed by different typical trace elements for each rock type. This choice will allow to compare the background input from rocks and from diffuse input from the atmosphere. The soil description and related abbreviations follow the recommendations proposed by the Food and Agriculture Organization of the United Nations (1988).

Rock weathering and soil-forming processes are undoubtedly the result of complex biologically catalyzed chemical reactions between water, plants, microorganisms and rocks. These processes involve elemental transfer from the top of a soil profile to the base and vice versa, various organic and inorganic ligands and sorption processes (Sposito, 1989; McBride, 1994). In the following discussion, we will only focus on the result of these processes, i.e. the concentrations of a given element in the different parts of the studied ecosystem. Detailed discussions of the inferred soil forming processes can usually be found in the original reference given for each example.

2.2. Case study 1 for typical natural background values: granitic Si-rich silicate rocks from the Mont Blanc area (Argentière, France)

Fig. 8A,B shows the general geographic-geological context of this site, which is located at an altitude of 1950 m, north-west of Mont Blanc. The site is situated in a pasture area above the tree limit. It is characterized by a soil cover of about half a meter on local till material, formed by granitic gneisses (Atteia, 1992; Dalla Piazza, 1996). Fig. 8D shows the complete profile which has the typical characteristics of an evolved silicate-dominated soil (podzol with its characteristic "eluvial" ash-grey E-horizon). The same figure shows the pH values of the local rain, the soil, the percolation and local spring waters. Complete analyses can be found in Appendix A.1. To the right of the soil column, some of the chemical variation of the solid material and the percolation waters of the six different soil horizons and the parent rock (granitic gneiss) are portrayed.

Soil and soil water pH values are considerably acid and range from 3.7 to 4.9 (Fig. 8D). The granitic parent rock is typically rich in Si, K, Na and Al but poor in Fe–Mg and in most trace elements, except for F, Zr, Rb, Pb and Ba (Appendix A.1). This type of rock is usually among the richest in U (cf. Table 1), due to the presence of

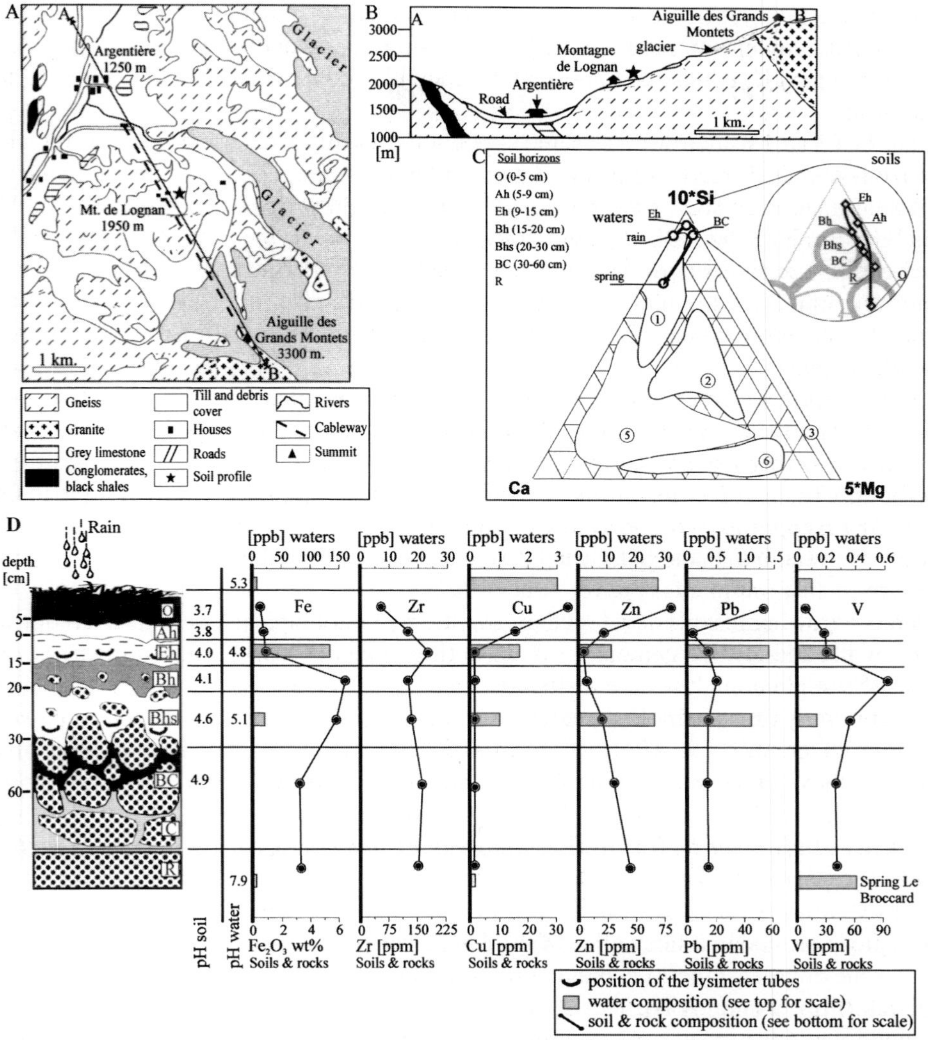

Figure 8. Case study 1 for background values: granitic Si-rich silicate rocks from the Mont Blanc area (Argentière, France). (A) Geographic-geologic situation of the soil profile studied. The studied spring is situated outside the area shown – Le Broccard, Martigny area, Valais, Switzerland (Dubois, 1992). (B) Vertical section through the area (location see line A–B in (A). (C) Major element changes of the related waters (rain, soil percolation and spring), soil horizons (O, Ah, E, Bh, Bhs, BC) and the parent rock R. Nos. 1–6: see Table 1. See (D) for details on the soil profile and trace elements. (D) Soil composition and waters on granitic parent rocks of the Mont Blanc area (case study 1), data from Atteia (1992); Dalla Piazza (1996). On the left, the soil profile with its different horizons is portrayed and the fresh parent rock is schematically indicated at the bottom. The changes in water composition throughout the soil profile have been monitored by sampling local rain water (see top row of the diagram) and putting a series of free flow lysimeter tubes on different soil levels. These data are then compared with the chemical signature of a typical spring, representative of the ground water of the area (bottom of the diagram). The general trend in soil composition and the parent rock is shown with circles linked by a solid line. The complete set of data is given in Appendix A.1.

pitchblende (UO_2), zircon ($ZrSiO_4$) and other U-bearing minerals in the rock. In this case, the U content of the parent material is low (3 ppm U), but the spring water contains 9 μg/l U, compared to normal waters which contain less than 1 μg/l (Appendix A.1).

Fe, Ti, Cr and V are low in the three upper horizons, reach a maximum in the two Bh horizons and decrease towards the bottom (rock substrate). Zr, contained in the very resistant mineral $ZrSiO_4$, shows little variation in the lower part of the section. Si, Al, Mg, Ca, Ba, Y, Sr and Rb increase evenly from the top to the bottom of the profile. However Cu, Zn and Pb decrease with soil depth (Fig. 8D). Since these elements are of little importance in the rock, this trend, as well as the one in the percolation water, indicates a predominance of the atmospheric contribution (for Pb at least in the uppermost 20 cm).

In the spring water, except for U, Sr, B and F, trace element concentrations are low, usually lower than in the local rain water, suggesting that these elements were retained in the soil. On the Ca–Si–Mg diagram (Fig. 8C) the soil horizons of the aquifer, from the bottom to the top of the profile, become progressively enriched in Si relative to Ca and Mg (except the two uppermost organic horizons). The waters typically become enriched in Ca, Na, F and SO_4. The latter is usually attributed to the oxidation of FeS_2 present in small amounts (Pfeifer et al., 1992).

Plants growing in soils developed on granitic rocks are typically acidophilous due to the low pH and low Ca content – <1 wt.% (Landolt, 1977; 1992). Unfortunately, for this site, no plant analyses are available.

In summary, this example demonstrates, that except for U and F, the trace element input from granitic rocks to the surrounding ecosystems is very low. A case of a considerable U-enrichment of an ecosystem is discussed in Section 3.3.

2.3. Case study 2 for typical background values: basaltic Ca–Mg–Fe-rich silicate rocks (greenstones) from the Val d'Hérens, Valais/Switzerland

This site is located in a side valley of the Rhone catchment, in an approximately 2-km-wide zone of greenstones (i.e. metamorphic gabbros and basalts) at an altitude of about 2000 m (Figs. 1 and 9A,B). This zone represents a 120 million-years-old ocean floor. About 50% of the area is covered by local till material, on top of which a half meter thick podzol developed (Derron, 1999). The soil profile lies in a pasture area close to the upper tree limit (Lac Bleu area) and comprises four different horizons (A, E, B and C; Fig. 9D).

In this case, the parent rocks contain about 25% less SiO_2 than granites, but considerably more Mg, Fe and Ca. Except Pb, its trace element content is higher than in granites. This is particularly the case for Ni, Cr, V, Cu and Zn (complete analyses in Appendix A.2). The major element composition of the waters produced by this system follows the same general trend as granitic waters, but spring waters are richer in Ca and Mg relative to Si and thus correspond to the general composition of waters expected from basaltic rocks (Fig. 9C, Appendix A.2). pH values of both soil material and soil waters are 1 to 2 units higher than for granites (Fig. 9D, left side).

The trace element composition in the soil profile is 2–3-fold higher in Fe, Ni, Cr, Cu and Sr, but only half of Rb, Ba and Pb, compared to granitic rocks (Fig. 9D, Appendix

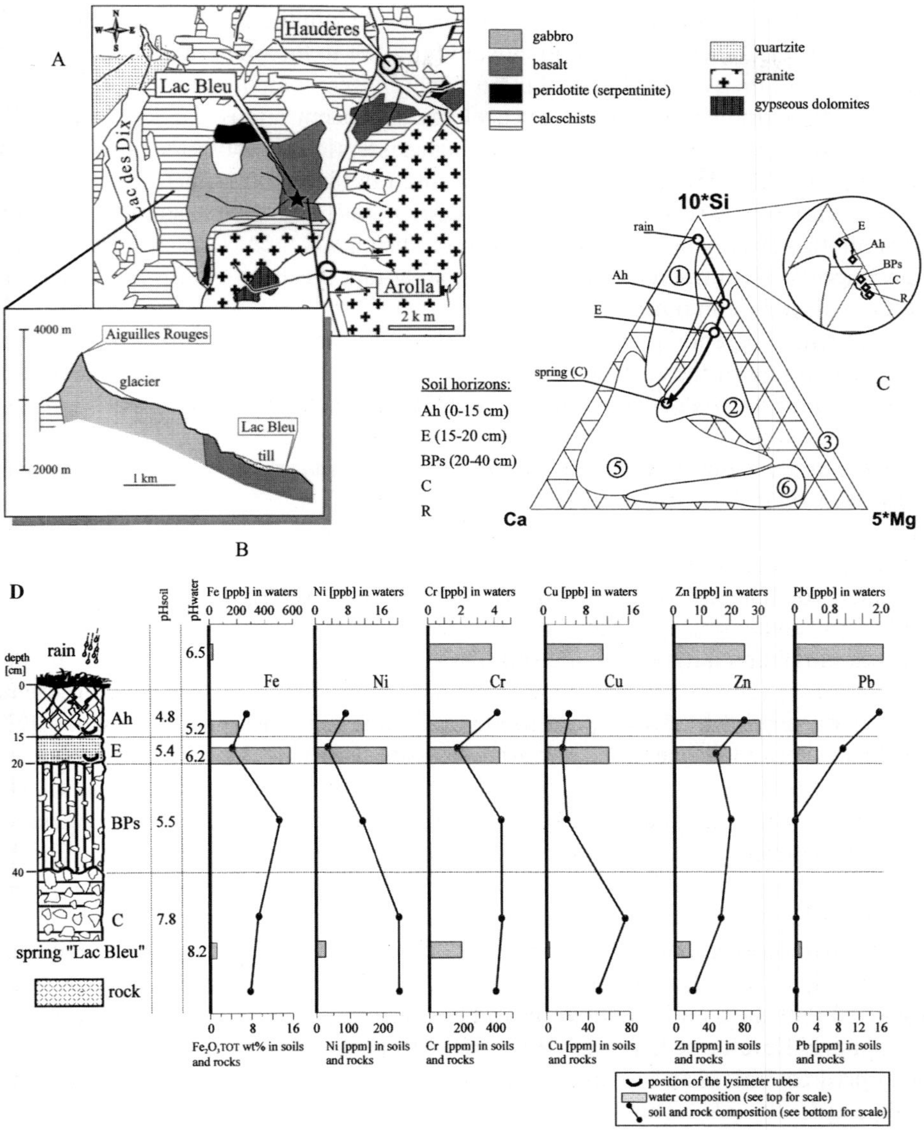

Figure 9. Case study 2: basaltic Ca–Mg-rich silicate rocks (greenstones) from the Val d'Hérens, Valais/Switzerland). (A) Geographic-geologic situation of the soil profile and the spring studied. Both sites lie just about 50 m apart. (B) Vertical section through the area. C: Major element changes of the related rain, soil percolation, spring waters and soil horizons (Ah, E, BPs, C) and the parent rock R. Nos. 1–6: see Table 1. (D) Soil composition and waters on basaltic rocks (case study 2). Data from Derron (1999). On the left, the soil profile with its different horizons is portrayed (Ah, E, BPs, C) and the fresh parent rock R is schematically indicated at the bottom. On the right, as in Fig. 8D, the changes in water composition throughout the soil profile from rain water down to the spring water are indicated with bars. The soil and rock composition is shown with circles linked by a solid line. The complete set of data is given in Appendix A.2.

A.2) while Zn and Al are about equal. However, except for Ni and Fe, concentrations of soil percolation waters are only slightly higher than in the local rain water. For Zn the situation found here is almost identical to the one in the granitic system discussed above. For Pb, the rock-soil contribution is negligible compared to the atmospheric input. In the spring water, most trace element concentrations are low due to the relatively high pH of 8.2 which enhances their adsorption on solids. Sr is, as expected for relatively Ca-rich rocks (12 wt.% CaO), high in the springwater.

Plants growing on soils influenced by basaltic rocks have often calciphilous tendencies, although usually no carbonate is present. The only weakly acid soil water and the abundance of calcium seems to be sufficient to attract this type of vegetation (Buszko and Giugni, pers. commun.).

In summary, depending on the elements, the trace metal input of basaltic rocks can be important, but at the same time relatively high Ca contents and pH values create chemical conditions resembling those found in carbonate dominated areas.

2.4. Case study 3 for typical background values: peridotitic Mg–Fe-rich silicate rocks from the Canavese area (Baldissero, Italy)

This site is located in the southern Alps at low altitude (400 m) on a prominent elongated south-west–north-east directed hill (Monte Pelati), west of the village of Baldissero-Canavese (Figs. 1 and 10A,B). The local climate is somewhat Mediterranean with high precipitations in the spring and the fall and a dry period in the summer and the winter. The main underlying rocks are peridotites, pyroxenites and serpentinites, characterized by Si–Ca-poor, but Mg–Fe-rich compositions (Schlegel, 1997). Mg-carbonate rocks (magnesite) occur in the southern part. Locally veins of amorphous silica (opale) are associated to the latter. The dominant soil of the area is a Mg-rich regosol with a maximum thickness of 20 cm. This only allows limited vegetation due to the deficiency in Ca (Brooks, 1987) and the relative dryness of the area. For this site, analytical results of several plants are available (Birch/*Betula pendula*, Ling Heather/*Calluna vulgaris*, Appendix A.3).

In the parent rock, only Co, Zn, V and particularly Cr and Ni are important trace elements (up to several thousand ppm for the latter two). The other trace elements are below the detection limit of most instruments (bottom of Fig. 10D, Appendix A.3). The soil profile described on Fig. 10D is only 10-cm-thick and lies on peridotites. The soil is acid (pH between 5.7 and 5.9), but the soil percolation and spring waters are alkaline (pH 7.6–7.9). This is a general tendency in soils developed on peridotitic rocks. In a comparable situation on serpentinites, but in a colder climate, Juchler (1988) describes percolation waters whose pH increased from 6.1 in the O-horizon to 7.2 in the BC-horizon and 8.1 in local springs. Such alkaline waters, buffered by various hydrous Mg-carbonates are typical for ground waters from Mg-rich rocks all over the world (Pfeifer, 1977).

As expected, these waters are rich in Mg, Ni and V. This is especially true for the soil percolation waters (Fig. 10C,D, Appendix A.3). However, their Cr-content is not much higher than in basaltic rocks.

The trace element concentrations of the soils forming on the rocks described here are two to three times richer in Zn and ten to 15 times richer in Ni, Cr and Co than in

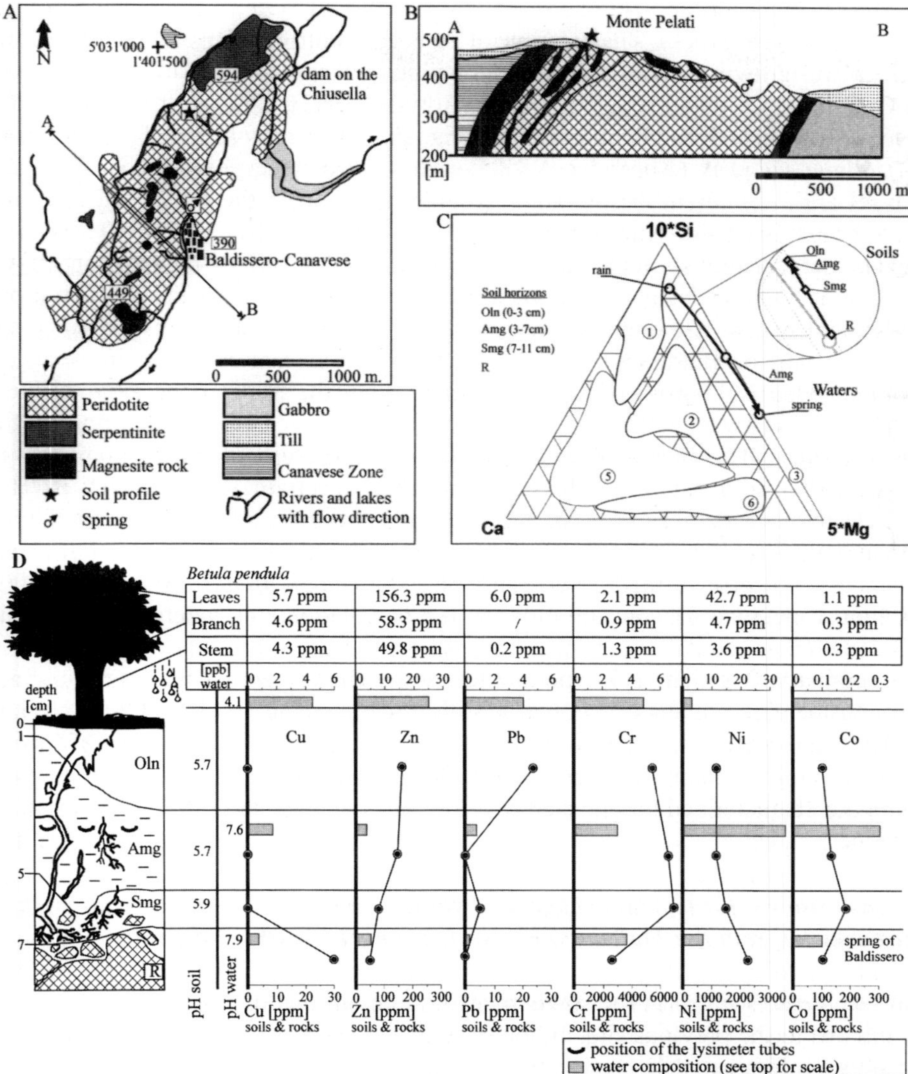

Figure 10. Case study 3 for typical background values: peridotitic Mg–Fe-rich silicate rocks from the Canavese area (Baldissero, Italy). (A) Geographic-geologic situation of the soil profile and the spring studied. (B) Section through the area showing rocks lying under the till material. (C) Major element changes of the related waters (rain, soil percolation and spring), soil horizons (Oln, Amg, Smg) and the parent rock R. Nos. 1–6: see Table 1. (D) Soil, plant and water compositions on peridotite rocks (case study 3). Data from Schlegel (1997). On the left, the soil profile with its different horizons (Oln, Amg, Smg, R) is portrayed. The trace elements of a young birch tree are shown. On the right, the changes in water composition throughout the soil profile from rain water down to the spring water are indicated with bars. The soil and rock compositions are shown with circles linked by a solid line. The complete data set is given in Appendix A.3.

those typical for basaltic rocks (Fig. 10D). Overall, this example and peridotitic rocks in general, represent the natural soil system with the highest trace element contents known for common rocks. The presence of Pb is again attributed to atmospheric input.

The plants growing on this soil are acidophilous and contain three to ten times more Zn, Cr, Ni and Co than average in plants (Adriano, 1992). This is especially true for the leaves of the analyzed *Betula pendula* (Fig. 10D, Appendix A.3) and the flowers of the *Calluna vulgaris* (Appendix A.3). For these two widespread plants only few analyses are available in the literature. In order to compare the results with Ni- and Cr-poor situations, the same plants were analyzed in the Malcantone area, where granites and micaschists dominate (Sections 2.5 and 3.2). This comparison allows some interesting conclusions about the relation between trace elements concentrations in soils and plants. Although Ni concentrations are up to 1000 times higher at the peridotitic site of Baldissero, the plant concentrations are just ten times higher (*Calluna vulgaris* 15 mg/kg Ni) than in those from the Malcantone area (*Calluna* 2.7 mg/kg Ni). However, there is no proportionality between Ni concentrations of the soils and plants. This indicates that the analyzed plants are not hyperaccumulators of Ni, but rather stop taking up this element when a certain level is reached. This certainly prevents damage. For Cr, although the peridotitic soil contains much more Cr than in the Malcantone area, plants from the peridotitic site contain about the same amount of Cr (*Calluna* 2 mg/kg Cr). This is in accordance with the low solubility of Cr and almost identical Cr concentrations in the percolation soil waters (cf. Appendix A.7).

In summary, Mg-rich rocks (peridotites, serpentinites) enrich their surroundings with considerable amounts of rather specific trace elements such as Cr, Ni, V and Co. These elevated concentrations of trace elements, together with the lack of Ca and the often very surficial water circulation, lead to restricted plant growth. Ground waters are characterized by elevated Mg- and Ni- contents and often relatively high pH values.

2.5. Case study 4 for typical background values: shale-type Al-rich silicate rocks (mica schists) from the Malcantone Area (Astano, Ticino/Switzerland)

This site is also located in the southern Alps at an altitude of 500 m. The climate is comparable to the previous site, although summers are less dry (Figs. 1 and 11A,B). The area is dominated by relatively Al- and Fe-rich silicate rocks (gneisses and micaschists) and almost completely covered by till material whose origin is not necessarily local, but can be regional (catchment of the Lisora and Magliasina rivers, cf. Fig. 14). Soil depths vary from 50 to 200 cm and are often of podzol or crypto-podzol type, especially in chestnut forest sites (Blaser, 1973; Richard and Lüscher, 1987). The locality described here is in a forested area, whose age is not known. It was perhaps a vineyard some 40 years ago (Rey, 1996).

The solid bedrock of the area has the typical chemical characteristics of a shale-type rock (cf. Table 1) and is rich in less common traces such as As, Cd, Pb, Zn and Sb related to sulfide minerals (bottom of Fig. 11D, Appendix A.4). The As content of the rock included in the Appendix A.4 of over 900 ppm seems high (Pfeifer et al. 1996), but is not unusual for this region. In addition, in the whole area, numerous natural

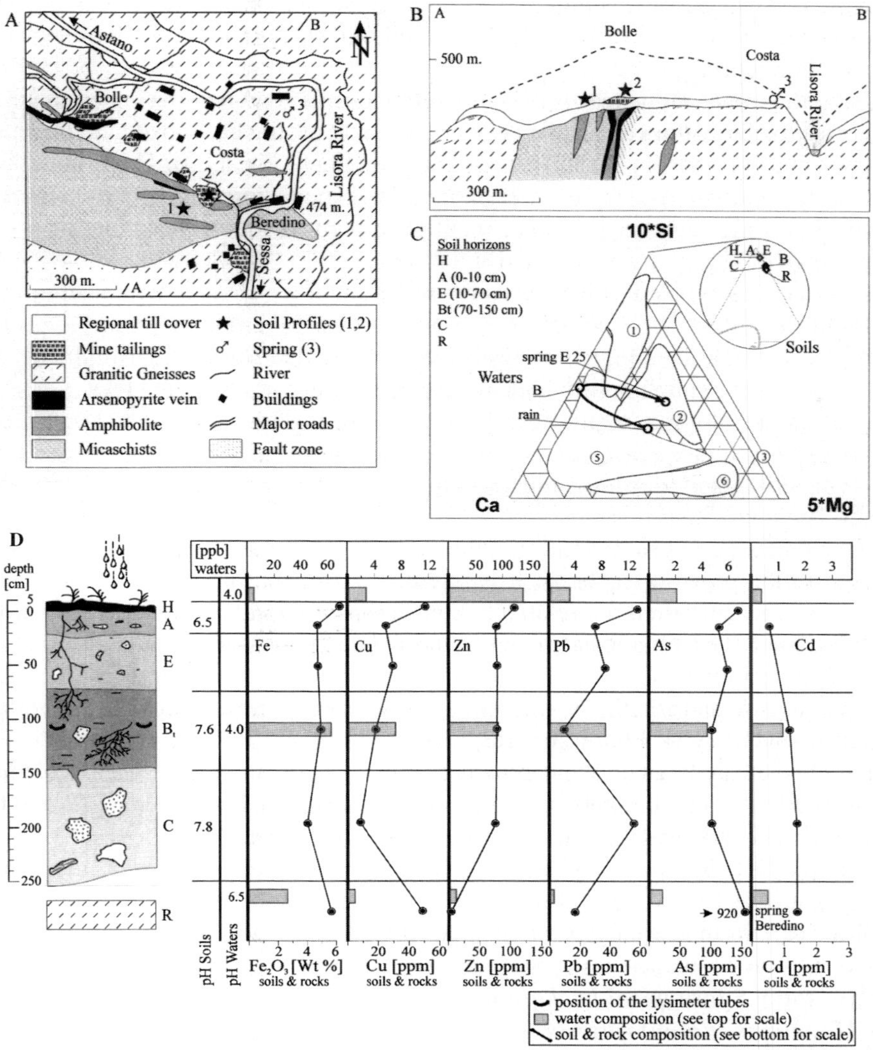

Figure 11. Case study 4 for typical background values: shale-type silicate rocks (mica schists) from the Malcantone area – Astano, Ticino/Switzerland (Rey, 1996). (A) Geographic-geologic situation of the soil profile (1) background situation, (2) on the arsenopyrite vein material and mine tailings) and the studied spring (3). Almost the whole area is covered by till material, the map shows the reconstructed distribution of the rocks underneath, based on detailed geological mapping. (B) Vertical profile through the area. (C) Major element changes of the related waters (rain, soil percolation and spring), soil horizons (H, A, E, Bt and C) and the parent rock R. Nos. 1–6: see Table 1. (D) Soil and water compositions on shale rocks (micaschists, case study 4). Data from Rey (1996); Bondietti et al. (1994). On the left, the parent rock and the soil profile with its different horizons (H, A, E, Bt and C) is portrayed. On the right, the changes in water composition throughout the soil profile from rain water down to the spring water are indicated with bars. The soil and rock compositions are shown with circles linked by a solid line. The complete set of data is given in Appendix A.4.

sulfide ore deposits occur (Section 3.2). One of the largest arsenopyrite-dominated ore deposit is located nearby (Fig. 11A). The environmental impact of this deposit is described in Section 3.

Outside the direct influence zone of the ore deposits, regional glacial till material often forms the basis on which soils develop. The soil profile described here in detail, comprises five horizons (H–C) and is more than 2 m thick (Fig. 11D). The E-horizon is only weakly bleached, as typical for cryptopodzols. The C-horizon is very clayey and corresponds to the regional till cover mentioned above. The soil percolation water profile shown in Fig. 11D was sampled at a distance of 200 m, where the whole profile was less deep and evolved. It has the same major and trace element composition, but shows soil pH values between 4 and 6, instead of 6.5 to almost 8, as in the profile shown in Fig. 11D. This higher soil pH seems to be due to the slight influence of nearby outcropping amphibolitic rocks of basaltic composition (Fig. 11A).

The different soil horizons are rich in arsenic (100–150 ppm), Cd (0.5–2 ppm), and Pb (20–60 ppm) and reach values which menace soil fertility (VBBo, 1998). Other traces typically linked to sulfide minerals, are not above average: Zn, Cu and Sb (Fig. 11D and Appendix A.4).

Trace element compositions in water remain within the limits reported for the soils discussed in the previous sections. It is interesting to note that in this natural case, heavy metal concentrations are relatively high compared to the arsenic concentrations (which are low). The situation is reversed in the heavily contaminated soil profile as discussed in Section 3.2 (cf. Fig. 16). We tie this to the different water pH values, acid in the natural case and alkaline in the contaminated case, which change the adsorption behavior (Bowell, 1994; Pfeifer and Rey, 1998; Halter and Pfeifer, 1999).

Some selected analyses of plants growing in this type of ecosystem, but sampled on a site situated 10 km to the south are given in Appendix A.4. Their average As and Cd contents are 1.5 and 1 mg/kg, respectively (Nawratil, 1998).

In summary, shale type rocks have a tendency to add a whole series of metals and metalloids such as Cd, Pb, Zn and As to the ecosystem, which approach concentrations typical for certain industrially contaminated sites.

2.6. Case study 5 for typical background values: limestones from Vallée de Joux (Jura Mountains, Vaud/Switzerland)

This site is located at an altitude of 1300 m in the eastern part of the southern Jura chain (Figs. 1 and 12A,B). It is characterized by a relatively humid climate (precipitation: around 180 cm/a) and partly covered with conifer forests. The underlying rock is a white marine limestone of the upper Jurassic period. As usual in a limestone area, there is almost no surficial water circulation. The high solubility of the carbonates is at the origin of widespread fissures and cavities, which favor drainage. Such "karst" systems give locally rise to springs with large discharge rates such as "la Lionne", in our case (Fig. 12A).

The soil profile (Fig. 12D) is situated under pasture, on top of a typical south-west–north-west oriented elongated hill. It is a brown soil, as commonly found in the Jura chain. The profile is composed of a thin organic upper horizon (less than 10 cm depth, 8 mg/kg C) and a deeper yellow-brown silty horizon. Detailed investigations showed

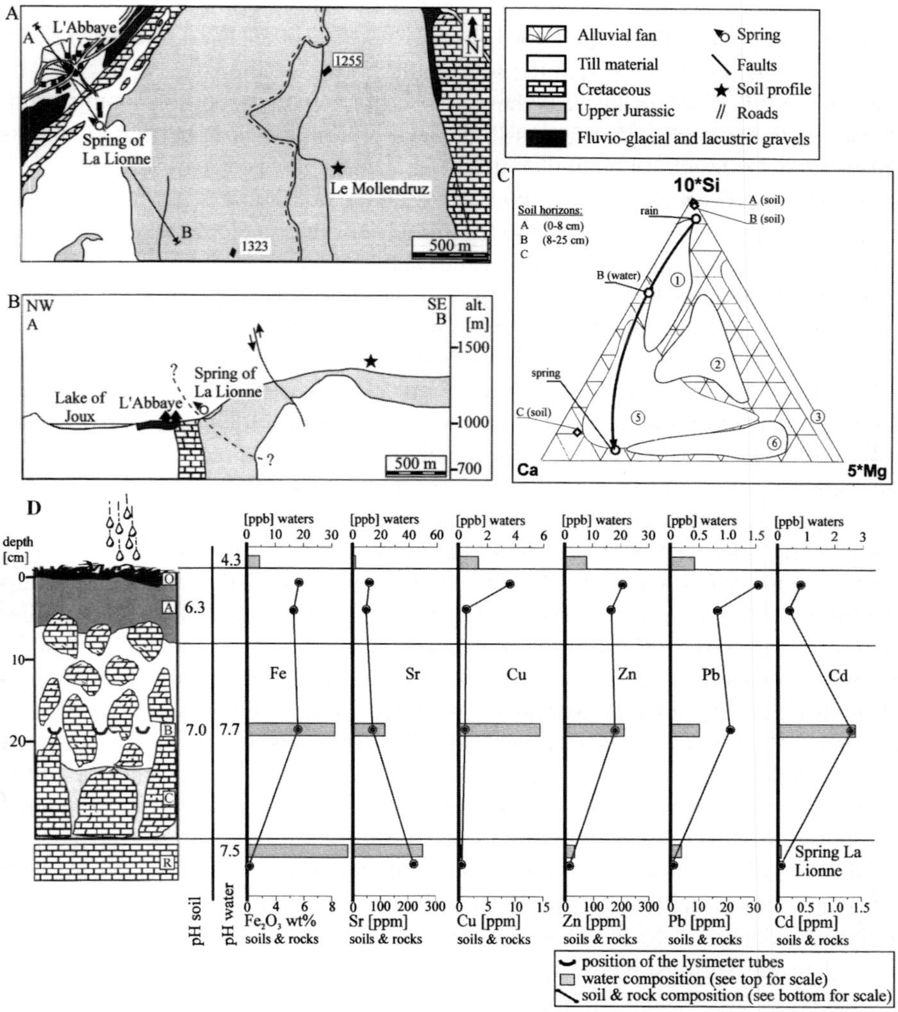

Figure 12. Case study 5 for typical background values: limestones from Vallée de Joux (Mollendruz, Jura mountains, Vaud/Switzerland). (A) Geographic-geologic situation of the soil profile and the spring (La Lionne, E of the village L'Abbaye). (B) Vertical profile through the area. (C) Major element changes of the related waters (rain, soil percolation and spring), soil horizons (O, A, B and C) and the rock substrate R. Nos. 1–6: see Table 1. (D) Soil and water compositions on limestones (case study 5). Data from Atteia (1992); Dalla Piazza (1996). On the left, the rock substrate R and the soil profile with its different horizons is portrayed (O, A, B and C). On the right, the changes in water composition throughout the soil profile from rain water down to the spring water are indicated with bars. The soil and rock compositions are shown with circles linked by a solid line. The complete set of data is given in Appendix A.5.

an important discontinuity of mineralogical composition and particle size distribution between the horizon in contact with the calcareous bedrock and the insoluble residue of the bedrock itself. On one hand, the soil was enriched with typical silicate minerals of the Alps, such as chlorite and feldspars and which are not present in the residue of the decarbonatation of the calcareous bedrocks. On the other hand, the large proportion of coarse material, present in the soil, could not be attributed completely to rock dissolution. Since Pochon (1978), these two criteria have been commonly used to demonstrate the presence of silicate till material of alpine origin, deriving from post-glacial eolian deposits and characteristic of externally derived material in soils. The data presented in Fig 12D and in Appendix A.5 were taken from Atteia (1992); Dalla Piazza (1996), except those relating to Cd and U. These two elements were measured in a similar profile, located in a north-eastern site (Atteia et al., 1995). Trace elements such as Zr, Rb, V and Pb, which are typical of granitic parent materials, predominate in accordance with the soil mineralogy. With the exception of Sr, the trace element concentrations are larger in soil than in the underlying rock. The important Cd concentrations in soil (ranging typically from 0.5 to 5 ppm) was previously attributed to anthropogenic input (Atteia et al., 1995). A new more extended study carried out by Dubois et al. (1998) showed that rock weathering was the main Cd source in soil. Moreover, the authors confirmed that most of the Cd was bound to carbonate (in the lattice of calcite minerals) and did not come from eolian deposits.

Surbeck and Medici (1990) reported high radon (^{222}Rn) emissions from the underground in many places in the Jura area. In cavities up to 40 kBq/m^3 and in houses up to 5 kBq/m^3 have been detected, compared the average of 0.08 kBq/m^3 measured in houses in Switzerland. These emissions can be attributed to the high U and Th concentrations which were found in the limestones and marls of the Jura chain – 5 to 20 ppm U, 5–30 ppm Th, Appendix 5 – (Atteia et al., 1995; Hamel, 1998).

The trend of major elements in water and soil is illustrated in Fig. 12C. As expected, the spring waters are the most Ca-enriched waters of all the yet presented examples. Moreover, the two upper horizons (A and B), rich in Si, are very different from the Ca-rich C-horizon. Most of the elements present in soil solution (Fig. 12D, Appendix A.5) were dissolved in the upper soil horizon but, due to an increase of pH, their concentrations in solution decrease strongly as soon as the water gets in contact with the underlying calcareous rock. That explains the low trace element concentrations, measured in the springs. These decreases in concentration can be attributed to the precipitation of Al, Fe and Mn as oxy-hydroxide with which trace elements are co-precipitated or adsorbed. Some elements, such as Cd, Pb and Ni can also precipitate as carbonate or be adsorbed on calcite.

No plant analyses were available for the site. However, another study, not published yet, was carried out in a similar site of the Jura. The Cd contents in plants varied from 0.5 to 5 mg/kg and were slightly correlated with the exchangeable Cd (extracted with 0.1 M $NaNO_3$). However, no relationship was observed between the Cd content in plants and the Cd concentration in soils (extracted with 2.0 M HNO_3).

In summary, due to a relatively large proportion of externally derived material in soil, the trace element concentrations in the soil of the Swiss Jura are only partially representative for limestone-dominated systems. Nevertheless, many soils of the Jura mountains are rich in Cd of natural origin. It was demonstrated that this Cd derives from calcite dissolution and not from weathering of the insoluble clayey fraction of the rock. Moreover, unexpectedly high amounts of radon are released in this region.

2.7. Case study 6: dolomitic-sulfatic carbonate rocks from different areas

In the Alps, another type of carbonate rock is rather abundant: rocks containing large amounts of dolomite, i.e. $CaMg(CO_3)_2$ (Mandia, 1991; Kündig, 1997). It is often accompanied by Ca-sulfate (gypsum and anhydrite), rock salt (NaCl), Ca-fluoride and various sulfide-type minerals rich in Fe, Zn, Pb and As (Escher, 1933; Hein and Schneider, 1983; Schmidt and Amstutz, 1983; Schroll, 1983, Knill, 1996). These rocks belong to the Triassic period and correspond to marine sediments formed in a warm laguna some 230 million years ago (Amstutz and Fontboté, 1985).

Soils developed on such terrains are often thin, rich in organic material and acid. Nutrients are often leached out, restricting plant growth (Landolt, 1992). However, waters emerging from these parent material belong to the most mineralized ones. Their total salt content typically varies between 300 and 2500 mg/l (surface, respectively springwaters) and they are dominated by Ca–Mg and sulfate (cf. Fig. 6). Their pH ranges between 7 and 8.5.

Many bottled mineral waters from Switzerland stem from these aquifers (Högl, 1980). The most striking feature of these waters is their relatively high trace element content, such as Zn, Rb, Mn, Li, Cu and sometimes F and As. These trace concentrations are comparable to waters found in shales (Section 2.5). Examples of spring and river waters of this type are described by Mandia (1991), Scheder and Streiff (1997) and Pastorelli et al. (1999), however without detailed informations about the related soils. Appendix A.6 contains some typical water compositions related to gypseous dolomites. The high trace element concentrations in spring waters are possibly due to the fact that large amount of carbonate material are dissolved and small quantities of traces contained in the accessory clay and sulfide fraction are amplified.

A very interesting and unique rock-soil-water-plant system in gypseous dolomite rocks is situated in the Val Piora of the Gotthard area in Central Switzerland (Fig. 1). A prominent Mg- and sulfate-rich carbonate zone with a thickness of about 1.2 km forms the basement of an alpine valley lying at an altitude between 1800 and 2200 m, at the tree limit. Among the 20 small lakes of the area, three lie directly on the carbonate zone together with an adjacent bog area: Lake Cadagno, Lake Ritom and Lake Tom. Especially the 20 m deep Cadagno lake has a rather unique natural stratification ("meromixis"): a lower H_2S-dominated anoxic water layer is separated by a thin layer of red light-dependant anaerobic sulfur bacteria from the upper SO_4-rich and oxygenated water layer (Peduzzi et al., 1998). The larger Ritom lake had the same vertical stratification up to 1915, when a dam was built and the water was used to produce electricity. The anoxic lake sediments of Lake Cadagno contain relatively high amounts of Zn and Cd (400 and 2.5 mg/kg, respectively). The uppermost part seems to anthropogenic (Birch et al., 1996), but elevated trace element concentrations in the lower part of the sedimentary column appear to be natural (Zn, 100mg/kg; Cd, 0.5 mg/kg). In the gas phase above the bog areas rare gases such as trimethylated arsenic and antimony species and even elemental mercury have been detected (Wickenheiser et al., 1998).

Plants growing in the dolomite-rich soil are mainly calciphilous, but no trace element analyses are available. All this phenomena are clearly the result of a unique water-rock-microorganism-plant-interaction system with the Mg-carbonates as parent material. Its particularity are the surprisingly high trace element content and the presence of sulfur which often leads to reducing conditions.

3. Examples of naturally contaminated situations

3.1. Introduction

This chapter is devoted to situations in which trace elements are found as natural contaminants, in concentrations which considerably exceed natural background values discussed above. Such situations are typical for areas rich in metal ore deposits. Their potential environmental impact depends on their location within a catchment. If they are situated in the upper part, a large area and many different environmental compartments will be affected (Fig. 13). Local contamination of soils, plants, sediments and water occurs almost in every case, the degree to which a whole catchment is affected by contaminated waters and glacial or alluvial sediments depends on the size of the ore deposit, the slope and the climate of a given area.

In the following section, we present two cases of metal contamination from the Swiss Alps, for which relatively complete data for soils, plants, water and sediments exist. Two elements are considered: As and U. Both are quite soluble in water. Arsenic occurs in surface and shallow ground waters either as the anionic species AsO_4^{3-} (As V, arsenate) and more rarely as AsO_3^{3-} (As III, arsenite). Due to its anionic nature, it is sorbed under acid conditions (pH $<$ 7) and liberated under alkaline conditions

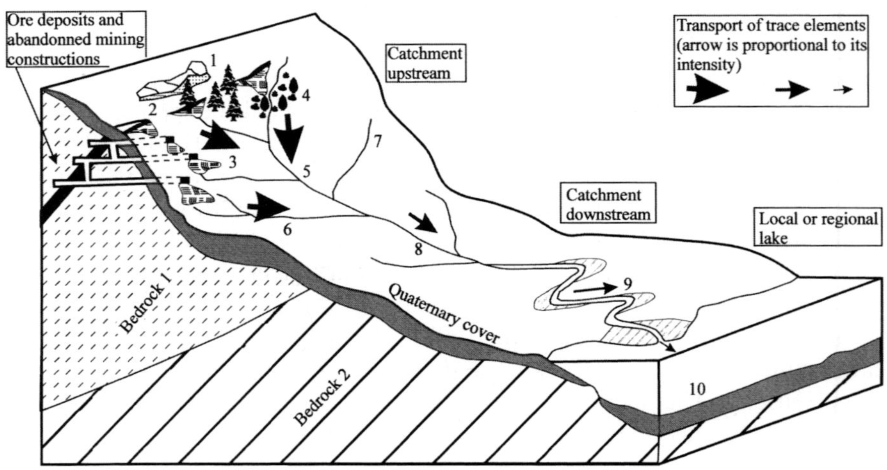

Figure 13. Overview of contamination processes induced by an ore deposit in a river catchment of a mountainous area. The different places potentially affected by the dispersed chemical elements are indicated with numbers: (1) natural outcrop, (2) open pit, (3) mine dump, (4) local vegetation, (5) local surface water, (6) river draining the area of the ore deposit, (7) uncontaminated control area, (8) major river water, possibly carrying trace elements from the ore deposit, (9) river sediment and water plants (e.g. mosses), possibly enriched in trace elements, (10) lake sediments, possibly enriched in trace elements.

(Bowell, 1994; Halter and Pfeifer, 1998). Uranium is very soluble under oxidizing conditions and occurs as UO_2^{2+} in water (Hsi and Langmuir, 1985).

3.2. Case study 1 for contaminated situations: arsenic contamination in the Malcantone area, Southern Switzerland

The Malcantone area is located in the Southern Alps at an altitude varying from 300 to 1600 m. It comprises two separate catchments: the Magliasina and the Lisora basin (Fig. 14). The geological substrate is formed by various granitic and Al-rich silicate

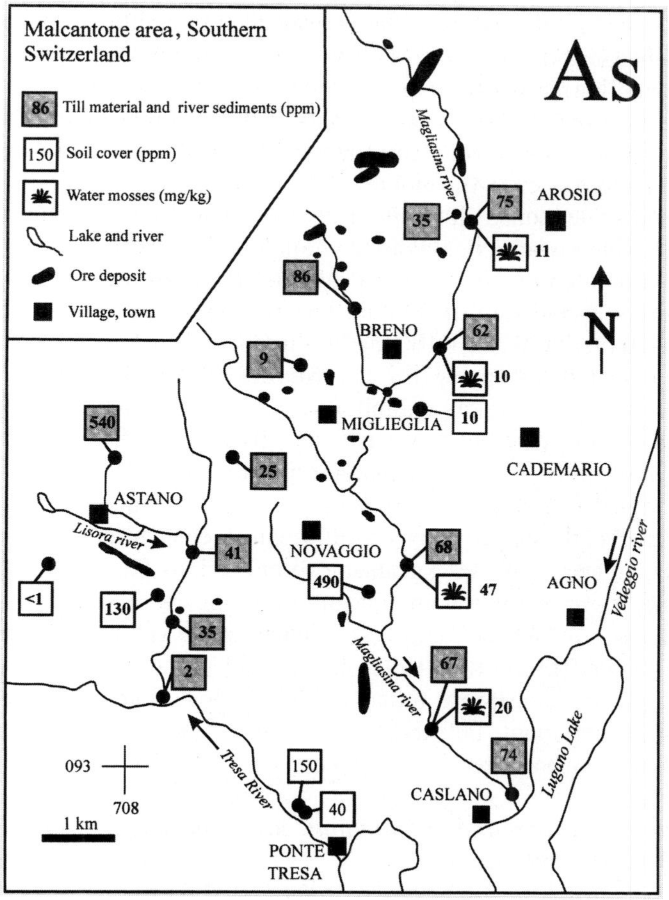

Figure 14. Case study 1: contaminated situations. Geographic distribution of arsenic contents in different environmental compartments of the Malcantone area, Ticino /Switzerland: soils, river sediments and aquatic mosses. Data from this study, State Laboratory, Canton Ticino, and Martini and Pillet (1996).The whole area is naturally contaminated with arsenic originating from numerous smaller and larger ore deposits dominated by Fe, As, Zn, Pb and Sb (indicated with black spots). The general background of As in the till material forming the aquifers varies between 50 and 150 ppm, but locally values up to 500 ppm occur.

rocks (gneisses and micaschists). They contain abundant small metal and metalloid sulfide ore deposits dominated by Fe, As, Pb, Zn, Cd and Sb and were partly mined for gold (Köppel, 1964; Wenger et al., 1990; Martini and Pillet, 1996). The recent discovery of unexpectedly elevated As contents in drinking waters of several villages are at the origin of a region-wide environmental monitoring campaign (Figs. 14 and 15).

Fig. 16A shows the typical situation in the immediate vicinity of an ore deposit (cf. Fig. 11A,B) (Rey, 1996): a soil cover which contains the ore material releases waters considerably enriched in As, SO_4 and Sr to the environment (As, up to 80 μg/l; SO_4, 120 mg/l). The ore material contains the minerals Ca–Mg-carbonate, arsenopyrite, zinc blende and galena. Soil percolation and local seepage waters exhibit the typical major element composition of carbonate rocks (Fig. 13B). However, as mentioned already in Section 2.5, in contrast to the site outside the ore deposit, metal traces are relatively low because of the high pH (Fig. 16A, Appendix A.7). Plants (birch, ling heather and dandelion) growing on this substrate are at least two to ten times enriched in Zn, Cd and As compared to the same plants from the Canavese region (cf. Section 2.4 and Appendix A.3) and mean plant values (Adriano, 1992). The highest values are found in leaves and flowers (Appendix A.7).

On a regional scale, soils, till material and river sediments exhibit trace element contents comparable to the example discussed in Section 2.5, but have up to 100 times higher values compared to the typical worldwide As-background of 7 ppm (Figs. 14 and 17). Aquatic mosses from the Magliasina river show the same trend as the river sediments (10–20 mg/kg As; 120–180 mg/kg Zn), that is much more than mosses from a carbonate dominated catchment (Fig.17) (Martini and Pillet, 1996).

The surface and spring waters exhibit also a strong regional anomaly in arsenic: typical surface waters contain between 6 and 9 μg/l As. Spring waters, used as drinking water in most villages, contain typically between 20 and 80 μg/l As (WHO and US-EPA limiting value, 10 μg/l As). Considering that the average discharge of the Magliasina river is 1.5 m^3/s with a concentration of 9 μg/l As, an average annual quantity of 400 kg arsenic brought into the Lugano Lake can be calculated. The origin of these high concentrations clearly is the till material which forms the major aquifers of the region. The highest value yet measured in glacial tills was 540 ppm As (Pfeifer and Rey, 1998). In one place, erratic blocks containing large amount of sulfide minerals were found (Köppel, 1964), indicating that during the glacial period an important, but not preserved ore deposit must have been eroded and dispersed in the whole area.

3.3. Case study 2 for contaminated situations: uranium contamination in the Rhone catchment (Western Switzerland)

The upper part of the Rhone catchment in southwestern Switzerland contains abundant U-mineralizations. They are at the origin of an important flux of dissolved uranium towards the Lake of Geneva (around 13 ton/year) (Dominik et al., 1992). Some complementary studies were carried out by Woodtli et al. (1985) on the sediment contents, by Baertschi and Keil (1992) on the surface and ground waters and by Pfeifer et al. (1994) on soil and plant contamination in the Martigny area. These studies allow to reconstruct the natural dispersion process of uranium in some detail (Figs. 18 and 19).

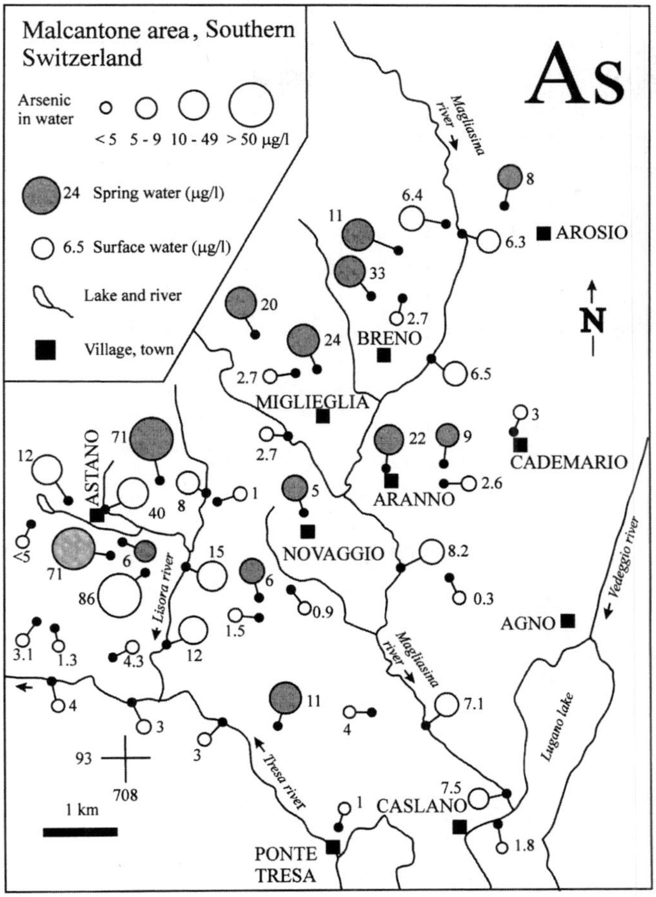

Figure 15. Malcantone area, Ticino /Switzerland: arsenic contents of spring and river waters. Data from this study, State Laboratory Canton Ticino, and Martini and Pillet (1996). Arsenic in waters ranges from 6 µg/l (river water, Lake of Lugano) up to 80 µg/l (Astano area, in part clearly outside the direct influence area of the arsenopyrite vein, cf. Fig. 12).

The example discussed here is situated on a steep slope between the villages of Les Marécottes and La Creusaz above Martigny. The area is dominated by granitic rocks, in part covered by weathered rock debris and till material (Fig. 19). Locally, UO_2-bearing (pitchblende) zones of few centimeters in width and several meters in length occur in cliffs. Weathered and eroded material from these zones is deposited below the cliffs and slowly moves downhill by gravity sliding and pushed by occasional avalanches. This leads locally to U accumulation in soils and plants, such as grass, blueberry and moss (up to 2500 ppm U in the soil and 3–100 mg/kg in the plants, Appendix A.8). Outside the direct weathering and erosion zone of the U-rich rock material, uranium contents of soils correspond to classical granitic values (around 3 ppm; see Section 2.2). No local soil percolation water data are available for this specific site, but the overall regional effect of these processes on the water composition

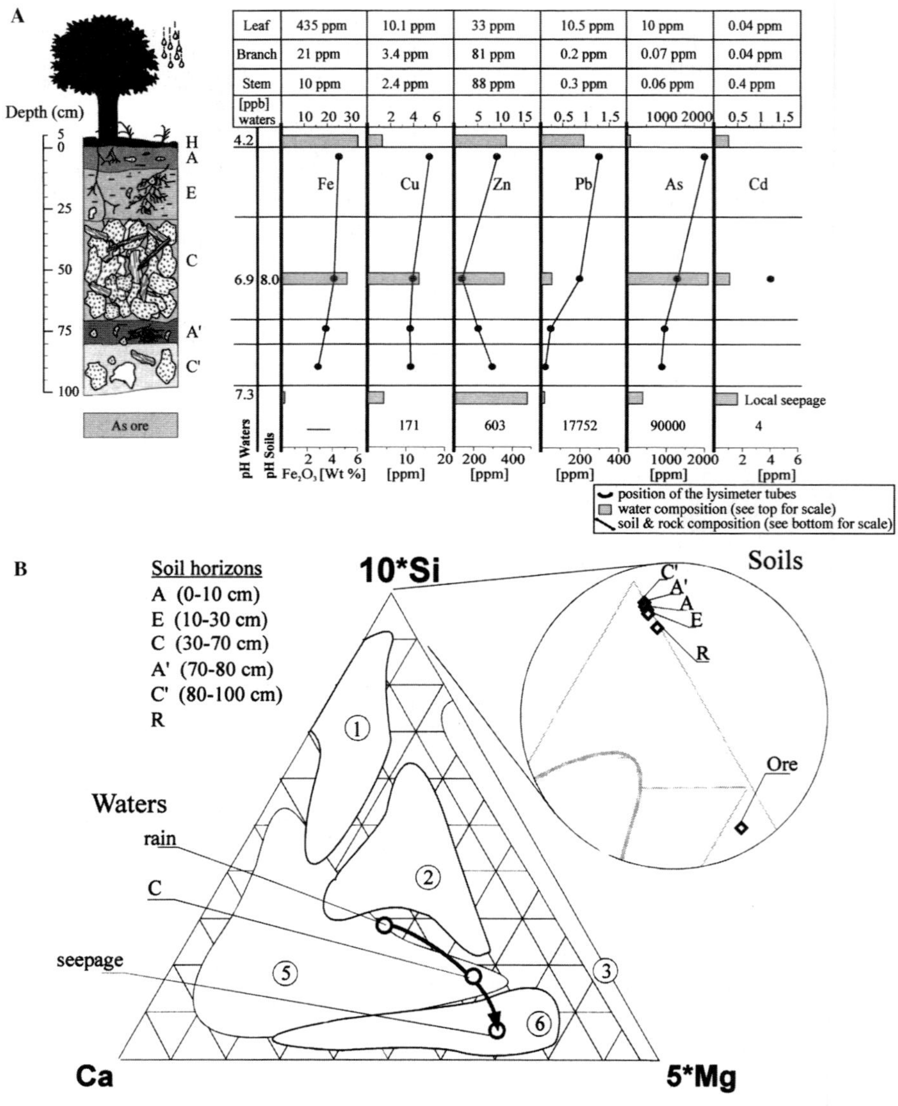

Figure 16. (A) Case study 1, contaminated situations: soil, plant (*Betula pendula*) and water compositions from a soil profile situated on an arsenopyrite deposit and its mine dump at Astano, Malcantone/Ticino. Data from Bondietti et al. (1994) and Rey (1996). The exact location of this site (no. 2) is shown on Fig. 11A. On the left, the soil profile with its different horizons is portrayed (H, A, E, C: soil on fallen and excavated material rich in metals and carbonate; A′, C′: older soil profile underneath). On the right, the changes in water composition throughout the soil profile from rain water down to the spring water are indicated with bars. The soil composition is shown with circles linked by a solid line and the ore composition is indicated with numbers. The complete data set is given in Appendix A.7. (B) Major element compositions of waters (rain, percolation and spring) and soils (H, A, C, A′, C′, R = As-ore) related to the case study 1, contaminated situations (soil profile no. 2, Astano, Malcantone, Switzerland). Nos. 1–6: see Table 1.

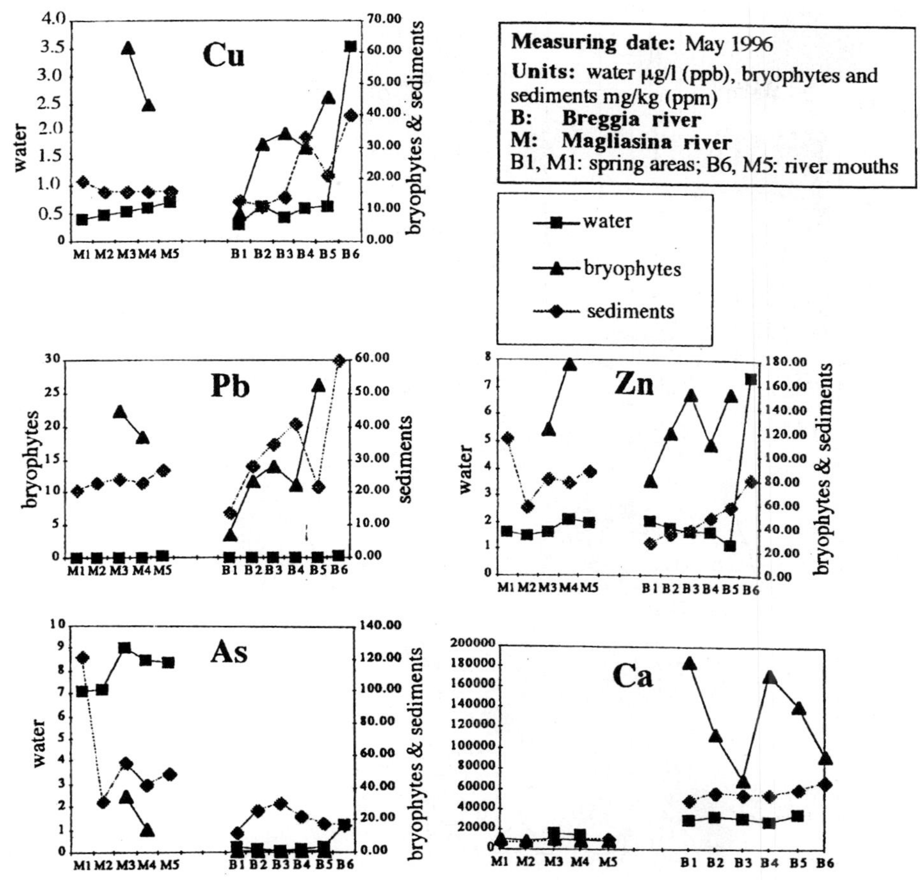

Figure 17. Typical water, sediment and moss compositions from the Magliasina river (major river of the Malcantone) compared to the Breggia river (Ticino, Switzerland), which lies in a limestone area with low natural metal contents. Data from Martini and Pillet (1996).

is illustrated in Fig. 18: local surface and spring waters contain between 5 and 30 µg/l U and river sediments contain up to 60 ppm U. Obviously, further down in the catchment towards the Lake of Geneva, U concentrations of waters and sediments decrease gradually, due to dilution by waters and sediments, containing typical background values of less than 1 µg/l in the water and below 6 ppm U in the sediments.

4. Discussion and conclusions

4.1. General aspects

This interdisciplinary research mainly dealt with the fractionation of the different chemical elements present in different rock types at the outset of the various weath-

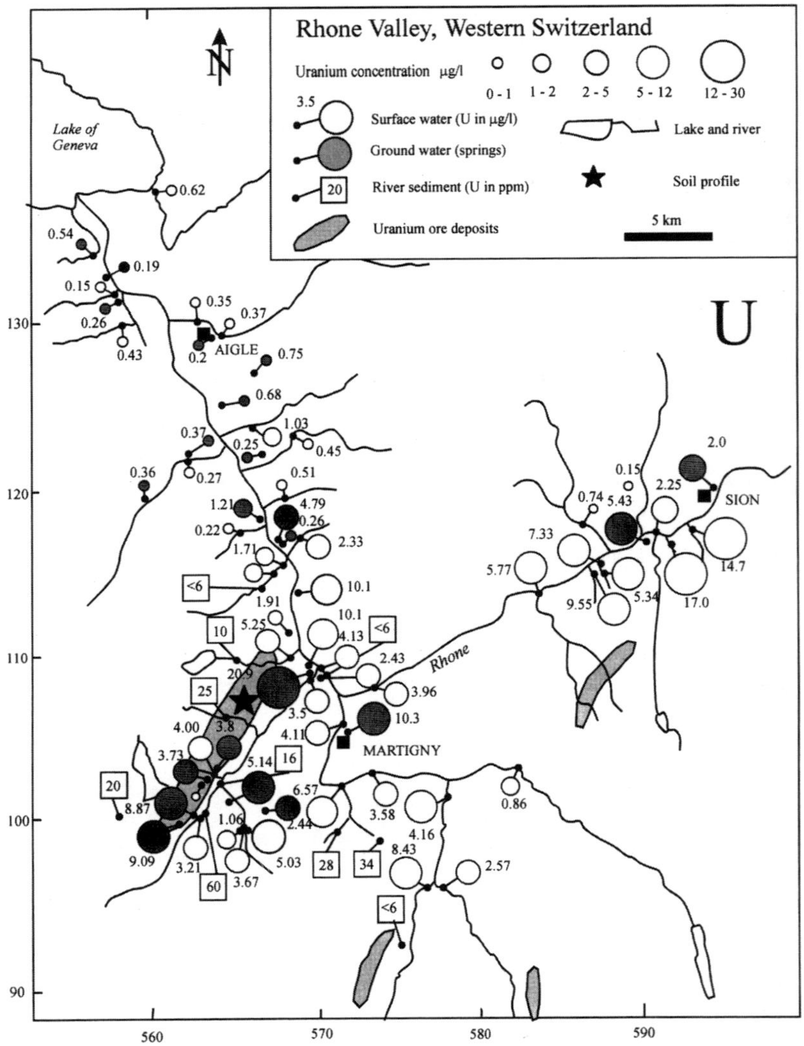

Figure 18. Case study 2, contaminated situations: geographic distribution of uranium in waters and sediments in the Martigny area (Rhone catchment, Western Switzerland). Sediment data from Woodtli et al. (1985) and water data from Baertschi and Keil (1992). Obviously the widespread uranium ore deposits (pitchblende/UO_2 veins) are at the origin of a widespread regional enrichment. The site described in detail in Fig. 19 (La Creusa-Les Marécottes) is marked with an asterisk.

ering and dispersion processes. The complex details of why a given element increases or decreases in the soil column has not been discussed in detail, in order not to overload the general message. The main question was the nature and extent of the chemical contribution of natural rocks to the different compartment of the environment. Table 3 gives an overview of the fingerprint of the different rocks in the various compartments dealt with in this paper. In the following discussion we treat each of them separately.

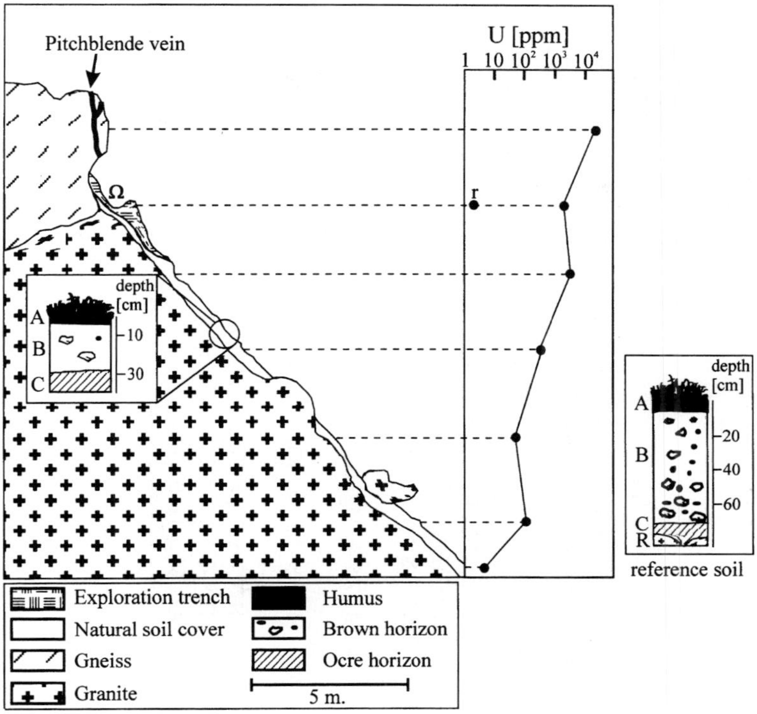

Figure 19. Typical example of how uranium is released to the environment (redrawn and completed after Pfeifer et al. (1994)). By weathering and erosion U-rich ore quality material (pitchblende veins in granitic gneiss) is deposited on the slope beneath a cliff. The soil and plants developing on it are enriched in uranium (up to 2500 ppm in the soils vs. 3 ppm in the reference soil outside the touched area, 50 m to the south-east, marked with r; 100 mg/kg U in contaminated plants, – see Appendix A.8). The waters from this type of environment contain between 10 and 30 μg/l U (cf. Fig. 18, Section 2.2 and Appendix A.1).

4.2. Soils

This study has confirmed that soils, especially not highly evolved soils, as they are typical for mountainous areas under temperate climatic conditions, are a good mirror of the total chemical characteristics of the parent rock. However, this is only strictly true for the B- and C-horizons which contain little organic material. The organic top horizons often turned out to be influenced by atmospheric input, certainly often of anthropogenic origin. The solid soil phases usually concentrate most trace elements together with less soluble major elements such as Fe, Al and Si. They do this more or less in the proportions present in the parent rock.

Likewise, the soil percolation waters differ from one rock-soil system to the other. These waters are especially rich in Fe and Al, certainly in colloidal as well as in dissolved form. If a certain trace element is present in elevated amounts in the parent rock, it is usually again present in elevated concentrations in the soil water. But in

Table 3. Summary of important chemical parameters for soils, soil percolation waters, plants and sediments as a function of different parent rocks, based on case studies in the Central Alps and the Jura. The list of elements indicated for the river sediment compartment is not exhaustive

Rock type	Soil	Percolating water	Springwater	Plants	River sediment
Granite	pH 4–5, Si, Na, K, Rb, Zr, Pb	pH 4.5–5.5, Rb, Pb, Al	pH 6.8–7.5, Si, SO_4, K, F, U, Mo	Acidophilous	U, As
Basalte	pH 5–8, Ti, Al, Fe, Ca, Mg, Cu, Zn, V, Cr, Ni	pH 5–6.5, Fe, Ca, Ni, Sr	pH 7–8, Ca, Mg, Sr	Calciphilous	–
Peridotite	pH 4–6, Mg, Fe, V, Co, Ni, Cr	pH 5.5–7.5, Ni, V,	pH 7.5–9. Mg, B	Acidophilous, Ni, Co	Cr, Ni
Shale	pH 4–5, Si, Al, Mg, As, Cd, Zn, Pb	pH 4.5–5.5, Cu, Zn, Pb, Al	pH 6.8–7.5, Si, SO_4	Acidophilous	As, Sb, Cd
Limestone	pH 5–8, Ca, Sr	pH 6–8, Ca, Cd	pH 7.5–8.5, Ca, Sr	Calciphilous, Cd	–
Gypsous dolomites	pH 6–8, Mg, Ca, S, Na	pH 6–8.5	pH 8–8.5, Ca, Mg, SO_4, Mn, As, Ba, Cl	Calciphilous	–

several cases, the trace element concentrations present in the rain water were preserved throughout the soil column.

4.3. Spring and surface waters

When resurging to the surface, the picture seen in the water radically changes. Whereas the major elements mirror the parent rocks amazingly well, the trace metals exhibit a completely different behavior and often have similar trace concentrations, especially metal traces. When we compare trace-rich basaltic rocks with trace-poor gypseous-dolomite rocks this phenomena is especially striking. Therefore, a uniforming process must exist in the water-unsaturated zone of the aquifers (soils and rock fissures above the ground water level), which levels out quite different initial trace element patterns, but which are not completely elucidated (Tessier, 1992; Salbu and Steinnes, 1995). Along the path from rain water entering a soil and resurging as spring water, pH values usually increase due to the proton consumption during mineral dissolution and the limited amount of air- or root CO_2 present. In certain cases, this can contribute to the fixation of trace elements on solid soil particles (including colloids) and channel walls by adsorption (McCarthy and Degueldre, 1993). Solid Fe and Mn phases probably play a key role in these processes (Jenne, 1968). Their stability depends, in addition to pH, on the oxidation potential, which certainly decreases along the path towards the spring, but which is extremely difficult to measure correctly. A certain leveling effect is conceivably also due to the relatively low solubility of the trace-bearing minerals in silicate rocks and the increased solubility of pure carbonate and associated minerals in lime- and dolostones.

4.4. Plants

Apart from the metal rich zones in the neighborhood of ore deposits and other rocks especially rich in particular trace elements such as peridotites and serpentinites, from central Europe, little data exist yet on the dependence of trace element behavior in common rocks. For several elements, such as Ni and As, upper limits of the trace uptake of plants seem to exist. As soil and plant concentrations are concerned, the uptake is normally dependant on the "available" concentrations determined with extraction methods such as 0.1 N $NaNO_3$. Whereas data on higher plants are relatively abundant, more data on naturally influenced trace elements in mosses and lichens would certainly be useful.

4.5. Sediments

The natural trace element composition of glacial and river sediments is usually close to its parent rock. However, since weathering intensity increases from the upper part to the lower part of a catchment, lake sediments often reflect the same differentiation between well soluble and less soluble elements that we discussed for soils in Section 1.3.3 and can represent an important sink of trace elements. In Switzerland little data is yet available on this aspect.

In the case of high trace element concentrations related to ore deposits, the important mass flow related to weathering and glacial or river sedimentation is at the origin of frequent cases of local or regional natural contamination. In the case of the Malcantone area discussed, there are the till sediments and not the original ore deposits which are at the origin of the high concentrations of arsenic of the spring and surface waters.

4.6. Synthesis

In many places of the world, rocks are locally and regionally at the origin of elevated trace element contents in soils, sediments, waters and plants. Especially in terms of soil, plant and sediment composition, the parent rock plays a key role. However, a single rock type is usually only of local importance. In general, different rock types contribute to the overall chemical composition of an area. In that sense, the cases of natural background presented in this paper will help to identify the main source rocks responsible for the chemical signature of an area. As a typical example we can cite the large areas in Switzerland, situated on till material derived from areas rich in peridotitic and basaltic rocks, which are systematically enriched in Cr and Ni (Desaules and Studer, 1993; Tuchschmid, 1995). The most marking rock-systems in terms of trace elements are peridotites, shales and any kind of ore deposits. Important trace element contributions of these rocks to a given ecosystem could easily be mistaken for an industrial contamination.

Acknowledgements

Most rocks and soils have been analyzed under the supervision of J.-C. Lavanchy (University of Lausanne), soil and spring waters under J.-D. Teuscher (Ecole Polytechnique Féderale – Lausanne) and P.-Y. Favarger (University of Geneva). The critical remarks on contents and English of an earlier version of this paper by W. Halter were very constructive and are thankfully acknowledged. We also thank M. Jäggli (Canton Ticino) for the permission to use unpublished data on water composition of the Malcantone area. The senior author would like to thank the following diploma students to allow him to use their data: M. Hamel, R. Martini, C. Nawratil, M. Pillet, M. Scheder and A. Streiff. He also benefited considerably from field visits and discussions on soils of the Southern Alps with P. Blaser, P. De Pury and G. Gini. J. Buszko and G. Giugni gave useful hints regarding soil-plant relations. This research has been partially financed by grants nos. 20-40756.94, 20-50531.97 and 21-52758.97 of Swiss National Science foundation.

Appendix A

This section contains the chemical data for soils, rocks, waters and plants presented in the case studies as a function of rock type (Appendices A1–A6) and for two naturally contaminated river catchments (Appendices A7 and A8). Legend: <3, detection limits; –, not measured; LOI, loss of ignition (weight loss determined during an 1 h heating at 1050°C, corresponds to $H_2O + CO_2 + S - O_2$ (taken up from the air to oxidize Fe II)). For analytical methods and uncertainties see Section 1.2.

A.1. Mont Blanc area (Argentiére, France): granitic rocks (Atteia, 1992; Dalla Piazza, 1996)

Soils and rocks

Horizon	Depth (cm)	SiO_2 (wt.%)	TiO_2 (wt.%)	Al_2O_3 (wt.%)	Fe_2O_3 (wt.%)	MnO (wt.%)	MgO (wt.%)	CaO (wt.%)	Na_2O (wt.%)	K_2O (wt.%)
O	0–5	9.26	0.12	2.34	0.59	0.01	0.17	0.47	0.22	0.42
Ah	5–9	32.30	0.33	6.76	0.82	0.01	0.25	0.36	1.02	1.47
Eh	9–15	72.95	0.38	12.73	0.95	0.02	0.31	0.50	2.82	3.40
Bh	15–20	53.91	0.67	12.13	6.38	0.03	0.47	0.81	2.04	2.70
Bhs	20–30	58.83	0.46	13.59	5.99	0.04	0.62	1.09	2.35	2.99
BC	30–40	68.89	0.42	14.76	3.13	0.05	0.82	1.35	2.71	3.45
R	–	71.67	0.42	14.70	3.35	0.05	1.10	1.22	2.70	3.47

Horizon	LOI (wt.%)	Ba (ppm)	Ce (ppm)	Co (ppm)	Cr (ppm)	Cu (ppm)	La (ppm)	Nd (ppm)	Ni (ppm)	Pb (ppm)
O	86.41	<3	39	7	18	34	16	16	10	51
Ah	56.67	<3	55	<2	18	12	29	23	2	3
Eh	5.93	256	47	<2	9	<4	19	16	<2	14
Bh	20.87	206	64	15	30	<4	31	25	<2	20
Bhs	14.06	318	67	12	32	<4	30	31	<2	12
BC	4.44	427	68	5	22	<4	30	30	5	12
R	1.32	481	52	5	26	<4	22	26	7	13

Horizon	Rb (ppm)	S (ppm)	Sr (ppm)	U (ppm)	V (ppm)	Y (ppm)	Zn (ppm)	Zr (ppm)
O	33	852	37	–	9	8	77	64
Ah	69	674	57	–	28	11	21	139
Eh	129	47	79	–	29	15	3	182
Bh	107	258	78	–	92	19	9	134
Bhs	122	167	93	–	56	25	20	140
BC	151	24	121	–	41	30	32	170
R	173	51	105	3	43	32	44	151

Waters

Type	pH	Ca (mg/l)	K mg/l)	Mg (mg/l)	Na (mg/l)	NH_4 (mg/l)	Cl (mg/l)	F (mg/l)	HCO_3 (mg/l)	NO_3 (mg/l)
Rain	5.3	0.28	0.11	0.02	0.26	0.06	0.34	–	–	1.15
Eh (soil)	4.8	0.70	0.17	0.15	0.45	0.03	0.42	–	–	0.66
Bhs (soil)	5.1	0.48	0.16	0.07	0.29	0.04	0.45	–	–	1.24
Spring	7.5	12.59	1.05	0.60	1.63	–	0.16	0.50	32.90	0.93

Type	SO_4 (mg/l)	Al (mg/l)	B (μg/l)	Ba (μg/l)	Co (μg/l)	Cr (μg/l)	Cu (μg/l)	Fe (μg/l)	Mn (μg/l)	Ni (μg/l)
Rain	0.81	<0.20	2.9	2.4	–	0.10	3.0	9.2	3.90	0.5
Eh (soil)	1.53	0.40	3.8	20.7	–	0.09	1.6	131.0	2.30	1.9
Bhs (soil)	1.20	0.28	4.3	15.4	–	<0.05	1.0	21.6	9.80	0.8
Spring	8.52	–	12.2	0.7	0.03	<0.05	0.1	<3.0	0.05	<0.3

Type	Pb (μg/l)	Rb (μg/l)	SiO_2 (μg/l)	Sr (μg/l)	U (μg/l)	V (μg/l)	Zn (μg/l)
Rain	1.1	0.1	0.43	3.1	–	0.10	27.5
Eh (soil)	1.4	1.3	4.73	2.2	–	0.25	11.3
Bhs(soil)	1.1	1.1	2.76	1.9	–	0.12	26.1
Spring	<0.2	0.5	7.31	20.0	9.0	0.40	<0.5

A.2. Val d'Hérens (Valais, Switzerland): basaltic rocks (Derron, 1999)

Soils and rocks

Horizon	Depth (cm)	SiO_2 (wt.%)	TiO_2 (wt.%)	Al_2O_3 (wt.%)	Fe_2O_3 (wt.%)	FeO (wt.%)	MnO (wt.%)	MgO (wt.%)	CaO (wt.%)
Ah	0–15	44.19	1.11	13.18	0.76	5.24	0.10	4.69	5.47
E	15–20	55.85	1.37	16.49	2.92	2.07	0.08	3.72	6.42
BPs	20–40	43.35	1.57	16.18	4.79	6.53	0.13	6.95	6.95
C		48.52	1.33	16.19	4.50	4.32	0.15	7.96	8.92
R		47.32	0.28	17.56	1.61	2.58	0.09	10.13	12.50

Horizon	Na_2O (wt.%)	K_2O (wt.%)	P_2O_5 (wt.%)	H_2O (wt.%)	CO_2 (wt.%)	Corg. (wt.%)	Ba (ppm)	Ce (ppm)	Co (ppm)
Ah	2.65	0.78	0.17	9.56	3.86	6.88	126	34	53
E	3.63	1.24	0.07	2.73	1.48	1.14	191	52	40
BPs	3.00	0.52	0.15	5.00	2.69	1.18	<3	16	38
C	3.12	0.38	0.13	3.61	0.61	0.06	<3	16	37
R	2.20	0.05	0.02	4.36	0.15	–	<3	<6	29

Horizon	Cr (ppm)	Cu (ppm)	La (ppm)	Nb (ppm)	Nd (ppm)	Ni (ppm)	Pb (ppm)	Rb (ppm)	S (ppm)
Ah	380	20	13	10	16	82	16	42	595
E	202	16	13	9	24	43	9	65	39
BPs	388	<4	12	<5	13	126	<2	16	54
C	440	75	8	<5	9	248	<2	12	22
R	936	53	<4	<5	4	246	<2	<1	<10

Horizon	Sr (ppm)	Th (ppm)	U (ppm)	V (ppm)	Y (ppm)	Zn (ppm)	Zr (ppm)
Ah	138	5	3	240	29	82	149
E	189	5	3	260	44	49	227
BPs	139	<1	<2	262	22	66	129
C	170	<1	<2	223	24	55	104
R	115	<1	<2	92	8	21	40

Waters

Type	pH	Ca (mg/l)	K (mg/l)	Mg (mg/l)	Na (mg/l)	Cl (mg/l)	HCO_3 (mg/l)	NO_3 (mg/l)	SO_4 (mg/l)
Rain	6.3	0.8	0.2	0.1	1.4	1.4	<0.1	0.9	4.2
Ah(soil)	5.2	2.0	0.7	0.6	0.6	1.5	4.9	5.7	2.5
E (soil)	6.2	3.1	0.4	0.7	0.7	1.6	2.4	3.6	3.7
Spring	8.2	6.5	0.2	1.5	0.9	1.4	58.6	0.6	25.6

Type	Al (mg/l)	As (μg/l)	B (μg/l)	Ba (μg/l)	Cr (μg/l)	Cu (μg/l)	Fe (μg/l)	Mn (μg/l)	Ni (μg/l)
Rain	<0.01	0.2	–	1–3	3.7	1–5	7–27	2–17	<1
Ah (soil)	0.14	1.0	2.0	2.0	2.6	9.0	210.0	13.0	12
E (soil)	0.21	0.3	8.0	2.0	4.3	12.0	580.0	4.0	17
Spring	0.02	0.2	4.0	1.0	2.0	<1.0	40.0	1.0	2

Type	Pb (μg/l)	Rb (μg/l)	SiO_2 (μg/l)	Sr (μg/l)	V (μg/l)	Zn (μg/l)
Rain	0.5–8	–	–	1–7	–	5–90
Ah (soil)	0.5	0.4	1.1	8.0	–	30.0
E (soil)	0.5	0.5	1.1	10.0	–	19.0
Spring	0.1	0.2	1.5	60.0	0.5	4.0

A.3. Canavese area (Baldissero-Canavese, Torino, Italy): peridotitic rocks (Schlegel, 1997)

Soils and rocks

Horizon	SiO_2 (wt.%)	TiO_2 (wt.%)	Al_2O_3 (wt.%)	Fe_2O_3 (wt.%)	FeO (wt.%)	MnO (wt.%)	MgO (wt.%)	CaO (wt.%)	Na_2O (wt.%)	K_2O (wt.%)
Oln	44.31	0.29	7.79	2.44	4.80	0.14	17.83	3.13	0.45	0.21
Amg	47.88	0.30	8.64	4.54	3.49	0.17	20.51	3.33	0.46	0.19
Smg	46.94	0.21	7.90	7.58	1.69	0.23	25.55	2.79	0.26	0.08
R	43.56	0.08	3.58	0.74	7.39	0.12	39.66	2.74	0.11	0.00

Horizon	LOI (wt.%)	O_2 (wt.%)	H_2O (wt.%)	CO_2 (wt.%)	Corg. (wt.%)	Ba (ppm)	Co (ppm)	Cr (ppm)	Cu (ppm)	Ni (ppm)
Oln	16.86	0.53	8.21	3.66	5.52	33	104	5483	<4	1225
Amg	8.87	0.39	4.34	2.35	2.57	55	127	6449	<4	1220
Smg	5.64	0.19	3.59	1.61	0.63	35	186	6477	<4	1501
R	0.64	–	–	–	–	16	100	2741	30.00	2318

Horizon	Pb (ppm)	Rb (ppm)	S (ppm)	Sr (ppm)	V (ppm)	Zn (ppm)	Zr (ppm)
Oln	23	12	196	28	130	163	81
Amg	<2	11	96	27	146	138	85
Smg	<2	5	42	10	153	84	30
R	<2	<1	40	<1	75	46	<2

Waters

Type	Date	pH	T (°C)	Cond. (S/cm)	Ca (mg/l)	Li (mg/l)	K (mg/l)	Mg (mg/l)	Na (mg/l)	NH_4 (mg/l)
Rain	22.02.96	4.1	5.0	26	<0.50	<0.20	0.03	0.20	0.29	<0.50
Amg (soil)	22.02.96	7.6	5.0	89	0.57	<0.20	0.02	11.35	0.48	–
Spring	22.02.96	7.9	5.0	371	8.34	<0.20	0.03	43.52	0.86	–

Type	Cl (mg/l)	F (mg/l)	HCO_3 (mg/l)	NO_3 (mg/l)	SO_4 (mg/l)	Al (μg/l)	As (μg/l)	Ba (μg/l)	Co (μg/l)	Cr (μg/l)
Rain	2720	0.30	12.00	2.72	2.05	–	0.4	2.6	0.2	4.8
Amg (soil)	0.29	<0.20	46.00	3.52	3.09	9.0	1.3	0.3	2.0	
Spring	0.67	<0.20	193.00	4.56	7.43	18.0	–	1.2	0.1	3.7

Type	Cu (μg/l)	Mn (μg/l)	Ni (μg/l)	Pb (μg/l)	SiO_2 (μg/l)	Sr (μg/l)	V (μg/l)	Zn (μg/l)
Rain	4.3	3.1	1.1	3.9	0.3	0.7	0.1	25.1
Amg (soil)	1.6	2.0	35.0	0.31	8.8	1.3	–	1.9
Spring	0.6	–	6.4	0.1	33.6	10.0	1.0	4.2

Calluna vulgaris (Ling heather)

Material	Al (mg/kg)	As (mg/kg)	B (mg/kg)	Ba (mg/kg)	Ca (mg/kg)	Co (mg/kg)	Cr (mg/kg)	Cu (mg/kg)	Fe (mg/kg)	K (mg/kg)
Root	39.7	<0.2	4.3	16.4	937.6	2.4	2.7	4.2	203.0	1100.9
Stem	19.3	<0.2	4.5	5.9	557.1	0.5	0.8	2.0	29.7	645.2
Leaf	168.7	<0.2	39.3	14.5	3745.9	0.4	3.5	4.9	197.0	1338.3
Flower	63.2	<0.2	18.4	11.4	2282.2	<0.2	2.2	5.5	61.9	3054.6

Material	Mg (mg/kg)	Mn (mg/kg)	Ni (mg/kg)	Na (mg/kg)	Pb (mg/kg)	Rb (mg/kg)	Si (mg/kg)	Sr (mg/kg)	Zn (mg/kg)
Root	1273.5	476.2	30.7	172.4	1.4	2.4	267.8	3.4	15.1
Stem	613.0	300.4	5.8	70.7	0.7	1.5	57.2	1.8	6.0
Leaf	3732.8	218.4	28.0	85.1	4.7	2.9	1791.6	4.7	20.4
Flower	2395.8	85.8	15.7	120.4	1.0	8.6	934.3	2.5	12.9

Betula pendula (Birch)

Material	Al (mg/kg)	As (mg/kg)	B (mg/kg)	Ba (mg/kg)	Ca (mg/kg)	Co (mg/kg)	Cr (mg/kg)	Cu (mg/kg)	Fe (mg/kg)	K (mg/kg)
Root	41.7	<0.2	8.6	14.4	382.3	0.4	1.7	6.3	45.2	753.0
Stem	20.9	<0.2	8.6	12.6	615.9	0.3	1.3	4.3	31.4	766.2
Branch	19.1	<0.2	12.0	11.4	733.6	0.3	1.0	4.6	15.9	841.1
Leaf	235.9	<0.2	41.6	5.8	5606.6	.1	2.1	5.7	140.4	673.2

Material	Mg (mg/kg)	Mn (mg/kg)	Na (mg/kg)	Ni (mg/kg)	Pb (mg/kg)	Rb (mg/kg)	Si (mg/kg)	Sr (mg/kg)	V (mg/kg)	Zn (mg/kg)
Root	877.7	54.2	65.5	4.3	0.3	2.8	42.8	2.9	<0.3	25.7
Stem	859.9	90.4	45.9	3.6	0.2	2.0	34.0	3.4	<0.3	49.8
Branch	1049.5	90.2	53.8	4.7	<0.2	2.4	<12.0	3.7	<0.3	58.3
Leaf	12772.2	520.3	86.8	42.8	6.0	1.6	516.2	7.1	0.5	156.3

A.4. Malcantone area (Astano, Ticino, Switzerland): micaschists, outside of the Arsenopyrite vein area (Rey, 1996)

Soils and rocks

Horizon	SiO_2 (wt.%)	TiO_2 (wt.%)	Al_2O_3 (wt.%)	Fe_2O_3 (wt.%)	FeO (wt.%)	MnO (wt.%)	MgO (wt.%)	CaO (wt.%)	Na_2O (wt.%)	K_2O (wt.%)
H	57.85	0.71	18.89	6.35	–	0.12	2.37	1.55	1.87	2.15
A	71.74	0.63	16.04	5.3	–	0.09	1.87	1.54	2.54	2.11
E	56.40	0.63	18.27	5.57	–	0.10	2.27	1.60	1.99	2.00
Bt	75.94	0.66	16.89	5.58	–	0.09	1.96	1.91	2.66	2.18
C	72.56	0.42	13.14	3.44	–	0.05	1.01	1.15	2.59	2.11
R	65.00	0.32	16.44	2.67	2.24	0.01	0.26	0.20	5.12	2.52

Horizon	P_2O_5 (wt.%)	H_2O (wt.%)	As (wt.%)	Ba (wt.%)	Cd (wt.%)	Co (wt.%)	Cr (wt.%)	Cu (wt.%)	Ni (wt.%)	Pb (wt.%)
H	0.28	7.42	142	504	–	19	101	47	39	63
A	0.13	3.24	110	466	0.5	45	83	28	38	27
E	0.19	10.6	134	512	–	22	99	34	38	38
Bt	0.13	3.56	108	467	1.2	78	89	21	42	14
C	0.08	1.23	105	382	1.0	44	36	9	22	59
R	0.05	3.27	920	289	–	44	3	49	<2	18

Horizon	Rb (ppm)	S (ppm)	Sb (ppm)	Sr (ppm)	V (ppm)	Y (ppm)	Zn (ppm)	Zr (ppm)
H	134	621	<2	181	147	31	139	227
A	110	851	3	165	135	36	85	175
E	127	281	<2	196	134	33	90	217
Bt	112	701	<2	185	147	37	86	194
C	81	69	3	158	67	28	69	143
R	93	3295	3	76	9	24	<2	654

Waters

Type	Date	pH	T (°C)	Cond. (μS/cm)	Ca (mg/l)	K (mg/l)	Mg (mg/l)	Na (mg/l)	NH_4 (mg/l)	Cl
Rain	6.95	4.0	25.0	–	1.49	0.41	0.18	0.60	1.17	0.98
Bt (soil)	6.95	4.0	15.0	–	3.69	1.53	0.00	2.31	0.46	3.55
Spring	6.95	6.5	8.5	89	10.53	0.46	1.68	2.70	0.00	4.42

Type	F (mg/l)	HCO_3 (mg/l)	NO_3 (mg/l)	SO_4 (mg/l)	Al (μg/l)	As (μg/l)	Ba (μg/l)	Cd (μg/l)	Co (μg/l)	Cr (μg/l)
Rain	0.04	0.00	3.49	6.38	0.1	1.7	14.3	0.1	0.2	1.3
Bt (soil)	0.04	0.00	5.93	7.31	6772.5	3.6	102.7	1.1	5.4	2.4
Spring	0.07	18.30	4.98	10.02	8.0	0.8	11.1	0.5	0.1	2.4

Type	Cu (μg/l)	Mn (μg/l)	Ni (μg/l)	Pb (μg/l)	SiO_2 (μg/l)	Sr (μg/l)	V (μg/l)	Zn (μg/l)
Rain	3.3	14.3	0.8	2.5	0.18	5.8	1.9	110.3
Bt (soil)	7.0	523.6	7.2	8.7	0.47	24.3	0.5	84.7
Spring	0.6	0.4	1.0	0.8	2.40	36.5	0.3	7.5

Plants (Nawratil, 1998)

Material	Al (mg/kg)	As (mg/kg)	Ca (mg/kg)	Cd (mg/kg)	Co (mg/kg)	Cr (mg/kg)	Cu (mg/kg)	Fe (mg/kg)	K (mg/kg)	Li (mg/kg)
Plantain	1868.0	4.4	15048.0	0.3	1.3	5.1	24.0	2283.0	20647.0	2.3
Dandelion	2603.0	8.1	6502.0	0.2	1.9	6.8	34.0	3434.0	34145.0	3.3
Dock	409.0	0.7	4862.0	0.6	0.4	1.9	13.0	528.0	35838.0	0.7

Material[a]	Mg (mg/kg)	Mn (mg/kg)	Na (mg/kg)	Ni (mg/kg)	Pb (mg/kg)	Si (mg/kg)	V (mg/kg)	Zn (mg/kg)
Plantain	3393.0	80.0	203.0	4.7	7.3	460.0	5.7	57.0
Dandelion	2529.0	114.0	205.0	5.0	10.6	394.0	5.1	70.0
Dock	4247.0	34.0	206.0	2.7	3.5	705.0	0.7	96.0

[a] Plantain, *Plantago*; Dandelion, *Taraxacum officinale*; Dock, *Rumex*.

A.5. Vallée de Joux area (Jura mountains, Vaud/Switzerland): limestones (Dalla Piazza, 1996)

Soils and rocks

Horizon	Depth (cm)	SiO_2 (wt.%)	TiO_2 (wt.%)	Al_2O_3 (wt.%)	Fe_2O_3 (wt.%)	MnO (wt.%)	MgO (wt.%)	CaO (wt.%)	Na_2O (wt.%)	K_2O (wt.%)	P_2O_5 (wt.%)
O	0–5	56.91	0.95	13.25	5.52	0.14	1.01	1.23	0.54	1.41	–
A	5–15	56.69	0.90	13.08	5.33	0.15	0.99	1.62	0.56	1.46	–
B	15–25	57.76	0.94	13.02	5.56	0.16	1.03	2.74	0.57	1.49	–
R	–	0.74	0.02	0.31	0.11	0.01	0.68	54.79	0.00	0.01	0.01

Horizon	LOI (wt.%)	Ba (ppm)	Ce (ppm)	Cd (ppm)	Co (ppm)	Cr (ppm)	Cu (ppm)	La (ppm)	Nd (ppm)	Ni (ppm)	Pb (ppm)
O	19.04	110	118	0.4[a]	18	111	9	79	61	46	31
A	19.22	85	117	0.2[a]	17	110	<4	84	64	46	16
B	16.73	140	110	1.3[a]	18	112	<4	77	55	46	21
R	43.61	<3	13	0.04[a]	11	14	<4	5	12	<2	<2

[a] Values taken from La Chaux-de-Fonds area (Atteia et al., 1995).

Horizon	Rb (ppm)	S (ppm)	Sr (ppm)	Th (ppm)	U (ppm)	V (ppm)	Y (ppm)	Zn (ppm)	Zr (ppm)
O	102	283	66	33[a]	22[a]	141	54	190	335
A	99	218	63	19[a]	8[a]	134	55	167	315
B	109	263	70	12[a]	2[a]	142	57	176	336
R	5	722	218	3[b]	23[b]	3	<1	16	<2

[a] Values taken from La Chaux-de-Fonds area (Atteia et al., 1995).
[b] Values taken from Hamel (1998).

Waters

Type	Date	pH	Ca (mg/l)	K (mg/l)	Mg (mg/l)	Na (mg/l)	Cl (mg/l)	F (mg/l)	NO_3 (mg/l)	SO_4 (mg/l)	Al (μg/l)
Rain	18.11.94	4.3	0.33	0.12	<0.05	0.22	0.20	0.20	2.15	3.60	10.3
B (soil)	18.11.94	7.7	36.53	0.06	0.32	0.41	0.30	0.00	0.89	3.89	20.0
Spring	05.11.92	7.5	62.59	–	2.43	0.64	1.22	0.02	2.66	7.10	<50.0

Type	B (μg/l)	Ba (μg/l)	Cd (μg/l)	Cr (μg/l)	Cu (μg/l)	Fe (μg/l)	Mn (μg/l)	Ni (μg/l)	Pb (μg/l)	Rb (μg/l)	SiO_2 (mg/l)
Rain	3.8	1.2	–	0.2	1.3	4.3	8.3	9.1	0.4	0.4	0.7
B (soil)	5.9	14.2	2.8	0.7	5.7	30.4	0.7	3.6	0.5	0.4	14.2
Spring	1.5	2.0	0.1	0.4	<0.2	<50.0	0.2	1.2	<0.2	0.1	0.5

Type	Sr (μg/l)	V (μg/l)	Zn (μg/l)
Rain	0.8	<0.2	7.7
B (soil)	22.2	0.3	20.5
Spring	48.9	<0.2	3.0

A.6. Waters from gypseous dolomites (coordinates of localities see Appendix A.9) (Scheder and Streiff, 1997)

Data marked with an asterisk are from Mandia (1991).

Waters

Type	Date	pH	T °(C)	Cond. (μS/cm)	Ca (mg/l)	K (mg/l)	Mg (mg/l)	Li (μg/l)	Na (mg/l)
Rain Bramois	01.04.97	5.5	–	10	1.30	1.30	0.10	–	1.30
Spring Evolène	06.04.97	7.3	8.9	1650	350.00	1.00	62.00	–	1.00
Creek Nax	05.04.97	8.6	5.9	648	70.00	6.00	45.00	40.00	9.00
Average Switzerland*	–	–	8.7	1626	419.00	2.70	66.60	27.10	5.80

Type	NH_4 (mg/l)	F (μg/l)	Cl (mg/l)	HCO_3 (mg/l)	NO_3 (mg/l)	SO_4 (mg/l)	Al (μg/l)	As (μg/l)	Ba (μg/l)
Rain Bramois	0.50	–	1.800	0.00	4.4	4.50	–	0.2	1.9
Spring Evolène	–	0.700 149.00	–	982.00	–	0.3	29.0		
Creek Nax	–	–	22.00	329.00	10.00	47.00	72.0	21.0	51.0
Average Switzerland*	–	129.40	7.00	251.30	3.00	1041.00	16.4	0.6	10.6

Type	Cd (μg/l)	Co (μg/l)	Cr (μg/l)	Cu (μg/l)	Mn (μg/l)	Ni (μg/l)	Pb (μg/l)	Rb (μg/l)	SiO_2 (mg/l)
Rain Bramois	–	–	3.7	43.3	3.4	–	3.6	–	0.3
Spring Evolène	–	1.0	2.0	1.2	5.0	17.0	0.5	–	5.8
Creek Nax	–	–	–	4.0	2.0	3.0	0.1	–	11.1
Average Switzerland*	0.2	–	0.3	3.1	12.3	3.8	18.3	6.8	

Type	Sr (μg/l)	V (μg/l)	Zn (mg/l)
Rain Bramois	20.5	–	99.4
Spring Evolene	7500.0	–	25.0
Creek Nax	1950.0	–	7.0
Average Switzerland*	7.9	0.5	7.1

A.7. Malcantone area (Astano, Ticino, Switzerland): soil on sulfide ore deposit of Astano (Rey, 1996)

Soils and rocks

Horizon	SiO_2 (wt.%)	TiO_2 (wt.%)	Al_2O_3 (wt.%)	Fe_2O_3 (wt.%)	MnO (wt.%)	MgO (wt.%)	CaO (wt.%)	Na_2O (wt.%)	K_2O (wt.%)	P_2O_5 (wt.%)
A	67.93	0.57	14.9	5.06	0.11	1.22	1.10	1.96	2.71	0.14
C	62.69	0.55	13.96	4.52	0.09	1.19	1.13	2.01	2.30	0.14
A′	67.83	0.52	14.03	4.10	0.06	1.05	1.00	2.00	2.31	0.10
C′	70.69	0.53	14.06	3.94	0.06	1.03	0.91	2.32	2.54	0.08
R	–	–	–	–	–	–	–	–	–	–

Horizon	H_2O (wt.%)	Ag (ppm)	As (ppm)	Ba (ppm)	Cd (ppm)	Co (ppm)	Cr (ppm)	Cu (ppm)	Ni (ppm)	Pb (ppm)
A	3.66	0	2218	451	–	48	62	19	30	361
C	3.38	8	1337	412	5	81	64	13	30	247
A′	2.81	<2	797	440	–	64	55	14	29	57
C′	1.78	<2	794	543	–	51	44	13	27	17
R	–	130	89000	442	4	104	134	111	<2	18000

Horizon (ppm)	Rb (ppm)	S (ppm)	Sb (ppm)	Sn (ppm)	Sr (ppm)	V (ppm)	Y (ppm)	Zn (ppm)	Zr (ppm)
A	123	1095	3	<5	126	118	39	330	185
C	105	978	2	<5	129	98	35	175	185
A′	101	450	4	<5	127	92	34	245	185
C′	105	<10	2	<5	138	80	34	282	185
R	151	10500	73	14	74	125	35	603	251

Waters

Type	Date	pH	T °(C)	Ca (mg/l)	K (mg/l)	Mg (mg/l)	Na (mg/l)	NH4 (mg/l)	Cl (mg/l)	F (mg/l)
Rain	9.95	4.2	25.0	0.30	0.03	0.03	0.19	0.50	0.16	0.12
C (soil)	7.95	6.9	15.0	26.97	0.92	7.56	0.12	<0.50	0.34	0.04
Spring	1.94	7.3	12.3	56.29	2.20	16.39	3.01	<0.50	2.47	0.01

Type	HCO_3 (mg/l)	NO_3 (mg/l)	SO_4 (mg/l)	Al (μg/l)	As (μg/l)	Ba (μg/l)	Cd (μg/l)	Co (μg/l)	Cr (μg/l)	Cu (μg/l)
Rain	0.00	1.66	2.58	13.8	0.6	2.0	0.2	0.1	2.0	1.4
C (soil)	115.90	0.86	19.27	10.2	838.0	12.4	0.5	0.4	5.6	4.8
Spring	103.70	9.24	126.40	11.6	85.9	14.3	0.3	0.4	0.7	1.7

Type	Mn (μg/l)	Ni (μg/l)	Pb (μg/l)	SiO_2 (mg/l)	Sr (μg/l)	V (μg/l)	Zn (μg/l)
Rain	7.1	0.6	0.8	–	1.1	0.7	11.8
C (soil)	1.2	5.4	0.2	–	127.4	0.6	11.4
Spring	1.2	4.4	0.0	2.49	165.2	0.3	20.9

Calluna vulgaris (Ling heather)

Material	Al (mg/kg)	As (mg/kg)	B (mg/kg)	Ba (mg/kg)	Ca (mg/kg)	Co (mg/kg)	Cr (mg/kg)	Cu (mg/kg)	Fe (mg/kg)	K (mg/kg)
Root	14.3	<0.2	4.0	22.5	610.0	<0.2	0.6	2.6	8.0	1551.0
Stem	7.4	<0.2	3.0	19.3	538.0	<0.2	0.8	3.6	12.0	1434.0
Leaf	116.2	0.6	43.0	40.6	4693.0	<0.2	2.1	6.7	112.0	2907.0
Flower	88.6	0.5	16.0	26.3	2826.0	<0.2	1.4	8.1	62.0	919.0

Material	Mg (mg/kg)	Mn (mg/kg)	Ni (mg/kg)	Na (mg/kg)	Pb (mg/kg)	Rb (mg/kg)	Si (mg/kg)	Sr (mg/kg)	Zn (mg/kg)
Root	232.0	214.0	1.5	119.4	2.0	4.7	13.0	1.6	11.3
Stem	227.0	122.0	1.8	59.3	0.9	3.8	14.0	1.4	4.9
Leaf	1449.0	221.0	5.3	91.0	3.0	6.8	1619.0	5.3	36.3
Flower	769.0	88.0	2.7	247.0	2.5	3.5	867.0	3.9	21.7

Betula pendula (Birch)

Material	Al (mg/kg)	As (mg/kg)	B (mg/kg)	Ba (mg/kg)	Ca (mg/kg)	Co (mg/kg)	Cr (mg/kg)	Cu (mg/kg)	Fe (mg/kg)	K (mg/kg)
Stem	11.4	<0.2	4.2	2.1	958	<0.2	0.8	2.4	10.3	1297.5
Branch	13.5	<0.2	9.4	3.7	1531.7	<0.2	0.8	2.1	21.2	1181.4
Leaf	541.3	10.6	35.2	12.9	19752	0.3	2.3	10.1	435.2	3443.4

Material	Mg (mg/kg)	Mn (mg/kg)	Na (mg/kg)	Ni (mg/kg)	Pb (mg/kg)	Rb (mg/kg)	Si (mg/kg)	Sr (mg/kg)	V (mg/kg)	Zn (mg/kg)
Stem	415.4	8.3	8.1	0.6	0.3	2.1	<12.0	3.8	<0.3	88.6
Branch	361.5	2.4	38.7	0.4	1.2	2.2	<12.0	16.6	<0.3	14.7
Leaf	4657.4	20.7	153.6	3.9	10.5	5	980.2	103.9	1	32.7

Taraxacum officinale (Dandelion) (Pfeifer, this chapter)

Material	As (mg/kg)	Ba (mg/kg)	Ca (mg/kg)	Cr (mg/kg)	Cu (mg/kg)	Fe (mg/kg)	Mn (mg/kg)	Ni (mg/kg)	Pb (mg/kg)	Si (mg/kg)	Sr (mg/kg)	Zn (mg/kg)
Leaves	3.7	42	12	3.4	28.9	519	53.1	2.2	5.1	626	40.6	110.2

A.8. Martigny area (La Creusaz): site rich in uranium (Pfeifer et al., 1994)

Soils (pit 4) and rocks

Horizon	SiO_2 (wt.%)	TiO_2 (wt.%)	Al_2O_3 (wt.%)	Fe_2O_3 (wt.%)	P_2O_5 (wt.%)	MnO (wt.%)	MgO (wt.%)	CaO (wt.%)	Na_2O (wt.%)	K_2O (wt.%)
B	55.66	0.76	13.92	6.3	0.21	0.15	2.05	0.87	1.27	.44
C	59.71	0.74	15.34	6.4	0.2	0.12	1.6	0.71	1.42	2.77
R (U-vein)	73.91	0.97	8.73	6.09	0.26	0.01	0.42	0.5	2.13	2.64

Horizon	H_2O (wt.%)	CO_2 (wt.%)	Corg. (wt.%)	Ba (ppm)	Ce (ppm)	Co (ppm)	Cr (ppm)	Cu (ppm)	La (ppm)	Ni (ppm)
B	7.41	3.57	4.61	415	87	31	72	45	36	16
C	4.9	3.24	1.92	435	86	44	71	27	35	16
R (U-vein)	2.3	–	–	<3	143	204	32	125	28	16

Horizon	Pb (ppm)	Rb (ppm)	S (ppm)	Sr (ppm)	Th (ppm)	U (ppm)	V (ppm)	Y (ppm)	Zn (ppm)	Zr (ppm)
B	69	190	152	107	29	465	102	12	130	170
C	23	206	77	105	16	81	99	15	104	178
R (U-vein)	1217	316	11000	103	436	13830	63	<3	97	158

Plants

Type of plant	U (mg/kg)
Grass (*Agrostis schraderiana*)	
Fresh green	51
Dry yellow	12.3
Green uncontaminated (ref)	5.6
Mosses	
Polytrichum sp. (small)	93.1
Rhytidiadeiphus triquetrus	6.31
Polytrichumformosum	3.14
Pleurizium schreberi	10.7
Hylocomium splendens	7.41

(*continued*)

Type of plant	U (mg/kg)
Shrubs (Blueberry)	
Vaccinium vitis-idaea, leaf	2.76
Vaccinium vitis-idaea, stalk	2.36
Vaccinium myrtillus, leaf	2
Vaccinium myrtillus, stalk	2.05

A.9. Exact location of the different case studies discussed in this paper

Locality	Country	Coordinates[a]	Dominant rock type
Mont Blanc, Argentière	France	561.250/92.000	Granitic gneisses
Val d'Hérens, Valais	Switzerland		
Lac Bleu		603.150/99.850	Greenstones
Evolene spring		604.725/107.525	Gypsous dolomites
Nax, creek		598.525/119.100	The same
Bramois, rain water		596.900/119.500	
Canavese, Baldissero	Italy	7°44′53″E/ 45°25′22″N[b]	Peridotites, serpentinites
Malcantone, Ticino	Switzerland		
Costa background site		707.350/095.750	Micaschistes, granitic gneisses
Costa sulfide ore deposit		707.380/095.800	Arsenopyrite, galena, sphalerite
Vallée de Joux, Jura mts, Le Mollendruz	Switzerland	516.100/166.400	Limestones
Val Piora, Ticino, Lake Cadagno	Switzerland	697.500/156.200	Gypsous dolomites

[a] Swiss coordinate system in km; Bern $x = 600.000$, $y = 200.000$.
[b] Longitude/latitude.

References

Adriano, D.O. (Ed.), 1992. Biochemistry of Trace Metals. Lewis, Boca Raton, FL, p. 513.

Amstutz, G.C., Fontboté, L., 1985. Correlative observations on the lead-zinc-(barite-fluorite) occurrences in the Germanic Triassic lithofacies of Central Europe. Geol. J. D70, 179–211.

Appelo, C.A.J., Postma, D., 1993. Geochemistry, Groundwater and Pollution. Balkema, Rotterdam, p. 536.

Atteia, O., 1992. Rôle du sol dans le transfert des éléments traces en solution -application à l'étude de quelques écosystèmes d'altitude. Ph.D. Thesis No. 1031. EPF, Lausanne.

Atteia, O., Dubois, J.-P., Thélin, P., Hunziker, J., Pfeifer, H.-R., 1995. A search for cadmium in soils of the Jura, near La-Chaux-de-Fonds. Geoderma 68, 149–172.

Baize, D., Jabiol, B., 1995. Guide pour la Description des Sols. Inra Ed., Paris, p. 375.

Baertschi, P., Keil, R., 1992. Urangehalte von Oberflächen-. Quell- und Grundwässern der Schweiz. NAGRA, Techn. Ber. 91–26, 51.

Birch, L., Hanselmann, K., Bachofen, R., 1996. Heavy metal conservation in lake Cadagno sediments: historical records of anthropogenic emissions in a meromictic alpine lake. Wat. Res. 30, 679–687.

Blaser, P., 1973. Die Bodenbildung auf Silikatgestein im südlichen Tessin. Eidg. Anst. forstl. Versuchsw., Mitt. 49, 253–340.

Bowell, R.J., 1994. Sorption of arsenic by iron oxides and oxyhydroxides in soils. Appl. Geochem. 9, 279–286.

Bondietti, G., Gex, P., Gini, G.F., Hansen, J., Hunziker, J., Pfeifer, H.-R., 1994. Heavy metal contamination around the As-Pb-Zn-Au-mine at Astano (Malcantone, Ticino). Eclogae Geol. Helv. 87, 487–490.

Bourg, A.C.M., 1988. Metals in aquatic and terrestrial systems: sorption, speciation and mobilization. In: Salomons, W., Förstner, U. (Eds.), Chemistry and Biology of Solid Waste. Springer, Berlin, pp. 3–32.

Bourg, A.C.M., 1995. Speciation of heavy metals in soils and groundwater and implications for their natural and provoked mobility. In: Salomons, W., Förstner, U., Mader, P. (Eds.), Heavy Metals. Springer, Berlin, pp. 19–31.

Bourg, A.C.M., Loch, J.P.G., 1995. Mobilization of heavy metals as affected by pH and redox conditions. In: Salomons, W., Stigliani, W.M. (Eds.), Biogeodynamics of Pollutants in Soils and Sediments. Springer, Berlin, pp. 87–102.

Brooks, R.R., 1972. Geobotany and Biogeochemistry in Mineral Exploration. Harper & Row, New York. p. 290.

Brooks, R.R. (Ed.), 1987. Serpentine and its Vegetation. A Multidisciplinary Approach. Dioscorides Press, Portland, OR, p. 454.

Carmichael, I., Turner, F.J., Verhoogen, J., 1974. Igneous Petrology. MacGraw-Hill, New York, p. 739.

Chester, R., Nimmo, M., Keyse, S., 1996. The influence of the Saharan and Middle Eastern desert-derived dust on the trace metal composition of Mediterranean aerosols and rainwaters: an overview. In: Guerzoni, S., Chester, R. (Eds.), The Impact of Desert Dust Across the Mediterranean. Kluwer, Dordrecht, The Netherlands, pp. 253–273.

Dalla Piazza, R., 1996. Géochimie des altérations dans trois écosystèmes sol tempérés application à l'acquisition des caractéristiques chimiques des solutés. Ph.D. Thesis no. 1483, EPF, Lausanne.

De Caritat, P., Saether, O.M., 1997. Chemical changes attending water cycling trough a catchment- an overview. In: Saether, O.M., De Cartiat, P. (Eds), Geochemical Processes, Weathering and Groundwater Recharge in Catchments. Balkema, Rotterdam, pp. 381–391.

Delarze, R., Gonseth, Y., Galland, P., 1998. Guide des Milieux Naturels de Suisse, Écologie, Menaces, Espèces Caractéristiques. Delachaux et Niestlé, Lausanne, p. 413.

Derron, M.-H., 1999. L'interaction eau-roche-sol à basse température dans les Alpes. Ph.D. Thesis, University of Lausanne.

Derron, M.-H., Pfeifer, H.-R., 1998. Influence of rock composition on some trace metals in spring waters from alpine watersheds. Goldschmidt Conference Toulouse 1998. Min. Mag. 62A, 373–374.

Desaules, A., Studer, K., 1993. NABO, réseau national d'observation de sols, période d'observation 1985–91. Cahier de l'environnement no. 200. OFEFP, Bern.

Dominik, J., Cuccodoro, S., Gourcy, L., Santiago, S., 1992. Uranium enrichment in the surface and ground waters of the Alpine Rhone watershed. In: Vernet J.-P. (Ed.), Impact of Heavy Metals on the Environment. Elsevier, Amsterdam, pp. 397–416.

Drever, J.I., Clow, D.W., 1995. Weathering rates in catchments. In: White, A.F., Brantley, S.L. (Eds.), Chemical Weathering Rates of Silicate Minerals. Reviews in Mineral. 31. Mineral. Society of America, Washington, DC, pp. 463–483.

Dubois, J.-D., 1992. Typologie des aquifères du cristallin: exemple des massifs des Aiguilles Rouges et du Mont-Blanc (France, Italie et Suisse) Ph.D. Thesis no. 950. 324 pp.

Dubois, J.-P., Okopnik, F., Benitez, N., Védy, J.-C., 1998. Origin and spatial variability of Cd in some soils of the Swiss Jura. Proceedings of the World Congress of Soil Science, Montpellier, August 20–26.

Ellenberg, H., Weber, H.E., Duell, R., Wirth, V., Werner, W., Paulissen, D., 1991. Zeigerwerte von Pflanzen in Mitteleuropa. Erich Goltze, Göttingen.

Ernst, W.H.O, 1988. Response of plants and vegetation to mine tailings and dredged materials. In: Salomons, W., Förstner, U. (Eds.), Chemistry and Biology of Solid Waste, Springer, Berlin, pp. 54–69.

Escher, E., 1933. Erzlagerstätten und Bergbau im Schams. In: Mittelbünden und im Engadin. Beitr. Geol. Schweiz, Geotechn. Ser. 18. Kümmerly & Frey, Bern, 120 pp.

Food and Agriculture Organization of the United Nations, 1988. FAO/Unesco Soil Map of the World (revised legend). World Soil Resources Report 60. FAO, Rome, 119 pp.

Faure, G., 1992. Principles and Applications of Inorganic Geochemistry. Maxwell-Macmillan, New York, p. 626.

Förstner, U., Contaminated Sediments. Springer, Berlin, 157 pp.

Govet, G.I.S. (Ed.), 1983. Handbook of Exploration Chemistry. Six volumes, edited between 1983 and 1994. Elsevier, Amsterdam.

Halter, W., Pfeifer, H.-R., 1998. The effect of temperature on the surface properties of α-Al_2O_3 and AS(V) adsorption. Mineralogical Magazine, 64, 563–564.

Hamblin, W.K., 1992. Earth's Dynamic Systems. 6th ed., 646 pp.

Hamel, M., 1998. Minéralogie et sédimentologie du sidérolitique et de son substratum mésozoique à Vicques-Courcelon (bassin de Delémont, Jura). Diploma Thesis, Sciences de la Terre. University of Lausanne, 153 pp.

Hein, U.F., Schneider, H.-J., 1983. Fluorine anomalies accompanying the alpine Pb-Zn deposits compared to the geochemistry of their fluorites. In: Schneider, H.-J. (Ed.), Mineral Deposits of the Alps. Springer, Berlin, pp. 198–212.

Högl, O., 1980. Die Mineral- und Heilquellen der Schweiz. Verlag Paul Haupt, Bern, p. 302.

Hsi, C.K., Langmuir, D., 1985. Adsorption of uranyl onto ferric oxy-hydroxides: application to the surface complexation site-binding model. Geochim. Cosmochim. Acta 49, 1931–1941.

Huber, K., 1997. Altération minérale et transferts de matière dans les sols issus de moraines acides d'âge croissant (chronoséquence) des Alpes Suisses. Ph.D. Thesis no. 1714. EPF, Lausanne.

Jenne, E.A., 1968. Controls on Mn, Fe, Co, Ni, Cu and Zn concentrations in soils and water: the significant role of hydrous Mn and Fe oxides. In: Gould, R.F. (Ed.), Trace Inorganics in Water. Adv. Chem. series 73, American Chemical Society Publication, pp. 337–387.

Juchler, S.J., 1988. Die Böden auf Serpentinit in der subalpinen Stufe bei Davos. I. Bodenbildung. II. Nickel und Chromdynamik. Ph.D. Thesis no. 8716, ETH-Zurich, p. 174.

Knill, M.D., 1996. The Pb-Zn-As-Tl-Ba-deposit at Lengenbach, Binn Valley, Switzerland. Beitr. Geol. Schweiz, Geotechn. Ser. 90, p. 74.

Köppel, V., 1964. Die Vererzungen im insubrischen Kristallin des Malcantone. Beitr. Geol. Schweiz, Geotechn. Ser. 40, p. 123.

Kündig, R. (Ed.), 1997. Die mineralischen Rohstoffe der Schweiz. Schweiz. Geotechn. Komm. Zürich, p. 522.

Landolt, E., 1977. Oekologische Zeigerwerte zur Schweizer Flora (with an English summary). Veröff. Geobotan. Inst. ETH, Stiftung Rübel, 64, p. 208.

Landolt, E., Unsere Alpenflora, 6th ed. Fischer, Stuttgart, p. 318.

Lozet, J., Mathieu, C., 1990. Dictionnaire de Science du Sol (with English-French index). Technique et Docum. Lavoisier, Paris, p. 384.

Mandia, Y., 1991. Typologie des aquifères évaporitiques du Trias dans le bassin lémanique du Rhône (Alpes occidentales). Ph.D. Thesis no. 948, p. 345.

Markert, B. (Ed.), 1993. Plants as Biomonitors. Indicators for Heavy Metals in the Terrestrial Environments. VCH Publishers, Weinheim, p. 644.

Martini, R., Pillet, M., 1996. Etude biogéochimique comparative des bassins versants de la Magliasina (Malcantone) et de la Breggia (Val Muggio), Tessin. Diploma Thesis, Sciences de l'Environnement, University of Lausanne and University of Geneva.

McBride, M.B., 1994. Environmental Chemistry of Soils. Oxford University Press, Oxford.

McCarthy, J.F., Degueldre, C., 1993. Sampling and characterization of colloids and particles in groundwater for studying their role in contaminant transport. In: Buffle, J., van Leeuwen, H.P. (Eds.), Environmental Particles, vol. 2. Lewis, Boca Raton, FL, pp. 247–315.

Meyer, K., 1991. Bodenverschmutzung in der Schweiz. Themenbericht. Ed. Nationales Forschungsprogramm No. 22 "Boden", Liebefeld, Bern, 240 pp.

Mouvet, C., 1986. Métaux lourds et mousses aquatiques, synthèse méthodologique. Agence de l'eau Rhin-Meuse, Moulins-les Metz, 110 pp.

Nawratil, C., 1998. Contamination en métaux lourds des sols et plantes de la décharge de Croglio, Sottoceneri/TI. Diploma Thesis, Sciences de l'Environnement, University of Lausanne and University of Geneva.

Nesbitt, H.W., Young, G.M., 1982. Prediction of some weathering trends of plutonic and volcanic rocks based on thermodynamic and kinetic considerations. Geochim. Cosmochim. Acta 48, 1523–1534.

Parriaux, A., Dubois, J.D., Mandia, Y., Basabe, P., Bensimon, M., 1990. The AQUITYP-project: towards an aquifer typology in the alpine orogen. Mem. 22nd Congress of IAH, XXII. Lausanne, pp. 254–262.

Pastorelli, S., Marini, L., Hunziker, J., 1999. Water chemistry and isotope composition of the Acquarossa thermal system, Ticino, Switzerland. Geothermics 28, 75–93.

Peduzzi, R., Bachofen, R., Tonolla, M., 1998. Lake Cadagno: a meromictic alpine lake. Docum. Ist. Ital. Idrobiol. 63, 152.

Pfeifer, H.-R., 1977. A model for fluids in metamorphosed ultramafic rocks. I. Observations at surface and subsurface conditions (high pH spring waters). Schweiz. Mineral. Petrogr. Mitt. 57, 361–396.

Pfeifer, H.-R., Lavanchy, J.-C., Serneels, V., 1991. Bulk chemical analysis of geological and industrial materials by X-ray fluorescence, recent developments and application to material rich in iron oxide. J. Trace Microprobe Technol. 9, 127–147.

Pfeifer, H.-R., Sanchez, A., Degueldre, C., 1992. Thermal springs in granitic rocks from the Grimsel Pass (Swiss Alps): the late stage of a hydrothermal system related to the Alpine orogeny. In: Kharaka, Y., Maest, A. (Eds.), Water Rock Interaction, 7th International Symposium Water-Rock Interaction, Park City. Balkema, Rotterdam. pp. 1327–1331.

Pfeifer, H.-R., Vust, M., Meisser, N., Doppenberg, R., Croci Torti, R., Domergue, F.-L., Keller, C., Hunziker, J., 1994. Uranium enrichment of soils and plants in the vicinity of a pitchblende vein at La Creusaz/Les Marécottes (Wallis). Eclogae Geol. Helv. 87, 491–502.

Pfeifer, H.-R., Hansen, J., Hunziker, J., Rey, D., Schafer, M., Serneels, V., 1996. Arsenic in Swiss soils and waters and their relation to rock composition and mining activity. In: Prost, R. (Ed.), Contaminated Soils, 3rd Internatational Conference on Biogeochemistry of Trace Elements, Paris, May 15–19, D:/data/communic/050.PDF, Colloque 85, INRA ed., Paris.

Pfeifer, H.-R., Rey, D., 1998. Arsenic and heavy metal contents in soil percolation and surface water from the former mining area of Astano (Switzerland), a key to their sorption behavior. Proceedings of the Goldschmidt Conference, Toulouse, 1998. Min. Mag. 62A, 1167–1168.

Pochon, M., 1978. Origine et évolution des sols du Haut Jura suisse. Ph.D. Thesis. University of Neuchâtel. Mém. Soc. Helvétique Sci. Nat. XV, 190.

Rey, D., 1996. Arsenic dans les sols et eaux d'Astano (TI). Diploma Thesis. Sciences de la Terre, University of Lausanne.

Richard, F., Lüscher, P., 1987. Physikalische Eigenschaften von Böden der Schweiz. Vol. 4, No. 19: Lokalform "Cima Pianca" Humus(-Eisen) Podsol (Krypto-Humuspodsol). Eidg. Anst. forstl. Versuchsw., Birmensdorf.

Rose, A.W., Hawkes, H., Webb, J.S., 1979. Geochemistry in Mineral Exploration. 2nd ed., 657 pp.

Salbu, B., Steinnes, E. (Eds.), 1995. Trace Elements in Natural Waters. CRC Press, Boca Raton, FL, 302 pp.

Salomons, W., Förstner, U. (Eds.), 1984. Chemistry and Biology of Solid Waste. Springer, Berlin.

Scheder, M., Streiff, A., 1997. Etude biogéochimique du bassin versant de la Borgne (Val d'Hérens, Valais, Suisse). Diploma Thesis. Sciences de l'Environement, University of Lausanne and University of Geneva.

Scheder, M., Pfeifer, H.-R., Beatrizotti, G., Righetti, G., 1999. Sorgenti del Ticino: Composizione chimica e relazione con le rocche e il tipo di circulazione sottoterranea. Rapporto di Lavoro del Istituto di Scienze della Terra, Univ. Professionale, Ticino.

Schlegel, C., 1997. Contamination des sols, plantes et eaux en Ni, Cr, Co et V au voisinage de roches ultramafiques à Baldissero- Canavese (I). Diploma Thesis. Sciences de la Terre, University of Lausanne.

Schmidt, S. Th., Amstutz, G.C., 1983. Mineralogical investigations of strata-bound Pb-Zn-Ba-occurrences in the Triassic of the Central Alps (Switzerland). In: Schneider, H.-J. (Ed.), Mineral Deposits of the Alps. Springer, Berlin, pp. 117–127.

Schroll, E., 1983. Geochemical characterization of the Bleiberg type and other carbonate hosted lead-zinc mineralizations. In: Schneider, H.-J. (Ed.), Mineral Deposits of the Alps. Springer, Berlin, pp. 189–197.

Sposito, G., 1989. The Chemistry of Soils. Oxford University Press, Oxford, 227 pp.

Spreafico, M., Weingartner, R., Leibundgut, C., 1992. Hydrological Atlas of Switzerland. Ed. Landeshydrologie und -geologie, Bern.

Surbeck, H., Medici, F., 1990. Rn-22 transport from soil to karst caves by percolating water. Mem. 22nd Congress IAH, Lausanne, pp. 348–355.

Swoboda-Colberg, N., Drever, J.I., 1993. Mineral dissolution rates in plot-scale field and laboratory experiments. Chem. Geol. 105, 51–69.

Tessier, A., 1992. Sorption of trace elements on natural particles in oxic environments. In: Buffle, J., van Leeuwen, H.P. (Eds.), Environmental Particles, Vol. 1. Lewis, Boca Raton, FL, pp. 425–453.

Tuchschmid, M., 1995. Quantifizierung und Regionalisierung von Schwermetall-und Fluorgehalten bodenbildender Gesteine der Schweiz. Umwelt Materialien no. 32. Boden, Swiss Federal Office of the Environment (BUWAL), Bern.

VBBo, 1998. Ordinance relating to pollutants in soil of July 1, 1998. SR 814.12, Government of Switzerland, EDMZ, Bern.

Wenger, C., Steiger, R., Bianconi, F., Kündig, R., 1990. Karte der Vorkommen mineralischer Rohstoffe der Schweiz 1: 200000, Schweiz geotech. Komm., Zurich, 114 pp.

Wickenheiser, E.B., Michalke, K., Hensel, R., Dreschner, C., Hirner, A.V., Rutishauser, B., Bachofen, R., 1998. Volatile compounds in gases emitted from the wetland bogs near Lake Cadagno. In: Peduzzi, R. Bachofen, R., Tonolla, M. (Eds.), Lake Cadagno: a Meromictic Lake. Documenta Ist. ital. Idrobiol. 63, pp. 137–140.

Woodtli, R., Bugnon, C., Della Valle, G., Escher, A., Gex, P., Grebert, Y., Imfeld, N., Jaffe, F., Lavanchy, J.-C., Moritz, R., Pfeifer, H.-R., von Raumer, J., Sartori, M., Thierrin, J., 1985. Projet UROMINE, recherches minières exécutées au Valais par les universités de Lausanne, Fribourg et Genève. National Project no. 7. 488 pp.

Trace elements – Their distribution and effects in the environment
B. Markert and K. Friese, editors

Chapter 3

Cadmium in the environment of Central Europe

M. Ursínyová, V. Hladíková

Abstract

Problems with metal pollution of the environment still exist both on a local and regional scale. The most important sources of cadmium contamination in order of significance are air pollution, river sediments, sewage sludge, waste. Increases in environmental cadmium content result in an increase in cadmium levels in foods. This chapter presents the information about the environmental pollution and human exposure to cadmium in the Central European countries (Slovak Republic, Czech Republic, Poland, Hungary and Austria). The evaluations of available local data on cadmium concentrations in emissions, ambient air, drinking water, river water, sediments and food were done considering WHO guidelines. The cadmium content in human biological materials (placenta, human milk, blood and tissues) from these countries is presented. The survey indicates that the main sources of cadmium industrial emissions in this area are in the Upper Silesian Industrial Region in Poland.

Keywords: Cadmium, Air, Water, Sediments, Food, Human biological materials, Central Europe

1. Production and uses

Cadmium is recovered as a by-product of smelting zinc concentrates and as such its production is dependent upon the demand for zinc. The percentage of cadmium in zinc concentrates varies from mine to mine, ranging from 0.07 to 0.83% with an average of 0.23%. Since the zinc content of these concentrates is 55%, approximately 3 kg of cadmium will be produced for every ton of refined zinc.

Production of refined cadmium in the western world (except of Central and Eastern European countries) in 1993 was approximately 15 000 tons. Japan is the largest producer (2808 tons in 1993), treating concentrates from South Africa, Asia and Australia as well as from its own mines. Production in countries with economies in transition was estimated to be around 4500 tons in 1990, but fell to 3000 tons more recently with more than half produced in the former Soviet Union (OECD, 1994).

Cadmium is relatively easy to recycle after separation from other materials with fairly low energy expenditure. Cadmium is volatilized at the comparatively low temperatures, making it easy to separate from metals with higher melting points used in a high-temperature recovery process. Cadmium is also active chemically and electrochemically, and thus can be readily removed using hydrometallurgical techniques. Cadmium recycling in general has been developed in the major industrialized countries.

Secondary cadmium production is in the range of 1500–2000 tons per year, approximately 10–15% of total current production in western world (OECD, 1994).

Cadmium is used in a number of industrial processes, but for most of its uses there are alternatives of lower toxicity. The principal applications of cadmium fall into five categories: protective plating on steel, stabilizers for polyvinyl chloride, pigments in plastics and glasses, electrode material in nickel–cadmium batteries, and as a component of various alloys.

2. Sources of environmental contamination

Cadmium is widely distributed in the earth's crust at an average concentration of about 0.1–0.2 mg/kg and is commonly found in association with zinc. Volcanic activity is major natural source of cadmium release to the atmosphere. The annual global flux from this source has been estimated to be 820 tons. Deep sea volcanism is also a source of environmental cadmium release, but the role of this process in the global cadmium cycle remains to be quantified (OECD, 1994). About 10–15% of total airborne cadmium emissions arise from natural processes (WHO, 1992a).

The WHO reported a total world-wide emission of between 3350 and 14 640 tons per year and with emission from natural sources being between 150 and 2600 tons per year (WHO, 1992b).

Between 15 000 and 17 000 tons of cadmium per year is transformed into products belonging mainly to five categories mentioned above (OECD, 1994). At the global level, the smelting of non-ferrous metal ores has been estimated to be the largest human source of cadmium release to the environment (2660–8200 tons per year) accounting for about 75% of the total anthropogenic emissions world-wide but only 25% in the European Economic Community (WHO, 1992b).

3. Distribution between media and environmental effects

Anthropogenic emissions of heavy metals in Europe and North America have greatly increased the concentrations of especially cadmium, lead and mercury even in remote areas, particularly in the northern hemisphere. The regional atmospheric deposition pattern of cadmium as well as other metals (arsenic, copper, mercury, lead, zinc) exhibits large-scale gradients that can be related to long-range transport from anthropogenic source areas. In many cases concentrations of heavy metals in topsoils, bogs and lake sediments show regional gradients that are parallel to the atmospheric deposition patterns (Anon, 1994).

Metals are stable environmental contaminants, therefore, they tend to accumulate in the biosphere, being preferentially stored in soils and sediments (Anon, 1994).

Increases in soil cadmium content result in an increase in the uptake of cadmium by plants. The uptake by plants from soil is greater at low soil pH. Processes that acidify soil (e.g. acid rain) may therefore increase the average cadmium concentrations in foodstuffs (WHO, 1992b).

However, the uptake from soil is limited. Plants in locations with high-level exposure to cadmium (in the range of hundreds of mg/kg), have growth reduction as the major effect. Plants exposed to cadmium in the field for long periods can develop tolerance to the metal (WHO, 1992a).

Elisher (WHO, 1992a) concluded that there are at least six trends evident from the abundant data available for cadmium:

1. Marine organisms generally contain higher cadmium residues than their freshwater and terrestrial counterparts.
2. Cadmium tends to concentrate in the viscera of vertebrates, especially the liver and kidneys.
3. Cadmium concentrations are generally higher in older organisms.
4. Higher cadmium residues are generally associated with industrial and urban sources, although this does not apply to sea birds and sea mammals.
5. Cadmium residues in plants are normally less than 1 mg/kg. However, plants growing in soil amended with cadmium (e.g. from sewage sludge) may contain significantly higher levels.
6. The species analyzed, season of collection, ambient cadmium levels, and the sex of the organism probably all affect the residue level.

4. Cadmium intake

Human intake of cadmium occurs via inhalation and ingestion of food and water (WHO, 1992b).

Assuming a cadmium concentration of 10 ng/m^3 in both indoor and outdoor air and a daily inhalation rate of 15 m^3 for an adult, the average intake of cadmium from the atmosphere would be 0.15 μg, of which about 25% will be absorbed (OECD, 1994).

Cadmium intake from drinking water, based on daily consumption of 2 l, is usually less than 1 μg (OECD, 1994).

Tobacco is an important source of cadmium uptake in smokers, no matter whether passive or active. Smoking a pack of 20 cigarettes daily can result in the inhalation of 2–4 μg cadmium, the amount varying considerably according to the country or origin of the tobacco. Of this amount, 25–50% may be absorbed via the lungs, resulting in an uptake of 1–2 μg, a much larger amount than from air alone. Those who smoke two or more packs of cigarettes daily will absorb correspondingly greater quantities of cadmium (OECD, 1994).

The cadmium content of agricultural crops varies according to species, variety cultivated and season. It is evident that cadmium is a normal constituent of most foodstuffs. Meat, fish, and fruit generally contain similar cadmium levels and values of 5–10 μg/kg fresh weight are representative for these food classes. The cereals and root vegetables contain higher cadmium concentrations and value of 25 μg/kg fresh weight is considered representative for these food items. Adult animals and certain shellfish contain even higher concentration; values in excess of 50–100 μg/kg fresh weight are considered normal. Food preparation (milling, washing, peeling, cooking) results in a reduction of cadmium content (WHO, 1992a).

In order to prevent any further increase of cadmium in agricultural soils likely to increase the dietary intake of future generations, the WHO recommended that airborne levels of cadmium in rural areas should be as low as possible and not exceed 5 ng/m^3 and in urban or industrial areas levels should not exceed 10 ng/m^3 (WHO, 1994). The provisional tolerably weekly intake (PTWI) for the total cadmium intake was set at 7 μg/kg of body weight (WHO, 1989). A guideline value for cadmium in drinking water of 3 μg/l was established based on an allocation of 10% of the PTWI (WHO, 1993, 1996).

5. Evaluation of human health risks

The general population is exposed to cadmium through the contaminated food and tobacco smoke and in occupational settings by inhalation (Gerhardsson and Skerfving, 1996). At the inhalatory exposure, absorption is dependent upon the chemical speciation, particle size and solubility in biological fluids. Uptakes up to 50% of the inhaled doses have been reported (WHO, 1992b). Of the cadmium inhaled at smoking, about 25–50% is absorbed (OECD, 1994). Between 0.1 and 0.2 μg of cadmium intake per cigarette has been reported. Furthermore, more cadmium is found in the particulate phase than in the gaseous phase (Medeiros et al., 1997). The gastrointestinal absorption of cadmium is influenced by the type of diet and nutrition status. On average, 5% of the total oral intake of cadmium is absorbed, but individual values range from less than 1 to more 20% (WHO, 1992b).

The mechanisms of cadmium toxicity are not fully understood. However, it is well established that cadmium is bound to the sulfhydryl group of proteins. Cadmium is absorbed by blood plasma and penetrates the blood cells. Therefore, the cadmium concentration in erythrocytes is much higher than in plasma (Gerhardsson and Skerfving, 1996). Cadmium absorbed from the lungs or gastrointestinal tract is stored principally in the liver and kidneys where more than half of the cadmium body burden is deposited (WHO, 1992b). Both the liver and kidneys store cadmium as metallothionein complex, which serves not only for the transport of cadmium but also acts as a defense mechanism against the toxicity of the unbound cadmium ion. Ironically, it is the cadmium–metallothionein complex accumulated in kidneys that is partially responsible for toxic renal effect. Cadmium does not undergo the metabolic conversion in vivo (Agency for Toxic Substances and Disease Registry, 1990).

Cadmium is excreted from the body primarily via urine and the excretion is very slow. The daily excretion via faces and urine comprises only about 0.01–0.02% of total body burden of cadmium in human beings (Friberg et al., 1986). The mean concentration of urinary cadmium in unexposed people is $<$0.5–2.0 μg/l (WHO, 1992b). The excretion via urine is increased with age (Friberg et al., 1986) It was found that smokers have higher urinary excretion than non-smokers (WHO, 1992b). Whereas cadmium concentration in the blood reflects the recent exposure, urinary cadmium concentration more closely reflects the total body burden (Agency for Toxic Substances and Disease Registry, 1990).

In the exposed people suffering from some renal impairment, urinary excretion of cadmium is increased and, thus, the whole body half-time is shortened. The renal damage leads to losses of cadmium from the kidney, and the renal concentrations may be lower compared to people with similar exposure but without renal damage (WHO, 1992b).

In clinical evaluation, adverse effects of excessive cadmium exposure may include the following:

- acute exposure: gastroenteritis (ingestion only), bronchitis (inhalations only), interstitial pneumonitis (inhalation only), pulmonary edema (inhalation only);
- chronic exposure: proteinuria, osteomalacia (itai-itai disease), pulmonary fibrosis (inhalation only), liver damage, hypertension, lung cancer, prostatic cancer, mild anemia, yellow discoloration of front teeth, anosmia (Agency for Toxic Substances and Disease Registry, 1990).

There is some evidence that cadmium, following the inhalation, is carcinogenic, and IARC has encountered cadmium and its compounds to Group 2A. However, there is no evidence for cadmium carcinogenicity of the oral route, and there is no clear evidence that cadmium is genotoxic (WHO, 1996).

The total body burden of cadmium is less than 1 μg at birth, which gradually increases with age, up to 50 years. At this age, one who has essentially been unexposed to cadmium may accumulate 20–30 mg of cadmium in the body (Medeiros et al., 1997). The biological half-time has been estimated to be 15–30 years in human beings (Piscator, 1982). The placenta acts only as a partial barrier against the fetal exposure. Before the formation of the placenta, cadmium reaches the embryo. Later, cadmium is deposited in the placenta, and there is limited passage of cadmium into the fetus. Moreover, there is only minimal cadmium excretion in breast milk (Gerhardsson and Skerfving, 1996).

During a long-term exposure to low concentrations of cadmium, the highest cadmium concentration was found in the kidneys, especially in the renal cortex (the concentration of which is 1.25 times higher than in the whole kidney), followed by the liver, pancreas, and adrenals (Agency for Toxic Substances and Disease Registry, 1990; Gerhardsson and Skerfving, 1996). The lowest concentration was found in the brain, bone, and fat (WHO, 1992b).

At higher levels of cadmium exposure, a greater proportion of the body burden was found in the liver. After the development of severe cadmium-induced renal dysfunction, cadmium was found to be lost from the renal tissue (WHO, 1992b).

6. Summary and evaluation of data from Central European countries

The data on cadmium in the environment, food and biological materials used in this paper have been published by scientists and national ministries for the environment in countries of the Central Europe – Slovakia, Czech Republic, Poland, Hungary, and Austria.

6.1. Cadmium in the atmosphere

Emissions of cadmium to the atmosphere are an important transport pathway, leading to its precipitation and deposition onto soil, vegetation and surface waters.

Table 1. Emissions of cadmium in the Slovak Republic (t)

Activity	Cd	
	1992	1994
Primary and secondary iron and steel industry	0.199	0.214
Primary and secondary non-ferrous metal production	1.405	0.655
Inorganic chemical industry	0.0003	0.0002
Cement production	0.0009	0.003
Glass production	9.029	5.100
Combustion of fossil fuels	0.404	0.251
Magnesite production	0.015	0.007
Waste incineration	0.060	0.060
Traffic	0.529	0.569
Total	11.288	6.859

The results of the cadmium emission inventory in Slovakia carried out by the Slovak Hydrometerological Institute in 1992 and 1994 are shown in Table 1. Shutting down environmentally and economically ineffective productions have significantly reduced emissions of cadmium (Magulová et al., 1996).

National emission balance of heavy metals in the Czech Republic includes the emissions of cadmium, lead and mercury from all major pollution sources, for which the emission factors were available in the Atmospheric Emission Inventory Guidebook, published in other countries (Austria, Germany, Poland, Slovakia) or calculated from experimental measurements of projects of the Ministry of the Environment of the Czech Republic. The results of the cadmium emission balance in the Czech Republic in 1992 are presented in Table 2. The authors assume that the amount of emissions from the sources not included in this inventory do not exceed 5–10% of the total emissions (Fiala et al., 1998).

Cadmium emission in Poland in 1990 was 86.5 tons, coming mostly from fuel combustion for energy production purposes (80%). The Polish contribution to the European emission of cadmium is 8.4%. On the Polish list of main environmental pollutants heavy metals such as cadmium and lead are regarded as the leading ones. The Upper Silesian Industrial Region in the Katowice Province, situated in the south of Poland, has the biggest industry and population concentration in the whole country. There are about 200 industrial plants which have been recognized as particularly noxious to the environment. About 50% of national steel production, over 90% of coal and 100% of zinc lead ore processing, are located on this Province, especially in

Table 2. Emissions of cadmium in the Czech Republic, year 1992 (t)

Activity	Cd
Incineration	7.0
Industry	1.7
Traffic	1.2
Total	9.9

Table 3. Emissions of cadmium in Austria, year 1993

Activity	Cd	
	t	%
Primary and secondary iron and steel industry	1.373	35
Primary and secondary non-ferrous metal production	0.210	5
Cement production	0.196	5
Glass production	0.066	2
Combustion of fossil fuels	1.609	41
Waste incineration	0.008	<1
Traffic	0.457	12
Total	3.919	100

its central part. As a result the Upper Silesian Industrial Region is responsible for the emission of the largest quantity of heavy metals in Poland (Gzyl, 1997).

One of the first estimates of the total emission of heavy metals in Austria for the year 1992 yielded approximately 5 tons of cadmium, from which approximately 3 tons represented fine particulate matter relevant for long-range transport (Federal Ministry for Environment, Youth and Family, 1997). Table 3 shows the evaluation of cadmium air emissions in Austria in 1993 (Österreichische Akademie der Wissenschaften-Kommission für Reinhaltung der Luft, 1993).

The bad health status of the population in Hungary is – to a significant extent – related to the effect of air pollution. About 13% of the Hungarian territory, where 49% of the population lives is regarded as ''polluted''. Environment protecting authorities' efforts as well as decreasing industrial energy use as a result of economic recession and structural changes played an important role in the decrease of pollutant emission (Pintér, 1997). The trend of annual cadmium emission in Hungary is presented in Table 4 (Anon, 1998).

The results of earlier cadmium emission estimates within the German Federal Environmental Protection Agency, the available data from Paris Commission program (PARCOM), the European Soil and Sea Quality and Atmospheric Deposition of Selected Substances (ESQAD) data and emissions estimated by national experts in some countries are presented in Table 5 (Anon, 1994).

Air pollution problems may vary greatly with the geography, demography, and the socio-economic profile of a region. The climate and topography of a region will

Table 4. Trend of annual cadmium emission in Hungary (t)

Year	Cadmium
1980	5.3
1985	5.2
1990	3.9
1991	3.9
1992	3.4
1993	3.6
1994	3.5

Table 5. Emission estimates for cadmium in Europe (t)

Country	German UBA Project (published in 1988)	Pacyna 1993 (1989-data)	Other national data (reference year)	TNO data (1983-data)	ESQUAD Project (1990-data)
Austria	5.3	4.3	5.1 (1994)		4
Czechoslovakia (former)	21.6	21.8	11.2 (1992) (only Slovakia)		18
Hungary	4.4	4.4			5
Poland	180.4	177	255.5 (1990)	45	51

Table 6. Cadmium concentrations in ambient air in the Slovak Republic in 1996–19

Locality	Arithmetic mean (ng/m^3)	Geometric mean (ng/m^3)	Minimum (ng/m^3)	Maxim (ng/m^3)
Bratislava I	0.42	0.33	0.11	0.79
Bratislava II	0.61	0.52	0.20	1.37
Bratislava III	0.47	0.30	0.02	0.98
Bratislava IV	0.35	0.28	0.07	0.96
Bratislava V	0.46	0.29	0.05	1.86
Topol'níky	3.00	0.79	0.20	19.94
Hurbanovo	0.82	0.46	0.08	3.72
Mochovce	0.89	0.45	0.19	3.51
Žiar n. Hronom	0.55	0.41	0.14	1.24
Prievidza	0.38	0.29	0.11	1.24
Žilina	0.71	0.56	0.16	1.33
Ružomberok	0.73	0.59	0.18	1.39
Krompachy	5.80	4.34	1.09	13.48
Strážske	1.27	0.80	0.15	2.93
Starina	0.61	0.47	0.19	1.55
Košice I	3.09	1.94	0.82	13.98
Košice II	0.95	0.83	0.40	2.49
Košice III	3.34	1.56	0.22	14.9
Košice IV	5.71	1.22	0.26	39.79
Vel'ká Ida	10.27	5.27	0.81	27.94

influence the distribution and atmospheric processes of the pollutant and, ultimately, its effect on the environment and/or human health.

In the framework of a PHARE project in 1996–1997 cadmium concentrations in ambient air were analyzed in 20 localities of Slovakia (Table 6). In National Ambient Air Quality Standard defined by the Slovak legislation is the annual arithmetic mean of average 24-h concentrations of cadmium and is quoted 0.01 μg/m^3 (i.e. the same as the WHO guideline for Cd in urban or industrial areas). The results show that the arithmetic mean concentration in Vel'ká Ida exceeded the WHO guideline for Cd in urban areas. The average concentrations in Krompachy, Košice IV were higher than 5 ng/m^3. The highest means were associated with metallurgical industries (at Krompachy, Košice and Vel'ká Ida) but also in agricultural locality Topol'níky was daily value higher than recommendation of the WHO (Ursínyová and Hladíková, 1998a).

Monitoring of ambient air pollution in the Czech Republic in 1994 realized in 25 selected localities showed that the annual arithmetic concentrations for cadmium in these regions were below the National Ambient Air Quality Standard 0.01 μg/m^3 except for Ostrava and Prague 7 (Table 7) (National Institute of Public Health in Prague, 1995a). The Czech Hydrometeological Institute reported the long-term exceeding of the cadmium immision limit 0.01 μg/m^3 in the districts on the north of the Czech Republic (Liberec, Tanvald, Souš) not included in the monitoring mentioned above (Fiala et al., 1998).

There is little information about ambient air concentrations for heavy metals in Austria. In Table 8 cadmium concentrations in some Austrian cities are presented

Cadmium concentrations in ambient air in the Czech Republic in 1994

n	Arithmetic mean (ng/m^3)	Geometric mean (ng/m^3)
gue 1	2.37	0.57
ague 4	1.70	0.90
rague 5	1.43	0.64
Prague 6	1.02	0.48
Prague 7	11.1	0.74
Prague 10	1.49	0.54
Benešov	1.19	0.63
Kladno	4.70	0.73
Kolín	0.69	0.33
Klatovy	0.29	0.11
Plzeň	1.18	1.01
Sokolov	0.60	0.29
Děčín	0.93	0.55
Ústí nad Labem	0.43	0.15
Havlíčkův Brod	0.34	0.27
Hradec Králové	1.89	1.45
Svitavy	1.18	1.13
Ústí nad Orlicí	0.31	0.14
Brno	0.42	0.14
Hodonín	0.47	0.12
Jihlava	1.00	1.00
Kroměříž	1.20	0.73
Karviná	2.93	2.26
Olomouc	1.23	1.18
Ostrava	21.00	7.04

(Österreichische Akademie der Wissenschaften-Kommission für Reinhaltung der Luft, 1993). Due to zinc and lead ore processing Arnoldstein belongs to the Austrian areas with the highest heavy metal pollution.

In Poland cadmium in suspended dust does not exceed standard values (0.01 μg/m^3) except for some results obtained in the Upper Silesian Industrial Region, which is generally considered to be the most polluted region in Poland as well as in Europe (Andrzejewski and Bararnowski, 1993; Gzyl, 1997). The Regional Sanitary and Epidemiological Station, Katowice, has measured cadmium and lead in total suspended dust as 24 hour concentrations at 26 measurement stations in the Katowice Province. The average cadmium concentration was 0.003 μg/m^3 (range: 0.001–0.054 μg/m^3). On the other hand, average annual concentration of cadmium on the territory of the Upper Silesian Industrial Region in 1989 reported by the State Inspectorate for Environmental Protection and Environmental Information Centre GRID-Warsaw was much higher – 0.10 μg/m^3 with maximum concentrations in the Silesian cities reaching 0.80–1.16 μg/m^3. Apart from this region, in many other places in Poland there are local problems, often of great intensity. Examples of serious heavy metal pollution are the contaminations of air and soil near the coloured metals smelters in Legnica, Głogów and Miasteczko Śląskie. This local problems often significantly affect the

Table 8. Concentrations of cadmium in ambient air in Austria

Region	Cd (ng/m^3)
Arnoldstein, 1992	50
Linz Steyregg, 1984	1
Linz Urfahr, 1984	1
Linz Steyregg, 1982	3
Linz Urfahr, 1982	2
Hallein, 1982	7

health of local inhabitants, although in a country-wide scale, they are shadowed by general pollution problems (Andrzejewski and Bararnowski, 1993).

6.2. Cadmium in water and sediments

Ground and surface water contamination from industrial, communal, and agricultural sources has been identified as one of the three environmental priorities for Central and Eastern Europe. Recent information shows that the regulations introduced in response to the alarming concentrations of heavy metals in many European lakes and rivers in the 1970s have been effective. In general, concentrations today are below standards for drinking water; only cadmium and mercury exceed drinking-water standards in some rivers. There are, however, still some heavy-metal ''hot spots'' associated with mining areas and industries using large quantities of metals that pose a risk to human health (Anon, 1997).

The Water Quality Protection Working Group of the Hungarian-Slovak Transboundary Water Commission has elaborated, at its Tiszafüred meeting in 1989, a program for extended water quality measurements and assessment of Danube reach between Bratislava and Visegrád. The memorandum of LVIth Session of the Hungarian-Slovak Transboundary Water Commission requested to carry out an extended comparison of the results of investigations of Danube and its tributaries for the period 1989–1995. The objective of this evaluation is to identify the dynamics of water quality changes and their trends in Danube and in tributaries of the Hungarian and Slovak catchment area for the study period. The limit values stemmed from the 6-class quality classification system of the former COMECON and from the proposals of the European Economic Committee (Makovinská et al., 1997). Cadmium concentrations were substantially lower than 3 μg/l, the limit value of class I, in both the Danube and its tributaries. The highest measured value was 3.2 μg/l.

The pollution levels of suspended solids and bottom sediments were measured in the Danube station (Bratislava, Komárno-Komárom, Szob) from micropollutants that are attached to solid particles (Tables 9 and 10). The analyzed components included organic micropollutants and heavy metals. Bottom sediment samples were taken at the river bank, while suspended solids were sampled in the main stream. Cadmium concentrations remained below the Severe Effect Level (SEL) (10 mg/kg for cadmium) in both the suspended solids and the bottom sediments (Makovinská et al., 1997).

The concentrations of cadmium in surface water in the Czech Republic in 1996 are shown in Table 11. According to the cadmium contamination, the most of the

Table 9. Cadmium in suspended solids of the Danube (1989–1995)

Locality	Cd (mg/kg)		
	Min	Mean	Max
Bratislava	0.3	2.8	7.1
Komárno/Komárom	0.07	2.9	9.9
Szob	0.4	2.6	9.8

observed surface belongs to the purity class I except for Odra-Bohumín (class II) and Olše-Věřňovice (class III) (Ministry of the Environment of the Czech Republic, 1998).

In Slovakia in 1996 the median concentration of cadmium in drinking water was 0.06 μg/l. None of the 235 samples exceeded the limit value for cadmium (5 μg/l) (Food Research Institute, 1997).

Extensive monitoring of drinking water quality from the public water supply in the Czech Republic showed the cadmium concentrations up to 200 μg/l (geometric mean, 0.1 μg/l). More than 60% of the 785 analyzed samples of the drinking water was below the tenth of the limit value for cadmium (5 μg/l), and only 1.6% exceeded that value (National Institute of Public Health in Prague, 1997).

Cadmium concentrations in drinking water in Poland ranged between 1 and 4 μg/l, however, in the Katowice Province in a few cases higher concentration of cadmium were observed (Gzyl, 1997).

6.3. Cadmium in food

Since 1980, systematic efforts have been made by the Joint UNEP/FAO/WHO Food Contamination Monitoring and Assessment Programme (GEMS/Food) to collect information on dietary intake of various contaminants, including cadmium, lead and mercury, for evaluation of the risk to human health. Different foods contain cadmium in the following (increasing) order: milk, eggs, fruits, vegetables, cereals, meat, fish and internal organs. The cadmium uptake of plants depends on the type and species of soil and plant. The highest level of cadmium occurs in bran.

The concentrations of cadmium in foods in Slovakia purchased during the period 1993–1997 in retail stores in ten localities that each of them represented about 20 000

Table 10. Cadmium in sediments of the Danube (1989–1995)

Locality	Cd (mg/kg)		
	Min	Mean	Max
Bratislava left side	0.08	0.74	1.8
Bratislava right side	0.06	0.57	1.8
Komárno left side	0.05	0.35	0.8
Komárno right side	0.02	0.47	1.6
Szob left side	0.06	0.66	2.4
Szob right side	0.04	0.70	2.4

Table 11. Cadmium in surface water in the Czech Republic (1996)

Locality	Cd (μg/l)
Labe-Valy	0.2
Labe-Lysá nad Labem	0.2
Labe-Obříství	0.3
Labe-Děčín	0.3
Jizera-Nový Vestec	0.1
Berounka-Lahovice	0.4
Bílina-Ústí nad Labem	0.3
Odra-Bohumín	2.6
Olše-Věřňovice	4.0
Svratka-Vranovice	0.2

inhabitants and different types of settlements are shown in Table 12 (Food Research Institute, 1997).

The comparison of cadmium intakes from two types of total diet (ovo-lacto-vegetarian diet and non-vegetarian standard diet) collected from the kindergartens in Bratislava (capital of Slovakia) was studied between 1991 and 1994 (Table 13). The authors found statistical significantly higher weekly intake in vegetarian children. The relatively low median intake of cadmium was explained by the fact that the diet samples were not prepared from ingredients containing high concentrations of cadmium (i.e. entrails, mushrooms) (Ursínyová and Hladíková, 1998b). In adults the median weekly intake of cadmium from vegetarian diet (0.96 μg/kg body weight) was also statistical significantly higher than that from non-vegetarian diet (0.37 μg/kg body weight), however, in both groups the intakes were quite lower than those in children (Ursínyová et al., 1997).

The results of the project of National Institute of Public Health, Prague, "The Health Risk Assessment of Dietary Exposures to the Selected Chemical Substances (Total Diet Study), Czech Republic, 1994" are presented in Table 14 (National Institute of Public Health in Prague, 1995b). In this study 552 composite food samples were analyzed representing 1920 individual commodities from 12 sites of the Czech Republic. No statistically significant difference was found between individual sites with a different environmental pollution level. The estimated exposure dose for the Czech Republic ranges between 28 and 41% of the PTWI.

The content of cadmium in vegetables (carrot, red beet, parsley, celery) in Poland and in Katowice Province varied between 0.004 and 0.078 mg/kg fresh matter and 0.08 and 19.30 mg/kg dry matter, respectively. The results of contamination of plant with cadmium in Poland are shown in Table 15 (Gzyl, 1997).

Exposure to cadmium from vegetables was calculated in Katowice Province as a percentage of PTWI established by WHO (0.4–0.5 mg Cd/week). They ranged between 2.4–80.8% of PTWI, whereas in reference area these values amounted to 4.6% of PTWI. Cadmium (and lead) coming from carrot and red beet have the largest share in human exposure to heavy metals from vegetables. The elimination of these vegetables may reduce cadmium intake from 1.7 to 50.5% of PTWI (Gzyl, 1997).

Table 12. Contents of cadmium (μg/kg fresh weight) in foodstuffs in Slovakia

Foodstuff	Average	Median	% of PTWI
Potatoes	12.51	10	4.36
Apples	3.69	1	0.10
Subtropic fruits	3.69	2	0.16
Carrots	12.04	10	0.62
Cabbage	4.87	2	0.16
Tomatoes	9.67	7	0.21
Onion	16.27	10	0.48
Milk	2.07	1	0.71
Eggs	4.21	2	0.24
Rice	19.55	11	0.43
Vegetable oils	4.25	0	0.14
Beef	7.51	4	0.44
Pork	10.42	3	1.48
Meet products	7.66	6	0.77
Pork fat	5.35	2	0.14
Poultry	10.23	3.5	0.47
Cheese – soft	7.98	5.5	0.15
Butter	5.10	2	0.08
Yoghurts	4.47	2	0.43
Bread	16.45	12.5	3.25
Rolls and cakes	15.48	14	2.79
Lemonades	1.81	0	0.30
Beer	2.76	1	0.97
Wine	2.51	1	0.14
Total	–	–	19.02

The data submitting to GEMS/Food by investigators for the period 1981–1987 show that general population mean weekly dietary intake of cadmium in Poland ranged from 2.2 to 5.3 μg/kg body weight (i.e. about 31–76% of PTWI) and the intakes of those living in industrial areas were much higher than those living in a non-industrial area. In the case of young children the mean weekly intake was 2.7–7.5 μg/kg body weight and even exceeded the PTWI value (UNEP/FAO/WHO, 1992).

The mean weekly dietary intake of cadmium reported from Hungary for the similar period was 0.4–0.9 μg/kg body weight (i.e. only about 6–13% of PTWI) (UNEP/FAO/WHO, 1992).

Table 13. Cadmium dietary exposure in children in Slovakia

Diet	*n*	Concentration (μg/kg total diet)		Weekly intake (μg/kg body weight)	
		Median	Range	Median	Range
Vegetarian	120	5.35	0.49–25.12	3.70	0.34–18.74
Non-vegetarian	120	4.90	1.06–19.10	2.61	0.58–11.53

Table 14. Daily dietary intake of cadmium (μg/kg body weight) in the Czech Republic

Foodstuff	Median	% of PTWI
Potatoes	0.12	12
Apples	0.00032	0.032
Subtropic fruits	0.0004	0.04
Vegetable products	0.012	1.2
Stem vegetables	0.0029	0.29
Fruiting vegetables	0.014	1.4
Leafy vegetables	0.0031	0.31
Root vegetables	0.01	1
Milk	0.00027	0.027
Eggs	0.00037	0.037
Sugar	0.00088	0.088
Sweets	0.0085	0.85
Cereals	0.025	2.5
Rice	0.0023	0.23
Plant oils	0.00025	0.025
Plant oil butters	0.00053	0.053
Beef	0.00047	0.047
Pork	0.00097	0.097
Canned meats	0.00049	0.049
Fish products	0.0013	0.13
Poultry meats	0.00042	0.042
Cheeses	0.00028	0.028
Butter	0.00014	0.014
Bread and rolls	0.071	7.1
Legumes	0.0011	0.11
Non-alcoholic beverages	0.00024	0.024
Alcoholic beverages	0.00026	0.026
Total	–	29

Cadmium levels in Hungarian foods have not changed considerably over recent years. The following median values were reported in 1995: in cereals 15–18 μg/kg, in vegetables 4–150 μg/kg, in fruits 1–7 μg/kg (Table 16).

The results of contamination of meat with cadmium are shown in Table 17. The average cadmium content of pig kidneys has not changed since 1992, but that of cattle kidneys shows a continuous increase: 357, 485, and 375 μg/kg, in the years 1992, 1993 and 1994, respectively (Pintér, 1997).

Austria, also one of the countries that participated in GEMS/Food Programme, did not submitted to information on dietary intakes of cadmium.

6.4. Cadmium in human biological materials

The full-term placenta samples collected at maternity hospitals in four regions of different environmental pollutants and traffic density in Slovakia were examined for cadmium (and lead) contents. The average cadmium concentrations from these regions were in range 0.0120–0.0221 mg/kg and placental contamination with

Table 15. Contamination of plant with cadmium in Poland

Plant	Cadmium content (mg/kg fresh matter)
Vegetables	0.004–0.078
Fruits	0.001–0.028
Potatoes	0.010–0.018
Cereals	>0.1 (18% of samples)
Flour	<0.1

cadmium reflected the active/passive smoking of the mothers rather than environmental pollution (Reichrtová et al., 1998).

In the Slovak pilot study the cadmium content was investigated in human milk and reconstituted infant formula obtained from the Lactarium of the Pediatric Hospital in Bratislava (Table 18) (Ursínyová and Hladíková, 1997). The calculated median intakes of cadmium from human milk and infant formula were similar and they represent 8.0 and 9.6% of PTWI, respectively. The calculated intake of cadmium by maximum oral exposure scenario via neither human milk nor infant formula did not exceed the PTWI level.

In the framework of the Slovak monitoring programme ''Impact of environmental factors on population health'' in 1995, cadmium was determined in total blood of general adult population and in biological material from pathological department of hospitals localized in polluted and unpolluted regions of the Slovak Republic (Table 19) (Sigmundová et al., 1997).

Cadmium blood levels were also investigated in three groups of subjects on an alternative diet (54 semivegetarians, 82 ovo-lacto-vegetarians and 44 vegans) aged 19–63 years who were for an average period of 4.5 years on a vegetarian diet. The results were compared with a matched group of omnivores ($n = 80$) from the same region of Slovakia. Significantly higher cadmium blood values were recorded in alternative nutrition groups (1.72 vs. 0.42 μg/l in omnivores). The cadmium incidence over 5 μg/l was in 7.4% of vegetarians vs. 0% of omnivores (higher intake of whole grain food and germs – the cadmium accumulation is generally higher in the bran and germ fractions than in the endosperm). Low levels of zinc (the antagonist of cadmium) were recorded in 11% vegans (Kudláčková et al., 1998).

In the framework of the system of monitoring the environment impact on population health in the Czech Republic the concentrations of cadmium blood and urine were measured in adults ($n = 670$), children ($n = 599$) and umbilical blood ($n = 549$) using AAS. The median blood cadmium level in adults (0.90 μg/l,

Table 16. Contamination of vegetables with cadmium in Hungary

Food	Cadmium content (μg/kg)	
	Median	Maximum
Cereals	5–62	200–800
Vegetables	5–30	20–461
Fruit	1–16	5–90

Table 17. Contamination of meat with cadmium in Hungary

Year	Average content of cadmium (μg/kg)			
	Pork	Beef	Poultry	Game
1989	41	144	15	15
1990	42	57	27	75
1991	32	31	17	55
1992	23	22	19	27
1993	13	16	10	14
1994	11	12	9	13

i.e.0.008 μmol/l) depends on smoking habit (1.25 μg/l, i.e. 0.01 μmol/l in smokers vs. 0.65 μg/l, i.e. 0.005 μmol/l in non-smokers). The median blood cadmium level in children was 0.70 μg/l and in umbilical blood 0.60 μg/l. The median urine cadmium level was 0.585 μg/g creatinine (0.59 μmol/mol creatinine) in adults and 0.37 μg/g creatinine (0.37 μmol/mol creatinine) in children (Cerna et al., 1997).

In Poland there are many authors investigating the contamination of human biological material with heavy metals, especially in association with local pollution problems.

Bursa (1996) studied the population consisted of 40 pregnant women (living in the Silesia district over 10 years and non-professionally exposed to heavy metals) and their 40 newborns, who had perinatal period problems (mainly respiratory effort). The concentrations of cadmium in whole maternal blood and in cord blood of their newborns with different birth weight (BW) are shown in Table 20. The author concluded, that the placenta barrier between mothers and babies exist only in relation to cadmium transfer, independently of birth weight.

Baranowska (1995) measured the concentrations of lead and cadmium in human placenta and in maternal and neonatal (cord) blood with the aim to assess the influence of the strongly polluted environment on the content of metals in tissues and on permeability of placenta to cadmium (and lead). The author found the following mean concentrations of cadmium: 4.90 μg/l in venous blood, 0.11 mg/kg in human placenta, 1.13 μg/l in cord blood, and concluded, that the placenta is a better barrier for cadmium than for lead.

The effect of smoking during pregnancy in 170 mother-neonates pairs in Legnica-Glogow Copper Basin was examined by Zareba et al. (1996). Higher blood cadmium (and lead) levels were found in women who smoked during pregnancy than in non-

Table 18. Cadmium dietary exposure in infants in Slovakia

Milk	*n*	Concentration (μg/kg milk)		Weekly intake (μg/kg body weight)	
		Median	Range	Median	Range
Human milk	35	0.5	0.3–1.7	0.56	0.34–1.90
Infant formula	8	0.6	0.4–0.7	0.67	0.45–0.78

Table 19. Cadmium in human biological material in the Slovak Republic

Biological material	Localities					
	Polluted			Unpolluted		
	Median	Min	Max	Median	Min	Max
Kidney (mg/kg)	8.55	2.91	29.88	23.85	5.92	51.26
Liver (mg/kg)	1.23	0.15	2.9	0.52	0.15	2.41
Blood (mg/l)	0.0023	0.0002	0.0077	0.0022	0.0009	0.0042

smokers (1.41 ± 0.39 vs. 0.79 ± 0.27 μg/l, $P = 0.022$). Similar results were observed in their neonates (0.69 ± 0.22 vs. 0.34 ± 0.19 μg/l, $P = 0.001$). The authors have found no difference between passive smokers and non-smokers.

An assessment of school children's exposure to cadmium was carried in Cracow and in a rural area in the years 1987–1989. The average blood cadmium concentration (3.3 μg/l) in young residents of Nowa Huta quarter (modern part of Cracow close to the steel and iron plant) was found and it was significantly higher than that of the city center (2.9 μg/l). This was related to higher industrial cadmium pollution observed in Nowa Huta. Unexpectedly, the highest blood cadmium levels (3.9 μg/l) were found in the groups from the rural area. The blood cadmium in children did not exceed the value of 10 μg/l, accepted as an upper limit for the general population (Zielonka and Wodzien, 1993).

Andrzejak et al. (1993) measured cadmium concentration in blood of people ($n = 60$, including 28 children) living in a village 900 m from a copper smelter in Legnica. The average blood cadmium concentration 28 ± 7 μg/l (children 19 ± 4 μg/l) pointed to the contamination of the surroundings of the copper smelter.

Research concerning cadmium in human milk was carried out in 33 samples from the women living in Poznan (urban industrial area). The cadmium content of milk (0.62 ± 0.28 μg/l) was the source of 1/6 PTWI given by FAO/WHO for adults. The author concluded that, however, it can present the danger to babies' health (Rydzewska and Krol, 1996).

The concentrations of cadmium in blood of the 51 healthy Austrian mothers was 0.44 (SD: 0.4) μg/l and in umbilical cord blood 0.08 (SD: 0.16) μg/l. Urine excretion of cadmium (and lead) assayed was normal. The concentrations of cadmium in human milk was below limits of detection of AAS (Plockinger et al., 1993).

The study of cadmium concentrations in 60 autopsy tissue samples obtained from individuals (25–87 years of age) from Styria, a moderately industrialized region of

Table 20. Cadmium in whole maternal blood and cord blood in the Silesia District (μg/l)

Body weight of newborns	Mothers			Newborns				
	n	x	s_x	n	x	s_x	t	P
Total	46	12.5	3.7	46	5.6	2.1	11.02	<0.001**
BW ≤ 1500 g	16	12.0	3.2	16	5.6	2.1	6.38	<0.001**
BW > 1500 g	30	12.8	3.8	30	5.6	2.0	8.90	<0.0001***

Austria, revealed that median concentrations in adults showed no statistical significant age dependency (i.e. liver 7.6 nmol/g; kidney 59.8 nmol/g; thyroid 11.2 nmol/g) (Tiran et al., 1995).

7. Conclusion

In general, the information about the environmental pollution and human exposure to cadmium in the Central European countries is very limited.

The provisional results of this survey indicate that the main sources of cadmium industrial emission in the region of Central European countries – Slovakia, Czech Republic, Poland, Hungary, and Austria – are in Poland (Upper Silesian Industrial Region). The long-range transport of air pollutants from the Upper Silesian Industrial Region is a problem leading to contamination of different compartments of the environment and to the increasing intake of cadmium in the human body also in more distant urban and rural areas in Poland as well as in the neighboring countries. The highest levels of cadmium in human blood were also determined in the southern Poland, where proximal tubular damage could be expected in a part of the population.

There is a great need for more data on contamination of the environment and food as well as on human exposure. Careful quality control is necessary, especially, when analyzing trace metals in biologic media. The reduction of cadmium pollution from industrial and agricultural sources is desirable. The environmental and biological monitoring programs exist mostly at the national level, but transboundary pollution problems should result in establishing coordinated international monitoring programmes.

Acknowledgements

We wish to thank Eva Reichrtova, MD, DSc. and Professor Ladislav Rosival MD, DSc. for their kind assistance in revising this publication.

References

Agency for Toxic Substances and Disease Registry, 1990. Cadmium Toxicity. Case Studies in Environmental Medicine. USA.

Andrzejak, R., Antonowicz, J., Tomczyk, J., Lepetow, T., Smolik, R., 1993. Lead and cadmium concentrations in blood of people living near a copper smelter in Legnica. Poland. Sci. Total Environ. (Suppl. 1), S233–S236.

Andrzejewski, R., Bararnowski, M. (Eds.), 1993. State of Environment in Poland. State Inspectorate for Environmental Protection, Environmental Information Centre GRID, Warsaw.

Anon., 1994. Task Force on Heavy Metals Emissions. Economic Commission for Europe, Prague.

Anon., 1997. Global Environmental Outlook. United Nations Environment Programme, Oxford University Press, New York.

Anon., 1998. Immissziós határértékek. http://www.joboki.hu/levego/hat_ert.htm.

Baranowska, I., 1995. Lead and cadmium in human placentas and maternal and neonatal blood (in heavily polluted area) measured by graphite furnace atomic absorption spectrometry. Occup. Environ. Med. 52, 229–232.

Bursa, J., 1996. Lead and cadmium concentrations in whole maternal blood and in cord blood of their newborns with different birth weight ($\leq$1500 g; $>$1500 g). In: Abstracts of International Conference Environmental Pollution and Child Health: Critical Needs and Issues for Central and Eastern Europe, Sosnowiec, Poland.

Cerna, M., Spevackova, V., Cejchanova, M., Benes, B., Rosser, P., Bavorova, H., Ocadlikova, D., Smid, J., Kubinova, R., 1997. Population-based biomonitoring in the Czech Republic – the system and selected results. Sci. Total. Environ. 204, 263–270.

Federal Ministry for Environment, Youth and Family, 1997. State of the Environment in Austria. Vienna.

Fiala, J., Bartáčková, Z., Dvořáková, M., Kurfürst, J., Machálek, P., 1998. Emission, and deposits of heavy metals in the Czech Republic. In: Proceedings of the III. Conference Air '98, pp. 49–52 (in Czech).

Food Research Institute, 1997. Partial Monitoring System of Foreign Substances in Foods and in Feeds. Report from 1996. Bratislava (in Slovak).

Friberg, L., Kjellström, T., Nordberg, G.F., 1986. Cadmium. In: Friberg, L., Nordberg, G.F., Vouk, V.B. (Eds.), Handbook on the Toxicology of Metals, vol. II: Specific Metals. Elsevier, Amsterdam, pp. 130–184.

Gerhardsson, L., Skerfving, S., 1996. Concepts on biological markers and biomonitoring for metal toxicity. In: Chang, L.W. (Ed.), Toxicology of Metals. Lewis Publishers, CRC Press, Boca Raton, FL, pp. 81–107.

Gzyl, J., 1997. Assessment of Polish population exposure to lead and cadmium with special emphasis to the Katowice Province on the basis of metal concentrations in environmental compartments. Cent. Eur. J. Publ. Health 5, 93–96.

Kudláčková, M., Béderová, A., Šimončič, R., Ursínyová, M., Hladíková, V., Magálová, T., Klvanová, J., Brtková, A., 1998. Risks of vegetarian nutrition. Chem. Papers 52, 477.

Magulová, K., Sajtáková, E., Uhlík, J., Marinkovičová, J., Šedíková, D., 1996. Air pollution in selected non-attainment areas of the Slovak Republic by emissions from stationary sources. Enclosure to report: Occurrence of Selected Toxic and Carcinogenic Organic and Inorganic Substances in Ambient Air of Selected Locations in the Slovak Republic [Report of MoE SR supporting project for PHARE project EU/93/AIR/22b]. Institute of Preventive and Clinical Medicine, Bratislava.

Makovinská, J., László, F., Csányi, B., Pál, G., Hock, B., Németh, J., Szabó, S., Škoda, P., Topolská, J., 1997. Tendency and Dynamics of Water Quality Changes of the Danube River and its Tributaries (1989–1995). Water Research Institute, Stimul, Bratislava.

Medeiros, D.M., Wildman, R., Liebes, R., 1997. Metal metabolism and toxicities. In: Massaro, E.J. (Ed.), Handbook of Human Toxicology. CRC Press, Boca Raton, FL, pp. 149–188.

Ministry of the Environment of the Czech Republic, 1998. State of the Environment in the Czech Republic. http://www.ceu.cz/special/czechrep/rocenka97/b224tab0.htm.

National Institute of Public Health in Prague, 1995a. Monitoring the health status of inhabitants in relation to the outdoor and indoor air. System of Monitoring of Health Status of Inhabitants in Relation to the Environment. Prague (in Czech).

National Institute of Public Health in Prague, 1995b. The Health Risk Assessment of Dietary Exposures to the Selected Chemical Substances in the Czech Republic. Prague.

National Institute of Public Health in Prague, 1997. Health effect and risks due to contamination of drinking water. System of Monitoring of Health Status of Inhabitants in Relation to the Environment. Prague (in Czech).

OECD, 1994. Cadmium. Environment Monograph Series No.104. Risk reduction monograph No. 5. OECD, Paris.

Österreichische Akademie der Wissenschaften-Kommission für Reinhaltung der Luft, 1993. Umweltwissenschaftliche Grundlagen und Zielsetzungen im Rahmen des Nationalen Umweltplans für die Bereiche Klima, Luft, Geruch and Lärm, Vienna.

Pintér, A. (Ed.), 1997. Hungarian Environmental Health Action Programme. Prospektus Nyomda, Veszprém.

Piscator, M., 1982. Cadmium – exposure and effects in the general population and in occupationally exposed workers. In: Prasad, A.S. (Ed.), Clinical, Biochemical and Nutritional Aspects of Trace Elements. Alan R. Liss, NewYork, pp. 521–536.

Plockinger, B., Dadak, C., Meisinger, V., 1993. Lead, mercury and cadmium in newborn infants and their mothers. Z. Geburtshilfe Perinatol. 197, 104–107.

Reichrtová, E., Ursínyová, M., Palkovičová. L., Wsolová, L., 1998. Contents and localization of heavy metals in human placentae. Fresenius J. Anal. Chem. 361, 362–364.

Rydzewska, A., Krol, I., 1996. Contents of zinc, copper and cadmium in milk of women living in Poznan. Ginekol. Pol. 67, 125–128 (in Polish).

Sigmundová, V., Škripeňová, A., Borošová, B., Benasová, A., 1997. Content of toxic metals in biological materials in inhabitants from selected regions of the Slovak Republic. In: Proceedings of the Conference Health and the Environment, Milovy, Czech Republic, pp. 242–247 (in Slovak).

Tiran, B., Karpf, E., Tiran, A., 1995. Age dependency of selenium and cadmium content in human liver, kidney, and thyroid. Arch. Environ. Health 50, 242–246.

UNEP/FAO/WHO, 1992. Assessment of Dietary Intake of Chemical Contaminants, WHO/HPP/FOS/92.6.

Ursínyová, M., Hladíková, V., 1997. The intake of selected toxic elements from milk in infants. Fresenius Environ. Bull. 6, 627–632.

Ursínyová, M., Hladíková, V., 1998a. Toxic elements in ambient air in the Slovak Republic. In: Proceedings of the III. Conference Air'98, pp. 257–260 (in Slovak).

Ursínyová, M., Hladíková, V., 1998b. Dietary intake of cadmium, lead and mercury in vegetarian and non-vegetarian children. Fresenius Environ. Bull. 7, 585–592.

Ursínyová, M., Hladíková, V., Sokolovič, P., 1997. Intake of toxic elements from vegetarian and non-vegetarian diet. In: Proceedings of Conference Foreign Substances in Food, Tatranská Štrba, Slovak Republic, pp. 145–147 (in Slovak).

WHO, 1989. Evaluation of Certain Food Additives and Contaminants. Thirty-third Report of the Joint FAO/WHO Expert Committee on Food Additives, Technical Report Series No. 776. Geneva.

WHO, 1992a. Cadmium – Environmental Aspects. Environmental Health Criteria 135, WHO, Geneva.

WHO, 1992b. Cadmium. Environmental Health Criteria 134. WHO, Geneva.

WHO, 1993. Guidelines for Drinking-Water Quality, vol. I. Recommendations. WHO, Geneva.

WHO, 1994. Updating and Revision of the Air Quality Guidelines for Europe. Report on the WHO Working Group on Inorganic Air Pollutants, Regional Office for Europe, EUR/ICP//EHAZ 94 05/MT04, 24–27 October 1994, Düsseldorf, Germany.

WHO, 1996. Guidelines for Drinking-Water Quality. Health Criteria and Other Supporting Information, vol. 2. WHO, Geneva.

Zareba, A., Strugala-Stawik, H., Rudkowski, Z., Dembicka, D., Pastuszek, B., 1996. Blood levels of lead and cadmium in pregnant women, active and passive smokers and their neonates. In: Abstracts of International Conference Environmental Pollution and Child Health: Critical Needs and Issues for Central and Eastern Europe, Sosnowiec, Poland.

Zielonka, E., Wodzien, M., 1993. Blood cadmium and lead levels among children in Cracow. Folia. Med. Cracov 34, 85–96.

Trace elements – Their distribution and effects in the environment
B. Markert and K. Friese, editors

Chapter 4
Lead in the environment of Central Europe

M. Ursínyová, V. Hladíková

Abstract

Lead contamination was identified as one of the most significant environmental problems in Central Europe. Lead adversely affects many organ systems in the human body, especially the nervous system of children, manifesting itself as an impairment of cognitive functions. Therefore, the countries of the region introduced policies to reduce lead use in gasoline. Austria and the Slovak Republic phased out leaded gasoline completely by 1993 and 1995, respectively; the Czech Republic, Hungary and Poland have significantly reduced the permissible lead content of gasoline (to 0.15 g/l). These restrictions have resulted in a marked decrease in lead atmospheric emissions and, consequently, in a significant decrease of lead concentrations in ambient air. The reduction of lead emissions contributes to a significant decline in accumulation of lead in various environmental media, especially soil, dust and food, reducing long-term exposures through ingestion. Nevertheless, further measures for reduction of lead are necessary due to its higher local levels. Special attention should be paid to the Upper Silesian Industrial Region in Poland, where lead concentrations in particular environmental compartments are still very high, and the long-range transport of lead emissions from this industrial area contributes also to the contamination of different compartments of the environment and increased human exposure in other areas of Poland as well as in the neighbouring countries.

Keywords: Lead; Environmental contamination, Air, Water, Sediments, Food, Human biological materials, Central Europe

1. Production and uses

Primary lead from mining is often produced as a by-product with other metals such as zinc, copper and cadmium. In 1989, 155 mines in 35 countries produced 2.21 million tons of lead in concentrate. Since 1970, the world demand for lead's unique chemical, electrical and physical properties has increased by 25% to a record level of 5.627 million tons in 1990. OECD countries accounted for 65% of world demand in 1990, with the Central and Eastern European countries consuming 21%. Asia is now third largest lead-consuming region, representing 9% of world demand. Lead is also among the most recycled non-ferrous metals in the world (OECD, 1993).

Major uses are in batteries, cables, pigments, petrol (gasoline) additives, solder, steel products and various semimanufacturers (in ceramic glazes, crystal tableware, some traditional remedies). Lead and lead compounds are also used in solder applied to water distribution pipes and to seams of cans used to store foods (Tsuchiya, 1986).

New applications are being developed for lead and lead compounds in microelectronics, superconductors, asphalt, earthquake damping materials and radon gas shielding, and for retrievable storage or permanent disposal of nuclear waste (OECD, 1993).

2. Sources of environmental contamination

The level of lead in the earth's crust is about 20 mg/kg. Lead in the environment may drive from either natural or anthropogenic sources. Natural sources of atmospheric lead include geological weathering and volcanic emissions and have been estimated at 19 000 tons/year, compared to an estimate of 126 000 tons/year emitted to the air from mining, smelting and consumption of over three million tons of lead per year (WHO, 1989).

Sources of atmospheric lead emissions may be considered as belonging to one of two main categories: mobile source (use of lead fuels in automobiles) and stationary ones (refining, manufacturing and incineration). However, restrictions on lead levels in gasoline in many countries have resulted in a marked decrease in automotive emissions of lead into atmosphere. Most lead emissions are deposited near the source, although some particulate matter (<2 μm in diameter) is transported over long distances and results in the contamination of remote sites such as arctic glaciers (WHO, 1989).

3. Distribution between media and environmental effects

Human activities remove lead from the earth's crust, where it is relatively immobile, and transfer it to environmental media. Substantial human and ecosystem exposure to lead can occur in all environmental media.

Airborne lead can contribute to human exposures by the contamination of food, water and dust, as well as through direct inhalation (WHO, 1989).

Atmospheric deposition of lead is a major source of soil, water and sediment contamination. Depending on the particulate size, airborne suspended particles may have a long residence time in the atmosphere. Studies from remote areas indicate that lead-containing particles are transported over substantial distances, up to thousands of kilometres, by general weather systems. In rural areas of Europe and North America, the annual deposition rates (wet and dry) are typically in the range of 20–80 $\mu g/m^2$ per day. In large cities the deposition rates are about 1.5–10 times higher than in rural areas (OECD, 1993). Some data indicate that lead fallout in excess of 250 $\mu g/m^2$ per day will increase blood lead levels (WHO, 1994).

Concentrations of lead in soil range widely, depending on the nature of the parent material, land use and deposition from industrial, automobile and urban sources. The natural concentrations of lead in soil is in range of 2–200 mg/kg (Tsuchiya, 1986). The concentration of lead in soils near highways is related to traffic density, local meteorological conditions, vegetation and topography. Generally, the concentrations of lead in soil decreases exponentially downwind of a point source (OECD, 1993). Lead soil concentrations is greatest within 10 m of the road and within the top 1–5 cm (OECD,

1993). Transport within soil and the bioavailability of lead from soil are dependent upon many factors, including pH, mineral composition of the soil, and amount and type of organic material (WHO, 1995). Plant species differ widely in their ability to absorb, accumulate and tolerate lead. The percentage of accumulation of lead is higher in the leafy crops, lettuce and cabbage, than in the root crops, radish and carrot. A proportion, typically around 50% of the lead content of vegetables and fruit crops can be removed by washing (OECD, 1993).

Lead is natural, usually very minor constituent of surface and ground waters. Nriagu and Pacyna estimated for 1983, inputs of 97 000–180 000 tons per year of lead to aquatic ecosystem. The total annual input of lead to aquatic ecosystem has declined since 1983 (reduction of lead emissions from mobile sources). In general, lead decreases in concentration from rainwater (generally acidic; about 20 μg/l) to fresh-water (generally neutral; about 5 μg/l) to sea water (generally alkaline; below 1 μg/l) (OECD, 1993). Lead deposited in water partitions rapidly between sediment and aqueous phase, depending upon pH, salt content, and presence of organic chelating agent (WHO, 1989).

4. Lead intake

The total intake of lead by adults and children in the general populations varies greatly as to the relative contributions from individual sources (air, water, food, soil/ dust, and others) and is partly depended on life-style and socio-economic status. Additional intake of lead will take place in certain groups from the use of tobacco and alcoholic beverages (WHO, 1995). Occupational exposure to lead occurs in many countries, especially in those with developing industries and small home-based industries.

The major sources of lead for infants and children are dust and soil, chips of lead paint, water and food (Lockitch, 1993). According to the US EPA in 2-year-old infants the intake of lead is 47% from food, 1% from air, 45% from dust, 6% from water and 1% from soil. Deficit of calcium, phosphate, selenium or zinc may result in increased lead absorption. Iron and vitamin D also affect absorption of lead (WHO, 1995).

Concentrations of lead in air range from 7.6×10^{-5} $\mu g/m^3$ in remote areas to >10 $\mu g/m^3$ near lead smelters. Airborne concentrations of lead in urban areas are likely to be in range of 0.5–3 $\mu g/m^3$. Where leaded vehicle fuel is no longer used, concentrations are likely to fall to <0.2 $\mu g/m^3$. Levels of airborne lead in indoor shooting ranges have been shown to range from 2.7 to 90.5 $\mu g/m^3$ depending upon the location of sampling (WHO, 1995).

Lead is present in tobacco. The mean content of lead in filter-tipped cigarettes produced between 1960 and 1980 was 2.4 μg/g. Approximately 5% of this lead may be inhaled; the remainder occurs in the ash and side/stream smoke (WHO, 1995).

WHO recommended a maximum lead guideline value in ambient air of 0.5 $\mu g/m^3$. To prevent further increases of lead in soils and consequent increases in the exposure of future generations, the air lead levels should be kept as low as possible (WHO, 1994).

Lead concentrations in drinking water result primarily from lead leaching from water delivery system and from previously contaminated drinking water sources. In

1993 WHO guideline value of lead in drinking water was revised from 50 μg/l to 10 μg/l (WHO, 1995).

The concentrations of lead in foods are typically less than 100 μg/kg fresh weight. Foods with high surface areas (such as spinach) also tend to have higher concentrations of lead due to atmospheric deposition. Canned foods (especially if the food is at a low pH) often have much higher concentrations of lead than foods packed in another manner (OECD, 1993).

WHO and FAO establish the PTWI (Provisionally Tolerable Weekly Intake) for lead 25 μg/kg body weight (WHO, 1993).

5. Evaluation of human health risks

Inhalation and ingestion (of water, food, paint, soil and/or dust) are the primary routes of human exposure to lead (OECD, 1993). Percutaneous absorption is minimal in humans. The absorption of lead from air to blood involves two processes: the deposition of airborne particles in the respiratory tract; and the absorption and clearance from the respiratory tract into the circulation. The respiratory deposition of airborne lead is in the range of 30–50% and varies with particle size, chemical speciation, solubility in body fluids and ventilation rate. Higher deposition rates may occur with larger particles, but this deposition takes place in the upper respiratory tract. The smaller particles of inhaled lead (generated by automobile exhaust) are almost (>90%) completely absorbed after deposition in the lower respiratory tract (WHO, 1995).

The amount of lead absorbed from the gastrointestinal tract of adults is typically 10 to 15% of the ingested quantity; for pregnant women and children, the amount absorbed can increase to as much as 50% (Agency for Toxic Substances and Disease Registry, 1992).

In the adult, almost 90% of the total body lead burden is contained in bone. Accumulation of lead in bone begins in fetal life and continues into old age. The half-life in adult bone has been estimated in different models to be 10–20 years, but in rapidly growing children with active bone turnover, it may be much less. The bone lead reservoir is important in situations of increased bone turnover, such as pregnancy or osteoporosis. Prolonged immobilisation resulting in bone breakdown may increase blood lead levels. Congenital lead poisoning may result from release of lead from bone stores accumulated by the mother through lead poisoning in her childhood (Lockitch, 1993).

Rather little is known about the basic mechanism behind lead toxicity. However, it is well established that lead binds to the sulfydryl groups of proteins. If this occurs on an enzyme, its function may be inhibited, which may result in toxic effects. Also, lead alters calcium-mediated cellular processes (Gerhardsson and Skerfving, 1996).

Lead is eliminated from the body in both urine and faeces. Any dietary (including waterborne and airborne after swallowed) lead not absorbed in the gastrointestinal tract is excreted in faeces. The blood lead not retained in the body is either excreted by the kidneys or through biliary clearance into the gastrointestinal tract. In single-exposure studies with adults, lead has a half-life, in blood, of approximately 25 days; in soft tissue, about 40 days; and in the non-labile portion in bone, more than 25 years (Agency for Toxic Substances and Disease Registry, 1992).

A special form of excretion of endogenous lead is through breast milk. Studies of breast milk indicate that lead concentrations correlate with maternal blood lead concentrations; most studies reported that lead secreted from breast milk varies in concentration between 10 and 30% of maternal blood lead concentration (WHO, 1995).

In humans, lead can result in a wide range of biological effects depending upon the level and duration of exposure. Lead has long been known to be highly toxic in acute doses, but it is now becoming clear that chronic exposure to low levels of this element can cause a variety of behavioural and psychological problems, particularly in children. Children are much more sensitive to lead's toxic effects than the adults. Lead toxicity causes haematological, gastrointestinal, and neurological dysfunction in adults and children. Symptoms are usually noted with blood lead greater than 1.93 μmol/l. Severe or prolonged exposure may also cause anemia, CNS and peripheral nervous system damage (encephalopathy, peripheral lead neuropathy), nephropathy, renal tubular dysfunction and morphological renal changes, hypertension, and reproductive impairment. Lead inhibits enzymes; alters cellular calcium metabolism; stimulates synthesis of binding proteins in kidney, brain, and bone; and slows nerve conduction. Less severe exposure to lead, designated by blood lead levels of 0.48–0.96 μmol/l, has been implicated in poor pregnancy outcome, impaired neurobehavioral development, reduce stature in young children, and higher blood pressure in adults (Lockitch, 1993).

The lowest observable blood lead levels associated with specific health effects in chronically exposed children and adults are shown in Fig. 1 (Agency for Toxic Substances and Disease Registry, 1992).

Blood lead is the generally accepted indicator of lead exposure. A highly significant association was found between lead exposure and children's IQ. An increase in blood lead concentration from 10 to 25 μg/dl was associated with a decrease of IQ. Based on such findings the "Centers for Disease Control and Prevention" (CDC) in the USA have recently lowered the community action level for blood in children from 25 to 10 μg/dl (Table 1) (Brody et al., 1994).

Epidemiological studies of workers exposed to lead do not support a carcinogenic effect in humans. Still, on basis of animal data, IARC classified lead as "possibly carcinogenic to humans" (Group 2B) (International Agency for Research on Cancer, 1987).

6. Summary and evaluation data from Central European countries

The first pan-European conference of environmental ministers, held at Dobříš Castle, Czechoslovakia, in 1991, requested that a comprehensive state of the environment report be done for the whole Europe. Consequently, chemical risks were identified as one of the 12 most significant environmental problems of concern to Europe in this report (UNEP, 1997). Moreover, the next extensive review (Hertzman, 1995) of environmental concerns in Central and Eastern Europe came to the conclusion that

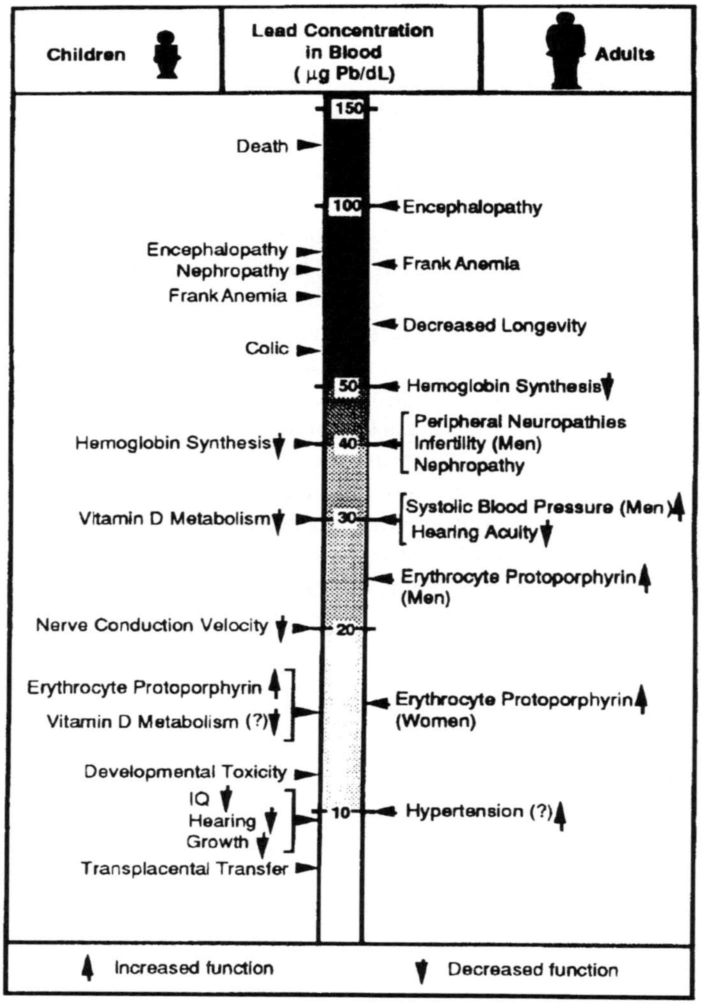

Figure 1. Lowest observed adverse effects levels of inorganic lead in children and adults.

the lead was one of the most serious and widespread environmental hazards in this region.

6.1. Lead in the atmosphere

According to the 1992 inventory of heavy metal emissions in the Slovak Republic, stationary sources played a significant role in lead emissions. Steel and iron production (Košice, Podbrezová) emitted the largest amounts of lead to the atmosphere (43%). The power sector (coal combustion) and non-ferrous metal production were also significant contributors (17 and 7%, respectively). Traffic was the second largest source of lead emissions in 1992, after iron and steel production, representing about

Table 1. Interpretation of blood lead test results and follow-up activities: class of child based on blood lead concentration

Class	Blood lead concentration (μg/dl)	Comment
I	≤9	A child in Class I is not considered to be lead-poisoned.
IIA	10–14	Many children (or a large proportion of children) with blood lead levels in this range should trigger community-wide childhood lead poisoning prevention activities. Children in this range may need to be screened more frequently.
IIB	15–19	A child in Class IIB should receive nutritional and educational interventions and more frequent screening. If the blood lead levels persist in this range, environmental investigation and intervention should be done.
III	20–44	A child in Class III should receive environmental evaluation and remediation and medical evaluation. Such a child may need pharmacologic treatment of lead poisoning.
IV	45–69	A child in Class IV will need both medical and environmental interventions, including chelation therapy.
V	≥70	A child with Class V lead poisoning is a medical emergency. Medical and environmental management must begin immediately.

29% of total emissions in Slovakia. Vehicular lead emissions have been declining, however, due to the introduction of unleaded gasoline in Slovakia in 1992. Since the beginning of 1995, only unleaded gasolines have been distributed in Slovakia. The Slovak Republic was one of the first four European countries (after Austria, Finland and Sweden) to use only unleaded gasoline (Violová et al., 1997). The results of the emission inventory of lead in Slovakia in 1992 and 1994 are presented in Table 2 (Magulová et al., 1996).

Table 2. Emissions of lead in the Slovak Republic (t)

Activity	Lead	
	1992	1994
Primary and secondary iron and steel industry	30.435	28.220
Primary and secondary non-ferrous metal production	14.775	9.197
Cement production	3.075	1.057
Glass production	14.668	11.494
Combustion of fossil fuels	10.141	6.234
Magnesite production	0.009	0.004
Waste incineration	1.724	1.724
Traffic	96.800	21.100
Total	171.627	79.030

Table 3. Emissions of lead in the Czech Republic, year 1992 (t)

Activity	Lead
Incineration	68.2
Industry	206.1
Traffic	245.0
Total	519.3

Lead releases in the Czech Republic in 1992 are presented in Table 3 (Fiala et al., 1998).

In Austria, in the last couple of years first steps have been set to the reduction of heavy metal emissions. These were specific measures such as the prohibition of leaded gasoline from 1993, but also non-specific measures for the reduction of particulate emissions in industry and power plants which simultaneously brought about a reduction in heavy metal emissions. Nevertheless, further measures for the reduction of heavy metal emissions are considered to be necessary due to their higher local incidence. Generally, lead represents the most important heavy metal emitted in Austria (Table 4) (Federal Ministry for Environment, Youth and Family, 1997).

The lead emission into the air from industrial sources in Hungary in 1992 was 5.015 t/year. Considering the county emissions, 85% the total industrial emission is stated to derive from three counties: Borsod-Abaúj-Zemplén, Heves and Fejér counties. The number of reported lead emitting sources is 731. As can be seen in Table 5, the trend of the total lead emission in Hungary has been decreasing, mostly due to the decrease of the lead content of gasoline (Rakics and Öri, 1996).

The available data of earlier lead emission estimates within the German Federal Environmental Protection Agency, the available data from PARCOM (Paris Commission program), the ESQAD (European Soil and Sea Quality and Atmospheric Deposition of Selected Substances) data and emissions estimated by national experts in some countries are shown in Table 6 (Anon, 1994).

The lead concentrations were investigated in ambient air in selected localities of the

Table 4. Emissions of lead in Austria, year 1993

Activity	Lead	
	(t/year)	(%)
Primary and secondary iron and steel industry	90.708	16
Primary and secondary non-ferrous metal production	11.422	2
Cement production	5.393	1
Glass production	5.269	1
Combustion of fossil fuels	61.865	11
Waste incineration	0.082	<1
Petrol-Pb	174.000	30
Mining	221.738	39
Total	570.477	100

Table 5. Trend of lead emission in Hungary (t/y)

Year	1980	1985	1990	1991	1992	1993	1994	1995	1996
Lead	634.2	578.9	695.3	522.8	241.6	217.7	189.6	153.7	126.4

Slovak Republic in the framework of the PHARE Project (1996–1997) (Table 7). Considering the situation in Slovakia, all the mean levels from all the sampling sites were below the annual National Ambient Air Quality Standards (0.5 $\mu g/m^3$) and guideline by WHO. Comparison with typical values for atmospheric lead levels in Europe indicates that, in general, the levels measured in Slovakia are at the lower end of typical European levels (Ursínyová and Hladíková, 1998a). Cities in Slovakia show the relationship between the decreasing use of lead in gasoline and the decline in airborne lead concentrations (Fig. 2) (Violová et al., 1997).

The annual arithmetic concentrations for lead in all the 27 studied localities of the Czech Republic were below the National Ambient Air Quality Standard 0.5 $\mu g/m^3$ (Table 8). The highest values were found in Ostrava, Karviná, Brno and Kladno - above 0.1 $\mu g/m^3$ (National Institute of Public Health in Prague, 1995a).

In Table 9 ambient air concentrations for lead in some Austrian cities are presented (Österreichische Akademie der Wissenschaften-Kommission für Reinhaltung der Luft, 1993).

Ambient atmospheric lead concentrations in urban areas are routinely monitored in Poland. They were found to be at moderate levels (rarely exceeding the WHO guideline of 0.5 $\mu g/m^3$), except for Silesia. Even in Silesia, a decline of atmospheric lead levels has also been registered during recent years. At some areas of Poland, however, small increases in ambient atmospheric lead levels have been observed. For example, in Warsaw, where the major sources of lead emissions are traffic and power plants, concentrations increased in five of the six administrative units between 1992 and 1993, although only in one of them they exceeded 1.0 $\mu g/m^3$. Unfortunately, a monitoring of these sites was discontinued in 1994 (Violová et al., 1997). Average and maximum annual ambient air concentrations for lead in the Upper Silesian Industrial District (Poland) in 1989 were 0.2 $\mu g/m^3$ and 1.30–1.9 $\mu g/m^3$, respectively (Andrzejewski and Bararnowski, 1993). Gzyl (1997) presented the average concentrations of lead in dustfall in this region 38 (8–616) mg/m^2 per year while the Polish standard value is 100 mg/m^2 per year. WHO refer, that lead dustfall in excess of 250 $\mu g/m^2$ per day (90 mg/m^2 per year) will increase blood lead levels (WHO, 1994).

The lead levels measured in ambient air and dust in Hungary are summarized in Table 10. The hygienic limit values of lead in Hungary are 0.3 $\mu g/m^3$ in air and 12.0 mg/m^2 per 30 days in dust (Rudnai and Horváth, 1996). An analysis of trend of lead concentrations in ambient air during 1991–1995 showed significant decreases in ten large cities. In Budapest, for example, the lead concentration in ambient air progressively decreased from the mean value of 1.7 $\mu g/m^3$ in 1991 to 0.54 $\mu g/m^3$ in 1994 (Violová et al., 1997).

6.2. Lead in water and sediments

In the framework of the monitoring programme elaborated by the Water Quality

Table 6. Emission estimates for lead in Europe (t)

Country	German UBA Project (published in 1988)	Pacyna 1993 (1989-data)	Other national data (reference year)	TNO data (1983-data)	ESQUAD Project (1990-data)
Austria	1128	660			292
Czechoslovakia (former)	1187	1170	176 (1992) (only Slovakia)		509
Hungary	604	670			764
Poland	2950	3000		2650	2303

Table 7. Lead concentrations in ambient air in the Slovak Republic in 1996–1997

Locality	Arithmetic mean (ng/m^3)	Geometric mean (ng/m^3)	Minimum (ng/m^3)	Maximum (ng/m^3)
Bratislava I	70.6	56.6	20.2	184.8
Bratislava II	47.4	41.7	9.6	77.8
Bratislava III	44.4	40.2	15.7	65.3
Bratislava IV	57.3	44.2	12.2	173.9
Bratislava V	39.6	33.5	16.4	100.9
Topol'níky	29.6	24.5	8.0	61.2
Hurbanovo	30.3	27.4	14.8	57.7
Mochovce	22.2	18.9	10.4	53.0
Žiar n. Hronom	22.3	16.9	6.7	58.0
Prievidza	28.2	17.7	4.8	106.4
Žilina	38.8	29.4	7.5	99.9
Ružomberok	35.5	28.2	10.3	90.8
Krompachy	297.3	252.0	106.1	625.5
Strážske	53.1	38.0	12.2	120.0
Starina	26.9	21.4	9.0	77.3
Košice I	207.8	156.4	66.7	583.9
Košice II	67.2	62.2	34.1	135.8
Košice III	104.5	95.7	38.3	189.0
Košice IV	84.4	63.2	11.5	189.6
Vel'ká Ida	315.5	189.9	32.5	777.7

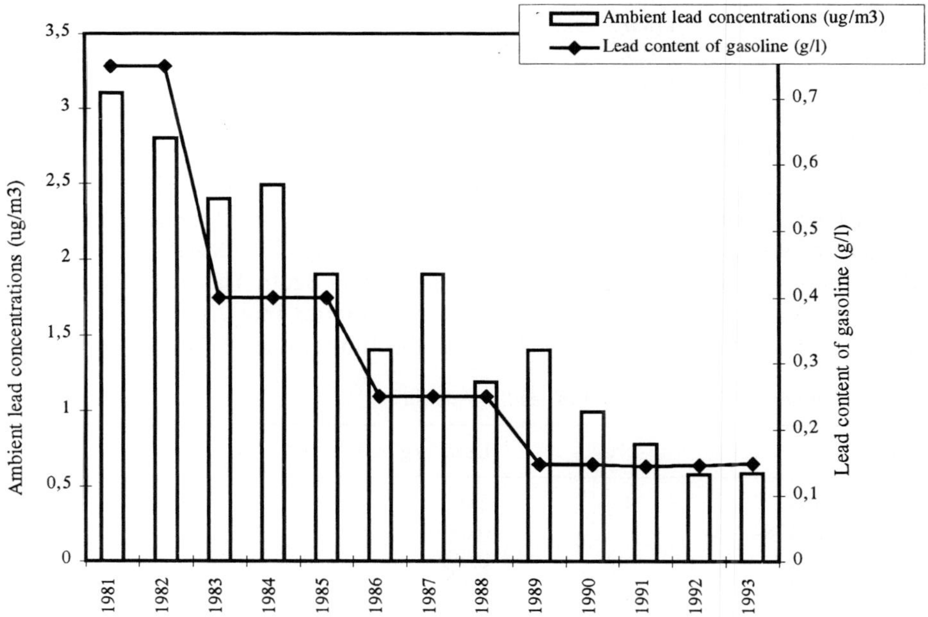

Figure 2. Lead content of gasoline and ambient atmospheric lead concentrations in Bratislava (capital of Slovakia), 1981–1993.

Table 8. Lead concentrations in ambient air in the Czech Republic in 1994

Region	Arithmetic mean (ng/m^3)	Geometric mean (ng/m^3)
Prague 1	63.7	46.4
Prague 4	97.4	53.1
Prague 5	87.2	62.1
Prague 6	46.9	35.2
Prague 7	51.9	44.9
Prague 10	48.6	41.6
Benešov	10.3	6.52
Kladno	114.0	32.4
Kolín	8.19	3.75
Příbram	75.7	58.5
Klatovy	84.1	59.1
Plzeň	56.6	48.6
Sokolov	44.1	32.9
Děčín	67.9	63.0
Ústí nad Labem	68.8	39.7
Havlíčkův Brod	1.03	0.90
Hradec Králové	54.8	52.0
Svitavy	59.5	57.0
Ústí nad Orlicí	56.9	31.1
Brno	107.0	80.1
Hodonín	44.6	41.3
Jihlava	62.8	60.7
Kroměříž	30.1	17.8
Karviná	118.0	106.0
Olomouc	70.9	68.0
Ostrava	118.0	82.4

protection Working Group (WG) of the Hungarian-Slovak Transboundary Water Commission, ten different metals and trace elements were investigated in the Danube and its tributaries for the period 1989–1995. Lead concentration exceeded 10 mg/l, the limit value of WG class I (according to the 6-class water quality classification system), in the Danube between Medved'ov/Vámosszabadi and Szob. The observed range of 10–15 mg/l corresponds to class II. Among the tributaries Ipoly/Ipel' was characterized with concentrations exceeding 20 mg/l, the limit value of WG class II. Lead

Table 9. Ambient air concentrations of lead in Austria

Region	Lead ($\mu g/m^3$)
Arnoldstein, 1992	2.0
Linz Steyregg, 1984	0.045
Linz Urfahr, 1984	0.082
Linz Steyregg, 1982	0.2
Linz Urfahr, 1982	0.57
Wien, 1980	0.43
Hallein, 1982	0.09

Table 10. Lead content of the environmental media in Hungary, 1985–86

Area	Source of lead	Lead in air (μg/m^3)	Lead in dust (mg/m^2 per 30 days)
Romhány	Ceramic industry	–	14.0
Budapest	Traffic	5.3	–
Bem.u.	Traffic	2.9	–
Csanády	Traffic (city centre)	2.0	–

remained below the SEL-Severe Effect Level value (250 mg/kg) in both the suspended solids and bottom sediments (Tables 11 and 12) (Makovinská et al., 1997).

The concentrations of lead in the river waters in the Czech Republic are shown in Table 13. According to the lead characteristic concentration (90th percentile), the most of the observed surface water in the Czech Republic belongs to the purity class I except for Sázava-Pikovice, Otava-Písek, Svratka-Vranovice (class II) and Bílina-Ústí nad Labem, Odra-Bohumín, Olše-Věřňovice (class III) (Ministry of the Environment of the Czech Republic, 1998).

The Hungarian Ministry of Environment and Regional Policy in 1990 initiated investigation of the lead content of the different types of water. The survey of unfettered surface water showed that 16 of the 23 samples contained lead in the range 1–5 μg/l, three samples in the range 5–10 μg/l, two samples in the range 10–25 μg/l, and two samples above 25 μg/l (max = 92 μg/l). Underground aquifers of operating waterworks showed a very low lead content: 9 of 25 samples were below 0.1 μg/l, 15 samples were in the range 0.1–1 μg/l, and only one exceeded 1 μg/l (Csanady, 1996).

In the Slovak Republic in 1996 the median concentration of lead in drinking water was 2.6 μg/l. None of the 235 samples exceeded the limit value for lead (Food Research Institute, 1997). In 1998 the national limit value for lead concentration in drinking water was lowered from 50 to 10 μg/l (Anon, 1998).

The results of the follow up of drinking water quality from the public water supply in the Czech Republic showed the geometric mean lead concentrations of 0.2 μg/l. More than 75% of the 779 analysed samples of the drinking water was below the tenth of the limit value for lead (50 μg/l), and only 1.7% exceeded that value (National Institute of Public Health in Prague, 1997).

Lead concentrations in drinking water in Poland were up to 50 μg/l (Gzyl, 1997).

Regarding lead in drinking water, the Hungarian National Environmental Health

Table 11. Lead in the suspended solids of the Danube (1989–1995)

Locality	Lead (mg/kg)		
	Min	Mean	Max
Bratislava	20	65	180
Komárno/Komárom	18	71	170
Szob	34	66	150

Table 12. Lead in the sediments of the Danube (1989–1995)

Locality	Lead (mg/kg)		
	Min	Mean	Max
Bratislava left side	4.6	24	91
Bratislava right side	4	23	93
Komárno left side	3.2	14	40
Komárno right side	3	14	37
Szob left side	7	24	68
Szob right side	<5	24	54

Programme elaborated after 1995 formed the following tasks at the national level of action: prohibiting production of PVC pipes with lead content or modification in technology; information campaigns to limit consumption lead as secondary pollutant (tea making) from heater tanks; and survey on lead content in water from PVC pipes and heater tanks (Pintér, 1997). The lead content of drinking water at the consumer's tap is usually small. From the data available, only 5% was at or above 4 μg/l (Csanady, 1996).

6.3. Lead in food

The aim of the Partial Monitoring System ''Contaminants and Additives in foods and feeds'' realised by the Slovak Ministry of Agriculture from 1993 is to receive objective data on contamination of foods and feeds in association with contamination of the

Table 13. Arithmetic mean lead concentrations in surface water in the Czech Republic (1996)

Locality	Lead (μg/l)
Labe-Valy	1.7
Labe-Lysá nad Labem	1.8
Labe-Obříství	2.6
Labe-Děčín	2.0
Jizerý-Nový Vestec	1.2
Vltava-České Budějovice	0.8
Vltava-Vrané	2.2
Vltava-Zelčín	2.2
Sázava-Pikovice	4.4
Berounka-Lahovice	4.5
Otava-Písek	1.2
Bílina-Ústí nad Labem	8.1
Odra-Bohumín	17.5
Olše-Věřňovice	22.5
Morava-Lanžhot	2.7
Bečva-Troubky	1.9
Svratka-Vranovice	6.2
Jihlava-Iváň	2.8

Table 14. Contents of lead (μg/kg fresh weight) in foodstuffs in the Slovak Republic

Foodstuff	Average	Median	% of PTWI
Potatoes	23.32	20	1.14
Apples	26.56	20	0.1
Subtropic fruits	23.10	20	0.19
Carrots	34.14	20	0.25
Cabbage	29.32	10	0.14
Tomatoes	37.29	20	0.12
Onion	28.40	10	0.12
Milk	21.17	20	1.01
Eggs	44.85	33.5	0.36
Rice	38.54	21	0.12
Vegetable oils	25.10	10	0.12
Beef	48.85	40	0.4
Pork	60.13	40	1.19
Meet products	82.16	70	1.16
Pork fat	36.46	22	0.13
Poultry	48.49	40	0.31
Cheese -soft	81.12	60	0.22
Butter	29.48	20	0.07
Yoghurts	37.13	30	0.5
Bread	56.43	25	1.56
Rolls and cakes	38.00	20	0.96
Lemonades	20.51	10	0.48
Beer	14.99	10	0.74
Wine	54.43	44	0.43

environment in the Slovak Republic on one hand and exposure of its inhabitants on the other hand. The concentrations of lead in foods in the Slovak Republic purchased during 1993–1997 in retail stores in 10 localities representing about 20 000 inhabitants and different types of settlements are shown in Table 14 (Food Research Institute, 1997).

The evaluation of lead intakes from diet was realised in the Slovak Republic in two types of total diet (ovo-lacto-vegetarian diet and non-vegetarian standard diet) collected from the kindergartens in Bratislava (capital of Slovakia) in 1991–1994 (Table 15). The results of median weekly intakes represented 27% of the PTWI in non-vegetarian children and 40% in vegetarian children. The authors did not find

Table 15. Lead dietary exposure in children in the Slovak Republic

Diet	*n*	Concentration (μg/kg total diet)		Weekly intake (μg/kg body weight)	
		Median	Range	Median	Range
Vegetarian	120	16.2	<3.6–83.7	9.9	<2.1–48.6
Non-vegetarian	120	13.2	<3.6–103.4	6.7	<2.0–57.0

statistical significantly difference between weekly intakes in these groups. The maximum weekly intake of lead from vegetarian diet was lower than that from non-vegetarian diet. On the other hand, the number of samples with the content of lead resulting in the intake exceeding the PTWI value was lower in non-vegetarian diet (5.2%) than that in vegetarian diet (11.8%). The relatively low median intake of lead was explained by the fact that according to instructions of the Slovak Ministry for Education the diet in kindergartens cannot be prepared from entrails, mushrooms (i.e. ingredients probably containing high concentrations of heavy metals) (Ursínyová and Hladíková, 1998b). In adults, the median weekly intake of lead from vegetarian diet (4.20 μg/kg body weight) was higher than that from non-vegetarian diet (2.47 μg/kg body weight), however, in both groups the median intakes were below the PTWI and lower than those in children (Ursínyová et al., 1997).

In the framework of the project of the National Institute of Public Health, Prague: "The Health Risk Assessment of Dietary Exposures to the Selected Chemical Substances (Total Diet Study), Czech Republic, 1994" altogether 552 composite samples representing 1920 individual food samples were analysed on lead content for evaluation of the human exposure (Table 16). The lead dietary exposure did not exceed on not a single one of the 12 controlled sites of the Czech Republic the PTWI level. Statistically significant difference was not found between individual sites with a different environmental pollution level. The estimated intake for the population of the Czech Republic ranges between 8 and 16% of the PTWI (National Institute of Public Health in Prague, 1995b).

Lead concentrations in different plants in Poland were up to 0.3 mg/kg fresh matter in vegetables (90% of samples), <0.1 mg/kg fresh matter in fruits (90% of samples), and >0.3 mg/kg fresh matter in cereals (25% of samples). In the Katowice Province, lead contents in vegetables (carrot, red beet, parsley, celery) varied between 0.50–280 mg/kg dry matter. Exposure to lead from vegetables was 1.5 mg/week per person. The results ranged between 4.8–44.0% of PTWI, whereas in the reference area these values amounted to 6.0% of PTWI for lead. The elimination of vegetables with high contents of lead in diet may reduce the dietary intake to 2.0–18.4% of PTWI (Gzyl, 1997).

The GEMS/Food Programme reported, that the mean weekly intake in adults from Poland ranged in the period 1982–1987 between 10.5–22.4 μg/kg body weight. Data indicate an increase in dietary intakes near industrial areas. The mean weekly intake for Polish children living in industrial areas amounted to 33 μg/kg body weight, and was double that of children living in non-industrial areas (UNEP/FAO/WHO, 1992).

In the recent years in Hungary, practically parallel with the decrease of the lead content of petrol, the decrease of lead content of foodstuffs started, too (Sohár, 1996). Significant differences in the lead level of food of plant origin have been found in different regions. Results of the surveys on the lead content in vegetables are shown in Table 17 (Pintér, 1997).

Soft drink and alcoholic beverages contain much less lead than fruit and vegetables juices. The lead concentration of Hungarian wines was always lower than literature values due to the fact that capsules containing lead were never permitted in Hungary. The median lead content of beverages (1990–1991) was <1–120 μg/kg and range of maximum contents was 11–800 μg/kg (Sohár, 1996).

The result of meat tests shows a decreasing trend in the lead content of food in

Table 16. Daily dietary intake of lead (μg/kg body weight) in the Czech Republic

Foodstuff	Median	% of PTWI
Potatoes	0.017	0.48
Apples	0.0015	0.04
Subtropic fruits	0.002	0.06
Vegetable products	0.013	0.36
Stem vegetables	0.0074	0.21
Fruiting vegetables	0.011	0.31
Root vegetables	0.0058	0.16
Milk	0.0083	0.23
Eggs	0.0037	0.1
Sugar	0.0068	0.19
Sweets	0.016	0.45
Cereals	0.026	0.73
Rice	0.0023	0.06
Plant oil butters	0.0035	0.1
Beef	0.0047	0.13
Pork	0.0049	0.14
Canned meats	0.0013	0.04
Fish products	0.0026	0.07
Poultry meats	0.0032	0.09
Sausages	0.0059	0.17
Cheeses	0.0031	0.09
Butter	0.0014	0.04
Bread and rolls	0.093	2.6
Legumes	0.0032	0.09
Coffee and tea	0.0066	0.18
Nonalcoholic beverages	0.0078	0.22
Alcoholic beverages	0.011	0.31
Total	–	8.6

Hungary (Table 18). The reduction is due to the stricter limit values in fodder introduced in 1990 (Pintér, 1997).

The assessment of the daily dietary intake of lead via maximum exposure scenario was 321 μg/person, i.e. 2247 μg/person per week. If one takes into account the other lead sources (PTWI value applies to lead from all sources), the exposure over the PTWI value is likely for some special population groups in Hungary (Sohár, 1996).

Table 17. Contamination of vegetables with lead in Hungary

Food	Lead content (μg/kg)	
	Median	Maximum
Cereals	50–600	110–2100
Vegetables	3–308	90–6400
Fruit	3–190	40–750

Table 18. Contamination of meat with lead in Hungary

Year	Average content of lead (μg/kg)			
	Pork	Beef	Poultry	Game
1989	430	500	310	360
1990	210	250	240	400
1991	190	190	200	460
1992	150	140	130	370
1993	70	70	60	11
1994	40	40	40	50

Data of the GEMS/Food Programme from Hungary on the mean weekly dietary intakes of lead in the period 1980–1984 were 11.3–14.0 μg/kg body weight in adults and 11.6 μg/kg body weight in infants (UNEP/FAO/WHO, 1992).

One of the highest mean weekly intakes in infants (118 μg/kg body weight) was reported from Austria in a study based on lead levels in selected foods. Higher intakes were noted after introduction of mixed feeding of adult foods such as vegetables at the age around 4 months (UNEP/FAO/WHO, 1992). But the results of this study come from 1981 and specification of the studied population is not available.

6.4. Lead in human biological materials

In the Slovak study investigating the average lead concentrations in the placental tissue (200 samples) collected from four regions of different environmental pollutants and traffic density the levels in range 0.0094–0.0324 mg/kg were found. This study revealed that placental contamination with lead depended on the traffic-related environmental pollution rather than the industrial sources. Among the observed contaminated structures in the placental tissue, the syncytiotrophoblast was the most frequent site for lead (and nickel) particulate deposits. Additionally, a transport of the lead particles from basal plate to chorionic plate was found (Reichrtová et al., 1998a,b).

The lead contents in human milk and reconstituted infant formula obtained from the Lactarium of the Paediatric Hospital in Bratislava are presented in Table 19. The calculated median intakes of lead from human milk and infant formula were similar and they represent 18.8 and 23.8% of PTWI, respectively. The calculated intake of lead by maximal oral exposure scenario via the human milk exceeded the PTWI level by 9.8%. In the investigated group of 35 mothers, no significant relation between lead

Table 19. Lead dietary exposure in infants in the Slovak Republic

Milk	*n*	Concentration (μg/kg milk)		Weekly intake (μg/kg body weight)	
		Median	Range	Median	Range
Human milk	35	4.2	<0.48–24.5	4.70	<0.54–27.44
Infant formula	8	5.3	0.9–7.6	5.94	1.01–8.51

Table 20. Lead in human biological material from general population of the Slovak Republic

	Localities					
	Polluted			Unpolluted		
	Median	Min	Max	Median	Min	Max
Kidney (mg/kg)	0.11	0.03	0.24	0.03	0.01	0.13
Liver (mg/kg)	0.29	0.15	0.61	0.14	0.05	0.37
Blood (mg/l)	0.077	0.0412	0.104	0.051	0.0078	0.256

levels in breast milk and the age, eating habits (lactovegetarians and non-vegetarians) and number of multiparae's was observed (Ursínyová and Hladíková, 1997).

The preliminary results of the Slovak monitoring programme ''Impact of environmental factors on population health'' regarding to lead concentrations in total blood and biological material of general adult population are shown in Table 20. The biological material was collected from pathological departments of hospitals localised in polluted and unpolluted regions of the Slovak Republic in 1995 (Sigmundová et al., 1997).

The influence of the environmental factors on blood lead concentrations was studied in 318 children aged 9–10 year living in different regions of the Slovak Republic. The arithmetic mean blood lead level was 3.9 μg/dl (SD = 1.98) and the median value was 3.41 μg/dl. The mean blood lead concentrations for girls (3.47 μg/dl) and boys (4.29 μg/dl) were statistical significantly different ($P < 0.001$). In relation to the selected environmental risk factors a statistically significant ($\alpha = 0.05$) increase in mean blood lead level for coal heating fuel at home ($P < 0.002$), consumption of vegetables and fruit from private gardens ($P < 0.001$), and consumption of water from the private wells ($P < 0.001$) was found (Hladíková et al., 1997).

The relation of lead levels in blood and neuropsychic changes was evaluated in the Slovak population of 9–10 year old children living permanently in industrial parts of the country near metallurgical plants (big city, $n = 87$; surrounding of this city, $n = 92$; small town, $n = 102$) and children living in relatively clean area, as a control group ($n = 130$). Based on the blood lead level was found: children living in polluted areas (surrounding of the big city, average of lead level 4.58 μg/dl; small town, average of lead level 3.89 μg/dl) were significantly worse in sensomotor performance, in short-term memory, in intelligence test and in general behaviour (unconcentration, unrest) than children from big city (average of lead level 2.81 μg/dl) and children from non-polluted area (average of lead level 3.04 μg/dl). The multifactorial analysis has shown that in children with higher blood lead concentration (≥4.0 μg/dl) were in significant interaction the worse social and living conditions (smoking habits in family, lower income, more children than two in family, and lower education of mothers) and neurobehavioral tests of performance. The authors concluded, that the load of body by low-level of lead in interaction with worse psychosocial status were the risk factors for psychic, somatic development and life quality of children (Šovčíková et al., 1997, 1998a,b).

Biological monitoring in the Czech Republic has been implemented in four districts

– two of them represent more industrial, the other two more agricultural and recreational living conditions. The median blood lead level for the overall group of adults (4.78 μg/dl) was significantly higher in men (5.15 μg/dl). Smoking significantly influenced the blood lead level in women. The median blood lead level in umbilical blood (2.77 μg/dl) was much lower than that in children (3.97 μg/dl). The 90th percentile in no group exceeded the value of 10 μg/dl (Cerna et al., 1997).

The average blood lead levels about 37.2 μg/dl were recorded in an elementary school population living in a neighbourhood close to the smelter in Příbram-Czech Republic (within 3 km from plant) in 1986–1990. In contrast to earlier, the recent investigation (1992–1994) has revealed among a comparable group of children significantly lower exposure to lead. The mean blood lead levels was 11.35 μg/dl (95% CI = (9.32; 13.82)). Besides blood lead, teeth lead was used to assess exposure among children. Statistically significant differences ($P < 0.05$) were observed between the geometric mean teeth lead level of 6.44 μg/g ($n = 13$, 95% CI = (3.95; 10.50) in the most contaminated zone and 1.43 μg/g ($n = 35$, 95% CI = (1.11; 1.84) in zones further away from the point source. Both biomarkers, blood and teeth lead levels, reflected a similar pattern of lead exposure in children (Cikrt et al., 1996).

As a part of the European Multicenter Study blood, hair, and tooth samples were collected in 1985 and 1986 from 7–9 year old children living in settlements with various levels and sources of environmental lead pollution in Hungary. The average lead concentrations were in the range 15.6–27.3 μg/dl in blood, 3.1–11.3 μg/g in hair, and 1.8–3.4 μg/g in teeth. The authors concluded that about 8–13% of the primary schoolchildren had blood lead levels higher than 10 μg/dl, and that the blood lead levels in pre-school children might be substantially higher than those measured in schoolchildren. Considering the impact of environmental lead exposure on the health and especially on the behaviour and mental performance of children, from a preventive point of view, the authors highly recommended that the pre-school children be more widely tested for blood lead levels, most evidently, as a part of the compulsory check-up before school admission (Rudnai and Horváth, 1996).

The results of the study on lead levels in blood of mothers, their new-borns and in mother milk, which was conducted in 1987–1988 in Salgótarján (industrialised Hungarian town with significant traffic) should be viewed with some caution, because the lead analyses were done under substandard technical conditions (Table 21) (Rudnai and Horváth, 1996).

Rudnai and Horváth (1996) conducted a survey in an area of Budapest which had formerly been heavily polluted by a lead-waste processing plant. At the time of the study, the plant had already been closed. The measured blood levels are presented in Table 22. The distance of residence from the plant, the length of residence time in the area, and in case of women age, were found to be significantly correlated with blood lead values.

As a result of reduction of the lead content of gasoline, the blood lead level decreased steeply in the early 1990's. In 1994 a mean blood lead level 6.9 μg/dl was found in school children living in Hungarian provincial towns, and 7.9% of the data exceeded the limit of 10 μg/dl. Recently a representative survey in pre-school children ($n = 1401$, age 0–4 years) showed a mean blood lead level of 5.7 μg/dl and only 4.7% exceeded the limit. The authors concluded, that despite this improvement

Table 21. Lead in biological samples from Salgótarján (Hungary)

Parameter	PbB (μg/dl) mothers	PbB (μg/dl) new-born babies	Pb (μg/dl) mothers milk
n	38	38	38
Mean values	21.3	26.5	6.4
SD	11.5	14.2	4.4

nation-wide condition is not satisfactory. Dangerous situation were created in 1996 by contamination of paprika powder with lead and battery dismantling without permission (Bitto et al., 1997).

The results of the first in Poland large-scale blood lead levels survey in children living in an industrial area were reported by Zejda et al. (1996). Six thousand, nine hundred and sixty-nine children aged 2–6 years were examined in three towns (Chorzów, Katowice, Sosnowiec) from 1993 to 1995. Based on the estimated size of the children population in the target age group the participation rate was 80% in Chorzów, 14% in Katowice and 53% in Sosnowiec. Geometric mean concentration of blood lead was slightly but not statistically significantly higher in boys (all age groups, all locations: 6.68 ± 1.51 μg/dl) than in girls (all age groups, all locations: 5.58 ± 1.54 μg/dl). Association of blood lead levels with family, household and environmental factors was analysed. The following variables explained variation in blood lead levels: town ($P = 0.0001$), age ($P = 0.005$), floor on which apartment was located ($P = 0.0001$), number of siblings ($P = 0.0001$), apartment quality ($P = 0.0001$), carpet in a child's room ($P = 0.0001$), consumption of locally grown vegetables ($P = 0.007$), frequent trips outside the region ($P = 0.0001$). The coefficient of variable expressing the location of apartment in the area of intense vehicle traffic did not reach the level of statistical significance ($P = 0.06$). The occurrence of blood lead concentration above 10 μg/dl was associated with floor on which apartment was located, number of siblings, apartment's quality, the presence of carpet in child's room and frequent trips outside the region. The occurrence of blood lead concentration above 15 μg/dl was associated with the same variables and, additionally, with the place of residence and intensity of vehicle traffic.

An assessment of 12-year old children's exposure to lead was carried out in Cracow ($n = 431$) and in a rural area ($n = 152$) in 1987–1989. About 70% of the results from the city centre of Cracow exceeded 20 μg/dl, reaching the mean value of 30.5 μg/dl. In the group from Nowa Huta (the modern part of Cracow close to the steel and iron plant) blood lead concentrations were markedly lower (mean: 20.4 μg/dl), correspond-

Table 22. Blood lead levels in an Hungarian area polluted by a lead-waste processing plant

Group	*n*	Mean PbB (μg/dl)	SE	Median	98% frequency
Men	64	12.9	0.79	12.1	28.1
Women	129	7.5	0.34	6.1	17.9
Pregnant women	30	5.6	0.63	4.7	17.9
Children (7–8 years)	421	7.8	0.26	6.7	19.4

Table 23. Lead in venous blood and hair in Polish children

Area	n	Lead in blood (μg/dl)			Lead in hair (μg/g)		
		$x \pm$ SD	Geom. mean	Min–max	$x \pm$ SD	Geom. mean	Min–max
Cracow	62	6.71 ± 2.19	6.40	2.43–16.06	2.38 ± 3.42	1.53	0.31–23.6
Krosno	44	5.46 ± 1.63	5.25	3.14–10.73	1.55 ± 1.15	1.16	0.23–5.08

ing to the degree of environmental pollution. The rural control group also had elevated blood lead levels (22.5 μg/dl), probable caused by transboundary emissions (Zielonka and Wodzien, 1993).

The later study in Cracow showed much lower lead exposure in children (Table 23). The study was carried out in 1995 on healthy schoolchildren (aged about 11, $n = 106$). The children lived near the steel works in Cracow and in the rural area Krosno region-Humniska. In Cracow the lead content in falling dust was 5.30 mg/m^2 per 3 months and in soil 0–50 μg/g. In rural area the lead in falling dust was 0.22 mg/m^2 per 1 month. The authors found significant correlation between lead in blood and hair ($P < 0.0001$, $r = 0.5786$). The lead content was significantly higher in the boys' hair ($P < 0.0001$). Children from the rural area had significantly less lead in their blood ($P < 0.01$). Only 5.66% of boys from the studied population had lead in the blood higher than 10 μg/dl (Zachwieja et al., 1996).

The contents of lead in whole maternal blood and cord blood were measured in 40 pregnant women living over ten years in Silesia District in Poland and their 40 new-borns (with different birth weight). The results are shown in Table 24. A new-borns with very low birth weight (BW $\leq$ 1500 g) can be recognised as children with high risk of lead intoxication (Bursa, 1996).

Zareba et al. (1996) measured the blood lead (and cadmium) levels in pregnant women-active and passive smokers and their neonates. Higher blood lead concentrations were found in women who smoked during pregnancy than in non-smokers (5.8 ± 2.2 μg/dl vs. 5.0 ± 1.6 μg/dl, $P = 0.07$). Similar results were observed in their neonates (4.5 ± 2.4 μg/dl vs. 4.2 ± 1.6 μg/dl, $P = 0.43$). The authors found no differences between passive smokers and non-smokers, and they concluded that smoking might play a synergistic role in lead toxicity to the fetus by enhancing penetration of lead through placenta.

Baranowska (1995) investigated the influence of the strongly polluted environment on the content of metal in human tissue. The mean lead concentration was 7.25 μg/dl in maternal venous blood, 3.83 μg/dl in cord blood, and 0.5 μg/g in placenta.

Table 24. Lead in whole maternal blood and cord blood (μg/dl) in Silesia District (Poland)

Body weight of new-borns	Mothers			New-borns				
	n	x	s_x	n	x	s_x	t	P
Total	46	4.96	2.05	46	3.97	2.00	2.34	<0.02
BW ≤ 1500 g	16	4.90	1.79	16	4.09	2.06	1.16	<0.25
BW > 1500 g	30	4.99	2.19	30	3.90	1.92	2.05	<0.05

In the study of the toxic element status of healthy Austrian women ($n = 51$) and their new-born babies a significant correlation between maternal and neonatal blood lead level ($P < 0.001$) was found. The mean concentration of lead was 3.7 (SD: 1.27) μg/dl in whole blood of the mothers, 2.63 (SD: 1.16) μg/dl in whole blood of the new-borns, and 35.8 (SD: 15.0) μg/l in breast milk (Plockinger et al., 1993).

7. Conclusion

In the Central European countries, the main sources of the environmental pollution and human exposure to lead are the exhaust of vehicles using leaded gasoline, industrial processes that utilize lead and lead compounds (ferrous and non-ferrous metal smelting and processing, and battery manufacturing and dismantling), and combustion sources.

Water contamination by pipes, that contain lead, and lead-containing paint do not appear to be a significant factor in human exposures in this region.

Data on air quality indicate that a considerable reduction of population exposure to lead occurred in the early 1990s. But there are indications that even in countries that took measures to reduce the amount of lead in gasoline, improvements have slowed down or stopped recently. Austria and the Slovak Republic phased out leaded gasoline completely by 1993 and 1995, respectively; whereas the Czech Republic, Hungary and Poland have significantly reduced the permissible lead content of gasoline (to 0.15 g/l), and introduced and promoted the use of unleaded gasoline during the late 1980s and early 1990s.

Although considerable progress has been made in the Central European countries in reducing human exposure to lead, much still needs to be done. Special attention should be paid to the Upper Silesian Industrial Region in Poland, where lead concentrations in particular environmental compartments are still very high, and the long-range transport of lead emissions from this industrial area contributes also to the contamination of different compartments of the environment and increased human exposure in other areas of Poland as well as in the neighbouring countries.

General strategies to eliminate lead poisoning include reducing sources of exposure, increasing safe and effective abatement programs, and identifying persons at risk. According to WHO, strategies for exposure reduction must not be limited to elimination of extreme cases of high pollution; instead, they should aim at decreasing average exposure levels throughout the population.

References

Agency for Toxic Substances and Disease Registry, 1992. Lead Toxicity. Case Studies in Environmental Medicine. USA.

Andrzejewski, R., Bararnowski, M. (Eds.), 1993. State of Environment in Poland. State Inspectorate for Environmental Protection, Environmental Information Centre GRID-Warsaw. Warsaw.

Anon., 1994. Task Force on Heavy Metals Emissions. Economic Commission for Europe, Prague.

Anon., 1998. Water quality. Drinking Water STN 75 7111, Slovak Republic (in Slovak).

Baranowska, I., 1995. Lead and cadmium in human placentas and maternal and neonatal blood (in heavily polluted area) measured by graphite furnace atomic absorption spectrometry. Occup. Environ. Med. 52, 229–232.

Bitto, A., Horvath, A., Sarkany, E., 1997. Monitoring of blood lead levels in Hungary. Cent. Eur. J. Public. Hlth. 5, 75–78.

Brody, D.J., Pirkle, J.L., Kramer, R.A., Flegal, K.M., Matte, T.D., Gunter, E.W., Paschal, D.C., 1994. Blood lead levels in the US Population, Phase 1 of the Third National Health and Nutrition Examination Survey (NHANES III, 1988 to 1991). JAMA 272, 277–283.

Bursa, J., 1996. Lead and cadmium concentrations in whole maternal blood and in cord blood of their newborns with different birth weight ($\leq$1500 g; $>$1500 g). In: Abstracts of International Conference Environmental Pollution and Child Health: Critical Needs and Issues for Central and Eastern Europe. Sosnowiec, Poland.

Cerna, M., Spevackova, V., Cejchanova, M., Benes, B., Rosser, P., Bavorova, H., Ocadlikova, D., Smid, J., Kubinova, R., 1997. Population-based biomonitoring in Czech Republic - the system and selected results. Sci. Total. Environ. 204, 263–270.

Cikrt, M., Smerhovsky, Z., Blaha, K., Nerudova, J., Sediva, V., Fornuskova, H., Knotkova, J., Roth, Z., Kodl, M., Fitzgerald, E., 1996. Biological monitoring of lead exposure among children in the Czech Republic. In: Abstracts of International Conference Environmental Pollution and Child Health: Critical Needs and Issues for Central and Eastern Europe. Sosnowiec, Poland.

Csanady, M., 1996. Lead content of the drinking water in Hungary. In: Füzesi, Z., Levy, B.S., Levenstein, Ch. (Eds.), From Science to Action: The Lead Hazard in Hungary - a Fact Report. Ferling Press, Pécs, pp. 152–156.

Federal Ministry for Environment, Youth and Family, 1997. State of the Environment in Austria. Vienna.

Fiala, J., Bartáčková, Z., Dvořáková, M., Kurfürst, J., Machálek, P., 1998. Emission, and deposits of heavy metals in the Czech Republic. In: Proceedings of the III. Conference Air'98, pp. 49–52, (in Czech).

Food Research Institute, 1997. Partial Monitoring System of Foreign Substances in Foods and in Feeds. Report from 1996. Bratislava, (in Slovak).

Gerhardsson, L., Skerfving, S., 1996. Concepts on biological markers and biomonitoring for metal toxicity. In: Chang, L.W. (Ed.), Toxicology of Metals. Lewis Publishers, CRC Press, Boca Raton, New York, pp. 81–107.

Gzyl, J., 1997. Assessment of Polish population exposure to lead and cadmium with special emphasis to the Katowice Province on the basis of metal concentrations in environmental compartments. Centr. Eur. J. Publ. Hlth. 5, 93–96.

Hertzman, C., 1995. Environment and Health in Central and Eastern Europe. A report for the Environmental Action Programme for Central and Eastern Europe, World Bank, Washington, DC.

Hladíková, V., Ursínyová, M., Šovčíková, E., Cviková, V., 1997. Children's blood lead levels in relation to the selected environmental risk factors. In: Proceedings of the International Symposium on Environmental Epidemiology in Central and Eastern Europe: Critical Issues for Improving Health. Smolenice, Slovak Republic, pp. 64–66.

International Agency for Research on Cancer, 1987. Monographs on the evaluation of carcinogenic risks to humans, overall evaluations of carcinogenicity; an updating of IARC Monographs, vols. 1–42 (Suppl. 7). Lyon, pp. 230–232.

Lockitch, G., 1993. Perspectives on lead toxicity. Clin. Biochem. 26, 371–381.

Magulová, K., Sajtáková, E., Uhlík, J., Marinkovičová, J., Šedíková, D., 1996. Air pollution in selected nonattainment areas of the Slovak Republic by emissions from stationary sources, Enclosure to report: Occurrence of Selected Toxic and Carcinogenic Organic and Inorganic Substances in Ambient Air of Selected Locations in the Slovak Republic [Report of MoE SR supporting project for PHARE project EU/93/AIR/22b], Institute of Preventive and Clinical Medicine, Bratislava.

Makovinská, J., László, F., Csányi, B., Pál, G., Hock, B., Németh, J., Szabó, S., Škoda, P., Topolská, J., 1997. Tendency and Dynamics of Water Quality Changes of the Danube River and its Tributaries (1989–1995). Water Research Institute, Stimul, Bratislava.

Ministry of the Environment of the Czech Republic (8 Oct 1998) State of the Environment in the Czech Republic. http://www.ceu.cz/special/czechrep/rocenka97/b224tab0.htm

National Institute of Public Health in Prague, 1995a. Monitoring the health status of inhabitants in relation to the outdoor and indoor air. System of Monitoring of Health Status of Inhabitants in Relation to the Environment, Prague, (in Czech).

National Institute of Public Health in Prague, 1995b. The Health Risk Assessment of Dietary Exposures to the Selected Chemical Substances in the Czech Republic. Prague.

National Institute of Public Health in Prague, 1997. Health effect and risks due to contamination of drinking water. System of Monitoring of Health Status of Inhabitants in Relation to the Environment, Prague, (in Czech).

OECD, 1993. Lead - Background and National Experience with Reducing Risk. Risk reduction monograph No. 1. Paris.

Österreichische Akademie der Wissenschaften-Kommission für Reinhaltung der Luft, 1993. Umweltwissenschaftliche Grundlagen und Zielsetzungen im Rahmen des Nationalen Umweltplans für die Bereiche Klima, Luft. Geruch and Lärm, Vienna.

Pintér, A. (Ed.), 1997. Hungarian Environmental Health Action Programme. Prospektus Nyomda, Veszprém.

Plockinger, B., Dadak, C., Meisinger, V., 1993. Lead, mercury and cadmium in new-born infants and their mothers, Z. Geburtshilfe Perinatol. 197, 104–107.

Rakics, R., Öri, I., 1996. Sources and uses of lead in Hungary. In: Füzesi, Z., Levy, B.S., Levenstein, Ch. (Eds.), From Science to Action: The Lead Hazard in Hungary - a Fact Report. Ferling Press, Pécs, pp. 221–231.

Reichrtová, E., Ursínyová, M., Palkovičová, L., Wsolová, L., 1998a. Contents and localization of heavy metals in human placentae. Fresenius J. Anal. Chem. 361, 362–364.

Reichrtová, E., Dorociak, F., Palkovičová, L., 1998b. Sites of lead and nickel accumulation in the placental tissue. Human Environm. Toxicol. 17, 176–181.

Rudnai, P., Horváth, A., 1996. Biological monitoring of environmental lead exposure in Hungary. In: Füzesi, Z., Levy, B.S., Levenstein, Ch. (Eds.), From Science to Action: The Lead Hazard in Hungary - a Fact Report. Ferling Press, Pécs, pp. 232–238.

Sigmundová, V., Škripeňová, A., Borošová, B., Benasová, A., 1997. Content of toxic metals in biological materials in inhabitants from selected regions of the Slovak Republic. In: Proceeding of Conference Health and the Environment. Milovy, Czech Republic, pp. 242–247, (in Slovak).

Sohár, J., 1996. Lead content of foodstuffs in Hungary. In: Füzesi, Z., Levy, B.S., Levenstein, Ch. (Eds.), From Science to Action: The Lead Hazard in Hungary - a Fact Report. Ferling Press, Pécs, pp. 239–251.

Šovčíková, E., Ursínyová, M., Wsolová, L., Rao, V., Lustik, M., 1997. Effects on the mental and motor abilities of children exposed to low levels of lead in Bratislava, Toxic Substance Mech. 16, 221–236.

Šovčíková, E., Wsólová, L., Ursínyová, M., Hladíková, V., Cviková, V., 1998a. Rural-urban neuropsychic and blood lead differences in Slovak children, Epidemiology 9(Suppl.), S139.

Šovčíková, E., Wsolová, L., Ursínyová, M., 1998b. Cognitive changes in lead exposed children. In: Abstract book of the 12th Conference of the European Health Psychology Society, Prevention and Intervention. Vienna, Austria, 89 pp.

Tsuchiya, K., 1986. Lead. In: Friberg, L., Nordberg, G.F., Vouk, V.B. (Eds.), Handbook on the Toxicology of Metals, vol. 2: Specific Metals. Elsevier, Amsterdam, pp. 298–353.

UNEP, 1997. Global Environmental Outlook, Oxford University Press, New York, Oxford.

UNEP/FAO/WHO, 1992. Assessment of Dietary Intake of Chemical Contaminants. WHO/HPP/FOS/92.6.

Ursínyová, M., Hladíková, V., 1997. The intake of selected toxic elements from milk in infants. Fresenius Environ. Bull. 6, 627–632.

Ursínyová, M., Hladíková, V., 1998a. Toxic elements in ambient air in the Slovak Republic. In: Proceedings of the III. Conference Air'98, pp. 257–260, (in Slovak).

Ursínyová, M., Hladíková, V., 1998b. Dietary intake of cadmium, lead and mercury in vegetarian and non-vegetarian children. Fresenius Environ. Bull. 7, 585–592.

Ursínyová, M., Hladíková, V., Sokolovič, P., 1997. Intake of toxic elements from vegetarian and non-vegetarian diet. In: Proceeding of Conference Foreign Substances in Food. Tatranská Štrba, Slovak Republic, pp. 145–147, (in Slovak).

Violová, A., Bratský, D., Šovčíková, E., Ursínyová, M., 1997. Complete phase-out of leaded gasoline: Policies and implementation in the Slovak Republic. In: Lovei, M. (Ed.), Phasing out Lead from Gasoline in Central and Eastern Europe. The World Bank, USA, pp. 49–65.

WHO, 1989. Lead – Environmental Aspects. Environmental Health Criteria 85. Geneva.

WHO, 1993. Evaluation of Certain Food Additives and Contaminants. Forty-first Report of the Joint FAO/WHO Expert Committee on Food Additives, Technical Report Series No. 837. Geneva.

WHO, 1994. Updating and Revision of the Air Quality Guidelines for Europe. Report on the WHO Working Group on Inorganic Air Pollutants, Regional Office for Europe, EUR/ICP//EHAZ 94 05/MT04, 24–27 October 1994, Düsseldorf, Germany.

WHO, 1995. Inorganic Lead. Environmental Health Criteria 165. Geneva.

Zachwieja, Z., Schlegel-Zawadzka, M., Chlopicka, J., Bartoń, H., Pitrzyk, J.J., Huzior-Balajewicz, A., 1996. Lead in hair and blood of children living near a steel works and in a rural area. In: Abstracts of International Conference Environmental Pollution and Child Health: Critical Needs and Issues for Central and Eastern Europe. Sosnowiec, Poland.

Zareba, A., Strugala-Stawik, H., Rudkowski, Z., Dembicka, D., Pastuszek, B., 1996. Blood levels of lead and cadmium in pregnant women, active and passive smokers and their neonates. In: Abstracts of International Conference Environmental Pollution and Child Health: Critical Needs and Issues for Central and Eastern Europe. Sosnowiec, Poland.

Zejda, J.E., Grabecki, J., Król, B., Panasiuk, Z., Jedrzejczak, A., Jarkowski, M., 1996. Blood lead levels in urban children of Katowice Voivodship, Poland: Results of the population-based biomonitoring and surveillance program. In: Abstracts of International Conference Environmental Pollution and Child Health: Critical Needs and Issues for Central and Eastern Europe. Sosnowiec, Poland.

Zielonka, E., Wodzien, M., 1993. Blood cadmium and lead levels among children in Cracow, Folia Med. Cracov. 34, 85–96.

II

Analytical techniques

Trace elements – Their distribution and effects in the environment
B. Markert and K. Friese, editors

Chapter 5
Instrumental techniques for trace analysis

R. Djingova, I. Kuleff

Abstract

Trace analysis of environmental materials is a complex process including not only the analysis itself but also sampling, sample preparation, physical and chemical pretreatment and data evaluation. Therefore the basic rules for sampling of environmental materials and methods for preconcentration are presented. The possibilities and potential of the most significant methods for environmental trace analysis (X-ray fluorescence (XRF), inductively coupled plasma-mass spectrometry (ICP-MS), inductively coupled plasma-atomic emission spectroscopy (ICP-AES), atomic absorption spectrometry (AAS), instrumental neutron activation analysis (INAA), electrochemical methods and methods measuring radionuclides) are discussed and comparative evaluation is done. The methods for speciation of metals in environmental matrices are briefly presented. Since there is not a single method that can solve all tasks in environmental investigations a combination of at least two methods might be recommended.

Keywords: Trace elements; Radionuclides; Environment; Water; Soil; Aerosol; Plant; Neutron activation analysis; Atomic absorption spectrometry; Inductively coupled plasma-atomic emission spectroscopy; Inductively coupled plasma-mass spectrometry; X-ray fluorescence; Electrochemical methods; Speciation

1. Introduction

Production of metals has been among the first human "industrial" activities, and thus since ancient times man influences the state of metals in nature. Thanks to human activity metals are redistributed in ways different from natural ones and new metal compounds are obtained and released into the environment. Thus the natural cycle of metals is disturbed. Since metals and metalloids are distributed everywhere in nature and cover a large concentration interval between mg/kg and more to ng/kg and less, and participate in a number of important geochemical and biochemical processes this disturbance leads to serious impact on living organisms. This is one of the reasons that determination of metals and metalloids in environmental materials is an object of constant interest.

Although most analytical techniques have reached maturity and their possibilities are well defined still the analysis of environmental materials cannot be considered without problems due mainly to two reasons:

It is already recognized that the quality of analysis and the justification of final conclusions depend not only on the method for analysis but on all steps before, such as sampling, sample preparation etc. Therefore the planning step is decisive when a proper strategy has to be developed where the task, sampling and sample preparation should be planned with the idea what method for analysis is to be used afterwards and vice versa. The connection between sampling, sample preparation and analysis is so close that each step can totally geopardize the preceding and following ones. It is recognized that incorrect sampling may introduce up to 1000% error (Sansoni, 1986; Markert, 1989a; Wagner, 1995), and sample treatment is a source of a number of possible losses. Although a number of contributions have appeared discussing these problems (Paris, 1987; Stoeppler, 1991, 1992; Markert, 1996) still investigations on sampling and analysis remain somehow separate fields and in many papers presenting results in the field of environmental analytical chemistry little or no attention is paid to correct description of sampling or storage parameters. In this way the results cannot be used further by other investigators and remain of importance only to the author of the respective paper. Even more dangerous is to use such data for general conclusions when they actually might not be representative for the purpose.

The second reason is connected to the analysis itself. As Nuernberg (1984) stated: "Many institutions without any or with only insufficient analytical expertise start buying expensive instrumentation and performing analysis of chemical pollutants. In this manner the already significant amount of false data could be increased dramatically by a wave of irrelevant results...". This statement is a good explanation for the impending need of quality assurance and quality control. This is well known to the analytical community but it is still not well enough understood by people with other education working in the field of environmental investigations who take it for granted that analytical techniques have reached a level of guaranteed high accuracy, precision and detection limits. Actually the real possibilities are far from the theoretical "instrumental" ones, specified by the producers of equipment, and each separate case should be controlled. The results from quality control have to be presented to the public together with the results from environmental analysis. This will help for a further realistic evaluation of the role of analytical techniques in environmental analysis in respect to the introduction of meaningful detection limits, accuracy, precision and overall possibilities in the analysis of different environmental materials.

Another question which is rarely discussed is the variety of determined elements. Generally it seems that the type of determined elements depends more on the method used in the investigation than on the real need of determination. Elements such as Pb, Cd, Zn, Cu are among the very important pollutants but they are also easily determined by atomic absorption spectrometry (AAS), a very popular method. Therefore a considerable number (if not the majority) of environmental investigations deal with these elements only. The elements that remain outside the scope of popular analytical techniques are actually those that still have unknown functions in the organisms, and the necessity of their determination came forward quite recently. A good example in this respect are rare earth elements and platinum metals in plants.

The aim of the present review is to stress on some of the important factors in the investigation of metals and some metalloids in environmental materials by instrumental techniques. Basic rules for sampling, storage and treatment will be presented and the realistic possibilities of popular analytical techniques for trace analysis discussed.

The enormous amount of literature in this field renders an extensive review impossible within such a limited volume. Therefore more general contributions, review articles and papers containing results from analysis of standard reference materials (SRM) are cited with the clear idea that many interesting papers in this field remain outside.

2. Sampling, storage and sample treatment

2.1. Atmospheric aerosols

The term atmospheric particulate matter is used to describe any substance in liquid or solid form in the atmosphere under normal conditions and with dimensions roughly in the micrometer range. The most investigated representative of this group are the aerosols and very often the two are used as synonyms though under particulate matter dust, fog, fume, haze etc. are also included.

The representative sampling of atmospheric aerosols is influenced by several natural factors which should be considered carefully in order to prepare a good sampling plan. Meteorological conditions and particle size play very decisive role in this respect. Among the meteorological conditions: temperature, humidity, precipitations, wind speed etc., the wind direction is the most important factor (Keith, 1991) since its change influences strongly the imminent distribution pattern of the pollutants. It should be kept in mind also that wind direction may be changed also by topographic factors such as mountains, hills, seas etc. (Keith, 1991). The proper knowledge of meteorological situation and topography of the region enables sampling that is representative for the investigated area under well described and reproducible conditions.

Particle size determines more or less the origin, the chemical and optical properties, the mechanism of removal from the atmosphere and therefore it plays an important role in the sampling strategy planning. Generally particles with diameter less than 2.5 μm are referred to as fine, and those with greater diameter are referred to as coarse (aerosol particle size analysis is described in Griffits et al. (1998)).

Sea spray, volcanic eruptions, soil dust, as well as some industries (cement manufacturing) produce the so called primary aerosols, i.e. the material is emitted directly in particulate state (Klockow, 1982), and they are both fine and coarse. Secondary aerosols are produced in the atmosphere usually by condensation after emission from high temperature sources, and they are fine as a rule. Considering the difference in the chemical composition it is recognized that the major components of the fine aerosols are toxic substances of anthropogenic origin such as As, Cd, Pb, Se, Zn etc. while the course aerosols are enriched in elements like Ca, Fe, Si coming from erosion, sea aerosols and other natural sources.

The factors very briefly mentioned above determine representative sampling of aerosols as a very difficult step. Several approaches may be used for collecting of aerosols (Alian and Sansoni, 1985, 1990):

- gravimetric settling (for $d \geq 20$–50 μm)
- wet impingement (100% efficiency for particles of ≥1 μm)
- thermal precipitation ($d \geq 0.01$–10 μm)

- impaction
- filtration

Sampling by impaction permits collection of aerosols according particle size and is especially useful in the study of distribution of the concentrations of pollutants with particle size. Different types of impactors are used for this purpose.

The most widely used technique for sampling of aerosols is filtration. It permits to collect any desired volume of air, size of particles (by choosing appropriate filter size and medium), needs relatively simple equipment, permits parallelly to sampling concentration of the pollutants and last but not least analysis may be performed either directly using the filter or after extraction with different media (Klockow, 1982; Alian and Sansoni, 1985, 1990; Watson and Chow, 1994). The choice of an appropriate filter is determined by the aim of analysis and the method used. Suitable for the majority of instrumental methods are Teflon membrane, cellulose fibre, glass fibre and mixed quartz fibre filters. The parameters that have to be considered when choosing a filter are its mechanical and chemical stability, sample efficiency, loading capacity and blank values. As a rule of thumb filters should be analyzed before sampling. Often the blank values may be commensurable to the aerosol concentrations to be determined (Alian and Sansoni, 1985, 1990; Watson and Chow, 1994). This is especially important for methods using directly the filter for analysis such as instrumental neutron activation analysis (INAA) and XRFA. Certain types of filters may need pretreatment before sampling. Quartz fibre filters for example are usually heated to about 900°C for minimum 3 h and afterwards kept in freezers.

During the last decade a number of investigations have been directed towards direct analysis of air particulate matter without using sampling on filters. This approach includes either direct injection of air or use of impactor directly in the furnace to be applied as atomizer in electrothermal atomic absorption spectrometry (ETAAS) or electrostatic precipitation of metals in aerosols on a tungsten collecting electrode which afterwards is put in the electrothermal atomizer (Cresser et al., 1988; 1992).

Some instrumental methods (INAA, XRFA) analyse directly the aerosol filter or portions of filters. Recently such approach is intensively tested for other methods especially ETAAS. For the majority of instrumental methods however treatment of the filter is required leading to a liquid extract. Wet ashing with different acidic mixtures is very popular though dry ashing in a furnace, followed by acid dissolution is also used. The choice of the method depends not only on the measurement technique and the range of substances to be analyzed but also on the chemical form of the element (Kneip and Kleinman, 1982). For example high temperature secondary aerosols may contain refractory oxides or silicates and the treatment requires fusion or hydrofluoric acid dissolution.

Dry ashing may result in losses due to volatilization. Glass fibre filters after ashing at 700°C give up to 80% losses of aerosol. In this case for example fusion with Na_2O_2 is recommended. A more appropriate technique is ashing at low temperatures using oxygen plasma (Alian and Sansoni, 1985). A dry ashing procedure for perchlorovinyl air filters in quartz microfurnace under a stream of oxygen is described by Cresser et al. (1992).

Acid treatment is performed using different acidic mixtures: HCl + HF; HF +

$HClO_4 + HNO_3$; $H_2SO_4 + HNO_3 + H_2O_2$ (Alian and Sansoni, 1985), HNO_3 (Katz, 1977; NIOSH, 1977). Perchloric acid extraction from inert (quartz and glass fibre) filters is also practised though this procedure is not to be used for ETAAS (Kneip and Kleinman, 1982).

Ultrasonic vibration is utilized to promote acid extraction (e.g. with 5% HNO_3 at 70°C) as well as for suspension of fly ash as slurry directly onto the furnace of an autosampler (Cresser et al., 1992).

The use of solvent extraction is beneficial in some cases. For example for the separation of Au, Pt, Pd, Ag and Ir (Schutyser et al., 1977) proposed a combination of precipitation scavenging exchange–solvent extraction procedure. Very low detection limit for Be in fly ash (ng/cm^3) has been reached after extraction with *N*-benzoyl-*N*-phenylhydroxylamine in isobutyl methylketone (IBMK) and measurement by flame atomic absorption (FAAS) (Cresser et al., 1992).

2.2. Water

Thanks to the movement of water and the diffusion of dissolved substances different geochemical processes take place in the Earth's crust. To a great extent they are responsible for the redistribution of chemical elements in soil and further on to the biosphere. The anthropogenic activity leads to disposal of a lot of wastes to water basins which slowly accumulate in the oceans. This is one of the reasons that makes the determination of metals and toxic elements in water as one of the important tasks in environmental pollution control.

There are different classifications of natural waters. Depending on the salt content waters are divided into fresh water (salt content <0.5 g/dm^3 – rivers and lakes), brackish water (<10 g/dm^3 – estuarine), and salt water (about 35 g/dm^3 – sea water, although some closed seas like the Baltic and the Black Sea have lower salt content). Depending on their origin, waters may de classified as river and lake, precipitation (snow and rain), sea water, waste water and underground water.

In comparison with other environmental materials water has some very specific features which are extremely important for sampling and analysis. Some of its properties change extremely quickly with the time. Among these are acidity, dissolved gases, temperature as well as the chemical form of some elements and substances. Therefore sampling and analysis should be performed under special care, as quickly as possible and for some parameters – in the fields directly after sampling. Besides, the possibilities for losses and contamination during the storage and transport are greater than with other materials. Therefore the sampling strategy should be planned so as to account for:

- aim of investigation
- type of water basin (full hydrological characteristic inclusive major inflows and outflows)
- geochemical features of the region
- climatic conditions (a month prior to sampling)

The information on these items determines the frequency of sampling, replications, sampling sites and depth, type of samples and sampling devices, the procedures for initial treatment, and the parameters that must be determined in the fields (temperature, pH, colour etc.). The sampling protocol must include an exact description of all sampling parameters characterizing the sampling site, climatic conditions, devices, filtration procedure, acidity etc.

Sampling of snow and rain is perhaps most difficult for several reasons:

- significant change of the chemical composition with the duration of precipitation and standing;
- difficulties in isolation of dry anthropogenic deposition.

Therefore sampling is performed usually during storms using two possible techniques – "open rain" or "wet only" sampling (Schutyiser, 1978). Sampling collectors of Teflon may be used covered with nylon metting (Cowgill, 1994). Containers with automatically opening lids with the first rain drop and closing after rainfall are better.

Sampling from rivers, lakes, seas etc. may be performed either 0.5 m below the surface or at different depths depending on the aim of investigation. Recommendable for stratified bodies (depth >5 m) is to take five samples (0.5 m below surface; middle of the epilimnium, 1 m above and below the thermocline, and about 0.5 m above the bottom). Samples are collected by pumping or with polyethene buckets. Sub-surface samples can be taken either by pumping or by automatic samplers opening and closing at a desired depth. Automatic samplers are preferable to manual sampling. Different types of peristaltic, vacuum and bladder pumps are used on the samplers (Dick, 1994). Different aspects of sampling have been presented in (Cowgill, 1988; Ferguson et al., 1976; 1977; ISO, 1980; Knoechel and Prange, 1981; ASTM, 1982; Bond, 1982; EPA-600/4-82-029, 1982; Sturgeon and Berman, 1987; Whitworth et al., 1998).

Sampling containers may be plastic or glass, depending on the investigated parameters. They have to be as inert as possible to the compounds of interest and not cause leaching or adsorption. The containers should be cleaned before sampling. For trace metal analysis cleaning is usually performed by treating the vessels with hot HNO_3 or HCl at least for 2–3 h followed by rinsing with bidistilled water. Treatment with different complexing agents is also practised. Before sampling vessels are rinsed several times with the water to be sampled.

Natural waters always contain suspended particles. These are inorganic substances like alumosilicates, carbonates, quartz as well as organic colloidal matter. Their concentration varies in a large interval: 10–100 mg/dm^3 for river water; 10–20 mg/dm^3 for coastal waters; >1 mg/dm^3 for deep sea water and the open ocean. Suspended particles should be isolated from the water after sampling otherwise processes of adsorption, desorption etc. may take place which lead to change in the original content of the sample. There are three ways to isolate suspended matter: sedimentation, filtration and centrifugation.

The gravitational method permits the sedimentation of all particles with $d > 10$ μm after 10 days stay in a special container, but this is very rarely used.

Filtration is the most widely used method. Standardly a 0.45 μm membrane filters are used. Filtration may be performed either at high pressure with nitrogen or at low pressure. High pressure filtration is preferred since it permits direct filtering of the sample in inert atmosphere and thus the possibilities of contamination are reduced. Filtration has to be performed immediately after sampling in the fields. A lot of devices, specially constructed for different investigations exist (Knoechel and Prange, 1981; Sturgeon and Berman, 1987; Whitworth et al., 1998). Despite the long experience in filtration as a preconcentration step for water samples still it may be a source of serious analytical errors (Gardner and Comber, 1997).

Isolation of suspended matter by centrifugation is realized using flow centrifuges by varying the revolutions and the stream velocity. Thus suspended particles with different dimensions may be isolated. At 10 000 rev./min and at a stream velocity of 1.2 l/min, for example, the effect is the same as with filtration (Calmano and Lieser, 1981; Das et al., 1983).

The quantity of isolated suspended matter should be exactly determined and the concentration of suspended particles is used as a characterization of the investigated water basin. The concentration of metals in the filtered water is actually only the quantity of dissolved metals therefore separate analysis of the suspended matter may be recommended for evaluation of the total content.

The preservation of the water samples during transport and storage before analysis has to prevent sorption of metals on the container walls. This depends on the concentration of the elements, pH, duration of contact, container material, presence of complexing agents. Usually samples are preserved by acidifying with c. HNO_3 (ultrapure) to pH 1.5–1.8. About 1.5 cm^3 are added per litre sample. Storage temperature of 4°C is recommended to reduce microbial activity. There are of course exceptions to this rule. For example: preservation of Hg is achieved for about 3 months after addition of strong oxidizing agent, Cr^{6+} at pH 9, presence of carbonate buffer and EDTA. When speciation is to be performed no treatment should be done, and analysis has to be immediately after sampling. If this is absolutely impossible then freezing of the sample at -196°C (liquid nitrogen) may be done and storage at -20°C. Filters with suspended matter are also to be kept at low temperature.

The rules for preservation and storage given above are not finalized as yet. One of the reasons is that most experiments leading to these conclusions have been and are performed using model solutions and not real samples.

2.3. Soils

The processes of rock weathering lead to the formation of soils where additionally the climate, biological activity, relief etc. play very important role. Therefore, nowadays the diversity of soils is so great that a complete classification is hardly possible. Besides soil composition is very complex. It includes solid and liquid phase (soil solution) as well as gaseous phase. The solid phase is a polydisperse system consisting of larger ($d > 0.01$ mm) and fine ($d < 1$ μm) particles. Besides minerals and inorganic substances (about 99%) organic substances are also present. The mineral part is very heterogenous reflecting not only the composition of parent rocks but also is influenced by secondary minerals. One of the most important soil features is that it includes living organisms which amount to about 200–300 kg/ha in the upper 0–20 cm layer.

Keeping in mind these features soil can be defined as “The upper layer of the Earth’s crust composed of mineral particles, organic matter, water, air and organisms” (DIN, 1991).

Trace metals in soils are existing in several forms:

- located in the lattice of primary minerals;
- located in the lattice of clay and other secondary minerals;
- chemically bound to organic matter;
- physically or chemically sorbed, or readily exchangeable ions on mineral, or organic exchange active surfaces;
- in water soluble form in the soil solution.

The last two forms include up to 2% of the total metal content but they are considered as most important for plant nutrition. Until recently in agrochemical, nutrient and ecological investigations attention has been paid mostly to these two forms and analysis has been performed after extraction with water, buffer solutions or other weak extraction media. However the anthropogenic activity leads among other things to increase in soil pH. The result is slow transition from the silicate to the aluminium buffer region where heavy and toxic metals are released from the minerals. Therefore it is very useful to know more about the content of soils than the usual available quantities.

Sampling strategies of soils depend mostly on the purpose of the investigation and a lot of sampling procedures have been standardized (Fraenzle and Kuhnt, 1983; Lewis, 1987; Fraenzle, 1990, 1994; DIN, 1991; ISO, 1991; Wagner and Sprengart, 1991) and used (Davis et al., 1988; Ho and Tai, 1988; Bajo et al., 1989; Steinnes et al., 1989; Hewitt and Candy, 1990; Berggren, 1992a,b; Severson et al., 1992; Wagner, 1993a; Bloemen et al., 1995). A very good discussion on development of international standards for soil sampling is presented in (Paetz and Groessmann, 1994). On this basis some general rules may be formulated:

A. When sampling is planned complete information about the type of soil, colour, consistency, texture, sampling horizon, drainage conditions, type of vegetation, weather conditions and human activity should be available.

B. For monitoring and agrochemical purposes usually the top 0–20 cm layer is considered, because most of the pollutants remain there after deposition on the one hand and on the other the root uptake even that of larger trees occurs there. The lower horizons however are also significant because the concentration there may give useful information about the spatial variability, behaviour of elements and probable influence on deeper roots. The representative sampling needs the construction of a good sampling grid in conformity of the region. The maximal single sector should not exceed 4 ha. The minimum number of field replicates can be calculated if the variance is known, if not – then 5–10 is the obligatory number, though 20 is considered to lead to more representative results. Care should be taken in respect to soil type. Peats, sandy or stony soils are hard to be taken in the usual ways, and higher soil volume may be required.

C. Sampling from surface soils usually is performed using steel trowels, and for depth studies – soil corers are preferable. Contamination from sampling equipment is a real possibility especially when metals are to be determined. Though stainless steel equipment gives good results, polythene trowels are recommended or coating the samplers with polyurethane for example may be advantageous.

D. Sampling directly after agrochemical treatment of soils (herbicides, pesticides, fertilizers, lime etc.) should not be performed.

E. Sampling from locations where fertilizers, lime etc. have been stored or fire has been made is to be avoided.

F. If the purpose of the investigation is not the study of roadside pollution or transport influence, sampling should be performed at least 200 m from highways, metal fences or cable conduits. Otherwise the content of Pb, As, Sb, Br, Cd may be influenced near highways and of Cu and Zn near cables.

Storage of soil samples is to be done carefully depending on the purpose of the investigation. It is recommended that transport and storage time should be minimal in the case of speciation analysis. In the general case it is advisable to keep the transport and storage temperatures of samples below ambient air temperature and with as low as possible aeration to prevent microbial activity. After performing certain obligatory determinations (pH, etc.) the laboratory samples should be dried at temperatures up to 110°C (not for speciation analysis).

Reduction of bulk sample size is performed either manually by successive quartering the bulk sample, by increment sampling or by using rotation sampling devices.

Laboratory treatment of soil samples starts with removal of stones, roots, small living organisms. After crushing lightly the soil is sieved through 2 mm sieve. The fraction greater than 2 mm is collected and estimated separately. In some special cases this procedure is not followed (Wagner and Sprengart, 1991; Wagner, 1993b).

The determination of total metal levels in soils by INAA and XRFA needs no special additional treatment of the laboratory sample. All other methods however demand certain treatment no matter whether total or extractable content is to be determined.

The variety of procedures is enormous (Berrow and Mitchell, 1980; Berrow and Stein, 1983; Berrow and Reaves, 1984, 1985; Ho and Tai, 1988; Bajo et al., 1989; Steinnes et al., 1989; Eriksson et al., 1990; Hewitt and Candy, 1990; Berggren, 1992a,b; Bloemen et al., 1995; Haapala et al., 1996; Garcia and Millan, 1998) though the general principles are quite clear nowadays. The determination of the total metal content demands complete dissolution of the sample which practically is possible only after alkaline fusion (with Na_2CO_3; NaOH etc.) followed by dissolution of the melt in acid (e.g. HCl) or after hydrofluoric/perchloric digestion. Since both approaches require extremely good analytical experience and Teflon, or platinum (nickel) crucibles, are time-consuming nowadays they are rarely used for monitoring and agrochemical purposes. The general practice at present is to extract metals with c. HNO_3 or mixture of $HNO_3 + HClO_4$ or aqua regia. The remaining metals are so strongly bound in the undissolved silicates that they are practically unimportant for monitoring, agrochemical and nutrient investigations. Microwave or Teflon bomb assisted acid digestion is also practised as well as ultrasonic agitation for slurry introduction in ETAAS.

The amount of water soluble, readily exchangeable and organically bound metals in soils may be evaluated after different extractions. Various procedures have been proposed using extractions with acetic acid, calcium chloride, ammonium acetate, EDTA, diluted nitric acid etc. Good results are obtained by using sequence of extractions for the different forms, e.g. 1 M ammonium acetate (pH 7), 0.1 M hydroxilaminhydrochlorid +0.01 M nitric acid (pH 2); 30% hydrogen peroxide + nitric acid (pH 2) (Foerstner et al., 1981). A complete procedure for sequencial extractions is summarized in (Davies, 1994).

2.4. Plants

Plants occupy a special place among the ecological materials usually investigated. The reason is that plants are living organisms which thanks to photosynthesis are responsible for oxygen production. Besides they are an extremely important link in the food chain collecting from the soil, air and water and transmitting further to man and animals. The elemental composition of plants usually reflects the geochemical features of the environment where they grow. That is why some of them are successfully used as biomonitors. Their composition however is influenced by the processes and specific features of the living organism resulting in intra- and inter-species differences, differences due to stage of development etc.

Metals in plants can generally be divided into three groups depending on the role in the living system. Essential for all or some plant species are: Ca, Co, Cu, Fe, K, Mg, Mn, Mo, Na, Ni, Se, V and Zn. Among the toxic metals are: Al, As, Au, Cd, Cr, Hg, Pb, Pd, Pt, Sb, Te, Tl and U. Of course each essential metal can be toxic as well depending on the amount. The rest of the metals in the periodic system are with still unknown functions in the plant organism.

Planning representative sampling of plants has to include a number of abiotic and biotic factors: genetic variability, climatic, edaphic conditions etc., which determine the difference among species, population, stands, organs etc. irrespective of the aim of investigation: nutrient, physiological, ecotoxicological or environmental monitoring. A number of excellent contributions have discussed different aspects of representative sampling of plants (Martin and Couphtrey, 1982; Kuleff and Djingova, 1984; Markert, 1987, 1988, 1989b,c,d, 1995a, 1996; Wagner, 1987, 1990, 1993a; Ernst, 1990, 1994; Djingova and Kuleff, 1994, 1999; Turcsanyi et al., 1994; Djingova et al., 1996). It is accepted that the sampling error may exceed orders of magnitude the analytical error and the error due to physical or chemical treatment of plant material may reach 300%. Due to the heterogeneous distribution of plants in ecosystems and sites on the one hand and the peculiarities and physiological functions of the living organism on the other hand the minimizing of sampling error (taking a representative sample) needs a lot of preliminary investigations prior to real sampling. After the choice of a specific plant to be studied in order to plan sampling one has to have a very good estimate on:

- biological variation, including individual variations, distributions among organs, influence of stage of development etc.;
- seasonal variation in the metal concentration;
- location specific differences,

as well as the procedure for initial treatment of the samples after sampling.

The assessment of biological variation includes the determination of the minimal number of individuals to be sampled so as to have minimal error due to individual variation. In the general case the individual plants should not be less than 10, and better 25 in the case of herbacious species, bushes etc. In the case of trees the age of the tree, sampling height, age of twigs and leaves, position of leaves etc. should be studied and standardized (Wagner, 1987).

In monitoring investigations mainly leaves are used however other parts of the plants are also important. The pattern of distribution and redistribution among different organs of the plant should be studied so as to have information about the degree of exposure, the ways pollutants enter the plant and about the direction and extent of the physiological processes. Usually the concentrations increase in the direction seed $<$ flower $<$ stem $<$ leaves $<$ root, but there are exceptions to this rule.

Seasonal variations are considered greater than location specific variations and may reach up to 200%. The studies of the seasonal behaviour of metals in plants have lead to some more general conclusions:

- metal concentrations in deciduous trees and shrubs reach maximum in September–October, and a possible minimum in the summer months. Sampling is recommended before senescence:
- conifers have a continuous age gradient;
- some mosses exhibit highest values in winter months, and minimum in summer;
- grasses may reach maximum in winter; herbacious species reach highest values in different periods but sampling is to be performed shortly after blossoming when the concentrations are relatively stable (Djingova and Kuleff, 1999).

The behaviour of each plant should be studied separately. It is also necessary to perform sampling not at the time of highest concentrations but during an interval of relative stability in the concentrations.

Location-specific differences are dependant on the edaphic and microclimatic conditions and for some elements and plants may reach up to 100%. To account for this variation a very good sampling grid must be constructed and all parameters: e.g. soil characteristics, climatic conditions etc. recorded carefully in the sampling protocol. The biological variation due to location should be determined as well.

Physical treatment of the samples prior to analysis is a parameter belonging to the sampling procedure. While in food chain investigations cleaning is not performed it is strongly recommended in all other studies. There are two major reasons:

- to avoid soil contamination (extremely important for grass, herbage and decisive when roots are investigated);
- to reduce the effect of precipitation leading to wash out of the leaves, and the aim is to remove reproducibly the mechanically deposited dust and/or aerosols without leaching elements out of the leaves.

A number of cleaning procedures have been used and studied: wiping with cloth, blowing with nitrogen jet, rinsing or washing with tap and/or bidistilled water, rinsing with tap water and detergent, treatment with acid solution, and complexing agents etc. In each case additionally temperature and duration of treatment play significant role. Summarizing these investigations the following may be derived:

- most reproducible results for conifers are obtained after treatment with chloroform for 1 min and the solution is investigated separately (Wyttenbach et al., 1985; Krivan and Schaldach, 1986);
- for deciduous trees, shrubs, grass, herbage different procedures have been used. Reproducible results have been obtained by cleaning of Ailanthus leaves with 1% Alconox and 0.01 M EDTA (Porter, 1986; Markert, 1992). The comparison of

several treatments (energetic 1 min washing with tap water/bidistilled water; rinsing with distilled water; 1 min treatment with 0.01 M HCl; 1 min with 0.05 M EDTA, wiping with cloth) for three plants (*Populus nigra*, *Taraxacum officinale* and *Plantago major*) proved best results for the treatment tap/bidistilled water. The procedure leads to reproducible removal of Pb, Cd, Fe, Cu, and Zn and no leaching was established for the investigated Ca, K, Mg. Comparison with untreated leaves sampled before and after raining from the same locations indicated that the rain effect is similar to the removal with tap/bidistilled water (Djingova and Kuleff, 1994, 1999). It is worth mentioning that especially for decidious trees, grass and herbage cleaning should be done on fresh leaves or practically in the fields. Therefore washing with tap/bidistilled water may be the easiest approach to standardize for different leaf surfaces in the fields.

Standardization of all above steps is a time consuming, tedious work but it ensures successful results from the analysis otherwise the analytical work is useless to be done. There are not many examples of good standardized sampling procedures for higher plants, and among them the procedures for *Populus nigra* 'Italica' (Wagner, 1987), *Picea abies* (Knabe, 1982), *Taraxacum officinale* (Kuleff and Djingova, 1984; Djingova et al., 1986, 1993; Djingova and Kuleff, 1993, 1999) may be cited.

Drying and homogenization are the following steps that are to be performed after the samples have been transported to the laboratory and prior to analysis. Usually drying is done at temperatures not exceeding 80°C. Gentle drying at 40–50°C in vacuum has also been a successfully used approach. The theoretical and practical approaches of homogenization are discussed in Markert (1995b,c) in detail. Homogenization is usually performed using different types of mills. The most important aspect is that care should be taken to avoid contamination from the construction material.

Generally as in the case of soils two approaches may be used for sampling preparation: dry ashing and acid digestion.

Although dry ashing at 450–900°C is used for plant materials (Kabata-Pendias and Dudka, 1990, 1991; Cresser et al., 1992) this approach is not recommended unless special precautions are taken. Even then many volatile elements of interest (Pb, Cd, As etc.) may be lost. For the determination of rare earth and other non-volatile elements however it may be a successful procedure before acid treatment.

Digestion with different acid cocktails is a more convenient approach for dissolution of plant materials. HNO_3 is very attractive for analysts but a variety of mixtures are used: HNO_3 + $HClO_4$; HF + $HCLO_4$; HNO_3 + HCl etc. (Cresser et al., 1988, 1992; Cresser, 1994). A successful mixture is HNO_3 + H_2O_2 even in open systems when dissolution only with nitric acid is not complete. Microwave-assisted digestion, sometimes with predigestion step or high asher pressures are recommended for dissolution of plants. In these systems the use only of nitric acid usually leads to complete dissolution. In all cases high purity acids (even additionally purified by distillation) should be used as well as quartz or Teflon vessels.

2.5. Preconcentration

The direct determination of many heavy and toxic elements is not possible in various environmental materials due to insufficient detection limits of analytical methods and the very low concentrations of most of the elements of interest. The sensitivity of analytical methods may be increased if before measurement concentration of the element(s) is performed. During this stage the concentration of the element is increased and very often the matrix is eliminated which further may reduce spectral interferences (Hoste et al., 1971; Das et al., 1983; Toelgyessy and Kyrs, 1989; Stoeppler, 1992). The preconcentration step is performed when the quantity of the analyte element in the aliquot is below the limit of determination and there are no possibilities for increase of volume or mass of the sample. The detection and determination limits of most of the analytical methods indicate that a considerable number of elements mostly in atmospheric air and natural water cannot be determined without preconcentration.

The following techniques are usually used for preconcentration:

- evaporation
- lyophilization
- dry ashing
- coprecipitaion
- adsorption
- liquid–liquid extraction
- ion exchange

Evaporation is a method for preconcentration used in the analysis of water samples. The enrichment factor is between 10^2 and 10^3. This approach however may lead to losses of certain elements due to adsorption on the walls of the vessels. That is why evaporation nowadays is practically used in the preconcentrations of radionuclides from waters since the carriers prevent adsorption (Das et al., 1983; Mizuike, 1983; Toelgyessy and Kyrs, 1989).

Lyophilization is mainly used for the sample treatment of plants (and biological materials) having undefined quantity of water (fruits, vegetables) as well as when volatile elements are to be determined. Water is removed under low temperature and pressure. The freezing temperatures do not influence the mass of the final product. The enrichment factor depends on the quantity of water and is in the range 4–20 for fruits and vegetables.

Dry ashing is not highly recommended due to the very probable and irreproducible losses of volatile elements. It is possible to perform it at low temperatures or by oxidation with KNO_3 and slow increase of temperature (Stoeppler, 1992; Tschoepel, 1992). The enrichment factors are about 100–600 when plant materials are treated.

Coprecipitation is a method for separation and preconcentration based on the formation of mixed crystals thanks to isomorphic exchange or adsorption of microcomponents on the surface of ionic crystals. Microelements in solutions in concentrations below ng/cm^3 can hardly be isolated by direct precipitation, therefore different reagent carriers are used (Hoste et al., 1971; Das et al., 1983; Mizuike, 1983; Toelgyessy and Kyrs, 1989; Stoeppler, 1992; Nickson et al., 1995). The off-line approach is slow, unprecise yet enrichment factors are well over 1000. When reagent carriers are chosen correctly recovery may reach up to 99%. This method is widely used in the analysis of different radionuclides (Das et al., 1983; IAEA, 1989).

Ion exchange is used to eliminate matrix, major and some interfering trace elements and concentrate the trace element(s) of interest (Hoste et al., 1971; Das et al., 1983; Fang, 1993; Mizuike, 1983; Toelgyessy and Kyrs, 1989; Stoeppler, 1992; Nickson et al., 1995). Ion exchange ensures high decontamination factors of elements such as Na, Cl and Br and is very useful in the analysis of sea water. Four types of exchange resins are used: inorganic exchangers (e.g. hydrated antimony pentoxide (HAP)), chelating resins (e.g. Chelex 100), anionic exchangers, and chemically modified exchangers (chelating agents-loaded resins) (Hoste et al., 1971; Mizuike, 1983; Stoeppler, 1992; Fang, 1993; Nickson et al., 1995). Ion exchange permits automation and continuous performance of the process. After the separation, elution with different acids is performed.

Sorption: Ions or their complexes may be collected in microcolumns on activated carbon, hydrous iron(III) oxide or reversed phased silica based sorbents (Das et al., 1983; Mizuike, 1983; Fang, 1993). The techniques are easily automated. Either elution or direct analysis may be performed. However the problem with blank values (by direct analysis of the sorbent) deserves serious attention.

Liquid–liquid extraction: The solvent extraction with chelating agents is a well established method for preconcentration of some ultra-trace elements (Hoste et al., 1971; Das et al., 1983; Stoeppler, 1992). Derivatives of dithiocarbamic acid - sodium pirrolidindithiocarbamate, ammonium pirrolidindithiocarbamate are widely used as chelating agents. The complexes are afterwards easily extractable by different organic solvents or using solid-phase extraction (Hoste et al., 1971; Poole and Poole, 1996). By variation of pH a selective extraction might be performed.

3. Instrumental techniques for analysis

3.1. X-Ray fluorescence methods

The excitation of atoms with suitable radiation causes elimination of electrons from the inner shell (K, L, M). The electrons from the outer shells drop to the free positions and the process is accompanied by emission of electromagnetic radiation with energy up to 120 KeV (X-rays), characteristic for the respective atom (element). The excitation can be induced by X-ray, gamma ray, or accelerated particles (protones, electrons etc.). The emitted radiation can be used for analysis either after wavelength or energy dispersion. These possibilities determine the existence of several versions of X-ray fluorescence methods (XRF): wavelength, energy dispersive, proton induced, total reflection and synchrotron XRF.

The general advantages of XRF methods are (Hahn-Weinheimer et al., 1984; Sansoni, 1987; Toelg, 1987; Stoeppler, 1991):

- non-destructive in principle and multi-element;
- suitable for elements with $Z > 6$ in concentration interval % – mg/kg;

- good selectivity;
- relatively simple spectrum with less spectral interferences than emission methods;
- low cost, easy operation;
- high sample throughput.

Disadvantages of XRF methods are:

- high self absorption of emitted radiation limiting the analysis to thin films or surface layers;
- dependence on particle size and homogenity;
- strong matrix effects demanding exact matching of samples and standards;
- conventional XRF is generally not sensitive enough for background determinations in various environmental matrices.

Wavelength XRF (WDXRF) is a sequential technique, more time-consuming than the other XRF methods and with worse sensitivity. Detection limits lie normally in the higher mg/kg range. WDXRF has been successfully applied for the determination of Al, Ca, Fe, K, Mg, Mn, Na, P, Si in plant samples and of Cr, Cu, Fe, Ni, Pb, and Zn also in vegetation (Hahn-Weinheimer et al., 1984).

Energy dispersive XRF (EDXRF) has the advantage of being a typically multi-element, quick method with sensitivities in the lower mg/kg range. It has been applied for analysis of a number of environmental materials: aerosols and fly ash, soils, plants (Beitz et al., 1980; Hahn-Weinheimer et al., 1984; Van Dyck et al., 1986; Krumpen et al., 1989; Toelgyessy et al., 1990; Bandhu et al., 1996; Osan et al., 1996; Alfani et al., 1997).

In aerosol samples 12 elements (K, Ca, Ti, V, Mn, Fe, Cu, Zn, Br, Rb, Sr and Pb) have been determined in concentration range 10–700 ng/m^3 (Bandhu et al., 1996). In fly ash where the concentrations are higher the elements Al, Si, K, Ca, Ti, Mn, Fe, Ni, Cu, Zn, As, Rb, Sr, Y, Pb and Th are determined with precision below 10% in an concentration interval 15 mg/kg (Th) – 13% (Si). The accuracy evaluated by analysis of NIST-SRM 1633a is very good for all elements but As (Van Dyck et al., 1986). The good analytical parameters have been achieved after introduction of a secondary fluorescence correction for medium thickness samples. The same approach has lead to the successful determination of 20 elements in soils (IAEA Soil 5) with excellent accuracy and precision, and of 12 elements in plant matrices (NIST-SRM 1571).

Osan et al. (1996) determined K, Ca, Mn, Fe, Cu, Zn, Br, Rb and Sr in plant samples in the concentration range 0.6–280 mg/kg. Detection limit was not good enough to allow determination of Cr, As and Ni.

Analysis of four plant certified materials (NIST-SRM-1571, -1573, -1570, and -1575) has proved the reliable determination of Cr, Mn, Fe, Cu, Zn and Pb. The accuracy for Ni was unsatisfactory (Beitz et al., 1980). In all these examples Mo-tube or Mo/W anode tube with Mo secondary target have been used for excitation.

EDXRF with *radioisotope sources* is attractive due most of all to its very low cost and stable intensity of the radiation, depending only on the half-life of the radio-nuclide. The conventional radioisotope sources that are used are ^{241}Am, ^{109}Cd, ^{55}Fe, ^{228}Pu etc. (Hoffmann, 1986; Toelgyessy et al., 1990).

Using radioisotope EDXRF Kumar et al. (1989) determine S, Cl, K, Ca, Ti, V, Fe,

Zn, Br, Rb, Sr, Zn and Pb in aerosol samples in the concentration range 0.6–100 ng/m^3. Due to poor peak-to-background ratio the elements Br, Rb, Sr and Zn are determined with unsatisfactory accuracy. In the same paper results for Ca, Fe, Zn and Sr in ground water are presented in concentrations 0.07–7.4 μg/dm^3.

The use of ^{241}Am permitted the determination of As, Ba, Fe, Pb, Rb, Sr and Zn in IAEA-Soil 7 and IAEA SDN1/2 with excellent accuracy and precision well below 10%. Detection limits vary between 1 and 10 mg/kg (Ivanova et al., 1998). In the same paper four plant standard reference materials are investigated and Co, Cr, Fe, Mn, Ni and Pb are determined with very good analytical parameters. The values for the detection limits (0.1–10 mg/kg) however indicate that the method is suitable only for samples from highly polluted regions.

Theoretically the combination of several radioisotope sources each one suitable for excitation of elements with different Z should lead to considerable improvement of the possibilities of the method. The determination of 11 elements has been reported by Belakova et al. (1995) in medicinal plants using ^{238}Pu, ^{241}Am/Ag and ^{125}I as excitation sources. The detection limit of 0.5 mg/kg has not been enough in the case of Hg and Se, and no analysis of standard reference materials is performed.

Using a combination of ^{241}Am with Dy secondary target, ^{55}Fe and ^{238}Pu, Djingova et al. (1998) investigate seven plant standard reference materials. Radionuclide EDXRF permitted the reliable determination of Ca, K, Mn, Fe and Zn and the detection limits or accuracy are not good enough to determine any of the ecotoxicologically important elements.

Particle-induced XRF (PIXE) has many advantages over conventional XRF. Irradiations are performed in ion accelarators (Muminov and Haidarov, 1980). It ensures up to 100 times better detection limits and values of 0.1–1 mg/kg can be obtained even in routine analysis. Accuracy and precision normally lie in the 1–5% range. The proton beam can be focused to a extremely small spot ($d \sim 1$ μm) which permits microdistribution analysis of elements with $Z = 25 \div 80$. In this interval lies the best sensitivity of the method. Using this approach over 20 elements have been determined in air particulate matter (Kumar et al., 1989). It is considered very suitable for investigation of aerosol samples (Johansson and Campbell, 1988).

The target in PIXE may be either filter (aerosol, water), thin carbon foil or microscopic cuts of biological materials.

After preconcentration with chelating agents 13 elements are determined in drinking water (Bi, Cd, Co, Cu, Fe, Hg, In, Mn, Ni, Pb, Pd, Tl, Zn). The detection limits are in the range 0.2–3.5 μg/dm^3. Comparison with AAS for Zn, Fe, Ni, Cu and Mn proved very good agreement (Gonsior and Roth, 1985).

A disadvantage of PIXE is the limitation of sensitivity due to bremsstrahlung of secondary electrons produced by the proton beam. The most serious handicap for the wide use of the method however is the necessity to perform the analysis in laboratories where accelerators or cyclotrones are available.

Total reflection XRF (TXRF) is already a competitive method for environmental trace analysis. The principle is that the sample is irradiated in grazing incidence by a narrow collimator X-ray beam. The sample has to be a very thin film on a polished quartz or Plexiglass carrier. The beam is totally reflected by the sample carrier and the sample is passed and excited. The advantages of this method are (Sansoni, 1987;

Klockenkaemper and von Bohlen, 1992, 1996; Prange and Schwenke, 1992; Toelg and Klockenkaemper, 1993):

1. detection limits up to 1000 times better than conventional XRF. Most of the elements with $Z > 11$ are detectable to ng/g levels in 100 μl solution;
2. elimination of matrix effects thanks to the positioning of the sample as a very thin liquid layer on a carrier. Therefore the method is independent on the matrix;
3. quantification by internal standardization. A single standard element added to the total sample or after the sample has been placed on the carrier is used for multi-element determination;
4. due to the very thin sample self absorption effects are eliminated.

Restriction of the method is that the sample must have mass between 1 and 100 μg and a thickness of 10 nm–10 μm to prevent matrix effect and compensate for count rate limitations.

The original application of TXRF has been solutions therefore other matrices, especially complex ones have to be digested and the matrix eventually separated. Hence TXRF is extremely good for water analysis. Different types of natural waters have been investigated: rain, drinking, river, and sea water (Knoechel and Prange, 1980, 1981; Klockenkaemper and von Bohlen, 1992; Reus et al., 1993). Approximately 50–100 μ of water are used for the determination of S, Cl, Ca, Fe, Zn, Mn, Tl, V, Ni, Cu, Pb, Br, Sr, Rb and U in a concentration range of 1–230 ng/cm^3 with precision about 3% (accuracy is about 4%). After appropriate preconcentration (chemical matrix separation) the detection limit may reach 3–20 pg/cm^3.

In the analysis of NIST-SRM-1643c, Markert et al. (1997) determined As, Cd, Co, Cr, Cu, Fe, Mn, Ni, Pb and Zn. with very good accuracy. The result for Cr is rather low however agreeing very well with inductively coupled plasma-mass spectrometry (ICP-MS).

The detection limits for river and sea water may be improved further after freeze drying and subsequent digestion with nitric acid (Klockenkaemper and von Bohlen, 1996).

The analysis of air particulate matter by TXRF permits the determination of K, Ca, Cr, Mn, Fe, Cu, Zn, Sr and Pb with detection limits of 0.1 ng/m^3. Either the particles may be removed from the filter into an acidic solution or the sample carrier may be used as a plate in impactors for air particulate collection (Schmelling et al., 1997a) (the precision varies between 5 and 20%). Markert (1996) reports the determination of 34 elements in NIST-SRM-1633a with very good analytical parameters.

TXRF has been applied for certification of IAEA-SRM lichen 336. The concentrations of Ca, Cu, Fe, Mn, Pb, Rb, Sr and Zn have been determined after acid digestion. The concentration of determined elements is in the mg/kg range, and precision varies between 3.5 and 7%. All the results are within the range of accepted values (Schmelling et al., 1997b).

Markert (1996) reports analysis of NIST-SRM-1573 (Tomato leaves) and presents data about 18 elements. The results for Ni, Mo, Cd and Pb agree not very well with certified values.

Cyclotrone XRF (SYRF) demands the use of cyclotrone to induce X-ray fluorescence of an element. This leads to the following advantages (Ketelsen et al., 1985; Sansoni, 1987):

1. thanks to the high particle flux density and good collimation the irradiated surface is very small;
2. the excitation of elements is more uniform thanks to the white spectrum up to 35 KeV which enables the use of K-fluorescence to $Z = 60$;
3. the high degree of polarization of the cyclotrone radiation and its possible total reflection on the sample holder leads to low background and good detection limits.

A decisive disadvantage is that cyclotrones are hardly available therefore the method has not found wide application.

Results from the analysis of NIST-SRM- 1575 regarding Mn, Fe, Zn, As, Rb, Sr, Zr, Pb, have been reported (Ketelsen et al., 1985). The agreement between experimental and certified values is very good with exception for As and Pb.

Summarizing these examples on XRF analysis of environmental matrices the methods may be arranged in the following order: WDXRF $<$ EDXRF (radioisotope) $<$ EDXRF $<$ PIXE $<$ TXRF in respect to overall possibilities. This order may be directly derived from theoretical considerations however there are some differences in the predicted application of different XRF methods to the environmental matrices. In the case of aerosols and fly ash samples all methods permit the reliable determination of about 12 elements. PIXE has the advantage of reported 20 elements. The only positive impressive exception is the determination of 34 elements in Markert (1996).

For natural waters TXRF should definitely be the choice though PIXE also gives good possibilities. The number of reliably determined elements does not exceed 16.

For plants and soils all methods permit the determination of five to 15 elements, radionuclide EDXRF having the lowest possibilities.

Among the determined elements dominate are the alkaline earth elements and alkali metals (K, Ca, Rb, Sr) and the transition metals Mn, Fe, Cu, Zn etc. The reliable determination of some ecologically important elements like Hg, Se, As, Sb has not been proven. Obviously the determination of rare earth elements, platinum metals and heavy elements is a problem although theoretically XRF is a good method for their analysis. Values for La and Nb by TXRF have been reported in standard reference materials only in Markert (1996).

3.2. Mass spectrometry

Theoretically mass spectrometry (MS) is a dream method allowing the principle determination of all elements in the periodic system down to the ng/g range. The most important mass spectrometry methods are: spark source-MS (SSMS), laser ionization-MS (LIMS), glow discharge-MS (GDMS), secondary ion-MS (SIMS), thermal ionization-MS (TIMS) and ICP-MS. The application of the first five methods which are solid state spectrometric methods is mainly directed for trace and ultra trace analysis of metals, alloys, semi- and non-conductors and various high purity materials, though a very spectacular result has been reported with GDMS (Duckworth et al., 1993) where 51 elements have been determined in concentrations 0.1 mg/kg-%. The reported detection limits were in the ng/g range. Very good review on these methods may be found in (Bacon et al., 1996; Becker and Dietze, 1998). The most environmentally oriented method is ICP-MS.

The principle of ICP-MS is atomization of the sample in inductively coupled argon plasma and high degree ionization, after which the positively-charged ions are extracted from the plasma into the mass spectrometer where the ions are separated according their mass to charge and detected by photomultiplier. The advantages of ICP-MS are (Gray, 1985; Barnes, 1996; Becker and Dietze, 1998):

1. more than 90% of the elements can be detected in the 0.01–1 ng/g range;
2. about 50 elements are detectable below 10 pg/ml level;
3. with ICP quadropole MS the precision is 0.1–2%;
4. the dynamic range is six to seven orders of magnitude;
5. high sample throughput.

Some of the general limitations of the method are:

1. necessity for dissolution of samples which poses the decisive question about blank values;
2. matrix effects as well as problems with dissolved solids, mass resolution and spectral inteferences are serious for earlier equipment and are diminished to a great extent in third generation instruments;
3. high cost of instrumentation.

When analysing environmental materials and depending on the instrument the following problems appear to a different extent.

(a) Isobaric species are not resolved. Polyatomic and molecular ions can be formed in the plasma and interfere with coincident analyte ions. When environmental and biological materials are analysed then the polyatomic ions include Cl, C, Ca, P, K, Na etc. and limit the determination of Cr, Fe, Mn, Ni, etc. A number of approaches have been proposed to overcome these difficulties:

1. sample preparation methods
2. multivariate corrections
3. introduction of high resolution spectrometers

Barnes (1996) stresses on these problems in the analysis of environmental samples because they limit seriously the number of determined elements and worsen the results for some elements like B, Al, Mg.

(b) Another problem is the so called memory effect which can be overcome after extensive clean out washing periods with different media depending on the determined element. For example for B acidified solution of NaF is used, and for Au in plant samples – 8% aqua regia (Barnes, 1996).

ICP-MS has been used for the determination of 40 ultratrace elements in water in the concentration range 15 pg/dm^3–20 μg/dm^3. The detection limits reached 0.5 pg/dm^3 for the lanthanides (Yamasaki et al., 1994).

Twelve heavy elements are determined in polar snow in the pg/g region. Reported detection limits vary between 0.009 (U) pg/g to 0.9 μg/g (Zn) (Barbante et al., 1997). However results from analysis of standard reference materials have not been presented.

McLaren et al. (1995) analyzed water samples for production of standard reference materials and determine 16 elements with detection limits in the subng/g range. Garbe-Schoenberg (1993), Guo and Lichte (1995) and Guenther et al. (1997) determine a number of elements in aerosols, soils and rock standard reference materials. Detection limits for lanthanides reached 0.002 μg/g and were 0.01 μg/g for the rest of the elements.

Fifteen to 18 elements are determined by Markert (1996) in plant standard reference materials (NIST-SRM 1572,1575 and BCR-SRM-281). In NIST-SRM-1572 the accuracy for some of the elements (As, Pb, Mo) is not satisfactory. In the analysis of NIST-SRM-1643a (Markert et al., 1997) determine 12 elements with excellent analytical characteristics in the μg/g and ng/g regions. Forty-five elements are determined by Madeddu and Rivoldini (1996) in plants and standard reference materials reaching 0.2 ng/g detection limits, and McLaren et al. (1995) use ICP-MS for certification of plant reference materials in the sub-nanogram range for 16 elements.

Further extension of the possibilities of the method is the introduction of isotope dilution to MS. It is applicable to elements with at least two stable isotopes or to monoisotopic elements having a long lived radioactive isotope (e.g. ^{230}Th). IDA-ICP-MS has been applied to analysis of various standard reference materials: sediments, waters etc. (Barnes, 1996). Analytical precision of 2% has been reached and even down to 1% in the case of flow injection IDA-ICP-MS and determination of Cu, Cd and Pb in river water.

In summarizing this very brief survey, it is quite clear that ICP-MS seems to have an enormous potential in environmental trace analysis. The method is still developing therefore any final conclusions will be premature. Obviously however it is able to cover the determination of a large number of elements including very heavy elements, rare earth elements and platinum metals in a variety of environmental matrices. It is perhaps worth mentioning that when results from analysis of SRMs are presented the spectacular number of reliably determined elements decreases to 15–25 which indicates that still a lot of investigations have to be performed in order the method to occupy a realistic place among other instrumental techniques for environmental trace analysis.

3.3. Atomic emission spectrometry

By absorbing external energy one or several electrons are excited to higher energy levels. During spontaneous returning to the ground state, each energy transition emits a spectral line and its energy or wavelength are characteristic for the element (type of electron shell). The intensity is more or less proportional to the concentration of atoms. The larger the number of electrons in the shell system desto greater is the number of possible transitions. Therefore the heavier elements are more suitable for atomic absorption spectroscopy (AES) determinations. Classical AES has good power of detection in the entire μg/g region and is a very good method for quick semi-quantitative analysis. The usual mode of excitation is by arc, flame or spark sources.

The introduction of ICP and DCP increased the possibilities of AES in the environmental analysis thanks to their relative freedom from interferences and wide applicability (Keliher et al., 1986; Sansoni, 1987; Toelg, 1987; Stoeppler, 1991; Markert, 1996).

The main advantages of ICP-AES are:

- ability to detect up to 50 elements
- precision varies between 0.2–5%
- relatively low chemical interferences
- sensitivity lies in the ng/g–μg/g range
- low detection limits for refractory elements
- high sample throughput
- dynamic range about five orders of magnitude

Disadvantages are:

- spectral interferences similar to arc and spark AES
- dependence on blank values
- dependence on salt and acid content

ICP-AES is an analytical technique already well established among the methods for environmental analysis with well known characteristics (Stoeppler, 1991; Jackson and Mahmood, 1994).The relative lack of papers dealing with special development of the method in respect to environmental analysis is even explained by this fact (Cresser et al., 1992). The attention nowadays is mostly directed towards development of different systems for sample introduction (Marshall et al., 1996).

In the analysis of natural waters Speer et al. (1986) discuss the influence of salt content on the reliability of results. Li et al. (1996) determined Co, Cu, Mn, Fe, V, Ti and Mn in waters after third phase preconcentration and reach detection limits of 0.02–0.6 μg/cm^3 (the reported recovery is 95–100%). Analysis of standard reference materials produced by national Research Council Canada: SRLS-3 (river water) and NASS-3 (sea water) proved the reliable determination of V, Mn and Cu in river water and Mo in sea water. For the rest of the elements sensitivity was not good enough (Li et al., 1996). Very good accuracy has been obtained in the determination of Ca, Co, Cu, Fe, Mn and Zn (Ebdon et al., 1987).

Reported analysis of soil standard reference materials (IAEA Soil 5, IAEA SL-1 and BCR-141) indicate that about 16–20 elements may be determined with very good analytical characteristics in sediments and soils after introduction of background and spectral corrections (Liese, 1985a,b). Severson et al. (1992) reported the determination of 26 elements in soils. The accuracy and precision have been proven by analysis of NIST-SRMs. The concentration range varies between 0.1 (Se) μg/g to %. Seventeen elements were reliably determined in soil and sediment standard reference materials without introducing any corrections (Kanda and Taira, 1988).

Between five and 15 elements can be determined in aerosol filters, including some rare earth elements and Zr in concentrations down to 0.17 ng/g (Cresser et al., 1988, 1992; Ebdon et al., 1987).

Severson et al. (1992) report the determination of 22 elements in moss samples at background levels in the interval 0.5 μg/g to % following the procedure which has been developed for soils (see above). Liese (1985b) analyses NIST-SRM-1571 and determines 18 elements with precision well below 10%. The accuracy for Na, Cd and Cr was unacceptable. Cd, Co, Cr, Cu, Fe, Mn, Mo, Ni, Pb and Zn were determined in NIST-SRM-1575 (Kos et al., 1996) with precision better than 10% but for Cr (20%). Detection limits were between 0.06 (Mn) and 1.8 (Pb) μg/g. After appropriate digestion, concentration and instrumental corrections 36 elements have been determined in certified environmental materials (Que Hee and Boyle, 1988).

The determination of 26 elements in several plant reference materials including rare earth elements after appropriate separation and preconcentration procedure is reported in (Markert, 1996). Djingova et al. (1998) analyzed seven plant standard reference materials. The content of Al, Ba, Ca, Cd, Co, Cr, Cu, Fe, K, Mg, Ni, Pb, Sr and Zn was determined with detection limits varying between 0.3 and 7 μg/g. The detection limit for Pb was not good enough to determine lower than 1 μg/g.

The position of ICP-AES among other instrumental techniques for trace analysis is well established. Although instrumentation is still developing it is more in respect to reduce laborious introductions of different types of corrections than increase the quality of analysis. As an interesting example in this respect a comparison of ICP-AES analysis of soil and sediment standard reference materials done in 1985 (Liese, 1985a) and 1998 (Leivuori, 1998) may be done. In 1985 the content of 21 elements in IAEA soil, five and 19 elements in IAEA SL-1 has been determined after introduction of two type of corrections. The accuracy was 0.1–18% and precision below 10%. In 1998 in a sediment standard reference material (NIST-SRM-2704 - Buffalo River Sediment) ten elements were determined (for the rest of the elements in this study ETAAS is preferred) with accuracy of 0.1–16% and precision below 10%. The same result is obtained if a comparison of analysis of plant standard reference materials is done (Liese, 1985a; Kos et al., 1996; Djingova et al., 1998).

The position of ICP-AES in respect to sensitivity is between FAAS and ETAAS. The multielement determination of about 15 elements is relatively easy to perform in all types of environmental matrices. However the direct determination of some ecotoxicologically important elements (Pb, As, Hg, Cd, Sb etc.) is not possible in natural waters, plants and aerosol samples. Therefore a preconcentration step is required. After preconcentration the determination of lanthanides in plants has been achieved (Markert, 1996) with excellent accuracy.

3.4. Atomic absorption analysis

The principle of atomic absorption analysis (AAS) is based on the fact that atoms absorb radiation at the same wavelength at which they emit. The sample is atomized in the light path of a radiation source emitting the atomic spectrum of the analysed element, and the extent of absorbed radiation (absorbance) is proportional to the concentration of the element.

The general advantages of AAS are:

- high sensitivity and low detection limits down to the sub-nanogram range (ETAAS)
- good accuracy and precision
- well established methodology
- high sample throughput
- relatively low cost

The disadvantages are:

- single element technique
- narrow dynamic range
- works predominantly with solutions

AAS is perhaps the most popular, thoroughly tested and well established method in environmental research, and therefore regular major advances cannot be expected any more (Cresser et al., 1992). Annually appear an enormous number of contributions using AAS in environmental investigations. In the period 1983–1992 more than 2000 papers were published on AAS in the environmental health practice (Tsalev, 1994). A survey on the annual reviews in *J. Anal. At. Spectrom.* on Atomic Spectroscopy Update – Environmental Analysis indicates that the articles using AAS exceed tens of times the contributions using other analytical techniques. A number of excellent reviews discuss the state of the art and perspectives of AAS in environmental research (Welz, 1985; Broekaert and Toelg, 1987; Toelg, 1987; Stoeppler, 1991; Cresser, 1994; Tsalev, 1994, 1995, 1998; Halls, 1995).

Depending on the sample introduction and atomization several versions of AAS are distinguished.

FAAS is the oldest version of AAS. It works with liquid samples which after nebulization are mixed with acetylene and introduced in a flame atomizer burner with air-acetylene or N_2O-acetylene flame. A single measurement can be completed within 10 s. Theoretically the method is applicable to 60–70 elements and due to its low cost, selectivity and simple operation is preferred whenever the concentration of the determined elements is within its possibilities. The sensitivity of FAAS is of the same order of magnitude as of ICP-AES and for some elements even worse which in recent years has lead to the replacement of FAAS with the multielement ICP-AES. To increase the sensitivity of FAAS usually preconcentration procedures are applied before measurement. Otherwise FAAS permits direct measurements in the low mg/kg range of a number of metals in soils and sediments, six to ten elements in plants and in natural waters. It is sensitive enough for the direct determination of Al., Ba, Ca, Cu, Fe, K, Na, Mg, Mn and Zn in different environmental materials (alkaline metals are determined by flame atomic emission spectrometry). Due to the narrow dynamic range problems with the accuracy appear (e.g. Djingova et al., 1991) and very often dilutions are necessary which decreases relative sensitivity and increases the possibilities for errors.

ETAAS. In ETAAS atomization takes place in an electrothermal atomizer which is heated to the appropriate temperature programme. The detection limits of the method are about two or three orders of magnitude better than FAAS. It is applicable to about 40 elements but generally for about 20 elements detection limits at the ng and pg level can be reached. Commensurable or better sensitivities have only INAA, ICP-MS and stripping voltammetry. Therefore ETAAS is widely used for environmental analysis. However the method suffers from serious interferences leading to systematic errors due to thermochemical processes in the atomizer. Background absorption is also a potential source for systematic errors. Spectral interferences are additive and cannot be corrected by the popular standard addition method. ETAAS is also not free of memory effects for refractory elements.

To compensate for the above effects background correction systems are introduced such as the use of Zeeman effect. Zeeman AAS uses splitting of atomic spectral lines into several components under the influence of a magnetic field which are either influenced or not by background absorption.

To minimize the interferences, chemical modifiers such as Pd, Ni, $Mg(NO_3)_2$, $(NH_4)_2HPO_3$, etc. are used. Their effect is thermal stabilization of highly volatile analytes; increasing volatility of interfering components; and some modifiers enhance ashing of organic matrices ($Mg(NO_3)_2$, HNO_3, H_2O_2 etc.) (Tsalev and Slaveykova, 1998).

ETAAS is a very suitable method for analysis of natural waters especially with low total dissolved solids (river, drinking water). About 20 elements may be determined in such samples and after preconcentration the number of elements increases to about 30. In waters with high total solid content (marine, estuarine waters) directly about ten to 12 elements are determined. Of course after appropriate chemical treatment and preconcentration their number increases significantly.

ETAAS is used to determine about 15–20 elements in aerosol filters, usually after acid extraction. Preconcentration adds about five to six more elements. In soils and plants the method permits the direct determination of about ten to 15 elements after dissolution of the samples.

Sample dissolution for ETAAS is preferably done using HNO_3 or HNO_3–H_2O_2. The method does not tolerate presence of perchloric and hydrofluoric acids. The presence of other acids in the final digests should also be avoided. Otherwise interferences may appear, sensitivity is reduced as well as the lifetime of the atomizer is shortened. Therefore the introduction of solid samples has been an object of permanent interest in ETAAS.

In the case of soils and plants direct introduction of solids or suspensions is already practised. Although automatic solid sample systems are already available the slurry technique is preferable since it is more independent on inhomogenities. Using solid or slurry introduction about five to 12 elements may be directly determined in soils and plants.

Hydride generation AAS (HGAAS) and *cold vapour AAS* (CVAAS) are special combinations of chemical separation and enrichment with AAS. In HGAAS the analyte is transformed to a volatile hydride, stripped off by an inert gas and atomized in a quartz tube, flame-in tube etc. About ten elements (As, Se, Bi, Sb etc.) can be determined by this technique. The accuracy and detection limits depend on the proper isolation of the hydride. CVAAS is the universally acknowledged most sensitive method for determination of Hg. The generation of elemental mercury vapour is similar to the hydride generation however the quartz cell may not be heated and this gives the name of the method.

The above presentation indicates that ETAAS may be considered a powerful method in environmental trace analysis. About 25–30 elements may be determined in environmental materials, including platinum metals. CVAAS is the best method for Hg and HGAAS for the hydride-forming elements. It is however not the general practice to determine all these elements in a single study. The usual number of routinely determined elements are about ten to 12. An evidence in that respect may be found in the reports on standardization of various reference materials. For example in the process of standardization of IAEA SL-1 (La Brecque et al., 1987) acceptable AAS results have been reported for Ag, Ca, Cd, As, Al., Fe, Hg, Mn, Ni, Pb, Cu, Rb and Zn.

About 30% of all reported results have been obtained by the different modes of AAS during the standardization of IAEA-SRM-V-10 (hay powder) (Pazonieki and Hanna, 1985), 10% of which are outliers. The elements for which results have been reported by AAS are: Ag, As, Ba, Ca, Cd, Cr, Cu, Fe, Hg, Mg, Zn, Mn, Ni, Pb, Se. In the production of NIES vehicle exhaust particles 14 elements have been standardized with the participation of AAS (Okamoto, 1987). Recent investigations on the routine use of ETAAS in the analysis of soil and plant reference materials (Djingova et al., 1998) also indicate that about ten to 12 elements are determined with good accuracy and precision.

Obviously in routine determinations AAS yields reliable results for about ten to 15 elements in all possible environmental matrices. Extensions of these possibilities need more efforts so as to reach the number of predicted determinable elements (about 30).

3.5. Activation analysis

If a sample is irradiated with particles or radiation then the non-radiactive nuclei take part in nuclear reactions. As a result of which radioactive nuclides are obtained. Decaying they emit radiation which is proportional to the quantity of atoms and its energy is characteristic for the respective nucleus. Depending on the type of irradiation and emission several versions of activation analysis (AA) can be distinguished, some of which are rather exotic though very effective (Hoste et al., 1971; De Soete et al., 1972; Das et al., 1983; Erdtmann and Petri, 1986; Alfassi, 1990).

Charged particle activation analysis, where activation is performed after irradiation with protones, deuterons, tritons, ^{3}He or ^{4}He (Hoste et al., 1971). A large variety of nuclear reactions take place. With 5–15 MeV protones the dominant reaction is (p,n) and for many elements detection limits reach 50 ng/g. *Protone AA* with energy of the irradiating protones up to 25 MeV (accelarators) permits also the determination of light elements with detection limits down to 0.01 mg/kg. Among the advantages is that favourable detection limits are obtained for the light elements C, N, O, B etc.

Gamma activation analysis (Hoste et al., 1971; Kushelevsky, 1990) carried out with bremstrahlung from an electron beam of 50 μA at 27 MeV allows determination of:

- Pb, Si (detection limit 10^{-5}–10^{-6} g)
- Cr, Fe, Pt, S (10^{-6}–10^{-7} g)
- C, Cl, Cu, F, K, N, O, P, Zn, Zr (10^{-7}–10^{-8} g)
- Mn, Ni, Sb (10^{-8}–10^{-9} g)
- As, Cd, Co (10^{-9}–10^{-10} g)

Prompt activation analysis (Erdtmann and Petri, 1986; Alfassi, 1990) uses the prompt radiation accompanying a nuclear reaction for determining elemental or isotopic concentrations. The variety of prompt methods is large because a sample can be irradiated with various particles - neutrons, charged particles or gamma-rays. Prompt activation analysis permits the determination of several elements – about 17 elements in environmental matrices (Germani et al., 1980) – but most analysis are used for the determination of light elements (H, He, Li, B, C, N, Si, S, Cl) as well of Cd and Gd.

Delayed neutron counting is an excellent method for determination of U and Th

(Amiel, 1981; Erdtmann and Petri, 1986; Alfassi, 1990). A great number of samples (more than 10 000 per year) could be analyzed with detection limits of 2×10^{-8} g/g for uranium and 2×10^{-6} g/g for thorium. Using the following reaction sequence: $^{6}Li(n_{th},\alpha)^{3}H \rightarrow {}^{18}O(^{3}H,\alpha)^{17}N$ and counting of delayed neutrons emmited by ^{17}N, the quantitative assay of lithium in natural waters is possible.

Undoubtedly the method with widest application is *neutron activation analysis* (NAA), more often in its instrumental non-destructive form (INAA) but also after radiochemical separation or concentration (RNAA) (Hoste et al., 1971; De Soete et al., 1972; Das et al., 1983; Erdtmann and Petri, 1986; Alfassi, 1990; Toelgyessy and Klehr, 1987). The irradiation of the samples in NAA is performed in experimental nuclear reactors with neutron fluxes usually in the order of 10^{11}–10^{14}/cm^2 per s. Thanks to nuclear reactions – most often (n, γ), a lot of elements which radionuclides are formed may be determined by measuring the induced radioactivity which is proportional to the number of atoms (concentration of elements). By variation of the irradiation time in dependence on the half life of the radionuclides optimal conditions for the analysis are chosen. The irradiation may be performed with neutrons with different energies: pile (whole reactor spectrum), thermal ($E < 0.1$ eV), resonance (epithermal, epicadmium – 0.4 eV $< E >$ 1 keV) and 14 MeV neutrons (nuclear generators). Depending on this different versions of the method exist, though pile NAA is normally used (Hoste et al., 1971; De Soete et al., 1972; Alfassi, 1990). INAA is a typical multielement method, and it may be performed either as a multistandard method (standards for each element) or as a monostandard (single comparator, K_o-method) (Kim, 1981; Alian et al., 1984, 1985; Erdtmann and Petri, 1986; Sansoni, 1987).

General advantages of INAA are:

- high selectivity and sensitivity
- very good precision and accuracy
- very large dynamic range
- suitable for the determination of 60 – 70 elements
- high sensitivity for rare earth elements, platinum metals and heavy elements

RNAA usually is used when due to insufficient detection limit or matrix interference the element of interest cannot be determined. Then chemical group or individual separations are performed preferably after irradiation which introduces several more advantages:

- total independence on reagent blanks
- possibility for exact determination of yield which permits to apply non-quantitative separations
- possibility to avoid difficult separations in the ng range and below by addition of inactive carrier

Disadvantages of INAA are:

- the necessity of a nuclear facility for irradiation
- rather long time of analysis, imposed by the long waiting (cooling) periods for reduction of short-lived radioactivity

- generally liquids are to be avoided
- difficulties in the determination of some environmentally important elements, e.g. Pb

INAA has been used for routine determination of about 30–48 elements in coal., soils, fly ash with excellent accuracy and precision proved by analysis of NIST and IAEA standard reference materials (Germani et al., 1980; Alian and Sansoni, 1990; Shtangeeva, 1994). Combination of pile and epithermal irradiations increases the precision and accuracy and detection limits for a number of elements (Kostadinov and Djingova, 1980, 1981a). Detection limits are in the low mg/kg and μg/kg range and the precision is 1–8%. Analysis of soil EDTA and acetic acid extracts has been performed in (Bajo et al., 1989) and the content of 29 elements is determined with a precision of 4–10%. For some of the elements however (Tb, Ti and Zn) accuracy was not good (20–40%).

An excellent review on the use of INAA in the analysis of aerosols can be found in (Alian and Sansoni, 1985, 1990). Over 25 elements are easily and reliably determined in different aerosol filters. The use of epithermal and thermal irradiations increases the analytical parameters of the analysis for some of the elements (Alian et al., 1984, 1985). Forty-one elements have been successfully determined in combustion effluents from power plants (James and Acevedo, 1993).

In the analysis of natural waters by INAA usually very long irradiation times are necessary due to very low concentrations of determined elements which is not recommendable. The reason for avoiding direct INAA of liquids in a nuclear reactor are: the significant variation of the neutron flux over large samples; the danger of excessive gas formation during irradiation due to radiolysis of water; bad counting geometry configuration for measurement and efficiency of semiconductor detectors drops with increasing of volume. That is why although direct analysis is practised (e.g. Lavi et al., 1992) water samples are preferably subjected to treatment (van der Sloot, 1976; Das et al., 1983; Tanizaki, 1990). Usually such techniques are recommended which lead to formation of a solid phase, e.g. ion exchange, co-precipitaion, sorption, evaporation. Besides during these procedures the spectral interfering elements Na, Mg, P, S, Cl, K, Ca, Br may be successfully eliminated. A very effective procedure for preconcentration of trace elements from water has been developed by van der Sloot (1976). It is based on adsorption of trace elements on active carbon (van der Sloot, 1976; Das et al., 1983) and permits the reliable determination of 25–30 elements. However attention should be paid to blank values in the active carbon since preconcentration is performed before irradiation.

INAA is a very suitable method for analysis of plant samples, though in rare cases high sodium content may produce difficulties (the same problem can be encountered with soils). Up to 30 elements have been determined in plant standard reference materials either after pile irradiation (Djingova et al., 1998) or by combination of pile and epithermal irradiation (Kostadinov and Djingova, 1981a) or using thermal monostandard NAA (Alian et al., 1984). The accuracy is excellent, precision varies between 1 and 10% and detection limits are in the ng/g and sub-nanogram range. The determination of REE (La, Lu, Sm, Yb) and of U and Th has been done in NIST-SRM-1571 and Bowen's kale after two irradiations – epithermal (18 h) and pile (12 h) neutrons and three measurements (Kostadinov and Djingova, 1981a). ENAA is beneficial in the determination of As, Sb, Ba, Rb, Sr, U, Th, Ta.

INAA has been used in the standardization of numerous reference materials. INAA results have been reported for 45 elements in IAEA SL-1 (La Brecque et al., 1987). INAA is the only method providing data for nine of them (Br, Ce, Cl, Cs, Lu, Sc, Ta, Tb, W) and the majority of data for La, Yb, and Th have also been supplied by INAA.

In the standardization of IAEA V-10 (Pazonieki and Hanna, 1985) values for 32 elements are supplied by INAA 30% of all mean values reported by participating laboratories are by INAA. The determination of Sm, Th, Sc, Eu, Hf and I has only been possible by INAA while the determination of La, Sb, Ce and Cl has been predominantly by INAA.

Fifteen out of 16 elements, certified in NIES Vehicle Exhaust particles are certified with the participation of INAA (Okamoto, 1987).

To summarize the discussion above, INAA is a very useful method in the analysis of a variety of environmental materials. The element that cannot be determined is Pb and the sensitivity for Cd and Cu is not always enough (Djingova et al., 1998). Otherwise, routinely, 30–40 elements are determined in soils, sediments, plants and aerosols. This number may be increased to about 50 if short time irradiations are used (which however demands work at a laboratory in the reactor facility) and/or irradiation with neutrons with different energies is applied. These analytical parameters are obtained by simple optimization of irradiation conditions, cooling and measurement times. Thus the method is without any doubts the least labour consuming.

As the most serious shortcomings of INAA its cost have been pointed out as well as the necessity for nuclear reactor for irradiation. It is really a pity that a number of experimental nuclear reactors have already been closed however there are still a lot working in the world. Whenever there is a nuclear facility available, then the cost of analysis is commensurable to the cost of FAAS (the cost of instrumentation is also not higher).

3.6. Electrochemical methods

The electrochemical methods are based on oxidation or reduction of a substance in an electrolyte solution on an electrode. If the electrode is dropping mercury then the method is called *polarography*. All other approaches using current/potential relations are classified under the name *voltammetry*. The methods use the law of Faraday according to which the reaction of 1 mol substance on an electrode is equivalent to the transfer of a large electrical charge of $n \times 96\ 500$ C through the interface electrode/solution. Usually for heavy metals $n = 2$ (number of transferred electrons). The current/potential curves exhibit peaks characteristic for each present species reduced or oxidized at a constant redox potential. The peak height corresponds to the concentration.

To perform voltammetric determinations the following conditions should be fulfilled and limitations met (Kalvoda, 1984; Nuernberg, 1985; Sansoni, 1987; Stoeppler, 1991, 1992):

- samples should be in solution
- solutions should be without any traces of organic substances

- solutions must have sufficient conductivity, therefore usually acids or buffers are added
- dissolved oxygen interferes the determination, therefore deaeration with ultrapure nitrogen is recommended
- voltammetry is an oligo element method. It permits the sequential determination of 5 – 8 elements in one solution

Differential pulse polarography can determine about 25 elements (five to six in a group). The detection limits reach 50–500 μg/dm^3 and direct determinations are possible.

If the metal concentrations are below this range then enrichment has to be performed using the methods of *stripping voltammetry* (Wang, 1985). The preconcentration step is within the same experimental system. Metals (such as Cd, Cu, Pb etc.) are concentrated by cathodic electrolytic deposition on an hanging mercury drop electrode. The amalgam is anodally oxidized – *anodic stripping voltammetry*. In ultratrace determinations instead of mercury drop, mercury films are used. Some metals are better anodally accummulated and cathodally reduced and the approach is called *cathodic stripping voltammetry*. In *absorptive stripping voltammetry* metal organic complexes are absorbed on the surface of a hanging mercury drop electrode.

Stripping voltammetric methods have phenomenal detection limits (below 100 ppt) for metals: Zn, Cd, Pb, Bi, Cu, Sn, Tl, As, Se. The precision in pure solutions is 5–10%. Time for complete (six to eight elements) determination is up to 40 min. The method is an absolute favourite in the analysis of marine and river waters, rain and snow (Golimovsky et al., 1985; Nuernberg, 1985; Van den Berg, 1986; Donat and Bruland, 1988). The determination is independent of salt content Therefore the technique is very useful in marine water analysis. For all other environmental samples where digestion is necessary (soil, plants, sediments, aerosols) the fantastic detection limits are reduced to the normal μg/kg level (Adeluji et al., 1985; Ostapczuk et al., 1988). This is due to the necessity to obtain interference-free solutions and connected to the problem with reagent blanks.

Therefore electrochemical methods which undoubtedly are the most sensitive for the determination of some metals (Pb, Cd, Zn, Bi, Sn, As, Tl) in natural waters and precipitation are hardly competitive for other environmental matrices. In this respect, as Stoeppler (1991) concludes, they are complementary to rather than competing to the other instrumental methods.

3.7. Methods for determination of radionuclides

Besides NAA which deals with radioactivity but "starts" analysis with a non-radioactive sample the other methods for radioactivity determination are rarely included in reviews on instrumental trace analysis – one positive exception is Sansoni (1987). The reason probably is that relatively few people from the analytical community are acquainted with these methods. Their use however enables the understanding of the ecotoxicological behaviour of elements which are considered either neutral or non-toxic to living systems on the one hand and on the other, permits the determination of some heavy and super heavy elements. Metals such as Cs, Sr, La, Ba etc. are usually not an object of special environmental interest and some of them are very rarely determined. The situation however changes drastically if they are radioactive. The radioactive isotopes such as ^{60}Co, ^{90}Sr, ^{137}Cs, and ^{239}Pu, ^{241}Am etc. are amongst the most dangerous products of human activity. Starting with the nuclear weapon tests in the early 1960s, the rapid development of nuclear power production, nuclear medicine and nuclear scientific research until now a number of radioactive nuclides are being released to the environment under well controlled conditions or as a result of accidents. Though normally in very low concentrations their content, distribution and behaviour should be carefully studied and controlled. The methods for radioactivity measurements besides permit the determination of naturally existing radioactive nuclides of ecotoxicological importance such as ^{226}Ra, ^{222} Rn etc. Additionally they enable the easy and reliable determination of U and Th which normally present a problem for the majority of methods as well as of the actinide elements. Last but not least these methods are used as measuring techniques in radiotracer experiments.

The extremely low concentrations of natural and artificial radionuclides in environmental materials (soil, water, plants) demand radiochemical isolation and preconcentration prior to measurement of radioactivity (Das et al., 1983; Toelgyessy and Kyrs, 1989). Some of the schemes for the isolation of radiolead, radiostrontium, Ac, Am, Cm, Pu, Ra, Th, U, etc. using extraction chromatography deserve special attention (Brunett and Cable, 1995; Pin, 1996).

Depending on the emitted radiation and measurement the methods that are used are *gamma spectrometry*, *alpha spectrometry* and measurements of β-activity (e.g. *liquid scintillation counting*).

Gamma spectrometry is a method used as a measurement technique in NAA (Hoste et al., 1971; De Soete et al., 1972; Erdtmann and Petri, 1986; Alfassi, 1990). When used independently it is applicable for the single or simultaneous determination of gamma-emitters such as ^{137}Cs, ^{60}Co, ^{226}Ra and a number of daughter products from the natural U and Th radioactive series. The method is very sensitive and when appropriate shielding of the detector (normally semiconductors are used like HP Ge, or Ge/Li) is done to reduce cosmic radiation then concentration levels down to parts of Bq/kg may be determined (corresponding to approx. ppt depending on the nuclide). Solid samples need no special treatment except grinding and homogenization. Liquid samples are not preferred though their measurement is possible under well standardized conditions. The analysis of standard reference materials (NIST-SRM-4353, NIST-SRM-4355, IAEA-SD-N-1/2, NBL-75A, NBL-76B, NBL-77A, NBL-80A, NBL-81A and NBL-82A) has provided excellent accuracy (below 5%) and precision (about 2%) in the determination of U, Th, and ^{226}Ra, ^{214}Pb, ^{214}Bi, ^{208}Tl, ^{228}Ac, ^{234}Th etc. (Djingova, unpublished data). Comparison with other laboratories in the determination of ^{60}Co, ^{40}K and ^{137}Cs in the leaves of trees, herbages and in fruits and vegetables proved the reliability of analysis (Djingova, unpublished data). A number of investigations appeared, especially after the Chernobyl accident, using gamma-spectrometry for the determination of a variety of radionuclides, emitted to the environment in soils, sediments, aerosols water samples (IAEA, 1995).

Alpha spectrometry can be used for analysis of every alpha-emitting material. From environmental standpoint especially interesting is the possibility to determine U, Th, ^{238,239,240}Pu, ^{241}Am and a number of natural and artificially produced alpha emitters. The method is independent on natural background (in contrast to gamma-spectrometry) therefore special shielding of detectors is not necessary and lower backgrounds may be achieved. The most serious problem is connected to the strong self absorption of α particles in the sample, which is much more significant than the similar problems in XRF (Helmbold, 1985; Sansoni, 1987; Toelgyessy and Kyrs, 1989). In this way only the upper layer of the sample is analyzed and even only the most upper atomic layers are emitting undisturbed alpha particles. Therefore the sample preparation is a very important step. Usually after chemical separation of the matrix carrier-free electrodeposition is done on a steel disk by electroplating. Measurements are done either by a Si charged particle detector (surface barrier detectors) or a grid ionization chamber. Thanks to the 100% efficiency of detectors and the large mass of alpha particles it is not a problem to determine the absolute activity of an alpha source. Alpha spectrometry may be considered as the best method for the determination of actinides and has been widely used for investigation of environmental materials, especially after the Chernobyl accident.

Among the methods for radioactivity measurements several others should be at least mentioned. The ecotoxicollogically important radon (^{222}Rn), thoron (^{230}Rn) and actinon (^{219}Rn) and their direct parents (e.g. ^{226}Ra) can be reliably determined by *emanation analysis* (Toelgyessy and Kyrs, 1989). It demands the quantitative removal of the inert gas from the sample before its activity is measured. The method is easily applicable for waters but solids have to be melted either directly or by fusion.

^{90}Sr deserves special attention since it is released by nuclear power facilities, accumulates in organisms and replaces Ca (mainly in bones) and has a long half life. ^{90}Sr is a pure beta emitter and before measurement (liqiud scintillation counting, proportional counters, etc.) a laborious and time-consuming procedure for radiochemical separation has to be performed (IAEA, 1989). This is practically the best way for reliable determination of this radionuclide (IAEA, 1989).

Liquid scintillation counting is a preferable method for determination of low energetic beta emitters such as ^{3}H, ^{14}C and ^{63}Ni as well as a number of alpha emitters such as Ra, U, Th, Pu, Cm (Dacheux and Aupiais, 1997, 1998; Aupiais et al., 1998).

The methods for determination of radionuclides in the environment occupy a special place among instrumental methods. They are not competing with the other methods. These methods provide very valuable information which may be used as an indication for other environmental studies (besides radioactivity measurements) because in the general case radioactive and stable nuclides of the same element have similar behaviour (exceptions to this rule are well known and described).

4. Comparative evaluation of the possibilities of instrumental methods for environmental analysis

Figs. 1–3 show a comparison of the accuracy and number of determined elements of several instrumental methods is presented. Results from the analysis of coal fly ash NIST-SRM 1633a (Fig. 1), IAEA Soil 5 (Fig. 2) and plant SRMs (Fig. 3) are given as ratio $C_{method}/C_{certified}$. The value 1 is that of full coincidence of experimental to certified results. The data are transformed from values published in (Djingova, 1982; Liese, 1985a; Van Dyck et al., 1986; Kumar et al., 1989; Djingova et al., 1991, 1998; Lavi et

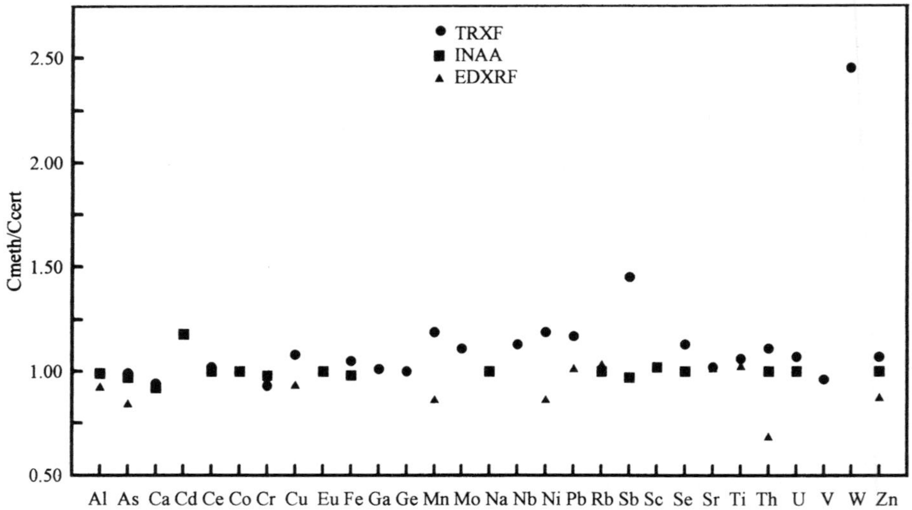

Figure 1. Comparison of several methods in the analysis of NIST-SRM-1633a.

al., 1992; Bandhu et al., 1996; Markert, 1996; Kos et al., 1996; Markert et al., 1997; Leivuori, 1998).

The comparison between TXRF, EDXRF and INAA in the analysis of NIST-SRM-1633a (Fig. 1) indicates that by TXRF 22 elements are determined, 17 by INAA and 12 by EDXRF (TXRF was applied for the determination of 32 elements, INAA for 27 and EDXRF for 14, but not all of them are certified in the material) The accuracy achieved by TXRF for Sb (45%), W (245%) and Mn (20%) is not acceptable and for

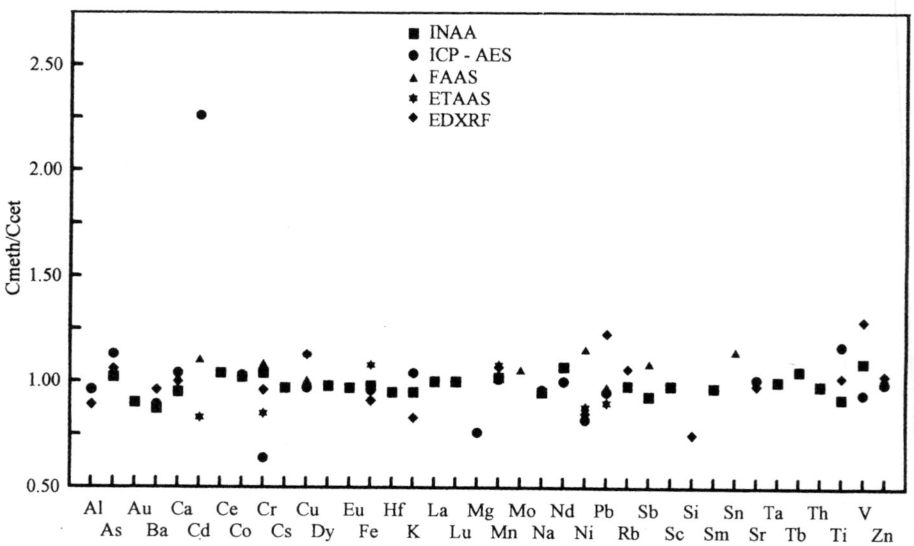

Figure 2. Comparison of several methods in the analysis of IAEA Soil 5.

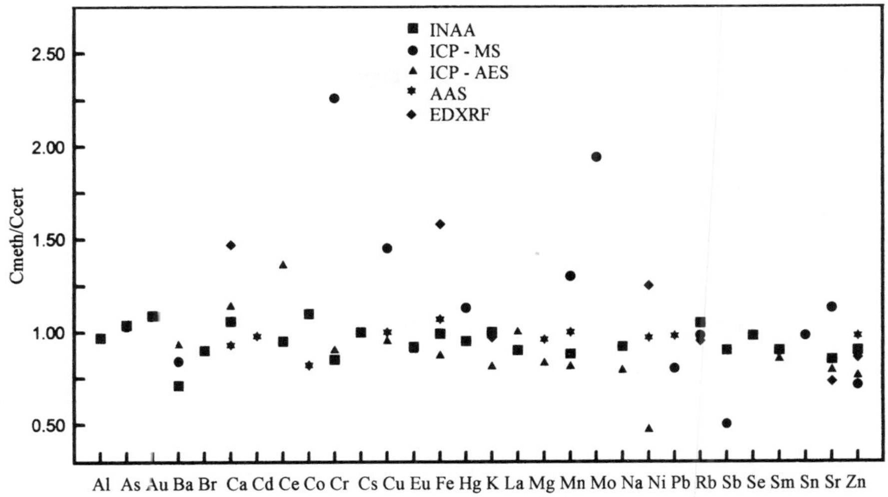

Figure 3. Comparison of several methods in the analysis of plant SRMs.

Pb (17%) is rather poor. For the rest of the elements excellent results are obtained by this method. All the results by INAA are with accuracy below 5% but for Cd (18%). EDXRF has unsatisfactory accuracy for As and especially for Th (32%). The determination for Ni is a problem for all methods. Values for P, S, Ga, Ge, Nb, and Mo are obtained only by TXRF while INAA permits the determination of Cd, Au, Cu, Sm, Cs, Yb, Ta, Eu, Na and Co. The common elements determined by the three methods are K, Mn, Fe, Zn, As and Rb.

The results from the analysis of IAEA-Soil 5 are presented in Fig. 2. The data of FAAS are obtained under more special conditions: Sb, Sn and Mo are determined with flame N_2O/air, Cd, Cr, Pb, Ni – after extraction with methyl isobutilketone (IBMK) and Cu and Mn – directly in flame acetylene/air (S. Arpadjan, pers. commun.). The number of determined elements is 27 by INAA (only by INAA - Ce, Cs, Dy, Eu, Hf, La, Lu, Sc, Ta, Th, Au and Tb), 20 elements by ICP-AES, 9 by FAAS (Sn and Mo only by this method), 7 by ETAAS and 17 by EDXRF. Fig. 2 indicates unsatisfactory accuracy for Cd, Cr, and Mg for ICP-AES and for V and Pb by EDXRF. Problems in the determination of Ni have all methods as in the analysis of fly ash.

Fig. 3 compares the accuracy of INAA, ICP-MS, ETAAS, ICP-AES and EDXRF in the analysis of plant reference materials (NIST-SRM-1575 Pine needles, NIST-SRM-1572 Citrus leaves and Bowen's kale). Twenty-three elements are determined by INAA, 13 by ICP-MS, 17 by ICP-AES – lanthanides are determined after chemical separation and preconcentration (Markert, 1996) – 10 by ETAAS and 9 by EDXRF. The method displaying greatest problems with accuracy is ICP-MS where accuracy for Cr is 126%, for Cu 45%, for Mn 30%, for Mo 94% and for Sb 50%. EDXRF faces difficulties in the determination of Ca and Fe and ICP-AES for Ce and Na. ETAAS is the only method permitting determination of Cd in plants at background levels and again as in the other cases Ni is problematic for all methods.

To demonstrate the potential of the methods measuring radioactivity a comparison

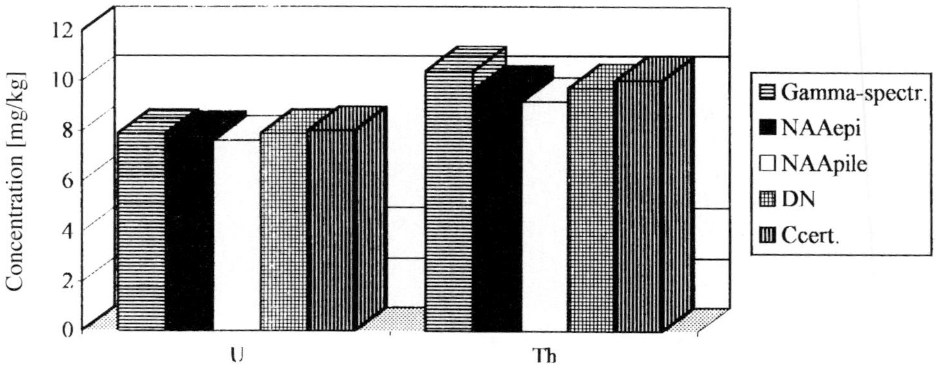

Figure 4. Analysis of U and Th in geological SRMs.

between gamma spectrometry, INAA with epithermal neutron irradiation (very beneficial in the determination of U), INAA with pile irradiation and delayed neutron counting is presented in Fig. 4 for the determination of U and Th in geological SRMs (BNL 82A and NIST-SRM-435). The results for U indicate that all methods give accurate results. Serious systematic errors are not observed for Th as well. These results illustrate that a method easily performed and relatively cheap as gamma spectrometry yields as good results as the three other methods (the best in the determination of U and Th) which practically use the same measurement equipment but need irradiation facilities.

A summary of the above results is presented in Table 1 where recommended

Table 1. Recommended methods for determination of selected elements at background levels in environmental materials

Element	Recommended method
Al	INAA[a], ICP-AES, ETAAS, EDXRF
Am	α Spectrometry, γ spectrometry
As	INAA, ICP-MS, HGETAAS, TXRF
Au	INAA
Ba	ICP-MS, ICP-AES, EDXRF, INAA
Br	INAA
Ca	INAA, ICP-MS, ICP-AES, ETAAS, FAAS, EDXRF, TXRF
Cd	ETAAS
Ce	INAA, ICP-AES[b]
Co	INAA, ICP-AES (^{60}Co-γ spectrometry)
Cm	α-Spectrometry
Cr	INAA, ICP-AES, ETAAS
Cs	INAA, TXRF
Cu	FAAS, ETAAS, ICP-AES, ICP-MS, TXRF
Dy	INAA[a], ICP-AES[b]
Er	ICP-AES[b] (?)
Eu	INAA, ICP-AES[b]

Table 1 (*continued*)

Element	Recommended method
Fe	INAA, ICP-MS, ICP-AES, ETAAS, FAAS, EDXRF, TXRF
Hf	INAA
Hg	CVAAS, ICP-MS, INAA
Ho	ICP-AES[b] (?)
K	INAA, ICP-MS, ICP-AES, ETAAS, FAAS, EDXRF, TXRF
La	INAA, ICP-AES[b], TXRF
Li	ICP-MS
Lu	INAA, ICP-AES[b]
Mg	INAA, ICP-MS, ICP-AES, ETAAS, FAAS, EDXRF, TXRF
Mn	INAA, ICP-MS, ICP-AES, ETAAS, FAAS, EDXRF, TXRF (^{54}Mn-γ spectrometry)
Mo	ICP-MS, AAS, TXRF
Na	INAA, ICP-AES, EDXRF
Nb	ICP-MS (?) (^{95}Nb-γ spectrometry)
Nd	INAA
Ni	ICP-AES, ETAAS, TXRF (^{63}Ni-β liquid scintilation)
P	ICP-AES, EDXRF
Pb	ETAAS, ICP-MS, TXRF; (^{210}Pb-β counting)
Pd	ETAAS[c], INAA[c]
Po	α Spectrometry
Pr	ICP-AES[b]
Pt	ETAAS[c], INAA[c]
Pu	α Spectrometry
Ra	α Spectrometry, γ spectrometry
Rb	INAA, ICP-MS, ICP-AES, TXRF, EDXRF
Sb	INAA (^{125}Sb-γ spectrometry)
Sc	INAA
Se	ETAAS, INAA
Si	EDXRF
Sm	INAA, ICP-AES
Sn	ETAAS, ICP-MS
Sr	INAA, ICP-MS, ICP-AES, ETAAS, FAAS, EDXRF, TXRF ($^{89,\ 90}$Sr-β counting)
Ta	INAA
Tb	INAA, ICP-AES[b]
Th	INAA, delayed neutron counting, α spectrometry, γ spectrometry
Ti	INAA[a], ICP-AES
U	INAA, delayed neutron counting, α spectrometry, γ spectrometry
V	INAA[a], ICP-AES
Yb	INAA, ICP-AES[b]
Zn	INAA, ICP-MS, ICP-AES, ETAAS, FAAS, EDXRF, TXRF
Zr	EDXRF (^{95}Zr-γ spectrometry)

[a] Short irradiation at the reactor site.
[b] ICP-AES after chemical separation and preconcentration (Markert, 1996).
[c] After chemical separation and preconcentrations.

Table 2. Recommended methods for analysis of different environmental matrices

Matrix	Recommended method
Water	TXRF, ETAAS, stripping voltammetry, ICP-MS
Soil	INAA, ICP-AES, EDXRF, ICP-MS
Aerosols	INAA, ICP-AES, ETAAS, TXRF, ICP-MS
Plants	INAA, ETAAS, ICP-MS, ICP-AES

methods for the determination of 57 elements in environmental matrices are given. Table 2 recommends the most suitable methods for analysis depending on the matrix keeping in mind the features and potential of the methods already discussed.

5. Speciation

The term *speciation* refers to the determination of the chemical form of metals and metalloids and according to IUPAC it is the process yielding evidence of the atomic or molecular form and oxidation state of an analyte. Included in this definition are the organometallic compounds and those of N, P, S and Si. The determination of organic substances is not referred to.

Speciation is of great importance in the analysis of environmental materials and it is this field of environmental research which attracts an increasing attention in the last decade. The reason is that the knowledge of the chemical form under which an element is in a specific environmental compartment leads to revealing the essence of different geochemical and biochemical processes. Besides the toxicity to metals depends to a great extent on their form and the dependence is not always predictable. For example mercury organic compounds are more toxic than inorganic mercury but for arsenic the opposite is true.

If the goal of an investigation is speciation then all analytical steps discussed up till now (sampling, storage and measurement techniques) have to undergo certain specific changes (Lund, 1990; Baeyens, 1992; Ceulemans and Adams, 1996; Quevauviller, 1998).

During the sampling and storage steps care should be taken so as to prevent any changes in the original composition and if possible to separate the existing species.

Methods have been developed for separation of mercury and arsenic species during sampling of aerosols. A sampling procedure for mercury species is based on passing the air through selective absorption tubes: glass wool filter (for particulate Hg); Chromosorb W, coated with 3% SE-30 (for $HgCl_2$); Chromosorb W, treated with 0.05M NaOH (for CH_3HgCl); Ag coated glass beads (for Hg^0) and Au coated glass beads (for $(CH_3)_2Hg$). Tetra-, tri- and dibutyltin are absorbed on Chromosorb 102. Organic arsenic compounds may be determined after trapping on silver glass beads.

The sampling of water and soil for speciation analysis is not significantly different than for total metal content determinations. The distribution of metals ranges from truly dissolved to particulate-bound species. This includes inorganic complexes or complexes with organic molecules, surface-bound metals (organic and inorganic colloids and particles) and absorbed inside microorganisms. All the species can coexist and may not be in thermodynamic equilibrium. That is why the storage step is of vital importance. No treatment and conservation of water samples is usually permitted. This means that analysis should be done immediately after sampling. Keeping the samples at very low temperature (−196°C, or at least −20°C after initial liquid nitrogen freezing) may be helpful to prevent changes.

The use of a separation, preconcentration and measuring procedure depends to a great extent on the species that is to be determined (Lund, 1990; Baeyens, 1992; Szpunar-Lobinska et al., 1995; Mendez Garcia et al., 1996; Buffle et al., 1997).

Stable anions and cations are usually determined by ion-chromatography and/or voltammetry. Spectrophotometry with selective colour forming reagents has been also used.

Some metals and metalloids (Ge, Sn, Pb, Sb, As, Se) easily can form stable alkyl element compounds which are of special interest in environmental matrices. Usually methylation is a natural process (not for Pb) while higher alkyls are products of anthropogenic activity. Gas chromatography may be used for determination of non-polar compounds (polar compounds have to be derivatized first).

The determination of prophyrin complexes of Ni, V, and Fe is important due to their geochemical importance as well as their poisonous effect on catalysts during refinery operations. Capillary gas chromatography might be successfully used for their determination.

Detection of metals in proteins and especially of those able to form cysteine thiol groups (Zn, Hg, Cd, Bi, Cu, Au) is a modern field in speciation analysis. Much attention is paid to metallothioneins which are very stable to proteolytic digestion and thermal denaturation, have high metal content and occur in all living organisms. Metallothioneins attract attention thanks to the fact that they exhibit detoxication effect, and are preferably synthesized when certain organisms are poisoned by heavy metals. Size-exclusion chromatography is a favourite among methods for the determination of metals in proteins. Reverse-phase chromatography, capillary zone electrophoresis and gel electrophoresis techniques have also been successfully applied.

The methods used for speciation analysis have to include a separation and preconcentration step and very sensitive detection. These demands are usually attainable after combination of different chromatographic techniques to sensitive and selective element specific detectors which has lead to the emerging of modern hyphenated techniques either with on-line or off-line preconcentration. Besides the coupling of gas chromatography (GC) and high performance liquid chromatography (HPLC) to ETAAS, ICP-AES and ICP-MS a lot of more complicated systems have been proposed. Among them are vapour generation-cryogen trapping-gas chromatography – AAS (VG-CT-GC-AAS), flow injection-microwave digestion-cold vapour AAS (FI-MWD-CVAAS), flow injection-UV photooxidation-hydride generation AAS (FI-UV-HGAAS), HPLC-UV-VGAAS, HPLC-MWD-VGAAS, HPLC-MWD-ICP-AES, continuous flow-UV photooxidation-cold vapour AAS (CF-UV-CVAAS). This field is developing rapidly and therefore only some major approaches will be mentioned here (Lund, 1990; Donard and Martin, 1992; Puk and Weber, 1994; Baxter and Frech, 1995; Szpunar-Lobinska et al., 1995; Lobinski, 1996; Tsalev, 1999).

Flow injection is a method using on-line discrimination chemical reactions. Direct selective preconcentration is performed on microcolumns, the analytes being concentrated up to three to four orders of magnitude and injected into the detector. Usually as separation techniques ion-exchange, liquid–liquid extraction, gas diffusion are used. Flow injection systems are connected either to ETAAS or ICP-AES detection systems. These methods have been used for determinations of redox species such as Cr(III)/Cr(IV), Fe(II)/Fe(III), As(III)/As(V), Se(IV)/Se(VI) in soil extracts and water samples.

VG-CT-GC-AAS combines very successfully several techniques. It is applicable to hydride forming, ethylated or otherwise derivatized inorganic or organoelement species of As, Bi, Sb, Sn, Hg etc. The detection limits are in the ng/dm^3 range.

Separations by GC are very popular for non-polar organic compounds. Prior to the separation however extraction from the sample and preconcentration have to be performed as well as derivatization. The last step is usually done by addition of Grignard reagent (RMgX) and its excess after the process is removed by acid addition. As detectors after GC separation successfully are used: ICP-AES (for N, P, S, C, Br, Cl, Hg, Zn, Pb) with detection limit in the pg range; atomic fluorescence and mass spectrometry.

HPLC has several significant advantages over GC and besides does not need derivatization. HPLC-ETAAS systems are in use for determination of organoarsenic and organotin compounds. Speciation done by HPLC coupled to ICP-AES and ICP-MS provides better detection limits and progress is expected from coupling HPLC systems with microconcentric nebulizer to ICP-MS.

Among the other non-chromatographic techniques for speciation analysis the following may be mentioned:

Electroanalysis: Direct potentiometry with ion selective electrodes may be used in speciation studies of Cu, Pb and Cd. However the sensitivity is not enough and is in the μmol/dm^3 to mmol/dm^3 range. Differential pulse polarography may be useful for determination of free metal ions and corresponding complexes and differentiate between different redox states of some elements. Anodic stripping voltammetry is determining concentrations of metal ions in the presence of metal complexes with naturally occurring ligands.

Very promising techniques for speciation of metals in water are: diffusion in gels; integrated diffusion in gel and square wave anodic stripping voltammetry and integrated supported liquid membranes and voltammetry (Buffle et al., 1997). A thin layer of complexing resin ($\sim$100 μm) is placed on a Plexiglass plate and is covered by a layer of polyacrylamide gel (500–1000 μm). The plate is placed in a water column and trace metals diffuse through the gel and are caught in the resin. This preconcentration technique is specific for mobile species. The same technique in combination with voltammetric cells of microelectrode arrays permits sensitivity limit of $\sim 10^{-11}$M. Using a similar technique, supported liquid membrane as preconcentration step and voltammetric microelectrode cell as measurement device, the concentration of free copper (Cu^{2+}) and lead (Pb^{2+}) ions in river and drinking water is determined (Buffle et al., 1997).

Isotope methods: Isoptope exchange and substoichiometric isotope dilution have been successfully used for determination of dimethyl and diphenyl mercury in waters and biological liquid samples (Kostadinov and Djingova, 1981b; 1982). The reactions R_2HgCl–^{131}I and $(C_6H_5)_2Hg$–^{203}Hg have been used. The detection limit reached 0.0005 ng.

Speciation in water is most simple and widespread due to the fact that water samples do not need dissolution steps. A scheme of separation methods, aiming to discriminate according to size, charge or extractability are combined with ASV, AAS or ICP methods. A popular separation scheme (Batley and Florence scheme) uses separation through filtration, passage through a Chelex 100 column, irradiation with UV and final Chelex 100 separation. At all stages measurement by ASV is done (Lund, 1990). Organometallic compounds in waters are determined dominantly by GC or HPLC.

The studies of speciation in air involve among other things enrichment of heavy metals with particle size. A scheme for separation and determination of different metal species in dust is very similar to procedures for soils:

- extraction with $MgCl_2$ for exchangable metals
- extraction with acetic acid buffer - for carbonates
- treatment with hydroxylamine at 96°C – for metals bound to Fe–Mn oxides
- treatment with H_2O_2 at 85°C – for organic bound metals
- $HF/HClO_4$ digestion for residual metals

Selective sampling of organometallic compounds in air has already been mentioned. It is usually followed by hydride generation AAS or CT-GC-AAS measurements.

The speciation of metals in soils (Rubio and Ure, 1993) and sediments is done by sequential extraction procedures which ensure discrimination between different metal species. Extractions are performed using buffer solutions, weak acids, complexing agents, redox reagents etc. Some examples are already mentioned in the section for sampling and sample preparation of soils. The outcome of the procedures depends to a great extent on the temperature, volume and extraction times and are by no means standardized. These extractions permit the separation and determination of the inorganic forms and generally organic bound ions. Determination of organometallic compounds in soils is a very difficult task. Organoarsenic and organolead species might be determined by GC-HGAAS but organotin compounds are not easily extractable and practically the determination is hardly possible.

The most complicated speciation analysis is in plants and biological samples. Besides the compounds and ions already mentioned metals in living systems participate in a lot more complicated species. They may be bound to amino acids, proteins, peptides and the separation methods may influence their original distribution. For the determination of organo-arsenic, selenium, lead and tin hydride generation combined with different preconcentration steps may be used (Dedina and Tsalev, 1995).

6. Presentation of analytical data and environmental interpretation of results

After the measurement stage has been completed the concentrations of determined elements are calculated using calibration curves, multielement standards, standard reference materials, standard addition methods and single comparator approaches.

Presenting the single value to the public (as regrettably is often done even in otherwise very valuable contributions) makes no sense and brings no information. Firstly it is statistically incorrect to present a value from single determination and secondly if the respective value is a mean the number of parallel determinations and the respective standard deviation should be given as well which permits the direct evaluation of the error in the determination of each element. Of course it is permissible to give a percentage error but it is absolutely not enough to present only the measurement error. It is only a small additive to the overall uncertainty of the result.

Presenting standard deviations (with respective number of determinations) permits also the use of the results by other investigators for comparisons, applying standard statistical criteria (Doerffel, 1990).

It is extremely important (as mentioned several times already in this contribution) to characterize the accuracy of the final results. The best way is to present data from the analysis of SRMs with a similar matrix. If this is not possible then comparison with a basically independent method should be reported.

Ecological investigations usually aim the evaluation of pollution levels, distribution patterns of elements, geochemical behaviour etc. For such purposes the methods of multivariate statistics are more helpful than standard statistical approaches. Using methods like *cluster analysis* (Everitt, 1980; Massart and Kaufman, 1989), *principal component analysis*, *discriminant analysis* etc. the spatial distribution of elements over large territories may be investigated (Djingova et al., 1993, 1995a,b, 1999), comparisons of the pollution among regions may be done (Djingova et al., 1996), patterns of distribution and probable origin of pollutants may be revealed as well as correlations between elements (Djingova et al., 1987), inter-species behaviour may be followed (Djingova and Kuleff, 1993). A number of computer programs is available e.g. BMDP (Dixon et al., 1981; Bollinger et al., 1983), BNL Program Package (Sayre, 1988), etc. One must keep in mind however that good understanding of the meaning of mathematical operations is necessary. The results from the mathematical operations should be critically evaluated and checked in combination with all additional information for the investigated system. Otherwise mathematically sound but totally erroneous conclusions can be made, explicable only by the low expertise of the investigator.

7. Conclusion

The instrumental methods discussed in the present contribution may be conditionally divided in two groups: well established techniques which are not expected to undergo significant increase of possibilities and developing methods. Among the first are INAA, EDXRF, FAAS, electrochemical methods and ICP-AES. In the second group are ICP-MS and TXRF. ETAAS occupies an intermediate position not because the method itself is developing but it is a part of many hyphenated techniques and thus is an object of methodological and instrumental development.

According to present possibilities in respect to number of determined elements, accuracy, precision and detection limits INAA is still the leading method, followed by ETAAS and ICP-AES. According to potential and perspectives ICP-MS is undoubtedly the most promising technique. However it must be clearly stated that there is not a single method that can solve all tasks in environmental analysis. Therefore a combination of at least two methods may be strongly recommended. Promising combinations might be INAA/ETAAS, INAA/ICP-AES, ICP-AES/ICP-MS, EDXRF/ETAAS etc.

Obviously the field in environmental trace analysis which is rapidly developing is speciation. In this respect increase in the role of electrochemical methods in determination mostly of inorganic species might be expected while for the determination of both inorganic and organometallic species the combinations between highly selective chromatographic or/and extraction techniques with sensitive methods like ICP-MS, AAS, NAA are most promising.

Acknowledgements

The authors will like to express their gratitude to Associate Professor Dr.Sc. Dr. D. Tsalev for his constant interest, helpful discussions and suggestions and to Dr G. Gencheva for her support and graphic presentation.

References

Adeluji, S.B., Bond, A., Briggs, M., 1985. Multielement determinations in biological materials by differential pulse voltammetry. Anal. Chem. 17, 1286–1390.

Alfani, A., Arpalad, C., Cafiero, G., 1997. Assessing trace metals in leaves of *Quercus ilex* L. by energy dispersive X-ray spectrometry. J. Trace Elements Med. Biol. 11, 188–190.

Alfassi, Z. (Ed.), 1990. Activation Analysis, CRC Press, Boca Raton, FL, p. 617.

Alian, A., Sansoni, B., 1985. A review on activation analysis of air particulate matter. J. Radioanal. Nucl. Chem 89, 191–275.

Alian, A., Sansoni, B., 1990. Activation analysis of air particulate matter. In: Alfassi, Z. (Ed.), Activation Analysis, vol. 2. CRC Press, Boca Raton, FL, pp. 503–565.

Alian, A., Djingova, R., Kroener, B., Sansoni, B., 1984. The monostandard method in thermal neutron activation analysis of geological, biological and environmental materials. Fresenius Z. Anal. Chem. 319, 47–53.

Alian, A., Djingova, R., Sansoni, B., 1985. Comparison of different neutron activation analysis methods for multielement analysis of geological materials. In: Sansoni, B. (Ed.), Instrumentelle Multielementanalyse. VCH Publishers, Weinheim, pp. 123–132.

Amiel, S., 1981. Neutron counting in activation analysis. In: Amiel, S. (Ed.), Nondestructive Activation Analysis. Elsevier, Amsterdam, pp. 43–52.

ASTM D3370, 1982. Standard Practices for Sampling Water. American Society for Testing Materials, Philadelphia, PA.

Aupiais, J., Fayolle, C., Gilbert, P., Dacheux, N., 1998. Determination of ^{226}Ra in mineral drinking waters by α liquid scintillation with rejection of β-γ emitters. Anal. Chem. 70, 2353–2359.

Bacon, J., Crain, J., Williams, J., 1996. Atomic spectrometry update – Atomic mass spectrometry. J. Anal. At. Spectrom. 11, 355R–393R.

Baeyens, W., 1992. Speciation of mercury in different compartments of the environment. Trends Anal. Chem. 11 (7), 245–254.

Bajo, S., Wyttenbach, A., Tobler, L., Conradin, H., 1989. Multielement determination in soil extracts by instrumental neutron activation analysis. J. Radioanal. Nucl. Chem. 134, 181–191.

Bandhu, H., Puri, S., Shahi, J., Mehta, D., Garg, M., Singh, N., Mangal, P., Suri, S., Swietlicki, E., Trehan, P., 1996. An evaluation of the sources of air pollution in the city of Chandigarh, India – A study using EDXRF technique. NIM-B 114, 341–344.

Barbante, C., Belloni, T., Mezzardi, G., Cescon, P., Scarponi, G., Morel, C., Jay, S., van de Velde, K., Ferrari, C., Boutrou, C., 1997. Direct determination of heavy metals at picogram per gram levels in Greenland and

Antarctic snow by double focusing inductively coupled plasma mass spectrometry. J. Anal. At. Spectrom. 12, 925–931.

Barnes, R., 1996. Analytical plasma source mass spectrometry in biomedical research. Fresenius Z. Anal. Chem. 355, 433–441.

Baxter, D., Frech, W., 1995. Speciation of lead in environmental and biological samples. Pure Appl. Chem. 67, 615–648.

Becker, J.S., Dietze, H.-J., 1998. Inorganic trace analysis by mass spectrometry. Spectrochim. Acta B in press.

Beitz, L., Plesd, R., Rethfeld, H., 1980. Roentgenanalyse auf Schwermetalle in pflanzliche Producten. Landwirtsch. Forsch. 33, 5–11.

Belakova, M., Havranek, E., Bumbalova, A., 1995. Heavy metals and some other elements in medicinal plants, determined by X-ray fluorescence analysis. J. Radioanal. Nucl. Chem. 201, 431–437.

Berggren, D., 1992a. Speciation of copper in soil solution from podzols and cambisols of S. Sweden. Water Air Soil Pollut. 62, 111–123.

Berggren, D., 1992b. Speciation and mobilization of aluminium and cadmium in podzols and cambisols in S. Sweden. Water Air Soil Pollut. 62, 125–156.

Berrow, M., Mitchell, R., 1980. Location of trace elements in soil profiles: total and extractable contents of individual horizons. Trans. R. Soc. Edinburgh Earth Sci. 71, 103–121.

Berrow, M., Reaves, G., 1984. Background levels of trace elements in soils. Proceedings of the International Conference on Environmental Contamination, London, July 1984. CEP Consultants Ltd., Edinburgh, pp. 333–340.

Berrow, M., Reaves, G., 1985. Extractable copper concentrations in Scottish soils. J. Soil Sci. 36, 31–43.

Berrow, M., Stein, W., 1983. Extraction of metals from soils and sewage sludges by refluxing with aqua regia. Analyst 108, 277–285.

Bloemen, M.-L., Markert, B., Lieth, H., 1995. The distribution of Cd, Cu, Pb and Zn in topsoils of Osnabrueck in relation to land use. Sci. Total Environ. 166, 137–148.

Bollinger, G., Herrmann, A., Moentmann, V., 1983. BMDP – Statistikprogramme fuer die Bio-Human- and Sozialwissenschaften. Gustav Fischer Verlag, Stuttgart, p. 431.

Bond, R. (Ed.), 1982. Surveillance Field Procedures Review for the Great Lakes: International Surveillance Plan. International Joint Commission, Windsor, Ontario.

Broekaert, J., Toelg, G., 1987. Recent developments in atomic spectrometry methods for elemental trace determinations. Fresenius Z. Anal. Chem. 326, 495–509.

Brunett, W., Cable, P., 1995. Determination of ^{228}Ra in natural waters using extraction chromatographic resins. Radioact. Radiochem. 6, 36–44.

Buffle, J., Wilkinson, J., Tercier, M.-L., Parthasarathy, N., 1997. In situ monitoring and speciation of trace metals in natural waters. In: Palmissano, F., Sabbatini, L., Zambonin, P. (Eds.), Reviews on Analytical Chemistry Euroanalysis IX, Bologna 1–7 September, 1996. Societa Chimica Italiana, Roma, pp. 67–82.

Calmano, W., Lieser, K., 1981. Trace element determination in suspended matter and in sediments by INAA. J. Radioanal. Chem. 63, 335–343.

Ceulemans, M., Adams, F., 1996. Integrated sample preparation and speciation analysis for simultaneous determination of methylated species of tin, lead and mercury in water by purge-and-trap injection-capillary gas chromatography – Atomic emission spectrometry. J. Anal. Atom. Spectrom. 11, 201–206.

Cowgill, U., 1988. Sampling waters: the impact of sample variability on planning and confidence levels. In: Keith, L. (Ed.), Principles of Environmental Sampling. American Chemical Society, Washington, DC, pp. 171–203.

Cowgill, U., 1994. Sampling of freshwaters for estimation of all detectable elements. In: Markert, B. (Ed.), Environmental Sampling for Trace Analysis. VCH Publishers, Weinheim, pp. 187–202.

Cresser, M., 1994. Flame Spectrometry in Environmental Chemical Analysis: a Practical Guide. Royal Society of Chemistry, Cambridge, UK, p. 108.

Cresser, M., Ebdon, L., Dean, J., 1988. Atomic spectrometry update – Environmental analysis. J. Anal. Atom. Spectrom. 3, 1R–44R.

Cresser, M., Armstrong, J., Dean, J., Watkins, P., Cave, M., 1992. Atomic spectrometry update – Environmental analysis. J. Anal. Atom. Spectrom. 7, 1R–52R.

Dacheux, N., Aupiais, J., 1997. Determination of uranium, thorium, plutonium, americium, and curium ultratraces by photon electron rejecting alpha liquid scintilation. Anal. Chem. 69, 2275–2282.

Dacheux, N., Aupiais, J., 1998. Determination of americium and curium utratraces by photon/electron rejecting alpha liquid scintilation. Anal. Chim. Acta 363, 279–294.

Das, H., Faanhof, A., van der Sloot, H., 1983. Environmental Radioanalysis. Elsevier, Amsterdam, pp. 296.

Davies, B., 1994. Chemistry and bioavailability with special reference to trace elements In: Farago, M. (Ed.), Plants and the Chemical Elements. VCH Publishers, Weinheim, pp. 1–30.

Davis, R., Carlton-Smith, C., Stark, J., Campbell, J., 1988. Distribution of metals in grassland soils following surface applications of sewage sludge. Environ. Pollut. 49, 99–115.

De Soete, D., Gijbels, R., Hoste, J., 1972. Neutron Activation Analysis. Wiley-Interscience, New York, pp. 836.

Dedina, J., Tsalev, D., 1995. Hydride Generation Atomic Absorption Spectrometry. Wiley, Chichester, pp. 544.

Dick, E., 1994. Water and wastewater sampling for environmental analysis. In: Markert, B. (Ed.), Environmental Sampling for Trace Analysis. VCH Publishers, Weinheim, pp. 255–278.

DIN Deutsches Institut fuer Normung e.V., 1991. Probenahmestrategien fuer Kulturboeden und Boeden mit oekologischen Funktionen. (Draft ISO/TC 190) Berlin.

Dixon, W., Brown, M., Engelman, L., Frane, J., Hill, M., Jennrich, R., Toporek, J., 1981. BMDP-Biomedical Computer Programs. University of California Press, Berkeley, pp. 431.

Djingova, R., 1982. Neutron Activation Analysis of Mercury and Other Heavy Metals in Coal and Biomedical Samples. PhD Thesis, University of Sofia, Bulgaria, pp. 179.

Djingova, R., Kostadinov, K., 1982. A method for determination of diphenylmercury in biological liquid samples. Radiochem. Radioanal. Lett. 51, 399–402.

Djingova, R., Kuleff, I., 1993. Monitoring of heavy metal pollution by *Taraxacum officinale*. In: Markert, B. (Ed.), Plants as Biomonitors. VCH Publishers, Weinheim, pp. 435–465.

Djingova, R., Kuleff, I., 1994. On the sampling of vascular plants for monitoring of heavy metal pollution. In: Markert, B. (Ed.), Environmental Sampling for Trace Analysis. VCH Publishers, Weinheim, pp. 395–414.

Djingova, R., Kuleff, I., 1999. Seasonal variations in the metal concentration in the leaves of *Taraxacum officinale*. *Plantago major* and *Plantago lanceolata*. Chem. Ecol. in press.

Djingova, R., Kuleff, I., Penev, I., Sansoni, B., 1986. Bromine, copper, manganese, and lead content of the leaves of *Taraxacum officinale* (dandelion). Sci. Total Environ. 50, 197–208.

Djingova, R., Kuleff, I., Arpadjan, S., Alexandrov, S., Voulgaropoulos, A., Sawidis, T., 1987. Neutron activation and atomic absorption of *Ulva lactuca* and *Gracilaria verrucosa* from Thermaikos gulf. Greece. Toxicol. Environ. Chem. 15, 149–158.

Djingova, R., Arpadjan, S., Kuleff, I., 1991. INAA and flame AAS of various vegetable reference materials. Fresenius J. Anal. Chem. 339, 181–186.

Djingova, R., Kuleff, I., Andreev, N., 1993. Comparison of the ability of several vascular plants to reflect environmental pollution. Chemosphere 27, 1385–1396.

Djingova, R., Kuleff, I., Voulgaropoulos, A., 1995a. Heavy metal pollution in Sofia (Bulgaria) and Thessaloniki (Greece) using *Taraxacum officinale* as bioindicator. Ecol. Chem. (St. Petersburg) 4, 241–248.

Djingova, R., Wagner, G., Peshev, D., 1995b. Heavy metal distribution in Bulgaria using *Populus nigra* 'Italica' as a biomonitor. Sci. Total Environ. 172, 151–158.

Djingova, R., Wagner, G., Kuleff, I., Peshev, D., 1996. Investigation of the time-dependant variations in metal concentration in the leaves of *Populus nigra* 'Italica'. Sci. Total Environ. 184, 197–203.

Djingova, R., Ivanova, J., Kuleff, I., 1998. Comparative evaluation of the possibilities of INAA, EDXRF, ICP-AES and AAS in the analysis of plants. J. Radioanal. Nucl. Chem. 237, 25–34.

Djingova, R., Wagner, G., Kuleff, I., 1999. Screening of heavy metal pollution in Bulgaria using *Populus nigra* 'Italica'. Sci. Total Environ. in press.

Doerffel, K., 1990. Statistik in der analytischen Chemie. Deutscher Verlag fuer Grundstoffindustrie, Leipzig, pp. 256.

Donard, O., Martin, F., 1992. Hyphenated techniques applied to environmental speciation studies. Trends Anal. Chem. 11, 17–26.

Donat, J., Bruland, K., 1988. Direct determination of dissolved cobalt and nickel in seawater by differential pulse cationic stripping voltammetry preceded by adsorptive collection of cyclohexane-1-2-dione dioxime complexes. Anal. Chem. 60, 240–244.

Duckworth, D., Barshick, C., Smith, D., 1993. Analysis of soils by glow-discharge mass spectrometry. J. Anal. Atom. Spectrom. 8, 875–879.

Ebdon, L., Cresser, M., McLeod, C., 1987. Atomic spectrometry update – Environmental analysis. J. Anal. Atom. Spectrom. 2, 1R–28R.

EPA-600/4-82-029, 1982. Handbook for sampling and sample preservation of water and wastewater. US Department of Commerce, Springfield, IL.

Erdtmann, G., Petri, H., 1986. Nuclear activation analysis: fundamentals and techniques. In: Elving, P. (Ed.), Treatise on Analytical Chemistry, part 1, Vol. 14, Wiley, Inc., New York, pp. 419–643.

Eriksson, J., Andersson, A., Wenblad, A., 1990. Cd, Ni and Zn contents of oat grain as related to soil factors and precipitation. Swed. J. Agric. Res. 20, 81–87.

Ernst, W.H.O., 1990. Element allocation and (re)translocation in plants and its impact on representative sampling. In: Lieth, H., Markert, B. (Eds.), Element Concentration Cadasters in Ecosystems. VCH Publishers, Weinheim, pp. 17–40.

Ernst, W.H.O., 1994. Sampling of plants for environmental trace analysis in terrestrial, semiterrestrial and aquatic environments. In: Markert, B. (Ed.), Environmental Sampling for Trace Analysis. VCH Publishers, Weinheim, pp. 381–394.

Everitt, B. 1980, Cluster Analysis. Halsted Press, Wiley, New York, pp. 136.

Fang, Z., 1993. Flow-injection Separation and Preconcentration. VCH Publishers, Weinheim, pp. 259.

Ferguson, R., Price, V., Baucom, E., 1976. Field Manual for Stream Sediment Reconnaissance. Savannah River Laboratory, Aiken, SC, DPST-76-358.

Ferguson, R., Price, V., Baucom, E., 1977. Field Manual for Stream Sediment Reconnaissance. Savannah River Laboratory, Aiken, SC, DPST-76-416.

Foerstner, U., Rath, V., Schoer, J., Mueller, G., 1981. Chemische Extraktionsversuche an metallkontaminierten Sedimenten aus dem Neckar und aus dessen Zufluessen. Chemische Zeitung 105 (6), 175–179.

Fraenzle, O., 1990. Representative sampling of soils in the Federal Republic of Germany and the EC countries. In: Lieth, H., Markert, B., (Eds.), Element Concentration Cadasters in Ecosystems. VCH Publishers, Weinheim, pp. 63–72.

Fraenzle, O., 1994. Representative soil sampling. In: Markert, B. (Ed.), Environmental Sampling for Trace Analysis. VCH Publishers, Weinheim, pp. 305–320.

Fraenzle, O., Kuhnt, G., 1983. Regional repraesentative Auswahl der Boeden fuer eine UPB. (BU-FB 106 05 028), Kiel.

Garbe-Schoenberg, C.-D., 1993. Simultaneous determination of thirty seven trace elements in twenty eight international rock standards by ICP-MS. Geostand. Newslett. 12, 81–97.

Garcia, R., Millan, E., 1998. Assessment of Cd, Pb and Zn contamination in roadside soils and grasses from Gipuzkoa (Spain). Chemosphere 37, 1615–1625.

Gardner, M., Comber, S., 1997. Sample filtration as a source of error in the determination of trace metals in marine waters. Analyst 122, 1029–1032.

Germani, M., Gokmen, I., Sigello, A., Kowalczyk, G., Olmez, I., Small, A., Anderson, D., Failey, M., Gulovall, M., Choquette, C., Lepel, E., Gordon, G., Zoller, W., 1980. Concentrations of elements in the National Bureau of Standards' bituminous and subbituminous coal standard reference materials. Anal. Chem. 52, 240–245.

Golimovsky, J., Valenta, P., Nuernberg, M., 1985. Trace determination of chromium in various water types by adsorption differntial pulse voltammetry. Fresenius Z. Anal. Chem. 322, 315–322.

Gonsior, B., Roth, M., 1985. PIXE analysis – Physical basis and examples of application. In: Sansoni, B. (Ed.), Instrumentelle Multielementanalyse. VCH Publishers, Weinheim, pp. 291–300.

Gray, A., 1985. The potential of ICP-source mass spectrometry. In: Sansoni, B. (Ed.), Instrumentelle Multielementanalyse. VCH Publishers, Weinheim, pp. 227–237.

Griffits, W., Mark, D., Marshall, I., Nichols, A., 1998. Aerosol Particle Size Analysis. Royal Society of Chemistry, Cambridge, UK, pp. 116.

Guenther, D., Cousin, H., Magyar, B., Leopold, I., 1997. Calibration studies on dried aerosol for laser-ablation-inductively coupled plasma mass spectrometry. J. Anal. Atom. Spectrom. 12, 165–170.

Guo, X., Lichte, F., 1995. Analysis of rocks, soils and sediments for the chalkophile elements by laser ablation-inductively coupled plasma mass spectrometry. Analyst 129, 2707–2711.

Haapala, H., Goltsova, N., Pitulko, V., Lodenius, M., 1996. The effect of simultaneous large acidic and alkaline airborne pollutants on forest soil. Environ. Pollut. 94, 159–168.

Hahn-Weinheimer, P., Hirner, A., Weber-Diefenbach, K., 1984. Grundlagen und praktische Anwendung der Roentgenfluoreszenzanalyse (RFA). Friedreich Vieweg, Braunschweig/Wiesbaden, pp. 253.

Halls, D., 1995. Analytical minimalism applied to the determination of trace elements by atomic spectrometry. J. Anal. Atom. Spectrom. 10, 169–175.

Helmbold, M., 1985. Multielement analysis by means of alpha particle spectrometry. In: Sansoni, B. (Ed.), Instrumentelle Multielementanalyse. VCH Publishers, Weinheim, pp. 98–108.

Hewitt, C.N., Candy, G.B.B., 1990. Soil and street dust heavy metal concentrations in and around Cuenca. Ecuador. Environ. Pollut. 63, 129–136.

Ho, Y.B., Tai, K.M., 1988. Elevated levels of lead and other metals in roadside soil and grass and their use to monitor aerial metal depositions in Hong Kong. Environ. Pollut. 49, 37–51.

Hoffmann, P., 1986. Application of radionuclide sources for excitation in energy-dispersive X-ray fluorescence analysis. Fresenius Z. Anal. Chem. 323, 801–806.

Hoste, J., Op de Beeck, J., Gijbels, R., Adams, F., Van den Winkel, P., De Soete, D., 1971. Activation Analysis. CRC, Butterworths, London, pp. 148.

IAEA 1989. Measurement of Radionuclides in Food and the Environment. IAEA Techn. Rep. Ser. No. 295, Vienna, p. 165.

IAEA, 1995. Proceedings of the Symposium Environmental Impact of Radioactive Releases, Vienna, 8–12 May 1995. IAEA, Vienna, p. 874.

ISO, 1980. International Standard Organization. Part 1. Guidance on the Design of Sampling Programs. Part 2. Water Quality, Sampling, Guidance on Sampling Techniques, ISO-5667/1.

ISO, 1991. International Standard Organization, Soil Quality – Sampling. (Draft ISO/CD 10381/1.2).

Ivanova, J., Djingova, R., Kuleff, I., 1998. Determination of some heavy and toxic elements in plants and soils using ^{241}Am as excitation source. J. Radioanal. Nucl. Chem 238, 29–32.

Jackson, K., Mahmood, T., 1994. Atomic absorption, atomic emission, and flame emission spectrometry. Anal. Chem. 66, 252R–279R.

James, W., Acevedo, L., 1993. Trace element partitioning in Texas lignite combustion. J. Radioanal. Nucl. Chem. 171, 287–302.

Johansson, S.A.E., Campbell, J. (Eds.), 1988. PIXE: a Newer Technique for Elemental Analysis. Wiley, New York.

Kabata-Pendias, A. Dudka, S., 1990. Evaluating baseline data for cadmium in soils and plants in Poland. In: Lieth, H., Markert, B. (Eds.), Element Concentration Cadasters in Ecosystems. VCH Publishers, Weinheim, pp. 265–280.

Kabata-Pendias, A., Dudka, S., 1991. Trace metal contents of *Taraxacum officinale* (dandelion) as a convenient environmental indicator. Environ. Geochem. Health 13, 108–113.

Kalvoda, R., 1984. Electrochemical analytical methods used in environmental analysis. Sci. Total Environ. 37, 3–7.

Kanda, V., Taira, M., 1988. Sequential multi-element analysis of sediments and soils by inductively coupled plasma atomic emission spectroscopy with a computer controlled rapid scanning echelle monochromator. Anal. Chem. 207, 269–281.

Katz, M. (Ed.), 1977. Methods of Air Sampling and Analysis. Intersociety Committee. American Public Health Association, Washington, DC.

Keith, L., 1991. Environmental Sampling and Analysis: a Practical Guide. Lewis Publishers, Chelsea, MI, p. 127.

Keliher, P., Boyko, W., Clifford, R., Snyder, J., Zhu, S., 1986. Emission spectroscopy. Anal. Chem. 58, 335R–356R.

Ketelsen, P., Knoechel, A., Petersen, W., Tolkiehn, G., 1985. Roentgenfluoreszenzanalyse mit Hilfe Synchrotronstrahlung (SYRFA). In: Sansoni, B. (Ed.), Instrumentelle Multielementanalyse. VCH Publishers, Weinheim, pp. 301–309.

Kim, J., 1981. Monostandard activation analysis: evaluation of the method and its accuracy. J. Radioanal. Chem. 63, 121–144.

Klockenkaemper, R., von Bohlen, A., 1992. Total reflection X-ray fluorescence – An effective method for microtrace and surface layer analysis. J. Anal. Atom. Spectrom. 7, 273–279.

Klockenkaemper, R., von Bohlen, A., 1996. Elemental analysis of environmental samples by total reflection X-ray fluorescence: a review. X-Ray Spectrom. 25, 156–162.

Klockow, D., 1982. Analytical chemistry of the atmospheric aerosol. In: Georgii, H., Jaeschke, W. (Eds.), Chemistry of Unpolluted and Polluted Troposphere. D. Reidel Publishing Company, Dordrecht, Holland, pp. 57–90.

Knabe, W., 1982. In: Steubing, L., Jaeger, H.-J. (Eds.), Monitoring of Air Pollutants by Plants, Jung, The Hague, pp. 45–72.

Kneip, T., Kleinman, M., 1982. Analysis of airborne particles in the workplace and ambient atmosphere. In: Cantle, J. (Ed.), Atomic Absorption Spectrometry. Elsevier, Amsterdam, pp. 123–137.

Knoechel, A., Prange, A., 1980. Analysis of trace elements in seawater. Part II. Determinaton of heavy metal traces in sea water by X-ray fluorescence analysis with totally reflecting sample holders. Microchim. Acta (Wien) II, 395–408.

Knoechel, A., Prange, A., 1981. Analytik von Elementspuren in Meerwasser. Part I. Tracestudien zur Entwicklung von spurenanalytischen Verfahren fuer die Multielementbestimmung von Schwermetallen in waessrigen Loesungen mit Hilfe von Varianten der energiedispersiven Roentgenfluorescenzanalyse. Fresenius Z. Anal. Chem. 306, 252–258.

Kos, V., Budic, B., Hudnik, V., Lobnik, F., Zupan, M., 1996. Determination of heavy metal concentration in plants exposed to different degrees of pollution using ICP-AES. Fresenius Z. Anal. Chem. 354, 648–652.

Kostadinov, K., Djingova, R., 1980. Trace element investigation of coal samples by thermal and epithermal neutron activation analysis. Radiochem. Radioanal. Lett. 45, 297–304.

Kostadinov, K., Djingova, R., 1981a. Trace element analysis of biological materials by thermal and epithermal neutron activation analysis. J. Radioanal. Chem 63, 5–12.

Kostadinov, K., Djingova, R., 1981b. Determination of dimethylmercury and diphenymercury in liquid biological samples. Radiochem. Radioanal. Lett. 48, 401–404.

Krivan, V., Schaldach, G., 1986. Untersuchungen zur Probenahme und -vorbehandlung von Baumnadeln zur Elementanalyse. Fresenius Z. Anal. Chem. 324, 158–167.

Krumpen, W., Freiburg, C., Markert, B., 1989. Instrumentelle Multielementanalyse von repraesentativen pflanzlichen Matrizen mit Roentgenfluoreszenz (RFA). Verh. Ges. Oekologie, XIX (I), 276–277.

Kuleff, I., Djingova, R., 1984. The dandelion (*Taraxacum officinale*) – A monitor for environmental pollution? Water Air Soil Pollut. 21, 77–85.

Kumar, S., Singh, S., Garg, M., Mehta, D., Singh, N., Mangal, P., Trehan, P., 1989. Elemental analysis of environmental samples using energy dispersive X-ray fluorescence technique. Indian J. Environ. Health 31, 8–16.

Kushelevsky, A., 1990. Photon activation analysis. In: Alfassi, Z. (Ed.), Activation Analysis. CRC Press, Boca Raton, FL, pp. 219–237.

La Brecque, J., Hanna, A., Abdel-Rassoul, A., Schelenz, R., 1987. An IAEA Intercomparison Study. Inorganic Costituents in Lake Sediments (IAEA/SL-3), IAEA/RL/143, IAEA, Vienna, p. 99.

Lavi, N., Neeman, E., Nir-El, Y., 1992. Neutron activation analysis for environmental trace element research. J. Radioanal. Nucl. Chem. 163, 307–312.

Leivuori, M., 1998. Heavy metal contamination in surface sediments in the gulf of Bothnia. Chemosphere 56, 43–59.

Lewis, R., 1987. Guidelines for Environmental Specimen Banking with Special Reference to the Federal Republic of Germany: Ecological and Marginal Aspects. US MAB Report No., 12.

Li, X.J., Schramel, P., Wang, H.Z., Grill, P., Kettrup, A., 1996. Determination of trace metal ions Co, Cu, Mo, Mn, Fe, Ti, V in reference seawater samples by inductively coupled plasma emission spectrometry combined with third phase preconcentration. Fresenius Z. Anal. Chem. 356, 52–56.

Liese, T., 1985a. Die Analytik von Elementen in Rahmen der Bestimmung von Transfaktoren Boden-Pflanze. Kernforschungszentrum, Karlsruhe, Kfk 3830, 43.

Liese, T., 1985b. Zur Bestimmung von Elementen in Pflanzen- und Bodenproben mittels sequentieller ICP-AES. Fresenius Z. Anal. Chem. 321, 37–44.

Lobinski, R., 1996. Hyphenated techniques for speciation analysis, status and trends. In: Welz, B. (Ed.), CANAS'95 Colloquium Analytische Atomspektroskopie, Bodenseewerk. Perkin–Elmer, Ueberlingen, pp. 389–397.

Lund, W., 1990. Speciation analysis – Why and how? Fresenius Z. Anal. Chem. 337, 557–564.

Madeddu, B., Rivoldini, A., 1996. Analysis of plant tissues by ICP-OES and ICP-MS using an improved microwave oven acid digestion. Atom. Spectrosc. 17, 148–154.

Markert, B., 1987. Ein Vergleich von analytischer und biologischer Varianz bei der Elementanalytik von Fe, Cu, Pb und Zn mit Hilfe der AAS. In: Welz, B. (Ed.), 4 Colloquium Atomspektrometrische Spurenanalytik, Bodenseewerk. Perkin Elmer, Ueberlingen, pp. 385–392.

Markert, B., 1988. The distribution of some chemical elements in different herds of *Vaccinium vitis-idaea* (red whortleberry), a contribution to the representative sampling procedure of terrestrial plants for multi-element analysis. Angew. Bot. 62, 343–353.

Markert, B., 1989a. Multi-element analysis in ecosystems: basic conditions for representative sampling of plant materials. Fresenius Z. Anal. Chem. 335, 562–565.

Markert, B., 1989b. Distribution of chemical elements in *Vaccinium myrtillus* (blueberry), basic problems for representative sampling of plants for multi-element analysis in ecosystems. Fresenius Z. Anal. Chem. 333, 11–14.

Markert, B., 1989c. Klassifizierung biologischer Varianzen im Elementhaushalt von Pflanzen. Verh. Ges. Oekologie (Goettingen XVII, 779–784.

Markert, B., 1989d. Multi-element analysis in ecosystems, basic conditions for representative sampling of plant materials. Fresenius Z. Anal. Chem. 335, 562–565.

Markert, B., 1992. Multi-element analysis in plant materials – Analytical tools and biological questions. In: Adriano, D. (Ed.), Biogeochemistry of Trace Metals. Lewis, Boca Raton, FL, pp. 401–408.

Markert, B., 1995a. Quality assurance of plant sampling and storage. In: Quevauviller, P. (Ed.), Quality assurance in environmental Monitoring – Sampling and Sample Preparation. VCH Publishers, Weinheim, pp. 215–254.

Markert, B., 1995b. Communution of Samples. Encyclopedia of Analytical Sciences. Academic Press, London, pp. 803–809.

Markert, B., 1995c. Sample preparation (cleaning, drying, homogenization) for trace element analysis in plant matrices. Sci. Total Environ. 176, 45–61.

Markert, B., 1996. Instrumental Element and Multi-Element Analysis of Plant Samples. Wiley, Chichester, pp. 295.

Markert, B., Pedrozo, F., Geller, W., Friese, K., Korhammer, S., Baffico, G., Diaz, M., Woelfl, S., 1997. A contribution to the study of the heavy metal and nutritional element status of some lakes in the Southern Andes of Patagonia (Argentina). Sci. Total Environ. 206, 1–15.

Marshall, J., Fisher, A., Chenery, S., Sparkes, S., 1996. Atomic spectrometry update – Atomic emission spectrometry. J. Anal. Atom. Spectrom. 11, 213R–238R.

Martin, M., Couphtrey, P., 1982. Biological Monitoring of Heavy Metal Pollution: Land and Air. Applied Science, London, pp. 475.

Massart, D., Kaufman, L., 1989. The Interpretation of Analytical Chemical Data by the Use of Cluster Analysis. Robert E. Krieger, Malabar, pp. 237.

McLaren, J.W., Methven, B.A.J., Lam, J.W.H., Berman, S.S., 1995. The use of inductively coupled plasma mass spectrometry in the production of environmental certified reference materials. Microchim. Acta 119, 287–295.

Mendez Garcia, A., Fernandez Sanchez, M., Sanchez Uria, J., Sanz Mandel, A., 1996. Speciation of mercury by continuous flow liquid-liquid extraction and inductively coupled plasma atomic emission spectrometry detection. Mikrochim. Acta 122, 157–166.

Mizuike, A., 1983. Enrichment Techniques for Inorganic Trace Analysis. Springer Verlag, Berlin, p. 152.

Muminov, V., Haidarov, R., 1980. Rentgenofluorescentnyi Analiz Vozbuzhdeniem Uskorennyimi Legkimi Ionami (X-Ray Fluorescence Analysis Using Excitation with Accelerated Ions). Fan, Tashkent, p. 171.

Nickson, R., Hill, S., Worsfold, P., 1995. Solid phase techniques for the preconcentration of trace metals from natural waters. Anal. Proc. Incl. Anal. Commun. 32, 387–395.

NIOSH Manual for Analytical Methods, 1977. US Department of Health, Education and Welfare, US Government Printing Office, Publ No. 77-157-A.

Nuernberg, H., 1984. Editorial to inorganic analysis in environmental research and protection. Fresenius Z. Anal. Chem. 317, 197–199.

Nuernberg, H., 1985. Potentialities of voltammetry in environmental oligoelement analysis of trace metals. In: Sansoni, B. (Ed.), Instrumentelle Multielementanalyse. VCH Publishers, Weinheim, pp. 415–443.

Okamoto, K., 1987. New certified reference material vehicle exhaust particulates. Anal. Sci. 3, 191–192.

Osan, J., Török, S., Török, K., Nemeth, L., Labar, J., 1996. Physiological effect of accidental fly ash deposition in plants and chemical study of the dusted plant leaves by XRF and EPMA. X-Ray Spectrom. 25, 167–172.

Ostapczuk, P., Stoeppler, M., Duebeck, H., 1988. Present potential of electrochemical methods for metal determinations in reference materials. Fresenius Z. Anal. Chem. 322, 662–665.

Paetz, A., Groessmann, G., 1994. Problems and results in the development of international standards for sampling and pretreatment of soils. In: Markert, B. (Ed.), Environmental Sampling for Trace Analysis. VCH Publishers, Weinheim, pp. 321–334.

Paris, I. (Ed.), 1987. New Results in the Research of Hardly Known Trace Elements and the Analytical Problems of Trace Element Research. University Press, Budapest, p. 309.

Pazonieki, L., Hanna, A., 1985. Report on Intercomparison IAEA V-10 of the Determination of Trace Elements in Hay Powder. IAEA, Vienna, p. 92.

Pin, C., 1996. Sequential separation of rare earth elements, thorium, and uranium by miniturized extraction chromatography: application to isotope analysis of silicate rocks. Anal. Chim. Acta 339, 79–89.

Poole, S., Poole, C., 1996. Trends in extraction of semivolatile compounds from water for environmental analysis. Anal. Commun. 33, 15H–17H.

Porter, J.R., 1986. Evaluation of washing procedures for pollution analysis of *Ailanthus altissima* leaves. Environ. Pollut. Ser. B. 12, 195–202.

Prange, A., Schwenke, H., 1992. Trace analysis using total reflection X-ray fluorescence spectrometry. Adv. X-Ray Anal. 35, 899–911.

Puk, R., Weber, J., 1994. Critical review of analytical methods for determination of inorganic mercury and methylmercury compounds. Appl. Organometal. Chem. 8, 293–302.

Que Hee, S., Boyle, J., 1988. Simultaneous multielement analysis of some environmental and biological samples by inductively coupled plasma atomic emission spectrometry. Anal. Chem. 60, 1033–1942.

Quevauviller, P., 1998. Method Performance Studies for Speciation Analysis. The Royal Society of Chemistry, Cambridge, UK, p. 272.

Reus, U., Markert, B., Hoffmeister, C., Spott, D., Guhr, H., 1993. Determination of trace metals in river water and suspended soils by TXRF spectrometry. Fresenius Z. Anal. Chem. 347, 430–435.

Rubio, R., Ure, A., 1993. Approaches to sampling and sample pretreatment for metal speciation in soils and sediments. Int. J. Environ. Anal. Chem. 51, 205–217.

Sansoni, B., 1986. Fortgeschrittener chemischer Analysendienst fuer Elemente. Radionuclide und Phasen das Juelicher Baukastensystem fuer Analysenschritte. Fresenius Z. Anal. Chem. 323, 573–600.

Sansoni, B., 1987. Multi-element analysis for environmental characterization. Pure Appl. Chem. 59, 579–610.

Sayre, E., 1988. Brookhaven Data Handling Programs, BNL.

Schmelling, M., Klockenkaemper, R., Klokow, D., 1997a. Application of total-reflection X-ray fluorescence spectrometry to the analysis of airborne particulate matter. Spectrochim. Acta, B 52, 985–994.

Schmelling, M., Alt, F., Klockenkaemper, R., Klokow, D., 1997b. Multielement analysis by total reflection X-ray fluorescence spectrometry for the certification of lichen research material. Fresenius Z. Anal. Chem. 357, 1042–1044.

Schutyiser, P., 1978. Instrumental neutron activation analysis of dry atmospheric fall-out and rain water. Anal. Chim. Acta 100, 75–85.

Schutyser, P., Govaerts, A., Dams, R., Hoste, J., 1977. Neutron activation analysis of platinum metals in airborne particulate matter. J. Radioanal. Nucl. Chem. 37, 651–660.

Severson, R., Gough, L., Van den Boom, G., 1992. Baseline concentrations in soils and plants. Wattenmeer National Park, North and East Frisian Islands, Federal Republic of Germany. Water Air Soil Pollut. 61, 169–184.

Shtangeeva, I., 1994. Variation of the elemental composition of plants and soils. J. Radianal. Nucl. Chem. 177, 381–391.

Speer, R., Hoffmann, P., Lieser, K., 1986. Problems in measuring water samples of different salt content by ICP-AES. Fresenius Z. Anal. Chem. 325, 558–563.

Steinnes, E., Solberg, W., Petersen, H., Wren, C., 1989. Heavy metal pollution by long range atmospheric transport in natural soils of Southern Norway. Water Air Soil Pollut. 45, 207–218.

Stoeppler, M., 1991. Analytical chemistry of metals and metal compounds, In: Merian, E. (Ed.), Metals and Their Compounds in the Environment. VCH Publishers, Weinheim, pp. 106–206.

Stoeppler, M. (Ed.), 1992. Hazardous Metals in the Environment. Elsevier, Amsterdam, p. 541.

Sturgeon, R., Berman, S., 1987. Sampling and storage of natural waters for trace metals. CRC Crit Rev. Anal. Chem. 18, 209–244.

Szpunar-Lobinska, J., Witte, C., Lobinski, R., Adams, F., 1995. Separation techniques in speciation analysis for organometallic species. Fresenius Z. Anal. Chem 351, 351–377.

Tanizaki, Y., 1990. Neutron activation analysis of water samples. In: Alfassi, Z. (Ed.), Activation Analysis, Vol. 2. CRC Press, Boca Raton, FL, pp. 378–405.

Toelg, G., 1987. Extreme trace analysis of the elements – The state of the art today and tomorrow. Analyst 112, 365–376.

Toelg, G., Klockenkaemper, R., 1993. The role of total-reflection X-ray fluorescence in atomic spectroscopy. Spectrochim. Acta 488, 111–127.

Toelgyessy, J., Klehr, E., 1987. Nuclear Environmental Chemical Analysis. Ellis Horwood, Wiley, New York, p. 180.

Toelgyessy, J., Kyrs, M., 1989. Radioanalytical Chemistry. Ellis Horwood, Chichester, p. 345.

Toelgyessy, J., Havranek, E., Dejmkova, E., 1990. Radionuclide X-Ray Fluorescence Analysis with Environmental Applications. Elsevier, Amsterdam, p., 254.

Tsalev, D., 1994. Electrothermal atomic absorption spectrometry in occupational and environmental health practice – A decade of progress and establishment. J. Anal. Atom. Spectrom. 9, 405–414.

Tsalev, D., 1995. Atomic Absorption Spectrometry in Occupational and Environmental Health Practice, III, Progress in Analytical Methodology. CRC Press, Boca Raton, FL, p. 349.

Tsalev, D., 1998. Electrothermal atomic absorption spectrometry. In: Meyers, R., (Ed.), Encyclopedia of Environmental Analysis and Remediation, Vol. 3. Wiley, New York, pp. 1583–1605.

Tsalev, D., 1999. Hyphenated vapour generation atomic absorption spectrometric techniques. J. Anal. Atom. Spectrom. 14, 147–162.

Tsalev, D., Slaveykova, V., 1998. Chemical modification in elektrothermal atomic absorption spectrometry. Adv. Atom. Spectrosc., Vol. 4, Sneddon, J. (Ed). JAI Press Inc. Greenwich, pp. 27–150.

Kostadinov, K., Djingova, R., 1981b. Determination of dimethylmercury and diphenymercury in liquid biological samples. Radiochem. Radioanal. Lett. 48, 401–404.

Tschoepel, P., 1992. Sample treatment. In: Stoeppler, M. (Ed.), Hazardous Metals in the Environment. Elsevier, Amsterdam, pp. 73–95.

Turcsanyi, G., Penksza, K., Siller, I., Fuehrer, E., Toth, S., Kovacs, M., Buettner, S., 1994. Sampling in the stemflow and throughfall area of forests. In: Markert, B. (Ed.), Environmental Sampling for Trace Analysis. VCH Publishers, Weinheim, pp. 449–464.

Van den Berg, C., 1986. The determination of trace metals in sea water using cationic stripping voltammetry. Anal. Chem. 57, 1532–1536.

Van der Sloot, H.A., 1976. Thesis: Stichting Energieonderzoek Centrum Nederland. Rep. ECN-1.

Van Dyck, P., Török, S., Grieken, R., 1986. Enhancement effect in X-ray fluorescence analysis of environmental samples of medium thickness. Anal. Chem. 58, 1761–1766.

Wagner, G., 1987. Entwicklung einer Methode zur grossraeumigen Ueberwachung mittels standartisierter Pappelblaetterproben von Pyramidenpappeln (*Populus nigra* 'Italica') am Beispiel von Blei, Cadmium und Zink. In: Stoeppler, M., Duerbeck, H.W.0 (Eds.), Beitraege zur Umweltprobenbank, Nr. 5, Juel. Spez., 412, p. 224.

Wagner, G., 1990. Variability in element concentrations in tree leaves depending on sampling parameters. In: Lieth, H., Markert, B. (Eds.), Element Concentration Cadasters in Ecosystems. VCH Publishers, Weinheim, pp. 41–54.

Wagner, G., 1993a. Large scale screening of heavy metal burdens in higher plants. In: Markert, B. (Ed.), Plants as Biomonitors. VCH Publishers, Weinheim, pp. 425–434.

Wagner, G., 1993b. Plants and soils as specimen types from terrestrial ecosystems in the environmental specimen banking program of the Federal Republic of Germany. Sci. Total Environ. 139/140, 213–224.

Wagner, G., 1995. Basic approaches and methods for quality assurance and quality control in sample collection and storage for environmental monitoring. Sci. Total Environ. 176, 63–71.

Wagner, G., Sprengart, J., 1991. Richtlinie zur Probenahme und Probenbearbeitung fuer Boeden, ESB-Draft, Saarbruecken.

Wang, J., 1985. Stripping Analysis, Principles, Instrumentation and Applications. VCH Publishers, Weinheim.

Watson, J., Chow, J., 1994. Particle and gas measurements on filters. In: Markert, B. (Ed.), Environmental Sampling for Trace Analysis. VCH Publishers, Weinheim, pp. 125–161.

Welz, B., 1985. Atomic Absorption Spectrometry. VCH Publishers, Weinheim, p., 506.

Whitworth, D., Achterberg, E., Nimmo, M., Worsfold, P., 1998. Validation and in situ application of an automated dissolved nickel monitor for estuarine studies. Anal. Chim. Acta 371, 235–246.

Wyttenbach, A., Bajo, S., Tobler, I., Keller, T., 1985. Major and trace element concentrations in needles of *Picea abies*; levels, distribution functions, correlations and environmental, correlations and environmental influences. Plant Soil 85, 313–325.

Yamasaki, S.-I., Tsumura, A., Takaku, Y., 1994. Ultratrace elements in terrestrial waters as determined by high-resolution ICP-MS. Microchem. J. 49, 305–318.

Trace elements – Their distribution and effects in the environment
B. Markert and K. Friese, editors

Chapter 6

Nuclear analytical techniques in atmospheric trace element studies in Portugal

M.C. Freitas, M.A. Reis, L.C. Alves, H.Th. Wolterbeek

Abstract

In this work, the nuclear techniques INAA and PIXE were used in a lichen survey using Parmelia sulcata Taylor held in Portugal in 1993. The possibilities of the two techniques are discussed, and the elements determined by both are compared. It is presented the concentration patterns of 46 chemical elements and the assignment to pollution sources attempted and compared with results on factor analysis previously published. Results are also compared with other European surveys.

Keywords: Nuclear analytical techniques; Atmospheric trace elements; Instrumental neutron activation analysis; Proton induced X-ray emission; Biomonitoring

1. Introduction

Nuclear analytical techniques (NATs) and related isotopic tracer methods are well established as important tools in a wide variety of different kinds of environmental studies. They provide a wealth of information on sources, pathways and effects of many elements of environmental and health-related interest. Apart from being regarded as of particular strength in analytical quality assurance (IAEA, 1997), nuclear and related techniques cover studies of air particulates, solid waste products, sediments, food, water, human tissues, biomonitors and other kinds of environmental samples.

Key features of most nuclear techniques are the principal absence of effects by element chemical forms, and the absence of necessary sample digestion procedures (De Goeij and Bode, 1997). Apart from being an analytical advantage, the first property should be regarded, as a drawback where information on a particular form of an element is required (viz. methylmercury rather than total mercury), the second only applies for nuclear signals of high penetrative power. However, the intrinsic accuracy of most NATs makes that they are in regular use in validation procedures of analytical methods, the development of new reference materials, and the set-up of base-line elemental information in a variety of (health-related) environmental issues.

Typical examples of the NAT family are neutron activation analysis (NAA) and proton induced X-ray emission (PIXE). NAA is one of the most robust analytical methods available in a vast range of applied sciences (Trahey, 1996; Bode and De

Goeij, 1998), hardly asks for any extensive sample preparation, but shows up to 2–4 weeks analytical turn-around time, depending on the half-lives of the involved radionuclides. PIXE is a fast analytical method, but asks for a strict sample homogeneity down to the sub-mg level, which implies laborious sample preparation procedures (IAEA/AL/095, 1996; Tadic et al., 1997). The methods are intrinsically accurate, multi-elemental and partly complement each other with respect to elements of environmental interest.

Both instrumental NAA (INAA) and PIXE are available at the Instituto Tecnológico e Nuclear (ITN), Sacavém, Portugal, and have been applied in a national study on atmospheric trace elements, using lichens as biomonitor materials. In the present paper, this survey is used to illustrate the potential of NATs in environmental studies.

2. Multi-elemental techniques versus single element techniques

The selection of elements to be analyzed and/or used in data interpretation should be closely linked to the objectives of the study (Wolterbeek and Bode, 1995). A survey may be dedicated to a single or only a limited number of elements, but it may also be set up to gather information about sources/effects based on broader points of view. The latter set-up may be regarded as most effective, because, first, the analysis of a large number of elements may generally increase the modes for interpretation, may permit a more reliable recognition of source finger-prints, and may show effects which are not a-priori anticipated (Bode and Wolterbeek, 1990; Wolterbeek et al., 1996), while, second, the resources needed for the field-work will not or hardly depend on the number of elements of interest. This indicates that the selection of a large number of elements *principally* emerges from many survey's goals: clear and unequivocal interpretation of data on specific elements may largely depend on the simultaneous presence of data on various other elements.

The above can be illustrated by data taken from several multi-element air pollution biomonitor surveys carried out at IRI: The air pollution surveys commonly include a number of soil-associated elements (e.g. Al, Fe, Sc, Cr, Th) and several rare earth elements (Kuik et al., 1993a,b). In the Factor Analysis interpretation of the data on all selected 20 elements, the soil indicator elements serve to extract a "soil-factor" (De Bruin and Wolterbeek, 1984), based on which, for *all* individual elements, site-specific soil-associated fractions of the total concentrations can be calculated (Kuik et al., 1993a,b).

The selection of a relatively large number of elements occurring in high temperature processes (e.g. V, Sb, Se, As) helped to discriminate between processes such as waste incineration, coal combustion, and other high-T sources (Kuik et al., 1993a,b), while the simultaneous analysis of elements like Br, I, and Na suggested a sea-aerosol associated and long-distance origin of As, for about 25% of its total occurrence in mosses in a 1992 survey (Kuik and Wolterbeek, 1995).

One of the most clear and simple examples can be given for the Zn smelting industries in the south-eastern part (Kempen area) of The Netherlands: here, Cd occurs as a by-product in a Zinc ore at a characteristic relative abundancy of 1 Cd:200 Zn. The analysis of a large number of elements, combined with the application of Factor

Analysis, yielded a well-defined finger-print for these Zn smelting industries: they could be characterized by a Zn/Cd factor, in which Cd and Zn were obtained in a 1:212 relative abundancy (De Bruin et al., 1987).

The data suggest that a careful selection of elements facilitates the interpretation of results on each of the individual elements; the larger the number of relevant elements involved in the eventual analysis, the more detailed information may be present in the data-set. Thus, the principal choice may be the multi-elemental analysis: the problem here is how to extract the wealth of information from the set, which may contain thousands of analytical data. A fast and functional approach may be found by the application of Factor Analysis techniques (Kuik et al., 1993a,b).

3. Nuclear analytical techniques in environmental research

Although no sharp lines can be drawn between nuclear and non-nuclear techniques (see De Goeij and Bode, 1997 for a review), the principle of the *nuclear* technique says that the analytical information on element and concentration originates from the nucleus and not from the atom. As such, chemical binding, chemical compound or matrix composition has no essential influence on the accuracy of the results (Bode and Wolterbeek, 1990; De Goeij and Bode, 1997). It should be noted here that although techniques such as particle/proton induced X-ray emission (PIXE) and X-ray fluorescence spectrometry (XRF) are basically derived from the behaviour of inner orbital electrons rather than the nucleus itself, they are often counted as a nuclear technique, primarily because inner orbital electrons do not predominate in the characteristics of the atom's chemical behaviour (but see also De Goeij and Bode, 1997 for NMR and Mössbauer techniques).

Nuclear analytical techniques such as Instrumental Neutron Activation Analysis (γ-rays oriented INAA, see Bode and De Goeij, 1998), PIXE, or XRF are all *multi-element* techniques, and *non-destructive*. Advantages of the first characteristic are discussed elsewhere in this chapter. The latter makes them different from a large number of other widely used (non-nuclear) analytical techniques, such as atomic absorption spectroscopy (AAS), inductively coupled plasma spectrometry (ICP), or mass spectrometry (MS). It also makes that INAA, PIXE or XRF are principally well suited for the routine analysis of the solid samples often encountered in environmental research: it is not necessary to bring the sample into solution, with all the associated problems ranging from incomplete digestion (elemental losses) to impurities in the applied chemicals (elemental contamination) (Bode and Wolterbeek, 1990; De Goeij and Bode, 1995). It may be clear that the absence of effects of chemical forms and the non-destructive character of the nuclear technique pays off particularly in (environmental) base-line surveys comprising large numbers of samples and/or strongly varying sample matrices (e.g. ecosystem research, including samples of soils, plants, animal tissues). It should be noted, however, that nuclear techniques may be successfully applied also in mechanistic (process) dynamic environmental studies, due to the isotope-specific responses in nuclear analytical approaches (De Goeij and Bode, 1997).

4. Nuclear analysis by INAA and PIXE

INAA stands out by non-destructivity, multi-element capability and adequate limits of detection for the majority of elements of environmental interest. The absence of digestion steps and the independence of chemical forms make that a high level of accuracy can be obtained (De Goeij and Bode, 1994, 1997). The intrinsic technique characteristics imply that this high accuracy can be maintained over a large dynamic range, from ppb to % level (Bode and Wolterbeek, 1990).

Gamma-radiation is the radiation of choice in neutron activation analysis, since it is mono-energetic and in most cases characteristic for the emitting nucleus. The other advantage of γ-radiation is that it has a high penetrating power, so that it is hardly adsorbed in the radioactive material itself (Bode and De Goeij, 1998). INAA is routinely used in a mg to g mass range of samples (IAEA/AL/095, 1996), and only when the sample matrix has a high overall atomic number Z, and/or when large samples are being analyzed, problems may arise due to self-attenuation of the induced γ-radiation, and, even more exceptional, self-shielding of the neutrons during irradiation. These phenomena are well-understood, and proper corrections can be applied, thereby making INAA applicable in a mg to kg sample mass range (Bode and Overwater, 1993; Overwater et al., 1993; Bode and De Goeij, 1998).

Although INAA may be calibrated for a large number of elements (up to 70 elements calibrated at IRI, Delft, The Netherlands, see Bode and Wolterbeek (1990)), INAA is not capable of determining low-Z elements of environmental interest, such as Be, B, Li, or high-Z elements such as Pb, Bi and Tl. Here, complementary techniques should be applied.

PIXE and INAA overlap and partly complement each other with respect to elements. The main difference between INAA and PIXE is that X-ray energies associated with PIXE are much smaller than the energies of emitted γ-rays used in routine INAA. These differences come out in differences in self-absorption (absorption of rays within the sample), which makes that PIXE should be practically regarded as a "thin-layer" (surface-related) technique, with energy-related depth-profiles, also depending on general sample matrix characteristics. This means that for all sample materials, including environmental ones such as bio-organisms, biomonitor materials, air particulate matter etc., key steps in PIXE analyses are the sample homogenization and pelleting: eventual elemental determinations are generally carried out on basis of very small (lower mg range) sample masses.

Both INAA and PIXE are intrinsically accurate, but for PIXE the small X-ray energies involved, taken together with the small sample masses in actual analysis, implies a strong dependency on both sample matrix characteristics and bulk sample homogeneity. This makes that for PIXE particular difficulties may be encountered in quantitative calibration procedures. An associated problem is that existing certified reference materials (CRMs) are generally certified for much larger sample masses: they are mostly inadequate for quality control in PIXE (IAEA/AL/095, 1996). Therefore, in both sample and CRM preparations, much effort is devoted to increase the number of sample particles in analyzed sub-samples by grinding into smaller individual particle sizes and avoiding so called "nuggets" (IAEA/AL/095, 1996).

Although high-(energy)-resolution detectors are in use in both PIXE and INAA, peak overlaps are encountered in both techniques. The resulting doublets or multiplets can often be resolved mathematically without too many difficulties (Bode and De Goeij, 1998), but in both techniques, however, unsolvable multiplets remain, for which parallel information should be available (e.g. Si–Na–Al–Mg and P–Al in INAA, or the As (K_α), Pb (L_α) interferences in PIXE) (Gonsior et al., 1982). Here, too, both techniques may complement each other in resolving overlaps (Knoll, 1989).

5. Survey

5.1. Introduction

In Portugal, data of heavy metals in airborne particulate matter was scarce till the middle of the ninety-decade. Since then, we have published a few papers (Freitas et al., 1996, 1997, 1999a,b; Reis et al., 1996; Freitas and Nobre, 1997; Alves et al., 1998a,b) and some papers were published by the Universities of Aveiro (Pio and Feliciano, 1996; Harrison et al., 1997; Pio and Lopes, 1998, among others), Lisbon (Ruhling, 1992; Sérgio et al., 1993, among others) and Oporto (Tavares et al., 1993; Soares and Vasconcelos, 1995; Vasconcelos and Tavares, 1998, among others). The interest of the heavy metal monitoring is internationally recognised. Although it is not expected severe atmospheric pollution in heavy metals in Portugal due to the geographical position in Europe and due to the lower industrialisation, it is of the most importance to have a reference data basis. This importance is for the country itself but also for the international organisations of the atmosphere, within globalisation programs. The recognition of the problem of lack of data led our Institute to undertake some projects in this area since 1993, supported by Electricity of Portugal, International Atomic Energy Agency, and Portuguese Ministry of the Environment.

The sensibility of lichens to atmospheric pollution was demonstrated in the last two decades in various international and national publications. Recent studies showed that the survey of a large territory with lichens turns identification of atmospheric pollution sources possible (Sloof, 1993). The use of lichens is the only possibility to go through a large scale monitoring, which otherwise would be done with an enormous amount of air samplers and samples. Response of lichens to the environment is being studied through modelling very recently in our Institute (Reis et al., 1999). This is made with lichen transplants whose response to airborne particulate matter, and total deposition is followed experimentally.

5.2. Methods

5.2.1. Sampling

A lichen collection campaign was held in Portugal during the months of July and August 1993. Due to the high industrial asymmetry between the coast and the interior of the country, two different grid sizes were used for sampling site definition. Near the Atlantic coast and up to 30 to 80 km from the coast line, a 10 × 10 km grid was used; in the interior a 50 × 50 km grid was selected. A total of 228 sampling squares were thus

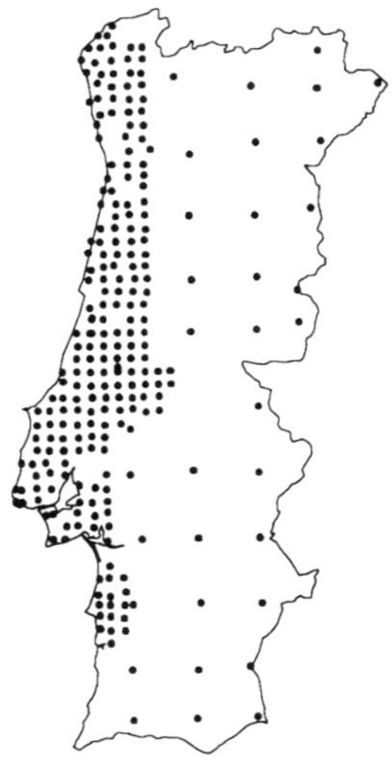

Figure 1. Grid 10 × 10 km at the Atlantic coast and 50 × 50 km at the interior.

defined; within each square and the nearest possible to the centre lichen samples were collected. The chosen lichen was *Parmelia sulcata Taylor* and it was collected from olive tree bark. In each place, this lichen was sampled from at least five different olive trees and all around the trees, at a height of 1–2 m above the ground level. Lichen samples of the same place were mixed in one bag, except in 10% of the places where they were kept separated for local variation determination. In Fig. 1 it is shown the sampling places. Spots missing are squares where it was not possible to find *P. sulcata* or olive tree.

5.2.2. Laboratory preparation

In the laboratory, remains of the tree bark and other lichen species where removed from *P. sulcata* transplants. Then each sample was put in a nylon sieve and rinsed in 18 MΩ water ejected for 30 s, freeze dried and ground in a Teflon® (balls and capsule) mill.

5.2.2.1. Instrumental neutron activation analysis (INAA) Two different size pellets were made: 500 mg was used for long irradiation (5 h) and 50–200 mg for short irradiation (30 s). The long irradiations were done in the Portuguese Research Reactor at ITN at a thermal neutron flux of 1.2×10^{13} cm^{-2} s^{-1} and the short irradiations in the Interfaculty Reactor Institute (Delft University) and the Hahn–Meitner Institut reactors at thermal neutron fluxes of 1×10^{12}–1×10^{16} cm^{-2} s^{-1}. Gamma-spectra of the irradiated samples were collected using hyperpure germanium detectors of high resolution, after 4 days and 4 weeks for long irradiation, and after 10 min. for short irradiation. Gold and zinc were used as comparators. Concentrations were determined using Gent University software for k_0-standardization method. In addition to the lichen samples, reference materials were prepared and analysed similarly to the lichen samples, for quality control (IAEA 336 lichen material, NIST Tomato leaves, Pine needles and Citrus leaves) (Freitas and Nobre, 1997). Contents for the following elements were determined: sodium, **chlorine**, magnesium, aluminium, **potassium**, **calcium**, scandium, **titanium, vanadium**, **chromium**, **manganese**, **iron**, cobalt, **copper**, **zinc, gallium**, **arsenic**, selenium, **bromine**, **rubidium**, **strontium**, molybdenum, silver, antimony, iodine, caesium, **barium**, lanthanum, cerium, neodymium, samarium, europium, terbium, lutetium, hafnium, tantalum, tungsten, mercury, thorium, and uranium (the elements which are determined by both techniques are shown in bold type).

5.2.2.2. Proton induced X-rays emission (PIXE) Samples were pelletized inserted in a boric acid cap. Samples were irradiated with a 2.0 MeV proton beam for filtered spectra and with a 1.2 MeV for filterless conditions. Spectra were obtained with a 200 eV Si(Li) X-ray detector and analysed with AXIL computer code. Concentrations were then calculated with the DATTPIXE program. Quality control was made by analysing in similar conditions to lichen samples the IAEA 336 lichen material and the CTA-OTL-1 tobacco leaves. Contents for the following elements were determined: silicon, phosphorous, sulphur, **chlorine**, **potassium**, **calcium**, **titanium**, **vanadium**, **chromium**, **manganese**, **iron**, nickel, **copper**, **zinc**, **gallium**, **arsenic**, **bromine**, **rubidium**, **strontium**, zirconium, **barium**, and lead (the elements which are determined by both techniques are shown in bold type).

The two multi-elementary techniques determine 46 elements (16 by both techniques) in *P. sulcata*.

5.3. Results and discussion

The local variation of *P. sulcata* was published somewhere else (Freitas and Nobre, 1997) and found to be on the average ±20%.

In Tables 1 and 2 it is shown the results obtained by PIXE and INAA, respectively. There it is mentioned the number of analysed samples, the maximum and minimum values found for each element, the mean and the median of each elemental set. It is observed that the elements Cu, Ga, Mo, Ag, Nd, and W, are determined by INAA only in a few samples, although Cu and Ga are determined by PIXE in most of the samples. The elements Ca, Sr, and Ba are not determined by INAA in 50 to 100 samples, although Ca is determined by PIXE in all the samples.

In Table 3 it is presented the mean and median differences (in %) between the results obtained by both techniques, and the maximum and the minimum of these differences. In general there is a normal distribution around the mean (which is quite similar to the median) for all the elements. The value of the mean gives an idea of the existence of a systematic error. For K, Ca, V, Fe, Zn, Br, and Sr the means are lower than ±10%, and between ±10 and ±30% for Cl, Ti, Mn, and Ba, for the others the means are larger than ±30% and go as far as ±88% (Cr, Cu, Ga, and As). The elemental heterogeneity of the samples may be the main cause for the differences. It is reminded that PIXE uses very small amounts of the sample (estimated less than 1 mg) while INAA uses 500 mg at least. In the further paragraphs, it will be discussed element by element.

Table 1. Results of PIXE (mg/kg) for the 1993 lichen survey

	Si	P	S	Cl	K	Ca	Ti	V	Cr	Mn	Fe
No. of samples	299	300	300	299	300	300	300	259	266	300	300
Maximum	3.24×10^4	2.92×10^3	3.96×10^3	2.65×10^1	8.05×10^3	3.13×10^4	7.77×10^2	1.30×10^2	3.40×10^1	1.43×10^2	6.39×10^3
Minimum	1.21×10^3	3.00×10^2	9.75×10^2	4.72×10^22	2.65×10^3	7.42×10^2	4.22×10^1	1.83×10^0	2.32×10^{-1}	1.56×10^1	3.57×10^2
Mean	8.99×10^3	1.34×10^3	2.03×10^3	1.08×10^3	4.65×10^3	6.18×10^3	2.42×10^2	1.45×10^1	2.17×10^0	3.96×10^1	1.94×10^3
Median	7.84×10^3	1.31×10^3	1.96×10^3	1.02×10^3	4.52×10^3	5.51×10^3	2.18×10^2	1.09×10^1	1.62×10^0	3.61×10^1	1.76×10^3
	Ni	**Cu**	**Zn**	**Ga**	**As**	**Br**	**Rb**	**Sr**	**Zr**	**Ba**	**Pb**
No. of samples	299	300	299	250	251	300	300	300	253	170	299
Maximum	3.31×10^1	2.65×10^2	5.08×10^2	3.94×10^0	1.87×10^1	5.73×10^1	7.64×10^1	1.03×10^2	8.75×10^1	7.99×10^1	1.42×10^2
Minimum	5.23×10^{-1}	3.30×10^0	2.21×10^1	3.52×10^{-1}	6.97×10^{-1}	8.71×10^0	4.71×10^0	2.84×10^0	2.52×10^0	2.01×10^0	2.03×10^0
Mean	3.92×10^0	2.88×10^1	6.91×10^1	1.13×10^0	3.82×10^0	2.29×10^1	1.75×10^1	2.29×10^1	1.32×10^1	2.29×10^1	1.84×10^1
Median	3.00×10^0	1.49×10^1	5.57×10^1	1.01×10^0	2.86×10^0	2.15×10^1	1.52×10^1	2.03×10^1	1.11×10^1	1.96×10^1	1.41×10^1

Table 2. Results of INAA (mg/kg) for the 1993 lichen survey

	Na	Mg	Al	Cl	K	Ca	Sc	Ti	V	Cr
No. of samples	301	295	297	297	301	255	297	297	297	301
Maximum	2.70×10^{3}	4.71×10^{3}	2.88×10^{4}	3.81×10^{3}	1.09×10^{4}	2.63×10^{4}	2.83×10^{0}	1.04×10^{3}	1.45×10^{2}	3.77×10^{1}
Minimum	1.31×10^{2}	1.90×10^{2}	2.17×10^{0}	2.48×10^{2}	2.10×10^{3}	1.15×10^{3}	1.35×10^{-1}	5.44×10^{1}	2.26×10^{0}	1.44×10^{0}
Mean	5.85×10^{2}	1.91×10^{3}	5.40×10^{3}	1.37×10^{3}	5.24×10^{3}	6.49×10^{3}	6.53×10^{-1}	3.20×10^{2}	1.57×10^{1}	5.45×10^{0}
Median	4.90×10^{2}	1.76×10^{3}	4.63×10^{3}	1.27×10^{3}	5.01×10^{3}	5.62×10^{-1}	2.48×10^{2}	1.14×10^{1}	4.69×10^{0}	

	Mn	Fe	Co	Cu	Zn	Ga	As	Se	Br	Rb
No. of samples	296	297	297	117	301	92	301	299	300	300
Maximum	1.47×10^{2}	7.57×10^{3}	2.90×10^{0}	3.45×10^{2}	449×10^{2}	7.89×10^{0}	3.10×10^{1}	1.17×10^{0}	9.31×10^{0}	+ 01
Minimum	1.87×10^{1}	3.90×10^{2}	1.36×10^{-1}	2.54×10^{0}	1.60×10^{1}	2.80×10^{-1}	4.35×10^{-1}	1.32×10^{-1}	7.08×10^{0}	3.97×10^{0}
Mean	5.02×10^{1}	2.10×10^{3}	7.75×10^{-1}	6.32×10^{1}	7.12×10^{1}	2.018×10^{0}	2.13×10^{0}	4.02×10^{-1}	2.25×10^{1}	1.59×10^{1}
Median	4.63×10^{1}	1.85×10^{3}	6.68×10^{-1}	4.72×10^{1}	5.61×10^{1}	1.78×10^{0}	1.34×10^{0}	3.73×10^{-1}	1.98×10^{1}	1.36×10^{1}

	Sr	Mo	Ag	Sb	I	Cs	Ba	La	Ce	Nd
No. of samples	219	41	11	291	297	299	250	301	301	90
Maximum	1.09×10^{2}	2.77×10^{0}	4.80×10^{-1}	6.73×10^{0}	1.16×10^{2}	3.95×10^{0}	1.60×10^{2}	1.34×10^{1}	2.62×10^{1}	1.34×10^{1}
Minimum	7.59×10^{0}	3.67×10^{-1}	5.17×10^{-2}	7.49×10^{-2}	1.05×10^{0}	1.03×10^{-1}	4.01×10^{0}	4.75×10^{-1}11	1.06×10^{0}	1.42×10^{0}
Mean	2.35×10^{1}	9.41×10^{-1}	1.89×10^{-1}	3.42×10^{-1}	7.57×10^{0}	6.40×10^{-1}	3.35×10^{1}	3.00×10^{0}	6.56×10^{0}	4.28×10^{0}
Median	2.06×10^{1}	8.28×10^{-1}	1.68×10^{-1}	2.20×10^{-1}	6.15×10^{0}	5.12×10^{-1}	2.96×10^{1}	2.49×10^{0}	5.70×10^{0}	3.72×10^{0}

	Sm	Eu	Tb	Lu	Hf	Ta	W	Hg	Th	U
No. of samples	301	289	297	292	301	301	126	296	301	200
Maximum	1.85×10^{0}	7.47×10^{-1}	2.44×10^{-1}	1.14×10^{-1}	1.83×10^{0}	3.93×10^{-1}	1.28×10^{1}	1.75×10^{0}	6.60×10^{0}	1.81×10^{0}
Minimum	8.89×10^{-3}	3.25×10^{-2}	7.26×10^{-3}	4.44×10^{-3}	5.53×10^{-2}	9.92×10^{-3}	8.97×10^{-2}	9.17×10^{-2}	8.66×10^{-2}	7.87×10^{-2}
Mean	4.47×10^{-1}	1.84×10^{-1}	6.09×10^{-2}	2.61×10^{-2}	4.13×10^{-1}	7.51×10^{-2}	5.03×10^{-1}	3.02×10^{-1}	9.26×10^{-1}	2.71×10^{-1}
Median	3.90×10^{-1}	1.62×10^{-1}	5.12×10^{-2}	2.22×10^{-2}	3.45×10^{-1}	5.99×10^{-2}	2.62×10^{-1}	2.78×10^{-1}	6.94×10^{-1}	2.07×10^{-1}

Table 3. Comparison of INAA and PIXE results (%) for the 1993 lichen survey

	Cl	K	Ca	Ti	V	Cr	Mn	Fe
No. of samples	296.0	300.0	255.0	296.0	259.0	266.0	296.0	297.0
Maximum	143.8	105.7	177.1	164.6	183.7	188.4	146.5	144.4
Minimum	−125.1	−79.8	−131.7	−132.3	−168.3	−135.1	−97.0	−142.5
Mean	20.7	9.8	0.0	24.6	2.2	87.8	23.3	5.1
Median	21.4	8.9	4.9	27.5	2.4	94.7	25.0	6.6
	Cu	Zn	Ga	As	Br	Rb	Sr	Ba
No. of samples	117.0	299.0	86.0	251.0	300.0	300.0	219.0	170.0
Maximum	191.2	154.1	159.2	179.8	126.0	141.2	170.3	178.1
Minimum	−145.3	−136.4	−155.1	−177.8	−120.4	−136.3	−143.5	−147.1
Mean	54.6	1.9	33.4	−60.8	−6.0	−11.5	0.5	27.5
Median	58.8	−2.1	36.1	−78.0	−10.0	−13.7	−2.9	25.4

The results obtained by INAA and PIXE for the different elements were mapped as concentration data patterns for the whole country by making use of an extinction rule of $1/r^3$ (r is the distance between plot point and sampling point) without any cut-off distance artificially introduced. It means that for every grid cell of the map a plot value was calculated from the concentrations found for all sampling points as the average of these concentrations weighted with the 3rd power of the distance between plot point and sampling point (Sloof, 1993). The computer code used was Surfer®. In Figs. 2–13 it is shown the maps so-obtained for all the elements determined. The discussion is made on basis of the pollution sources associated to elements as defined by Nriagu (Nriagu, 1989).

5.3.1. Coal-fired power stations

The elements associated to the coal-fired power stations are arsenic, sulphur, and selenium. Their pattern maps are shown in Fig. 2.

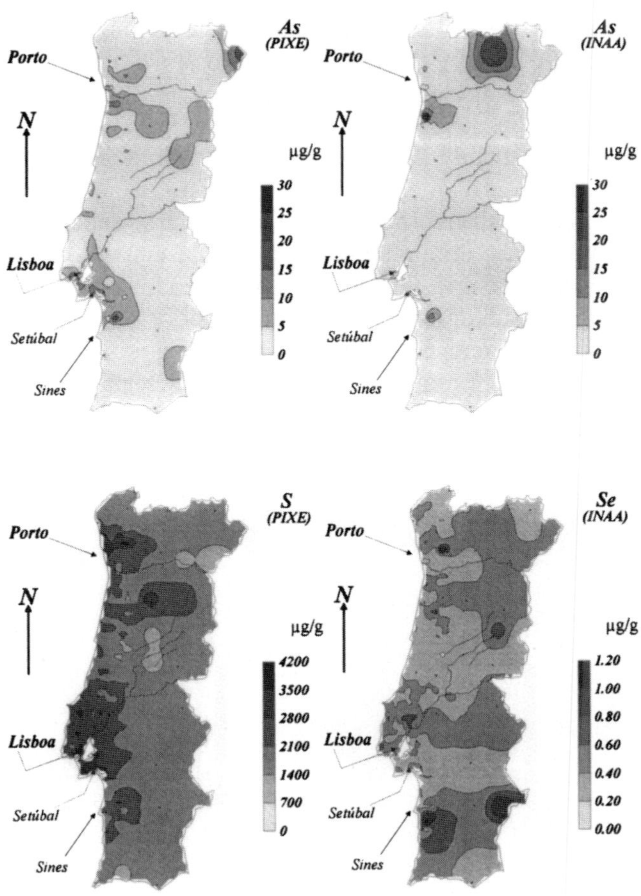

Figure 2. 1993 Survey using *P. sulcata*.

The arsenic data obtained by PIXE and INAA present quite different patterns. Whenever a large concentration of lead is present in the samples arsenic spectra peaks obtained by PIXE can be interfered, and also if arsenic is present in large quantities lead data can be also interfered. Comparing Fig. 2 with Fig. 5 we observe that in the south of the country around Lisbon the Pb-As pattern is quite similar. In the other hand Fig. 5 presents large concentrations of Pb in the north of the country which is there quite similar to the As pattern given by INAA. Therefore we are taken to conclude that the As-pattern given by PIXE should not be selected for this kind of samples. The difference between the two techniques shown in Table 3 is now explained. Therefore INAA data should be selected.

The correlation of As, S, and Se shown in Fig. 2 is hardly observed. In the north of the country, As-pattern resembles the pattern of the elements related to soil contribution and shown in Fig. 8, therefore there arsenic is predominantly originated on the ultrabasic complexes of that area, as already referred by Prudêncio et al. (1997). There is some similarity for As, S, and Se at south of Lisbon and near Oporto which are, respectively, in the neighbourhood of the Sines coal-fired power station and the Tapada do Outeiro coal-oil fired power station. Reis et al. (1996) and Freitas et al. (1997) did not identify clearly these elements to coal combustion in their factor analysis. Rahn and Huang (1999) define a coal source with a high peak of As and smaller peaks of S and Se, besides other elements.

5.3.2. Fuel-fired power stations

The elements associated to the fuel-fired power stations are vanadium, nickel, and rare earth elements (lanthanum, cerium, neodymium, samarium, europium, and terbium). Their pattern maps are shown in Fig. 3. Vanadium patterns obtained by both techniques are very similar which agrees with the similarity of values obtained for V and shown in Table 3 (mean difference between both techniques: 2.2%). Therefore INAA and PIXE are good techniques for V determination in lichens.

Comparing the figures we observe that the largest concentrations of vanadium are

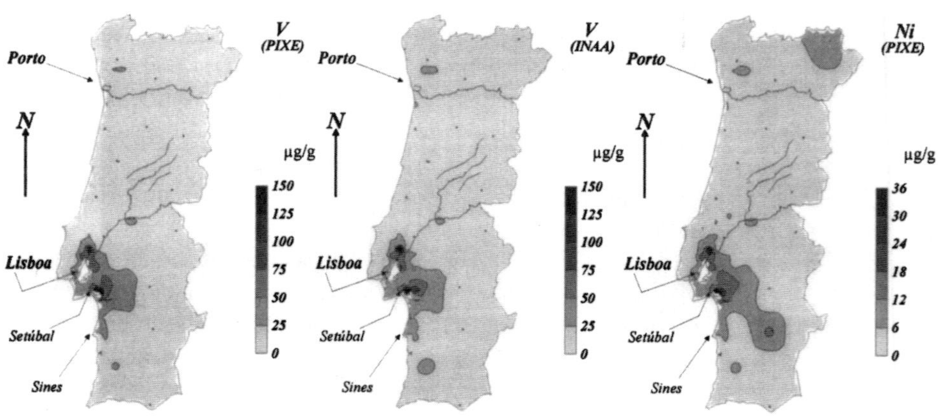

Figure 3. 1993 Survey using *P. sulcata*.

found in the same area of nickel, both indicating the oil-fired power stations of Carregado and Setúbal around Lisbon and the coal-oil power station of Tapada do Outeiro as the main sources (confirmed by factor analysis made by Reis et al. (1996), Freitas et al. (1997) and Rahn and Huang (1999)). The spot at north-east of Lisbon is not identified. Nickel has still a maximum in the north-east of the country in the same place as the elements correlated to soil contribution, which is referred by Prudêncio et al., 1997 as soil based in ultrabasic complexes.

When we look at the rare earth elements shown in Fig. 4, we conclude that there is no correlation with Ni and V. They are more similar to the elements discussed further in the soil contribution part, so we delay the discussion till that moment.

5.3.3. Mobile sources

The elements associated to mobile sources are bromine and lead. Their pattern maps are shown in Fig. 5. INAA and PIXE patterns of bromine are similar showing both a predominance of bromine at the Atlantic coast. The results obtained by the two techniques are similar as shown also in Table 3 (mean difference between the two techniques: −6%). Both maps show no correlation with Pb pattern and Br seems to have a marine origin. As referred before lead determined by PIXE may have influence on arsenic data and it was concluded that in north of Portugal lead shows a maximum (large spot in the middle of interior north) because there is a maximum of arsenic there. So we consider that spot as arsenic not lead. The largest concentrations of lead coincide with the regions of larger traffic in Portugal so we consider the mobile sources as the main origin of Pb. The spot at south-east should be considered as soil contribution since it also appears for the soil correlated elements.

5.3.4. Incinerators

The elements associated to incinerators are silver, zinc, antimony, cadmium, tin, and lead. Of these elements Zn, Sb, and Pb were determined in this work. Their pattern maps are shown in Fig. 6 (Fig. 5 for Pb). Zn obtained by both techniques gives similar information for the same large areas (near Oporto and near Lisbon), although the patterns are not so similar. In Table 3 it is presented a mean difference between the two techniques of 1.9%. The patterns of Zn, Sb, and Pb show larger concentrations of these elements in the same large areas: axis Lisbon-Setúbal-Sines and in the vicinity of Oporto, therefore it might be consequence of incineration. Officially there are no incinerators in our country, the first one started operation in February, 1999. We think that these elements appear associated due to incineration in the open air at small scale. Rahn and Huang (1999) applied the positive matrix factorisation to our data and found one factor which he identified as industry where antimony and lead are the largest components, and zinc is also present. Freitas et al. (1997) included it in one factor associated to general oil powered industries.

5.3.5. Cement production

The elements associated to cement production are calcium and magnesium. Their

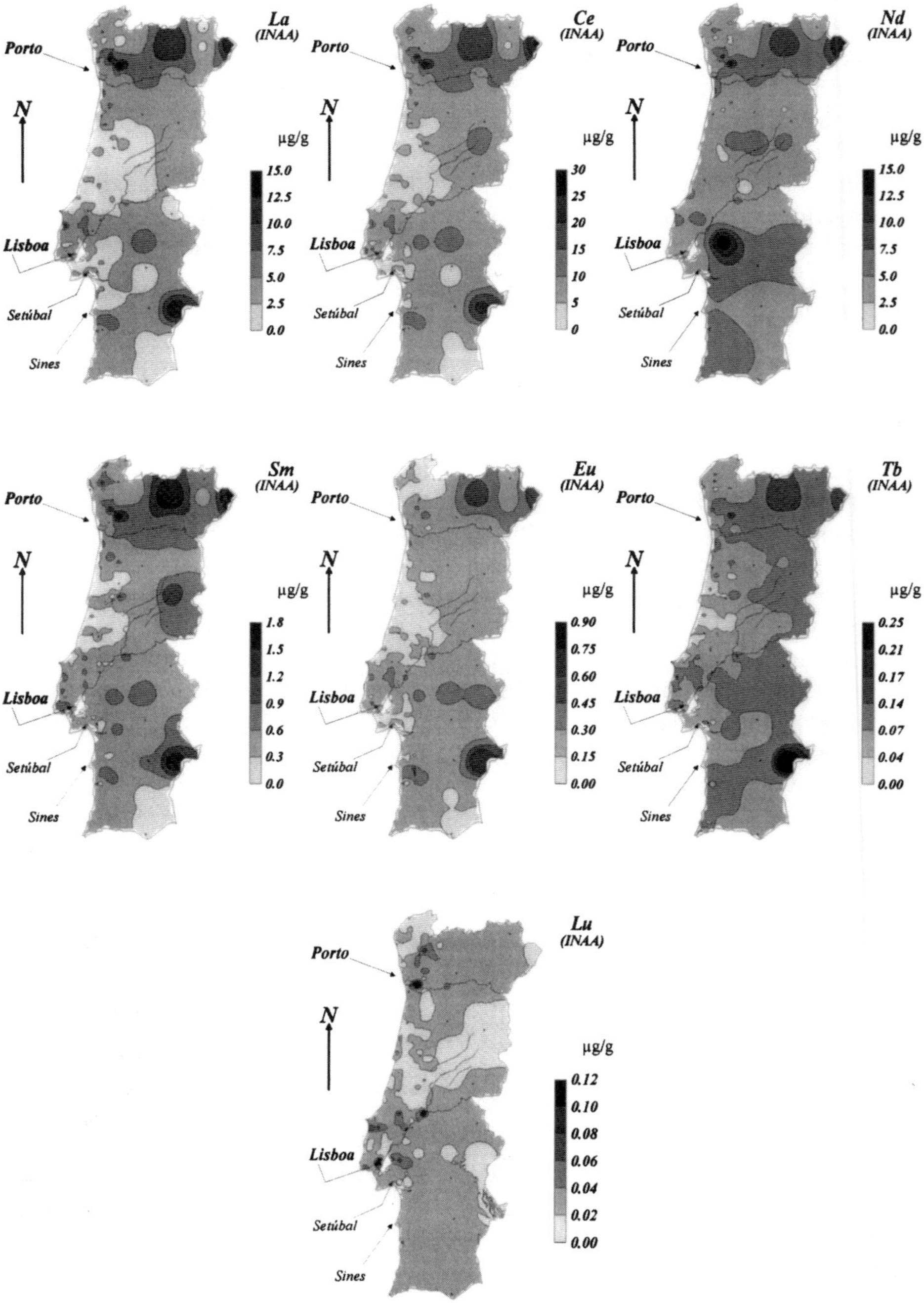

Figure 4. 1993 Survey using *P. sulcata*.

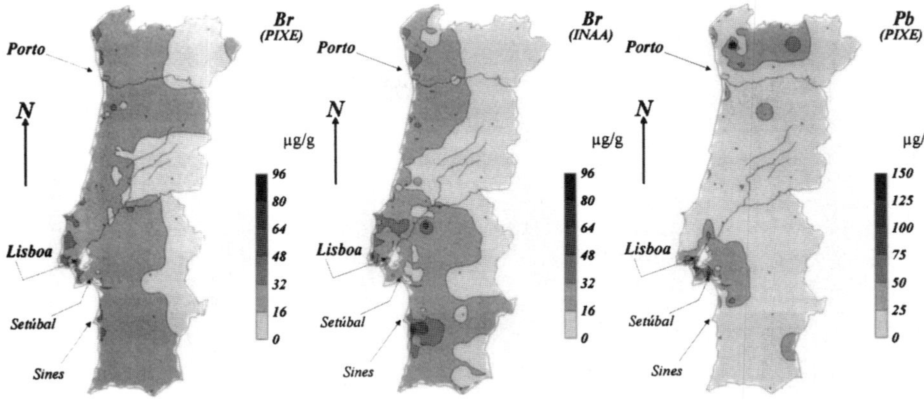

Figure 5. 1993 Survey using *P. sulcata*.

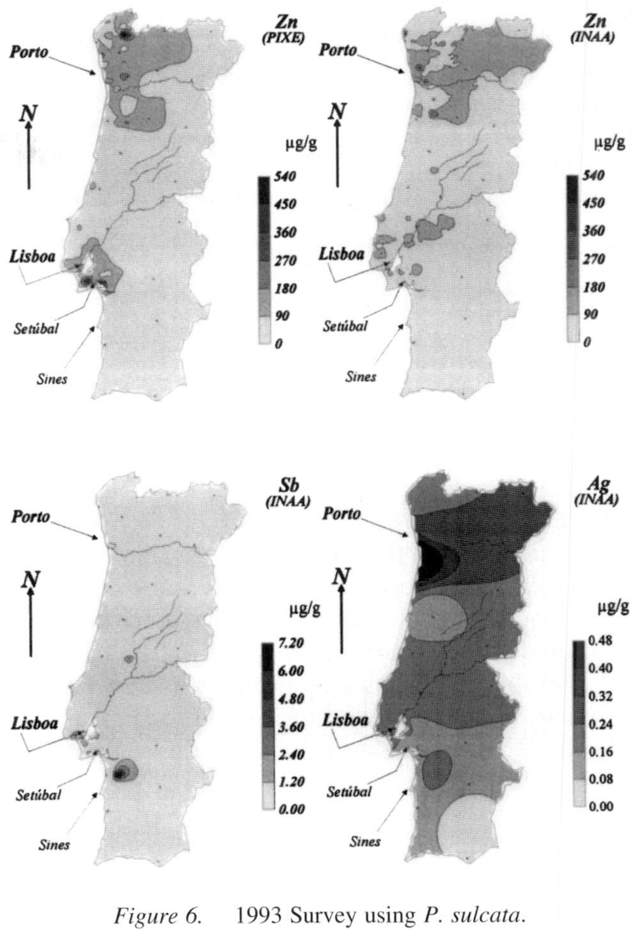

Figure 6. 1993 Survey using *P. sulcata*.

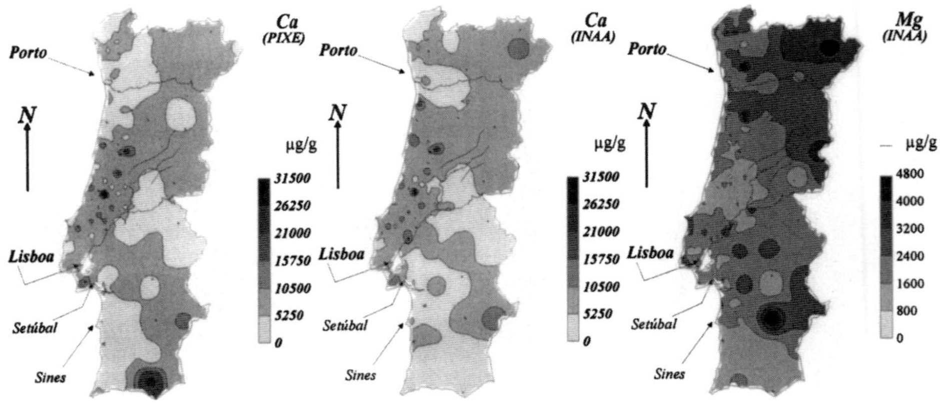

Figure 7. 1993 Survey using *P. sulcata.*

pattern maps are shown in Fig. 7. Calcium patterns determined by both techniques are similar with a few spots well pointed out, which in Freitas et al. (1997) factor analysis identified as cement sources (also Rahn (1999) identified this source after a more detailed factorisation). Ca values of INAA do not show the large spot in the south of the country as PIXE, because INAA Ca values were not measured. This is the reason why INAA values for Ca are only 255. In terms of the values itself the mean difference is 0% which is quite good (see Table 3). Magnesium is not associated to calcium and the pattern is more of one soil component (see further).

5.3.6. Soil contribution

The elements associated to soil contribution are, according to Nriagu (1989), manganese, aluminium, scandium, silicon, iron, and titanium. Their pattern maps are shown in Fig. 8. Manganese and titanium values determined by INAA are larger than the PIXE values. The mean difference found for both elements is about 20%. However, for each element the patterns are similar and the largest concentrations are found in the same areas. Iron data obtained by both techniques agree each other and also the patterns. The mean difference for Fe is 5% (see Table 3). For all the elements, the maps show high concentrations in the north-east of the country, due to the ultrabasic composition of the soils as referred previously for arsenic, nickel, the rare earth elements, lead, and magnesium. No explanation was found for the fact that only a few of the elements in discussion have heavy spots in south-east of the country. The patterns are similar for these elements demonstrating its correlation with soil contribution. The rare earth elements shown in Fig. 4 have similar patterns to the soil contribution elements. This is not surprising since the composition of soil in rare earths is soil tracer.

5.3.7. Wood burning

The elements associated to wood burning are potassium and carbon. Unfortunately

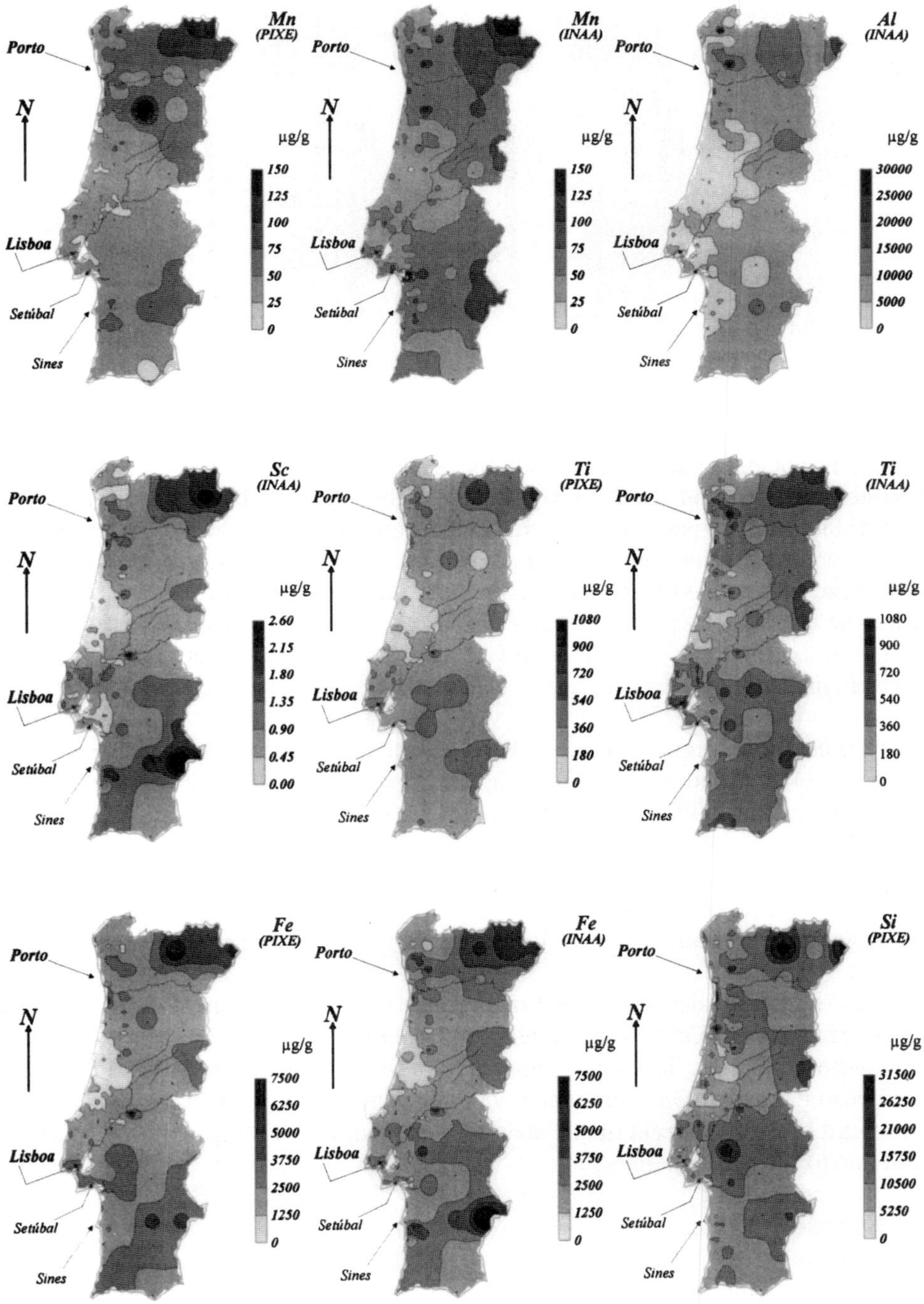

Figure 8. 1993 Survey using *P. sulcata*.

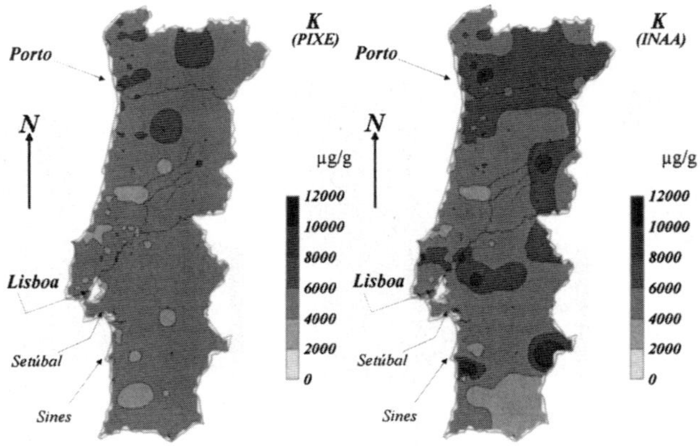

Figure 9. 1993 Survey using *P. sulcata.*

carbon is not determined by any of the techniques. Fig. 9 shows the potassium obtained by PIXE and INAA. The mean difference between both techniques is 10% (see Table 3). The patterns are similar although the PIXE values are lower. The regions with larger concentrations of potassium coincide with the areas of the country with more percentage of fires during summer, but we would need carbon to demonstrate the correlation.

5.3.8. Refineries

The rare earth elements are shown in Fig. 4. It was said previously that these elements are more soil contribution than anything else. Looking more in detail we observe that neodymium has a different pattern (except in the northern region) and samarium shows large concentrations in the centre east. The spot south of Sines might be related to the Sines refinery, as well as the small spot near Oporto.

5.3.9. Sulphide smelters

The elements associated to sulphide smelters are indium, cadmium, arsenic, selenium, and sulphur. In and Cd were not determined because of their high limits of detection with these techniques. The other elements were shown in Fig. 2 in association to coal-fired power stations. One spot which was not explained there was the one which is north-east of Sines, present in As- and S-patterns but not in Se-pattern. Might be that spot due to sulphide smelters?

5.3.10. Other elements

A few more elements were determined which are presented in Figs. 10–12. We will start to comment the similarity of the elements determined by both INAA and PIXE.

For Cl, INAA values are larger than the PIXE values and the difference is 20% on the average (see Table 3). In general the patterns indicate both more Cl near the

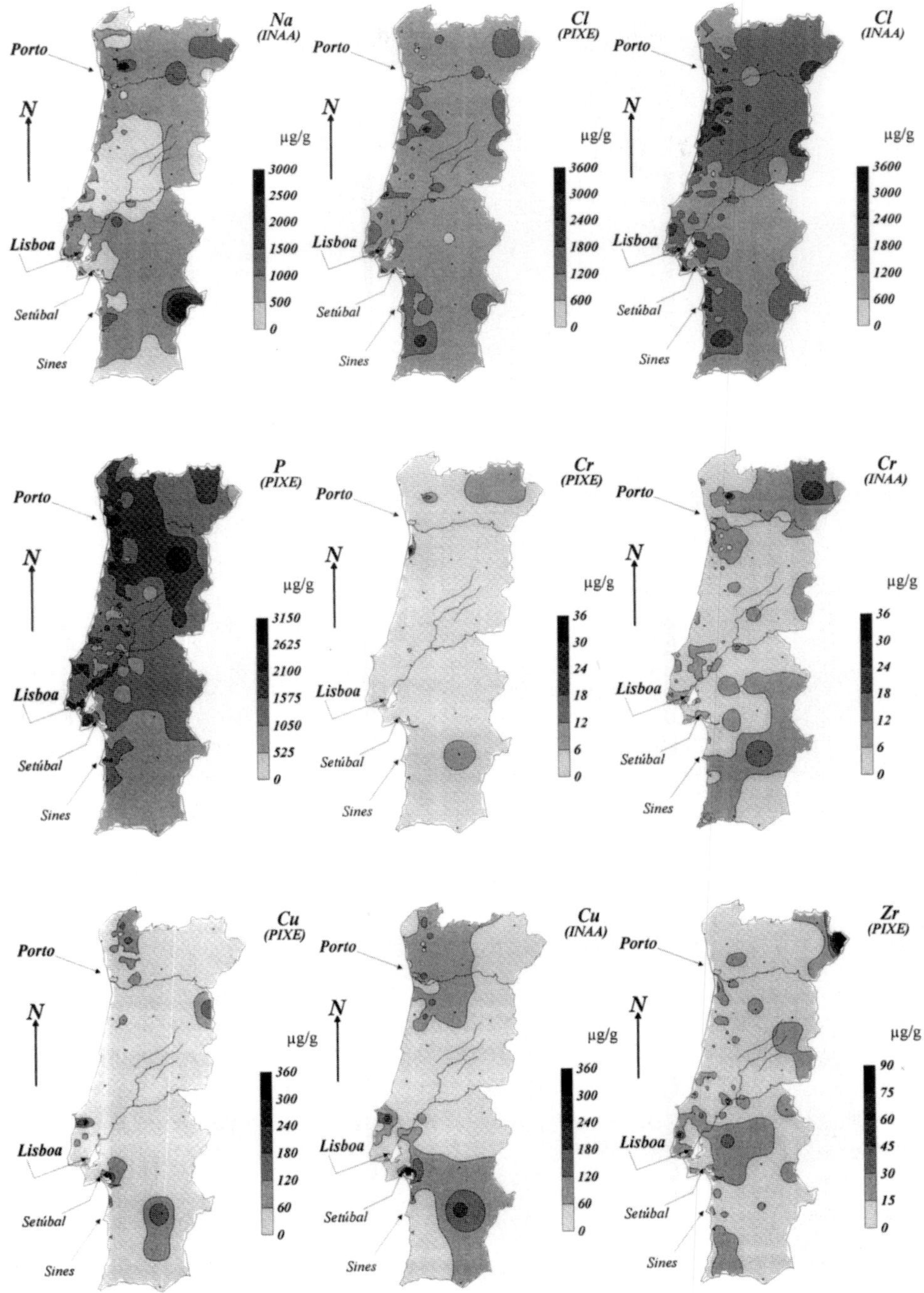

Figure 10. 1993 Survey using *P. sulcata*.

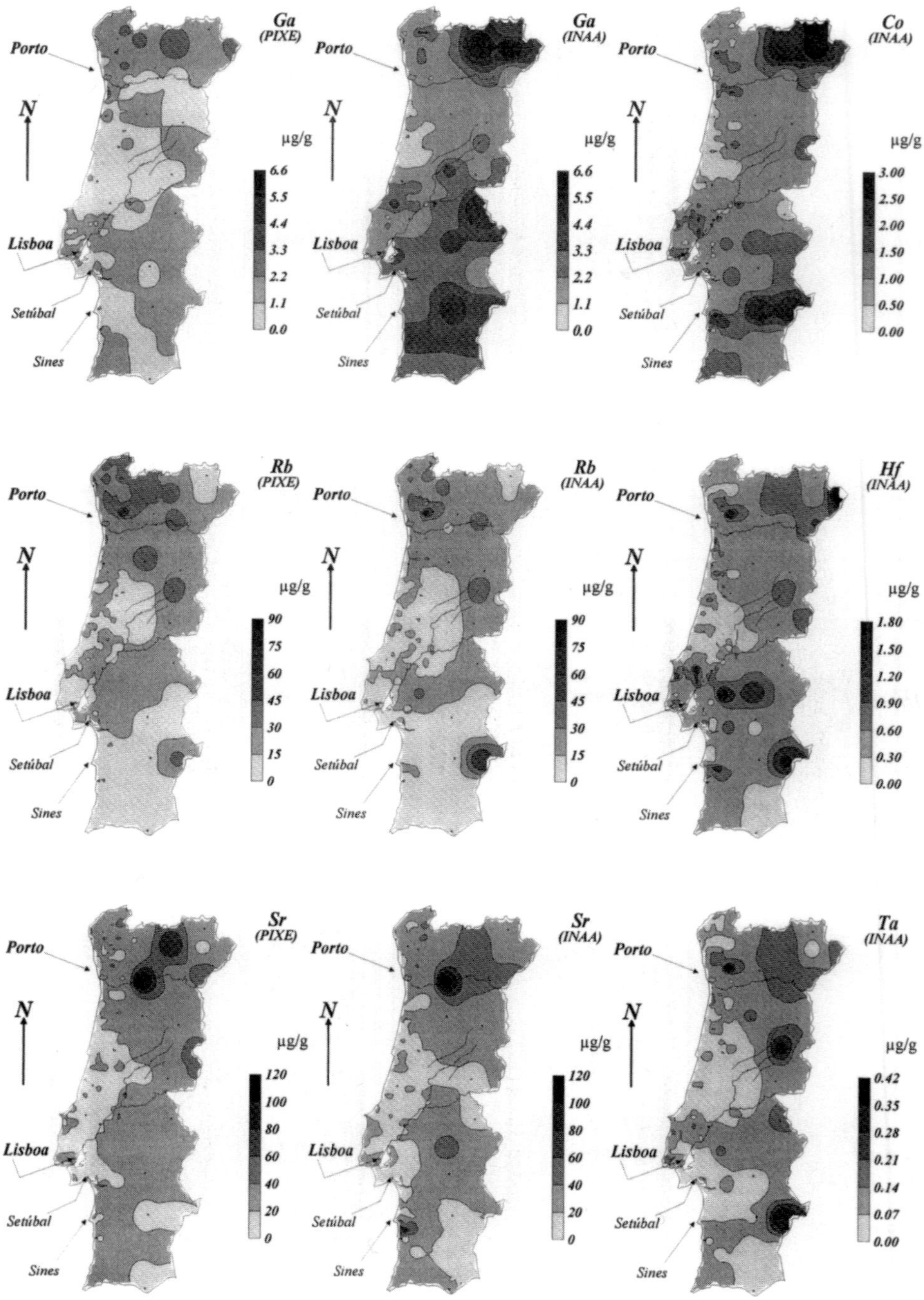

Figure 11. 1993 Survey using *P. sulcata*.

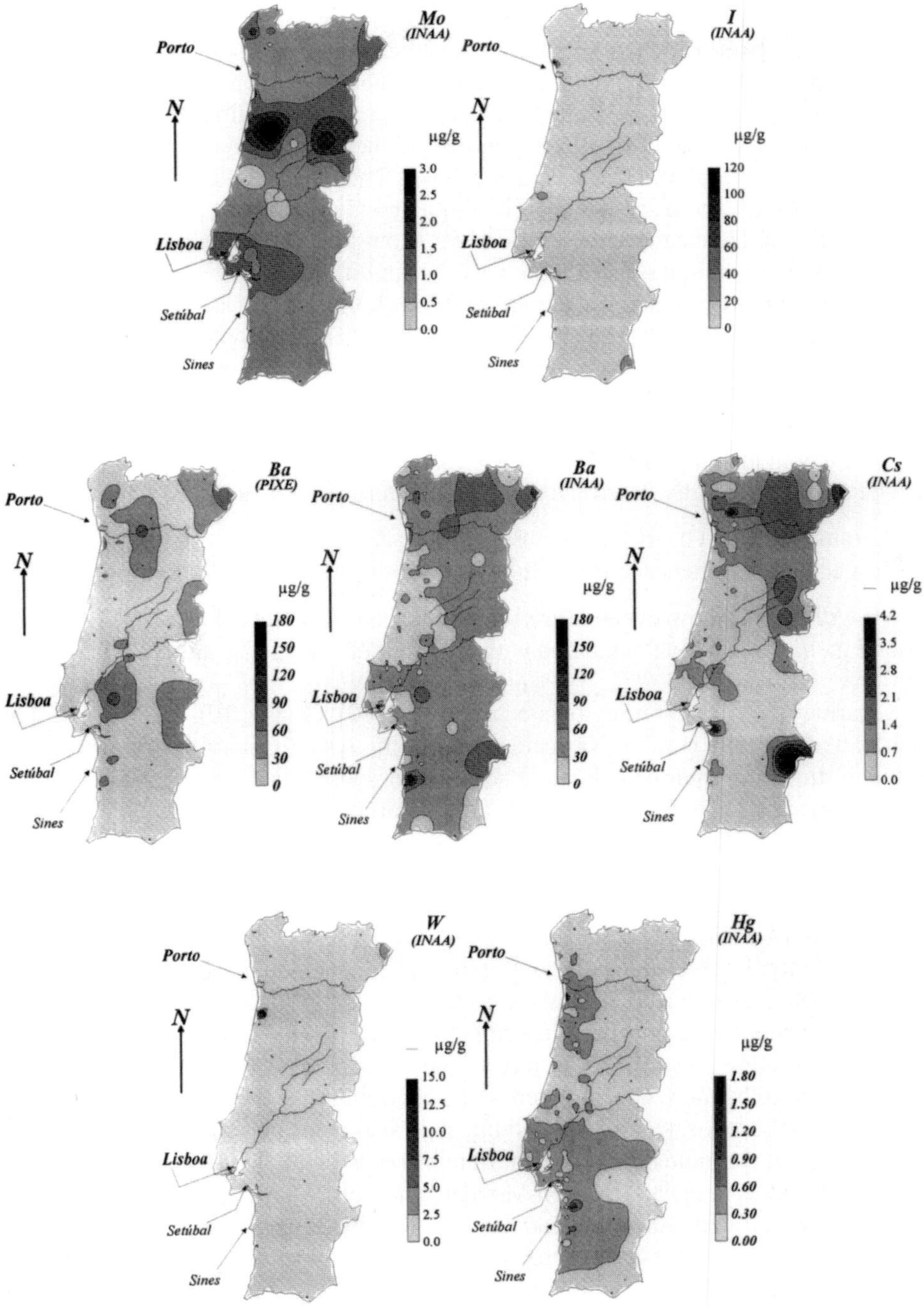

Figure 12. 1993 Survey using *P. sulcata*.

Atlantic coast as expected given the proximity of the ocean. For Cr, the mean difference (see Table 3) is quite large (88%). The INAA values are larger than the PIXE values and we think it might be due to interference of the sum peak of Sc-47 peak originated in Ca-47 ($2 \times 160 = 320$ keV). We conclude that Cr values obtained by PIXE are probably more accurate. INAA determined copper only in 117 samples in a total of 301 due to the worse detection limit. Also the mean difference is quite large (55%, see Table 3). We think PIXE is more reliable than INAA in Cu determination. For Ga, very few results were obtained by INAA. Therefore the PIXE values should be preferred. For Rb, both techniques give similar values (mean difference: 12%, see Table 3) and similar patterns. Therefore we should adopt both Rb sets. For Sr, the values of both techniques agree quite well (mean difference: 0.5%, see Table 3) and also the patterns are similar. The drawback for INAA is that Sr values of INAA are just 219 instead of 301. For Ba, the mean difference is 28% (see Table 3) and the patterns are not so similar. Ba values are interfered with U fission and correction is sometimes quite important. That may be the reason why INAA results are larger than the PIXE ones. In the other hand the PIXE results reach only 170 in a total of 300 samples. Therefore both techniques have drawbacks and it is suggested to accept both. In short, for the elements just discussed the following patterns will be adopted in Figs. 10–12.

- Both techniques: Cl, Rb, Sr, and Ba;
- Cr, Cu, and Ga: the selection will be done after patterns analysis.

We will discuss now eventual associations of the elements shown in Figs. 10–13 among themselves in one hand, and with the other elements discussed in the previous sections.

A marine source might be associated to Na (Fig. 10), Cl (Fig. 10), and I (Fig. 12) but their patterns are not similar. Na accumulates at the Atlantic coast but Cl also spreads to the north-centre interior indicating some other source than the ocean spray. Phosphorus concentrates more over the agricultural part of Portugal (north and centre) and copper in the north-west and south-west of the country might originate in copper sulphate used in the vineyards. Co (Fig. 11), Cs (Fig. 12), Hf (Fig. 11), and Ta (Fig. 11) have patterns similar to the soil elements. The values of Cr (Fig. 10), Ba (Fig. 12), and Ga (Fig. 11) obtained by INAA also have soil patterns but not the PIXE ones. In general these elements are indeed soil components therefore the INAA patterns look more plausible than the PIXE ones. Zr (Fig. 10) has a very specific pattern with spot in the north-east much probably originated in the soil. Mo (Fig. 12) shows two strong spots near the centre of the country of unknown source. Wolfram was expected to be more concentrated in the centre interior where some important W (Fig. 12) mines are located, but instead the main spot is located south of Oporto where important chemical industry is located. In this area we find also Hg (Fig. 12) contamination, as well as in the axis Lisbon-Setúbal-Sines axis where the chemical industries are also concentrated. As to Th and U (Fig. 13) the patterns are very similar with concentrated spots in the centre interior where the U mines of Urgeiriça are located. The spots in the north-east should be soil originated and the spots in the north-west are of unknown source.

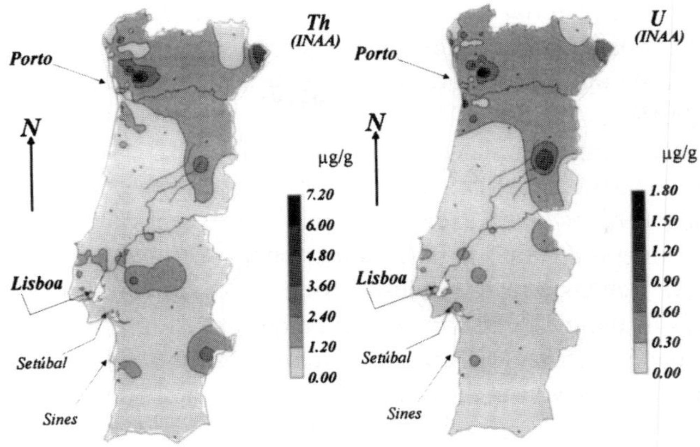

Figure 13. 1993 Survey using *P. sulcata*.

5.4. Comparison with other surveys

In Ruhling (1992) Portuguese mosses were collected from 1990 to 1992 and analysed for the elements Cd, Cr, Cu, Fe, Pb, Ni, and Zn. Comparison with Cr, Pb, and Ni was made in Freitas et al. (1999). It was found that in Portugal, surveys based on *P. sulcata* and the mosses *Hypnum cupressiforme* and *Scleropodium touretii* give similar information for these elements. We will compare now the elements Cu, Fe, and Zn. For Cu, similar large concentrations are found in the north-west and south-centre of the country. The large concentrations found in the interior centre are not visible in our survey probably because of the different grids adopted in the two surveys, finer with mosses.

Table 4. Results obtained in The Netherlands (mg/kg) in 1982/83 and 1986/87 reprinted from Sloof (1993)

Element	No. of samples analysed	Minimum	Maximum	Mean
Arsenic	110/329	1.7/0.5	17/17	5.7/5.7
Bromine	110/337	17/15	120/170	44/56
Chromium	110/337	7.6/4.0	58/270	22/26
Caesium	108/322	0.3/0.2	21/5.3	0.9/0.8
Iron	110/337	1100/720	15000/30000	4100/5800
Mercury	99/316	0.1/0.1	3.7/36	0.4/0.5
Lanthanum	99/316	0.8/0.8	16/64	4.1/6.2
Nickel	69/201	3.0/1.9	53/51	14/16
Lead	/232	/3.1	/367	/147
Selenium	98/332	0.4/0.4	4.0/7.5	1.4/1.8
Antimony	110/334	0.6/0.3	10/12	3.0/3.3
Vanadium	102/99	8.8/5.7	130/99	29/32
Wolfram	93/248	0.1/0.1	2.0/10	0.5/0.9
Zinc	110/327	80/61	630/1100	90/210

For Fe, the results are similar for both surveys: larger concentrations in the interior north and south. For Zn, the largest spots are in both surveys around Lisbon and around Oporto, although the Zn areas in the north of the country are larger for the lichen survey.

In The Netherlands similar survey was made in 1982/83 and 1986/87 using also *P. sulcata* (Sloof, 1993). In Freitas et al., (1999) comparison was made for arsenic, chromium, mercury, nickel, lead, antimony, selenium, and vanadium. It was concluded that when compared to The Netherlands, Portugal has regions with much lower contents in lichens, but the industrialised areas are as much polluted. Table 4 shows the levels of the elements which can be compared in this paper.

We observe that Br, Cr, Fe, Hg, and Zn contents in Portuguese lichens are quite lower than the Dutch lichens; V and W have similar values; Ni and Sb have lower minima; Cs, La, Pb, and Se lower maxima; and As has larger maxima. Comparing the means we observe that except for W all the elements in comparison are much lower in Portugal, for that contributing the not industrialised interior of the country. Therefore we may say that in Portugal pollutants can be as high as in The Netherlands in some regions of the country (near the Atlantic coast) but on the average the contents are still low.

6. Conclusions

In this work PIXE and INAA nuclear techniques were applied to a 1993 lichen survey held in Portugal. The two techniques determined 46 elements with 16 common elements. The common elements were compared and conclusions were attempted to select the best values, based on the concentration patterns obtained. It was concluded that:

- Ca, Cu, Ga, Sr, Mo, Ag, La, Nd, and W were determined by INAA but their detection limits are not good enough to give results in all the samples; Ca, Cu, and Ga are determined by PIXE in all the samples.
- On average: 1) K, Ca, V, Fe, Zn, Br, and Sr agree each other within $\pm 10\%$; 2) Cl, Ti, Mn, and Ba between $\pm 10\%$ and $\pm 30\%$; 3) Cr, Cu, Ga, and As between $\pm 30\%$ and $\pm 90\%$; the heterogeneity of the samples may be the main cause of the differences.
- In PIXE results whenever As and Pb are present in the samples, they interfere each other, and the values should be observed carefully.
- Some pollution sources could be regionally identified: 1) coal-fired power stations (As, Se, and S) at south of Lisbon and near Oporto; 2) fuel-fired power stations (V and Ni) in the axis Lisbon-Setúbal and near Oporto; 3) mobile sources (Pb) in regions of larger traffic (north-west and Lisbon and south Lisbon); 4) cement production (Ca); 5) soil contribution: Mn, Al, Sc, Si, Fe, and Ti, also Mg, REE, Co, Cs, Hf, and Ta (Cr, Ba, and Ga).
- It was observed two regions of large concentrations of metals in north-east of the country pointed out as originated in the ultrabasic complexes, main constituents of the soil of that region.
- The chemical complex of Estarreja (south of Oporto) is put in evidence by the Hg contents in lichens.

- Except for a few spots our results for Cr, Pb, Ni, Cu, Fe, and Zn agree quite well with the results obtained with a moss survey in a 1990/92 campaign made in Portugal using non-nuclear techniques.
- In general the pollutants determined in this work in the industrialised areas of the country near the Atlantic coast can be as highly concentrated as in The Netherlands (1982/83 and 1986/87 Dutch campaigns with the same lichen).There are regions of the country mainly located in the interior which have very low concentrations as compared with the Dutch values. Therefore on the average the elemental contents in the Portuguese lichens are still low as compared with the Dutch ones.

References

Alves, L.C., Reis, M.A., Freitas, M.C., 1998a. Air particulate matter characterisation of a rural area in Portugal. Nucl. Instrum. Methods Phys. Res. B136–138, 941–947.

Alves, L.C., Reis, M.A., Freitas, M.C., Gouveia, M.A., 1998b. Elemental analysis of particulate matter and source profile in Lisbon. X-Ray Spectrom. 27, 313–320.

Bode, P., De Goeij, J.J.M. 1998. Activation Analysis, In: R.A. Meyers (Ed.), Encyclopaedia of Environmental Analysis and Remediation, Wiley, New York, pp. 68–84, ISBN 0-471-11708-0.

Bode, P., Overwater, R.M.W., 1993. Trace element determinations in very large samples: a new challenge for neutron activation analysis. J. Radioanal. Nucl. Chem. 167, 169–176.

Bode, P., Wolterbeek, H.Th., 1990. Environmental research and instrumental neutron activation analysis: aspects of high accuracy and multi-element capability. J. Trace Microprobe Technol. 8, 121–138.

De Bruin, M., Wolterbeek, H.Th. 1984. Identification of sources of heavy metals in the Dutch atmosphere using air filter and lichen analysis. Proceedings of the Fifth International Conference on Nuclear Methods Environment Energy Resources, University of Missouri, Columbia, MO, pp. 266–276.

De Bruin, M., Van Wijk, P.M., Van Assema, R., De Roos, C., 1987. The use of multi-element concentration data sets obtained by INAA in the identification of sources of environmental pollutants. J. Radioanal. Nucl. Chem. 112, 199–213.

De Goeij, J.J.M., Bode, P., 1994. How to make nuclear methods the choice of a new generation of scientists. J. Radioanal. Nucl. Chem. 1179, 7–12.

De Goeij, J.J.M., Bode, P. 1995. Neutron Activation Analysis: Trends in Developments and Applications. Proceedings of International Conference on Neutrons and their Applications, Crete, Greece, 12–18 June 1994, SPIE-Vol. 2339, pp. 436–447.

De Goeij, J.J.M., Bode, P. 1997. Nuclear Analytical Techniques. Strong and weak points, obstacles and opportunities. In: Proceedings of the Symposium on Harmonization of Health related Environmental Measurements using Nuclear and Isotopic Techniques, Hyderabad, India, 4–7 November 1996, pp. 3–17, IAEA Proc. Series, Vienna, STI/PUB/1006, ISBN 92-0-103697-3.

Freitas, M.C., Nobre, A.S., 1997. Bioaccumulation of heavy metals using Parmelia sulcata e Parmelia caperata for air pollution studies. J. Radioanal. Nucl. Chem. 217 (1), 17–20.

Freitas, M.C., Reis, M.A., Alves, L.C., Wolterbeek, H.Th., Verdurg, T., Gouveia, M.A., 1996. Elemental accumulation in lichen transplants in the neighbourhood of thermal power stations. ANST/Nuclear Methods Environ. Res. 74, 117–118.

Freitas, M.C., Reis, M.A., Alves, L.C., Wolterbeek, H.Th., Verdurg, T., Gouveia, M.A., 1997. Monitoring of trace-element air pollution in Portugal: qualitative survey. J. Radioanal. Nucl. Chem. 217 (1), 21–30.

Freitas, M.C., Reis, M.A., Alves, L.C., Wolterbeek, H.Th., 1999a. Response of lichen Parmelia sulcata to environmental pollutants in Portugal. Environ. Pollut. 106, 229–235.

Freitas, M.C., Reis, M.A., Alves, L.C., Marques, A.P., Costa, C., 1999b. Environmental assessment in a industrial area of Portugal. Biol. Trace Elem. Res. 71/72, 471–479.

Gonsior, B., Bischof, W., Raith, B., Stratman, A., Wilde, H.R. 1982. Investigation of trace element distributions in biological structures using PIXE. In: Cesareo, R. (Ed.), X-ray Fluorescence (XRF and PIXE) in Medicine Acta Medica, ROMA, Italy, pp. 125–153, ISBN 88-7084-002-6.

Harrison, R.M., Smith, D.J., Pio, C.A., Castro, L.M., 1997. Comparative receptor modelling study of airborne particulate pollutants in Birmingham (UK). Coimbra (Portugal) and Lahore (Pakistan), Atmos. Environ. 31, 3309–3321.

IAEA 1997. Proceedings of the Symposium on the Harmonization of Health-related Environmental Measurements using Nuclear and Isotopic Techniques, Hyderabad, India, 4–7 November 1996. IAEA Proc. Series STI/PUB/1006, ISBN 92-0-103697-3, pp. 663.

IAEA/AL/095 1996. Report of the Second Research Co-ordination Meeting on Reference Materials for Micro-analytical Nuclear Techniques; Mexico, DF, Mexico, May–June 1996, pp. 143.

Knoll, G.F. 1989. Radiation Detection and Measurement. 2nd ed. Wiley, New York, ISBN 0-471-81504-7.

Kuik, P., Wolterbeek, H.Th., 1995. Factor analysis of atmospheric trace-element deposition data in The Netherlands obtained by moss monitoring. Water, Air, Soil Pollut. 83, 323–346.

Kuik, P., Sloof, J.E., Wolterbeek, H.Th., 1993a. Application of Monte-Carlo-assisted factor analysis to large sets of environmental pollution data. Atmos. Environ. 27A, 1975–1983.

Kuik, P., Blaauw, M., Sloof, J.E., Wolterbeek, H.Th., 1993b. The use of Monte Carlo methods in factor analysis. Atmos. Environ. 27A, 1967–1974.

Nriagu, J.O. 1989. Natural versus anthropogenic emissions of trace metals to the atmosphere. In: J.M. Pacyna, B. Offar (Eds.), Control and Fate of Atmospheric Trace Metals, NATO ASI Series, Kluwer Academic Publishers, Dordrecht.

Overwater, R.M.W., Bode, P., De Goeij, J.J.M., 1993. Gamma-ray spectroscopy of voluminous sources. Corrections for source geometry and self-attenuation. Nucl. Instrum. Methods A324, 209–218.

Pio, C.A., Feliciano, M.S., 1996. Dry deposition of sulphur dioxide and ozone over low vegetation in moderate southern European weather conditions. Measurements and modelling. Ann. Geophys. Part II (Suppl. II, Col. 14), C468.

Pio, C.A., Lopes, D., 1998. Chlorine loss from marine aerosol in a coastal atmosphere. J. Geophys. Res. 103, 25263–25269.

Prudêncio, M.A., Gouveia, M.A., Freitas, M.C., Chaves, L., Marques, A.P., 1997. Soil versus lichen analysis on elemental dispersion studies (north of Portugal). IAEA TECDOC in press.

Rahn, K.A., Huang, S. 1999. A graphical technique for distinguishing plant material in biomonitors from soil and atmospheric deposition. Sci. Total Environ. 232, 79–104.

Reis, M.A., Alves, L.C., Freitas, M.C., Wolterbeek, H.Th., Verburg, T., Gouveia, M.A., 1996. Main atmospheric heavy metal sources in Portugal by biomonitor analysis. Nucl. Instrum. Methods Phys. Res. B109/110, 493–497.

Reis, M.A., Alves, L.C., Freitas, M.C., Van Os, B., Wolterbeek, H.Th., 1999. Lichens (*Parmelia sulcata*) time response model to environmental availability. Sci. Total Environ. 232, 105–115.

Ruhling, A. 1992. Atmospheric heavy metal deposition in Europe - estimations based on moss analysis, Nord 1994:9, Copenhagen, Denmark.

Sérgio, C., Sim-Sim, M., Figueira, R. 1993. Quantificação da deposição de metais pesados em Portugal através da análise de briófitos, DGA, Lisbon.

Sloof, J.E. 1993. Environmental Lichenology Biomonitoring Trace Element Air Pollution. Ph.D. Thesis, Delft University of Technology, The Netherlands.

Soares, H.M.V.M., Vasconcelos, M.T.S.D., 1995. Application of potentiometric stripping analysis for speciation of copper complexes with adsorbable ligands on the mercury electrode. Anal. Chim. Acta 314, 241–249.

Tadic, T., Jaksic, M., Bogdanovic, I., Fazinic, S., Dujmic, D. 1997. Proton micro-PIXE control of standard reference materials for PIXE environmental applications. In: Proceedings of the Symposium on the Harmonization of Health-related Environmental Measurements using Nuclear and Isotopic Techniques, Hyderabad, India, 4–7 November 1996. IAEA Proc. Series STI/PUB/1006, ISBN 92-0-103697-3, pp. 251–264.

Tavares, H.M.F., Vasconcelos, M.T.S.D., Machado, A.A.S.C., Silva, P.A.P., 1993. Application of hydride generation atomic absorption spectrometry to the determination of lead collected on air filters and sphagnum moss. Analyst 118, 1433–1439.

Trahey, N.M. 1996. NIST Standard Reference Materials Catalog, 1995–1996, NIST Special Publication 260, NIST, Gaithersburg, MD, pp. 84.

Vasconcelos, M.T.S.D., Tavares, H.M.F., 1998. Atmospheric metal pollution (Ca, Cr, Cu, Fe, Mn, Ni, Pb and Zn) in Oporto city derived from results for low-volume aerosol samplers and for moss sphagnum auricularem bioindicator. Sci. Total Environ. 212, 11–20.

Wolterbeek, H.Th., Bode, P., 1995. Strategies in sampling and sample handling in the context of large-scaled plant biomonitoring surveys of trace element air pollution. Sci. Total Environ. 176, 33–43.

Wolterbeek, H.T.H., Bode, P., Verburg, T.G., 1996. Assessing the quality of biomonitoring via signal-to-noise ratio analysis. Sci. Total Environ. 180, 107–116.

Trace elements – Their distribution and effects in the environment
B. Markert and K. Friese, editors

Chapter 7

Chemical analysis of organoarsenic based chemical warfare agents in the environment

F.-A. Pitten, K. Thurow, A. Koch, A. Kramer

Abstract

This report summarizes important features of the environmental hazard, toxicological aspects and the problems of the chemical analysis of organoarsenic based warfare agents. Since most other warfare agents including mustard gas are degraded almost completely to less or even non-toxic products other warfare agents are not considered. The report consists of three different parts: In the beginning some general aspects of arsenic in the environment are presented to characterize the typical properties of these contaminants. After these data the specific problem "Löcknitz", an area with an unique contamination in Germany, is described. Finally detailed information about the speciation analysis of organo-arsenic warfare agents and their metabolites is provided.

Keywords: Arsenic; Chemical warfare agents; Environment

1. Introduction

Since the production, storage and military usage of any kind of chemical warfare agents (CWAs) is prohibited by international law threatening by these agents becomes more and more unlikely. Apart from terrorism presently the dominant problem with chemical weapons is the environmental hazard to the biosphere. In Germany for instance approximately 10 000 t of chemical warfare agents were dumped into the Baltic Sea, causing occasional accidents if fishermen collect rusty barrels with doubtful contents in their nets. At present little information is available if countries of the former USSR or Russia itself have problems with contaminated sites. However, in Germany a very specific site is the former Heeresmunitionsanstalt I and II at Löcknitz close to the German–Polish border. Prior to the rehabilitation of this area various investigations have been carried out to estimate the environmental hazard and the risks for people living close to this site. Finally, a definite risk assessment is required to protect people working at the site during the rehabilitation project itself.

2. Arsenic in the inanimate environment

Arsenic is an ubiquitous element showing metallic and non-metallic properties with concentrations ranging from 0.1 to 100 ppm or more depending on the types of rocks under consideration (Cullen and Reimer, 1989). The worldwide median soil concentration is 6.0 ppm, the concentration in uncontaminated soil rarely exceeds 40 ppm (Tanaka, 1988; Tamaki and Frankenberger, 1992). Seawater concentrations of arsenic range from 1.5 to 5.0 μg/l, in freshwater 1.0 – 10 μg/l may be found (Sanders, 1980). In oxygenated water arsenate predominates, whereas under reduced conditions arsenite is found more frequently. The WHO standard recommends a threshold limit value of 0.05 mg As mg/l in drinking water, the German standard 0.01 mg/l correspondingly. Atmospheric concentrations of arsenic range from 1.8×10^{-5} $\mu g/m^3$ (southern hemisphere) to 6.0×10^{-4} $\mu g/m^3$ (North Atlantic) over oceans and 1.0×10^{-3} $\mu g/m^3$ (southern hemisphere) to 2.8×10^{-3} $\mu g/m^3$ (northern hemisphere) over land masses (Chilvers and Peterson, 1987).

The main natural sources of arsenic are weathering of arsenic-containing rocks which liberates inorganic arsenic compounds such as arsenic trioxide, arsenite and arsenate, precipitation of the atmosphere, and volcanism (Tamaki and Frankenberger, 1992). The anthropogenic atmospheric input is dominated by combustion of fossil fuels, particularly coal, and smelting operations. Some organic arsenicals are still used as growth promoters for swine and poultry, but they are expected to be replaced by antibiotics (which may pose even a major environmental hazard).

Another important anthropogenic source in the past was the usage of arsenic derivatives in agriculture as herbicides (mono or disodium salts of arsenic acid), insecticides (calcium arsenate, lead arsenate), fungicides (sodium arsenite) or defoliants (dimethylarsinic acid or "cacodylic" acid) (Tamaki and Frankenberger, 1992). At present the usage of arsenic for wood preservation purposes is increasing, ammoniacal copper arsenate and chromated copper arsenate being the most common formulations (Hindmarsh and McCurdy, 1986).

3. Arsenic in living organisms

The reduced species (arsenite) is more toxic than the oxidized species (arsenate) since it may react with sulfhydryl groups of cysteine in structure and enzyme proteins. Therefore oxidation of arsenite in (micro)organisms is an important protective mechanism which is described for various species of the genus *Pseudomonas*, *Xanthomonas* and *Achromobacter*. However, some bacteria are able to reduce arsenate to the more toxic form arsenite. This process is described for some strains of *Pseudomonas fluorescens* and *Anabaena oscillaroides* (Cullen and Reimer, 1989).

Bacteria, fungi and yeasts are able to form volatile methylated arsenic compounds under aerobic and anaerobic conditions. Bacteria produce dimethylarsine while fungi synthesize trimethylarsine. Soil bacteria are responsible for the demethylation of both compounds (Tamaki and Frankenberger, 1992). One of the most popular arsenic intoxications due to indoor exposure was caused by the mold *Scopulariopsis brevicaulis* which produced trimethylarsine when growing on wallpaper colored with an arsenic-containing pigment (e.g. Scheele's green). The contaminated air had a characteristic garlic-like odor and resulted in a number of deaths before the specific correlation between the wallpaper pigment and the presence of the mold was recog-

Table 1. Organoarsenicals isolated from marine animals[a]

Animal	Organ	Wet weight based As concentration (mg/kg)
Western rock lobster (*Panulirus cygnus*)	Tail	20
	Hepatopancreas	39
Shark (*Prionace glaucus)*	Flesh	37
	Liver	28
Octopus (*Paroctopus dufleini)*	Muscle	49
Shrimp (*Sergestes lucens)*	Meat	6
Giant clam (*Tridacna maxima)*	Kidney	200
Haddock *(Melanogrammus aeglefinus)*	Muscle	6[b]
Herring *(Clupea harengus)*	Muscle	1[b]
Halibut *(Hippoglossus hippoglossus)*	Muscle	2[b]
Cod *(Gadus morrhua)*	Muscle	5[b]

[a] Reprinted from Cullen and Reimer (1989).
[b] Dry-weight basis.

nized. The metabolism of arsenic and various (organo)arsenic compounds by bacteria, fungi, marine plants and algae is very intricate and shall not be discussed in this report.

The dominant organoarsenic compounds in aquatic systems are $(CH_3)\ _3AS^+CH_{2-}COO^-$ (arsenobetaine) and $(CH_3)_3AS^+CH_2CH_2OH$ (arsenocholine). The arsenic concentration of fish and marine invertebrates differs widely depending on the species and the organ under consideration (Table 1).

Arsenic may be taken up by estuarine as well as terrestrial plants. Usually arsenate is preferred compared to arsenite in view to plant uptake (Aller et al., 1990). Table 2 presents the total arsenic content of some vegetables.

The uptake of arsenic by plants is determined by the content of arsenic *and* phosphorus in the soil and differs between the specific plant species. The velvetgrass *Holcus lanatus* for example strongly prefers phosphate compared to arsenate. Even though the ratio of available As:P could be as great as 200:1, the ratio of plant As:P does not exceed 1:2 (Meharg et al., 1994). Generally the presence of phosphate inhibits the uptake of arsenate while arsenate exerts only limited influence on the phosphate uptake (Tamaki and Frankenberger, 1992). However, this is not true for every plant species. The uptake of arsenic by the stinging nettle *Urtica dioica* for instance is negatively correlated with the arsenic concentration and positively with the phosphate concentration in the soil (Otte et al., 1990). Usually the arsenic concentra-

Table 2. Arsenic content of vegetables. Reprinted from Aller et al. (1990)

Plant	Mean value of arsenic content in μg/kg = ppb dry weight
Spinach leaves (*Spinacia oleraceae*)	200–1500
Lettuce leaves (*Lactuca sativa*)	20–250
Onion bulbs (*Allium cepa*)	50–200
Potato tubers (*Solanum tuberosum*)	30–200

tion in the roots is higher than in the above-ground organs of terrestrial plants. Above-ground organ contamination may also be due to atmospheric deposition which is described for lettuce and kale grown close to a wood preservation factory (Larsen et al., 1992). Therefore mosses have been used as biological indicators to assess the air pollution with metals including As (Markert et al., 1996). In this survey the As values of all moss samples ranged from 0.08 to 15.9 ppm.

In contrast to heavy metals arsenic accumulation is observed in very few mushrooms such as wild *Agaricus spp.*, where concentrations up to 14 ppm have been described (Vetter, 1994). The average arsenic concentration of edible mushrooms does not exceed 0.5 ppm usually (Stijve et al., 1990).

4. Toxicity of arsenic

Arsenic is frequently associated with poison by the public. However, since the toxicity of arsenic depends on the specific form of the arsenic compound inorganic and organic arsenicals have to be distinguished.

4.1. Inorganic arsenicals

Several physico-chemical properties of arsenic are similar to those of phosphorus, which is due to the similarity of the oxidation states and electron orbitals of both elements. Arsenic can occur in $+5$, $+3$, 0 and -3 states.

Inorganic arsenicals, particularly if they are water soluble (e.g. sodium arsenite), are absorbed to a high extent after oral application. The main toxicological mechanism of trivalent arsenic is the breakdown of the tricarboxylic acid cycle by inhibiting the oxidative decarboxylation of pyruvate and alpha-ketoglutarate due to its combination with the dithiols of lipoic acid (Hindmarsh and McCurdy, 1986). A different mechanism is claimed for the pentavalent form of arsenic. Arsenate replaces phosphate in high-energy compounds which brings about a loss of energy since these compounds are readily hydrolyzed. The result is an uncoupling of the oxidative phosphorylation from the mitochondrial respiratory chain. Another effect is the synthesis of 1-arseno-3-phosphoglycerate instead of 1:3 diphosphoglycerate which again hydrolyzes spontaneously without forming ATP (Hindmarsh and McCurdy, 1986). Generally, the trivalent arsenic seems to be more toxic than the pentavalent form and some authors argue that the main toxic systemic effects occur after reduction of the pentavalent form (Jonnalagadda and Prasada Rao, 1993).

The lethal dosage for man is approximately 100–150 mg arsenic trioxide resulting in severe vomiting, blood-stained diarrhoea, peripheral circulatory collapse and death (Hindmarsh and McCurdy, 1986). Chronic arsenic poisoning induces manifold symptoms such as general debility, weakness, loss of weight, appetite, and energy. Dermatologic manifestations including pigmentation, hyperkeratosis (palms of the hands and soles of the feet in particular) and, typically, white bands crossing the nails. These bands are called Mees' lines and appear at the same location on all fingers (Hindmarsh and McCurdy, 1986). Various peripheral vascular lesions have been observed in vineyard workers and people drinking arsenic-contaminated water for many years.

The latter have been reported to suffer from the so-called Blackfoot disease, a gangrene of the toes and feet described in Taiwan (Hindmarsh and McCurdy, 1986).

4.2. Naturally occurring organic arsenicals

The main organic arsenicals in marine fish and crustaceans, arsenobetaine and arsenocholine, are absorbed to a high extent and rapidly excreted via urine. Due to the fast renal elimination they are supposed to be of little toxicity. However, it is almost impossible to make general statements about the toxicity of organic arsenicals because each compound has to be examined separately. In fact there is a considerable lack of data concerning the specific toxicological data about organoarsenic compounds.

5. Carcinogenicity of arsenic

5.1. Inorganic arsenicals

There is strong evidence concerning the carcinogenicity of inorganic arsenic from epidemiological data. Occupational respiratory exposure as well as chronic ingestion of contaminated drinking water are reported to cause cancer in various studies. A compact review is given by Hayes (1997). The predominant malignancies are skin cancer (Bowen's disease and squamous cell carcinoma) and lung cancer. It is still under discussion if arsenic acts as a tumor promotor that needs other (co)carcinogens to induce cancer or rather than a tumor initiator. The main problem is the lack of a suitable animal model that allows to study cancer development after different application ways and dosages with different arsenic compounds. A recent report demonstrates that inorganic arsenic may induce chromosomal aberrations and sister chromatid exchange (Hayes, 1997). However, the IARC lists arsenic and different arsenic compounds among those metals which carcinogenicity to humans is beyond any doubt (Gibb and Farland, 1992).

5.2. Organic arsenicals

The toxicity and carcinogenic potential of seafood arsenic, arsenobetaine in particular, seems to be almost negligible compared to that of inorganic arsenic (Sabbioni et al., 1991). In contrast to these findings, other investigators recently demonstrated that organic arsenicals such as arsenobetaine and arsenocholine, which are common in marine seafood, may induce chromosomal aberrations in cultured human fibroblasts (Oya Ohta et al., 1996).

6. Arsenic-based chemical warfare agents

The most important arsenic-based chemical warfare agents are Chlorodiphenylarsine (Clark I), Cyanodiphenylarsine (Clark II), 2-chloro-ethenyldichloroarsine (Lewisite) and 10-chloro-9-10-dihydrophenarsazine (Adamsite). Detailed information about the synthesis and chemistry is provided by Jackson (1935) (Fig. 1).

Extensive experimental research including animal and human (!) trials has been carried out before Word War II in Germany. Flury summarizes detailed studies about Clark I, Clark II, Arsinetrichloride and other organic arsenicals (Flury, 1921). Clark I and Clark II belong to the so-called "Blaukreuzkampfstoffe", indicating that their main target is the respiratory tractus where they cause severe irritation, coughing and vomiting. Once inhaled at minimal dosages the exposed person has to discharge the gas mask which results in unprotected exposure to the substance. Therefore the Germans called these warfare agents "Maskenbrecher" meaning an agent that compels the discharge of the gas mask. Flury examined which concentration of different arsenic and non-arsenic compounds in the air makes it impossible to stay longer than 1 min in an experimental chamber. Due to his trials with healthy volunteers he classifies concentrations of Clark II $\geq$ 0.25 mg/m^3, Clark I $\geq$ 1 mg/m^3 and Phenyl arsine dichloride $\geq$10 mg/m^3 as "unbearable" (Flury, 1921). In contrast to these data very little is published about the toxicological effects of Lewisite. A recent review about this warfare agent, which had not been used in World War I but in an Iraqi attack against Iranian soldiers in, 1988, states that Lewisite was considered to be the "best" arsenical war gas (Goldman and Dacre, 1989). Among other reasons this is due to its specific property of skin irritation. Lewisite is not only poisonous after inhalation, but it irritates skin and eyes with stinging sensations and almost immediate pain. The LD50 for man after skin contact is estimated with $\geq$40 mg/kg, the lethal inhalation dosage is estimated 0.5 mg/l for 5 min (Goldman and Dacre, 1989). According to this review there is no experimental evidence for the mutagenicity of arsenic.

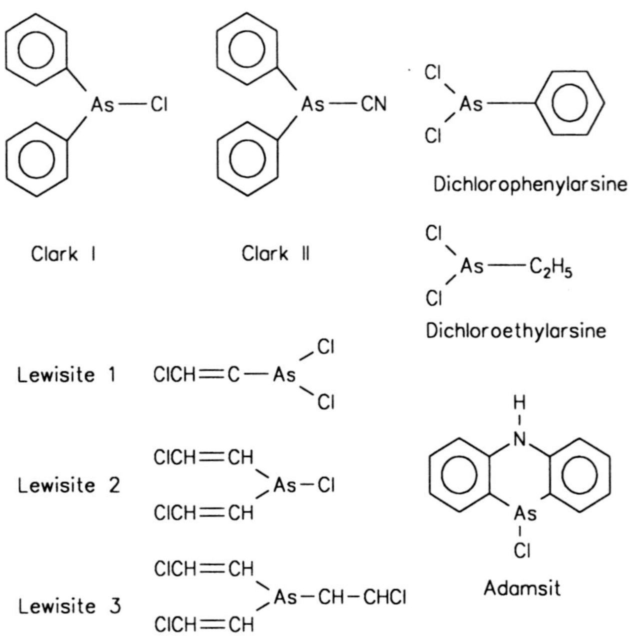

Figure 1. Typical arsenic-based chemical warfare agents.

A Swedish study compared the toxicity of Clark I, Clark II, Adamsite, and Lewisite with that of inorganic arsenic (As_2O_5) using an in vitro model (human leucozytes) and an in vivo model (dietary exposed juvenile three-spined sticklebacks). They concluded that the examined four warfare agents exert a toxicity more potent than arsenic itself (Henriksson et al., 1996).

7. Special problem Löcknitz

One of the vastest ammunition dumps of the Third' Reich was situated close to Löcknitz, a little town nearby the present German–Polish border (Fig. 2).

The so-called Heeresmunitionsanstalt Löcknitz consisted of two different parts, one for the storage of conventional ammunition (Area I, approximately 300 ha), the other for chemical weapons (Area II, 100 ha). At Area II Clark I, S-Lost (2,2-dichloroldiethylsulfide) and so-called "Arsine oil" which consisted of arsenic III chloride, chlorodiphenylarsine and triphenylarsine, were stored and decanted in grenades. The Heeresmunitionsanstalt was captured by the Soviet Army in April, 1945. At this time the tanks, many of them being subterranean, were completely filled with a total of approximately 3000 t of CWAs. The Soviets poured these compounds into concrete basins and burnt them after coverage with chlorinated lime. Concrete shelters, pillboxes, pipelines etc. were blasted and most buildings were demolished almost completely. Finally entry was prohibited and the whole area was secured by a fence being controlled by the police and armed forces. This status was maintained until the reunification of the German Democratic Republic with the Federal Republic of Germany in 1990.

Figure 2. Location of Löcknitz.

To prepare the definite rehabilitation of the area different governmental authorities, particularly the Bundesministerium für Bildung und Forschung (BMBF) and the Staatliches Amt für Umwelt und Natur (STAUN), cooperate since 1992. Up to now various investigations have been carried out including geological specification of the area and chemical analysis of hundreds of soil and groundwater samples. The total arsenic content was used as a marker for contaminated sites being very helpful to identify so-called hot spots. These hot spots are very small areas with very high, sometimes extreme, contamination. At some places the total arsenic of the soil exceeded 250 g/kg (!) and 1300 mg/kg Clark I. An analysis of the arsenic uptake by terrestrial plants was carried out to determine the entry of arsenic into the biosphere and to estimate the risk of a contamination of the fodder chain (Pitten et al., 1999).

8. On-site determination of total arsenic content

A portable X-Ray Fluorescence Spectrace 9000 (TN Technologies) has been used as an on-site screening tool for the determination of total arsenic content and establishing concentration profiles. The instrument has three radiation sources (Fe, Cd, Am) which provides different energy ranges (EPA, 1992). Due to the arsenic excitation energy of 10.532 eV the Cd-source is most suitable for the detection. The measurements were done without further sample preparation directly in-situ on the ground surface. Additional measurements were performed on similar places in different depths to determine the arsenic distribution in dependence on the ground depth.

The determined arsenic concentration varied between 40 ppm (MDL) and 20 000 ppm. Fig. 3 shows the arsenic concentration profile of a selected area (Schneider et al., 1999).

For the verification of the in-situ measured data complementary in vitro measurements using Spectrace 9000 and a laboratory XRF (QuanX, Noran Instruments Inc.) were performed. 40 soil samples have been investigated. For comparison of the results the arsenic concentration have been verified by classical AAS method (Thurow et al., 1998b).

The results of in-vitro XRF and AAS investigations show a systematic difference which resulted from incomplete sample preparation in the case of AAS measurements (Fig. 4). It is known from literature that classical sample preparation with HNO_3 or H_2SO_4/H_2O_2 does not lead to a total breakdown of arsenic compounds. The recorded systematical difference of 20% for the results of AAS in comparison to in-vitro XRF can easily be adapted to the complementary measuring technique because of the linearity over several magnitudes.

The MDL values were determined with 20 ppm (XRF with Spectrace 9000) and 1 ppm (AAS investigations) respectively.

9. Uptake of arsenic by plants

A total of 186 terrestrial plants grown on contaminated sites of the former Heeresmunitionsanstalt II was collected in summer, 1996 (Pitten et al., 1998a,b). The above-

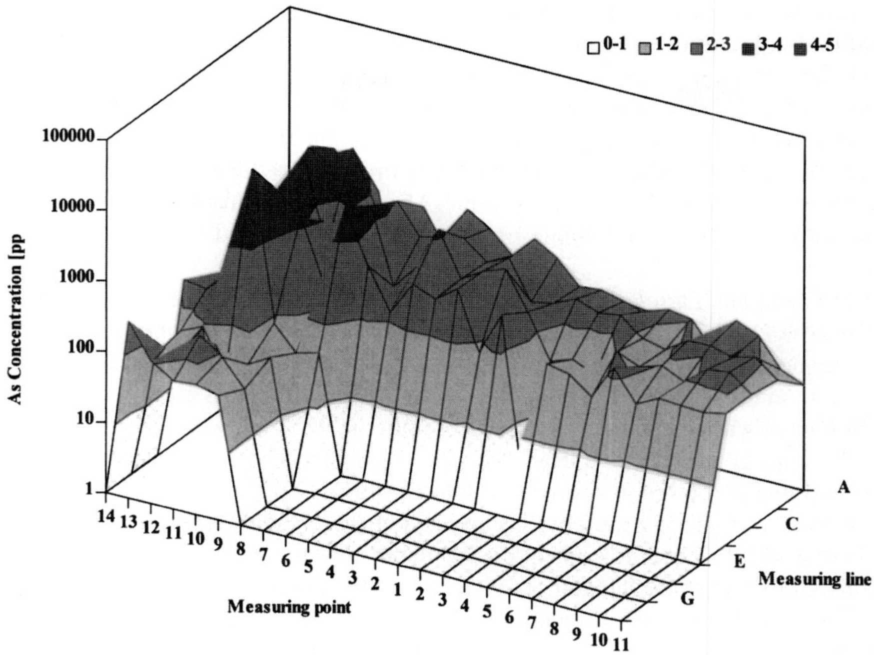

Figure 3. 3D Arsenic concentration profile of a selected area.

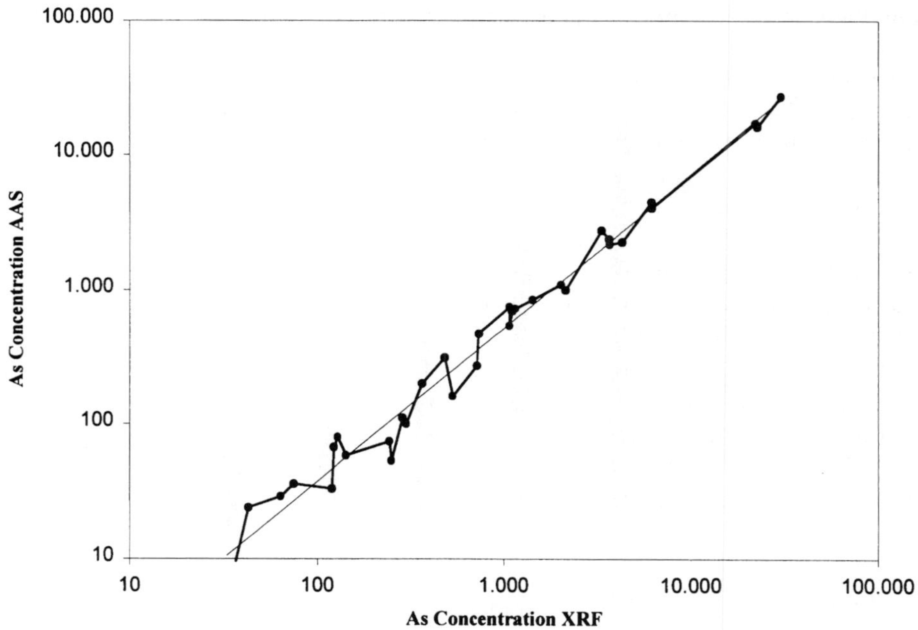

Figure 4. Comparison of in-vitro XRF and AAS results (double logarithmic) ($y = 0.79x$; $R^2 = 0.979$).

ground organs of typical fodder plants were preferred since the question of the entry of arsenic into the nutrition chain was in the center of our interest.

Therefore the sampled plants were not cleaned from adhering soil and dust particles, because animals harvesting these plants as fodder do not clean them either. This procedure is recommended if the entry of contaminants into the nutrition chain is under consideration (Markert, 1995). After a microwave digestion the samples were analyzed using Graphite Furnace Atomic Absorption Spectrophotometry (GFAAS). Approximately 9% of the sampled plants were contaminated with arsenic >2 ppm (Fig. 5).

Carex hirta and *Deschampsia cespitosa* showed the highest arsenic contents with a median of 0.7 and 1.0 ppm, respectively. The velvet grass *Holcus lanatus* was sampled most frequently. Its arsenic content was 4.3 ppm (mean value) with a median of 0.7 ppm. Soil samples taken close to the places where the plants were harvested were contaminated with arsenic ranging from 71 ppm to 2915 ppm with 840 ppm as median and 923 ppm as mean value. The differences between the arsenic content of monocotyledon and dicotyledon plants were not significant (Fig. 6).

However, considering the obvious and at some places extreme arsenic contamination of the soil, the arsenic content of the plants appears to be surprisingly moderate. As a result of this study the authors concluded that a dangerous contamination of the nutrition chain by arsenic via plant uptake was not observed.

10. Speciation of phenylarsenic compounds

Since only few speciation methods for phenylarsenic compounds are available the

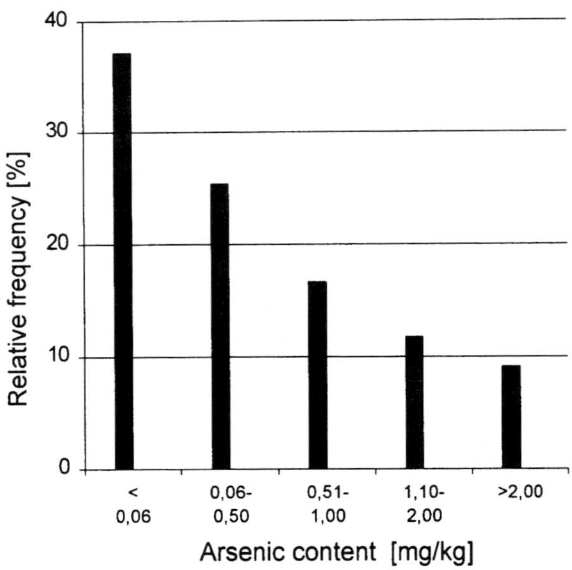

Figure 5. Distribution of arsenic-contaminated samples among 186 (=100%) plant samples. Reprinted from Pitten et al., 1998a.

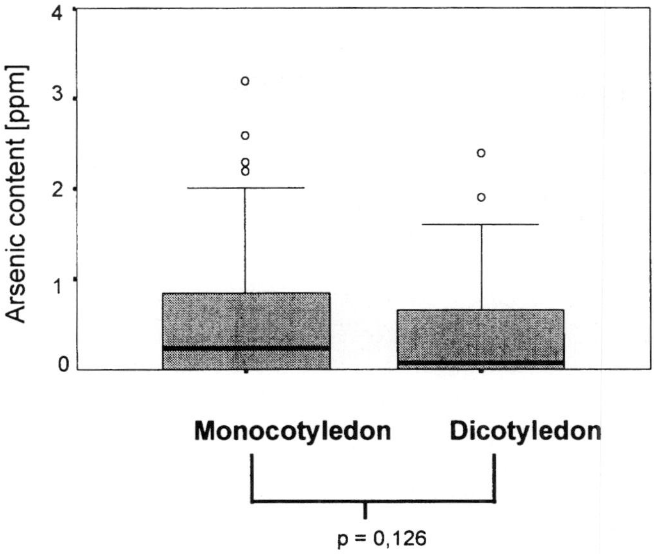

Figure 6. Box-and Whisker-Plot indicating median and 25%-percentiles of the arsenic content of mono-and dicotyledon plants. No statistical significant difference in the *U*-test according to Mann and Whitney. Reprinted from Pitten et al. (1998a).

development of suitable method combinations was necessary. The general concept for the analysis of arsenic containing CW agents includes sample preparation procedures as well as mass spectrometric techniques and atomic emission detection for structure elucidation. Speciation analyses were done for a total of 50 soil and plant samples from the Löcknitz area. The samples were freeze dried and homogenized prior to use.

10.1. Sample preparation

10.1.1. Classical extraction

The extraction of phenylarsenic compounds from soil was first described by Schoene (1993).

Different classical extraction methods (shaking, sonication, soxhlet) have been tested for the analysis of the phenylarsenic compounds. Extraction efficiency have been determined with standard soil spiked with 100 ppm triphenylarsine; the total arsenic content of the standards was verified using XRF measurements. The sample amount extracted varied between 10 g (shaking, ultrasonic) and 30 g (Soxleth). The recovery rates for triphenylarsine varied between 15 and 40% with standard deviations <5% depending on the extraction method (Fig. 7). Best results were achieved for soxhlet extraction (40%) whereas degradation of triphenylarsine was noticed using sonication as extraction methods. Schoene et al. described similar low recovery rates for Chlorodiphenylarsine and Bis(diphenylarsine)oxide according to the hydrolysis of these compounds in soil (Schoene et al., 1995). Since triphenylarsine does not show hydrolytic reactions in the soil, the

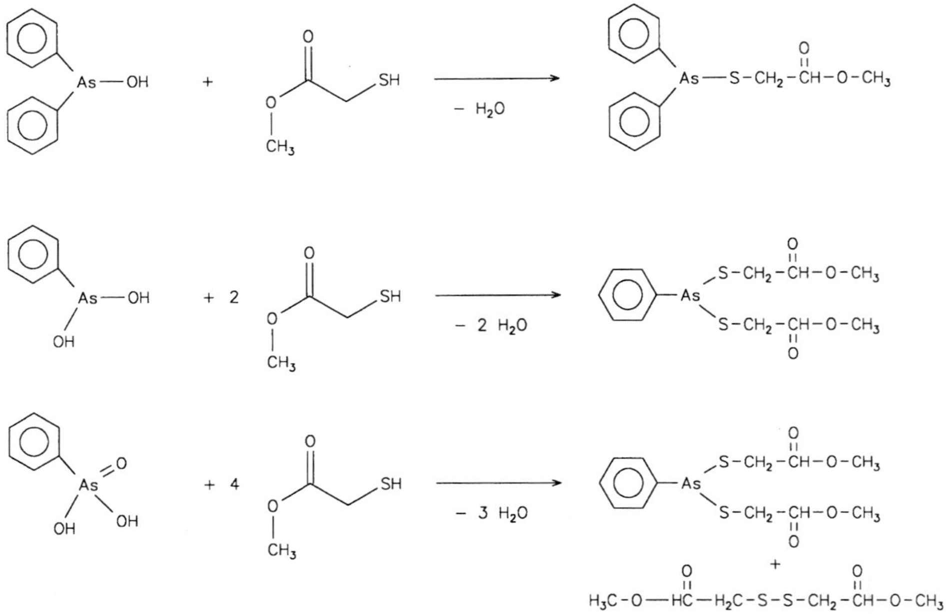

Figure 7. Hydrolysis and derivatization using TGM.

reasons for the low recoveries are strong matrix effects with bonding of arsenic compounds to organic soil matter.

The different extraction techniques were applied for the sample preparation of real soil samples, too. Differences were determined depending on the type of extraction methods. Ultrasonic extraction showed smaller results because of partial degradation of substances during the extraction process. The soxleth extraction with dichloromethane gave lower recovery rates for Chlorodiphenylarsine, Bis(diphenylarsine)oxide and Bis(diphenylarsine) because of decompositions probably occurring within the extraction as already described by Schoene et al.

For the determination of hydrophilic and non volatile compounds alcaline extractions followed by derivatization have been carried out (Thurow et al., 1999). For the derivatization silylation with MTBSTFA was used as well as thiomethylation with TGM.

Thiomethylation only gives derivatives of arsenic-(III) whereas pentavalent arsenic compounds are reduced by the thiol with formation of the disulfide $(SGM)_2$. The yields for the TGM-derivatization depend on the number of introduced thiomethyl groups. The determined yields and recovery rates are in accordance with the publication of Schoene (Schoene et al., 1995). The derivatization is subjected to sometimes strong matrix influence which can be corrected by determination of the yield of proper reference compounds. Multiple derivatized compounds show often thermal instability, thus the use of cold on-column injection or temperature programmed cold injection systems is required.

Silylation with MTBSTFA as silylating agent is a suitable method for derivatiz-

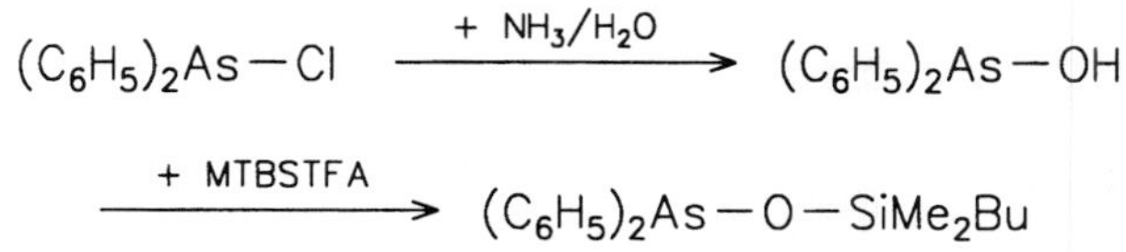

Figure 8. Hydrolysis and derivatization using MTBSTFA.

ation, too (Fig. 8). Due to the temperature instability of the silylethers the use of a cold injection system or on-column injection is necessary for further investigation.

10.1.2. Supercritical fluid extraction

An alternative extraction method is supercritical fluid extraction with CO_2, which has been widely used in different fields of environmental analysis. The development of a suitable extraction method for organoarsenic compounds requires a systematic optimization of the parameters chamber temperature, pressure, extraction time, number of extraction steps, modifier and rinse solvent, which could influence the extraction result crucially (Tena et al., 1993).

With respect to the solubility of the arsenic compounds methanol has been used as rinse solvent. Since organic solvents as modifiers can improve the extraction conditions (raise up the polarity of the extraction fluid) a number of modifiers has been tested. The combination of methanol with 15% dichloromethane as additional modifier gave optimum extraction yields and reproducibility.

The increase of extraction yields through several extraction steps has been described in literature (Bowadt and Hawthorne, 1995). In opposite to this Wenclawiak and Krah received no significant increase of extraction yields for inorganic and methylated arsenic compounds which could be confirmed for phenylarsenic compounds in the realized investigation (Wenclawiak and Krah, 1995).

In order to obtain optimum extraction yields with SFE the chamber temperature is another important parameter. Generally the extraction yields increase with increasing temperature because of a better solubility of the analyzed material in the supercritical fluid. The same behaviour has been reported for inorganic and methylated arsenic compounds by Krah (Wenclawiak and Krah, 1995). In opposite to this the extraction of phenylarsenic compounds showed decreasing efficiencies with increasing temperature (see Fig. 9).

As another important parameter the pressure in the extraction chamber has been varied because some chemical compounds may be disintegrated by high pressure. As reported in the literature the proportion between extraction yield and pressure depends on the specific properties of the current material and on the temperature. Best results have been found for 200 bar; Fig. 10 shows the influence of the pressure on the extraction results for triphenylarsine with a chamber temperature of 40°C.

The influence of extraction time showed a decrease of extraction efficiency of about 10–15% by decreasing the extraction time from 40 to 20 min.

Maximum extraction efficiency has been found to be about 40% regarding to possibly strong sorption interaction of arsenicals with the soil.

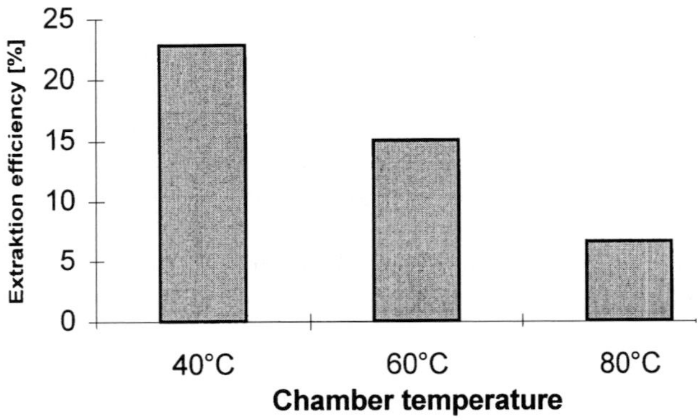

Figure 9. Influence of the chamber temperature on the extraction efficiency.

10.1.3. Headspace and thermal desorption

Headspace and thermal desorption are thermal extraction methods which can be directly connected to gas chromatography and do not need additional sample preparation. Usually both methods are applied for the determination of volatile compounds in air and water. Only few applications are known for the direct treatment of soil samples. The investigations for analysis of phenylarsenic compounds were carried out with an Headspace Sampler HS40 (Perkin–Elmer Inc.) and a Thermal desorption system TDS 2 (Gerstel GmbH, Germany).

Headspace: Results of headspace analysis are highly influenced by variation of thermostating temperature and time as well as by the sample amount. Additional methods for accelerating the equilibrium in the headspace above the sample and increase efficiency and reproducibility are saturation with inorganic salts or homogenization through shaking (Kolb, 1988).

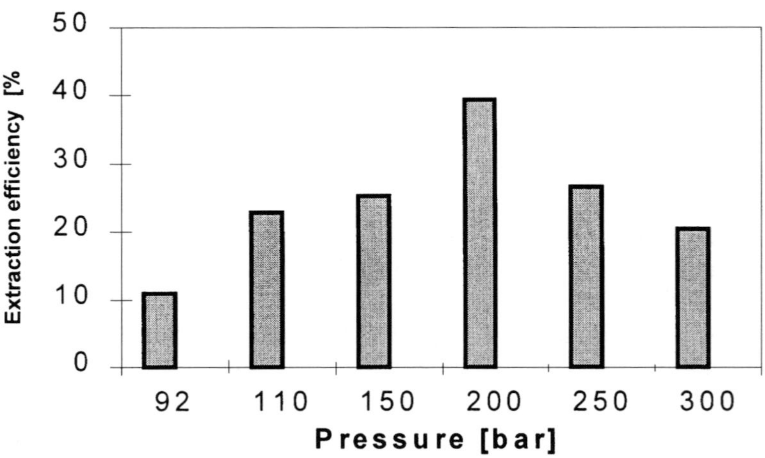

Figure 10. Influence of pressure on the extraction efficiency.

No peaks could be detected for samples with arsenic concentrations below 100 ppm (Thurow et al., 1998), since the concentration of the analyzed compound in the headspace is to low due to low vapor pressure. Thus headspace analysis is suitable for rapid screening procedures but can not be used for trace analysis. The reproducibility of the headspace results was about 2%.

Thermal desorption: Method development was done with sample amounts of 50–100 mg. The use of larger sample amounts can overload the injection system and the capillary column. Drying of the sample prior to thermal extraction is necessary because problems with humidity content of the soil and other matrix effects can happen. Variations of the parameters desorption time and desorption temperature lead to an optimized method. The calibration curve in the range from 1–100 ppm shows a good linearity with relative standard deviations of about 5–7% (Fig. 11). These values are higher compared to liquid extraction due to the inhomogeneity of the soil samples. An additional homogenization of the samples reduced the relative standard deviation to about 2–3%.

Repeated measurements with spiked and real soil samples indicated an extraction efficiency of nearly 100% for the phenylarsenic species. On the basis of characteristic ion traces it could be established, that the arsenic species are completely extracted from the soil in the first desorption step (Thurow et al., 1998).

10.2. Gas chromatographic investigations

10.2.1. GC/MS investigations

Mass spectrometry is today the method used most frequently for identification and structure elucidation of unknown compounds. GC/MS investigations have been used in the combination with thermal desorption as extraction procedure, for the analysis of the dichloromethane extract and for the derivatization products (Thurow et al., 1997). Investigations have been carried out with an HP 5973 MSD (Hewlett Packard Inc.) in combination with an cold injection system CIS 3 (Gerstel GmbH, Germany).

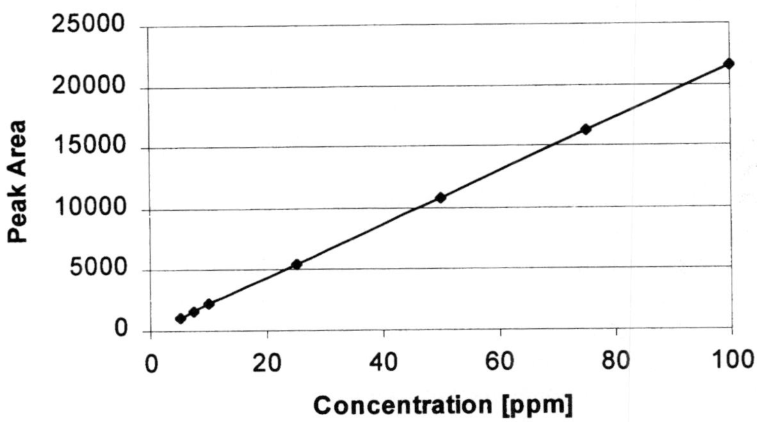

Figure 11. Calibration curve for Thermal desorption investigations of triphenylarsine.

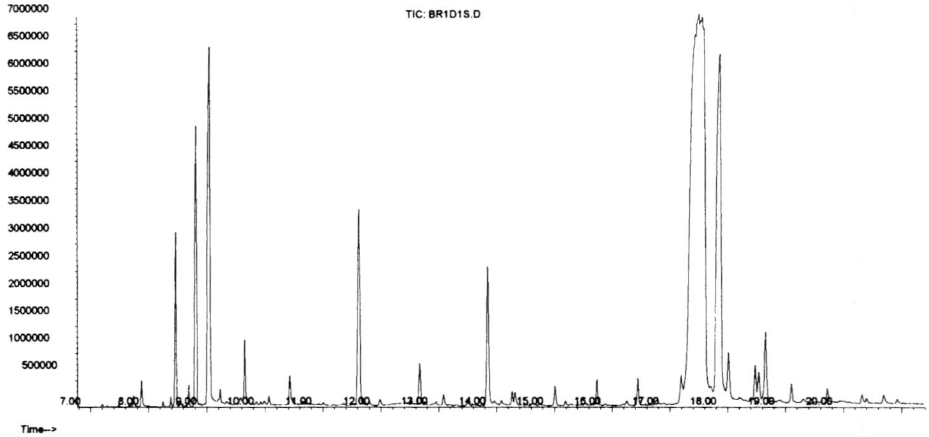

Figure 12. Typical total ion chromatogram for a real solid sample from Löcknitz.

Dichloromethane extracts and thermal desorption: The chromatograms of the samples (Fig. 12) showed main peaks at 9.04, 11.61, 13.85, 17.45 and 17.87 min. Chlorodiphenylarsine (Clark I, 9.04 min), Triphenylarsine (11.61 min) and Bis(diphenylarsine)oxide (17.45 min, main degradation product of Clark I) could be identified using library searches. Bis(diphenylarsine) (17.87 min) has been identified by means of its mass spectra according to Schoene et al. (1995). Isotopic pattern calculation indicated the peak at retention times 13.85 min to contain sulfur. By comparison of the mass spectrum the substance could be identified as Diphenylthiophenylarsine as proposed by Schoene et al. (1995).

Library searches gave additional prove of Methyldiphenylarsine at 8.26 min. At total 24 compounds have been detected in the mass spectra containing typical fragment ions $m/z = 152$, 229 and 227, which have been correlated to the fragments [PhAs], [Ph_2As] and [Ph_2As-2H] (see Table 3). Tentative identification of some species was done by means of typical fragments in the mass spectra. Diphenylthiomethylarsine was expected because of typical isotopic pattern for sulfur containing compounds. The compounds with $m/z = 340$ and 508 seems to contain chlorine because of isotopic pattern. Because of monoisotopy of arsenic an identification of arsenic in the compounds by isotopic pattern calculation is not possible. Thus for a validation of arsenic in the compounds the use of other detection principles is required.

The GC/MS speciation investigation of plants (after thermal desorption) showed typical total ion chromatograms for all measured above ground samples. The arsenic species were detected with the typical ions $m/z = 227$ and 152. The main contaminant in all above-ground samples was Triphenylarsine with about 95–97% of total organoarsenic content. Beside this a compound with molecular ion $m/z = 304$ and Methyldiphenylarsine could be detected with amounts of 3–5% and $< 0.1\%$ in these samples. None of the other arsenic compounds according to Table 3 were detected in the above-ground samples. An additional arsenic compound which could be identified as Ethyldiphenylarsine was found in the root samples. This compound was not detected in any of the soil samples investigated.

Table 3. Results of GC/MS-investigations

RT (min)	Fragments, *m/z*	MZ, *m/z*	Identification
7.88	152, 230	230	Ph_2AsH[c]
8.26	154, 227, 244	244	Diphenylarsine[a]
8.47	260, 245, 229, 152	260	Ph_2AsOMe[c]
8.69	274, 245, 229, 152	274	Ph_2AsOEt[c]
8.82	288, 245, 229, 152	288	$Ph_2AsOProp$[c]
9.04	154, 227, 264	264	Chlorodiphenylarsine[a]
9.65	278, 229, 1678, 152	(278)	?
9.93	276, 261, 227, 152	276	Ph_2AsSMe[c]
10.46	275, 167, 152	(275)	?
11.613	152, 227, 306	306	Triphenylarsine[a]
12.671	152, 227, 304	304	?
12.884			
12.927			
13.087	152, 227, 261, 340	(340)	Ph_2AsPh-Cl[c]
13.852	152, 229, 338	338	Diphenylthiophenylarsine[b]
15.433			
16.797			
20.695	152, 227, 382	(382)	Ph_2AsPh-Ph[c]
17.45	152, 227, 306, 474	474	Bis(diphenylarsen)oxid[a]
17.87	152, 229, 458	458	Bis(diphenylarsen)[b]
18.02	152, 227, 471, 548	(548)	?
18.48			
18.55			
18.66	152, 227, 263, 306, 340, 508	(508)	Ph_2As-O-AsPhPh-Cl[c]

[a] Identification by library search.
[b] Identification according to Schoene (Schoene et al., 1995).
[c] Tentative identification by means of mass spectra.

The total amount of organoarsenic in comparison to the total arsenic content varied between 30 and 50% (above ground) and 20–60% (soil).

Alcaline extracts after derivatization with TGM and MTBSTFA: The chromatograms of the alcaline extracts after derivatization with TGM or MTBSTFA were analyzed by GC/MS. In the chromatograms of TGM derivatization the typical oxidation product $(SGM)_2$ was identified by molecular mass and characteristic fragments. The proof of the derivatization products of the diphenyl compounds was possible with the characteristic masses at $m/z = 334$ ($[M]^+$), 261 ($[Ph_2AsS]^+$), 183 ($[PhAsS]^+$) and 152 ($[PhAs]^+$). Derivatives of monophenylarsenic compounds showed the expected molecular mass $m/z = 362$ amu. Additional signals were detected at $m/z = 257$, 225, 183 and 153 amu, which were correlated to the fragments $[M\text{-}SGM]^+$, $[PhAsC_3H_5O_2]^+$, $[PhAsS]^+$ and $[PhAs]^+$.

The chromatograms of the silylated samples showed the silylated derivatives of Mono- and Diphenylarsenic compounds. No derivatives of inorganic arsenic (As^{3+}) have been detected by this way.

10.2.2. GC/AES investigations

The combination of gas chromatography and atomic emission spectroscopy has been used for element selective investigations. The measurements were carried out with an HP G2350A Atomic emission detector (Hewlett Packard Inc.). For structural elucidation the channels carbon (C), hydrogen (H), arsenic (As), chlorine (Cl), oxygen (O) and sulfur (S) have been investigated. Spectral background correction was performed for elimination of interferences.

In order to reach high sensitivity and sufficient detection limits for the determination of the arsenic species the arsenic emission lines at 189.04, 193.7, 200.33, and 234.94 nm were studied using triphenylarsine as reference substance. The most sensitive arsenic line at, 193.7 nm is interfered by a strong carbon line and thus the wavelength at 189 nm has been used for arsenic identification according to Mothes and Wennrich (1997). A total of 24 arsenic containing compounds have been detected in the dichloromethane extracts and thermal desorption investigation, respectively (Thurow et al., 1999). The presence of arsenic was confirmed by the five characteristic lines between 180 and 210 nm (see Fig. 13).

Higher sensitivity for the specific detection of arsenic can be achieved by an increase of make up flow. In correlation with literature the optimum flow rate was found to be 150–190 ml/min. Hydrogen was applied as reagent gas for arsenic detection since it supports the dissociation of organoarsenic compounds and the excitation. Furthermore it protects the surface of the discharge tube from deposition of arsenic. Optimum results were achieved with an hydrogen pressure between 280 and 350 kPa.

The expected sulfur containing compounds at 9.93 and 13.85 min could be confirmed by GC/AES measurements by means of the sulfur channel. The chlorine channel showed seven peaks at retention times correlating to MS results. Furthermore different oxygen containing compounds have been found.

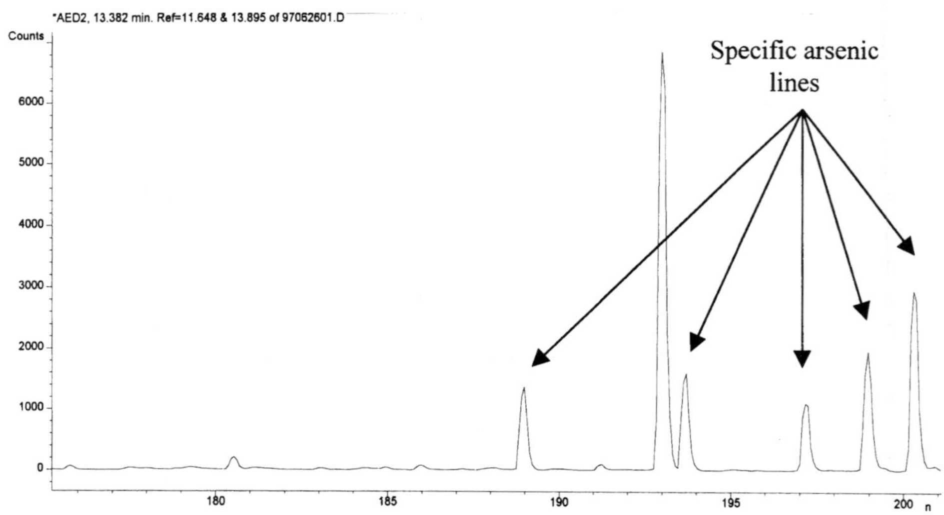

Figure 13. Snapshot of arsenic lines using triphenylarsine as reference compound.

Beside the qualitative identification of elements the atomic emission detector provides the possibility of determination of elemental composition and empirical formulas with compound independent calibration (Wylie et al., 1990; Quimby et al., 1995). This possibility enables an identification independent from retention times and does not require authentic reference standards for calibration. For structural elucidation the element ratios have been calculated using the method of relative response factors. Triphenylarsine was used as standard. The results have been confirmed by complementary high resolution mass spectrometric investigations. Table 4 summarizes the results of selected compounds.

Tentative identifications resulting from GC/MS investigations could be confirmed by means of empirical formula determination. The peaks at retention times 9.65 min and 10.46 min could not be identified due to the very low concentrations. The determinated elemental compositions varied about $\pm 5\%$ from the theoretical values. Greatest variations have been found in the detection of hydrogen according to literature (Pedersen-Bjergaard and Greibrokk, 1996). In order to achieve good results calibrants and unknown compounds should show similar concentration ranges in the investigated solutions.

10.3. HPLC/MS and HPLC/MS/MS investigations

To avoid derivatization reactions which influence the composition of the sample LC separation in combination with mass spectrometric detection was used for the analysis of water extracts. The measurements were carried out with an Quattro II (Micromass Inc.) in combination with a HP 1100 series HPLC (Hewlett Packard Inc.).

Electrospray and ApCI investigations of the standard compound triphenylarsine indicated that oxidation processes occur during the ionization process. The resulting mass spectra showed the oxidized species at $m/z = 323$ amu which correlates to Ph_3AsO. Phenylarsonic acid is preferential detected in negative ion mode because of their structure and properties.

The investigation of the extracts of real soil samples gave very complex mass spectra. On basis of their retention times triphenylarsine (14.10 min) and triphenylarsine oxide (8.43 min) could be identified. Additional arsenicals have been found by ion traces $m/z = 152$ and $m/z = 227/229$ which are known from GC/MS investigations. The corresponding mass spectra showed a high background. For reduction of the chemical noise and structural elucidation MS/MS investigations were carried out. Table 5 summarizes the results of different daughter and parent ion scans (Edens et al., 1998; Thurow et al., 1999).

The mass spectra at retention time 7.2 min showed an intensive molecule peak at $m/z = 263$. In the parent ion scans signals at $m/z = 229$, 245, 263 and 507 amu were found which could be correlated to the composition $Ph_2As(O)OH$ by the fragments $[Ph_2As]^+$ ($m/z = 229$) and $[Ph_2AsO]^+$ ($m/z = 245$). The fragment at $m/z = 507$ amu was identified as the dimeric condensation product $Ph_2As(O)$-O-$As(O)Ph_2$ that is formed during the ApCI process (Fig. 14).

In the parent ion scan of compound at retention time 10.28 min intensive signals were found at $m/z = 399$, 323, 307 and 229 amu. The corresponding daughter ion spectra of mass 399 have signals at $m/z = 229$, 227, 168 and 152 amu. From the

Table 4. Determination of elemental compositions and empirical formulas

RT (min)	Element ratio C:H:As	Additional elements	Molecular mass *m/z*	Empirical formula	Identification
8.264	1:1:0.077	–	244	$C_{13}H_{13}As$	Methyldiphenylarsine
9.037	1:0.87:0.079	Cl	264	$C_{12}H_{11}As_{1.1}Cl_{0.9}$	Chlorodiphenylarsine
13.087	1:0.72:0.056	Cl	340	$C_{18}H_{13}AsCl_{1.1}$	(Chlorophenyl)diphenylarsine
13.852	1:0.81:0.056	S	338	$C_{18}H_{15}AsS$	Diphenylthiophenylarsine
15.43	1:0.84:0.038	–	382	$C_{24}H_{20}As_{0.9}$	(Biphenyl)diphenylarsine
16.79	1:0.79:0.035	–	382	$C_{24}H_{19}As_{0.8}$	(Biphenyl)diphenylarsine
20.95	1:0.75:0.045	–	382	$C_{24}H_{18}As_{1.1}$	(Biphenyl)diphenylarsine
17.45	1:0.87:0.079	O	474	$C_{24}H_{21}As_{1.9}O$	Bis(diphenylarsine)oxide
17.87	1:0.91:0.079	–	458	$C_{24}H_{22}As_{1.9}$	Bis(diphenyl)arsine
18.48	1:0.81:0.08	Cl	508	$C_{24}H_{19}As_{1.9}O_{0.9}Cl$	Halogenated derivative of Bis(diphenylarsine)oxide
18.55	1:0.81:0.082	Cl	508	$C_{24}H_{20}As_{2}O_{0.9}Cl_{0.9}$	Halogenated derivative of Bis(diphenylarsine)oxide
18.66	1:0.81:0.79	Cl	508	$C_{24}H_{20}As_{1.9}O_{0.9}Cl$	Halogenated derivative of Bis(diphenylarsine)oxide

Table 5. Results of the MS/MS investigations

RT (min)	Parents of $m/z = 152$	Parent mass m/z	Daughters m/z	Identification
7.2	229, 245, 263, 507	507, 263	245, 152	$Ph_2As(O)OH$
8.43	323, 364	323	77, 91, 152, 154, 169, 227	$Ph_3As(O)$
9.25	229, 323	357	77, 91, 152, 154, 188, 227	$Ph_2As(O)PhCl$
10.28	227, 229, 323, 398	399	91, 152, 154, 168, 227, 229	$Ph\text{-}PhAs(O)Ph_2$
10.89	229, 307, 323, 399, 433	433	91, 152, 188, 227, 264	(PhCl)PhAs(O)Ph-Ph
12.12	339	339	77, 152, 154, 183, 227	Ph_3AsS
14.10	307	307	152, 154, 227	Ph_3As
24.6	383	383	152, 154, 227, 307	$Ph_2AsPh\text{-}Ph$

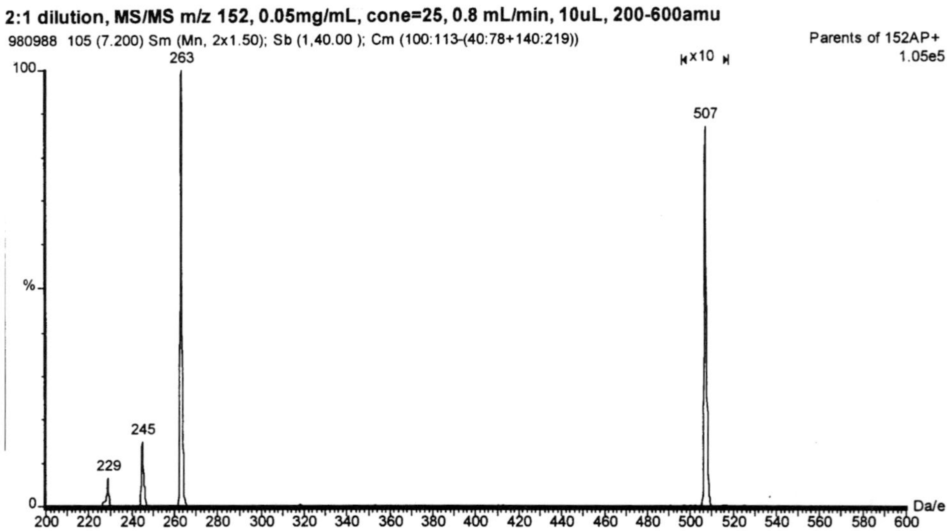

Figure 14. Mass spectra of $Ph_2As(O)OH$.

obtained data the substance could be identified as $Ph_2As(O)Ph$-Ph ($m/z = 338$ amu) with the fragments $[Ph_2As]^+$ ($m/z = 229$ amu), $[Ph_2As\text{-}2H]^+$ ($m/z = 227$ amu), $[PhAsO]^+$ ($m/z = 168$ amu) and $[PhAs]^+$ ($m/z = 152$). Identification of $Ph_2As(O)PhCl$ ($m/z = 357$), (PhCl)PhAs(O)Ph-Ph ($m/z = 433$ amu) and Ph_3AsS ($m/z = 338$ amu) was achieved by the same method.

10.4. High resolution mass spectrometry (HRMS)

High resolution mass spectrometry has been used as an additional tool for structural elucidation since this method provides exact mass determinations and therefore the possibility of the determination of elemental composition.

Table 6 summarizes the results of the HRMS investigations for the signals $m/z = 246$, 304, 338, 340, 382, 508, 534 and 550 PFK as reference substance (Grassert et al., 1998). The elemental compositions were detected by a special algorithm which includes the evaluation of all possible formulas as well as a verification of chemical correctness by calculating double bond equivalents (DBE) (Pellegrin, 1983). Since all compounds were proved to be arsenicals by using GC/AES the number of solutions resulting from HRMS could be decreased significantly from 31 to 18. For $m/z = 246$, 304 and 382 resulted only one elemental composition which could be related to the structures according to Fig. 15 by combining HRMS and LRMS data.

Furthermore the compounds with molecule peak $m/z = 340$ were proved to contain chlorine and could thus be identified as $Ph_2AsPhCl$. The structure is confirmed by the mass spectra given signals at $m/z = 186/188$, which show the typical chlorine isotopic pattern and could be correlated to the fragment $[C_6H_4ClAs]^+$.

The compound with molecular mass $m/z = 508$ amu could be identified as an chlorinated derivative of Bis(diphenylarsine)oxide by means of double bonding

Table 6. Results of the high resolution measurements for the investigated mass numbers

Nominal mass	Exact mass (measured)	Number of solutions	Number of As solutions	Possible formulas	DBE
246	246.00265	5	1	$C_{12}H_{11}AsO$	8
304	304.03158	2	1	$C_{18}H_{13}As$	13
338	338.0105	9	6	$C_{18}H_{15}AsS$	13
				$C_{21}H_{11}AS$	18
				$C_{11}H_{24}As_2S$	3
				$C_{15}H_{19}AsS_2$	11
				$C_{14}H_{20}As_2$	6
				$C_{11}H_{24}As_2O_2$	1
340	340.0137	5	3	$C_{14}H_{22}As_2$	5
				$C_{21}H_{13}As$	16
				$C_{18}H_{14}AsCl$	12
382	382.08346	2	1	$C_{24}H_{19}As$	16
508	507.98404	2	2	$C_{24}H_{19}As_2OCl$	16
				$C_{31}H_{10}AsOCl$	27
550	550.03178	3	2	$C_{37}H_{15}AsO$	31
				$C_{30}H_{24}As_2O$	20

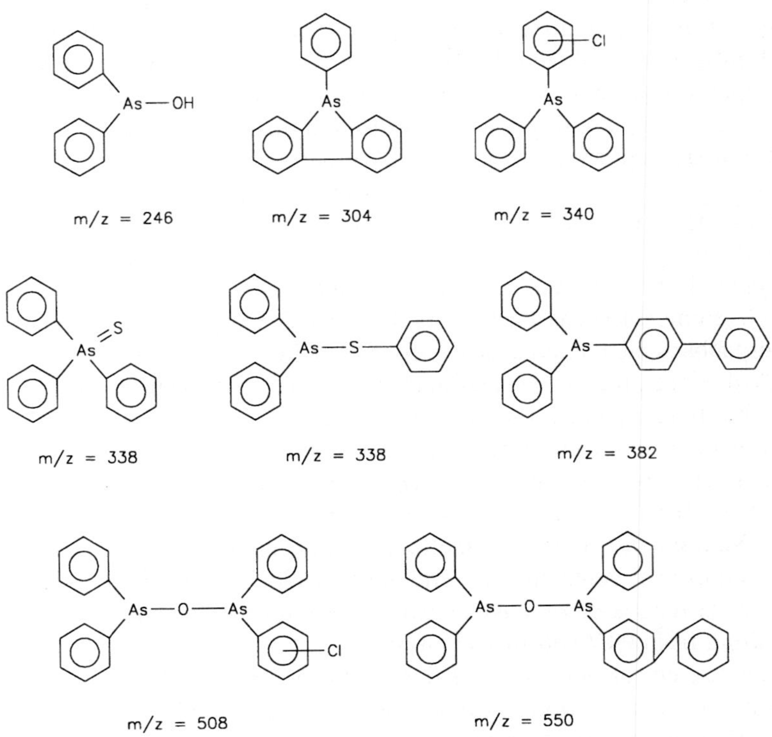

Figure 15. Phenylarsenic compounds identified by high resolution-MS.

equivalents and calculating the elemental ratios C/As using GC/AES investigations (see above). The structure could be verified by corresponding fragments $m/z = 152$ ($[PhAs]^+$), 263 ($[PhAsPhCl]^+$) and 279 ($[Ph(PhCl)AsO]^+$).

The same procedure has been used for the masses $m/z = 338$ and 550 amu as indicated in Table 6 and Fig. 15.

11. Summary and discussion of results

Suitable methods for speciation analysis of phenylarsenic compounds in soil and plant samples have been developed. In spite of the considerable contamination of the soil, the arsenic content of terrestrial plants is apparently modest.

The in-situ XRF investigations allow a rapid screening on arsenic contaminations in the field and thus a dedicated sampling from the resulting hotspots. The comparison of XRF and complementary AAS data showed a general difference resulting from incomplete sample preparation required for AAS investigations.

Different extraction methods have been tested for the sample preparation of arsenic contaminated samples. Best results were achieved using thermal desorption as extraction methods with extraction efficiencies of nearly 100%. Real soil or plant samples provide high matrix effects in the chromatograms. In this case the use of selective detectors such as MSD or AED is necessary for an easy determination of the arsenic compounds. Measuring of the arsenic channel in AES or selected ion monitoring provide quick information on the content of phenylarsenic compounds (Thurow et al., 1999b). Sonication should not be used at all since degradation of some species occurs. Supercritical fluid extraction is an suitable method too since it provides short extraction times, less use of organic solvents and matrix free extracts.

The use of derivatization methods for the analysis of organoarsenic compounds is suitable to generate GC-capable products, but changes significantly the composition of the sample. After derivatization it is often not possible to differentiate between compounds giving the same derivatization products. Furthermore it is necessary to pay attention on the influence of the matrix which interact the extraction and derivatization yield crucially. The further development in analysis of hydrophilic and non volatile organoarsenic compounds should thus include methods without any derivatization reactions like the direct measurement of water extracts using HPLC in combination with different detection principles.

A total of 22 arsenic containing compounds could be identified in the soil and plant samples using combination of mass spectrometric and atomic emission spectroscopic detection (Fig. 16). The soil samples showed a variety of arsenic species with Clark I, triphenylarsine and bis(diphenylarsine)oxide (which is a typical breakdown product of Clark I) as the main contaminants. The identification of numerous compounds was possible by a combination of mass spectrometric information and elemental compositions resulting from GC/AES investigations.

Additional arsenicals were identified using LC/MS and LC/MS/MS investigations. By this way the identification of triphenylarsine oxide and phenylarsine oxide in the samples was possible. Oxidation has been found to occur during ionization process for some species. Thus it is not fully clear whether the oxidized forms $Ph_2As(O)OH$,

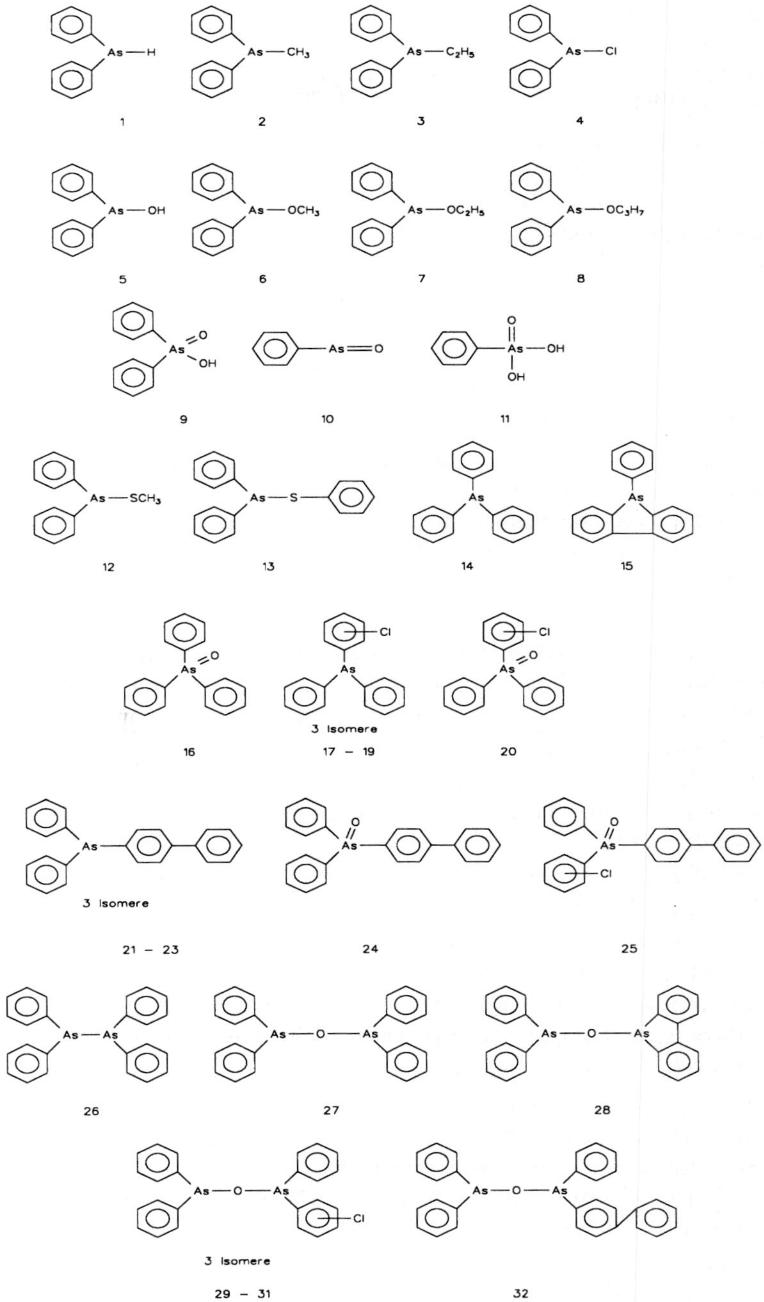

Figure 16. Summary of identified phenylarsenic species.

$Ph_2As(O)PhCl$ and (PhCl)PhAs(O)Ph-Ph are really sample constituents or only formed within the measurements.

Acknowledgements

The determination of the total arsenic content of the plants has been supported by the Landesamt für Umwelt und Natur Mecklenburg-Vorpommern, Germany. The analytical investigations have been supported by Hewlett Packard GmbH (Germany) and Gerstel GmbH (Germany). Further more we wish to thank Dr. Carol Haney from North Carolina State University Raleigh (USA) for support with the LC/MS and LC/MS/MS investigations and Dr. John Schneider from Argonne National Laboratory for support with portable XRF measurements.

References

Aller, A.J., Bernal, J.L., del Nozal, M.J., Deban, L., 1990. Effects of selected trace elements on plant growth. J. Sci. Food Agric. 51, 447–479.

Bowadt, S., Hawthorne, S.B., 1995. Supercritical fluid extraction in environmental analysis. J. Chromatogr. 703, 549–571.

Chilvers, D.C., Peterson, P.J., 1987. Global cycling of arsenic. In: Hutchinson, T.C., Meema, K.M. (Eds.), Lead, Mercury, Cadmium, and Arsenic in the Environment, Wiley, New York, 1987, pp. 279–301.

Cullen, W.R., Reimer, K.J., 1989. Arsenic speciation in the environment. Chem. Rev. 89, 713–764.

Edens, S.B., Haney, C.A., Nold, M.J., Thurow, K., 1998. HPLC/MS identification of organoarsenic compounds in soil. Proceedings South-East Regional Analytical Chemical Symposium, Raleigh.

EPA, 1992. Spectrace 9000 Field Portable X-ray Fluorescence Operating Procedure. SOP 1713 (Revision 1.0), US Environmental Protection Agency, Washington, D.C.

Flury, F., 1921. Über Kampfgasvergiftungen. Lokal reizende arsenverbindungen. Z Ges Exp Medizin 13, 523–579.

Gibb, H.J., Farland, W.H., 1992. Differences in animal and human responses to carcinogenic metals. Prog. Clin. Biol. Res. 374, 367–379.

Goldman, M., Dacre, J.C., 1989. Lewisite: its chemistry, toxicology, and biological effects. Rev. Environ. Contam. Toxicol. 110, 75.

Grassert, F., Stoll, N., Thurow, K., Baudisch, H., 1998. High resolution mass spectrometry for structure elucidation of chemical warfare agents. Proceedings I. International Symposium on Environmental Technologies. Rostock, pp. 36–37.

Hayes, R.B., 1997. The carcinogenicity of metals in humans. Cancer Causes Control 8, 371–385.

Henriksson, J., Johannisson, A., Bergqvist, P.A., Norrgren, L., 1996. The toxicity of organoarsenic-based warfare agents: in vitro and in vivo studies. Arch. Environ. Contam. Toxicol. 30, 213–219.

Hindmarsh, J.T., McCurdy, R.F., 1986. Clinical and environmental aspects of arsenic toxicity. Crit. Rev. Clin. Lab. Sci. 23, 315–347.

Jackson, K.E., 1935. Sternutators. Chem. Rev. 17, 251–292.

Jonnalagadda, S.B., Prasada Rao, P.V.V., 1993. Toxicity, bioavailability and metal speciation. Comp. Biochem. Physiol. 106c, 585–595.

Kolb, B., 1988. Applied Headspace Gas Chromatography. Wiley, New York.

Larsen, E.H., Moseholm, L., Nielsen, M.M., 1992. Atmospheric deposition of trace elements around point sources and human health risk assessment. II: Uptake of arsenic and chromium by vegetables grown near a wood preservation factory. Sci. Total Environ. 126, 263–275.

Markert, B., 1995. Sample preparation (cleaning, drying, homogenization) for trace element analysis in plant matrices. Sci. Total Environ. 176, 45–61.

Markert, B., Herpin, U., Siewers, U., Berlekamp, J., Lieth, H., 1996. The German heavy metal survey by means of mosses. Sci. Total. Environ. 182, 159–168.

Meharg, A.A., Naylor, J., Macnair, M.R., 1994. Phosphorus nutrition of arsenate-tolerant and non-tolerant phenotypes of velvetgrass. J. Environ. Qual. 23, 234–238.

Mothes, S., Wennrich, R., 1997. Capabilities for the direct determination of triphenyl- and triethylarsine in water by SPME followed by GC-AED. Am. Environ. Lab. 9, 10–12.

Otte, M.L., Rozema, J., Beek, M.A., Kater, B.J., Broekman, R.A., 1990. Uptake of arsenic by estuarine plants and interactions with phosphate, in the field (Rhine estuary) and under outdoor experimental conditions. Sci. Total Environ. 97/98, 839–845.

Oya Ohta, Y., Kaise, T., Ochi, T., 1996. Induction of chromosomal aberrations in cultured human fibroblasts by inorganic and organic arsenic compounds and the different roles of glutathione in such induction. Mutat. Res. 357, 123–129.

Pedersen-Bjergaard, S., Greibrokk, T., 1996. Environmental applications of capillary gas chromatography coupled with atomic emission detection - a review. J. High Res. Chrom. 19, 597–607.

Pellegrin, V., 1983. Molecular formulas of organic compounds: the nitrogen rule and degree of unsaturation. J. Chem. Educat. 60, 626–629.

Pitten, F.A., Müller, G., König, P., Schmidt, D., Kramer, A., 1998a. Arsenaufnahme von wildpflanzen auf einem kampfstoffkontaminierten rüstungsaltlastenstandort (Uptake of arsenic by plants grown at a former military base contaminated with compounds of chemical weapons). Z. Umweltchem. Ökotox. 10, 75–80.

Pitten, F.A., Müller, G., König, P., Schmidt, D., Kramer, A., 1998b. Risikoabschätzung der arsen-belastung von futterpflanzen, die auf einem rüstungsaltlastenstandort vom chemiewaffentyp wachsen. Zbl. Hyg. Umweltmed. 201, 44–45.

Pitten, F.A., Müller, G., König, P., Schmidt, D., Thuow, K., Kramer, A., 1999. Risk assessment of a former military base contaminated with organoarsenic-based warfare agents. Uptake of arsenic by terrestrial plants. Sci. Total Environ. 226, 237–245.

Quimby, B.D., Sullivan, J.J., Dryden, P.C., 1995. Automated Element Ratio and Quantitation Methods for Screening Unknowns Using GC/AED. Application Note 228–318. Hewlett Packard Company, Palo Alto.

Sabbioni, E., Fischbach, M., Pozzi, G., Pietra, R., Gallorini, M., Piette, J.L., 1991. Cellular retention, toxicity and carcinogenic potential of seafood arsenic. I. Lack of cytotoxicity and transforming activity of arsenobetaine in the BALB/3T3 cell line. Carcinogenesis 12, 1287–1291.

Sanders, J.G., 1980. Arsenic cycling in marine systems. Mar. Environ. Res. 3, 257–266.

Schneider, J.F., Johnson, D.O., Stoll, N., Thurow, K., Thurow, K., 1999. Portable X-ray fluorescence spectrometry characterization of a German military site for arsenic contamination in soil. J. Field Anal. Chem. 4, 12–17.

Schoene, K., 1993. Arsenkampfstoffe und ihre Abbauprodukte: erfassung, identifizierung und quantifizierung. In: Altlasten, Deponietechnik, Kompostierung. Hrsg. L. Schimmelpfeng, Akademia Verlag, St. Augustin.

Schoene, K., Bruckert, H.-J., Steinhanses, J., 1995. Analytik Kampfstoff-kontaminierter Rüstungsaltlasten. Erich Schmidt Verlag, Berlin.

Stijve, T., Vellinga, E.C., Herrmann, A., 1990. Arsenic accumulation in some higher fungi. Persoonia 14, 161–166.

Tamaki, S., Frankenberger, W.T., 1992. Environmental biochemistry of arsenic. Rev. Environ. Contamin. Toxicol. 124, 79–110.

Tanaka, T., 1988. Distribution of arsenic in the natural environment with emphasis on rocks and soils. Appl. Organomet. Chem. 2, 283.

Tena, M.T., Luque de Castro, M.D., Valcarcel, M., 1993. Systematic study of the influence of variables on the supercritical fluid extraction of polyaromatic hydrocarbons. Lab. Robot. Autom. 5, 255–262.

Thurow, K., Stoll, N., Thurow, K., 1997. Determination of organoarsenicals on military sites. Proceedings International Congress on Analytical Chemistry, Moskau, Russland. p. N-20.

Thurow, K., Koch, A., Stoll, N., 1998a. Direct determination of organoarsenic compounds in soil by thermal Desorption-GC and Headspace-GC. Proceedings Chemical Weapon Demilitarization Conference, Bournemouth, UK.

Thurow, K., Koch, A., Stoll, N., Thurow, K., Schneider, J.F., Johnson, D.O., 1998b. Screening of As contaminants in soils on-site with in-situ- and in-vitro XRF. Proceedings Chemical Weapon Demilitarization Conference, Bournemouth, UK.

Thurow, K., Haney, C.A., Nold, M.J., Koch, A., 1999a. Bestimmung von arsenverbindungen in rüstungsaltlasten mittels HPLC/MS und CE/MS. International Symposium on Instrumentalized Analytical Chemistry and Computer Technology, Düsseldorf.

Thurow, K., Koch, A., Stoll, N., 1999b. General approaches for the analysis of arsenic containing chemical warfare agents. Proceedings NATO Workshop: Environmental Aspects of Converting Former CW Facilities to Peaceful Purposes, Spiez.

Vetter, J., 1994. Data on arsenic and cadmium contents of some common mushrooms. Toxicon. 32, 11–15.

Wenclawiak, B.W., Krah, M., 1995. Reactive supercritical fluid extraction and chromatography of arsenic species. Fres. J. Anal. Chem. 351, 134–138.

Wylie, P.L., Sullivan, J.J., Quimby, B.D., 1990. An investigation of gas chromatography with atomic emission detection for the determination of empirical formulas. J. High Res. Chrom. 499, 506.

III
Botanical samples

Trace elements – Their distribution and effects in the environment
B. Markert and K. Friese, editors

Chapter 8

Environment and elemental content of lichens

J. Garty

Abstract

Of the large number of lichen species, only a few were applied as biomonitors of trace element deposition, in the frame of the global attempt to assess the rate of atmospheric contamination. This chapter represents a list of selected studies dealing with the elemental content of lichens, sampled in both populated and unpopulated regions, either in situ or transplanted. Research performed on this topic in the 1990s is at the focus of the present review which examines the following subjects: (1) accumulation of chemical elements in lichens around point-emission-sources: power plants, oil refineries, industrial plants, mines and refuse incinerators; (2) elemental content of lichens and vehicular activity in proportion to the reduced usage of lead as a fuel additive; (3) lichen-biomonitoring of airborne elements in large-scale surveys; (4) lichen biomonitoring of airborne elements in remote areas, including Arctic and Antarctic regions. Of the 171 references cited in the present chapter, 141 (82%) were published in the years 1990–2000.

Keywords: Lichens; Elemental content; Biomonitoring; Air pollution; Point emission sources; Vehicular activity; Large-scale surveys; Remote areas

1. Introduction

The capability of lichens to accumulate different elements, inclusive of trace metals, is a distinctive feature, extensively documented in the 1970s and 1980s (James, 1973; Nieboer et al., 1977, 1978; Puckett and Burton, 1981; Martin and Coughtrey, 1982; Brown and Beckett, 1984, 1985a,b; Lawrey, 1984; Arndt et al., 1987; Galun and Ronen, 1988; Puckett, 1988; Richardson, 1988; Nash, 1989; Tyler, 1989).

In the 1990s this characteristic of lichens was reviewed by the following authors (Brown, 1991; Brown and Brown, 1991; Garty, 1992, 1993; Richardson, 1992; Nash, 1996; Jacquiot and Daillant, 1997).

This chapter discusses selected articles dealing with the application of lichens as biomonitors of chemical elements in the environment. In the recent decade the relevant literature focused on the methods available to monitor effects of air pollution by means of in situ, resuspended or transplanted lichens in urban, industrial, rural and remote environments. These monitoring studies described spatial variations in the elemental content of lichens, relative to distance from natural or anthropogenic emission points apart from temporal fluctuations relative to climatic circumstances.

2. Accumulation of chemical elements in lichens around point emission sources

In the 1990s, point-emission-sources related to anthropogenic activities, were documented to a greater extent than natural emission sources of pollutants. Pollutants accumulated in lichens are emitted by diverse industrial plants. The elemental content and the correlation of certain pairs of mineral elements indicate definite sources of pollution.

2.1. Accumulation of chemical elements in lichens around power plants and oil refineries

Recent literature dealt to a large extent with airborne elements emitted by power plants using fossil fuel, such as coal, heavy fuel oil, oil shales and peat. The environmental impact of the emission of trace elements by geothermal power plants was dealt with to a lesser extent. The greater part of the investigations of power plant emissions, airborne pollutants and lichens as monitors were performed in temperate zones, whereas arid and semi-arid zones received little attention. Certain studies, related to arid and semi-arid zones, confined their focus of interest to the elemental content of lichens (Nash and Sommerfeld, 1981). Another study centered on physiological alterations in lichens: chlorophyll degradation and electrolyte leakage, but did not relate to the elemental content of the thallus (Benlap and Harper, 1990).

Coal-fired power plants are well-known sources of SO_2, NO_2, heavy metals and other elements (Fig. 1). Lichens applied as monitors near coal-fired power stations in Portugal were found to accumulate heavy metals such as Fe, Co, Cr, and Sb, originating in coal and ash particles drifting through the air and positioned on the thallus (Freitas, 1994).

In an additional study, Freitas (1995) analyzed the comparative accumulation of Cr, Fe, Co, Zn, Se, Sb and Hg in two vascular plants, *Cistus salvifolius* and *Inula viscosa* and in the epiphytic lichen *Parmelia sulcata* in an industrial region occupied by a thermal coal-fired power station, a chemical plant and an oil refinery. Of the three organisms, the lichen *P. sulcata* was found to be the most effective bioaccumulator, as the above-mentioned elements accumulated in large amounts in the lichen. Thus, *P. sulcata* was recommended as a reliable bioindicator of air pollution.

Another study, carried out in the Iberian Peninsula, reported on the response of epiphytic lichens affected by air pollution in the vicinity of a coal-fired power station in northern Spain (Terron Alfonso and Barreno Rodriguez, 1994). Certain parameters were developed to allow the differentiation of three isocontaminated areas. The dominant elements in the ash emitted by the power station were Si, Al, Fe, Ca and K. The ash contained additional elements in smaller amounts. However, no report was made of the elemental content of the lichens.

Passive biomonitoring was applied to estimate the air quality in the La Spezia district, Italy, with regard to a coal-fired power plant and other industrial activities (Nimis et al., 1990). The lichen *Parmelia caperata* collected from *Olea sativa* trees was meant to biomonitor the SO_2 pollution in the study area. The method, adopted by Nimis et al., was designed to avoid subjectivity, to begin with the strategy of sampling and end with the analysis of data. The applied index, based on frequency of species

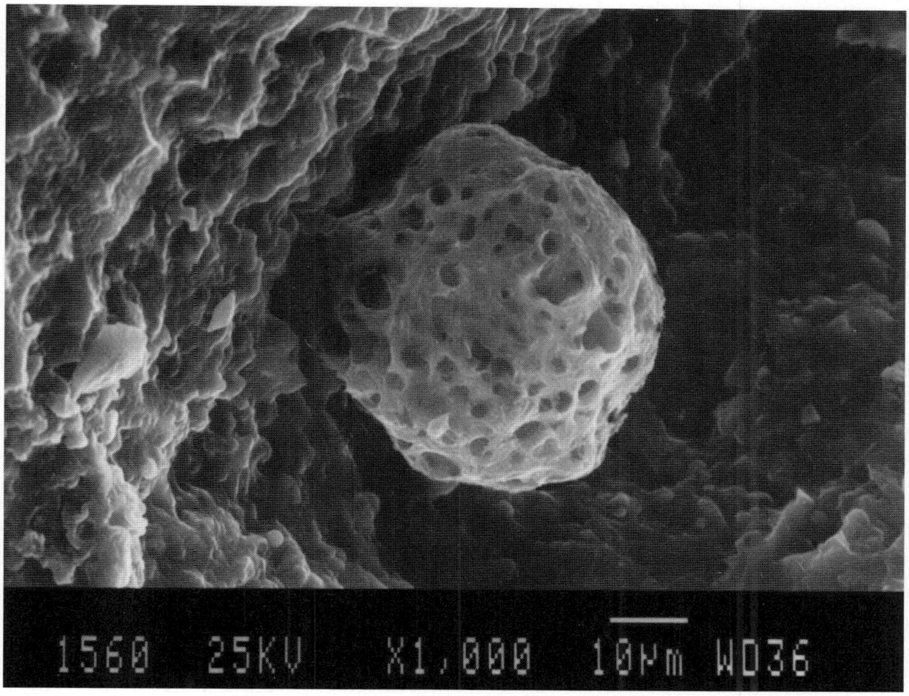

Figure 1. Large particle, derived from coal-combustion, detected on the thallus surface of in situ *Ramalina lacera* near the Oroth-Rabin coal-fired power plant near Hadera, Israel, in 1995. Scale bar, 10 µm (Garty, Kunin, Delarea, unpublished).

within a sampling grid, showed a high statistical correlation with pollution data measured by recording gauges. The distribution of the lichen *P. caperata* was found to relate best to the lichen index. *P. caperata* was used also as a bioindicator of heavy metal pollution in La Spezia (Nimis et al., 1993). Data on the amounts of 13 metals in lichen thalli in 30 stations were compared with data of lichens collected in other parts of Italy. No substantial phenomena of pollution related to Mn and Zn whereas extreme phenomena of pollution were found to relate to high levels of Pb and Cd in an area adjacent to an industrial zone. Within the industrial zone phenomena of evident deposition related to Al, As, Cr, Fe, Ni, Ti and V.

The La Spezia region was a subject of research for an additional group of researchers (Palmieri et al., 1997). Initial records made in 1989 detected enlarged amounts of metals in thalli of *P. caperata*. In some sites, deposits of Pb, Cd, Ni, As and Zn reached values which were five to ten times higher than the minimum values recorded in the surveyed area and than background values. The authors related this phenomenon to the metal content of the fly ash released from the power plant. The inaction of the power plant for a period up to two years led to a decrease of metal content, with the exception of Zn which is a constituent of organic compounds in local vineyards.

Vanadium is a frequent constituent of stack-emissions of oil-fired, power-generating stations. Juinchang et al. (1995) reported on the vanadium contamination of lichens in the vicinity of three oil-fired power plants in Eastern Canada. The V content of *Hypogymnia physodes, P. sulcata* and *Usnea strigosa*, sampled along ten transects in the vicinity of the plants, was found to decrease exponentially with the increase of distance from the plants. Lichens were detected as preferential biomonitors of environmental V-contamination near oil-fired power plants over and above tree foliage of spruce (*Picea glauca*), balsam fir (*Abies balsamea*) and red maple (*Acer rubrum*).

Transplants of the lichen *Punctelia subrudecta* in a polluted environment close to a thermal power station, fired by fuel oil and diesel oil, near Cordoba, Argentina, showed a significant variability of S content in accordance with industrial density and varying distance from the power station (Gonzales and Pignata, 1997). The authors observed significantly higher values for S content in samples located at a distance from the power station in sites with a low industrial density, probably as a result of the limited proportions of the study area and the height of the chimneys. In the immediate vicinity of the power station, however, the amounts of S increased significantly in accordance with high industrial density.

Six studies were designed to determine the environmental impact of pollutants emitted in the process of combustion of heavy fuel oil in the region of Ashdod, south-west Israel, by means of the fruticose lichen *Ramalina lacera,* defined in several publications as *Ramalina duriaei* (Garty et al., 1997a,b,c,d, 1998a,b). Lichen thalli growing on carob trees (*Ceratonia siliqua*) in the HaZorea Forest, north-east Israel, were transferred with their natural substrate and transplanted to the Ashdod region. It was confirmed that the power station and oil refineries were the major sources of SO_2 and metal-containing particles in this region. *R. lacera* transplants retrieved from different sites in the Ashdod region contained relative large amounts of S, Ni, and V. Nickel combined with V is considered to be a tracer of heavy fuel oil combustion in electricity-generating power stations (Ganor et al., 1988; Foner and Ganor, 1992). The Pearson correlation coefficients obtained for the amounts of Ni, V and S accumulated in lichen samples collected in HaZorea, resuspended in the same site or transplanted to the Ashdod region are presented in Table 1. High correlation coefficients suggest a common source of pollution, i.e., the combustion of heavy fuel oil. The consumption

Table 1. Pearson correlation coefficients of Ni, V and S values in *Ramalina lacera* collected in HaZorea, north-east Israel, resuspended in same site or transplanted to Ashdod region, south-west Israel

Period of exposure (and references)	Ni, V		Ni, S		V, S	
	r	*P*	*r*	*P*	*r*	*P*
March 1993–January 1994 (Garty et al., 1997c)	0.91	<0.001	0.57	<0.05	0.57	<0.05
February–November 1994 (Garty et al., 1997d)	0.93	≤0.005	0.83	≤0.005	0.75	≤0.05
February–November 1994 (Garty et al., 1998a)	0.94	<0.005	0.86	≤0.005	0.87	≤0.005
November 1994–August 1995 (Garty et al., 1997d)	0.96	≤0.001	0.75	≤0.01	0.70	≤0.05

of heavy fuel oil in the Eshkol Power Plant in 1993 was 1 092 716 tons and the calculated emission of SO_2 was 48 121 tons. The consumption of heavy fuel oil in the oil refineries in the same year was 223 047 tons and the calculated emission of SO_2 was 6693 tons.

In the majority of our studies using *R. lacera* transplants to monitor air pollution in the Ashdod region, we applied an additional parameter to provide further information on the nature of airborne elements in the study area. We have found previously (Garty et al., 1977) that coefficients of variation (CV) obtained by SD/mean, were higher for samples of the epilithic crustose lichen *Caloplaca aurantia* in polluted sites in the Tel Aviv area, than for samples located in rural sites. It is suggested that the dissimilarity of CVs indicates a difference between the dispersion capacity of metal particles. Low CVs indicate a low variation in the air due to an even suspension of minute particles; high CVs, on the other hand, indicate a limited, localized deposition of coarse particles. This assumption was adopted in previous studies (Garty, 1985; Garty and Ammann, 1987; Loppi et al., 1994a; Garty et al., 1995, 1996). Table 2 shows the CVs calculated for 21 mineral elements, obtained by SD and mean values, determined in different transplantation experiments (see refs. in Table 2). Table 2 presents in addition the grand averages of the mean CV values, calculated for three CV values or more. The mean CVs were the lowest for S and K and considerably higher for Pb, Ni, Al and Cr. The low CVs for S and K reflect their non-particulate form whereas the high CVs for Pb, Ni, Al and Cr reflect their particulate form, thus representing the impact of airborne particles deposited on the lichen thallus (Figs. 2–4).

Well-known sources of air pollutants in the eastern part of the Gulf of Finland are the energy-plants of Narva in Estonia near the Russian border which use oil-shales, and other power stations and industrial plants including oil refineries in the Leningrad region. Haapala et al. (1996) emphasized the ecological conditions, predominant in woodlands, in their study of the impact of pollutants in this region. At the margin of a lichen desert where no lichens or only small individuals of epiphytic lichens subsist, the authors detected the lichen *Xanthoria parietina*, which grows normally on *Populus tremula* as well as a red-colored green alga, *Trentepohlia umbrina* on pine trunks, under exceptional alkaline conditions. The emission of alkaline which was as large as the emission of sulfur, was assumed to neutralize the acidic effects, creating favorable conditions for extraordinary species to grow on pine trunks. The lichen *X. parietina* and the alga *T. umbrina* may therefore be regarded as indicators of alkaline pollution.

A peat-burning district-heating plant is a major point source of SO_2 pollution in Oulu, north Finland (Fig. 5). A sulfate pulp mill, a chemical factory and a heating plant belonging to the University Central Hospital, which burns heavy oil, wood chips and hospital waste, are additional producers of air contaminants (Kauppi and Halonen, 1992). In a study aimed to determine the extent to which the relatively sparse lichen flora on the trunks of Scots pines (*Pinus sylvestris*) can serve as an indicator of air purity, Kauppi and Halonen carried out a survey of percentage cover and conditions. Pine needles and thalli of *Hypogymnia physodes* were sampled for chemical analyses. The lichen species *Platismatia glauca* and *Hypogymnia tubulosa* differed in their tolerance of pollution as the latter was found to be the most sensitive of the macrolichens, and *H. physodes* and *Parmeliopsis ambigua* the most tolerant. With decreasing distance from the city center, the numbers of macrolichen species and the size and

Table 2. Mean coefficients of variation obtained by the SD/mean of elemental content in *Ramalina lacera* collected in HaZorea, north-east Israel, resuspended in same site or transplanted to Ashdod region, south-west Israel

	March–June 1993 (Garty et al., 1997a)	March 1993–January 1994 (Garty et al., 1997c)	March 1993–January 1994 (Garty et al., 1998b)	February–November 1994 (Garty et al., 1997d)	February–November 1994 (Garty et al., 1998a)	November 1994 –August 1995 (Garty et al., 1997b)	Grand average of the means (horizontal calculation)
S	0.09	0.09	0.12	0.10	0.12	0.07	0.10
Ni	0.41	0.18	–	0.27	0.26	0.32	0.29
V	–	0.19	–	0.18	0.18	0.17	0.18
Al	–	–	0.19	–	0.28	0.18	0.22
Cr	–	–	0.20	0.27	0.26	0.25	0.24
Fe	0.12	–	0.20	0.20	0.19	0.18	0.18
K	0.15	–	0.10	0.13	0.12	0.11	0.12
P	–	–	0.14	0.16	0.16	0.19	0.16
Ti	–	–	0.20	–	0.25	–	*
Zn	–	–	0.20	–	0.14	–	*
Cu	0.24	–	0.12	0.13	0.12	–	0.15
Pb	0.31	–	0.36	0.30	0.30	–	0.32
Si	–	–	0.24	–	0.27	–	*
Mg	0.11	–	0.10	0.18	0.18	–	0.14
B	–	–	0.25	–	0.20	–	*
Sr	–	–	–	–	0.30	–	*
Mn	0.10	–	–	0.17	0.16	–	0.14
Ca	0.22	–	–	–	0.19	–	*
Na	0.22	–	–	–	0.17	–	*
Ba	–	–	–	–	0.22	–	*
Cd	0.35	–	–	–	–	–	*

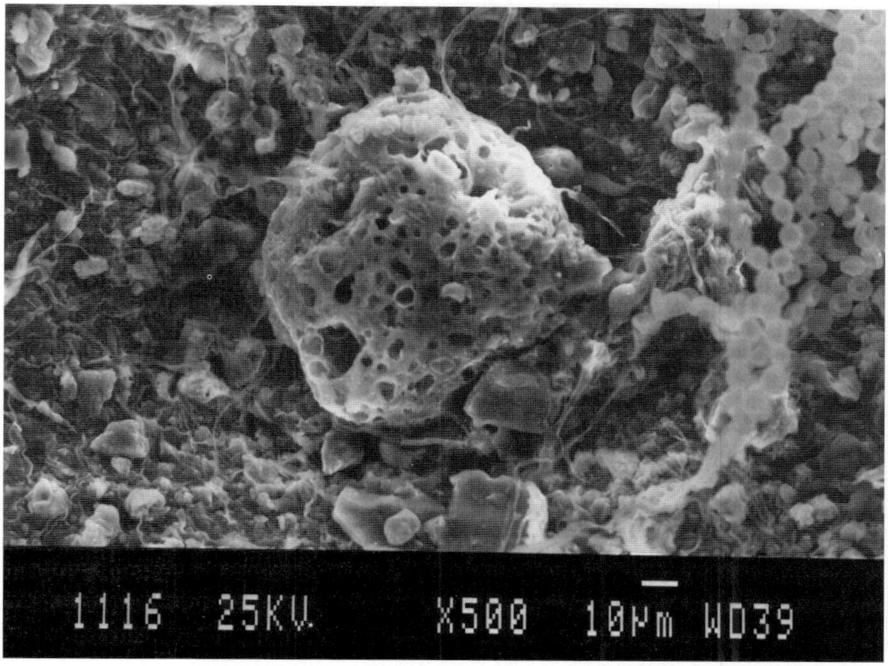

Figure 2. Large particle, derived from heavy-fuel combustion, detected on the thallus surface of in situ *Ramalina lacera* in the eastern part of the *Faidherbia (Acacia) albida* nature reserve north-east of Ashdod, south-west Israel, 1994. Scale bar, 10 µm (Garty, Kunin, Delarea, unpublished).

percentage-cover diminished, their external appearance changed markedly and the number of impurities increased. The values obtained for Cd in thalli of *H. physodes* in Oulu were similar to control values whereas S, Mg, Al, Fe and Ti reached high values relative to control values.

Lichens accumulate airborne trace elements in volcanic and geothermal areas (Barghigiani et al., 1990; Bargagli and Barghigiani, 1991; Bargagli et al., 1991; Dongara and Varrica, 1998). The emission of toxic trace elements from geothermal power plants is likely to pose a long-term environmental hazard. Certain researchers applied lichens as biomonitors of Hg emission and deposition in an area of geothermal and volcanic activity in Italy. Bargagli and Barghigiani (1991) found that the Hg content of the lichens *Physcia biziana* and *X. parietina* is related to their distance from the geothermal power plant. The Hg content of *X. parietina*, for example, which grows at a distance of 150–200 m from the geothermal power plant was 0.57 ± 0.13 $\mu g\ g^{-1}$ d.w. whereas samples of the same lichen species at a distance of 1000–5000 m from the power plant contained only 0.19 ± 0.03 $\mu g\ g^{-1}$ d.w. of Hg. Similarly, the authors found 3.08 ± 0.57 $\mu g\ g^{-1}$ d.w. Hg in samples of *P. sulcata* collected at a distance of 100–200 m from a geothermal power plant in Monte Amiata, and only 0.84 ± 0.28 $\mu g\ g^{-1}$ d.w. of Hg in samples at a distance of 1000–5000 m.

An additional study of the environmental impact of trace element emissions from geothermal power plants was performed by Loppi (1995). The author measured the

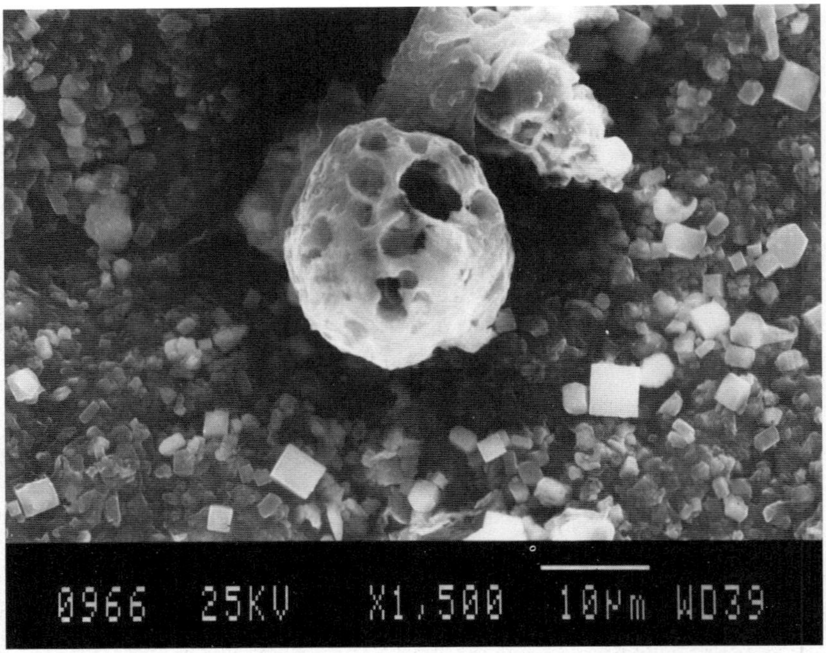

Figure 3. Large particle, derived from heavy-fuel combustion, detected on the thallus surface of *Ramalina lacera*, collected in the HaZorea Forest, north-east Israel, February 1994, transplanted to the industrial zone of Ashdod, south-west Israel and retrieved in November of the same year. Scale bar, 10 μm (Garty, Kunin, Delarea, unpublished).

amounts of 16 trace elements accumulated in thalli of the epiphytic lichen *P. caperata* collected in the Travale-Radicondoli geothermal field (central Italy). The amounts of Co, Mo and Sb were found to be higher in the geothermal area with respect to background values. At the above-mentioned site the average amounts of Al, As, B, Cd, Cu, Fe, Hg, Mn, Pb and Zn were high within 500 m from the geothermal power plant and decreased with distance. At a distance above 1 km from the power plant, the values were found to agree with background data except for As values. An additional study which applied the same lichen in the same study area revealed that B was the principal element of geothermal origin. Arsenic was assumed to originate from both power plants and adjacent thermal springs (Loppi and Bargagli, 1996).

Another study of *P. caperata* and oak leaves (*Quercus cerris*) detected greater amounts of As and B in the vicinity of the above-mentioned geothermal area than in remote areas (Loppi et al., 1997). As the average amounts of many trace elements such as As, Cd, Cu, Fe, Hg, Pb and Zn were higher in lichens than in oak leaves, the authors pointed to the preference of lichens in similar biomonitoring studies. It was suggested (Loppi et al., 1998) that H_2S is a major source of atmospheric sulfur and one of the pollutants responsible at large for the decline of lichens around the geothermal installations in the Travale-Radicondoli area. The joint effect of noxious pollutants like As and Hg was found to be detrimental to both lichens and crowns of oak trees. In a lichen bioindication survey of air quality in the Mount Amiata geothermal area,

Figure 4. Fly-ash particles (arrows) derived from heavy-fuel combustion, detected on the thallus surface of *Ramalina lacera* collected in the HaZorea Forest, north-east Israel in November 1994, transplanted to Kibbutz Hafetz Hayyim, 14 km east of the town of Ashdod, south-west Israel and retrieved in August 1995. Scale bar, 10 μm (Garty, Kunin, Delarea, unpublished).

Loppi and Nascimbene (1998) obtained the lowest values according to the Index of Atmospheric Purity (IAP) for an area which encompasses the geothermal power plant, up to a distance of about 500 m. The overall pattern of rising IAP values with the decrease of distance from geothermal power plants was definitely related to air contamination by geothermal installations. This pollution was the major cause of the observed zonation of lichen communities. H_2S was assumed to be the major contaminant responsible for lichen decline around geothermal power plants.

2.2. Elemental content of lichens near industrial plants, mines and refuse incinerators

Lichens were applied to delineate the distribution of airborne chemical elements emitted by industrial plants such as metal works and foundries (Belandria et al., 1991; Kansanen and Venetvaara, 1991; Manninen et al., 1991; Perkins, 1992; Zanini et al., 1992; Gailey and Lloyd, 1993; Caniglia et al., 1994; Jovanovic et al., 1995), chlor alkali plants (Makholm and Bennett, 1998), fertilizer manufacturers and strip mines (Palomäki et al., 1992; Tynnyrinen et al., 1992), mercury mines (Bargagli, 1990; Lupsina et al., 1992) and uranium mines (Fahselt et al., 1995; Jeran et al., 1995).

Figure 5. Peat used in the peat-burning plant in Oulu.

Figure 6. Twigs of *Pinus sylvestris* covered by *Hypogymnia physodes* and *Usnea hirta*, detached from trees in a natural forest in Peuhu, Oulunsalo, 15 km south-west of Oulu, N Finland, and resuspended in the same site as control samples, in the frame of a biomonitoring survey in the Oulu region (Garty et al., 1996).

Lichens were used to monitor air pollutants around cement works (Kortesharju et al., 1990), phosphate refineries (Dillman, 1996), cellulose factories (Bartok and Ostvath, 1990; Kytömaa et al., 1995) and areas liable to endure the negative impact of different industrial enterprises in addition to vehicular activities (Walther et al., 1990; Punning et al., 1991; Lackovicova et al., 1994; Takala et al., 1994; Bargagli et al., 1997; Carreras et al., 1998; Gonzales et al., 1998; Gonzales and Pignata, 1999). Table 3 presents a list of anthropogenic activities biomonitored by different lichen species as well as a list of chemical elements detected in the in situ/transplanted samples.

Lichens were applied to outline atmospheric emissions of hazardous gases and particulates around solid waste incinerators. A study related to this subject was carried out in the surroundings of Poggibonsi, central Italy. The establishment of floristic lists of lichens subsisting on different phorophytes, was the focus of this study apart from analyses of the chemo-physical characteristics of different phorophytes (Loppi and Putorti, 1995). An additional study performed in the above-mentioned area applied an IAP to confirm that the incinerator has no impact on air quality, at least at a radius of three km (Loppi et al., 1995b). However, Palmieri et al. (1997) in a study of lichens and mosses as bioindicators and bioaccumulators detected a high Pb content in samples close to a waste incinerator and a producer of Pb oxide in the La Spezia district, Italy. Epidemiological investigations proved that the Pb level in blood samples of the local population was rather high.

The environmental impact of the renovation of the Grenoble refuse incinerator was studied by means of the lichen *P. sulcata* and samples of a *Sphagnum* species, within a radius of 4 km, by Gombert and Asta (1997). The authors observed a significant decrease of chloride levels in in situ lichens in each direction to a distance as great as 1000 m from the incinerator, except for the south-west direction. Less than a year later, at the time of collection, a general decline was apparent, brought about by a decrease of chloride emissions. Transplants of *Sphagnum* mosses were found to be preferential over lichens as bioindicators of low chloride concentrations. Recently, Gombert and Asta (1998) used nine corticolous lichens to detect Pb and Cd emission of a municipal solid waste incinerator in Grenoble, France. This study involved lichens from unpolluted sites put through different environmental circumstances: the first batch was set up down-wind of the waste incinerator fumes, the second was set up in the ambient atmosphere of the plant, and the third was kept in a non-polluted site. After 1 month, lichens of the first and second batch did not exhibit any appreciable change. Pb and Cd were considered to be entrapped in the initial stage in a particulate form, as lichens accumulate metals in variable ways in accordance with species and conditions of pollution.

3. Elemental content of lichens and vehicular activity

Lichens, especially epiphytic lichens, are increasingly used to monitor air pollutants emitted by vehicles and by a wide range of industrial enterprises. Many publications showed that vehicular and industrial activities are the main source of heavy-metal air-pollution. Mosses, lichens, pine needles and tree bark and leaves are widely used to

Table 3. Anthropogenic activities biomonitored by different lichen species: selected publications in the 1990s

Anthropogenic activity	Lichen species	Chemical elements/ compounds	Reference
Ferrochrome and steel works	*Hypogymnia physodes*	Cr, Ni	Kansanen and Venetvaara (1991)
Steel works	*Xanthoria parietina*	Mn, Ni, Zn, Cu, Cr, Pb	Zanini et al. (1992)
Steel works	*Evernia prunastri*	Cr, Mn, Fe, Ni, Cu, Zn, Pb	Caniglia et al. (1994)
Steel works and oil refinery	*H. physodes*	S	Manninen et al. (1991)
Steel works	*H. physodes*, *Lecanora conizaeoides*	Fe, Mn, Zn, Pb, Cu, Cr, Ni, Cd	Gailey and Lloyd (1993)
Aluminum works	*Ramalina siliquosa*, *R. subfarinacea*	F	Perkins (1992)
Aluminum works	*Peltigera canina*, *Umbilicaria cylindrica*, *U. pustulata*, *Cladonia* gr. *pyxidata*, *Usnea muricata*, *H. physodes*, *Pseudevernia furfuracea*, *Parmelia conspersa*, *Bryoria fuscescens*	F	Belandria et al. (1991)
Aluminum works	*Hypogymnia caperata*, *Diploica canescens*, *Lecanora expallens*	33 Elements	Jovanovic et al. (1995)
Chlor alkali manufacturers	*H. physodes*	Hg	Makholm and Bennett (1998)
Fertilizer manufacturer, strip mine	*H. physodes*, *Bryoria capillaris*	F, S, N, P, K, Ca, Mg	Palomäki et al. (1992)
Fertilizer manufacturer, strip mine	*H. physodes*	F, S, N, P, K, Ca, Mg	Tynnyrinen et al. (1992)
Mercury mine	*Parmelia sulcata*	Hg, Al	Bargagli (1990)
Mercury mine	*H. physodes*	Hg, methyl mercury	Lupsina et al. (1992)

Uranium mine	*Cladina mitis*, *Cladina rangiferina*	Na, Sc, Cr, Zn, Rb, Cs, La, Ce, Sm, Th, U	Fahselt et al. (1995)
Uranium mine	*H. physodes*	Natural radionuclides	Jeran et al. (1995)
Cement works	*Cladina stellaris*	Ca, K, Mg, Na, N, P, Cu, Fe, Mn, Pb, Zn	Kortesharju et al. (1990)
Phosphate refineries	*Rhizoplaca melanophthalma*	Cd, Cr, Zn, Cu, P, Ca, Mg, K, Mn, B, Pb, Na, Al, Fe, Ni	Dillman (1996)
Cellulose manufacturer	*Xanthoria parietina*, *Physcia aipolia*	NH_4	Bartok and Osvath (1990)
Sulfite-cellulose manufacturer	*H. physodes*	Al, Fe, Zn, Cd, Hg	Kytömaa et al. (1995)
Industrial and urban sites	*Ramalina stenospora*, *Parmotrema praesorediosum*	Al, Cu, Fe, Zn, Pb	Walther et al. (1990)
Industrial and urban sites	*H. physodes*, *Pseudevernia furfuracea*, *Cladina* sp., *Peltigera aphtosa*	Fe, S, Ti	Takala et al. (1994)
Large industrial zone	*Cladina* spp., *H. physodes*	Zn, Pb, Cu, Cd	Punning et al. (1991)
Industrial zone	*H. physodes*	Hg, As, Cu, Pb, Cd, Zn, Mn	Lackovicova et al. (1994)
Industrial zone	*Parmelia caperata*, *P. sulcata*, *Xanthoria parietina*	Al, Cd, Cr, Cu, Fe, Hg, Mn, Ni, Pb, Zn	Bargagli et al. (1997)
Industrial zone + car manufacturers	*Usnea* sp.	S	Carreras et al. (1998)
Industrial zone	*Ramalina ecklonii*	S	Gonzales et al. (1998)
Industrial zone	*R. ecklonii*	S	Gonzales and Pignata (1999)

monitor trace element deposition. Various techniques of lichen biomonitoring in surveys of atmospheric heavy-metal deposition contributed considerably to our knowledge of anthropogenic air- contamination.

In recent times many investigations were carried out to estimate the air quality near roads, road intersections, highways and industrial plants, by means of either in situ or transplanted lichens. Emissions by motor vehicles include nitrogen oxides (NO_x), carbon monoxide (CO), sulfur dioxide (SO_2), hydrocarbons, volatile organic compounds (VOCs), suspended particulate matter and a group of attendant heavy metals (Campo et al., 1996). Of the heavy metals accumulated in lichens, Pb received the greatest attention as it is highly toxic to humans. The Pb content in lichens located near sites of heavy traffic may reach very high levels, e.g. 494 $\mu g\ g^{-1}$ detected in *P. caperata* in La Spezia, N Italy by Nimis et al. (1993). However, this heavy metal is not exclusive to emissions related to vehicular activity as it is detected as a major contaminant around metal smelters and refuse incinerators.

The combustion process in automobiles causes the release of elements such as Zn, Cd and Cu (Ormrod, 1984). Tire wear causes the release of Zn (Christensen and Guinn, 1979; Beckwith et al., 1985; Ward, 1989), Cd (Ormrod, 1984; Ward, 1989), Cu, Fe, Mn, Pb and Ni (Sadiq et al., 1989). Ward (1989) detected other elements such as Cd, Cu, V and Zn in lubricating oils, Cr, Ni and V in car metal plating and Mn, Cu and Zn in brake pads.

Studies related to heavy metals and lichens near motorways, carried out in the last decade, reveal that in certain species such as *Lecanora conizaeoides, Evernia prunastri* and *H. physodes,* growing in France, an equilibrium with the atmospheric Pb is reached very rapidly (Deruelle, 1992). Another study described the biomonitoring of SO_2 and metal pollution by in situ thalli of *P. caperata* in Savona, north Italy (Castello et al., 1994). Vehicular activity was indicated as a major source of Pb. The exponential fall down of Pb suggested that the lichen data were influenced to a great extent by their distance from heavily loaded roads. Campo et al. (1996) performed three seasonal surveys at eleven different extents from the edge of the main road (<10–100 m) in the higher tract of the Susa Valley, north-west Italy, to evaluate the motorway pollution in a mountain ecosystem. The largest amounts of Pb, Cr and Cd were found within 10 m from the edge of the main road. According to Semadi and Deruelle (1993) who transplanted samples of *Ramalina lacera* and *R. farinacea* from an unpolluted area to sites adjacent to three highways in Algeria, the accumulated Pb shows that vehicular pollution reaches an extent of 70 m off the road.

Castello et al. (1995) studied the SO_2 pollution and deposition pattern of Al, Cd, Cr, Cu, Fe, Mn, Ni, Pb and Zn in the province of Trieste (north-east Italy) by means of a lichen-based biomonitoring technique. An IAP based on frequency counts of epiphytic species was applied to record SO_2 contamination. The metal deposition in the area at the outskirts of the lichen desert was estimated by means of the lichen *P. caperata*. The most important phenomena of metal pollution were noted with regard to Cr, Pb and Ni, which showed exceptionally high background and maximum values, the latter being reached in a subset of stations located near high traffic roads. The authors assumed that Cr originated in industrial activity apart from vehicular activity. A moderate pollution by Cu and Zn could derive also from domestic heating.

Glenn et al. (1995) studied the metal content and community-structure of cryptogam bioindicators in relation to vehicular traffic in the Montsery biosphere reserve in Catalonia, Spain. The authors analyzed the metal content of the lichens *Parmelia subrudecta, P. sulcata* and *Anaptychia ciliaris*. The corticolous lichens and bryophytes detected in sites along paved roads subsisted in a poorly state of health relative to lichens found in sites accessible only by dirt roads. Roadside lichens showed Pb levels indicating that traffic along tourist roads had a measurable effect on vegetation. The results obtained in this study appeared to support the view that exhaust particles do not travel much beyond the roadsides, but do exert an apparent influence on vegetation at roadsides, whereas gaseous pollution arising from vehicular exhausts travels much further.

In Montecatini Terme (central north Italy), Loppi et al. (1995a) used in situ thalli of *P. caperata* to monitor heavy metals. The levels of Cd and Zn were found to be similar to levels detected in unpolluted areas, whereas Cu and especially Pb reached high values. The results of this study, which suggested that vehicular traffic is the major source of heavy metals, agreed with the incidence of motorized traffic in the area and the accompanying level of noise.

Monaci et al. (1997) performed a lichen-biomonitoring study in Siena by means of two different methods. The pattern of air quality in the study area was examined on the basis of the in situ frequency of different species of epiphytic lichens, i.e. using their species-specific sensitivity to the complex mixture of phytotoxic pollutants in the urban environment. The distribution of trace elements was evaluated quantitatively by an analysis of thalli of a tolerant species, *P. caperata*, known to be a reliable bioaccumulator of persistent atmospheric pollutants. The values obtained for Al, Ba, Cr, Cu, Fe, Pb and S were significantly higher in Sienese lichens over and above controls. Traffic was found to be the major source of atmospheric pollution. The pattern of trace-elemental deposition did not always coincide with air quality. IAP values were found to reflect essentially the emission of gaseous phytotoxic pollutants in the urban environment.

Bargagli et al. (1997) performed a study which applied a method of passive biomonitoring of trace element deposition in urban, industrial and reference areas in Italy, especially in Tuscany, Veneto and Friuli, by means of epiphytic lichens, mainly *P. caperata, P. sulcata* and *X. parietina*. Despite the fact that the Pb content of gasoline in Italy was lowered from 0.4 to 0.15 $g\,l^{-1}$ in 1991, samples of *P. caperata* collected in the urban environment of Siena still contained considerable amounts of Pb. The data presented in this study included average amounts of Al, Cd, Cr, Cu, Fe, Hg, Mn, Ni, Pb and Zn in lichens from five urban areas in northern and central Italy, from the entire region of Veneto, Tuscany and from remote areas in Calabria. It was shown that the pattern of urban pollution was characterized to a great extent by high values of Al, Cd, Cu, Ni and Pb. Traffic appeared to play a prominent role as these metals are known to be released by vehicular activity.

A passive, nationwide monitoring study of atmospheric heavy-metal deposition, performed in Finland by Lippo et al. (1995) detected the highest Pb content in the lichen *H. physodes* in southern Finland and around Oulu (northern Finland). The findings of Kubin (1990) agreed with this conclusion with regard to southern Finland. The lowest Pb values were distributed fairly even among Lapland and south-east Finland.

The region of Oulu was the subject of additional studies designed to obtain information on the accumulation of airborne elements of vehicular origin by means of transplanted lichens (Garty et al., 1996). The lichens *H. physodes* and *U. hirta* were collected in a relatively unpolluted site near Oulu and transplanted in the vicinity of streets of low volume and slow traffic and a highway in the city of Oulu for a period of 45 days. Relative large amounts of K, Fe, Mg, Zn, Mn, Pb and Cu were detected in thalli of *H. physodes* and of Mg, Zn, Pb, Cu and Cd in thalli of *U. hirta* in transplanted material in streets of low volume and slow traffic whereas the elemental content of thalli in the vicinity of the highway was much smaller. This may be explained by the higher rate of abrasion of car engines running idle near traffic lights and by the lesser ventilation in the close-clustered streets of the inner city. (Fig. 6)

The above-mentioned findings which showed that the Pb content in transplanted lichens was not in direct proportion to the density of traffic, agreed with the data compiled by Sawidis et al. (1995) who applied trees to monitor the heavy metal air pollution in Thessaloniki City (Greece). Two research sites were found to be the most severely contaminated by Pb and Cu. The increased Pb content in the leaves collected in these sites was assumed to be brought about by the density of traffic. Yet, the amount of Pb detected in the site located near the railway station was more than twice the amount detected in the other site whereas the load of traffic was reversed. The number of road junctions, which restrain the rapid progress of vehicles, should be taken into account. The described phenomenon with regard to Pb and to a lesser extent to Cu was interpreted as deriving from other factors than the density of traffic.

Certain studies which dealt with the biomonitoring of air pollutants of vehicular or industrial origin did not apply Pb content as an exclusive parameter to define the degree of pollution. Gonzales et al. (1996) found that in two sites in Cordoba, Argentina, transplants of *Ramalina ecklonii* exposed to heavy vehicular and medium industrial pollution accumulated the largest amounts of S, whereas the smallest S-content was found in freshly picked material. These findings obviate the necessity of research of additional chemical elements of vehicular origin in plant material. Monaci and Bargagli (1997) stated on the basis of a significant connection between Pb and Ba content in *Quercus ilex* leaves, that apart from the presence of the latter in diesel and unleaded gasoline, Ba should be considered a reliable tracer of vehicular emissions instead of Pb. Barium was indeed detected in samples of gasoline, unleaded gasoline and diesel oil in Siena and in emissions of diesel- and unleaded gasoline-powered vehicles in the USA by Que Hee (1994).

The recent reduction of Pb pollution is monitored at large by lichens. Lawrey (1993), for example, presented the values obtained for Pb and other elements (N, Al, Cd, Cr, Cu, Ni and Zn) in *Flavoparmelia baltimorensis* in Plummers Island and Bear Island, MD and on Stony Man Mountain, VA, USA. Measurements of elemental content of in situ thalli performed in 1988 and in 1992 revealed marked reductions of the entire range of nine elements except for Al in three stations. The findings were considered to provide limited but encouraging signs of an improved air quality in the region.

A monitoring study of a repeated, emission-related accumulation of elemental content in *H. physodes* in western Finland (Halonen et al., 1993) revealed a clear

Table 4. Concentrations of Pb in *Ramalina lacera,* in situ or resuspended in HaZorea, 1974–1997[a]

Date	In situ	Resuspended
January 1974	8.0[b]	–
May–December 1978	–	12.0[b]
February 1979	14.0[b]	–
February 1979–March 1980	–	13.0[b]
July 1980–July 1981	–	10.0 ± 4.0[c]
July 1981–July 1982	–	38.5 ± 12.0[d]
December 1981	22.6 ± 7.1[e]	–
December 1981–December 1982	–	20.6 ± 6.8[e]
January 1983–January 1984	–	20.5 ± 13.2[f]
July 1984	21.8 ± 6.7[g]	–
July 1984–July 1985	–	31.4 ± 14.7[g]
July 1985	22.1 ± 13.9[g]	–
June 1992	20.0 ± 8.0[h]	–
June 1992–March 1993	–	17.9 ± 2.4[h]
March 1993	24.7 ± 6.9[i]	–
June 1993	23.4 ± 3.8[i]	–
March–June 1993	–	22.5 ± 5.8[i]
March 1993–January 1994	–	12.0 ± 9.0[j]
February 1994	5.0 ± 3.0[k]	–
November 1994	15.0 ± 2.0[k]	–
February–November 1994	–	10.0 ± 3.0[k]
July 1996–May 1997	–	0.0–20.0[l]

[a] Pb values presented as $\mu g\ g^{-1}$ on dry weight basis.
[b] Garty and Fuchs, 1982; [c]Garty et al., 1985; [d]Garty, 1987; [e]Garty, 1988; [f]Garty and Hagemeyer, unpublished; [g]Garty et al., 1988; [h]Garty, unpublished; [i]Garty et al., 1997a; [j]Garty et al., 1998b; [k]Garty et al., 1997d; [l]Weissman, 1998, within the detection limit.

decrease of Pb amounts in the entire study area in 1989, in comparison with data obtained in 1974 and 1975, and in 1982 and 1983. The decreased Pb content was found to correspond with the data obtained by Rühling et al. (1992) for mosses in northern Europe.

Bennett (1995) made an attempt to detect slightly elevated levels of atmospheric pollution in Isle Royal National Park in northwestern Lake Superior, MI, USA, by means of different parameters in addition to elemental content. The elemental content of the lichens *H. physodes* and *Evernia mesomorpha* was ranked in comparison with the rank of elemental content of the earth's crust. As the rank of the elemental content of the lichens was found to be incompatible with the rank of elemental content of the earth's crust, it was concluded that the elevated level of certain elements derives from atmospheric pollutants. Both lichen species contained elevated levels of S, Zn, Pb, Cd and Se relative to the earth's crust. In the years 1983–1992 nine of sixteen elements increased in *H. physodes.* Pb, on the other hand, decreased at a percentage of −66% in *E. mesomorpha* and of −51% in *H. physodes.* The high Pb content which derived from the combustion of gasoline decreased substantially by 1992 but remained high enough to lead to significant physiological impairment.

Table 5. Total of motor vehicles in the years 1973–1996 in Israel[a]

Year	Total of motor vehicles
1973	370 895[b]
1974	408 280[b]
1975	419 389[c]
1976	432 499[c]
1977	450 415[c]
1978	473 773[d]
1979	524 950[d]
1980	532 430[e]
1981	592 916[e]
1982	662 671[f]
1983	729 567[f]
1984	760 213[f]
1985	776 217[g]
1986	819 102[h]
1987	881 999[i]
1988	952 786[j]
1989	983 774[j]
1990	1 015 404[j]
1991	1 075 459[j]
1992	1 175 876[k]
1993	1 261 004[k]
1994	1 373 089[l]
1995	1 459 018[m]
1996	1 542 870[n]

[a] Source: Statistical Abstract of Israel: [b]Table 18/22, p. 509, Vol. 26, 1975; [c]Table 18/25, p. 531, Vol. 30, 1979; [d]Table 18/21, p. 492, Vol. 31, 1980; [e]Table 18/20, p. 517, Vol. 33, 1982; [f]Table 18/20, p. 514, Vol. 36, 1985; [g]Table 18/18, p. 469, Vol. 37, 1986; [h]Table 18/18, p. 485, Vol. 38, 1987; [i]Table 18/18, p. 517, Vol. 39, 1988; [j]Table 18/18, p. 529, Vol. 43, 1992; [k]Table 18/17, p. 552, Vol. 45, 1994; [l]Table 18/17, p. 544, Vol. 46, 1995; [m]Table 18/15, p. 422, Vol. 47, 1996; [n]Table 18/15, p. 431, Vol. 48, 1997.

The above-mentioned findings were in accordance with data obtained by Bennett and Wetmore (1997) in a subsequent study of the elemental content of four lichen species on a transect leading into Voyageurs National Park, MN, USA. The amounts of 17 elements were measured in the lichens *Cladina rangiferina, E. mesomorpha, H. physodes* and *P. sulcata* sampled in ten sites throughout a period of three years. The highest value for Pb in *H. physodes* was measured at a distance of approximately 200 m from a moderately used highway. *Cladina*, a ground dwelling lichen, reached lower values for elemental content than the three epiphytic species. The elemental content decreased in three years from 1990 to 1992 at an average rate of 12%. The amount of Pb in thalli of *C. rangiferina* decreased at the end of the study at a rate of −43%, in *H. physodes* of −24% in *E. mesomorpha* of −47% and in *P. sulcata* of −19%.

A comparable decrease of Pb content was reported by Deruelle (1996) who studied the reliability of lichens as biomonitors of Pb pollution in the Fontainebleau forest in

France. In 1980 the Pb pollution produced by car exhausts on the adjacent motorway (A6) was assessed by means of several lichen species: *Cladonia portentosa, Cladonia chlorophaea, Lasallia pustulata, Umbilicaria grisea, H. physodes, P. caperata, Evernia prunastri, Lecanora conizaeoides* and *U. hirta*. In 1992 these lichens were found to contain less Pb than twelve years previously. The average figure for all measured species in all sites displayed a similar decrease: 3.2–3.5 times less. The decrease of Pb fallout along the motorway was ascribed to the consumption of unleaded gasoline.

In the years 1992–1997 we observed a rather uniform level of Pb in in situ thalli of *R. lacera* in the unpolluted HaZorea Forest, north-east Israel (Table 4). The same phenomenon related to resuspended thalli situated initially on carob trees and detached with their substrate.

A comparison of the total number of motor vehicles in Israel in the period 1973–1996 reveals that the specified amount increased progressively (Table 5). As the Pb content of the lichen did not follow the same trend it stands to reason that the introduction of unleaded gasoline in September 1990 is reflected to a noticeable extent.

Unlike the above-mentioned trend, the results of a long-term study (data of 1978 and 1993) performed by Loppi and De Dominicis (1996) in the village of Montieri, Tuscany, Italy, by means of the lichen *P. caperata* detected an unmodified level of Pb. The Pb content of the lichen in 1993 (32.68 $\mu g\ g^{-1}$) was quite similar to its content in 1978 (31.13 $\mu g\ g^{-1}$) thus indicating a consistent air quality in the study area.

Methylcyclopentadienyl manganese tricarbonyl (MMT) is an organic derivative of manganese (Mn) recently used as an antiknock agent in unleaded gasoline in North America (Loranger et al., 1994; Lytle et al., 1994). Lichens were indicated as apparent biomonitors of Mn along motorways. Lytle et al. (1994) found that the leaf tissue of aquatic, herbaceous and graminaceous plants accumulated larger amounts of Mn than of Pb along roads in UT, USA.

4. Lichen biomonitoring of airborne elements in large-scale surveys

Lichens were applied as biomonitors in the framework of large-scale surveys: biosphere reserves (Markert and Zhang, 1991; Markert and Wtorova, 1992; Vtorova and Markert, 1995), national parks (Frenzel et al., 1990; Sawicka-Kapusta and Rakowska, 1993), regional stations of an integrated background monitoring network (Skacel and Pekarek, 1992), extended rural sites (Loppi et al., 1994b; Bennett et al., 1996; Armstrong, 1997; Minger and Krähenbühl, 1997), and vast desert areas (Figs. 7–9).

Comprehensive surveys dealing with the chemical composition of either in situ or transplanted lichens were performed on a regional or nationwide basis apart from investigations which covered areas belonging to neighboring countries. Table 6 presents a short list of selected large-scale studies carried out in the 1990s. The lichen *H. physodes* was repeatedly used for this purpose. Bennett (2000) listed about 75 studies which reported on the elemental content of this lichen. This epiphytic lichen, singled out to determine levels of air pollution in boreal North America, may provide baseline data for future investigations (Pfeiffer and Barclay-Estrup, 1992) apart from substantial information relating to actual atmospheric contamination.

Figure 7. The epilithic desert lichen *Ramalina maciformis*, growing on a flintstone in the Negev Desert Highlands of Israel. This lichen previously used as a passive biomonitor (Garty, 1985) was applied in a study of the role of lichens in the cycling of metals in the Negev Desert (Garty et al., 1995a) and was recently used as an active biomonitor around the Ramat Hovav industrial region in the Negev.

Figure 8. The Ramat Hovav industrial region in the Negev Desert, Israel. Thalli of *Ramalina maciformis* transplanted in this region were found to contain higher Pb and Cu contents than thalli located 10–15 km away (Garty and Tamir, unpublished).

Figure 9. Sheep grazing in the Negev Desert Highlands. Sheep and goats graze on *Ramalina maciformis*, especially in summer, when no other plants are available. The food-chain which leads from lichens to animals and humans is endangered by the presence of heavy metals of industrial origin.

5. Lichen biomonitoring of airborne elements in remote areas, including Arctic and Antarctic regions

Nash and Gries (1995a) reviewed certain biomonitoring studies of atmospheric deposition, emphasizing arctic regions. This paper related for the greater part to publications made in the 1970s and 1980s. It focused on processes of deposition and accumulation of chemical elements in lichens and presented data obtained in urban/industrial and rural sites in the arctic region and the northern Taiga. In rural sites, the amounts of twelve accumulated elements were substantially lower than in other sites. Arctic lichens were found to be useful in the documentation of deposition patterns of persistent organic contaminants, particularly in relation to the lichen-caribou-human food chain, documented previously for radionuclides. The paper dealt with factors affecting the sensitivity of lichens to contaminants, listed relevant contaminants and described anatomical, morphological, reproductive, physiological and biochemical responses of lichens in the arctic. The effects of contaminants on arctic lichens should be considered a major concern, as lichen-dominated systems occur extensively in arctic regions (Nash and Gries, 1995b).

Jozwik (1990a) investigated the heavy metal content of lichens in the tundra of Spitsbergen. Fourteen lichen species were sampled in Calypsostranda, Lyellstranda and Chamberlindalen. The grand averages of the mean values obtained for the entire range of lichen species were as follows: 1.79–36.78 $\mu g\ g^{-1}$ for Cu, 3.91–244.47

Table 6. Regional and nationwide lichen-biomonitoring studies of elemental distribution in the 1990s

Lichen species	Chemical elements	Region/ country(ies)	Reference
Hypogymnia physodes	N, S	Norway	Bruteig (1993)
H. physodes	28 Elements	Slovenia	Jeran et al. (1996a)
H. physodes	28 Elements	Slovenia	Jeran et al. (1996b)
H. physodes	N	Sweden, Denmark	Sochting (1995)
H. physodes, *Pseudevernia furfuracea*, *Cladina* sp., *Peltigera aphtosa*	S	Finland	Takala et al. (1991)
H. physodes	16 Elements	Finland	Kubin (1990)
H. physodes	Cd, Cr, Cu, Fe, Ni, Pb, V, Zn	Finland	Lippo et al. (1995)
H. physodes, *P. furfuracea*, *Cladina* sp., *P. aphtosa*	Fe, S, Ti	Finland	Takala et al. (1994)
Cladonia convoluta, *C. rangiformis*, *Lecanora muralis*, *Neuphuscelia pulla*, *Xanthoparmelia tinctina*, *Xanthoria aureola*	^{137}Cs	Northern Greece	Sawidis and Heinrich (1992)
Parmelia sulcata	22 Elements	Portugal	Reis et al. (1996)
P. caperata	18 Elements	Portugal	Freitas and Nobre (1997)
P. sulcata	44 Elements	Portugal	Freitas et al. (1997)
P. sulcata	Co, Sc, Zn	The Netherlands	Sloof (1995)
P. sulcata	15 Elements	The Netherlands	Sloof and Wolterbeek (1991a)
P. sulcata	20 Elements	The Netherlands	Sloof and Wolterbeek (1991b)
P. sulcata, *Lecanora conizaeoides*	As, Br, Cd, Co, Cr, Cu, Fe, La, Sb, Se, V, W, Zn	The Netherlands	Sloof and Wolterbeek (1993)

μg g^{-1} for Mn, 3.40–68.22 μg g^{-1} for Zn, 2.99–52.75 μg g^{-1} for Pb and 0.16–6.13 μg g^{-1} for Cd. The Mn content of lichens, vascular plants and mosses in Spitsbergen was found to fall within the physiological limits of this element. Species was found to be the determinate factor for the accumulation of chemical elements in lichens and other plants in this remote area. The Pb content of *Xanthoria elegans* for example was 2.99 ± 0.25 μg g^{-1} and that of *Ochrolechia frigida* in the same area was 50.01 ± 1.29 μg g^{-1}. Copper, Zn, Pb and Cd occurred in certain Spitsbergen-tundra plants in amounts exceeding accepted levels of these elements (Jozwik, 1990b; 1991; Jozwik and Magierski, 1993; 1994). The above-mentioned findings for Pb in lichens collected in the Arctic match only partly data of Pb contents of five lichen species collected in 1990 in two sites on Ellesmere Island in the Canadian High Arctic by France and Coquery (1996). The average Pb content in the lichens was 2 μg g^{-1} with a range of 1.2–5.0 μg g^{-1}. As these data are lower than previously determined data for the same latitude, it was suggested that the decreased use of leaded gasoline in North America explains this discrepancy.

Geiser et al. (1994) established baseline values for 17 chemical elements in four lichen species for the Tongass National Forest in Alaska. A considerable difference between baseline and non-baseline values for certain elements around a significant point source indicated the outstanding sensitivity of the applied method, especially with regard to S and N. This study did not detect significant latitudinal or longitudinal trends. It was concluded that the established baseline should be valid for all parts of the study area. With the exception of Ca, Mg, K and Mn, whose higher values may be ascribed to marine influences, the entire range of S, N, P, Al, Fe, Na, Zn, Cu, B, Pb, Ni, Cr and Cd listed within expected background values for non-industrial locations and was comparable to or lower than values obtained for other national parks and forests.

Landers et al. (1995) examined US arctic ecosystems from 1990 to 1993 and obtained low values for total Hg in two lichens: *Cetraria cucullata* and *Masonhalea richardsoni*: 0.043 and 0.056 μg g^{-1}, respectively. Mercury values in lichens which showed a negative connection with distance from the marine coastline suggested the impact of natural sources, i.e. marine aerosols and lithologic sources associated with the coastal plain, in addition to anthropogenic sources: industrial emissions by petroleum and natural gas industries located near Prudoe Bay and Barrow in Alaska.

Certain studies carried out in remote sites in Canada detected limited amounts of heavy metals in lichens. Such were the findings of Crête et al. (1992) in northern Quebec and of Chiarenzelli et al. (1997) in the south-central District of Keewatin, Northwest Territories. However, the distribution of Cd, Pb, Hg and ^{137}Cs in lichens collected in northern Quebec was not uniform. The ^{137}Cs content was found to increase progressively with altitude. Hence a recommendation was issued that caribou should graze at low elevations in winter, in order to minimize the risk associated with cesium consumption by humans. In the middle altitude class (200–400 m) Pb levels were higher than in other altitude classes whereas the abundance of Hg in lichens growing in the tundra below 400 m was almost twice the abundance in the forest tundra and the boreal forest. Another study of Hg accumulations in transplanted thalli of *Cladonia rangiferina* in high elevation sites in Quebec (Evans and Hutchinson, 1996) detected a significant increase of Hg content after 12 months at the base and summit of a mountain and the summit of another mountain relative to controls. While enlarged inputs of Hg at summits may be ascribed to the effects of fog, the increase at base sites should be accounted for by processes of dry deposition.

Most of the studies of lichen elemental composition in the 1990s in the southern hemisphere were carried out in highly polluted areas. In Argentina, Gonzales and Pignata (1994, 1997, 1999), Levin and Pignata (1994), Gonzales et al. (1996, 1998) and Carreras et al. (1998) studied the deposition of S, in addition to Pb, Fe, Zn, Al, Cu and Mn (Gonzales and Pignata, 1994) whereas Saiki et al. (1995) studied 30 additional elements in Brazil.

Analyses of lichens growing in remote areas in the southern hemisphere yielded additional data relating to the accumulation of P, K, Ca, Mg, Zn, Cu, Pb, Cd and Al in *Nephroma antarticum* sampled in a remote *Nothofagus* forest in southern Chile (Wiersma et al., 1992). Most elements met quality assurance limits established for monitoring and exhibited low levels of spatial variability indicating their utility for baseline monitoring. The enrichment factor for Pb indicated the absence of atmospheric pollution. In accordance with these findings, the park was recommended as an appropriate site for integrated global baseline monitoring.

Another study was carried out in a pristine area at the west side of Nahuel Huapi National Park, Patagonia, Argentine by Ribeiro Guevara et al. (1995). Chemical analyses of 27 elements contained in five lichen species showed high intercorrelating levels of lithophile elements. The high rate of correlation among these elements led to the assumption that detritical material was entrapped within the thalli. All lichen species reflected the contribution of the eolic inputs of Pacific west winds, whereas the invariance of lithophile elements, which agreed with the elemental composition of local sediments suggested a single clastic source of trapped particulate material. An additional study carried out by Calvelo et al. (1997) in the above-mentioned study area, showed that foliose lichen species have the greater capability to entrap particles, over and above fruticose species.

Motivated by their wish to acquire basic knowledge on the elemental content of lichens in Antarctic regions and in Antarctica itself, a few authors collected and analyzed samples in different sites in and around this isolated continent. Bargagli et al. (1993) measured the Hg content of two species of macrolichens, *Umbilicaria decussata* and *Usnea antartica*, sampled on the coast of northern Victoria Land. These lichens contained higher Hg levels than lichens in other remote areas. It was suggested that a certain part of the airborne Hg, accumulated in the lichen, derived from local volcanic activity.

An additional investigation of elemental content of lichens sampled along the shore of glacial lakes around the Indian station Maitri in the Schirmacher Oasis, East Antarctica, was carried out by Kureishy et al. (1993). The name of the lichen species was not presented. This study detected low contents of Cd, Ni, Co and Pb (<0.05 $\mu g\ g^{-1}$ wet weight) and 14.0 $\mu g\ g^{-1}$ Cu.

According to Baeza et al. (1994), the radioactive contamination of sea and land in the Antarctic regions derives from the fallout of atmospheric atomic explosions executed since 1945, apart from emissions made by nuclear and radioactive facilities. The authors examined samples of *U. antartica* collected near the Spanish Antarctic Base, Juan Carlos I, situated on Livingston Island in the South Shetland archipelago in the Antarctic region. The ^{40}K content of the lichen was high in comparison with the content of a certain moss and of marine algae. The elemental content of the lichen was within the range constructed for lichens by different authors.

Upreti and Pandev (1994) determined the heavy metal content of the lichens *Umbilicaria aprina* and *U. decussata* in and around the Maitri Station. This study focused on the levels of Fe, Cu, Pb and Cr. The presence of Pb in all samples was attributed to anthropogenic activity, as polar expeditions transported large amounts of fuel into the region, including leaded-gasoline.

Poblet et al. (1997) recorded the trace-metal content of the Antarctic lichens *Usnea aurantiacoatra* and *U. antartica*, sampled in the vicinity of Argentina's scientific station "Jubany" on the "25 de Mayo" (King George) Island, in the Southern Shetland Archipelago (Antarctica). The average Fe content of the two lichens was similar: 699.05 ± 316.56 and 696.84 ± 279.23 $\mu g\ g^{-1}$. Although these values were lower than the greater part of the Fe values listed in a table of Fe content of natural or anthropogenic origin in different lichen species (Garty, 1993), they resemble the values recorded for *in situ* thalli from unpolluted sites in the UK, Canada, Italy and Finland (Garty, 1993). The levels of Mn, Zn, Cu, Pb and Cd were also lower or much lower than Fe levels. The atmospheric circulation of trace metals is assumed to extend to the 25 de Mayo Island and anthropogenic activity is the probable source: fossil fuel combustion, weathering of metallic structures of scientific stations, etc.

Acknowledgements

I thank Mrs Rachel Garty-Spitz for her valued contribution to the manuscript.

References

Armstrong, R.A., 1997. Are metal ions accumulated by saxicolous lichens growing in a rural environment?. Environ. Exp. Bot. 38, 73–79.

Arndt, U., Nobel, W., Schweizer, B., 1987. Bioindikatoren. Ulmer, Stuttgart.

Baeza, A., Miro, C., Paniaga, J.M., Navarro, E., Rodriguez, M.J., Sanchez, F., 1994. Natural and artificial radioactivity levels in Livingston Island (Antarctic regions). Bull. Environ. Contam. Toxicol. 52, 117–124.

Bargagli, R., 1990. Mercury emission in an abandoned mining area: assessment by epiphytic lichens. In: Cheremisinoff, P.N. (Ed.), Encyclopedia of Environmental Control Technology. Gulf Publishing, Houston, TX, pp. 613–640.

Bargagli, R., Barghigiani, C., 1991. Lichen biomonitoring of mercury emission and deposition in mining, geothermal and volcanic areas of Italy. Environ. Monitor. Assess. 16, 265–275.

Bargagli, R., Barghigiani, C., Siegel, B.Z., Siegel, S.M., 1991. Trace metal anomalies in surface soils and vegetation on two active island volcanoes: Stromboli and Vulcano (Italy). Sci. Total Environ. 102, 209–222.

Bargagli, R., Battisti, E., Focardi, S., Formichi, P., 1993. Preliminary data on environmental distribution of mercury in northern Victoria Land. Antarctica. Antarctic Sci. 5, 3–8.

Bargagli, R., Nimis, P.L., Monaci, F., 1997. Lichen biomonitoring of trace element deposition in urban, industrial and reference areas of Italy. J. Trace Elem. Med. Biol. 11, 173–175.

Barghigiani, C., Bargagli, R., Siegel, B.Z., Siegel, S.M., 1990. A comparative study of mercury distribution on the Aeolian volcanoes, Vulcano and Stromboli. Water Air Soil Pollut. 53, 179–188.

Bartok, K., Osvath, T., 1990. Influence of pollutants emanated from the Dej cellulose manufactures, over lichens. Contributii Botanice, Cluj-Napoca, 67–72.

Beckwith, P.R., Ellis, J.B., Revitt, D.M., 1985. Particles size distribution of Cu, Pb and Zn across a road surface. In: Lekkas, T.D. (Ed.), Heavy Metals in the Environment, Vol. I. CEP Consultants, Edinburgh, pp. 174–176.

Belandria, G., Asta, J., Garrec, J.P., 1991. Diminutions of fluorine contents in lichens due to a regression of pollution in an alpine valley (Maurienne, Savoie, France) from 1975 to 1985. Rev. Ecol. Alp. de Biologie alpine, Grenoble, 1, 45–58.

Benlap, J., Harper, K.T., 1990. Effect of a coal-fired power plant on the rock lichen *Rhizoplaca melanophtalma*: chlorophyll degradation and electrolyte leakage. Bryologist 93, 309–312.

Bennett, J.P., 1995. Abnormal chemical element concentrations in lichens of Isle Royale National Park. Environ. Exp. Bot. 35, 259–277.

Bennett, J.P., 2000. Statistical baseline values for chemical elements in the lichen *Hypogymnia physodes*. In: Agrawal, S.B., Agrawal, M. (Eds.), Environmental Pollution and Plant Responses. CRC/Lewis Publishers, Boca Raton, FL, pp. 343–353.

Bennett, J.P., Wetmore, C.M., 1997. Chemical element concentrations in four lichens on a transect entering Voyageurs National Park. Environ. Exp. Bot. 37, 173–185.

Bennett, J.P., Dibben, M.J., Lyman, K.L., 1996. Element concentrations in the lichen *Hypogymnia physodes* (L.) Nyl. after 3 years of transplanting along Lake Michigan. Environ. Exp. Bot. 36, 255–270.

Brown, D.H., 1991. Lichen mineral studies currently clarified or confused? Symbiosis 11, 207–223.

Brown, D.H., Beckett, R.P., 1984. Uptake and effect of cations on lichen metabolism. Lichenologist 16, 173–188.

Brown, D.H., Beckett, R.P., 1985a. Minerals and lichens: aquisition, localisation and effect. In: Vicente, C., Brown, D.H., Legaz, M.E. (Eds.), Surface Physiology of Lichens. Editorial de la Universidad Complutense de Madrid, Madrid, pp. 127–149.

Brown, D.H., Beckett, R.P., 1985b. The role of the cell wall in the intracellular uptake of cations by lichens. In: Brown, D.H. (Ed.), Lichen Physiology and Cell Biology. Plenum Press, New York, pp. 247–258.

Brown, D.H., Brown, R.M., 1991. Mineral cycling and lichens: the physiological basis. Lichenologist 23, 293–307.

Bruteig, I.E., 1993. The epiphytic lichen *Hypogymnia physodes* as a biomonitor of atmospheric nitrogen and sulphur deposition in Norway. Environ. Monitor. Assess. 26, 27–47.

Calvelo, S., Baccala, N., Arribere, M.A., Ribeiro Guevara, S., Bubach, D., 1997. Analytical and statistical analysis of elemental composition of lichens. J. Radioanal. Nucl. Chem. 22, 99–104.

Campo, G., Orsi, M., Badino, G., Giacomelli, R., Spezzano, P., 1996. Evaluation of motorway pollution in a mountain ecosystem. Pilot project: Susa Valley (Northwest Italy) years 1990–1994. Sci. Total Environ. 189/190, 161–166.

Caniglia, G., Calliari, I., Celin, L., Tollardo, A.M., 1994. Metal determination by EDXRF in lichens. A contribution to pollutants monitoring. Biol. Trace Elem. Res. 43-45, 213–221.

Carreras, H.A., Gudino, G.L., Pignata, M.L., 1998. Comparative biomonitoring of atmospheric quality in five zones of Cordoba city (Argentina) employing the transplanted lichen *Usnea* sp. Environ. Pollut. 103, 317–325.

Castello, M., Nimis, P.L., Alleteo, D., Bellio, M.G., 1994. Biomonitoring of SO_2 and metal pollution with lichens and barks in Savona (N Italy). Boll. Soc. Adriat. Sci. Trieste 75, 61–83.

Castello, M., Nimis, P.L., Cebulec, E., Mosca, R., 1995. Air quality assessment by lichens as bioindicators of SO_2 and bioaccumulators of heavy metals in the province of Trieste (NE Italy). Agric. Med., Special volume, 233–243.

Chiarenzelli, J.R., Aspler, L.B., Ozarko, D.L., Hall, G.E.M., Powis, K.B., Donaldson, J.A., 1997. Heavy metals in lichens, southern district of Keewatin, Northwest Territories, Canada. Chemosphere 35, 1329–1341.

Christensen, E.R., Guinn, V.P., 1979. Zinc from automobile tires in urban runoff. J. Environ. Eng. Divis. 105, 165–168.

Crête, M., Lefebvre, M.A., Zikovsky, L., Walsh, P., 1992. Cadmium, lead, mercury and ^{137}cesium in fruticose lichens of northern Quebec. Sci. Total Environ. 121, 217–230.

Deruelle, S., 1992. Accumulation du plomb par les lichens. Bull. Soc. Bot. de France, Actualités Bot. 139, 99–109.

Deruelle, S., 1996. La fiabilite des lichens comme bioindicateurs de la pollution plombique. Écologie 27, 285–290.

Dillman, K.L., 1996. Use of the lichen *Rhizoplaca melanophthalma* as a biomonitor in relation to phosphate refineries near Pocatello. Idaho. Environ. Pollut. 92, 91–96.

Dongara, G., Varrica, D., 1998. The presence of heavy metals in air particulate at Vulcano island (Italy). Sci. Total Environ. 212, 1–9.

Evans, C.A., Hutchinson, T.C., 1996. Mercury accumulation in transplanted moss and lichens at high elevation sites in Quebec. Water Air Soil Pollut. 90, 475–488.

Fahselt, D., Wu, T.-W., Mott, B., 1995. Trace element patterns in lichens following mine closures. Bryologist 98, 228–234.

Foner, H.A., Ganor, E., 1992. The chemical and mineralogical composition of some urban atmospheric aerosols in Israel. Atmosph. Environ., 26B, 125–133.

France, R., Coquery, M., 1996. Lead concentrations in lichens from the Canadian high arctic in relation to latitudinal pollution gradient. Water Air Soil Pollut. 90, 469–474.

Freitas, M.C., 1994. Heavy metals in *Parmelia sulcata* collected in the neighborhood of a coal-fired power station. Biol. Trace Elem. Res. 43-45, 207–212.

Freitas, M.C., 1995. Elemental bioaccumulators in air pollution studies. J. Radioanal. Nucl. Chem. 192, 171–181.

Freitas, M.C., Nobre, A.S., 1997. Bioaccumulation of heavy metals using *Parmelia sulcata* and *Parmelia caperata* for air pollution studies. J. Radioanal. Nucl. Chem. 217, 17–20.

Freitas, M.C., Reis, M.A., Alves, L.C., Wolterbeek, H.Th., Verburg, T., Gouveia, M.A., 1997. Bio-monitoring of trace-element air pollution in Portugal: qualitative survey. J. Radioanal. Nucl. Chem. 217, 21–30.

Frenzel, R.W., Witmer, G.W., Starkey, E.E., 1990. Heavy metal concentrations in a lichen of Mt. Rainier and Olympic National Parks, Washington, USA. Bull. Environ. Contam. Toxicol. 44, 158–164.

Gailey, F.A.Y., Lloyd, O.L., 1993. Spatial and temporal patterns of airborne metal pollution: the value of low technology sampling to an environmental epidemiology study. Sci. Total Environ. 133, 201–219.

Galun, M., Ronen, R., 1988. Interaction of lichens and pollutants. In: Galun, M. (Ed.), Handbook of Lichenology, Vol. 3. CRC Press, Boca Raton, FL, pp. 55–72.

Ganor, E., Altshuller, S., Foner, H.A., Brenner, S., Gabbay, J., 1988. Vanadium and nickel in dustfall as indicators of power plant pollution. Water Air Soil Pollut. 42, 241–252.

Garty, J., 1985. The amount of heavy metals in some lichens of the Negev Desert. Environ. Pollut. (Series B) 10, 287–300.

Garty, J., 1987. Metal amounts in the lichen *Ramalina duriaei* (De Not.) Bagl. transplanted at biomonitoring sites around a new coal-fired power station after 1 year of operation. Environ. Res. 43, 104–116.

Garty, J., 1988. Comparisons between the metal content of a transplanted lichen before and after the start-up of a coal-fired power station in Israel. Can. J. Bot. 66, 668–671.

Garty, J., 1992. Lichens and heavy metals in the environment. In: Vernet, J.-P. (Ed.), Impact of Heavy Metals on the Environment, Vol. 2. Elsevier, Amsterdan, pp. 55–131.

Garty, J., 1993. Lichens as biomonitors for heavy metal pollution. In: Markert, B. (Ed.), Plants as Biomonitors, Indicators for Heavy Metals in the Terrestrial Environment. VCH Publishers, Weinheim, pp. 193–263.

Garty, J., Ammann, K., 1987. The amounts of Ni, Cr, Zn, Pb, Cu, Fe and Mn in some lichens growing in Switzerland. Environ. Exp. Bot. 27, 127–138.

Garty, J., Fuchs, C., 1982. Heavy metals in the lichen *Ramalina duriaei* transplanted in biomonitoring stations. Water Air Soil Pollut. 17, 175–183.

Garty, J., Fuchs, C., Zisapel, N., Galun, M., 1977. Heavy metals in the lichen *Caloplaca aurantia* from urban, suburban and rural regions in Israel (a comparative study). Water Air Soil Pollut. 8, 171–188.

Garty, J., Ronen, R., Galun, M., 1985. Correlation between chlorophyll degradation and the amount of some elements in the lichen *Ramalina duriaei* (De Not.). Jatta. Environ. Exp. Bot. 25, 67–74.

Garty, J., Kardish, N., Hagemeyer, J., Ronen, R., 1988. Correlation between the concentration of adenosine triphosphate, chlorophyll degradation and the amounts of airborne heavy metals and sulphur in a transplanted lichen. Arch. Environ. Contam. Toxicol. 17, 601–611.

Garty, J., Harel, Y., Steinberger, Y., 1995. The role of lichens in the cycling of metals in the Negev Desert. Arch. Environ. Contam. Toxicol. 29, 247–253.

Garty, J., Kauppi, M., Kauppi, A., 1996. Accumulation of airborne elements from vehicles in transplanted lichens in urban sites. J. Environ. Qual. 25, 265–272.

Garty, J., Cohen, Y., Kloog, N., Karnieli, A., 1997a. Effects of air pollution on cell membrane integrity, spectral reflectance and metal and sulfur concentrations in lichens. Environ. Toxicol. Chem. 16, 1396–1402.

Garty, J., Karnieli, A., Wolfson, R., Kunin, P., Garty-Spitz, R., 1997b. Spectral reflectance and integrity of cell membranes and chlorophyll relative to the concentration of airborne mineral elements in a lichen. Physiol. Plant. 101, 257–264.

Garty, J., Kloog, N., Cohen, Y., Wolfson, R., Karnieli, A., 1997c. The effect of air pollution on the integrity of chlorophyll, spectral reflectance response, and on concentrations of nickel, vanadium and sulfur in the lichen *Ramalina duriaei* (De Not.). Bagl. Environ. Res. 74, 174–187.

Garty, J., Kloog, N., Wolfson, R., Cohen, Y., Karnieli, A., Avni, A., 1997d. The influence of air pollution on the concentration of mineral elements, on the spectral reflectance response and on the production of stress-ethylene in the lichen *Ramalina duriaei*. New Phytol. 137, 587–597.

Garty, J., Cohen, Y., Kloog, N., 1998a. Airborne elements, cell membranes, and chlorophyll in transplanted lichens. J. Environ. Qual. 27, 973–979.

Garty, J., Kloog, N., Cohen, Y., 1998b. Integrity of lichen cell membranes in relation to concentration of airborne elements. Arch. Environ. Contam. Toxicol. 34, 136–144.

Geiser, L.H., Derr, C.C., Dillman, K.L., 1994. Air quality monitoring on the Tongass National Forest. USDA-Forest Service, Petersburg, pp. 1–85, Appendices A1–C1.

Glenn, M.G., Gomez-Bolea, A., Lobello, R., 1995. Metal content and community structure of cryptogam bioindicators in relation to vehicular traffic in Montseny Biosphere Reserve (Catalonia, Spain). Lichenologist 27, 291–304.

Gombert, S., Asta, J., 1997. Étude du suivi de la pollution chlorée émise par une usine d'incineration d'ordures ménagères par l'utilisation de bioindicateurs végétaux: lichens et sphaignes. Écologie 28, 365–372.

Gombert, S., Asta, J., 1998. The effect of refuse incinerator fumes on the lead and cadmium content of experimentally exposed corticolous lichens. Water Air Soil Pollut. 104, 29–40.

Gonzales, C.M., Pignata, M.L., 1994. The influence of air pollution on soluble proteins, chlorophyll degradation, MDA, sulphur and heavy metals in a transplanted lichen. Chem. Ecol. 9, 105–113.

Gonzales, C.M., Pignata, M.L., 1997. Chemical response of the lichen *Punctelia subrudecta* (Nyl.) Krog transplanted close to a power station in an urban-industrial environment. Environ. Pollut. 97, 195–203.

Gonzales, C.M., Pignata, M.L., 1999. Effect of pollutants emitted by different urban-industrial sources on the chemical response of the transplanted *Ramalina ecklonii* (Spreng.) Mey and Flot. Toxicol. Environ. Chem. 69, 61–73

Gonzales, C.M., Casanovas, S.S., Pignata, M.L., 1996. Biomonitoring of air pollutants from traffic and industries employing *Ramalina ecklonii* (Spreng.) Mey and Flot. in Cordoba, Argentina. Environ. Pollut. 91, 269–277.

Gonzales, C.M., Orellana, L.C., Casanovas, S.S., Pignata, M.L., 1998. Environmental conditions and chemical response of a transplanted lichen to an urban area. J. Environ. Manag. 53, 73–81.

Haapala, H., Goltsova, N., Seppälä, R., Huttunen, S., Kouki, J., Lamppu, J., Popovichev, B., 1996. Ecological condition of forests around the eastern part of the gulf of Finland. Environ. Pollut. 91, 253–265.

Halonen, P., Hyvärinen, M., Kauppi, M., 1993. Emission related and repeated monitoring of element concentrations in the epiphytic lichen *Hypogymnia physodes* in a coastal area, W Finland. Ann. Bot. Fenn. 30, 251–261.

Jacquiot, L., Daillant, O., 1997. Bio-accumulation des métaux lourds et d'autres eléments traces par les lichen, revue bibliographique. Bull. de l'Observ. Mycol. 12, 2–31.

James, P.W., 1973. The effect of air pollutants other than hydrogen fluoride and sulphur dioxide on lichens. In: Ferry, B.W., Baddeley, M.S., Hawksworth, D.L. (Eds.), Air Pollution and Lichens. Athlone Press, London, pp.143–175.

Jeran, Z., Byrne, A.R., Batic, F., 1995. Transplanted epiphytic lichen as biomonitors of air-contamination by natural radionuclides around the Zirovski VRH uranium mine, Slovenia. Lichenologist 27, 375–385.

Jeran, Z., Jacimovic, R., Batic, F., Smodis, B., Wolterbeek, H.Th., 1996a. Atmospheric heavy metal pollution in Slovenia derived from results for epiphytic lichens. Fresenius J. Anal. Chem. 354, 681–687.

Jeran, Z., Jacimovic, R., Smodis, B., Batic, F., 1996b. The use of lichens in atmospheric trace element deposition studies in Slovenia. Phyton 36, 91–94.

Jovanovic, S., Carrot, F., Deschamps, C., Deschamps, N., Vukotic, P., 1995. A study of the air pollution in the surroundings of an aluminium smelter, using epiphytic and lithophytic lichens. J. Trace Microprobe Tech. 13, 463–471.

Jozwik, Z., 1990a. Heavy metals in tundra plants of Bellsund area. Spitsbergen. Pol. Polar Res. 11, 401–409.

Jozwik, Z., 1990b. Heavy metals in tundra plants of the Bellsund region, Spitsbergen. II. Blomlidalen, Tjörndalen, Dyrstadflya and Lognedalsflya. Wyprawy Geograficzne na Spitsbergen, UMCS, Lublin. pp. 159–165.

Jozwik, Z., 1991. Heavy metals in the tundra plants of the Bellsund region, Spitsbergen. III Chamberlindalen, Calypsostranda, Lyellstranda. Wyprawy Geograficzne na Spistbergen, UMCS, Lublin. pp. 163–170.

Jozwik, Z., Magierski, J., 1993. VII. Contents of Cu, Mn, Zn, Pb, and Cd in plants and soil on Western Spitsbergen-Bellsund region. Wyprawy Geograficzne na Spitsbergen, UMCS, Lublin. pp. 181–197.

Jozwik, Z., Magierski, J., 1994. VIII. Heavy metals in the plants and soil of Calypsostranda in the Bellsund region (West Spitsbergen), Wyprawy Geograficzne na Spitsbergen, UMCS, Lublin. pp. 179–189.

Juinchang, R., Freedman, B., Coles, C., Zwicker, B., Holzbecker, J., Chatt, A., 1995. Vanadium contamination of lichens and tree foliage in the vicinity of three oil-fired power plants in eastern Canada. J. Air Manag. Assoc. 45, 461–464.

Kansanen, P.H., Venetvaara, J., 1991. Comparison of biological collectors of airborne heavy metals near ferrochrome and steel works. Water Air Soil Pollut. 60, 337–359.

Kauppi, M., Halonen, P., 1992. Lichens as indicators of air pollution in Oulu, northern Finland. Ann. Bot. Fenn. 29, 1–9.

Kortesharju, J., Savonen, K., Säynätkari, T., 1990. Element contents of raw humus, forest moss and raindeer lichens around a cement works in northern Finland. Ann. Bot. Fenn. 27, 221–230.

Kubin, E., 1990. A survey of element concentrations in the epiphytic lichen *Hypogymnia physodes* in Finland in 1985–1986. In: Kauppi, P., Antilla, P., Kenttämies, K. (Eds.), Acidification in Finland. Springer-Verlag, Berlin, pp. 421–446.

Kureishy, T.W., Sen Gupta, R., Mesquita, A., Sanzgiry, S., 1993. Heavy metals in some parts of Antarctica and the southern Indian ocean. Marine Pollut. Bull. 26, 651–652.

Kytömaa, A., Nieminen, S., Thuneberg, P., Haapala, H., Nuorteva, P., 1995. Accumulation of aluminium in *Hypogymnia physodes* in the surroundings of a Finnish sulphite-cellulose factory. Water Air Soil Pollut. 81, 401–409.

Lackovicova, A., Martiny, E., Pisut, I., Stresko, V., 1994. Element content of the lichen *Hypogymnia physodes* and spruce needles in the industrial area of Rudnany and Krompachy (NE Slovakia). Ekologia 13, 415–423.

Landers, D.H., Ford, J., Gubala, C., Monetti, M., Lasorsa, B.K., Martinson, J., 1995. Mercury in vegetation and lake sediments from the U.S. arctic. Water Air Soil Pollut. 80, 591–601.

Lawrey, J.D., 1984. Biology of Lichenized Fungi. Praeger, New York.

Lawrey, J.D., 1993. Lichens as monitors of pollutant elements at permanent sites in Maryland and Virginia. Bryologist 96, 339–341.

Levin, A.G., Pignata, M.L., 1994. *Ramalina ecklonii* as a bioindicator of atmospheric pollution in Argentina. *Can. J. Bot.* 73, 1196–1202.

Lippo, H., Poikolainen, J., Kubin, E., 1995. The use of moss, lichen and pine bark in the nationwide monitoring of atmospheric heavy metal deposition in Finland. Water Air Soil Pollut. 85, 2241–2246.

Loppi, S., 1995. Trace element content of the lichen *Parmelia caperata* from a geothermal area of central Italy. Micol. Veg. Mediterranea 10, 67–72.

Loppi, S., Bargagli, R., 1996. Lichen biomonitoring of trace elements in a geothermal area (central Italy). Water Air Soil Pollut. 88, 177–187.

Loppi, S., De Dominicis, V., 1996. Lichens as long-term biomonitors of air quality in central Italy. Acta Bot. Neerland. 45, 563–570.

Loppi, S., Nascimbene, J., 1998. Lichen bioindication of air quality in the Mt. Amiata geothermal area (Tuscany, Italy). Geothermics 27, 295–304.

Loppi, S., Putorti, E., 1995. Lichen differentiation between lime and oak trees. Cryptogam. Bot. 5, 341–345.

Loppi, S., Chiti, F., Corsini, A., Bernardi, L., 1994a. Lichen biomonitoring of trace metals in the Pistoia area (central northern Italy). Environ. Monitor. Assess. 29, 17–27.

Loppi, S., Corsini, A., Bruscoli, C., 1994b. The amounts of Cd, Cr, Cu, Ni, Pb and Zn in the lichen *Parmelia caperata* growing in agricultural and suburban sites of the Pistoia area (central Italy). Micol. Veg. Mediterranea 9, 25–34.

Loppi, S., Corsini, A., Bruscoli, C., Rossetti, C., 1995a. Lichen biomonitoring of heavy metals in Montecatini Terme (central northern Italy). Micol. Veg. Mediterranea 10, 122–128.

Loppi, S., Putorti, E., De Dominicis, V., Barbaro, A., 1995b. Lichens as bioindicators of air quality near a municipal solid waste incinerator (central Italy). Allionia 33, 121–129.

Loppi, S., Cenni, E., Bussotti, F., Ferretti, M., 1997. Epiphytic lichens and tree leaves as biomonitors of trace elements released by geothermal power plants. Chem. Ecol. 14, 31–38.

Loppi, S., Cenni, E., Bussotti, F., Ferretti, M., 1998. Biomonitoring of geothermal air pollution by epiphytic lichens and forest trees. Chemosphere 36, 1079–1082.

Loranger, S., Demers, G., Kennedy, G., Forget, E., Zayed, J., 1994. The pigeon (*Columba livia*) as a monitor for manganese contamination from motor vehicles. Arch. Environ. Contam. Toxicol. 27, 311–317.

Lupsina, V., Horvat, M., Jeran, Z., Stegnar, P., 1992. Investigation of mercury speciation in lichens. Analyst 117, 673–675.

Lytle, C.M., McKinnon, C.Z., Smith, B.N., 1994. Manganese accumulation in roadside soil and plants. Naturwissenschaften 81, 509–510.

Makholm, M.M., Bennett, J.P., 1998. Mercury accumulation in transplanted *Hypogymnia physodes* lichens downwind of Wisconsin chlor-alkali plant. Water Air Soil Pollut. 102, 427–436.

Manninen, S., Huttunen, S., Torvela, H., 1991. Needle and lichen sulphur analyses on two industrial gradients. Water Air Soil Pollut. 59, 153–163.

Markert, B., Wtorova, W., 1992. Inorganic chemical investigations in the forest biosphere reserve near Kalinin, USSR. III. Comparison of the multielement budget with a forest ecosystem in Germany – aspects of rejection, indication and accumulation of chemical elements. Vegetatio 98, 43–58.

Markert, B., Zhang, d.L., 1991. Inorganic chemical investigations in the forest biosphere reserve near Kalinin, USSR. II. The distribution of lanthanide elements in the vegetation cover. Vegetatio 97, 57–62.

Martin, M.H., Coughtrey, P.J., 1982. Biological Monitoring of Heavy Metal Pollution. Applied Science Publishers, London.

Minger, A., Krähenbühl, U., 1997. Moss and lichen as biomonitors for heavy metals. Int. J. Environ. Anal. Chem. 67, 41–48.

Monaci, F., Bargagli, R., 1997. Barium and other trace metals as indicators of vehicle emissions. Water Air Soil Pollut. 100, 89–98.

Monaci, F., Bargagli, A., Gasparo, D., 1997. Air pollution monitoring by lichens in a small medieval town of central Italy. Acta Bot. Neerland. 46, 403–412.

Nash III, T.H., 1989. Metal tolerance in lichens. In: Shaw, A.J. (Ed.), Heavy Metal Tolerance in Plants: Evolutionary Aspects. CRC Press, Boca Raton, FL, pp. 119–131.

Nash III, T.H., 1996. Nutrients, elemental accumulation and mineral cycling. In: Nash III, T.H. (Ed.), Lichen Biology. Cambridge University Press, Cambridge, MA, pp. 136–153.

Nash III, T.H., Gries, C., 1995a. The use of lichens in atmospheric deposition studies with an emphasis on the Arctic. Sci. Total Environ. 160/161, 729–736.

Nash III, T.H., Gries, C., 1995b. The response of lichens to atmospheric deposition with an emphasis on the arctic. Sci. Total Environ. 160/161, 737–747.

Nash III, T.H., Sommerfeld, M.R., 1981. Elemental concentrations in lichens in the area of the Four Corners Power Plant. New Mexico. Environ. Exp. Bot. 21, 153–162.

Nieboer, E., Puckett, K.J., Richardson, D.H.S., Tomassini, F.D., Grace, B., 1977. Ecological and physicochemical aspects of the accumulation of heavy metals and sulphur in lichens. In: International Conference on Heavy Metals in the Environment, Toronto, Ontario, Canada, October 27–31, 1975, Symposium Proceedings, Vol. 2. pp. 331–352.

Nieboer, E., Richardson, D.H.S., Tomassini, F.D., 1978. Mineral uptake and release by lichens. Bryologist 81, 226–246.

Nimis, P.L., Castello, M., Perotti, M., 1990. Lichens as biomonitors of sulphur dioxide pollution in La Spezia (Northern Italy). Lichenologist 22, 333–344.

Nimis, P.L., Castello, M., Perotti, M., 1993. Lichens as bioindicators of heavy metal pollution: a case study at La Spezia (N Italy). In: Markert, B. (Ed.), Plants as Biomonitors, Indicators for Heavy Metals in the Terrestrial Environment. VCH, Weinheim, pp. 265–284.

Ormrod, D.P., 1984. Impact of trace element pollution on plants. In: Treshow, M. (Ed.), Air Pollution and Plant Life. Wiley, New York, pp. 291–319.

Palmieri, F., Neri, R., Benco, C., Serracca, L., 1997. Lichens and moss as bioindicators and bioaccumulators in air pollution monitoring. J. Environ. Pathol. Toxicol. Oncol. 16, 175–190.

Palomäki, V., Tynnyrinen, S., Holopainen, T., 1992. Lichen transplantation in monitoring fluoride and sulphur deposition in the surroundings of a fertilizer plant and a strip mine at Siilinjarvi. Ann. Bot. Fenn. 29, 25–34.

Perkins, D.F., 1992. Relationship between fluoride contents and loss of lichens near an aluminium works. Water Air Soil Pollut. 64, 503–510.

Pfeiffer, H.N., Barclay-Estrup, P., 1992. The use of a single species, *Hypogymnia physodes*, as an indicator of air quality in northwestern Ontario. Bryologist 95, 38–41.

Poblet, A., Andrade, S., Scagliola, M., Vodopivez, C., Curtosi, A., Pucci, A., Marcovecchio, J., 1997. The use of epilithic Antarctic lichens (*Usnea aurantiacoatra* and *U. antartica*) to determine deposition patterns of heavy metals in the Shetland Islands, Antarctica. Sci. Total Environ. 207, 187–194.

Puckett, K.J., 1988. Bryophytes and lichens as monitors of metal deposition. In: Nash III, T.H., Wirth, V. (Eds.), Lichens, Bryophytes and Air Quality. Bibliotheca Lichenologica 30. Cramer, Berlin, pp. 231–267.

Puckett, K.J., Burton, M.A.S., 1981. The effect of trace elements on lower plants. In: Lepp, N.W. (Ed.), Effect of Heavy Metal Pollution on Plants. Applied Science, London, pp. 213–238.

Punning, J.-M., Zobel, M., Lillemets, S., Trass, H., Roosma, A., Minayeva, O., Saare, L., Viitak, A., Hodrejarv, H., 1991. Heavy metals in snow and some indicator plants in Estonia. Proc. Eston. Acad. Sci. Ecol. 1, 65–71.

Que Hee, S.S., 1994. Availability of elements in leaded/unleaded automobile exhausts, a leaded paint, a soil, and some mixtures. Arch. Environ. Contam. Toxicol. 27, 145–153.

Reis, M.A., Alves, L.C., Wolterbeek, H.Th., Verburg, T., Freitas, M.C., Gouveia, A., 1996. Main atmospehric heavy metal sources in Portugal by biomonitor analysis. Nucl. Instrum. Meth. Phys. Res. B109/110, 493–497.

Ribeiro Guevara, S., Arribére, M.A., Calvelo, S., Roman Ross, G., 1995. Clastic contribution to the elemental composition of lichens at Nahuel Huapi National Park, Patagonia, Argentina. J. Radioanal. Nucl. Chem. 198, 437–448.

Richardson, D.H.S., 1988. Understanding the pollution sensitivity of lichens. Bot. J. Linnean Soc. 96, 31–43.

Richardson, D.H.S., 1992. Pollution Monitoring with Lichens. Richmond, Slough, UK.

Rühling, A., Brumelis, G., Goltsova, N., Kvietkus, K., Kubin, E., Lijv, S., Magnusson, S., Mäkinen, A., Pilegaard, K., Rasmussen, L., Sander, E., Steinnes, E., 1992. Atmospheric heavy metal deposition in northern Europe 1990. Nord 12, 1–41.

Sadiq, M., Alam, I., El-Mubarek, A., Al-Mohdar, H.M., 1989. Preliminary evaluation of metal pollution from wear of auto tires. Bull. Environ. Contam. Toxicol. 42, 743–748.

Saiki, M., Chaparro, C.G., Vasconcellos, M.B.A., Marcelli, M.P., 1995. Determination of trace elements in lichens by instrumental neutron activation analysis. In: 7th International Conference on Modern Trends in Activation Analysis, Seoul, Korea, September 24–30, 1995.

Sawicka-Kapusta, K., Rakowska, A., 1993. Heavy metal contamination in Polish national parks. Sci. Total Environ., Suppl., 161–166.

Sawidis, T., Heinrich, G., 1992. Cesium-137 monitoring using lichens and mosses from northern Greece. Can. J. Bot. 70, 140–144.

Sawidis, T., Marnasidis, A., Zachariadis, G., Stratis, J., 1995. A study of air pollution with heavy metals in Thessaloniki City (Greece) using trees as biological indicators. Arch. Environ. Contam. Toxicol. 28, 118–124.

Semadi, A., Deruelle, S., 1993. Détection de la pollution plombique à l'aide de transplant licheniques dans la region de Anaba (Algérie), Pollution Atmosphérique, Oct.–Dec., 86–102.

Skacel, F., Pekarek, J., 1992. Monitoring of lead, cadmium and mercury in environmental samples at the regional station of the integrated background monitoring network of GEMS in Czechoslovakia. Sci. Total Environ. 115, 261–276.

Sloof, J.E., 1995. Lichens as quantitative biomonitors for atmospheric trace-element deposition, using transplants. Atmosph. Environ. 29, 11–20.

Sloof, J.E., Wolterbeek, H.Th., 1991a. National trace-element air pollution monitoring survey using epiphytic lichens. Lichenologist 23, 139–165.

Sloof, J.E., Wolterbeek, H.Th., 1991b. Pattern in trace elements in lichens. Water Air Soil Pollut. 57–58, 785–795.

Sloof, J.E., Wolterbeek, H.Th., 1993. Interspecies comparison of lichens as biomonitors of trace-element air pollution. Environ. Monitor. Assess. 25, 149–157.

Sochting, U., 1995. Lichens as monitors of nitrogen deposition. Cryptogam. Bot. 5, 264–269.

Takala, K., Olkkonen, H., Krouse, H.R., 1991. Sulphur isotope composition of epiphytic and terricolous lichens and pine bark in Finland. Environ. Pollut. 69, 337–348.

Takala, K., Olkkonen, H., Salminen, R., 1994. Iron content and its relations to the sulphur and titanium contents of epiphytic and terricolous lichens and pine bark in Finland. Environ. Pollut. 84, 131–138.

Terron Alfonso, A., Barreno Rodriguez, E., 1994. Estimation of air pollution in the area of influence of the coal power station at La Robla (Leon, Northwest Spain) using epiphytic lichens as bioindicators. Cryptogam. Bryol. Lichenol. 15, 135–151.

Tyler, G., 1989. Uptake, retention and toxicity of heavy metals in lichens. Water Air Soil Pollut. 47, 321–333.

Tynnyrinen, S., Palomäki, V., Holopainen, T., Kärenlampi, L., 1992. Comparison of several bioindicator methods in monitoring the effects on forest of a fertilizer plant and a strip mine. Ann. Bot. Fenn. 29, 11–24.

Upreti, D.K., Pandev, V., 1994. Heavy metals of Antarctic lichens. 1. *Umbilicaria*. Feddes Repert. 105, 197–199.

Vtorova, V.N., Markert, B., 1995. Multi-element analysis of plants of the forest ecosystems in eastern Europe. Biol. Bull. 4, 447–454.

Walther, D.A., Ramelov, G.J., Beck, J.N., Young, J.C., Callahan, J.D., Marcon, M.F., 1990. Temporal changes in metal levels of the lichens *Parmotrema praesorediosum* and *Ramalina stenospora*, Southwest Louisiana. Water Air Soil Pollut. 53, 189–200.

Ward, N.I., 1989. Multielement contamination of British motorway environments. In: Vernet, J.-P. (Ed.), Heavy Metals in the Environment, Vol. I, CEP Consultants, Edinburgh, pp. 174–176.

Weissman, L., 1998. Factors affecting the production of endogenic ethylene, changes in spectral reflectance response, membranes and chlorophyll integrity and the composition of mineral elements in the lichen *Ramalina lacera* (With.) J.R. Laund., M.Sc. thesis, Tel Aviv University (in Hebrew).

Wiersma, G.B., Finely, K., Whitworth, C., Bruns, D.A., McAnulty, L., Boelcke, C., 1992. Elemental composition of lichens from a remote *Nothofagus* forest site in southern Chile. Chemosphere 24, 155–167.

Zanini, E., Bonifacio, E., Cuttica, G.C., 1992. Heavy metals in soils near a steel-making industry: a case study in a complex valley situation in Italy. J. Environ. Sci. Health A27, 2019–2036.

Trace elements – Their distribution and effects in the environment
B. Markert and K. Friese, editors

Chapter 9

Trace metals, other chemical elements and lichen physiology: research in the nineties

J. Garty

Abstract

Lichens are associations of two or more components: the alga and/or cyanobacterium, the photobiont partner, and the fungus, the mycobiont partner. The application of lichens as effective biomonitors of chemical contamination is based on investigations of stress-induced modifications of elemental content and physiological response. Stress derives from excessive amounts of chemical elements accumulated in the lichen, in addition to the leakage of essential nutrients. This chapter deals with the elemental content of the entire thallus. Physiological modifications on the other hand reflect the status of either photobiont, mycobiont, or both bionts. The present review focuses on the linkage of elemental content with lichen physiology and emphasizes research performed in the nineties. As such, this chapter does not report of many of the basic studies performed one to three decades ago, nor of recent studies which dealt with the physiological response to chemicals, e.g. fumigation studies but did not relate to the elemental content. Thus, the present chapter describes the following subjects: (1) the effect of chemical elements on the production of stress-ethylene in lichens; (2) elemental content and chlorophyll integrity in lichens; (3) elemental content and the spectral reflectance response of the lichen thallus; (4) chemical pollutants and changes in modulated chlorophyll a fluorescence: monitoring photosystem II (PSII) activities in lichens exposed to pollutants; (5) elemental content and degradation of cell membranes in lichens; (6) the accumulation of chemical elements and other physiological parameters: ATP (adenosine triphosphate), malondialdehyde, amino acids and protein biosynthesis; (7) heavy metals as substrate and as "lichen killers". Of the 142 references cited in the present chapter, 87 (61%) were published in the years 1990–1999.

Keywords: Lichens; Chemical elements; Ethylene; Chlorophyll; Spectral reflectance response; Photosystem II; Adenosine triphosphate; Malondialdehyde; Protein; Amino acids

1. The effect of chemical elements on the production of stress-ethylene in lichens

The phytohormone ethylene, a simple hydrocarbon gas, is an endogenous regulator of plant growth. Ethylene is produced in small quantities under normal conditions but it increases as living tissues are subject to a variety of stress factors, e.g. mechanical wounding, bruising, chilling, freezing, irradiation and attack by microorganisms (Lieberman, 1979). Chemicals, e.g. SO_2 gas, bisulfite ions or simulated acid rain were reported to have a stimulatory effect on the production of ethylene in different plants: cucumber (Bressan et al., 1979), paper birch (Kimmerer and Kozlowski, 1982), potato, radish, soybean (Arny and Pell, 1986), lettuce (Larsen, 1986), spruce (Hohendorff and Vogels, 1990) and sunflower (Zwoch et al., 1990).

The release of ethylene was shown to increase upon exposure of flowering plants to heavy-metal containing solutions, e.g. citrus fruit (Ben-Yehoshua and Biggs, 1970), beans (Abeles and Abeles, 1972; Fuhrer, 1982; Gora and Clijsters, 1989), *Spirodela oligorrhiza* (Mattoo et al., 1986), soybeans (Pennazio and Roggero, 1991), tomatoes (Bouzayen et al., 1991) and rice (Peng and Yamauchi, 1993; Kao, 1996).

Reports of ethylene production by microorganisms were made, e.g. by Ince and Knowles (1985) for bacteria, by Vanden Driessche et al. (1988) and Maillard et al. (1993) for algae, and by Axelrood-McCarthy and Linderman (1981), Rupp et al. (1989a,b) and Strzelczyk et al. (1994) for fungi.

The extensive literature relating to the production of ethylene by fungi refers to the presence of chemicals as indicative of alterations in the production of ethylene. Thus, we were led to test the probability of ethylene-production by lichens and investigate the impact of chemicals, either applied under controlled conditions or accumulated in the field. Changes in the mode of ethylene production would be indicative of the presence of stress-factors.

Lichens were proved to produce ethylene by Epstein et al. (1986), Lurie and Garty (1991), Ott and Zwoch (1992), Ott (1993), Ott and Schieleit (1994) and Schieleit and Ott (1996, 1997). The first indication of the impact of air pollution on the ethylene-producing system was provided by Epstein et al. (1986) who transplanted *Ramalina lacera*, defined as *R. duriaei* in many of the earlier studies (Fig. 1), originating in a relative clean air site, to a busy road intersection in Israel for a period of one year (Fig. 2).

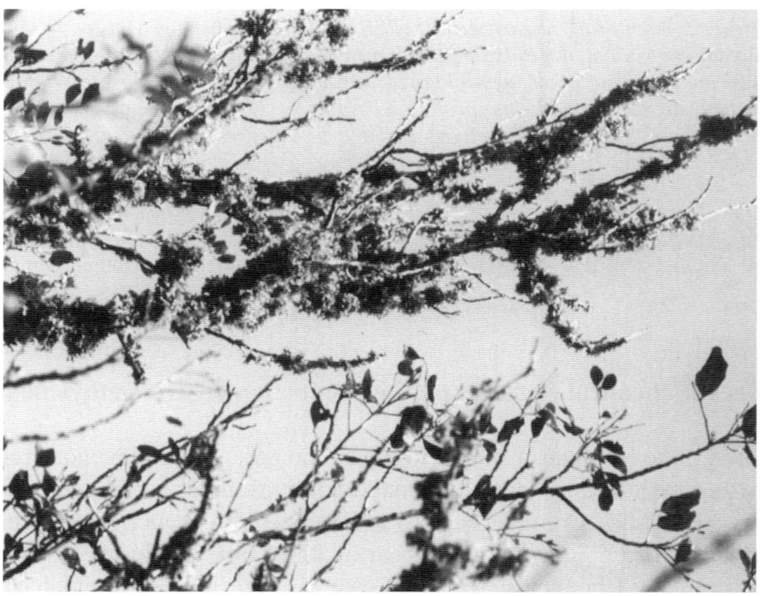

Figure 1. *Ramalina lacera* growing on twigs of carob trees (*Ceratonia siliqua*) in HaZorea, NE Israel.

Figure 2. Transplants of *Ramalina lacera* in a biomonitoring site in Israel.

The production of stress-ethylene was measured in thalli of both sites and found to be relatively high in the polluted site. In an additional study, transplanted lichens in an industrial region in Israel (Haifa-Bay, NW Israel) were found to produce greater amounts of ethylene than control lichens left in the unpolluted site (HaZorea) (Garty et al., 1993b). However, as no measurements were made of the content of airborne chemical elements in the entire thallus in the above-mentioned studies, we were unable at this stage to explain this alteration.

The production of stress-ethylene by thalli of *R. lacera* under controlled conditions was measured upon incubation for 30 min under low pH conditions: H_2O-treated thalli produced 1.38 nl g^{-1} h^{-1} ethylene; other thalli exposed to 20 mM of either $PbCl_2$, $ZnCl_2$, $MnCl_2$, $CuCl_2$ or $FeCl_2$, produced 3.12, 2.64, 4.12, 7.09 and 17.86 nl g^{-1} h^{-1} ethylene, respectively (Lurie and Garty, 1991). The high production of ethylene upon exposure to $FeCl_2$ was observed again upon exposure to various metal salts in solution, carried out to investigate the effect of both cations and anions. Table 1 shows the release of ethylene by *R. lacera* under the influence of different metals, combined with three kinds of anions in extreme acidic conditions (pH 2.0).

The presence of metal ions stimulated the release of ethylene by *R. lacera* according to the following sequence:

In the presence of Cl^-: Fe^{3+}, Fe^{2+} >> Cu > K; Mn, Zn, Pb.
In the presence of NO_3^-: Fe^{3+} >> Cu, K > Zn; Mn > Pb.
In the presence of SO_4^{2-}: Fe^{2+}, Cu > Mn, K, Zn.

At pH 2.0, the combination of Fe^{2+} or Fe^{3+} and Cl^- stimulated the release of

Table 1. Effect of K and heavy metal ions combined with different anions indicated by release of ethylene by *R. lacera* (Garty et al., 1995a)[a,b]

Solution	mM	pH	n	$X \pm SD$ (nl g^{-1} h^{-1})	
H_2O + KOH		7.0	15	1.2 ± 0.4	j
H_2O + HCl		2.0	15	4.6 ± 0.7	e
KCl	20	2.0	15	4.9 ± 0.6	de
KCl	40	2.0	15	4.4 ± 0.9	ef
KCl	60	2.0	15	4.4 ± 0.5	ef
KNO_3	20	2.0	15	4.5 ± 0.6	e
KNO_3	40	2.0	15	4.5 ± 0.7	e
KNO_3	60	2.0	15	4.5 ± 0.6	e
K_2SO_4	20	2.0	15	4.0 ± 0.5	fgh
$FeCl_3$	20	2.0	15	22.6 ± 4.7	a
$Fe(NO_3)_3$	20	2.0	15	20.1 ± 3.9	b
$FeCl_2$	20	2.0	15	21.8 ± 6.6	a
$FeSO_4$	20	2.0	15	8.8 ± 1.5	c
$CuCl_2$	20	2.0	15	6.4 ± 3.0	d
$Cu(NO_3)_2$	20	2.0	15	6.4 ± 2.6	de
$CuSO_4$	20	2.0	15	8.6 ± 3.9	c
$ZnCl_2$	20	2.0	15	3.2 ± 0.6	fghi
$Zn(NO_3)_2$	20	2.0	15	3.2 ± 0.8	fghi
$ZnSO_4$	20	2.0	15	2.4 ± 0.3	ghij
$MnCl_2$	20	2.0	15	4.1 ± 0.5	fgh
$Mn(NO_3)_2$	20	2.0	15	4.1 ± 0.5	fgh
$MnSO_4$	20	2.0	15	4.3 ± 0.5	fg
$PbCl_2$	20	2.0	15	2.3 ± 0.4	hij
$Pb(NO_3)_2$	20	2.0	15	2.1 ± 0.4	ij

[a] n = number of replicates; X = mean value; SD = standard deviation.
[b] Values of the vertical column followed by the same letter do not differ significantly at $P < 0.05$ by Duncan's multiple range test.

ethylene over and above the combination of NO_3^- and Fe^{3+} or SO_4^{2-} and Fe^{2+}. The results point to a synergistic effect of Fe^{3+}, Fe^{2+} and Cl^- ions. It was assumed that in most cases the release of ethylene is brought about by the injury endured by mycobiont and/or photobiont cells. The values obtained in this study were in general higher than the values obtained in two field studies (Epstein et al., 1986; Garty et al., 1993b).

Many observations were made of the decline of certain lichen species in correlation with the presence of acid rain and high rates of deposition of H^+, SO_4^{2-}, NH_4^+ and NO_3^-. In industrialized countries acid rain possibly accounts for the disappearance of certain lichen species (Gilbert, 1986; Hallingbäck and Kellner, 1992). In order to examine the immediate response, as measured by the production of stress-ethylene, of different lichen species to acidic solutions at pH 4.0 and lower, we conducted a study in Finland, using *Hypogymnia physodes, Usnea hirta, Cladina stellaris, Bryoria fremontii, Bryoria fuscescens* and *Alectoria sarmentosa* (Garty et al., 1995b). To determine the correlation of fluctuations in the amount of ethylene produced by lichens exposed to sulfur-containing solutions at a low pH, and the tolerance/sensitivity of these lichens to air pollution, we measured the amount of ethylene produced by thalli soaked in H_2SO_4 and $NaHSO_3$. The exposure of *Hypogymnia physodes, Cladina stellaris* and *Bryoria fuscescens* to H_2SO_4 at a pH ranging between 4.0 and 2.0 did not produce alterations in the amount of ethylene in comparison with samples wetted with H_2O at pH 6.8. The exposure of two pendulous lichens, *Usnea hirta* and *Alectoria sarmentosa*, to 1.0 and 5.0 mM H_2SO_4 at pH 2.7 and 2.0, respectively, made for a slight increase of ethylene, whereas another pendulous lichen, *Bryoria fremontii*, exposed to H_2SO_4 at pH 4.0–2.0 decreased its production of ethylene. The exposure of *H. physodes, U. hirta, C. stellaris,* and *A. sarmentosa* to $NaHSO_3$ at pH 4.0 gradually increased the production of ethylene. The exposure of *B. fremontii* and *B. fuscescens* to low $NaHSO_3$ concentrations depressed the production of ethylene.

The indifference of *H. physodes* to H_2SO_4 under strong acidic conditions correlated with its resistance to SO_2 in the air. This lichen is known to survive in cities having a concentration of 75 $\mu g\ m^{-3}$ of SO_2 in the air (Türk and Christ, 1986), of 70–100 $\mu g\ m^{-3}$ (Hawksworth and Rose, 1970) and of 125–150 $\mu g\ m^{-3}$ (Crespo et al., 1981). The lichen *U. hirta* was found to tolerate air pollutants better than other filamentous lichens in Oulu (Halonen et al., 1992), but was not observed to grow in the center of the city. According to Insarova et al. (1992) *U. hirta* is classified as fairly tolerant with respect to air pollution.

Two field studies performed in Oulu, tested the influence of air pollution on the amount of heavy metals and other elements and on the production of stress-ethylene in the lichens *Usnea hirta* (Garty et al., 1997c) and *Hypogymnia physodes* (Garty et al., 1997d). The accumulation of airborne elements by the two lichens was tested by means of transplantation, either in the vicinity of two streets with slow traffic, or in the vicinity of a highway in Oulu for a 45 d period. These studies analyzed the comparative production of stress-ethylene and the accumulated amounts of K, Ca, Na, Mg, Fe, Zn, Mn, Pb, Ni, Cu and Cd. Thalli of *U. hirta* accumulated more K, Ca, Na, Mg, Fe, Zn and Mn in samples exposed in one of the two streets with slow traffic in Oulu, than thalli left in a control site in a forest out of the city. Two of the observed elements, Pb and Ca, were in positive correlation with the amount of ethylene detected consecutive to the treatment of *U. hirta* transplants with H_2O. The amounts of stress-ethylene in *U. hirta* thalli retrieved from the streets with slow traffic and less cars (14 000 per d), soaked in solutions of 7.5 mM of $NaHSO_3$ (pH 4.0), were higher than those produced by thalli retrieved from a site near the highway with 25 000 cars per d and by thalli left in the forest. It is suggested that the streets with slow traffic and less cars are more polluted than the highway as a result of bad ventilation conditions and a slow rate of dispersion of airborne pollutants. *Hypogymnia physodes*, in comparison, contained more Fe and Mg in thalli exposed in either one or both streets with slow traffic in Oulu than thalli left in the control site. Two of the elements, Zn and Fe, exhibited a significant positive correlation with the amount of ethylene detected upon treatment with either H_2O or $NaHSO_3$. The Pb, Fe, Zn, Mn and Cu in *H. physodes* retrieved from one or both streets with slow traffic were higher than the amount of these metals in transplanted thalli near the highway after 45 d.

The macroelements K, Ca, Na, Mg and Fe detected in transplants of *U. hirta* and *H. physodes* derive mainly from geochemical sources and are important constituents of dust. Other elements such as Pb, Zn, Cu and Cd are linked for the greater part to anthropogenic activity, including car traffic.

Alterations in the production of stress-ethylene were compared with the content of sulfate-S, V, Ni, K, Pb, P, Cu, Cr, Fe, Mn and Mg in either in situ or transplanted thalli of the lichen *Ramalina lacera* in studies conducted in Israel (Garty et al., 1997f): The content of Cu, Mg, Fe, Ni, Pb, V, Cr, Mn and sulfate-S in in situ thalli collected in a nature reserve near an industrial town (Ashdod, SW Israel) was higher than that detected in in situ thalli collected in the HaZorea Forest (NE Israel). The high content of these chemical elements in *R. lacera* from the nature reserve coincided with a high rate of production of stress-ethylene. Lichens collected in the unpolluted forest in February 1994, either resuspended in the same site or transplanted in nine sites in the Ashdod region, and retrieved in November of the same year, contained more sulfate-S, Pb, Cu, Ni, V, Cr in the urban-industrial sites than in most of the rural sites. Statistical analysis revealed a positive correlation of the amount of ethylene and the V and Ni content. These two metals are emitted by the power station and the oil refineries in Ashdod and are well known tracers of the combustion of heavy fuel oil.

The role of heavy metals in the production of ethylene in lichens is not clearly defined, although it is evident that *R. lacera*, for example, produces more ethylene upon exposure to heavy metal solutions under acidic conditions, in comparison with H_2O (Lurie and Garty, 1991).

An additional lichen species, *Cladina stellaris*, produces large amounts of ethylene upon exposure to Fe salts at pH 3.5 and lesser amounts upon exposure to Cu and Zn salts (Tables 2–4, respectively).

Table 2. Production of ethylene (nl g^{-1} h^{-1}) by podetia of *Cladina stellaris* treated with Fe (II and III)-containing solutions; treatments with K_2SO_4 and KCl as controls to test influence of anions (Kauppi et al., 1998)[a,b]

	mM	pH	*n*	*X*	
H_2O		6.8	79	1.35	e
K_2SO_4	5.0	3.5	10	1.35	e
KCl	5.0	3.5	10	2.01	e
$FeSO_4$	0.5	3.5	13	6.56	c
	5.0	3.5	20	10.16	b
	10.0	3.5	9	10.48	b
$FeCl_2$	0.5	3.5	13	6.25	c
	5.0	3.5	26	9.89	b
	10.0	3.5	16	11.48	a
$Fe_2(SO_4)_3$	0.5	3.5	21	4.75	d
$FeCl_3$	0.5	3.5	21	4.69	d
$Fe(NO_3)_3$	0.5	3.5	33	4.36	d
ANOVA	*F* ratio		366.311		
	F probability		0.001		

[a] n = number of replicates; X = mean value.

[b] Values of the vertical column followed by the same letter do not differ significantly at $P < 0.05$ using one way ANOVA and Duncan's new multiple range test.

Table 3. Production of ethylene (nl g^{-1} h^{-1}) by podetia of *Cladina stellaris* treated with Cu^{2+}-containing solutions; treatments with K_2SO_4 and KCl as controls to test influence of anions (Kauppi et al., 1998)[a,b]

	mM	pH	*n*	*X*	
H_2O		6.8	79	1.35	f
K_2SO_4	5.0	3.5	10	1.35	f
KCl	5.0	3.5	10	2.01	de
$CuSO_4$	0.5	3.5	18	1.70	ef
	5.0	3.5	13	2.65	c
	10.0	3.5	22	3.14	b
$CuCl_2$	0.5	3.5	13	1.52	f
	5.0	3.5	17	3.07	b
	10.0	3.5	17	4.23	a
$Cu(NO_3)_2$	0.5	3.5	13	1.78	ef
	5.0	3.5	13	2.34	cd
	10.0	3.5	13	3.16	b
ANOVA	*F* ratio		54.022		
	F probability		0.001		

[a] n = number of replicates; X = mean value.
[b] Values of the vertical column followed by the same letter do not differ significantly at $P < 0.05$ using one way ANOVA and Duncan's new multiple range test.

Table 4. Production of ethylene (nl g^{-1} h^{-1}) by podetia of *Cladina stellaris* treated with Zn^{2+}-containing solutions; treatments with K_2SO_4 and KCl as controls to test influence of anions (Kauppi et al., 1998)[a,b]

	mM	pH	*n*	*X*	
H_2O		6.8	79	1.35	de
K_2SO_4	5.0	3.5	10	1.35	de
KCl	5.0	3.5	10	2.01	bc
$ZnSO_4$	0.5	3.5	12	1.15	ef
	5.0	3.5	13	1.17	ef
	10.0	3.5	13	1.82	bc
$ZnCl_2$	0.5	3.5	13	0.88	f
	5.0	3.5	13	2.16	b
	10.0	3.5	13	2.76	a
$Zn(NO_3)_2$	0.5	3.5	13	1.73	c
	5.0	3.5	13	1.67	cd
	10.0	3.5	13	1.90	bc
ANOVA	*F* ratio		21.173		
	F probability		0.000		

[a] n = number of replicates; X = mean value.
[b] Values of the vertical column followed by the same letter do not differ significantly at $P < 0.05$ using one way ANOVA and Duncan's new multiple range test.

The high ethylene production upon exposure of *C. stellaris* to Fe-containing solutions should be considered in the light of a study (Bouzayen et al., 1991) which showed that Fe is an essential cofactor in the conversion of 1-aminocyclopropane-1-carboxylic acid (ACC) to ethylene in a suspension-cultured tomato (*Lycopersicum esculentum*). The authors reported that the activity of the ethylene-forming enzyme (EFE) was almost completely abolished by the metal-chelating agent 1,10-phenanthroline, and that the subsequent addition of 0.4 mM $FeSO_4$ immediately reversed this inhibition. A partial reversion with 0.6 mM $CuSO_4$ and $ZnSO_4$ occurred probably as a consequence of the release of Fe ions from the 1,10-phenanthroline complex. It is uncertain whether the evolution of ethylene in lichens takes place by the conversion of ACC. Nevertheless, it is reasonable to link the high ethylene-production induced by the application of $FeCl_2$, $FeSO_4$, $Fe(NO_3)_3$, $Fe_2(SO_4)_3$ or $FeCl_3$ with the indispensability of Fe ions for the biosynthesis of ethylene via ACC. Our results with *C. stellaris* do not support the notion of a metabolic pathway proposed by Hottiger and Boller (1991) for *Fusarium oxysporum* f. sp. *tulipae*, where ethylene derives from C_3 or C_4 of glutamate or 2-oxoglutarate. The authors proved that Fe^{2+} ions stimulated markedly the rate of ethylene formation *in vivo*, whereas Fe^{3+}, Cu^{2+} and Zn^{2+} had little or no effect. These findings accord in part with early findings of Lynch (1974) who detected a stimulatory effect of Fe^{3+} in an extracellular preparation of *Mucor hiemalis*, a greater stimulatory effect of Fe^{2+}, an inhibitory effect of Cu^{2+} and Mn^{2+}, a stimulatory effect of acid and an inhibitory effect of alkali.

Our findings (Kauppi et al., 1998) suggest that *C. stellaris*, the dominant ground cover in subarctic woodlands (Roy-Arcand et al., 1989) will yield large amounts of ethylene if moistened by non-acidified rain water or melting snow. According to Ahti, 1977, the lichen phytomass can exceed 300 g m^{-2} in mature *C. stellaris* woodland. Based on the results of our study (Garty et al., 1995b), the calculated amount of ethylene released by the moist lichen may reach 140–220 ml ha^{-1} d^{-1} or 97 ± 25 ml ha^{-1} d^{-1} (Kauppi et al., 1998). These large amounts of ethylene produced by *C. stellaris* carpets upon exposure to acidic rain water, containing bisulfite ions and/or heavy metals, may confront us with the probability of a definite influence of ethylene as a phytohormone on the growth of epigeic lichens and other soil microorganisms, which may lead to the dichotomous branching of roots of trees as a result of inoculation with mycorrhizal fungi (Rupp et al., 1989b).

An additional subject which should be dealt with is the ethylene produced in great amounts by *Bryoria fremontii* and *B. fuscescens* under the influence of moderate acidity (Garty et al., 1995b) which could affect other epiphytic lichen and mosses. A quantitative estimation of the ethylene release of epiphytic lichens under field conditions is not under consideration at this stage of our research, as the biomass of *B. fremontii* and *B. fuscescens* in Finland is not documented yet. However, the gradient of the epiphytic biomass of three *Pseudotsuga-Tsuga* forests of different ages in western Oregon and Washington were studied by McCune (1993). The author applied the term "alectoroid" lichen to include a range of pendulous species: *Alectoria* (mainly *A. sarmentosa*), *Bryoria*, and *Usnea*. The biomass of these lichens ranged between 30 and 180 kg ha^{-1} in accordance with the age of the phorophyte, whereas the biomass of "other lichens", which included for the greater part *Platismatia* and *Hypogymnia* species, was estimated as 920 kg ha^{-1} on younger trees and 550 kg

Table 5. Production of ethylene (nl g^{-1} h^{-1}) by thalli of *Ramalina lacera* collected in HaZorea, February 1994, resuspended in same site or transplanted in Ashdod region, retrieved November 1994 (Garty et al., 1997f)[a,b]

Site description	H_2O, pH 5.6			$FeCl_2$, 5 mM, pH 3.7			$FeCl_2$, 10 mM, pH 3.7		
	n	$X \pm SD$		*n*	$X \pm SD$		*n*	$X \pm SD$	
1. Ashdod, Nature Reserve (center)	11	1.04 ± 0.32	bc	4	4.72 ± 0.75	c	4	4.46 ± 0.58	d
2. Ashdod, Nature Reserve (east)	12	0.79 ± 0.15	e	4	4.73 ± 1.55	c	4	5.71 ± 0.29	bc
3. Near chemical plant, Ashdod	9	0.99 ± 0.10	cd	6	6.84 ± 1.15	ab	6	6.76 ± 0.77	ab
4. Near power plant, Ashdod	5	1.23 ± 0.06	a		–			–	
5. Kevutzat Yavneh	12	0.85 ± 0.09	de	6	5.61 ± 0.96	bc	6	5.65 ± 0.74	bc
6. Sand dunes, south of Ashdod	11	0.79 ± 0.14	e	5	7.69 ± 1.68	a	6	7.81 ± 1.26	a
7. Nitzanim sand dunes	9	0.80 ± 0.16	e	6	7.12 ± 1.33	ab	6	6.81 ± 1.20	ab
8. Hafetz Hayyim Kibbutz	12	0.79 ± 0.17	e	6	6.08 ± 0.76	abc	6	5.58 ± 0.58	c
9. Nir Galim Village	8	1.19 ± 0.24	ab		–			–	
10. HaZorea Forest (control)	12	0.86 ± 0.10	de	5	6.13 ± 0.75	abc	6	6.17 ± 0.73	bc

[a] n = number of replicates; X = mean values; SD = standard deviation.
[b] Values of each vertical column followed by the same letter do not differ significantly at $P < 0.05$ using one-way ANOVA and Duncan's multiple range test.

Table 6. Production of ethylene (nl g^{-1} h^{-1}) by in situ thalli of *Ramalina lacera* collected in HaZorea, December 1989, and by thalli of *Ramalina lacera*, resuspended in same site or transplanted in Haifa Bay, collected September 1989, retrieved December 1989 (Garty et al., 1997f)[a,b]

	H_2O, pH 7.0			$FeCl_2$, 20 mM, pH 2.0		
Site description	*n*	$X \pm$ SD		*n*	$X \pm$ SD	
Frutarom Chemical Plant, Haifa Bay (transplanted material)	15	1.66 ± 0.31	a	15	13.33 ± 1.55	b
HaZorea Forest (resuspended material)	15	1.07 ± 0.46	b	15	14.82 ± 2.78	a
HaZorea Forest in situ thalli	15	0.98 ± 0.22	bc	15	14.92 ± 2.08	a
Oil Refineries, Haifa Bay (transplanted)	15	0.80 ± 0.26	bcd	15	11.52 ± 0.97	c
Kefar Massaryk Kibbutz, near steel smelter, Haifa Bay (transplanted material)	15	0.76 ± 0.14	cd	15	10.74 ± 0.95	c
Kefar Bialik, agricultural village, Haifa Bay (transplanted material)	16	0.61 ± 0.55	d	15	15.02 ± 1.72	a

[a] n = number of replicates; X = mean values; SD = standard deviation.
[b] Values of each vertical column followed by the same letter do not differ significantly at $P < 0.05$ using one-way ANOVA and Duncan's multiple range test.

ha^{-1} on older trees. Based on the above mentioned estimation, the production of ethylene by a biomass of 100 kg of *B. fremontii* wetted by non-acidified water may reach 14 ml ha^{-1} day^{-1}, whereas a similar biomass of this lichen exposed to rain water containing 0.1 mM $NaHSO_3$ at pH 4.0 may release about 60 ml ha^{-1} day^{-1} of ethylene.

The exposure of lichens to intense air contamination will affect the ethylene-producing system. Thus, lichens retrieved from very polluted sites will produce, under certain conditions, smaller amounts of ethylene than lichens exposed to moderate air contamination. A transplantation series of *Ramalina lacera* to polluted industrial sites and the subsequent exposure of the retrieved thalli to 5 or 10 mM $FeCl_2$ led to a lesser rate of ethylene- production than that of samples which did not endure injury of the ethylene-producing system (Tables 5 and 6).

2. Elemental content and chlorophyll integrity in lichens

Chlorophyll extraction and spectroscopic measurements are applied to assess the degree of integrity of the chlorophyll in photobiontic cells in lichens. The extraction of chlorophyll is usually performed by dimethyl sulfoxide (DMSO) or acetone. The exposure of lichen thalli to chemical pollutants, either under controlled conditions or under natural conditions in the presence of air pollutants, leads in many cases to a degradation of chlorophyll.

The advantage of DMSO as a solvent for the extraction of photosynthetic lichen pigments (Ronen and Galun, 1984) is that the extraction is simple, rapid, complete and that the extract is easily stored at low temperatures without degradation. An additional advantage is that small plant samples are sufficient since the procedure is not involved with any loss of extract and grinding or homogenizing are not required. Ronen and

Galun suggested the ratio of optical density at wavelengths 435 and 415 nm as a reliable parameter for an estimation of chlorophyll degradation. Its reliability derives from its high correlation with the acid factor used for the measurement of the ratio chlorophyll *a*/phaeophytin *a*.

An attempt to predict the combined effect of sulfur dioxide and nitrate on the green-algal lichen *Ramalina farinacea* was made by Balaguer et al. (1997) who measured the degree of phaeophytization and ultrastructural damage of a *Trebouxia* photobiont. The thalli were exposed to SO_2 and aqueous solutions of NO_3^-, for 6–14 days. The interactive effects of SO_2 and NO_3^- were found to be synergistic at slight effects of individual pollutants. The above-mentioned pattern reverted to additive effects, and sequentially to antagonistic effects as the detrimental impact of SO_2 intensified. Although variable results of the interactive impact of SO_2 and NO_3^- were obtained, the results are assumed to be predictable on the basis of the effects to be expected following exposure to each pollutant on itself.

In a study of the correlation of chlorophyll degradation and elemental content in transplanted *Ramalina lacera* in 13 biomonitoring sites in Israel (Garty et al., 1985) the values for the OD435 nm/OD415 nm ratio decreased with an increase of the amount of Br, Pb, Fe and Ti. The OD435 nm/OD415 nm ratio correlated positively with the amount of K. In another field study (Garty et al., 1988) the OD435 nm/OD415 nm ratio correlated inversely with the amounts of Pb, Zn and Cu in transplants of *R. lacera* in 12 biomonitoring sites. Thalli of *R. lacera* from the control site in HaZorea, transplanted for a period of ten months in the vicinity of oil refineries and a steel smelter in the Haifa Bay exhibited a definite degradation of chlorophyll (Garty et al., 1993a). The seasonal variation of Mg percentages detected by scanning electron microscopy (SEM) and energy dispersive X-ray analysis (EDX) on/in the cortical cells in the lichen was compared with changes in the chlorophyll integrity of *R. lacera* transplants. The amount of this element on/in the thallial cortex increased gradually in samples from the Haifa Bay whereas low Mg values were obtained for control thalli. The values obtained for Mg seem to represent a significant leakage from algal cells, indicated by a detectable accumulation on the lichen surface. As the control thalli exhibited a different pattern of Mg accumulation under similar climatic conditions we assume that the accumulation of this element on/in the cortical cells is associated with a leakage of ions from internal parts of the thallus, followed by a deposition on surface cells. Boonpragob and Nash (1991) detected a substantial decline of chlorophyll and net photosynthesis in summer in a polluted site in Los Angeles for *Ramalina menziesii*. The percentage of phaeophytins was found to increase in proportion. Leachable Mg, Ca, P, Na and K were found to increase in this lichen in summer (Boonpragob and Nash, 1990). The authors assumed that at least a certain part of the elevated, leachable Mg in *R. menziesii* derived from the degradation of chlorophyll in algal cells of this lichen in air-contaminated sites.

The bleaching of lichen thalli as a result of chlorophyll degradation is an obvious sign of damage to lichens affected by airborne pollutants (Nash, 1971; Puckett et al., 1973; Showman, 1975; Eversman, 1978). Our study (Garty et al., 1992) was initiated to determine the impact of different metal ions in the presence of certain anions under acidic conditions in order to explain the bleaching of transplanted *R. lacera* located around metal smelters, road intersections and industrial zones. Combined treatments of *R. lacera* with each of the metal ions Fe^{2+}, Fe^{3+}, Mn^{2+}, Cu^{2+}, Pb^{2+} and Zn^{2+} and the ions SO_4^{2-}, NO_3^- or Cl^- under acidic conditions were assumed to intensify the detrimental effect of heavily polluted sites. Treatments with Cu, Zn, Mn, Pb, Fe(II) and Fe(III) salts under acidic conditions were particularly effective. SO_4^{2-} combined with K^+ proved to be rather harmful to the photobiont chlorophyll over and above Cl^- anions; yet a $CuCl_2$-treatment led to a severe chlorophyll degradation. The following sequence lists the evaluated metal ions according to their relative impact:

(1) In the presence of chloride:	Cu > Zn > Mn; Fe^{2+} > Pb > Fe^{3+} > K.
(2) In the presence of nitrate:	Zn > Cu; Mn > Pb > Fe^{3+} > K.
(3) In the presence of sulfate:	Mn > Cu; Zn > K > Fe^{2+}.

The assumption that Cu is responsible for the reduced chlorophyll content of lichens growing in mining areas in which Cu, Zn and Pb are to be found in the soil was put forth by Chettri et al. (1998). This study tested the effect of these heavy metals on the chlorophyll content of the lichens *Cladonia convoluta* and *C. rangiformis* under controlled conditions. A Cu content up to 1600 $\mu g\ g^{-1}$ dry wt. had no effect on the total chlorophyll content of *C. rangiformis,* whereas Cu contents exceeding 175 $\mu g\ g^{-1}$ in *C. convoluta* led to a decrease of the total chlorophyll content. The Cu effects on chlorophyll were found to reduce in the presence of Pb and Zn in both lichens, but to a lesser extent in *C. rangiformis*. Metal cations appeared to be ionically bound within the cell wall in an exchangeable form with binding affinities of Pb > Cu > Zn. Of these cations, only Cu is taken up into photobiont cells. The higher sensitivity of *C. convoluta* relative to *C. rangiformis* is assumed to be related to an increased Cu-induced membrane permeability.

An additional study of the comparative sensitivity of different lichen species to chemical pollutants and of the impact of pollutants on chlorophyll integrity was performed by Riga-Karandinos and Karandinos (1998). Samples of *Anaptychia ciliaris, Lobaria pulmonaria* and *Ramalina farinacea* were collected in the vicinity of a lignite power plant in the plain of Megalopolis (Peloponesos) in southern Greece. The amounts of Ca, Cd, Cu, Fe, Mn, Pb, Zn and total S were determined. Multiple linear regressions showed that the amount of S and of the greater part of the metals depended on distance (negative effect) and elevation (positive effect) relative to the location of the power plant. Additional biochemical parameters such as thallus pH, total proteins, reducing sugars and the ratio of chlorophyll *a*/phaeophytin *a* examined in the framework of this study, found the lichen *L. pulmonaria* to be the most sensitive to SO_2, with the highest values for total S and reducing sugars, due probably to a breakdown of polysaccharides. The ratio of chlorophyll *a*/phaeophytin *a* in the three lichen species indicated a substantial degradation of chlorophylls. However, the absence of a consistent and significant correlation of the amount of total S and the degree of phaeophytization of chlorophyll suggests that although phaeophytinization should not be considered a reliable bioindicator of specific pollutants, it may be considered a reliable indicator of total environmental stress.

A three-year study on the impact of air pollution on chlorophyll degradation in the Ashdod region, SW Israel, used *Ramalina lacera* transplants (Garty et al., 1997a,b,e,

1998a). After 100 days, the transplants exhibited the greatest degree of chlorophyll degradation in the vicinity of a chemical plant producing herbicides and other pesticides (Table 7). The only significant inverse correlation for this short period of exposure was obtained for OD435 nm/OD415 nm ratios and sulfate-S (Table 8). However, longer periods of exposure led to an intensified degradation of chlorophyll near the above-mentioned chemical plant, apart from variable degrees of chlorophyll degradation in additional biomonitoring sites (Table 7). In those cases, inverse correlations were obtained for chlorophyll integrity and concentrations of several mineral elements such as Ni, V, B, Cr, Fe, Mn, Pb, Al and Ti (Table 8). The OD435 nm/OD415 nm ratio was found to correlate positively with the concentrations of both K and P, subsequent to long periods of exposure (Table 8).

The degradation of chlorophyll in *R. lacera* transplants in the Ashdod region often coincided with a severe plasmolysis of the algal cells in the lichen thallus. Fig. 3 shows a cross-section of a thallus of *R. lacera* sampled in the unpolluted control site at HaZorea in November 1994, resuspended in the same site and retrieved in August 1995. Fig. 4 shows a cross-section of the thallus of a transplanted lichen from an unpolluted site (Nitzanim sand dunes) sampled at the same time, whereas Fig. 5 shows the thallus of a lichen transplant sampled near a chemical plant in Ashdod. In the latter, the algal cells are deformed to a large extent.

Gonzales and Pignata (1994) introduced a new index of air-contamination (C.I.) based on an assessment of the status of the chlorophyll in lichens. For this purpose the researchers exposed transplants of *Punctelia subrudecta* in different biomonitoring sites in Cordoba, Argentine. At the end of a period of exposure of three months the total amount of Mn, Cu, Pb, Zn, Fe and Al in the thalli was measured. The elemental content was compared with chlorophyll degradation, amounts of S, MDA (malondialdehyde) and soluble proteins. The extraction of chlorophylls and phaeophytins was performed by acetone followed by an addition of oxalic acid-acetone. The results indicated a significant decrease in chlorophyll *a* relative to freshly picked control samples. Chlorophyll degradation expressed as phaeophytization ratio and S content correlated to a lesser extent than MDA and S content, and no correlation was detected for chlorophyll degradation and metal content. It is suggested that a short-term exposure is not sufficient to demonstrate the potential injury to chlorophylls due to the accumulation of metals. Our studies (Garty et al., 1988, 1997a) confirm that the degradation of chlorophyll in lichens is involved with a prolonged period of exposure to pollutants over and above other physiological injuries.

An additional study which used the lichen *Ramalina ecklonii* as a bioindicator of atmospheric pollution in Cordoba (Levin and Pignata, 1995) analyzed the content of S, chlorophylls, phaeophytins, soluble proteins and oxidation products measured as conjugated dienes in the lichen. This study, carried out in an urban polluted site, to evaluate the lichen as a bioindicator of atmospheric pollution, detected a significant degradation of chlorophyll and relative high S contents.

An additional study applying *R. ecklonii* (Gonzales et al., 1996) was designed to obtain information on the relative quality of the air in a certain section of Cordoba according to levels of vehicular and industrial activity. The highest chlorophyll degradation occurred in transplants exposed to heavy traffic and light industry. The highest S concentrations were obtained for sites with heavy traffic and a medium level of

Table 7. OD435 nm/OD415 nm ratios of *Ramalina lacera* collected with substrate in HaZorea, NE Israel, resuspended in same site or transplanted to Ashdod region, SW Israel, different periods[a,b]

Site description	Period of exposure							
	March–June 1993		March 1993–Jan. 1994		Feb.–Nov. 1994		Nov. 1994–Aug. 1995	
	n	*X* ± SD	*n*	*X* ± SD	*n*	*X* ± SD	*n*	*X* ± SD
1. Sulfur terminal, Ashdod (urban/industrial)	15	1.43 ± 0.02 ab	15	1.09 ± 0.21 c		–	12	0.95 ± 0.11 d
2. *Acacia albida* nature reserve (center), Ashdod	15	1.45 ± 0.01 a	15	1.43 ± 0.02 ab	10	1.42 ± 0.02 a	12	1.33 ± 0.05 b
3. *Acacia albida* nature reserve (east), Ashdod	14	1.44 ± 0.02 a	15	1.33 ± 0.16 b	9	1.43 ± 0.01 a	12	1.32 ± 0.06 b
4. Near chemical plant, Ashdod (urban/industrial)	15	1.33 ± 0.09 d	15	0.94 ± 0.26 d	10	0.68 ± 0.08 c	12	0.90 ± 0.19 d
5. Hafetz Hayyim Kibbutz (rural)	14	1.44 ± 0.03 a	14	1.42 ± 0.05 ab	10	1.43 ± 0.01 a	12	1.36 ± 0.04 ab
6. Near oil refineries, Ashdod (urban/industrial)	15	1.45 ± 0.01 a		–		–	12	1.04 ± 0.11 c
7. Near Ort School, Ashdod (urban)	14	1.41 ± 0.02 bc		–		–		–

8. Sede Uziyahu Village (rural)	14	1.44 ± 0.02 a	13	1.42 ± 0.04 ab		–	12	1.41 ± 0.03 a
9. Kevutzat Yavneh (rural)	15	1.41 ± 0.03 c	15	0.86 ± 0.22 d	10	1.37 ± 0.06 b		–
10. Nir Galim Village (rural)	16	1.45 ± 0.02 a	14	1.44 ± 0.08 ab		–	12	1.36 ± 0.04 ab
11.HaZorea Forest, control site (rural)	15	1.44 ± 0.02 ab	12	1.42 ± 0.02 ab	10	1.42 ± 0.03 a	12	1.38 ± 0.03 ab
12. Nitzanim Sand Dunes (rural)	12	1.44 ± 0.01 ab	12	1.42 ± 0.03 ab	10	1.43 ± 0.01 a	12	1.41 ± 0.04 a
13.Sand dunes, south of Ashdod (rural)	15	1.43 ± 0.01 ab	14	1.44 ± 0.02 ab	9	1.40 ± 0.03 ab	12	1.38 ± 0.03 ab
ANOVA F ratio	14.02		39.30		356.59		69.21	
ANOVA F probability	0.00		0.00		0.00		0.00	
Reference	Garty et al., 1997a		Garty et al., 1997e		Garty et al., 1998a		Garty et al., 1997b	

[a] n = number of replicates; X = mean value; SD = standard deviation.

[b] Values of each vertical column followed by the same letter do not differ significantly at $P < 0.05$ using one-way ANOVA and Duncan's multiple range test.

Table 8. Pearson correlation coefficients of OD435 nm/OD415 nm ratios and mineral elements in *Ramalina lacera* from HaZorea, resuspended controls and transplants in biomonitoring sites in Ashdod region, different periods

Period of exposure	Parameters	Pearson correlation coefficient	Reference	Remarks
March–June 1993	OD435 nm/OD415 nm, sulfate-S	−0.65*	Garty et al., 1997a	$*P < 0.05$
March 1993–January 1994	OD435 nm/OD415 nm, Ni	−0.70*	Garty et al., 1997e	$*P < 0.05$
	OD435 nm/OD415 nm, S	−0.73**		$**P < 0.01$
	OD435 nm/OD415 nm, V	−0.64*		
February–November 1994	OD435 nm/OD415 nm, B	−0.77*	Garty et al., 1998a	$*P \leq 0.05$
	OD435 nm/OD415 nm, Cr	−0.75*		
	OD435 nm/OD415 nm, Fe	−0.74*		
	OD435 nm/OD415 nm, K	0.89***		$***P < 0.005$
	OD435 nm/OD415 nm, Mn	−0.81*		
	OD435 nm/OD415 nm, Ni	−0.78*		
	OD435 nm/OD415 nm, Pb	−0.98***		
November 1994–August 1995	OD435 nm/OD415 nm, Al	−0.67*	Garty et al., 1997b	$*P \leq 0.05$
	OD435 nm/OD415 nm, Fe	−0.64*		
	OD435 nm/OD415 nm, K	0.97***		
	OD435 nm/OD415 nm, Ni	−0.72*		
	OD435 nm/OD415 nm, P	0.71*		
	OD435 nm/OD415 nm, sulfate-S	−0.65*		
	OD435 nm/OD415 nm, Ti	−0.83***		$***P < 0.001$
	OD435 nm/OD415 nm, V	−0.63*		

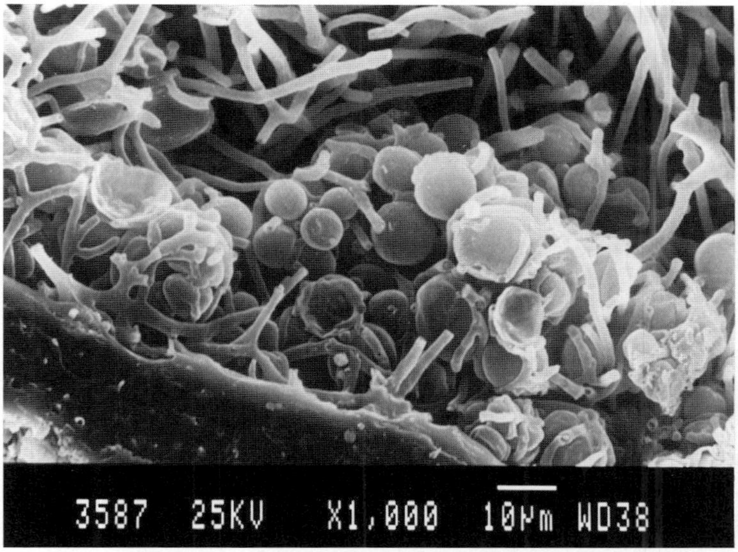

Figure 3. Cross-section of thallus of *Ramalina lacera* from HaZorea, November 1994, resuspended in the same site, retrieved in August 1995. Algal cells *(Trebouxia)* are in good condition. Scale bar, 10 μm (Garty, Kunin, Delarea, unpublished).

industrial activity, whereas in less busy sites the physiological status of the chlorophyll was better and the S concentrations were lower.

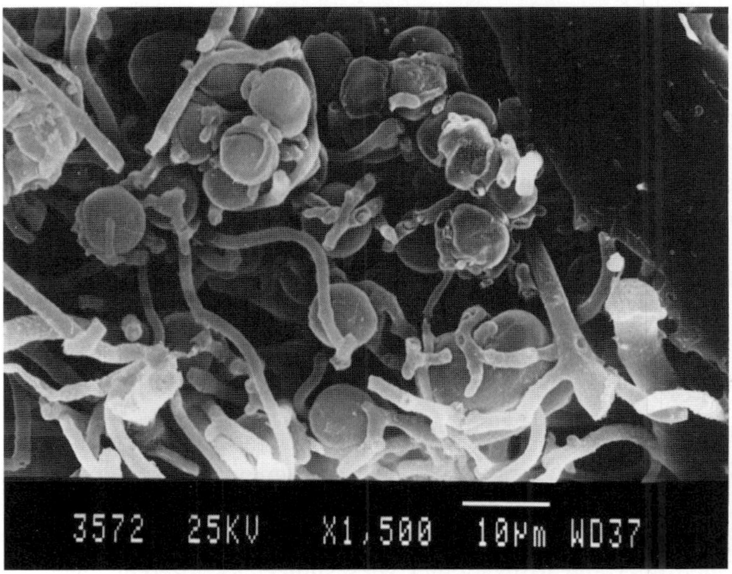

Figure 4. Cross-section of thallus of *R. lacera* from HaZorea, November 1994, transplanted to an unpolluted site in the Ashdod region (Nitzanim sand dunes), retrieved August 1995. Algal cells are in good condition following transplantation. Scale bar, 10 μm (Garty, Kunin, Delarea, unpublished).

Figure 5. Cross-section of thallus of *R. lacera* from HaZorea, November 1994, transplanted near a chemical plant in Ashdod, retrieved August 1995. Note shrunk *Trebouxia* cells. Scale bar, 10 μm (Garty, Kunin, Delarea, unpublished).

Gonzales and Pignata (1997) used transplants of *Punctelia subrudecta* to evaluate its specific response to pollution produced by traffic and/or industry in the surroundings of a power station and found greater amounts of S and MDA and higher PI (pollution index) values. This study detected insignificant differences for phaeophytization in different sites, apart from a certain relation of the ratio phaeophytin *a*/ chlorophyll *a* and the S content. In the same manner, Canas et al. (1997) investigated the chemical response of transplants of *Parmotrema austrosinense* and *P. conferendum* in non-polluted and urban sites in Cordoba. The authors examined the content of pigments, MDA, hydroperoxy-conjugated dienes and sulfur subsequent to varying periods of exposure. Significant higher levels of chlorophyll *a*, total chlorophyll and carotenoids were found for the fourth and sixth months of exposure relative to the first and second months. The S content was observed to increase in *P. austrosinense* only in the sixth month of exposure and it did not correlate either with a decrease of pigment content.

Canas and Pignata (1998) were able to show that the concentration of pigments and peroxidation products in lichens may be liable to temporal variation as exemplified by carotenoids, chlorophylls, hydroxy-conjugated dienes and MDA in *Parmotrema uruguense*. Thalli of this lichen from an unpolluted area were transplanted within the unpolluted area and to two other sites subject to different levels of vehicular activity. For most of the quantified parameters, the temporal variation throughout the period of exposure (July–November, 1994) depended on the site of transplantation. The lower content of carotenoids and chlorophyll *a* and the higher values for the ratio chlorophyll *b*/chlorophyll *a* and hydroperoxy-conjugated dienes detected in downtown Cordoba were attributed to the effect of atmospheric pollutants. The authors did not present, however, data of amounts of airborne pollutants in the lichen thallus, to link chemical contaminants with physiological damage.

The degradation of chlorophyll in epilithic desert lichens exposed to anthropogenic activity was hardly documented. Benlap and Harper (1990) however, observed chlorophyll degradation by measurements of the OD435 nm/OD415 nm ratio and electrolyte leakage for the umbilicate desert lichen *Rhizoplaca melanophthalma* in the vicinity of a coal-fired power plant near Page, Arizona. The data for both chlorophyll degradation and electrolyte leakage indicate a significant injury to chlorophyll and membranes in the immediate vicinity of the power plant. Highly significant correlations were obtained for both electrolyte leakage and chlorophyll degradation dependent on the distance from the power-plant.

Bartok et al. (1992) reported on the diminution of assimilation pigments and respiration as expressed by a decrease of dehydrogenase activity, in combination with signs of total exhaustion, dependent on the distance from a source of pollution, a chromite dryer. Transplants of *Xanthoria parietina* were placed around a refractory material plant in the Dej industrial area, Cluj county, Romania which released MgO, Al_2O_3, Fe_2O_3 and CrO_3. The physiological damage observed in the lichen samples coincided with an enlarged accumulation of Mg, Cr, Fe and Cd.

The influence of dust, originating in a limestone quarry, on the degradation of chlorophyll, was studied by Zaharopoulou et al. (1993) in Thessaloniki, N. Greece. The chlorophyll and phaeophytin extracted from the lichen *Physcia adscendens* were applied as criteria for pollution-induced injury. Chlorophyll concentrations decreased whereas phaeophytin percentages increased dependent on distance from the pollution source and consequently increasing load of dust. The presence of dust particulates from the limestone quarry led to the formation of a *Xanthorion* community, not normally found in inland sites, although representatives of this community are negatively affected by dust.

The greater part of those studies which dealt with high levels of pollution, showed significant degrees of chlorophyll degradation. Von Arb and co-workers (Von Arb and Brunold, 1990; Von Arb et al., 1990) on the other hand, detected a high content of chlorophyll in polluted sites in Biel, Switzerland, for thalli of *Parmelia sulcata*, related with low rates of sulfate uptake and protein synthesis.

Sochting (1995) investigated the uptake of nitrogen by lichens. The capability of lichens to absorb nitrogen is found to be extraordinary. As the process is rapid, the nitrogen content of lichen thalli reflects a deposition having taken place in the recent past. Sanchez-Hoyos and Manrique (1995) studied the effect of nitrate and ammonium ions on the pigment content (xanthophylls, carotenes and chlorophylls) of lichens. Thalli of *Ramalina capitata* were sprayed with deionized water, 10 mM solutions of KNO_3 or NH_4Cl, or maintained without any water supply for ten days under controlled conditions. Samples treated with nitrate solutions displayed a significant increase of all pigments except for antheraxanthin. Ammonium treatments, on the other hand, did not induce change in comparison with water, with the exception of antheraxanthin which decreased. Nitrate-treated lichens exhibited a significant increase of both chlorophyll *a* and chlorophyll *b*, whereas ammonium affected only chlorophyll *b*. The nitrogen-containing substances NO_3^- and NH_4^+ affect lichens such as *Ramalina capitata* in different ways. The presence of ammonium in the surroundings of lichen communities could stimulate a rapid modification of species. According to Kauppi (1980) and Kauppi and Mikkonen (1980) the presence of nutrient-rich dust, with nitrogen especially, can neutralize the effects of SO_2 and other acidic pollutants. This may explain the abundance of *Hypogymnia physodes* in some polluted sites in Finland and the occurrence of atypical species on pines, such as *Chaenotheca chrysocephala* and *Evernia prunastri* (Halonen et al., 1991; Kauppi and Halonen, 1992).

3. Elemental content and the spectral reflectance response of the lichen thallus

Measurements of spectroscopic features of plants have led to the construction of different vegetation indices (VI). Most indices are based on a combination of the ratio of two portions of the electromagnetic spectrum: the red band (R) at 600–700 nm, which corresponds to the region of maximum chlorophyll absorption and the near infrared band (NIR) at 700–1100 nm, which corresponds to the maximum reflectance of incident light by the living vegetation. The basic vegetation index, the Simple Ratio, is a ratio of the digital value of these two bands. The most widely used VI is known as the normalized difference vegetation index (NDVI; Rouse et al., 1974): $NDVI = (NIR\text{-}R)/(NIR + R)$.

The values obtained by this equation range from -1.0 to $+1.0$. The most important feature of NDVI is that healthy vegetation surfaces have higher index values than stressed vegetation surfaces as the former have a stronger 'greenness' signal. In consequence, NDVI has been empirically correlated with photosynthetic activity, fractional vegetation cover, green leaf biomass or leaf area, productivity, amount of standing biomass, etc. NDVI is considered to be appropriate for various applications such as image classification and detection of seasonal yield and cover in plants (Tucker, 1979; Curran, 1980; Holben et al., 1980; Asrar et al., 1984; Sellers, 1985; Prince, 1990).

The application of spectroscopic measurements to a study of the reflectance response of cryptogams yielded several interesting findings: Satterwhite et al. (1985) studied the impact of lichens on the reflectance of granitic rock surfaces and detected an interference of lichens in visible and near infrared reflectance spectra (400–1100 nm). Quantitative field surveys in the semiarid region of eastern Australia showed that microphytic crusts (lichens, mosses and cyanobacteria) contributed up to 27% of the ground cover (O'Neill, 1994). A record was made of the spectral reflectance of crusts on two types of soil, under different moisture conditions, at wavelengths between 0.4 and 2.5 μm. The presence of microphytic crusts on both types of soil decreased the reflectance throughout the spectrum. Karnieli and Tsoar (1995) measured the spectral reflectance of biogenic crusts on desert dune sand along the Israel-Egypt border. The crust was found to consist for the greater part of microphytes such as cyanobacteria. The higher vegetation cover was less than 30%. Therefore the soil itself was assumed to be the principal contributor to the overall spectral reflectance response. The spectral reflectance response of the dune vegetation was much lower than that of the active sand and of the biogenic crust. Consequently the dune vegetation signal was masked by the soil signal. The biogenic crust which covered most of the Israeli side of the border contributed a darker tone to the image. According to the authors, anthropogenic activities on the Egyptian side prevented the establishment of a biogenic crust, thus making for a brighter tone. Thus, the well known contrast of tone between Sinai and the Negev, is not, as would be expected, a result of the vegetative coverage but of the biogenic crust. In a subsequent study, Karnieli et al. (1996) showed that the levels of spectral reflectance of wet limestone covered with endolithic lichens were lower than those of dry limestone. The spectral response of the wet lichen-covered limestone was similar to that of higher plants.

Gouaux and Vincent (1990) performed a comparative analysis of polluted versus unpolluted areas by means of infrared color photography, digitalization, and data processing. The change in the spectral response indicated by the visible and infrared bands and by a decrease of the vegetation index (IR/R) was attributed to the action of pollutants. In the same manner, Cox et al. (1991) found that lichens exposed to Cu concentrations $>20\ \mu g\ g^{-1}$ exhibited a significant shift of 2–3% of the spectral response.

A series of transplantation studies were performed in Israel in order to estimate the degree of stress induced by exposure of lichens to pollutants, as expressed by changes in the spectral reflectance response of the thallus (Garty et al., 1997a,b,e,f). We compared the alteration of NDVI with the amount of mineral elements detected in thalli of *Ramalina lacera* collected in the HaZorea Forest (NE Israel) and transplanted to the Ashdod region (SW Israel) for different periods of time. The exposure of lichens to pollutants caused significant shifts in the spectral response of the lichen thallus (Table 9). Pearson correlation coefficients of NDVI and mineral elements accumulated in *R. lacera* in the Ashdod region (Table 10) indicated an inverse correlation of NDVI and total S, sulfate-S, Ni, V, Pb, Mn, Al, Cr, Fe and Ti and a direct correlation of NDVI, K and P.

Additional studies linked the elemental composition of lichens with the reflectance response expressed as NDVI and with the chlorophyll integrity of the photobiont expressed as OD435 nm/OD415 nm ratio. In a series of controlled experiments, thalli of *R. lacera* from HaZorea were treated either with double distilled water or with simulated acid rain (H_2SO_4:HNO_3, 2:1, pH 3.01 ± 0.07) for a five-months period (January–June 1996). Table 11 presents NDVI values, OD435 nm/OD415 nm ratios and elemental content at the conclusion of the experiment. Both physiological parameters indicate the degradation of chlorophyll upon exposure to simulated acid rain. In addition, it is demonstrated that the exposure of *R. lacera* thalli to simulated acid rain caused a significant decrease in the amount of Ca, K, Mg, Na, P, Mn, Sr and Ba, in comparison with the content of transplants left in HaZorea. On the other hand, the amounts of both total S and sulfate-S increased as expected in thalli treated with H_2SO_4:HNO_3 solutions.

4. Chemical pollutants and changes in modulated chlorophyll *a* fluorescence: monitoring photosystem II (PSII) activities in lichens exposed to pollutants

A non-invasive method to analyze photosynthetic parameters in lichens is the

Table 9. NDVI: changes in spectral reflectance of *Ramalina lacera* collected with substrate in HaZorea, NE Israel, resuspended in same site or transplanted to Ashdod region, SW Israel, for different periods[a,b]

	Period of exposure							
	March–June 1993		March 1993–Jan. 1994		Feb.–Nov. 1994		Nov. 1994–Aug. 1995	
	$X \pm SD$		$X \pm SD$		$X \pm SD$		$X \pm SD$	
1. Sulfur terminal, Ashdod (urban industrial)	0.28 ± 0.01	ef	0.11 ± 0.005	f	0.210 ± 0.000	g	0.135 ± 0.005	f
2. *Acacia albida* nature reserve (center), Ashdod (urban industrial)	0.31 ± 0.008	bc	0.25 ± 0.00	a	0.340 ± 0.000	g	0.332 ± 0.005	ab
3. *Acacia albida* nature reserve (east) Ashdod (suburban)	0.31 ± 0.009	ab	0.22 ± 0.00	c	0.372 ± 0.005	b	0.310 ± 0.014	c
4. Near chemical plant, Ashdod (urban/industrial)	0.15 ± 0.005	g	0.09 ± 0.00	g	0.160 ± 0.008	h	0.117 ± 0.009	g
5. Hafetz Hayyim Kibbutz (rural)	0.28 ± 0.009	def	0.20 ± 0.00	e	0.300 ± 0.008	f	0.330 ± 0.018	ab
6. Near oil refineries, Ashdod (urban/industrial)	0.29 ± 0.009	de	0.24 ± 0.005	a	–		0.155 ± 0.005	e
7. Near Ort School, Ashdod (urban)	0.31 ± 0.008	bc	–		–		–	

8. Sede Uziyahu Village (rural)	0.29 ± 0.008	cde	0.23 ± 0.00	b	–		0.330 ± 0.008	ab
9. Kevutzat Yavneh (rural)	0.27 ± 0.01	f	0.09 ± 0.00	g	0.320 ± 0.000	e	–	
10. Nir Galim Village (rural)	0.30 ± 0.008	bcd	0.21 ± 0.005	cd	0.327 ± 0.005	e	0.282 ± 0.009	d
11.HaZorea Forest, control site (rural)	0.32 ± 0.009	a	0.25 ± 0.00	a	0.412 ± 0.005	a	0.337 ± 0.005	a
12. Nitzanim Sand Dunes (rural)	0.28 ± 0.01	def	0.21 ± 0.006	d	0.357 ± 0.009	c	0.320 ± 0.021	abc
13.Sand dunes, south of Ashdod (rural)	0.28 ± 0.005	de	0.22 ± 0.00	c	0.367 ± 0.005	b	0.315 ± 0.012	bc
ANOVA F ratio	332.91		1654.21		606.24		219.95	
ANOVA F probability	0.00		0.00		0.00		0.00	
Reference	Garty et al., 1997a		Garty et al., 1997e		Garty et al., 1997f		Garty et al., 1997b	

[a] The number of replicates was four in all cases. X = mean value; SD = standard deviation.

[b] Values of each vertical column followed by the same letter do not differ significantly at $P < 0.05$ using one-way ANOVA and Duncan's multiple range test.

Table 10. Pearson correlation coefficients of NDVI and mineral elements in *Ramalina lacera* collected in HaZorea, NE Israel, resuspended in same site as control or transplanted to 9–12 biomonitoring sites in Ashdod region, SW Israel, for different periods

Period of Exposure	Parameters	Pearson correlation coefficients	Reference	Remarks
March–June 1993	NDVI, sulfate-S	−0.82***	Garty et al., 1997a	$^{***}P < 0.001$
March 1993–January 1994	NDVI, Ni	−0.71*	Garty et al., 1997e	$^{*}P < 0.05$
	NDVI, total S	−0.66*		
	NDVI, V	−0.59*		
February–November 1994	NDVI, sulfate-S	−0.72*	Garty et al., 1997f	$^{*}P \leq 0.05$
	NDVI, Ni	−0.68*		
	NDVI, K	0.84***		$^{***}P \leq 0.005$
	NDVI, Pb	−0.76*		
	NDVI, Mn	−0.72*		
November 1994–August 1995	NDVI, Al	−0.65*	Garty et al., 1997b	$^{*}P \leq 0.05$
	NDVI, Cr	−0.63*		
	NDVI, Fe	−0.70*		
	NDVI, K	0.92***		$^{***}P \leq 0.001$
	NDVI, Ni	−0.75**		$^{**}P \leq 0.01$
	NDVI, P	0.64*		
	NDVI, sulfate-S	−0.60*		
	NDVI, Ti	−0.79*		
	NDVI, V	−0.71*		

measurement of modulated chlorophyll *a* fluorescence (Demmig-Adams et al., 1990; Hovenden and Seppelt, 1995; Kappen et al., 1998). The above-mentioned studies, apart from other investigations not mentioned here, dealt with a wide range of physiological conditions including those prevailing under low temperatures. Less attention was paid to modifications of *in vivo* chlorophyll fluorescence of lichens exposed to pollutants. Gries et al. (1995) studied the effect of SO_2 fumigation on CO_2 gas exchange, chlorophyll fluorescence and chlorophyll degradation of different lichen species in western North America. The study aimed at a comparison of *Alectoria sarmentosa, Flavopunctelia praesignis, Parmotrema hypotropum, Ramalina menziesii* and *Usnea hirta*, in order to determine the relative sensitivity to SO_2 of major lichen species. The fluorescence response was indicated by the factors: F_r = maximum rate of the induced rise of fluorescence, F_v = variable fluorescence, F_m = maximal fluorescence yield when PSII reaction centers are closed, and F_v/F_m = photon yield of PSII photochemistry. The chlorophyll fluorescence parameters, applied as measurements of stress response, particularly F_r and F_v/F_m, exhibited a significant decline for fumigated lichens versus controls at 0.5–1.0 ppm SO_2. The gas exchange data of *R. menziesii* were found to correspond. For *A. sarmentosa* the fluorescence variables expressed a higher sensitivity than gas exchange data.

Table 11. NDVI, OD435 nm/OD415 nm ratios and amount of mineral elements (μg g^{-1} dry wt.) in thalli of *Ramalina lacera* collected in HaZorea, January 1996, resuspended in same site or transplanted to Botanical Gardens, Tel Aviv University. End of experiment: June 1996 (Source: Garty, Kunin, Cohen and Karnieli, unpublished)[a,b]

	HaZorea, resuspended material, untreated			Botanical Gardens, transplanted material, untreated			Botanical Gardens, transplanted material, H_2O-treated			Botanical Gardens, transplanted material, simulated acid rain			ANOVA	
	n	*X* ± SD		*n*	*X* ± SD		*n*	*X* ± SD		*n*	*X* ± SD		*F* ratio	*F* prob.
NDVI	4	0.318 ± 0.004	a	4	0.270 ± 0.000	c	4	0.290 ± 0.0006	b	4	0.245 ± 0.005	d	258.7143	0.0000
OD435 nm/OD415 nm	11	1.44 ± 0.03	a	11	1.43 ± 0.01	a	11	1.42 ± 0.02	a	11	1.29 ± 0.04	b	72.6020	0.0000
Ca	12	23383 ± 4913	a	12	22625 ± 4598	a	12	21876 ± 3909	ab	12	17844 ± 6119	b	2.9873	0.0412
K	12	2565 ± 198	a	12	1943 ± 164	b	12	1862 ± 330	b	12	1110 ± 164	c	84.4635	0.0000
Total-S	12	3108 ± 261	b	12	2614 ± 211	c	12	2564 ± 339	c	12	4790 ± 835	a	56.4923	0.0000
Sulfate-S	11	2032 ± 244	b	11	1903 ± 190	b	11	1967 ± 253	b	11	4074 ± 531	a	110.8066	0.0000
Mg	12	890 ± 100	a	12	726 ± 75	b	12	615 ± 103	c	12	330 ± 46	d	94.1472	0.0000
Na	12	582 ± 151	a	12	562 ± 99	a	12	189 ± 119	b	11	100 ± 36	b	59.0857	0.0000
P	12	1351 ± 180	a	12	839 ± 122	b	12	801 ± 248	b	12	817 ± 111	b	28.1095	0.0000
Mn	12	25 ± 4	a	12	21 ± 2	b	12	18 ± 3	c	12	12 ± 2	d	42.7986	0.0000
Sr	12	30 ± 6	a	12	36 ± 9	a	12	34 ± 10	a	12	21 ± 5	b	8.6760	0.0001
Ba	12	33 ± 6	a	12	32 ± 7	a	12	27 ± 8	ab	12	25 ± 6	b	3.5246	0.0225
Si	11	72 ± 17	c	12	165 ± 63	a	12	113 ± 19	b	12	80 ± 21	c	16.3915	0.0000
Fe	12	771 ± 133	b	12	945 ± 162	a	12	657 ± 154	b	12	690 ± 148	b	8.9009	0.0001
Ti	12	14 ± 2	b	12	21 ± 5	a	12	15 ± 2	b	12	14 ± 2	b	12.6926	0.0000
Zn	12	34 ± 7	bc	12	43 ± 8	a	12	39 ± 5	ab	12	29 ± 6	c	8.4935	0.0001
Ni	12	4 ± 2	b	11	11 ± 8	a	11	5 ± 2	b	12	4 ± 2	b	6.5519	0.0010
V	12	3 ± 1	bc	12	4 ± 1	a	12	3 ± 1	b	12	2 ± 1	c	9.3130	0.0001

[a] *n* = number of replicates; *X* = mean value; SD = standard deviation.
[b] Values of each horizontal line followed by the same letter do not differ significantly at $P < 0.05$ using one-way ANOVA and LSD test.

Scheidegger and Schroeter (1995) performed measurements of chlorophyll fluorescence to test the effect of ozone fumigation on the epiphytic macrolichens *Anaptychia ciliaris, Collema nigrescens, Evernia prunastri, Hypogymnia bitteri, Lobaria pulmonaria, Pseudevernia furfuracea* and *Usnea rigida s.l.* collected in Switzerland. The ratio F_v/F_m was found to decrease in all fumigated samples. The reduction was the greatest for *L. pulmonaria* with a fluorescence lowered by 43% whereas *E. prunastri* and *C. nigrescens* showed depressions of more than 15%. In *A. ciliaris* and *H. bitteri*, the reduction was less than 10%. *P. furfuracea* and *U. rigida*, according to this study, presented a slight, statistically insignificant reduction of F_v/F_m.

Additional studies of lichens exposed to chemicals under controlled conditions were carried out by Branquinho and co-workers (1997a,b) who investigated the impact of heavy metals on chlorophyll fluorescence, relative to membrane integrity in lichens. The uptake of Pb, particularly in cyanobiont lichens, caused a decrease of PSII photochemical reactions measured as a change in the F_v/F_m ratio (Branquinho et al., 1997a): for the lichens *Lobaria pulmonaria* and *Parmelia caperata*, the effect on PSII photochemical reactions could only be observed 48 h after the Pb uptake, thus confirming that a prolonged exposure to Pb leads to an additional decrease of F_v/F_m.

The fluorescence parameter F_v/F_m was shown by Branquinho et al. (1997b) to have an additional use in the observation of the sensitivity of lichens to the uptake of Cu. *Usnea* spp. were found to be the most sensitive to Cu uptake as shown by fluorescence studies: a total inhibition of PSII photochemical reactions occurred in *Ramalina fastigiata* when intercellular Cu concentrations exceeded a threshold of ca. 4.0 μmol g^{-1}, whereas an intracellular Cu concentration in *Usnea* spp. was already above this apparent threshold (ca. 4.0 μmol g^{-1}) even for the lowest Cu concentration supplied, hence Fv/Fm was equal to zero for all Cu concentrations supplied.

Field studies of modulated chlorophyll *a* fluorescence emissions and chlorophyll content are not always in accordance with studies carried out under controlled conditions: A study of *Parmelia quercina* in northern Castelon (Spain) detected an unaffected F_v/F_m ratio whereas the air pollution was found to be quite intense (Calatayud et al., 1996). On the other hand, significant changes in the maximal quantum yield of PSII were found for the lichen *Ramalina lacera* exposed in polluted sites in Israel and an application of heavy-metal containing solutions and simulated acid rain yielded an additional proof of change (Garty, Weissman, Tamir and Beer, unpublished).

5. Elemental content and degradation of cell membranes in lichens

The investigation of the impact of chemical contaminants on lichen cell membranes was characterized in the last three decades by an experimental approach.

In order to check the integrity of the plasma membrane which encloses lichen cells it is sufficient to place a piece of a lichen thallus in distilled deionized water for two or three minutes (Simon, 1974). This test was applied to assess the injury brought about by exposure to heavy metals (Puckett, 1976) or to SO_2 either in aqueous form (Puckett et al., 1977) or as a gas, either by an analysis of the K concentration of the leachate (Puckett et al., 1977) or by the measurement of its electric conductivity. Several studies (Pearson and Rodgers, 1982; Pearson, 1985; Hart et al., 1988; Alebic-Juretic and Arko-Pijevac, 1989) demonstrate that the disintegrating effect of SO_2 and other pollutants may be evaluated by this test which makes membrane leakage evident. Pearson and Henriksson (1981), for example, demonstrated the effect of SO_2 on membrane leakage in *Evernia prunastri*. The authors observed an increased leakage at all SO_2 levels from 0.32 ppm and above. The electric conductivity values obtained upon exposure of *E. prunastri* to 0.32, 100 and 300 ppm SO_2 was 10.04, 12.25 and 14.78 μmho g^{-1} ml^{-1}, respectively.

The cellular location of heavy metals and potassium in lichens and the uptake and effect of these cations on lichen physiology, were extensively studied in the seventies and the eighties and yielded a long list of publications. These studies established a strong base for further investigations of the relations between elemental content and the role of cell membranes in the control of the uptake, accumulation and release of elements in lichens (Brown, 1991; Brown and Brown, 1991). Additional information on the impact of chemicals on cell membrane integrity, based on research in the nineties, is given in the following parts of the present section.

In order to investigate the effect of Ca on the intracellular uptake of Cd in cyano-bacterial lichens, Brown and Avalos (1993) tested a number of inhibitors reported to have a considerable influence on the trans-membrane movement of Ca. The results of this study, in which lichens (*Peltigera membranacea*) were exposed to Cd-containing solutions, did not disclose unambiguous evidence to the notion that Ca is a direct competitive inhibitor of intracellular Cd uptake. In addition, this study suggested that the rate of intracellular Cd uptake correlates largely with the quantity of Cd bound to cell wall exchange sites. The authors conclude that the intracellular Cd uptake by *P. membranacea* does not appear to be directly and exclusively related to the Cd trans-membrane carrier system, as reported in an earlier study for the moss *Rhytidiadelphus squarrosus* (Wells and Brown, 1990).

In a study of the cellular location of Pb in *Cladonia portentosa*, Branquinho and Brown (1994) tested the capability of different displacing agents to release extracellular bound Pb from the lichen. It was found that Pb was displaced in extracellular exchange sites by other cations such as Cu, or thiol-rich chemicals, but at the same time these agents made for a considerable loss of intracellular K due to over-all membrane damage. The authors reported that disodium EDTA at pH 4.5 was found to be an effective displacing agent, although the presence of Na caused some release of intracellular K, without membrane damage. The authors suggested therefore the application of EDTA at a low pH to release effectively the entire amount of extracellular Pb in laboratory experiments.

As the K efflux from plant tissue demonstrates membrane disintegration in particular, Nifontova et al. (1995) investigated the permeability of lichen membranes as a result of gamma radiation. Thalli of different lichen species were irradiated for this purpose. The membrane permeability in irradiated thalli of *Cladonia arbuscula, Cetraria islandic*a and *Hypogymnia physodes* increased considerably in comparison with untreated control thalli. The K efflux from *H. physodes* thalli intensified with an increase of the radiation dose whereas no correlation was obtained for the K efflux and the radiation dose in *C. arbuscula* and *C. islandica*. The N content in *Peltigera aphtosa* appeared, however, to remain stable under acute gamma radiation.

Wells et al. (1995) performed a kinetic analysis of the Cd uptake of the lichen

Peltigera membranacea. The study, performed under laboratory conditions, revealed that this lichen which belongs to a Cd-tolerant population, showed a reduced intracellular Cd uptake in comparison with intolerant lichens collected in the field. A reduced intracellular uptake appears to be the result of unfavorable intracellular uptake kinetics in Cd-tolerant plant material. The authors were able to show, however, by means of a K pretreatment, which altered the balance of exchangeable ions on the cell wall, that a reduced Cd uptake by tolerant plant material may derive from prevailing extracellular ions at the time of collection. The study suggests that lichens are protected by their cell walls which bind substantial quantities of native mineral cations, thus providing an ionic atmosphere around the plasmalema which restricts the entry of toxic cations to the protoplast. The authors conclude that it is important to take the availability and abundance of native mineral ions into consideration in the investigation of mechanisms of metal tolerance in plant populations such as mosses and lichens.

Stress brought about by desiccation is proved not to influence the uptake of Cd by lichens. Beckett (1996) was able to show that the membranes of *Peltigera membranacea* were not damaged, even after a period of severe desiccation, and the ability of the lichen to take up Cd was not reduced. The author stressed the importance of membrane-transporting systems that may recover and take up ions following desiccation, for lichens subjected to wetting and drying cycles.

Branquinho and co-workers provided essential knowledge on the uptake of heavy metals and their effect on cell membrane integrity in lichens. One of these studies (Branquinho et al., 1997a) confirmed by means of exposure to Pb that this metal is capable of entering cells of lichens possessing different photosynthetic symbionts and various forms of thallus growth: *Lobaria pulmonaria, Ramalina farinacea, Parmelia caperata* and *Peltigera canina*. This study did not show, however, modifications of lichen membrane permeability, as measured by the loss of intracellular K and Mg. The effects of Pb were shown to be dependent solely on the nature of the photobiont. It was confirmed that Pb displaced extracellular exchangeable Ca and Mg. The authors concluded that the deficiency of Mg and Ca, if crucial to physiological processes, may become apparent upon long-term Pb contamination.

Branquinho and co-workers (1997b) analyzed *Usnea* spp. and *Ramalina fastigiata* to investigate the efficiency of Na_2-EDTA as a chelating agent for extracellular Cu supplied to lichens under laboratory conditions, without causing cellular membrane injury. The data provided by the authors support this hypothesis and show that a sequential elution procedure may be used to determine the cellular location of Cu in lichens. The patterns of extracellular uptake *vs.* time or concentration are anticipated by conventional kinetic studies of other organisms and heavy metals. The authors show that the Cu supplied in the laboratory replaced the naturally acquired extracellular Mg and Ca and induced alterations in the passage of K across the cell membrane. The intracellular uptake for *Usnea* spp. appeared however to saturate at low supplied Cu concentrations and was toxic, as indicated by another physiological parameter: chlorophyll fluorescence. *Usnea* spp. were found to be rather sensitive to the uptake of Cu relative to *Ramalina fastigiata*, as physiological change occurred at lower supplied Cu concentrations.

Ozone fumigation studies of lichens indicated a different sensitivity of lichen species. Tarhanen et al. (1997) used epiphytic lichens to demonstrate electrolyte leakage from ozone fumigated lichens. Both *Bryoria capillaris* and *B. fuscescens* showed a decreased K leakage, and *B. fuscescens* showed a decreased conductivity of the leachate, along four weeks of fumigation with high concentrations of ozone (150 and 300 ppb). In most lichens, the total concentration of K remained either unchanged or increased at 150 ppb O_3. Thus it is suggested that no leaching due to membrane injury occurred, whereas the uptake of K tended to increase in accordance with the concentration of O_3. The study suggests also that ozone supplied at natural concentrations (e.g. 40 ppb) barely affects the membrane permeability and the ultrastructure of the boreal lichens used in the experiment. Several lichens appear to tolerate higher concentrations of ozone than many vascular plants. Tarhanen et al. (1996) investigated the effect of industrial emissions on the permeability of membranes of *Hypogymnia physodes* and *Bryoria fuscescens* in northern Finland and in the Kola Peninsula. The K leakage from the lichens was found to be the most sensitive indicator of membrane injury. *B. fuscescens* had a higher rate of ion leakage than *H. physodes* whereas the overall response to pollution was similar for both species.

To evaluate the feasibility of the use of lichens as in situ biomonitors of atmospheric pollution in a semiarid climate, Rope and Pearson (1990) applied two techniques of evaluation in their study area in Idaho: trace element analysis and electrolyte leakage from cells. The lichen *Lecanora melanophthalma* accumulated for the most part higher levels of elemental content than vascular plants such as *Artemisia tridentata* and either comparable or lower levels than soil. The above-mentioned lichen was the only species to show a significant difference in electrolyte leakage which could be associated with the distance from a chemical processing plant.

In a study of the impact of air pollution on the integrity of cell membranes in *Ramalina lacera* transplanted to industrial sites in the Haifa Bay, NW Israel (Garty et al., 1993a), the rate of disintegration of cell membranes was found to be in accordance with the amount of S detected on the surface of the lichen thalli. This study applied the electric conductivity parameter, reflective of the integrity of lichen cell membranes, to express cellular injury in transplanted lichen thalli. Symptoms of disintegration of cell membranes were found to be detectable in *R. lacera* long before indications of damage became apparent in the photobiont chlorophyll. These findings are in accordance with the observations of Benlap and Harper (1990) of the effect of a coal-fired power plant on the rock lichen *Rhizoplaca melanophthalma* which led them to conclude that the parameter of electrolyte leakage as an indicator of injury is rather sensitive relative to the parameter of chlorophyll degradation.

The parameter of electrolyte leakage is applicable as a comparative tool of pollution-resistant and sensitive species. Silberstein et al. (1996a) tested the effects of air contaminants on the pollution-resistant *Xanthoria parietina* and the sensitive *Ramalina lacera* under controlled laboratory conditions and in field experiments. The presence of air contaminants, especially in the Tel Aviv area, Israel, and the exposure to bisulfite ions did not lead to a significant leakage of electrolytes from *X. parietina*, whereas a severe leakage was apparent in *R. lacera*. The authors did not detect a significant leakage of K in both control and bisulfite-treated thalli of *X. parietina*, whereas bisulfite caused leakage of K in *R. lacera*.

In order to assess the impact of anthropogenic activity, Garty and co-workers designed a series of transplantation experiments of *Ramalina lacera* in sites of industrial activity and heavy traffic, based on the assumption, validated by laboratory studies, that the status of cell membranes correlates with the elemental content. A significant correlation was obtained for electric conductivity values indicative of cell membrane integrity and contents of sulfate-S (Garty et al., 1997a), total S (Garty et al., 1998b), Mg (Garty et al., 1997a, 1998a), B (Garty et al., 1998a,b) Al, Cr, Fe, Si, Zn (Garty et al., 1998b), Ti (Garty et al., 1997b, 1998b), and V (Garty et al., 1997b). The K content of *R. lacera* on the other hand was in inverse correlation with electric conductivity values (Garty et al., 1998b).

Weissman (1998) used transplants of *Ramalina lacera* to investigate the possible connection of elemental content, electric conductivity of water in which lichen samples were soaked and elemental concentrations of leachates obtained upon soaking. The lichen was collected in the relative clean-air site in HaZorea, NE Israel, either resuspended in the same site or transplanted to the Haifa Bay industrial area and to Mount Carmel, NW Israel, for a period of eight months (April–November 1997). Tables 12–14 present the elemental content of the lichen, electric conductivity and elemental concentrations of the leachates. In the Haifa Bay region, transplants accumulated more Ba, Mg, P, S, V and Zn than transplants in the control site or in one site on the Carmel Mountain (Table 12). Table 13 shows higher values for electric conductivity in the Haifa Bay material, over and above two other sites, thus indicating a higher rate of disintegration of cell membranes. The Haifa Bay material contained much more K, Mg, P, S and V than material from two other sites (Table 14). The higher electric conductivity values obtained for thalli in the Haifa Bay region relative to the two other sites, is caused, partly at least, by the leakage of K, Mg, P, S and to a lesser extent - V, in the soaking period in the laboratory. The highest rate of leakage in the period of exposure under field conditions is that of K, as proved by its relative low contents in the Haifa Bay lichens.

6. The accumulation of chemical elements and other physiological parameters: recent research

6.1. ATP and elemental content

Subsequent to studies on alterations in ATP content and the accumulation of chemicals, carried out in the late eighties (Kardish et al., 1987; Garty et al., 1988), Silberstein and co-workers (1990) developed a new method for the extraction of ATP from lichens and applied it to compare the response of two lichen species to pollutants (Silberstein et al., 1996a). The authors compared the ATP content of *Xanthoria parietina* in polluted and non-polluted locations. Similar ATP contents coincided with dissimilar S contents, as samples from polluted locations contained elevated amounts of S relative to samples from non polluted locations (Silberstein et al., 1996b). Silberstein and co-workers (1996a) showed that the presence of bisulfite does not affect the ATP content in *Xanthoria parietina*. The decreased amount of ATP after exposure to 2 mM HSO_3^- was largely restored after a 20 h recovery, and may be regarded as temporary and reversible.

Table 12. Elemental content (μg g^{-1} on dry wt. basis) of *Ramalina lacera* collected in HaZorea, NE Israel, April 1997, resuspended in same site or transplanted to Haifa Bay–Mount Carmel region, NW Israel, retrieved November 1997 (Weissman, 1998)[a,b]

	HaZorea (control)			Yearot HaCarmel Hotel, Carmel Mountain			Oil refineries, Haifa Bay			Haifa power plant, Haifa Bay			ANOVA	
	n	*X* ± SD		*n*	*X* ± SD		*n*	*X* ± SD		*n*	*X* ± SD		*F* ratio	*F* probability
Al	10	894 ± 125	a	10	862 ± 66	ab	10	836 ± 59	ab	10	805 ± 76	b	1.9260	0.0000
Ba	10	14 ± 3	c	10	11 ± 1	d	10	19 ± 2	a	10	16 ± 2	b	32.9730	0.0000
Ca	10	19161 ± 1248	a	10	11369 ± 1079	c	10	16044 ± 1373	b	10	17008 ± 1241	b	70.3093	0.0000
Fe	10	892 ± 101	n.s.	10	918 ± 68	n.s.	10	917 ± 72	n.s.	10	863 ± 72	n.s.	1.0786	0.3704
K	10	2130 ± 184	b	10	2746 ± 113	a	10	1873 ± 111	c	10	1662 ± 134	c	114.7234	0.0000
Mg	10	731 ± 46	d	10	818 ± 68	c	10	1073 ± 60	a	10	1002 ± 56	b	100.6811	0.0000
Mn	10	21 ± 1	ab	10	20 ± 1	b	10	22 ± 1	a	10	21 ± 1	a	4.2734	0.0111
P	10	1008 ± 113	c	10	1685 ± 147	b	10	1801 ± 98	a	10	1840 ± 88	a	117.0702	0.0000
S	10	2562 ± 126	b	10	2325 ± 138	c	10	3103 ± 173	a	10	3024 ± 151	a	63.1876	0.0000
V	10	4.1 ± 0.8	c	10	3.4 ± 0.3	d	10	6.8 ± 0.8	b	10	5.8 ± 0.6	a	58.5772	0.0000
Zn	10	32 ± 4.3	b	10	24 ± 1.8	c	10	47 ± 6	a	10	50 ± 5	a	75.5775	0.0000

[a] *n* = Number of replicates; *X* = mean value; SD = standard deviation.
[b] Values of each horizontal line followed by the same letter do not differ significantly at $P < 0.05$ using one-way ANOVA and LSD test.

Table 13. Electric conductivity (mSm^{-1}) expressing electrolyte leakage from cells of *Ramalina lacera* collected in HaZorea, NE Israel, April 1997, resuspended in same site or transplanted to Haifa Bay - Mount Carmel region, NW Israel, retrieved November 1997 (Weissman, 1998)[a,b]

	n	$X \pm SD$	
HaZorea (control)	12	2.52 ± 0.18	c
Yearoth HaCarmel Hotel, Carmel Mountain	12	1.84 ± 0.20	d
Oil refineries, Haifa Bay	12	3.46 ± 0.30	b
Haifa power plant, Haifa Bay	6	4.12 ± 0.13	a
ANOVA F ratio	403.7433		
ANOVA F probability	0.0000		

[a] n = number of replicates; X = mean value; SD = standard deviation.
[b] Values of the vertical column followed by the same letter do not differ significantly at $P < 0.05$ using one-way ANOVA and LSD test.

6.2. Malondialdehyde (MDA) and pollutants

Malondialdehyde is a highly reactive aldehyde generally formed as a consequence of lipid peroxidation (Turton et al., 1997). A study which used thalli of *Hypogymnia physodes* transplanted in three locations with different levels of air pollution in Austria detected a positive correlation of atmospheric O_3 and MDA (Egger et al., 1994). An additional study using lichen transplants, performed by Gonzales and Pignata (1994) in Cordoba, Argentine, revealed a direct relation of S, Al, Pb and MDA content in thalli of *Punctelia subrudecta.* An additional study (Gonzales et al., 1996) observed that transplants of *Ramalina ecklonii* accumulated significantly different S contents in low, medium, or high-traffic sites, whereas MDA contents did not differ significantly. The S content differed significantly in high traffic and high industrial sites but no significant difference was observed for MDA contents. MDA reached significantly high values in transplants of *Punctelia subrudecta* near a power station relative to control material (Gonzales and Pignata, 1997). In this study the MDA values showed a certain relation with the S content. Additional studies using either *Parmotrema austrosinense, P. conferendum, Ramalina ecklonii* (Gonzales et al., 1998; Gonzales and Pignata, 1999) or *Usnea* sp. (Carreras et al., 1998) did not reveal such relations. Canas and Pignata (1998) could not detect significant modifications of MDA values in transplants of *Parmotrema uruguense* in different biomonitoring sites throughout the study period. After two and four months of exposure, sites differentiated significantly but no modification of MDA content could be attributed to the impact of air pollution in urban sites.

6.3. Amino-acid content, protein biosynthesis and chemical pollutants

The total amino-acid content of proline, alanine and arginine of pollution-resistant lichens may increase significantly in polluted sites. Silberstein and co-workers (1996b) found higher amounts of proline and arginine in samples of *Xanthoria parietina* in polluted sites, whereas the alanine content differed only a little in polluted and

Table 14. Elemental concentration (μg g^{-1} on air-dry basis) of 1 g *Ramalina lacera* in 95 ml leachates used for electric conductivity measurements. The lichen was collected in HaZorea, April 1997, resuspended in same site or transplanted to Haifa Bay - Mount Carmel region, retrieved November 1997 (Weissman, 1998)[a,b]

	HaZorea (control)			Yearoth HaCarmel Hotel, Carmel Mountain			Oil refineries, Haifa Bay			Haifa power plant, Haifa Bay			ANOVA	
	n	*X* ± SD		*n*	*X* ± SD		*n*	*X* ± SD		*n*	*X* ± SD		*F* ratio	*F* probability
Al	5	1.53 ± 0.19	ab	5	1.30 ± 0.27	b	5	1.72 ± 0.15	a	5	1.54 ± 0.14	ab	4.0047	0.0265
Ba	5	0.43 ± 0.08	a	5	0.22 ± 0.05	b	5	0.32 ± 0.17	ab	5	0.34 ± 0.06	ab	3.5104	0.0397
Ca	5	332.96 ± 39.78	a	5	184.31 ± 38.36	b	5	362.40 ± 37.43	a	5	336.62 ± 34.78	a	23.0883	0.0000
Fe	5	2.25 ± 0.36	a	5	1.01 ± 0.24	b	5	2.07 ± 0.30		5	1.24 ± 0.11	b	25.2098	0.0000
K	5	21.21 ± 9.16	c	5	36.55 ± 17.74	bc	5	48.14 ± 15.14	b	5	187.68 ± 5.19	a	180.8553	0.0000
Mg	5	49.12 ± 4.65	b	5	40.36 ± 2.24	b	5	74.67 ± 18.82	a	5	68.68 ± 14.96	a	8.6208	0.0012
Mn	5	0.39 ± 0.10	bc	5	0.31 ± 0.06	c	5	0.65 ± 0.05	a	5	0.49 ± 0.14	b	12.5673	0.0002
P	5	2.59 ± 1.61	c	5	10.26 ± 3.84	c	5	29.55 ± 12.44	b	5	49.44 ± 6.10	a	42.1122	0.0000
S	5	73.91 ± 11.73	c	5	66.84 ± 27.43	c	5	199.21 ± 29.86	b	5	253.89 ± 18.19	a	81.7610	0.0000
V	5	0.18 ± 0.04	c	5	0.12 ± 0.02	c	5	0.48 ± 0.04	b	5	0.71 ± 0.13	a	71.8008	0.0000
Zn	5	0.57 ± 0.29	b	5	0.13 ± 0.07	c	5	1.25 ± 0.22	a	5	0.82 ± 0.15	b	27.1043	0.0000

[a] *n* = Number of replicates; *X* = mean value; SD = standard deviation.
[b] Values of each horizontal line followed by the same letter do not differ significantly at $P < 0.05$ using one-way ANOVA and LSD test.

non-polluted sites. *Ramalina lacera*, a pollution-sensitive species, on the other hand, did not undergo an essential change of proline or arginine content, while alanine went through a nearly eight-fold elevation. The divergence of the amino acid composition of *X. parietina* coincided with a high sulfate content (2310 ± 577 $\mu g\ g^{-1}$) in polluted sites and a lower sulfate content (1519 ± 198 $\mu g\ g^{-1}$) in the unpolluted site. A bisulfite treatment led to a noticeable increase of alanine in *R. lacera*, and to a lesser extent in *X. parietina*. The authors suggest that the increase of amino-acids can be a result of either protein degradation, an increase in their synthesis or both. Indeed, in 17 biomonitoring stations in Switzerland the ^{35}S-protein content of *Parmelia sulcata* was found to correlate negatively with air pollution (Von Arb et al., 1990), *i.e.* the lowest values were obtained for the most polluted stations.

7. Heavy metals as substrate and as "lichen killers"

In their review of lichens in metal-enriched environments, Purvis and Halls (1996) present a list of lichens reported as rust-stained on iron-rich rocks, and an additional list of lichens typical of rocks rich in iron sulfide. Iron and steel serve sometimes as substrates for lichens (e.g. Brightman and Seaward, 1977; Garty, 1993). Crustose lichens are often observed to grow on old cannons (Figs. 6 and 7). The cannons located along the shore in Normandy, France were put in their present position less than 60 years ago (Fig. 7). Measurements of the lichen surface on these substrate may provide information on the rate of growth of certain crustose species.

Iron may produce a toxic effect. The toxic effect of Fe ions on lichens growing on rocks near the Svartisen, "Black Ice", glacier in the region of Mo i Rana, Northern

Figure 6. Crustose lichens on a cannon, Läckö, Sweden, September, 1992.

Figure 7. Crustose lichens on a cannon, "Omaha" Beach, Normandy, France, February, 1998.

Norway (Garty, 1993) is indicated by the absence of thalli below patches of a Fe-rich varnish. X-Ray spectra obtained of rock samples taken from the varnished area showed high percentages of Fe, 73.7 ± 7.1%, whereas a varnish-free lichen-covered

Figure 8. Epilithic crustose lichens on a stone wall (right side of the picture) in Florence, Italy, August, 1987. Fe ions prevent the growth of lichens and other lithobionts on the left side of the wall.

Figure 9. Lichens on heaps of mine spoil containing Fe and Pb near old silver mines, Sala, Sweden, September, 1992.

rock sample contained only 14.9 ± 7.3% of Fe. Another instance of the toxic effect of Fe was observed in the vicinity of a botanical garden in Florence, Italy. Epilithic crustose lichens were growing on one side of a wall on top of which an iron fence

Figure 10. The toxic effect of Zn on foliose and fruticose lichens as observed near the river Spean, Spean-Bridge, Scotland, September, 1991.

Figure 11. Roof pins reinforcing a lichen-covered asbestos-cement roof by means of zinc-galvanized plates. Note the 'lichen desert' (arrows) delineated by a water dripping pattern. Moshav Heruth, Israel, May, 1988.

Figure 12. A 'lichen desert' (arrows) produced by Zn-ions originating from galvanized screws on an electricity pole. Ganei-Am, Israel, winter, 1990.

Figure 13. Toxic effect of Zn-ions (arrows) on epilithic crustose lichens and mosses. The Zn ions derive from galvanized plates and objects fixed to the roof. Laerdalsoyri, Norway. Summer, 1992.

Figure 14. The toxic effect of Pb letters on a tombstone in Sydney, Australia. Lichens and other lithobionts cannot subsist underneath these letters as dissolved Pb ions leach out (see also Puckett and Burton, 1981), August, 1996.

Figure 15. Toxic effect of Cu ions on epilithic lichens growing on a tombstone. Val Rendena, Italy, August, 1998.

Figure 16. Toxic effect of Cu and additional heavy metals dissolved from screws reinforcing a plate on a tombstone. Linlithgow, Scotland, September, 1991.

was fastened. Iron and additional heavy metal ions dissolved in rainwater, prevented the establishment of visible lithobionts on the stony wall (Fig. 8).

Lichens may grow on Pb/Zn-rich rocks and associated soils. Purvis and Halls (1996) present a list of 61 terricolous lichen species characteristic of Pb/Zn-rich environments in Britain and Europe. Epilithic crustose lichens were observed also on heaps of mine spoil containing Fe and Pb near old silver mines in Sala, Sweden (Fig. 9).

The detrimental effect of both Zn and Pb on lichens is well known. Dissolved zinc ions from galvanized barbed wires (Seaward, 1974), roof fixtures (Garty et al., 1979) and fences (Plummer, 1982) are very toxic to lichens. Zinc ions lead to the elimination of lichens from different substrates such as wooden poles (Fig. 10), various constructions made of concrete or asbestos-cement (Figs. 11 and 12) and roof tiles (Fig. 13).

In cemeteries, the toxic effect of Pb (Fig. 14) and Cu (Fig. 15) is obviously indicated by the absence of lichen thalli below metal letters consisting of Pb or Cu and screws and other pieces of metal containing Cu and additional metals such as Ni or Cr (Fig. 16).

Acknowledgements

Many thanks to Mrs. Rachel Garty-Spitz for her valued contribution to the manuscript.

References

Abeles, A.L., Abeles, F.B., 1972. Biochemical pathway of stress-induced ethylene. Plant Physiol. 50, 496–498.

Ahti, T., 1977. Lichens of the boreal coniferous zone. In: Seaward, M.R.D. (Ed.). Lichen Ecology, Academic Press, London, pp. 145–181.

Alebic-Juretic, A., Arko-Pijevac, M., 1989. Air pollution damage to cell membranes in lichens - results of simple biological test applied in Rijeka. Yugoslavia. Water Air Soil Pollut. 47, 25–33.

Arny, C.J., Pell, E.J., 1986. Ethylene production by potato, radish and soybean leaf tissue treated with simulated acid rain. Environ. Exp. Bot. 26, 9–15.

Asrar, G., Fuchs, M., Kanemasu, E.T., Hatfield, J.I., 1984. Estimating absorbed photosynthetic radiation and leaf area index from spectral reflectance in wheat. Agron. J. 76, 300–306.

Axelrood-McCarthy, P.E., Linderman, R.G., 1981. Ethylene production by cultures of *Cylindrocladium floridanum* and *C. scoparium*. Phytopathology 71, 825–830.

Balaguer, L., Manrique, E., Ascaso, C., 1997. Predictability of the combined effects of sulphur dioxide and nitrate on the green-algal lichen *Ramalina farinacea*. Can. J. Bot, 75, 1836–1842.

Bartok, K., Nicoara, A., Victor, B., Tibor, O., 1992. Biological responses in the lichen *Xanthoria parietina* transplanted in biomonitoring stations. Rev. Roumaine Biol. - Ser. Biol. Vegetale 37, 135–142.

Beckett, R.P., 1996. Effect of desiccation stress on Cd uptake in the lichen *Peltigera membranacea*. S. Afr. J. Botany 62, 121–122.

Benlap, J., Harper, K.T., 1990. Effect of a coal-fired power plant on the rock lichen *Rhizoplaca melanophthalma*: chlorophyll degradation and electrolyte leakage. Bryologist 93, 309–312.

Ben-Yehoshua, S., Biggs, R.M., 1970. Effects of iron and copper ions in promotion of selective abscission and ethylene production by citrus fruit and the inactivation of indoleacetic acid. Plant Physiol. 45, 604–607.

Boonpragob, K., Nash III, T.H., 1990. Seasonal variation of elemental status in the lichen *Ramalina menziesii* Tayl. from two sites in southern California: evidence for dry deposition accumulation. Environ. Exp. Bot. 30, 415–428.

Boonpragob, K., Nash III, T.H., 1991. Physiological responses of the lichen *Ramalina menziesii* Tayl. to the Los Angeles urban environment. Environ. Exp. Bot. 31, 229–238.

Bouzayen, M., Latche, A., Pech, J.-C., Boller, T., 1991. Iron: an essential cofactor for the conversion of 1-aminocyclopropane-1-carboxylic acid to ethylene. Planta 184, 244–247.

Branquinho, C., Brown, D.H., 1994. A method for studying the cellular location of lead in lichens. Lichenologist 26, 83–90.

Branquinho, C., Brown, D.H., Maguas, C., Catarino, F., 1997a. Lead (Pb) uptake and its effects on membrane integrity and chlorophyll fluorescence in different lichen species. Environ. Exp. Bot. 37, 95–105.

Branquinho, C., Brown, D.H., Catarino, F., 1997b. The cellular location of Cu in lichens and its effects on membrane integrity and chlorophyll fluorescence. Environ. Exp. Bot. 38, 165–179.

Bressan, R.A., LeCureux, L., Wilson, L.G., 1979. Filner. Ph., Emission of ethylene and ethane by leaf tissue exposed to injurious concentrations of sulfur dioxide or bisulfite ion. Plant Physiol. 63, 924–930.

Brightman, F.H., Seaward, M.R.D., 1977. Lichens of man-made substrates. In: Seaward, M.R.D. (Ed.). Lichen Ecology, Academic Press, London, pp. 253–293.

Brown, D.H., 1991. Lichen mineral studies currently clarified or confused? Symbiosis 11, 207–223.

Brown, D.H., Avalos, A., 1993. The role of calcium in intracellular cadmium uptake by the lichen *Peltigera membranacea*. Ann. Bot. (Lond.) 71, 467–473.

Brown, D.H., Brown, R.M., 1991. Mineral cycling and lichens: the physiological basis. Lichenologist 23, 293–307.

Calatayud, A., Sanz, M.J., Calvo, E., Barreno, E., del Valle-Tascon, S., 1996. Chlorophyll *a* fluorescence and chlorophyll content in *Parmelia quercina* thalli from a polluted region of northern Castellon (Spain). Lichenologist 28, 49–65.

Canas, M.S., Pignata, M.L., 1998. Temporal variation of pigments and peroxidation products in the lichen *Parmotrema uruguense* (Kremplh.) Hale transplanted to urban and non-polluted environments. Symbiosis 24, 147–162.

Canas, M.S., Orellana, L., Pignata, M.L., 1997. Chemical response of the lichens *Parmotrema austrosinense* and *P. conferendum* transplanted to urban and non-polluted environments. Ann. Botanici Fenn. 34, 27–34.

Carreras, H.A., Gudino, G.L., Pignata, M.L., 1998. Comparative biomonitoring of atmospheric quality in five zones of Cordoba city (Argentina) employing the transplanted lichen *Usnea* sp. Environ. Pollut. 103, 317–325.

Chettri, M.K., Cook, C.M., Vardaka, E., Sawidis, T., Lanaras, T., 1998. The effect of Cu. Zn and Pb on the chlorophyll content of the lichens *Cladonia convoluta* and *Cladonia rangiformis*. Environ. Exp. Bot. 39, 1–10.

Cox, J.D., Beckett, P.J., Courtin, G.M., 1991. Factors affecting spectral responses from lichens. Proceedings of the Eighth Thematic Conference on Geologic Remote Sensing, vol. 3, Environmental Research Institute of Michigan. Ann Arbor, MI, pp. 1125–1137.

Crespo, A., Barreno, E., Sancho, L.G., Bueno, A.G., 1981. Establecimiento de una red de valoracion de pureza atmospherica en la provincia de la Coruna (Espana) mediante bioindicatores liquenicos. Lazaroa 3, 289–311.

Curran, P.J., 1980. Multispectral remote sensing of vegetation amount. Prog. Phys. Geog. 4, 315–341.

Demmig-Adams, B., Maguas, C., Adams III, W.W., Meyer, A., Killian, E., Lange, O.L., 1990. Effect of high light on the efficiency of photochemical energy conversion in a variety of lichen species with green and blue-green photobionts. Planta 180, 400–409.

Egger, R., Schlee, D., Türk, R., 1994. Changes of physiological and biochemical parameters in the lichen *Hypogymnia physodes* (L.) Nyl. due to the action of air pollutants - a field study. Phyton 34, 229–242.

Epstein, E., Sagee, O., Cohen, J.D., Garty, J., 1986. Endogenous auxin and ethylene in the lichen *Ramalina duriaei*. Plant Physiol. 82, 1122–1125.

Eversman, S., 1978. Effects of low-level SO_2 on *Usnea hirta* and *Parmelia chlorochroa*. Bryologist 81, 368–377.

Fuhrer, J., 1982. Ethylene biosynthesis and cadmium toxicity in leaf tissue of beans (*Phaseoulus vulgaris* L.). Plant Physiol. 70, 162–167.

Garty, J., 1993. Lichens as biomonitors for heavy metal pollution. In: Markert, B. (Ed.). Plants as Biomonitors, Indicators for Heavy Metals in the Terrestrial Environment, VCH Publishers, Weinheim, pp. 193–263.

Garty, J., Galun, M., Kessel, M., 1979. Localization of heavy metals and other elements accumulated in the lichen thallus. New Phytol. 82, 159–168.

Garty, J., Ronen, R., Galun, M., 1985. Correlation between chlorophyll degradation and amount of some elements in the lichen *Ramalina duriaei* (De Not.) Jatta. Environ. Exp. Bot. 25, 67–74.

Garty, J., Kardish, N., Hagemeyer, J., Ronen, R., 1988. Correlations between the concentration of adenosine tri phosphate, chlorophyll degradation and the amounts of airborne heavy metals and sulphur in a transplanted lichen. Arch. Env. Contam. Toxicol. 17, 601–611.

Garty, J., Karary, Y., Harel, J., 1992. Effect of low pH, heavy metals and anions on chlorophyll degradation in the lichen *Ramalina duriaei* (De Not.) Bagl. Environ. Exp. Bot. 32, 229–241.

Garty, J., Karary, Y., Harel, J., 1993a. The impact of air pollution on the integrity of cell membranes and chlorophyll in the lichen *Ramalina duriaei* (De Not.) Bagl. transplanted to industrial sites. Arch. Environ. Contam. Toxicol. 24, 455–460.

Garty, J., Karary, Y., Harel, J., Lurie, S., 1993b. Temporal and spatial fluctuations of ethylene production and concentrations of sulfur, sodium, chlorine and iron on/in the thallus cortex in the lichen *Ramalina duriaei* (De Not.) Bagl. Environ. Exp. Bot. 33, 553–563.

Garty, J., Karary, Y., Harel, J., Lurie, S., 1995a. The impact of heavy metals on lichens. In: Wilken, R.-D., Förstner, U., Knöchel, J. (Eds.). Heavy Metals in the Environment, International Conference, Hamburg, September 1995, vol. 1. CEP Consultants, Edinburgh, pp. 152–155.

Garty, J., Kauppi, M., Kauppi, A., 1995b. Differential responses of certain lichen species to sulfur-containing solutions under acidic conditions as expressed by the production of stress-ethylene. Environ. Res. 69, 132–143.

Garty, J., Cohen, Y., Kloog, N., Karnieli, A., 1997a. Effects of air pollution on cell membrane integrity, spectral reflectance and metal and sulfur concentrations in lichens. Environ. Toxicol. Chem. 16, 1396–1402.

Garty, J., Karnieli, A., Wolfson, R., Kunin, P., Garty-Spitz, R., 1997b. Spectral reflectance and integrity of cell membranes and chlorophyll relative to the concentration of airborne mineral elements in a lichen. Physiol. Plant. 101, 257–264.

Garty, J., Kauppi, M., Kauppi, A., 1997c. The influence of air pollution on the concentration of airborne elements and on the production of stress-ethylene in the lichen *Usnea hirta* (L.) Weber em. Mot. transplanted in urban sites in Oulu, N. Finland. Arch. Environ. Contam. Toxicol. 32, 285–290.

Garty, J., Kauppi, M., Kauppi, A., 1997d. The production of stress ethylene relative to the concentration of heavy metals and other elements in the lichen *Hypogymnia physodes*. Environ. Toxicol. Chem. 16, 2402–2408.

Garty, J., Kloog, N., Cohen, Y., Wolfson, R., Karnieli, A., 1997e. The effect of air pollution on the integrity of chlorophyll, spectral reflectance response, and on concentrations of nickel, vanadium, and sulfur in the lichen *Ramalina duriaei* (De Not.) Bagl. Environ. Res. 74, 174–187.

Garty, J., Kloog, N., Wolfson, R., Cohen, Y., Karnieli, A., Avni, A., 1997f. The influence of air pollution on the concentration of mineral elements, on the spectral reflectance response and on the production of stress-ethylene in the lichen *Ramalina duriaei*. New Phytol. 137, 587–597.

Garty, J., Cohen, Y., Kloog, N., 1998a. Airborne elements, cell membranes, and chlorophyll in transplanted lichens. J. Environ. Qual. 27, 973–979.

Garty, J., Kloog, N., Cohen, Y., 1998b. Integrity of lichen cell membranes in relation of airborne elements. Arch. Environ. Contam. Toxicol. 34, 136–144.

Gilbert, O.L., 1986. Field evidence for an acid rain effect on lichens. Environ. Pollut. Ser. A 40, 227–231.

Gonzales, C.M., Pignata, M.L., 1994. The influence of air pollution on soluble proteins, chlorophyll degradation, MDA, sulphur and heavy metals in a transplanted lichen. Chem. Ecol. 9, 105–113.

Gonzales, C.M., Pignata, M.L., 1997. Chemical response of the lichen *Punctelia subrudecta* (Nyl.) Krog transplanted close to a power station in an urban-industrial environment. Environ. Pollut. 97, 195–203.

Gonzales, C.M., Pignata, M.L., 1999. Effect of pollutants emitted by different urban-industrial sources on the chemical response of the transplanted *Ramalina ecklonii* (Spreng.) Mey and Flot. Toxicol. Environ. Chem., 69, 61–73.

Gonzales, C.M., Casanovas, S.S., Pignata, M.L., 1996. Biomonitoring of air pollutants from traffic and industries employing *Ramalina ecklonii* (Spreng.) Mey and Flot. in Cordoba, Argentina. Environ. Pollut. 91, 269–277.

Gonzales, C.M., Orellana, L.C., Casanovas, S.S., Pignata, M.L., 1998. Environmental conditions and chemical response of a transplanted lichen to an urban area. J. Environ. Manag. 53, 73–81.

Gora, L., Clijsters, H., 1989. Effect of copper and zinc on the ethylene metabolism in *Phaseolus vulgaris* L. In: Clijsters, H., DeProft, M. (Eds.). Biochemical and Physiological Aspects of Ethylene Production in Lower and Higher Plants, Kluwer Academic Publishers, Dordrecht, The Netherlands, pp. 219–228.

Gouaux, P., Vincent, J.P., 1990. Mise en evidence, controle de l'action de la pollution sur les lichens (*Peltigera canina*) par utilisation du film infrarouge couleurs. Sci. Total Environ. 95, 181–190.

Gries, C., Sanz, M.-J., Nash III, T.H., 1995. The effect of SO_2 fumigation on CO_2 gas exchange, chlorophyll fluorescence and chlorophyll degradation in different lichen species from western North America. Cryptogram Bot. 5, 239–246.

Hallingbäck, T., Kellner, O., 1992. Effects of simulated nitrogen rich and acid rain on the nitrogen-fixing lichen *Peltigera aphtosa* (L.) Willd. New Phytol. 120, 99–103.

Halonen, P., Hyvärinen, M., Kauppi, M., 1991. The epiphytic lichen flora on conifers in relation to climate in the Finnish middle boreal subzone. Lichenologist 23, 61–72.

Halonen, P., Väre, H., Soppela, K., 1992. Oulun puistojen epifyyttijäkälät. Lutukka 8, 77–83.

Hart, R., Webb, P.G., Hilton Biggs, R., Portier, K.M., 1988. The use of lichen fumigation studies to evaluate the effects of new emission sources on class I areas. J. Air Waste Manage. Assoc. 38, 144–147.

Hawksworth, D.L., Rose, L., 1970. Qualitative scale for estimating sulphur dioxide air pollution in England and Wales using epiphytic lichens. Nature 227, 145–148.

Hohendorff, U., Vogels, K., 1990. Einfluss von immisionen auf die ethenproduktion und den primärstoffwechsel von fichten (Picea abies [L.] Karst.). Eur. J. Forest Pathol. 20, 219–230.

Holben, B.N., Tucker, C.J., Fan, C.J., 1980. Assessing soybean leaf area and leaf biomass with spectral data. Photogrametric Eng. Remote Sens. 26, 651–656.

Hottiger, T., Boller, T., 1991. Ethylene biosynthesis in *Fusarium oxysporum* f. sp. *tulipae* proceeds from glutamate/2-oxoglutarate and requires oxygen and ferrous ions in vivo. Arch. Microbiol. 157, 18–22.

Hovenden, M.J., Seppelt, R.D., 1995. Utility of modulated fluorescence in measuring photosynthetic activity of antarctic plants, field and laboratory studies. Aust. J Plant Physiol. 22, 321–330.

Ince, J.A., Knowles, C.J., 1985. Ethylene formation by cultures of *Escherichia coli*. Arch. Microbiol. 141, 209–213.

Insarova, I.D., Insarov, G.E., Bråkenheim, S., Hutengen, S., Martinsson, P.-O., Semenov, S.M., 1992. Lichen sensitivity and air pollution - A review of literature data. Swedish Environmental Protection Agency Report 4007, Uppsala, Sweden.

Kao, C.H., 1996. Stimulation of ethylene production in detached rice leaves by vanadate. Plant Growth Regul. 18, 161–164.

Kappen, L., Schroeter, B., Green, T.G.A., Seppelt, R.D., 1998. Chlorophyll a fluorescence and CO_2 exchange of *Umbilicaria aprina* under extreme light stress in the cold. Oecologia 113, 325–331.

Kardish, N., Ronen, R., Bubrick, P., Garty, J., 1987. The influence of air pollution on the concentration of ATP and on chlorophyll degradation in the lichen *Ramalina duriaei* (De Not.) Bagl. New Phytol. 106, 697–706.

Karnieli, A., Tsoar, H., 1995. Spectral reflectance of biogenic crust developed on desert dune sand along the Israel-Egypt border. Int. J. Remote Sens. 16, 369–374.

Karnieli, A., Schachak, M., Tsoar, H., Zaady, E., Kaufman, Y., Danin, A., Porter, W., 1996. The effect of microphytes on the spectral reflectance of vegetation in semi-arid regions. Remote Sens. Environ. 57, 88–96.

Kauppi, M., 1980. The influence of nitrogen-rich pollution components on lichens. Acta Univ. Ouluensis, A. Sci. Rerum Nat. 101. Biologica 9, 1–25.

Kauppi, M., Halonen, P., 1992. Lichens as indicators of air pollution in Oulu, northern Finland. Ann. Bot. Fenn. 29, 1–9.

Kauppi, M., Mikkonen, A., 1980. Floristic versus single species analysis in the use of epiphytic lichens as indicators of air pollution in a boreal forest region. Northern Finland. Flora 169, 255–281.

Kauppi, M., Kauppi, A., Garty, J., 1998. Ethylene produced by the lichen Cladina stellaris exposed to sulphur and heavy-metal-containing solutions under acidic conditions. New Phytol. 139, 537–547.

Kimmerer, T.W., Kozlowski, T., 1982. Ethylene, ethane, acetaldehyde, and ethanol production by plants under stress. Plant Physiol. 69, 840–847.

Larsen, B.R., 1986. Synergistic effects of acid rain and ($\pm$)-2-(2,4-dichlorophenoxy)-propanoic acid (2,4-DP) on growth, mineral content and stress-induced ethylene in lettuce. Environ. Pollut. Ser. A 41, 179–196.

Levin, A.G., Pignata, M.L., 1995. *Ramalina ecklonii* as a bioindicator of atmospheric pollution in Argentina. Can. J. Bot. 73, 1196–1202.

Lieberman, M., 1979. Biosynthesis and action of ethylene. Ann. Rev. Plant Physiol. 77, 926–929.

Lurie, S., Garty, J., 1991. Ethylene production by the lichen *Ramalina duriaei*. Ann. Bot. (Lond.) 68, 317–319.

Lynch, J.M., 1974. Mode of ethylene formation by *Mucor hiemalis*. J. Gen. Microbiol. 83, 407–411.

Maillard, P., Thepenier, C., Gudin, C., 1993. Determination of an ethylene biosynthesis pathway in the unicellular green alga. *Haematococcus pluvialis*. Relationship between growth and ethylene production. J Appl. Phycol. 5, 93–98.

Mattoo, A.K., Baker, J.E., Moline, H.E., 1986. Induction by copper ions of ethylene production in *Spirodela oligorrhiza*: Evidence for a pathway independent of 1-aminocyclopropane-1-carboxylic acid. J. Plant Physiol. 123, 193–202.

McCune, B., 1993. Gradients in epiphytic biomass in three *Pseudotsuga-Tsuga* forests of different ages in Western Oregon and Washington. Bryologist 96, 405–411.

Nash III, T.H., 1971. Lichen sensitivity to hydrogen fluoride. Bull. Torrey Botanic. Club 98, 103–106.

Nifontova, M.G., Ravinskaya, A.P., Shapiro, I.A., 1995. Effect of acute gamma radiation on some physiological features of lichens. Lichenologist 27, 215–224.

O'Neill, A.L., 1994. Reflectance spectra of microphytic soil crusts in semi-arid Australia. Int. J. Remote Sens. 15, 675–681.

Ott, S., 1993. The influence of light on the ethylene production by lichens. In: Feige, G.B., Lumbsh, H.T. (Eds.), Phytochemistry and Chemotaxonomy of Lichenized Ascomycetes - A Festchrift in Honour of Siegfried Huneck. Bibl. Licheno. 53, Cramen, Berlin and Stuttgart, pp. 185–190.

Ott, S., Schieleit, P., 1994. Influence of exogenous factors on the ethylene production by lichens. I. Influence of water content and water status conditions on ethylene production. Symbiosis 16, 187–201.

Ott, S., Zwoch, I., 1992. Ethylene production by lichens. Lichenologist 24, 73–80.

Pearson, L.C., 1985. Air pollution damage to cell membranes in lichens. I. Development of a simple monitoring test. Atmos. Environ. 19, 209–212.

Pearson, L.C., Henriksson, E., 1981. Air pollution damage to cell membranes in lichens. II. Laboratory experiments. Bryologist 84, 515–520.

Pearson, L.C., Rodgers, G.A., 1982. Air pollution damage to cell membranes in lichens. III. Field experiments, Phyton 22, 329–337.

Peng, X.X., Yamauchi, M., 1993. Ethylene production in rice bronzing leaves induced by ferrous iron. Plant Soil 149, 227–234.

Pennazio, S., Roggero, P., 1991. Rapid ethylene production in soybean in response to cupric ion. Ann. Bot. (Lond.) 67, 247–249.

Plummer, G.L., 1982. Observations on lichens, granite rock outcrops and acid rain. Georgia J. Sci. 38, 202–210.

Prince, S.D., 1990. High temporal frequency remote sensing of primary production using NOAA AHVRR. In: Steven, M.D., Clark, J.A. (Eds.). Application of Remote Sensing in Agriculture, Butterworths, London, pp. 169–183.

Puckett, K.J., 1976. The effect of heavy metals on some aspects of lichen physiology. Can. J. Botany 54, 2695–2703.

Puckett, K.J., Burton, M.A.S., 1981. The effect of trace elements on lower plants. In: Lepp, N.W. (Ed.). Effect of Heavy Metal Pollution on Plants, Applied Science Publishers, London, pp. 213–238.

Puckett, K.J., Nieboer, E., Flora, W.P., Richardson, D.H.S., 1973. Sulphur dioxide: its effect on photosynthetic ^{14}C fixation in lichens and suggested mechanisms of phytotoxicity. New Phytol. 72, 141–154.

Puckett, K.J., Tomassini, F.D., Nieboer, E., Richardson, D.H.S., 1977. Potassium efflux by lichen thalli following exposure to aqueous sulphur dioxide. New Phytol. 79, 135–145.

Purvis, O.W., Halls, C., 1996. A review of lichens in metal-enriched environments. Lichenologist 28, 571–601.

Riga-Karandinos, A.N., Karandinos, M.G., 1998. Assessment of air pollution from a lignite power plant in the plain of Megalopolis (Greece) using as biomonitors three species of lichens: impacts on some biochemical parameters of lichens. Sci. Total Environ. 215, 167–183.

Ronen, R., Galun, M., 1984. Pigment extraction from lichens with dimethyl sulfoxide (DMSO) and estimation of chlorophyll degradation. Environ. Exp. Bot. 24, 239–245.

Rope, S.K., Pearson, L.C., 1990. Lichens as air pollution biomonitors in a semiarid environment in Idaho. Bryologist 93, 50–61.

Rouse, J.W., Haas, R.H., Schell, J.A., Deering, D.W., Harlan, J.C., 1974. Monitoring of the vernal advancements and retrogradation (green wave effect) of nature vegetation. NASA/GSFC Final Report. NASA, Greenbelt, MD, pp. 1–371.

Roy-Arcand, L., Delisle, C.E., Briere, F.G., 1989. Effects of simulated acid precipitation on the metabolic activity of *Cladina stellaris*. Can. J. Bot. 67, 1796–1802.

Rupp, L.A., DeVries, H.E., Mudge, K.W., 1989a. Effect of aminocyclopropane carboxylic acid and aminoethoxivinylglycine on ethylene production by ectomycorrhizal fungi. Can. J. Bot. 67, 483–485.

Rupp, L.A., Mudge, K.W., Negm, F.B., 1989b. Involvement of ethylene in ectomycorrhiza formation and dichotomous branching of roots of mugo pine seedlings. Can. J. Bot. 67, 477–482.

Sanchez-Hoyos, M.A., Manrique, E., 1995. Effect of nitrate and ammonium ions on the pigment content (xanthophylls, carotenes and chlorophylls) of *Ramalina capitata*. Lichenologist 27, 155–160.

Satterwhite, M.B., Ponder Henley, J., Carney, J.M., 1985. Effects of lichens on the reflectance spectra of granitic rock surfaces. Remote Sens. Environ. 18, 105–112.

Scheidegger, C., Schroeter, B., 1995. Effects of ozone fumigation on epiphytic macrolichens: ultrastructure. CO_2 gas exchange and chlorophyll fluorescence. Environ. Pollut. 88, 345–354.

Schieleit, P., Ott, S., 1996. Ethylene production and 1-aminocyclopropane-1-carboxylic acid content of lichen bionts. Symbiosis 2, 223–231.

Schieleit, P., Ott, S., 1997. Ethylene production in lichens with respect to possible bacterial contamination. Lichenologist 29, 492–495.

Seaward, M.R.D., 1974. Some observations on heavy metal toxicity and tolerance in lichens. Lichenologist 6, 158–164.

Sellers, P.J., 1985. Canopy reflectance, photosynthesis and transpiration. Int. J. Remote Sens. 6, 1335–1372.

Showman, R.E., 1975. Lichens as indicators of air quality around a coal-fired power generating plant. Bryologist 78, 1–6.

Silberstein, L., Siegel, S.M., Keller, P., Siegel, B.Z., Galun, M., 1990. A new method for the extraction of ATP from lichens. In: Jahns, M.H. (Ed.), Contribution to Lichenology, Vol. 38. In Honor of A. Henssen, Bibl. Licheno., Cramer, Berlin-Stuttgart, pp. 411–418.

Silberstein, L., Siegel, B.Z., Siegel, S.M., Mukhtar, A., Galun, M., 1996a. Comparative studies on *Xanthoria parietina*, a pollution-resistant lichen, and *Ramalina duriaei*, a sensitive species. I. Effects of air pollution on physiological processes. Lichenologist 28, 355–365.

Silberstein, L., Siegel, B.Z., Siegel, S.M., Mukhtar, A., Galun, M., 1996b. Comparative studies on *Xanthoria parietina*, a pollution-resistant lichen, and *Ramalina duriaei*, a sensitive species. II. Evaluation of possible air pollution-protection mechanism. Lichenologist 28, 367–383.

Simon, E.W., 1974. Phospholipids and plant membrane permeability. New Phytol. 73, 377–420.

Sochting U. 1995. Lichens as monitors of nitrogen deposition. Cryptogram Bot. 5, 264–269.

Strzelczyk, E., Kampert, M., Pachlewski, R., 1994. The influence of pH and temperature on ethylene production by mycorrhizal fungi of pine. Mycorrhiza 4, 193–196.

Tarhanen, S., Holopainen, T., Poikolainen, J., Oksanen, J., 1996. Effect of industrial emissions on membrane permeability of epiphytic lichens in northern Finland and the Kola Peninsula industrial areas. Water Air Soil Pollut. 88, 189–201.

Tarhanen, S., Holopainen, T., Oksanen, J., 1997. Ultrastructural changes and electrolyte leakage from ozone fumigated epiphytic lichens. Ann. Bot. 80, 611–621.

Tucker, C.J., 1979. Red and photographic infra-red linear combination for monitoring vegetation. Remote Sens. Environ. 8, 127–150.

Türk, R., Christ, R., 1986. Beiträg zur epiphytischen Flechtenflora im stadt-gebeit von Wien. Verh. Zool.-Bot. Ges. Oesterr. 124, 65–80.

Turton, H.E., Dawes, I.W., Grant, C.M., 1997. *Saccharomyces cerevisiae* exhibits a yAP-1-mediated adaptive response to malondialdehyde. J. Bacteriol. 179, 1096–1101.

Vanden Driessche, Th., Kevers, C., Collet, M., Gaspar, Th., 1988. *Acetabularia mediterranea* and ethylene: production in relation with development, circadian rhythms in emission, and response to external application. J. Plant Physiol. 133, 635–639.

Von Arb, C., Brunold, C., 1990. Lichen physiology and air pollution. I. Physiological responses of in situ *Parmelia sulcata* among air pollution zones within Biel, Switzerland. Can. J. Bot. 68, 35–42.

Von Arb, C., Mueller, C., Ammann, K., Brunold. Ch. 1990. Lichen Physiology and air pollution. II. Statistical analysis of the correlation between SO_2, NO_2, NO and O_3, and chlorophyll content, net photosynthesis, sulphate uptake and protein synthesis of *Parmelia sulcata* Taylor. New Phytol. 115, 431–437.

Weissman, L., 1998. Factors affecting the production of endogenic ethylene, changes in spectral reflectance response, membranes and chlorophyll integrity and the composition of mineral elements in the lichen *Ramalina lacera* ((With.) J.R. Laund), M.Sc. thesis, Tel Aviv University (in Hebrew).

Wells, J.M., Brown, D.H., 1990. Ionic control of intracellular and extracellular Cd uptake by the moss *Rhytidiadelphus squarrosus* (Hedw.) Warnst. New Phytol. 116, 541–553.

Wells, J.M., Brown, D.H., Beckett, R.P., 1995. Kinetic analysis of Cd uptake in Cd-tolerant and intolerant populations of the moss *Rhytidiadelphus squarrosus* (Hedw.) Warnst and the lichen *Peltigera membranacea* (Ach.) Nyl. New Phytol. 129, 477–486.

Zaharopoulou, A., Lanaras, T., Arianoutsou, M., 1993. Influence of dust from a limestone quarry on chlorophyll degradation of the lichen *Physcia adscendens* (Fr.) Oliv. Bull. Environ. Contam. Toxicol. 50, 852–855.

Zwoch, I., Knorre, U., Schaub, H., 1990. Influence of SO_2 and O_3, singly or in combination, on ethylene synthesis in sunflower. Environ. Exp. Bot. 30, 193–205.

Trace elements – Their distribution and effects in the environment
B. Markert and K. Friese, editors

Chapter 10

Assessment of metal pollution of aquatic systems with the water

moss *Fontinalis antipyretica* L. ex Hedw. – from fundamental investigations to physiological effects

I. Bruns, K. Friese, G. Krauss, G.-J. Krauss

Abstract

The water moss, Fontinalis antipyretica, was used for passive and active biomonitoring of metals in aquatic systems. For passive biomonitoring, plant material from different sites in the Harz region was collected. For active biomonitoring, plant material was exposed in locations of the river Elbe and of a former copper mining area in Central Germany (Mansfelder Land). The accumulation factors for the metals Mn, Fe, Cr, Ni, Co, As, Cd, Pb, Zn and Cu were calculated in relation to the background concentrations. In addition, Cd uptake was measured under laboratory conditions. The results indicate that F. antipyretica is a suitable biomonitor for metals over a wide range of concentrations.

Keywords: Biomonitoring; Metal accumulation; Elbe; *Fontinalis antipyretica*; Moss

1. Introduction

Biomonitoring represents a method of assessing metal pollution of aquatic and terrestrial ecosystems. Plants that accumulate these elements allow us to measure the metal input which is often below detection limit in water and air. Suitable plants for this procedure are mosses (Say et al., 1981; Say and Whitton, 1983; Mouvet, 1984; Markert and Weckert, 1993; Chovanec et al., 1994; Goncalves et al., 1994; Herpin et al., 1994, 1996; Markert et al., 1996; Siebert et al., 1996; Bruns et al., 1997). The lack of a fully developed cuticle enables the uptake of metals via the leaf surface. No translocation within the plants occurs, because mosses have no transport system. Also no uptake from the soil is detectable. The high accumulation rates for cations are a result of the large number of negatively charged cell walls. However, the extracellular adsorption of cations is followed by a migration to intracellular places. The physiological processes which enable the moss to cope with high intracellular metal concentrations are not yet completely understood.

One species suitable for biomonitoring in aquatic ecosystems is *Fontinalis antipyretica* L. ex Hedw (Say and Whitton, 1983; Mouvet, 1984; Lopez et al., 1994; Goncalves et al., 1994; Siebert et al., 1996; Bruns, 1997; Bruns et al., 1997). This water moss is widely spread in the northern hemisphere, which makes sampling over wide areas possible. *F. antipyretica* mainly grows in cold fast-flowing rivers and streams. In the cold seasons the biomass production is high and shoots can reach a length of up to 50 cm.

Two procedures of biomonitoring are available to estimate the pollution of aquatic locations: passive biomonitoring by sampling plant material from natural sites and active biomonitoring using plant material exposed to the sites.

In this contribution we will present our results from biomonitoring with *F. antipyretica*. For passive biomonitoring, plant material was collected from different streams of the Harz mountains in Central Germany. Active biomonitoring was done in the river Elbe and in an extremely polluted area of a former mining region in Saxony–Anhalt (Mansfelder Land). Experiments with extremely high concentrations of Cd were carried out under laboratory conditions.

2. Methods

2.1. Plant material and experimental procedures

Plant material was collected from different streams of the Harz region (Fig. 1) and cut into segments of 2–3 cm. For active biomonitoring approximately 10 g of the homogenous plant material was placed in plastic nets and fixed in the water as shown in Fig. 2. The moss bags were exposed at different sites of the river Elbe and in waters of a former copper mining district of Saxony–Anhalt according to Figs. 3 and 4, Tables 1 and 2. Passive biomonitoring was done at the sites presented in Fig. 1.

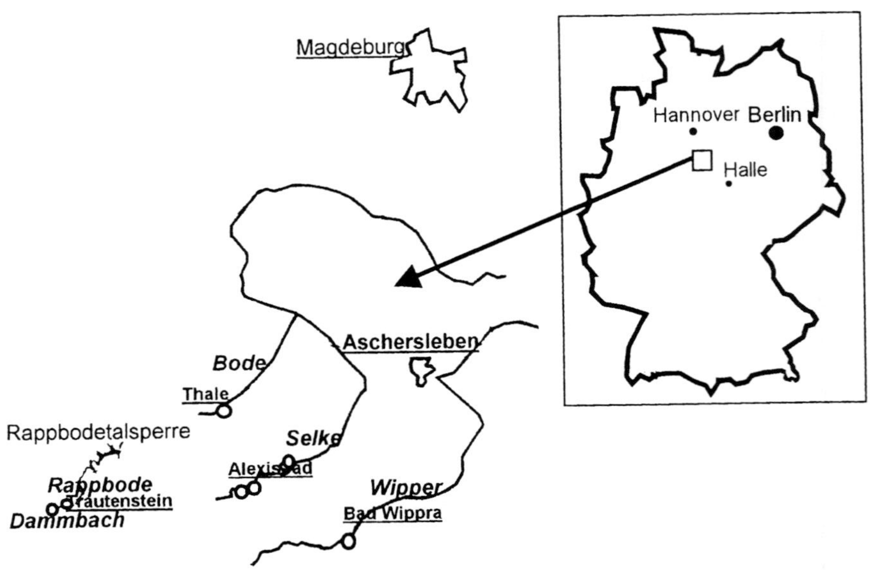

Figure 1. Passive biomonitoring with *F. antipyretica* in the Harz region.

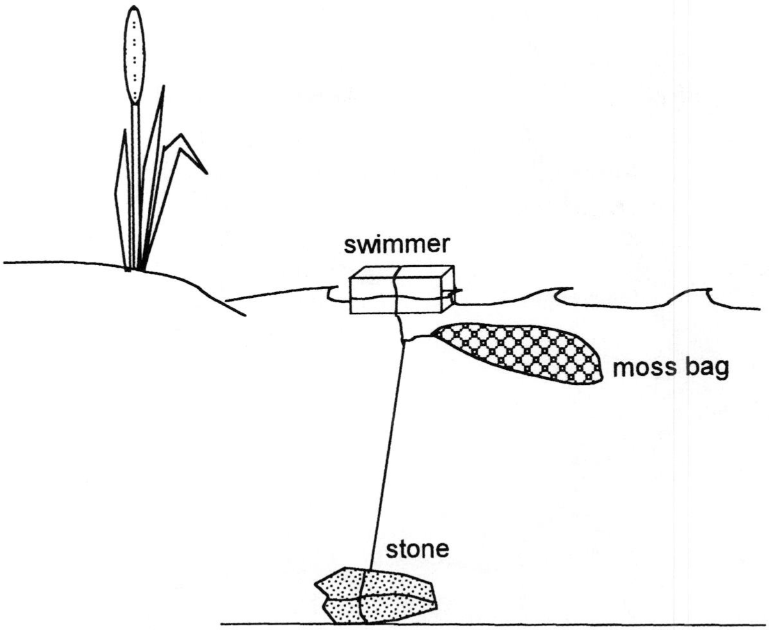

Figure 2. Exposure of plant material in active biomonitoring.

For the laboratory experiments, homogenous plant material was weighed into flasks containing 100 ml of Knop medium per g of fresh plant material. Cd, Zn, Pb and Cu were added as chloride salt.

Table 1. Exposure sites and duration of exposure of active biomonitoring in the river Elbe 1994 and 1995 with *F. antipyretica*

Exposure site	No.	Duration of exposure (days)	
		1994	1995
Tangermünde right	1	16	n.m.[a]
Tangermünde left	2	16	n.m.
Bittkau left	3	n.m.	14
Schönebeck right	4	16	14
Schönebeck left	5	16	14
Breitenhagen right	6	16	14
Breitenhagen left	7	16	14
Aken right	8	n.m.	15
Aken left	9	n.m.	15
Coswig right	10	16	15
Coswig left	11	16	15

[a] n.m., not measured.

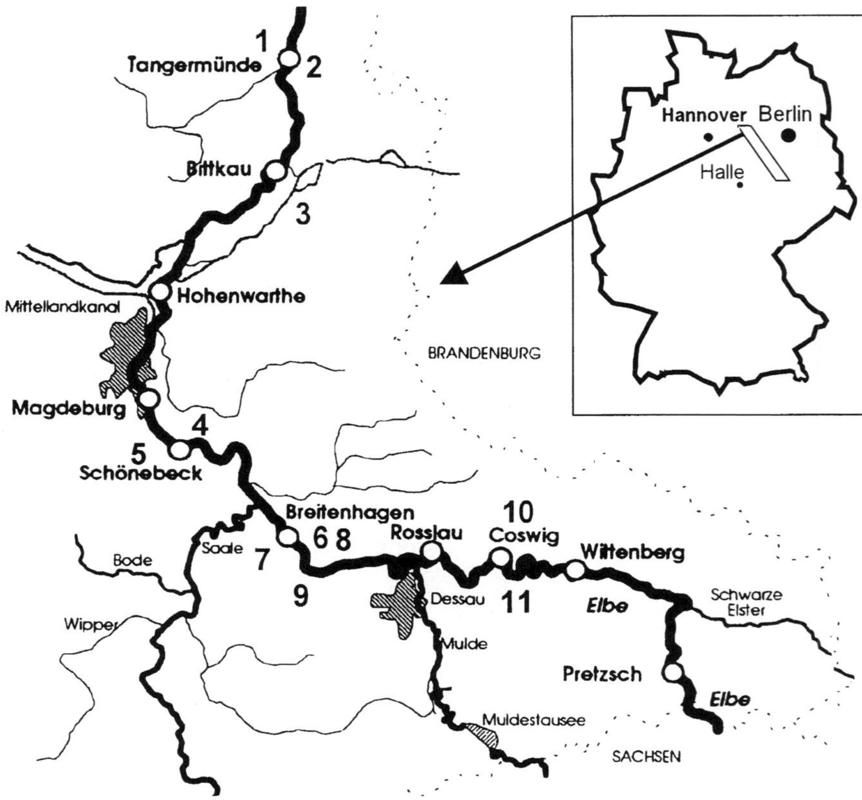

Figure 3. Active biomonitoring with *F. antipyretica* in the river Elbe

2.2. Digestion of plant material and preparation of water samples

After collecting from the streams the moss was washed and dried overnight in an oven at 75°C. The plant material was mortared with liquid N_2 and digested in a microwave system (CEM 2100) (Bruns et al., 1995, 1997). About 50 mg dw was weighed into Teflon vessels and 4 ml HNO_3 + 2 ml H_2O_2 (30% Perhydrol®) were added. After digestion the samples were filled with distilled water to a final volume of 50 ml. One

Table 2. Exposure sites in the region of Mansfelder Land

Sample	Exposure site
1	Helbra, collecting basin at Sanderschacht (stagnant water)
2	Stadtborn Hergisdorf (running water)
3	Sink Wormsleben (stagnant water)
4	Böse Sieben, bridge near Wormsleben (running water)
5	Böse Sieben, railroad bridge, Eisleben (running water)
6	Glume, Eisleben (running water)

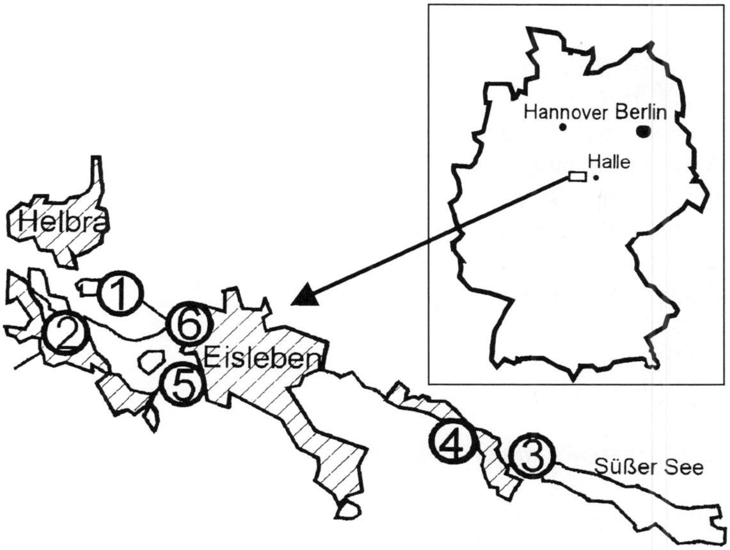

Figure 4. Active biomonitoring with *F. antipyretica* in the former copper mining district of Mansfelder Land (central Germany).

sample of BCR 61 (NIST, *Platihypnidium riparioides,* water moss) was included in each digestion passage.

Water samples were filtered through cellulose acetate filter (45 μm), acidified with 3% HNO_3 (65%, suprapure) and measured directly.

2.3. Analysis of metals

Concentration of the elements Mn, Fe, Cr, Ni, Co, As, Cd, Pb, Zn and Cu in water and plant samples were measured by inductively coupled plasma mass spectrometry (ICP-MS) (ELAN 5000, Perkin–Elmer). The instrument was calibrated by external standards of multi-element solutions for a range of 1–10 μg/kg for water samples and 0.5–20 μg/kg for plant material. Digested plant samples were diluted accordingly. Matrix effects and spectral interference were compensated by addition of internal standards (^{103}Rh and ^{232}Th).

In the case of moss samples exposed in the Mansfelder Land area, inductively coupled plasma atomic emission spectrometry (ICP-AES) was used for the determination of As, Cu, Fe, Mn, Ni, Pb, Zn in both aqueous and acidic solutions by standard addition technique. Atomic absorption techniques were applied for the determination of arsenic (hydride generation coupled with graphite furnace atomization), cadmium and lead (both by Zeeman-corrected graphite furnace AAS) (Krauss et al., 1998, 2000). Ion chromatography was used for the determination of sulphate.

2.4. Cd uptake in laboratory experiments

To investigate the Cd uptake in the laboratory, plant material was incubated in Knop medium (diluted 1:10, N and P reduced) containing 100 μM Cd for 10 days. After incubation the moss was washed three times with a 20 mM $NiCl_2$ solution (Brown and Wells, 1988). In this manner the surface bound Cd ions were replaced by Ni. The Cd content of the washing solutions was measured directly by atomic absorption spectroscopy (AAS) and the extracellular Cd content of the moss was calculated from these data. To measure the internal Cd concentration the moss was dried and digested as described above and Cd was measured by AAS.

3. Results and discussion

3.1. Analysis of reference material and recovery studies

Each digestion passage included one sample of reference material BCR 61 *Platihypnidium riparioides*. Table 3 presents the results of the element analyses of the reference material of different digestion passages. With the exception of Cd and Cu the recovery rates deviate about 5–15 % from the values given by the reference description. For most metals the recoveries are below the concentrations provided. This may be attributed to silicates, which are present in water mosses. Silicates probably bind some of the elements. Under the used digestion method the solution of silicates was not possible. Only the use of HF can dissolve the silicates totally (Bruns et al., 1995; Siebert et al., 1996). But for the routine analysis the handling of HF was not suitable. Only for Zn the recovery was significantly higher in all reference samples. Blanks were made and subtracted from the values. Minor contamination during sample preparation seemed to have occurred. For the aim of studying the metal pollution of different natural sites, the efficiency of recovery is acceptable.

3.2. Natural variability of metal contents of F. antipyretica

One sampling site (Wipper) was chosen to estimate the natural variability of the plant material. Five samples were collected within a 20-m long stream section. In this area the Wipper has a width of about 4 m. The metal concentrations are given in Table 4.

The natural variability was between 2 and 25 %. The highest values were observed for Mn, Cr and Cu. The lowest variations were found for Cd (2%). These data demonstrate a relative equal metal accumulation by *F. antipyretica* within one population at one sampling site. The small variability corresponds with a natural deviation for metal accumulation by water mosses of less than 10 % as reported by Mouvet (1985).

3.3. Passive biomonitoring in the Harz region

Plant material from different natural habitats in the Harz mountains was analyzed for its metal content. Collecting was repeated several times within three years. However,

Table 3. Metal contents and recoveries using reference material BCR 61 *Platihypnidium riparioides* (water moss) from different digestion passages (μg/g dw)

Sample	Mn	Fe	Cr	Ni	Co	As	Cd	Pb	Zn	Cu
A	3437	7524	441	391	38	6.7	1.0	60	629	706
B	3753	8098	483	424	39	8.2	1.2	61	657	748
C	3571	7887	446	386	35	4.6	1.2	55	650	699
D	3570	6985	414	387	33	4.1	1.1	55	636	698
Mean	3583 ± 112	7624 ± 422	446 ± 25	397 ± 15	36 ± 2.4	5.9 ± 1.7	1.1 ± 0.1	57 ± 2.8	643 ± 11	713 ± 21
Certified values	3771 ± 78	9092[a]	532[a]	420[a]	43[a]	7[a]	1.07 ± 0.08	64.4 ± 3.5	566 ± 13	720 ± 31
Recovery (%)	95.0	83.8	83.9	94.6	83.9	83.9	106.7	89.8	113.6	99.0

[a] Not certified.

Table 4. Biological variability of metal concentrations of five samples of *F. antipyretica* from an area of about 20 × 4 m of the river Wipper (Harz)

Sample	Mn	Fe	Cr	Ni	Co	As	Cd	Pb	Zn	Cu
A	8159	4382	9.4	12.5	5.8	3.4	0.57	12	170	15.0
B	16026	3301	5.6	8.8	6.8	4.0	0.57	10	129	9.0
C	18789	3017	5.5	9.3	8.4	4.0	0.54	10	137	10.6
D	13739	4291	8.0	12.4	8.1	3.9	0.56	12	161	14.6
E	16742	3315	5.7	9.4	8.0	4.0	0.57	11	148	12.4
Mean	14691	3661	6.8	10.5	7.4	3.9	0.56	10.9	149	12.3
SD	3643	562	1.6	1.6	1.0	0.2	0.01	1.1	15	2.3
SD %	24.8	15.4	23.0	15.4	12.9	5.6	1.9	10.0	10.0	18.6

at some places plant material was rare and continuous sampling impossible. The metal contents of the plants from two samplings in 1995 and one sampling in 1996 are given in Tables 5–7.

At the river Selke, moss was taken from three different sites within an area of 1 km along the stream. In comparison to the streams in the Harz mountains in July 1995 and July 1996 samples from the region of Rheingau (Hessen) were also analyzed.

The metal concentrations of the water were mostly below the detection limit (data are not presented). From the metal concentrations of the moss it is evident that the sampling sites at the river Selke are more contaminated than those of the other sites. The sites Selke 1 and Selke 2 are upstream of the village Alexisbad, whereas the site Selke 3 is downstream from this settlement. Waste from disused mining tunnels is dumped into the Selke as it passes the village. Moss from the Wipper contained significantly less metals. This sampling site lies in an area without settlement and surrounded by meadows. The interpretation of the data from the other streams is difficult. The accumulation by mosses is influenced by various parameters and metal concentrations of the water can not be calculated from the concentrations of the moss. The presence of competing ions like Ca^{2+} (Wells and Brown, 1987, 1990), the pH values and the amount of suspended matter may affect the quantity of free metal ions and the enrichment by the moss.

Besides the metal content, differences in the morphology of the plant material were obvious at the different places. The biomass production of *F. antipyretica* is highest in cold seasons and in late summer mainly leafless plants were found. But at some places

Table 5. Metal contents of *F. antipyretica* from different rivers of the Harz region, April 1995 (μg/g dw)

Site	Mn	Fe	Cr	Ni	Co	As	Cd	Pb	Zn	Cu
Wipper	787	4833	9.5	9.5	2.5	3.4	0.4	13	142	10
Selke (1)	9205	17664	13.6	34.5	35.9	53.3	5.6	408	1003	86
Selke (2)	2833	7096	9.5	20.3	17.3	26.0	2.4	188	428	49
Rappbode	973	7443	8.7	9.9	2.9	2.0	1.3	13	0	11
Dammbach	4149	14318	16.2	29.1	11.1	4.6	1.8	17	249	22

Table 6. Metal contents of *F. antipyretica* from different sampling sites of the Harz region and one site in the region Rheingau (Wisper), July 1995 (μg/g dw)

Site	Mn	Fe	Cr	Ni	Co	As	Cd	Pb	Zn	Cu
Wipper	18670	3098	4.1	13.9	4.6	3.8	1.1	12	416	12
Selke (2)	2982	7546	8.8	18.7	16.0	29.5	2.7	222	508	45
Selke (3)	27556	6874	9.2	75.6	64.8	61.1	16.8	186	3466	102
Rappbode	17114	3453	0.0	15.7	8.9	1.8	1.8	8	256	10
Dammbach	23370	12051	5.2	25.0	12.3	3.7	2.6	8	246	10
Wisper	756	1524	3.9	8.3	1.9	0.0	0.6	9	429	12

plants were green throughout the year. So the age and phenotype of the moss can differ strongly between the populations. The age is an important factor for the accumulation capacity of the moss. As shown in earlier studies, older parts of the plants have higher concentrations of metals than young apical tips (Siebert et al., 1996).

Thus, the distinct metal contents are at least partly caused by the different ages and phenotypes. Varying metal concentrations of the plant material are not only a result of differences in water pollution. Using *F. antipyretica* for passive biomonitoring, the variation of age and phenotype has to be considered. Repeated measurements throughout the seasons can confirm the results and give an impression of the metal pollution of the sampling sites.

3.4. Active biomonitoring in the river Elbe

Another procedure of biomonitoring with aquatic moss is the exposure of plant material. In contrast to passive biomonitoring, this method has some advantages. For active biomonitoring homogenous plant material from one place is used and therefore biological variability of phenotypes, age and accumulation capacity is minimized. But it should be expected that, like in passive biomonitoring, environmental parameters may also affect the metal uptake by the moss.

Active biomonitoring can be carried out independently of the natural moss vegetation. Also *F. antipyretica* is relatively tolerant against alterations of environmental conditions, provided the water temperature is below 20°C.

The uptake of metals can be calculated over a defined time period. This allows us to

Table 7. Metal contents of *F. antipyretica* from different sampling sites of the Harz region and one sampling site in the region of Rheingau (Wisper), July 1996 (μg/g dw)

Site	Mn	Fe	Cr	Ni	Co	As	Cd	Pb	Zn	Cu
Wipper	1790	3360	5.6	6.9	3.5	2.4	0.2	10	44	9
Selke (1)	832	4370	6.0	20.0	14.0	15.0	1.8	98	265	32
Rappbode	1660	5960	6.6	7.8	3.7	1.6	0.5	10	89	11
Dammbach	6740	21200	16.0	27.0	13.0	4.5	1.7	16	153	20
Bode	3050	2080	2.7	5.8	1.6	1.5	0.3	5.0	54	12
Wisper	1020	4860	13.0	14	3.1	1.5	0.3	6.9	121	13

estimate short time pollution, too. Accumulation factors in relation to the background concentrations can be calculated, which makes interpretation of the data easier.

The metal concentrations of the plant material exposed to the river Elbe 1994 and 1995 are presented in Tables 8 and 9. Compared to the control for most elements a significant accumulation was measured. The accumulation factors for 1994 are given in Table 8.

The metal concentration of the water on the first and last day of the exposure time are presented in Table 10.

The high Mn contents of the samples result from background concentration of the moss. Only at site 5 was Mn significantly enriched. Fe was accumulated at all sites, especially at site 6. A lower concentration of Cr was measured in the moss of sites 1, 2 and 5. The highest accumulation was observed at site 6. The enrichment of Ni was relatively equal over all sites. Co was highly accumulated at the most sites, up to 16 fold of the control at site 10. As concentrations were low at sites 2 and 5. Cd was mainly accumulated at sites 4, 6 and 7 and 9. No significant enrichment was found for sites 1, 2 and 5. Pb was accumulated at most of the sites with a maximum of 5.5-fold of the control. With the exception of sites 1, 2 and 5, Zn was enriched to

Table 8. Metal contents of *F. antipyretica* after exposure to the river Elbe in 1994 (μg/g dw)

Site	Mn	Fe	Cr	Ni	Co	As	Cd	Pb	Zn	Cu
1	1915	3675	6.8	35.2	35.2	12.4	0.9	14.9	312	23
AF[a]	1.2	2.2	0.7	2.3	9.0	4.1	1.0	2.7	1.1	1.2
2	1899	3201	5.6	22.9	5.6	2.5	1.1	6.5	372	21.0
AF	1.2	1.9	0.6	1.5	1.4	0.8	1.2	1.2	1.3	1.1
3	n.m.[b]	n.m.	n.m.	n.m.	n.m.	n.m.	n.m.	n.m.	n.m.	n.m.
AF	–	–	–	–	–	–	–	–	–	–
4[c]	1739	4764	22.3	47.8	43.6	9.0	4.6	18.1	1082	45.8
AF	1.1	2.8	2.2	3.1	11.2	3.0	4.9	3.3	3.7	2.3
5	4179	2233	4.2	30.5	16.3	3.4	0.9	9.6	562	21.0
AF	2.6	1.3	0.4	2.0	4.2	1.1	0.9	1.7	1.9	1.1
6	1939	10458	32.7	39.2	29.4	13.9	3.6	30.5	946	54.5
AF	1.2	6.2	3.3	2.5	7.5	4.7	3.8	5.5	3.3	2.7
7	2011	4207	10.0	52.2	29.3	8.7	4.1	14.1	1011	41.3
AF	1.2	2.5	1.0	3.4	7.5	2.9	4.4	2.6	3.5	2.1
8	n.m.	n.m.	n.m.	n.m.	n.m.	n.m.	n.m.	n.m.	n.m.	n.m.
AF	–	–	–	–	–	–	–	–	–	–
9	n.m.	n.m.	n.m.	n.m.	n.m.	n.m.	n.m.	n.m.	n.m.	n.m.
AF	–	–	–	–	–	–	–	–	–	–
10	1840	5400	15.0	52.0	62.0	10.0	3.0	19.0	1170	52.0
AF	1.1	3.2	1.5	3.4	15.9	3.3	3.2	3.5	4.0	2.6
11	1901	4149	11.6	45.2	42.0	7.8	2.5	15.8	982	45.2
AF	1.2	2.5	1.2	2.9	10.8	2.6	2.7	2.9	3.4	2.3
Control	1633	1679	10.0	15.4	3.9	3.0	0.9	5.5	289	19.9

[a] AF, accumulation factor (metal content sample/metal content control).
[b] n.m., not measured.
[c] Means of three samples.

the same range at all sites. Similar to Pb, Cu was equally distributed at most moss samples of the sites.

Active biomonitoring in the river Elbe was repeated in 1995. These results are presented in Tables 9 and 11. As in 1994, no significant Mn accumulation in *F. antipyretica* occurred. The accumulation factors for Fe were between 1.2 and 3.1. Cr was accumulated strongly at sites 8 and 9.

Ni was only moderately enriched at all sites. The enrichment of Co was up to 6.5-fold of the control at site 9.

As, like Ni, accumulated to a low extent but nearly equally over all exposure sites. Cd accumulated to a greater extent and reached 9.3-fold of the control at site 9. No significant accumulation of Pb could be observed. Zn was accumulated between 2.7- and 4.8-fold of the control. The enrichment of Cu was weak and did not differ significantly between the various sites.

For 1994 the order of accumulation factors (compared to the controls from the river Wipper) was Mn < Cu < Cr < Ni < Zn < As < Cd < Pb < Fe < Co. In 1995 the accumulation was Pb < Mn < As < Cu < Ni < Fe < Cr < Zn < Co < Cd. With the exception of Cd and Zn the accumulation factors were higher in 1994 for all elements.

Table 9. Metal contents of *F. antipyretica* after exposure to the river Elbe in 1995 (μg/g dw)

Site	Mn	Fe	Cr	Ni	Co	As	Cd	Pb	Zn	Cu
1	n.m.[a]	n.m.	n.m.	n.m.	n.m.	n.m.	n.m.	n.m.	n.m.	n.m.
AF[b]	–	–	–	–	–	–	–	–	–	–
2	n.m.	n.m.	n.m.	n.m.	n.m.	n.m.	n.m.	n.m.	n.m.	n.m.
AF	–	–	–	–	–	–	–	–	–	–
3	20985	4745	15.5	56.6	40.1	15.5	6.4	17.3	985	31.9
AF	1.2	1.5	1.2	1.6	4.8	1.3	5.0	1.2	4.8	1.7
4	25837	6433	15.7	66.9	25.1	19.9	6.9	17.8	1046	34.5
AF	1.5	2.1	1.2	1.9	3.0	1.7	5.4	1.3	5.1	1.8
5	17560	3789	16.6	59.1	29.6	12.9	3.5	12.9	828	28.7
AF	1.0	1.2	1.3	1.7	3.5	1.1	2.7	0.9	4.1	1.5
6	18592	4275	17.9	50.4	24.2	16.8	5.5	11.6	772	29.4
AF	1.1	1.4	1.4	1.4	2.9	1.4	4.2	0.8	3.8	1.6
7	16216	3874	14.4	55.9	19.8	13.5	10.8	9.9	861	27.9
AF	1.0	1.3	1.1	1.6	2.4	1.1	8.4	0.7	4.2	1.5
8	13889	9444	28.9	44.4	15.6	20.0	4.7	20.0	540	33.3
AF	0.8	3.1	2.2	1.2	1.9	1.7	3.6	1.4	2.7	1.8
9	19337	6077	28.5	60.8	54.3	18.4	12.0	15.7	1041	31.3
AF	1.1	2.0	2.2	1.7	6.5	1.5	9.3	1.1	5.1	1.7
10	19960	5788	24.0	46.9	24.0	16.0	2.9	15.0	608	29.9
AF	1.2	1.9	1.9	1.3	2.9	1.3	2.2	1.1	3.0	1.6
11[c]	20253	5397	16.2	48.6	24.3	16.9	2.7	13.5	591	29.0
AF	1.2	1.8	1.3	1.4	2.9	1.4	2.1	1.0	2.9	1.5
Control	16865	3075	12.9	35.7	8.3	11.9	1.3	13.9	203	18.8

[a] n.m., not measured.
[b] AF, accumulation factor (metal content sample/metal content control).
[c] Means of three digestions of one sample.

But after subtracting the background metal concentration of the controls, the absolute accumulation (μg/g dw) of Mn, Cr, Cd, Pb and Cu was higher in 1995. The plant material reflects a cumulative metal content, which allows us to assess the metal pollution over a defined period, even if no correlation between the element concentrations in moss samples and water was found. During both exposure periods the element concentrations of the water samples were higher at the beginning of the exposure time (Tables 10 and 11). This might be due to changes in the metal concentrations of the water. The water samples were taken only twice, at the beginning and at the end of the exposure time. It is likely that the metal concentration of the water changed during the exposure time.

3.5. Active biomonitoring in a former mining area

Active biomonitoring with *F. antipyretica* was also done in an extremely polluted area of a former copper mining region in Saxony–Anhalt.

The district of Mansfelder Land has a long history in copper shale mining and smelting (Schreck, 1998). After the closure of a major metallurgical plant in 1978,

Table 10. Metal content of water samples from the river Elbe taken at the beginning and end of the exposure time in 1994

No.[a]	Mn	Fe	Cr	Ni	Co	As	Cd	Pb	Zn 1	Cu
1a	340	−4.7	<2.00	2.7	0.8	3.9	<0.20	<2.00	−24.0	<2.00
1b	156	46.7	<2.00	2.9	0.9	4.9	<0.20	<2.00	−16.4	<2.00
2a	439	6.0	<2.00	<2.00	0.5	<1.00	<0.20	<2.00	−22.2	<2.00
2b	296	14.7	<2.00	<2.00	0.3	1.1	<0.2	<2.00	−18.3	<2.00
3a	n.m.[b]	n.m.	n.m.	n.m.	n.m.	n.m.	n.m.	n.m.	n.m.	n.m.
3b	n.m.	n.m.	n.m.	n.m.	n.m.	n.m.	n.m.	n.m.	n.m.	n.m.
4a	70	−6.6[c]	<2.00	4.3	0.6	3.4	0.20	<2.00	7.9	2.5
4b	36	−12.4[c]	<2.00	4.0	0.5	3.3	<0.20	<2.00	12.7	3.7
5a	86	55.5	<2.00	7.5	0.7	4.1	0.36	<2.00	65.1	3.4
5b	64	11.8	7	6.4	0.6	4.2	0.37	<2.00	53.5	3.1
6a	58	−4.7[c]	<2.00	3.7	0.5	3.6	<0.20	<2.00	−6.7[c]	2.3
6b	26	−3.7[c]	<2.00	4.0	0.4	3.5	<0.20	<2.00	2.0	2.8
7a	86	10.8	<2.00	4.9	0.6	3.7	<0.20	<2.00	2.0	5.4
7b	42	−0.8[c]	<2.00	4.9	0.4	3.7	<0.20	<2.00	11.7	2.7
8a	n.m.	n.m.	n.m.	n.m.	n.m.	n.m.	n.m.	n.m.	n.m.	n.m.
8a	n.m.	n.m.	n.m.	n.m.	n.m.	n.m.	n.m.	n.m.	n.m.	n.m.
9a	n.m.	n.m.	n.m.	n.m.	n.m.	n.m.	n.m.	n.m.	n.m.	n.m.
9a	n.m.	n.m.	n.m.	n.m.	n.m.	n.m.	n.m.	n.m.	n.m.	n.m.
10a	93	−0.8[c]	<2.00	3.9	0.9	4.1	<0.20	<2.00	0.1	2.4
10b	38	51.6	<2.00	3.9	0.6	3.7	0.29	<2.00	21.4	4.6
11a	62	−19.2[c]	<2.00	3.1	0.6	3.5	<0.20	<2.00	−8.6[c]	2.0
11b	36	44.8	<2.00	3.7	0.4	3.7	0.21	<2.00	1.1	4.1

[a] a, beginning; b, end (μg/l).
[b] n.m., not measured.
[c] Negative values are caused by the subtraction of the blank value.

Table 11. Metal content of water samples from the river Elbe taken at the beginning and end of the exposure time in 1995

Site[a]	Mn	Fe	Cr	Ni	Co	As	Cd	Pb	Zn	Cu
1a	n.m.[b]	n.m.	n.m.	n.m.	n.m.	n.m.	n.m.	n.m.	n.m.	n.m.
1b	n.m.	n.m.	n.m.	n.m.	n.m.	n.m.	n.m.	n.m.	n.m.	n.m.
2a	n.m.	n.m.	n.m.	n.m.	n.m.	n.m.	n.m.	n.m.	n.m.	n.m.
2b	n.m.	n.m.	n.m.	n.m.	n.m.	n.m.	n.m.	n.m.	n.m.	n.m.
3a	82	107	1.9	5.4	0.7	3.3	0.29	<2.0	60	3.8
3b	20	34	<2.0	4.9	0.5	2.7	<0.2	<2.0	30	4.3
4a	32	83	2.7	4.7	0.6	3.1	0.31	<2.0	46	4.8
4b	41	43	<2.0	5.6	0.7	2.6	<0.2	<2.0	29	4.4
5a	99	107	2.3	6.8	1.2	3.2	0.35	<2.0	87	4.6
5b	37	29	<2.0	5.9	0.7	2.5	<0.2	<2.0	24	3.5
6a	54	90	<2.0	4.2	0.6	3.1	<0.2	<2.0	41	2.9
6b	31	28	<2.0	4.5	0.6	2.7	<0.2	<2.0	24	3.3
7a	72	88	<2.0	4.9	0.7	2.9	0.35	<2.0	49	2.9
7b	165	42	2.0	4.9	0.8	2.6	<0.2	<2.0	24	4.1
8a	138	70	2.7	4.0	0.8	2.9	<0.2	<2.0	55	4.7
8b	33	44	<2.0	4.1	0.4	2.7	<0.2	<2.0	21	4.5
9a	80	97	2.1	5.3	0.8	2.7	0.39	<2.0	60	3.5
9b	52	36	<2.0	5.4	0.7	2.5	0.32	<2.0	39	3.7
10a	n.m.	n.m.	n.m.	n.m.	n.m.	n.m.	n.m.	n.m.	n.m.	n.m.
10b	42	20	<2.0	3.3	0.6	2.9	<0.2	<2.0	16	3.6
11a	65	81	2.3	3.7	0.8	3.3	<0.2	<2.0	36	3.7
11b	37	41	<2.0	4.1	0.7	2.9	<0.2	<2.0	19	3.0

[a] a, beginning; b, end (μg/l).
[b] n.m., not measured.

about 300 000 tons of solidified flue dust were deposited in open basins on mining and smelting waste heaps. Leachates from flue dust deposits occur around these heaps and pollute the surface and the groundwater. They contain up to 3000 mg/l Zn and exhibit elevated contents of Cd, Cu, Pb (Table 12), PAHs (Krauss et al., 1998, 2000) and visible mineral precipitation – glaucocerinite – with high Zn and Al content (Witzke, 1997).

The water moss was exposed for 3 weeks at the various highly contaminated sites (Table 12). The plant material was treated in the same way as described for the Elbe exposure. At the end of the exposure plant samples showed no obvious loss of vitality as determined by chlorophyll a/b ratio (not shown). The metal concentrations of samples are given in Table 13.

Compared to the moss exposed to the river Elbe, the accumulation factors were higher. The highest were found at site 4, where Cu was accumulated up to 263-fold, reflecting the strong pollution of this site. However, the higher background concentrations of the moss compared to those used for the Elbe exposure have to be considered.

The results of both active biomonitorings with *F. antipyretica* demonstrate that the moss accumulates metals over a wide range of concentration. Concerning the different accumulation factors and the vitality, this moss can be used as a monitor in weakly to

Table 12. Heavy metal content (average amounts from three samples) at six sites in the copper shale mining district of Mansfelder Land, water (mg/l), sediment (in brackets) (mg/kg)

Site	1	2	3	4	5	6
Sulphate (mg/l)	1362	6753	479	569	393	850
Mn	2.30 (540)	19.10 (180)	0.10 (500)	0.10 (770)	0.40 (880)	0.40 (420)
Fe	0.25 (7.80)	0.05 (2.10)	0.61 (2.20)	0.17 (2.60)	0.80 (0.37)	0.80 (3.60)
Cr	n.d. (149)	n.d. (75)	n.d. (54)	n.d. (45)	n.d. (81)	<0.05 (64)
Ni	<0.09 (76)	2.30 (400)	<0.08 (19)	<0.08 (26)	<0.08 (29)	<0.08 (720)
Co	n.d. (57)	n.d. (21)	n.d. (8)	n.d. (14)	n.d. (23)	<0.05 (14)
As	0.007 (5000)	0.003 (6800)	0.001 (33)	0.001 (64)	0.004 (19)	0.006 (290)
Cd	0.13 (63)	2.80 (29)	<0.05 (7)	<0.05 (6)	<0.05 (6)	<0.05 (9)
Pb	0.80 (49000)	1.90 (24500)	<0.10 (224)	<0.10 (372)	0.15 (540)	<0.10 (720)
Zn	56.10 (25960)	2630 (16660)	0.30 (1910)	0.40 (2060)	0.60 (1500)	1.00 (3100)
Cu	0.27 (10900)	13.25 (50220)	0.03 (320)	<0.02 (740)	0.08 (760)	<0.02 (460)

Table 13. Metal contents and accumulation factors (metal sample/metal control) of *F. antipyretica* after exposure at highly polluted sites of a former copper mining area in Saxony–Anhalt (μg/g dw)

Exposure site	Mn	Fe	Cr	Ni	Co	As	Cd	Pb	Zn	Cu
1	13900	6537	7.9	40	78	374	25	14500	17699	1169
AF	1.1	0.7	0.4	1.1	2.3	8.5	8.3	67.8	20.4	14.8
2	28317	4502	5.2	66.3	188	140	36.6	9109	32375	20778
AF	2.3	0.5	0.3	1.7	5.5	3.2	12.2	42.6	37.4	263.0
3	4334	8507	7.7	42.9	25	44	9.8	196	1940	76
AF	0.4	1.0	0.4	1.1	0.7	1.0	3.3	0.9	2.2	1.0
4	31061	11025	17.3	55.2	64	61	12.2	476	6882	347
AF	2.5	1.3	0.9	1.5	1.9	1.4	4.1	2.2	7.9	4.4
5	4058	8666	10.8	31.4	29	34	7	380	2986	169
AF	0.3	1.0	0.5	0.8	0.8	0.8	2.3	1.8	3.4	2.1
6	28222	7841	10.3	57.8	48	267	36.7	967	7910	439
AF	2.3	0.9	0.5	1.5	1.4	6.1	12.2	4.5	9.1	5.6
Control	12353	8749	20	38	34	44	3	214	866	79

moderately polluted systems like the river Elbe, as well as in extremely polluted surface water representing leachates from the flue dust deposits.

3.6. Cd uptake by F. antipyretica under laboratory conditions

In addition to the field experiments, the Cd uptake by *F. antipyretica* was tested under laboratory conditions. Concentrations of 1.12 mg/100 ml were added to the culture medium (Knop medium, modified) and the accumulation monitored over 10 days. The results are presented in Table 14. The data confirm that *F. antipyretica* accumulates heavy metals like Cd to a large extent. The enrichment is a time-dependent process, starting with a fast equilibration followed by several days of slower Cd uptake. There is a slow increase of the intracellular Cd content, which is a result of migration from extracellular binding sites into the cell.

Table 14. Contents of intra- and extracellular bound Cd in *F. antipyretica* (approx. 1 g fw) during 10 days of incubation in 100 ml Cd-containing medium (1.12 mg/100 ml)

Day	Intracellular (μg/g dw)	Extracellular (μg/g dw)	Total (μg/g dw)
1	1.53	4.69	6.22
2	1.40	3.73	5.12
3	2.58	4.58	7.16
4	1.31	2.61	3.92
5	2.52	3.83	6.35
6	3.15	4.25	7.40
7	3.04	4.15	7.19
8	2.02	2.99	5.01
9	3.04	3.26	6.31
10	3.17	3.46	6.62

4. Conclusions

The water moss *F. antipyretica* accumulates metals in aquatic systems. This property can be used to assess metal pollution of natural aquatic ecosystems. The biological variability of the plant material may have a significant influence on the accumulation capacity. Phenotype and age of the plants influence the metal uptake. Therefore active biomonitoring by exposing homogenous plant material has.

Acknowledgements

Our work was done with the financial support of GKSS Research Centre Gesthacht, UFZ-Centre for Environmental Research Leipzig-Halle, Department Inland Water Research Magdeburg, the Ministerium für Forschung und Wissenschaft in Saxony-Anhalt, the Ministerium für Umwelt, Naturschutz und Raumordnung. We thank Mrs. E. Pueschel and Mrs. C. Hoffmeister for technical assistance.

References

Brown, D.H., Wells, J.M., 1988. Sequential elution technique for determining the cellular location of cations. In: Glime, J.M. (Ed.), Methods in Bryology. Proc. Bryol. Meth. Workshop, Mainz. Hattori Bot. Lab., Nichinan, pp. 227–233.

Bruns, I., Siebert, A., Baumbach, R., Miersch, J., Günther, D., Markert, B., Krauss, G.-J., 1995. Analysis of heavy metals and sulphur-rich compounds in the water moss *Fontinalis antipyretica* L. ex Hedw. Fresenius J. Anal. Chem. 353, 101–104.

Bruns, I., 1997. Induktion thiolhaltiger Peptide im Wassermoos *Fontinalis antipyretica* L. ex Hedw. unter Schwermetalleinfluß und deren Nutzung als Biomarker für Schwermetallbelastungen aquatischer Systeme. Thesis, Department Biochemistry/Biotechnology, University of Halle-Wittenberg, Halle.

Bruns, I., Friese, K., Markert, B., Krauss, G.-J., 1997. The use of *Fontinalis antipyretica* L. ex Hedw. as a bioindicator for heavy metals, 2. Heavy metal accumulation and physiological reaction of *Fontinalis antipyretica* L. ex Hedw. in active biomonitoring in the river Elbe. Sci. Total Environ. 204, 161–176.

Chovanec, A., Vogel, W.R., Lorbeer, G., Hanus-Illnar, A., Seif, P., 1994. Chlorinated organic compounds, PAHs, and heavy metals in sediment and aquatic mosses of two upper Austrian rivers. Chemosphere 29, 2117–2133.

Goncalves, E.P.R., Soares, H.M.V.M., Boaventura, R.A.R., Machado, A.A.S.C., Esteves da Silva, J.C.G., 1994. Seasonal variations of heavy metals in sediment and aquatic mosses from the Cavado river basin (Portugal). Sci. Total Environ. 142, 143–156.

Herpin, U., Markert, B., Siewers, U., Lieth, H., 1994. Monitoring der Schwermetallbelastung in der Bundesrepublik Deutschland mit Hilfe von Moosanalysen. Forschungsbericht 10802087, 94–125.

Herpin, U., Berlekamp, J., Markert, B., Wolterbeek, B., Grodzinska, K., Siewers, U., Lieth, H., Weckert, V., 1996. The distribution of heavy metals in a transect of the three states The Netherlands, Germany and Poland, determined with the aid of moss monitoring. Sci. Total Environ. 187, 185–198.

Krauss, G. Bärlocher, F., Schreck, P., Kranich, W., Miersch, J., Dermietzel, J., Wennrich, R., Krauss, G.-J. 1998. Aquatic hyphomycetes at extremely polluted sites in the Mansfelder Land area. In: Becker, P.M. (Ed.), Microbiology of Polluted Aquatic Ecosystems. UFZ-Series, 10. UFZ Centre for Environmental Research, Leipzig-Halle, Germany, pp. 76–83.

Krauss, G., Baerlocher, F., Schreck, P., Wennrich, R., Glässer, W., Krauss, G.-J., 2000. Aquatic hyphomycetes occur in hyperpolluted waters in Central Germany. Sci. Total. Environ., submitted.

Lopez, J., Vazquez, D.M., Carballeira, A., 1994. Stress responses and metal exchange kinetics following transplant of the aquatic moss *Fontinalis antipyretica*. Freshwater Biol. 32, 185–198.

Markert, B., Weckert, V., 1993. Time- and site integrated long-term biomonitoring of chemicals by means of mosses. Toxicol. Environ. Chem. 40, 43–56.

Markert, B., Herpin, U., Siewers, U., Berlekamp, J., Lieth, H., 1996. The German heavy metal survey by means of mosses. Sci. Total Environ. 182, 159–168.

Mouvet, C., 1984. Accumulation of chromium and copper by the aquatic moss *Fontinalis antipyretica* L. ex Hedw. transplanted in a metal contaminated river. Environ. Technol. Lett. 5, 541–548.

Mouvet, C., 1985. The use of aquatic bryophytes to monitor heavy metal pollution of freshwaters as illustrated by case studies. Verh. Int. Verein. Limnol. 22, 2420–2425.

Say, P.J., Whitton, B.A., 1983. Accumulation of heavy metals by aquatic mosses. 1, *Fontinalis antipyretica* Hedw. Hydrobiologia 100, 245–260.

Say, P.J., Harding, J.P.C., Whitton, B.A., 1981. Aquatic mosses as monitors of the heavy metal contamination of the River Etherow, Great Britain. Environ. Pollut. 2, 295–307.

Schreck, P., 1998. Environmental impact of uncontrolled waste disposal in mining and industrial areas in Central Germany. Environ. Geol. 35, 66–72.

Siebert, A., Bruns, I., Krauss, G.-J., Miersch, J., Markert, B., 1996. The use of *Fontinalis antipyretica* L. ex Hedw. as a bioindicator for heavy metals. 1. Fundamental investigations into heavy metal accumulation in *Fontinalis antipyretica* L. ex Hedw. Sci. Total Environ. 177, 137–144.

Wells, J.M., Brown, D.H., 1987. Factors affecting the kinetic of intra- and extracellular cadmium uptake by the moss *Rhytidiadelphus squarrosus*. New Phytol. 105, 123–137.

Wells, J.M., Brown, D.H., 1990. Ionic control of intracellular and extracellular Cd uptake by the moss *Rhytidiadelphus squarrosus* (Hedw.). Warnst. New Phytol. 116, 541–553.

Witzke, T., 1997. Sekundärmineralbildungen in Kupferschlacke als Indikator für Schwermetallmobilisierung und -fixierung. In: Schreck, P., Glässer, W. (Eds.), Reststoffe der Kupferschieferverhüttung, Part 1: Mansfelder Kupferschlacken., UFZ Berichte, 23. UFZ Centre for Environmental Research, Leipzig-Halle, Germany, pp. 25–29.

Trace elements – Their distribution and effects in the environment
B. Markert and K. Friese, editors

Chapter 11

Antimony contents in plant species growing in an Sb-mining district (Tuscany, Italy)

F. Baroni, A. Boscagli, G. Protano, F. Riccobono

Abstract

We analyzed the antimony contents in 59 plant species (a total of 67 samples) from six sites in an Sb-mining area, each with a different Sb content in the soil. The plants were both cultivated and wild species, mostly herbaceous and tap-rooted, Dicotyledones, composites, grasses and legumes. The Sb concentration in plants is highly correlated with the concentration of total, soluble and extractable Sb in the soil, while the ability of the species to concentrate antimony is correlated only with the concentration of extractable Sb. Accumulation in roots is most responsive to the increase of available antimony in soils passing from the cultivated sampling sites to the mineral processing tailing pond. The most common site of Sb accumulation in the plant shifts from the roots to the leaves as the Sb content in the soil increases. Nevertheless, it appears to be largely species-specific. Antimony levels are high in plants from the tailing pond, above 1000 mg/kg in some wild species (Achillea ageratum, Silene vulgaris and Plantago lanceolata), while they are very low in crops and vegetables. Plant species traits, such as root morphology, life span, habitus and systematic rank, are important, although their influence is not statistically significant. All the plant species with the highest concentrations of antimony are herbaceous, almost all Dicotyledones, tap-rooted and mostly perennials, composites and legumes. The potential risk of the entry of antimony into food chains is discussed. The possible effects of plants on the environmental distribution of the element and the possible exploitation of some plant species as phytoremediators are also considered.

Keywords: Antimony; Mining waste; Environmental pollution; Soil-plant relationships; Biological accumulation; Translocation coefficient; *Achillea ageratum*; *Plantago lanceolata*; *Silene vulgaris*

1. Introduction

Antimony is a rather rare element in nature. Its abundance in the earth's crust is about 0.2 ppm and the highest values among rocks are found in clays (1.5 ppm). There is usually a higher concentration of antimony in soils than in the parent materials (Kabata-Pendias and Pendias, 1984). According to Ainsworth et al. (1990a, 1991), it is a largely immobile element. The latter study also reported that close to an antimony smelter 80–90% of the total Sb contained in the soil was in an immobile chemical form which accumulated in a thin (<5 cm) shallow horizon. Only a small fraction moved downward after being converted to a mobile form.

Antimony present in soluble forms can be readily absorbed by plants, in which it is concentrated in older leaves, lower stems and roots (Coughtrey et al., 1983). Hence, dead plant tissues can be richer in antimony than living ones. A low degree of Sb bioaccumulation in the food chain was found by Ainsworth et al. (1990b, 1991), both in contaminated grasslands and in experiments with diets enriched in antimony.

Little is known about the biological role of Sb. In plants, where it is not essential, antimony can be phytotoxic by competing with essential metabolites (Foy et al., 1978; Bowen, 1979). However, antimony phytotoxicity, like many other aspects concerning this element, has been poorly investigated. In mature leaf tissue, Kabata-Pendias and Pendias (1984) considered a content of 150 mg/kg to be toxic. Background Sb contents in terrestrial plants have been reported as 50 μg/kg (Brooks, 1972), from 2 to 30 μg/kg (Bowen, 1979) and from 0.2 μg/kg to 5 mg/kg (Coughtrey et al., 1983). Decidedly higher contents (7–50 mg/kg) were found in shrubs and trees growing in mineralized areas of Alaska (Shacklette et al., 1978) and in grasses (110–900 mg/kg) growing around Pb and Sb smelters (Ragaini et al., 1977; Istas et al., 1982; Ainsworth et al., 1991).

Important sources of Sb release into the environment are mining and smelting, emissions from coal firing, incineration of refuse, application of chemical fertilizers or other soil amendments, such as sewage sludge and fly-ash. Antimony released by these activities or also by the soil surface can become associated with fine particulate matter and widely dispersed. If ignored, the airborne element deposited on plant surfaces can bias the estimation of Sb uptake by the root system. In fact, particular care is necessary in choosing an appropriate washing procedure, as highlighted by Istas et al. (1982).

In many regions of the world where ores of toxic elements were exploited and smelted for long periods, the environmental impact of wastes is often an unresolved problem, mostly owing to the lack of adequate reclamation.

In southern Tuscany, important Sb ores were mined in the past and several roasting plants were active for a long time. In this paper, we focus on the Monti Romani area (where mining and smelting of antimony sulfide ores occurred until the recent past) in a survey of the impact of Sb diffusion on plant species growing in the environs of the mining sites. We will also try to elucidate the relationships between the Sb contents in soils and the plant species, the character of the Sb allocation within the plant species and the effect of some plant attributes on the Sb content and allocation.

2. Study area

The study area lies in the southernmost part of Tuscany, near the border with Latium, in the Monti Romani coastal range (Fig. 1). In this hill system the geological situation is rather complex, from both the stratigraphic and tectonic points of view. Indeed, a variety of terrains, ranging in age from the Paleozoic to the Pliocene, crop out, showing at least two main unconformity surfaces. The most important structural feature of the area was imparted by the extensional Mio-Pliocene tectonics, through the formation of a series of rift basins bordered by Apennine trending faults. Upper Triassic terrains, mostly consisting of highly fractured dolomitic limestones (Calcare Cavernoso fm.), host several ore-grade stibnite deposits in their top zone (Dessau et al., 1972; Dehm et al., 1983), which have been intensively exploited until recently.

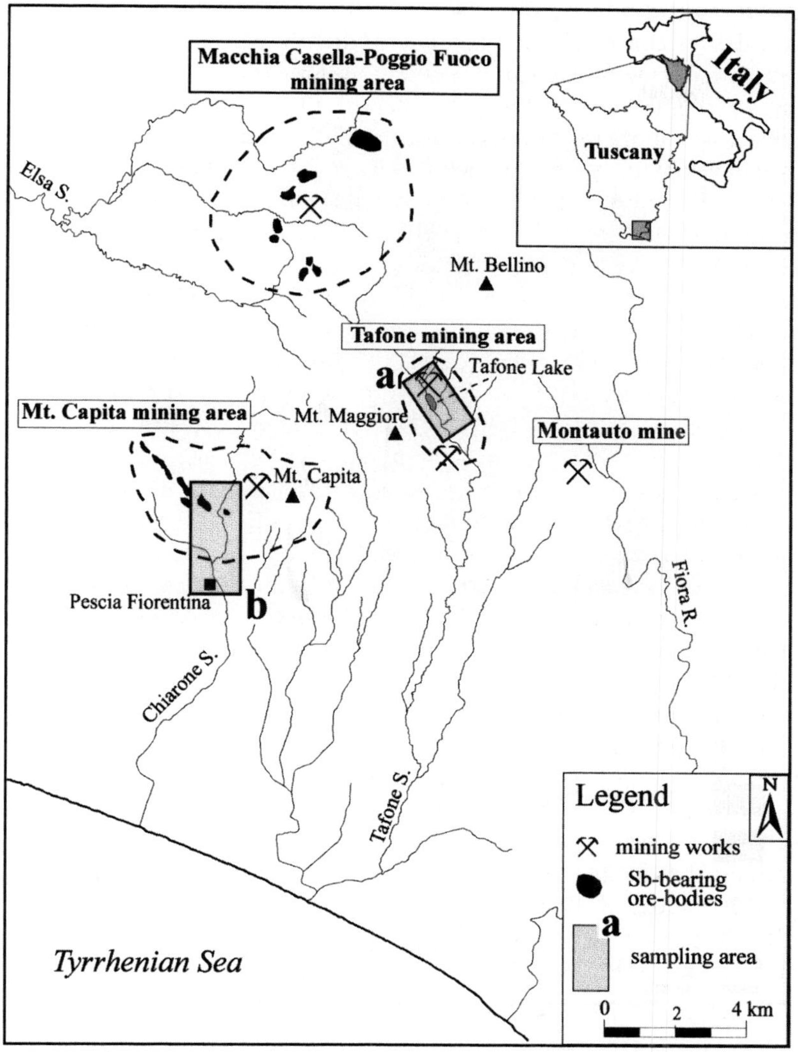

Figure 1. Study area.

The Sb-bearing ores of this area provided an essential contribution to the national antimony production. Between the 1960s and early 1970s, Sb production exceeded the Italian domestic consumption and this commodity was exported to other countries. These intensive mining and smelting activities also produced huge quantities of waste materials which are still releasing Sb and other toxic elements, such as As and Hg, into the surrounding environment (Zanzari et al., 1995; Protano and Riccobono, 1997). The high resolution (sampling density of 4 samples/km^2) geochemical map of Fig. 2, based on stream sediments, highlights the existence of rather pronounced and widespread geochemical anomalies for antimony closely linked to the mining sites of the area.

According to Thornthwaite's classification (Thornthwaite, 1948), the climate of the

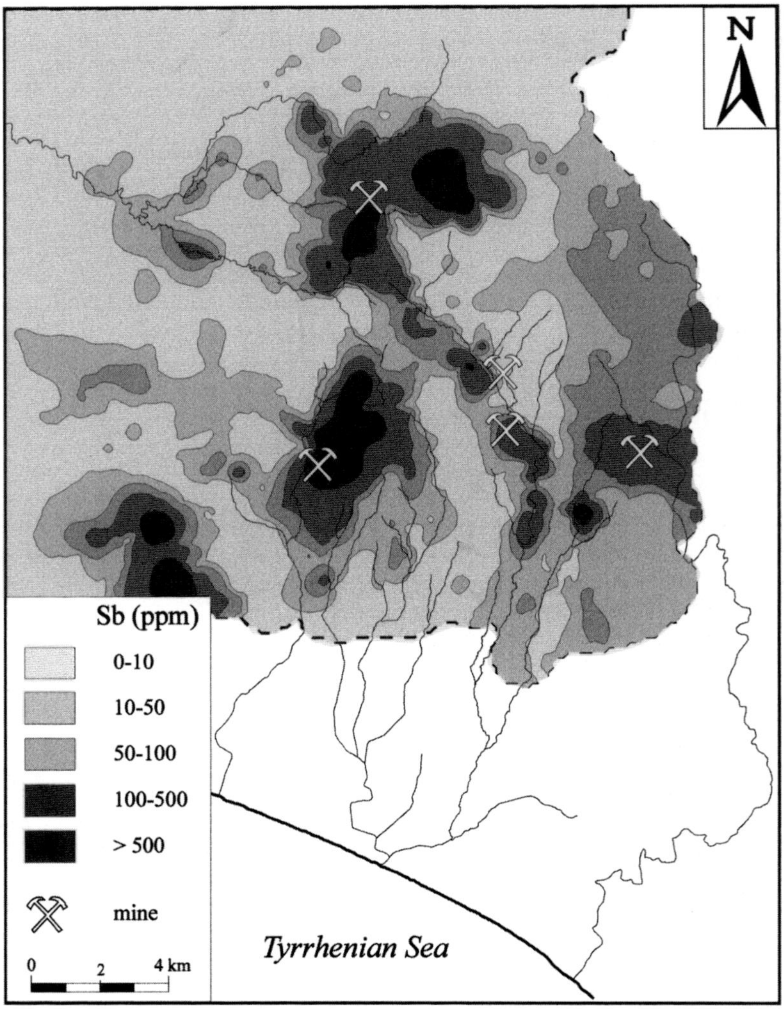

Figure 2. Geochemical map of the antimony distribution in stream sediments of the Monti Romani area. Reprinted from Protano et al. (1998).

area is of type $C_2\,B'_2\,s\,b'_4$, i.e. subhumid, mesothermic, with a moderate summer water deficit and thermal efficiency concentrated in the summer. Rainfall averages ca. 900 mm/y with two seasonal maxima: a primary peak in autumn and a secondary one in spring. Yearly mean temperature is 14–15°C; January is the coldest month (mean 7–8°C) and July the warmest (23–24°C) (Barazzuoli et al., 1993).

From a phytogeographical point of view, the study area is located at the boundary between the evergreen Mediterranean and the deciduous sub-Mediterranean belts. Thus the former vegetation type occurs under the warmest local conditions, while the latter occurs under the coldest ones. Under intermediate conditions, the two vegetation types overlap.

In the areas of the oldest mining works, shrubs like *Spartium junceum* L. and *Rubus ulmifolius* Scott dominate, while in the youngest ones, a colonizer like *Inula viscosa* (L.) Aiton dominates. Herbs, mainly ruderal ones, are often the most abundant plants. Chamaephytes (e.g. *Helichrysum italicum* (Roth) Don. and *Dorycnium hirsutum* (L.) Ser.) are also very common.

3. Sampling surveys

Plant and soil samples were collected from two areas located in the central part of the mining district (Fig. 1). Sampling was carried out especially near the old Tafone Mine open pit (now a small lake) and along a tract of the Chiarone stream, where mining works were more intensive (Fig. 3).

Sampling sites S1 and S2 were located in cultivated fields; sampling site S3 in a kitchen garden; sampling site S4 in an old field; sampling site S5 on mine dumps and site S6 at a mineral processing tailing pond. S1, S2 and S3 were used to investigate Sb contents in some vegetables and crops, S4 in some weeds and wild plants, S5 and S6 in several wild species.

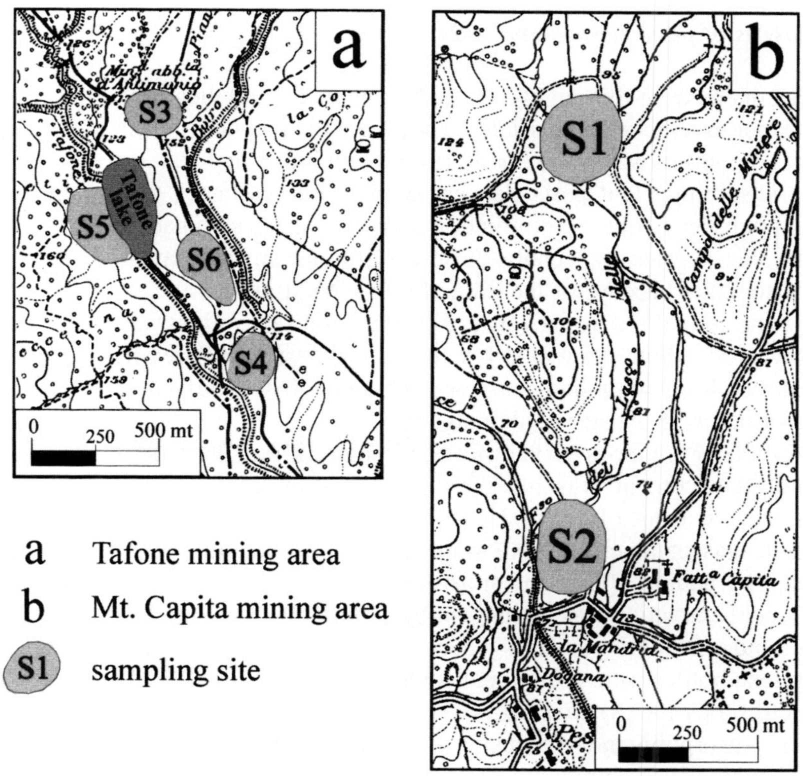

Figure 3. Location of the sampling sites in the environs of the two major mining works in the Monti Romani ore-field.

4. Materials and methods

4.1. Collection of plant and soil samples

Plant and soil samples were collected in 1996 and 1997. Cultivated plant species were selected to represent the most commonly used ones. Wild species were selected in consideration of their potential use as phytoremediators, according to the following guidelines:

- adaptation to adverse edaphic conditions (very common on mine tailings and dumps);
- large biomass allocation in the aerial part;
- fast maximization of epigeal biomass;
- frequency in the field.

For each selected species, individual specimens were randomly collected for analysis. In species with a low removable root mass, only the epigeal parts were collected.

Soil samples (representative of the upper 20 cm) were randomly collected. In some cases, the plant-soil couple was taken.

The nomenclature of the plant species is according to Pignatti (1982).

4.2. Analysis of plant tissues and soils

Plant specimens were thoroughly washed in tap water and then processed in an ultrasonic cleaner to remove soil particles. They were then repeatedly rinsed with acidified (HCl 3%) deionized water and finally rinsed with bidistilled water. Absorbing (roots) and non-absorbing plant parts (leaves, shoots, inflorescences, etc.) were generally separated in order to obtain information about the species' capacity for Sb translocation. In some species, especially those of Gramineae, the leaves and shoots were analyzed together because of the difficulty in separating them.

Plant parts were dried in an oven at 40°C until constant weight was attained and then were powdered. About 0.5 g of dry matter was digested in Pyrex flasks for 1 h using the Friederich's apparatus and a mixture of 5 ml of HNO_3 and 5 ml of H_2O_2. Digestion was performed with Aristar grade reagents.

Soil samples were air-dried and sieved through a 2-mm mesh before analysis. Their total antimony content was measured by digestion of 0.5 g with the procedure described above using a 1:1 (v:v) HCl + HNO_3 mixture. The estimated phytoavailable Sb content was obtained by both water and acid extraction. Water extraction was performed by adding 100 ml of bidistilled water to 50 g of soil and shaking it for 24 h; the acid extraction was carried out by adding 200 ml of 0.43 mol acetic acid to 5 g of soil and shaking it for 16 h – according to the procedure of Ure et al. (1993). Polyethylene bottles were used to collect and store the extracts before atomic absorption spectroscopy (AAS) analysis.

Atomic absorption measurements were made with a Perkin–Elmer 5000 AAS equipped with FIAS employing the hydride method. Working standards for AAS analysis were prepared from BDH stock solutions. Reference standards were the S-VM soil from the Institute of Radioecology and Applied Nuclear Techniques

(IRANT) of Kosice (CK) and the BCR 60 aquatic plant from the Community Bureau of Reference. The verified accuracy of analyses was within 10%.

4.3. Data analysis

Statistical tests were applied to the analytical results to investigate the following aspects: (i) the relationships between total, soluble and extractable antimony in the soil; (ii) the relationships between soil and plant Sb contents; (iii) the relationships between the Sb content in soil and the two plant coefficients: biological absorption coefficient (BAC) and plant concentration coefficient (PCC). The former is the ratio of antimony in the plant to total antimony in the soil (Edwards et al., 1998) and the latter is the ratio of antimony in the plant to mobile antimony (soluble and extractable) in the soil; (iv) the ability of the plant species to translocate antimony from the roots to the epigeal parts; (v) the effects of some plant traits on the Sb content in the plant as well as on the Sb translocation to epigeal parts. The plant traits considered are: root morphology (tap-rooted, hair-rooted and intermediate-rooted), life span (annual, biennial or perennial), habitus (herb, shrub and tree) and taxonomic rank (class: Mono- and Dicotyledones, and family).

Relationships were tested with the non-parametric Spearman rank correlation coefficient, while the translocation capacity was tested with the ratio C_E/C_R, where C_E and C_R are the antimony contents in an epigeal compartment (shoots, leaves, etc.) and in the roots, respectively. The effects of plant traits were tested with the Mann–Whitney *U*-test in each sampling site for which a sample size no smaller than four observations was possible. The null hypothesis was not rejected when an overlap of observations between small samples occurred.

5. Results

5.1. Antimony contents in the soil

Average Sb contents of soils collected from the six sampling sites are reported in Table 1. Total, soluble and extractable antimony reach very high levels in topsoils from the mineral processing tailing pond. However, there is a clear decrease in Sb content through soils collected on the mine dumps, fields and kitchen garden, in

Table 1. Average Sb contents in soils (mg/kg)

Sampling sites	Total	Soluble in H_2O	Extractable in $C_2H_4O_2$
S1 (cultivated field; $n = 4$)	145.40	0.07	0.71
S2 (cultivated field; $n = 4$)	28.60	0.05	0.18
S3 (kitchen garden; $n = 4$)	114.80	0.10	1.59
S4 (old field; $n = 4$)	27.74	0.08	0.18
S5 (mine waste dumps; $n = 7$)	1314.83	0.48	3.12
S6 (mineral processing tailing pond; $n = 5$)	16388.75	15.76	798.38

which the antimony concentration is lower by more than two or three orders of magnitude.

The ratio of the soluble or extractable Sb fraction to the total content does not follow the same trend. The soluble forms average 2.9‰ of total antimony in S4, but quite different values (from 0.4 to 1.7‰) are found elsewhere. The extractable forms are highest in S6 (4.87%), but range from 0.24 to 1.38% in the other sampling sites.

From the soil data from all sampling sites, it can be seen that the total, soluble and extractable antimony contents are highly correlated with each other (r from 0.873 to 0.926 and $P < 0.001$). However, rather different situations are found among the soils from the cultivated sites (S1, S2 and S3), the mine dumps (S5) and the mineral processing tailing pond (S6). In the first sites, only the total and extractable Sb are correlated ($r = 0.786$, $P < 0.05$); in the second, only the total and soluble Sb are correlated ($r = 0.857$, $P < 0.05$); in the last, both the soluble and extractable antimony are correlated with the total Sb ($r = 0.900$, $P < 0.05$).

5.2. Sampled plant species and their Sb content

The 59 plant species collected at the six sampling sites are listed in Appendix A along with their attributes. In total, 67 samples were analyzed (two in S1, two in S2, five in S3, 15 in S4, 34 in S5 and nine in S6; Tables 2–5), since *Achillea ageratum*, *Coleostephus myconis*, *D. hirsutum*, *I. viscosa*, *Medicago sativa*, *Plantago lanceolata* and *Silene vulgaris* were collected at two or three different sites. As shown in Table 6, the species are mostly Dicotyledones, Compositae, Leguminosae, Gramineae, tap-rooted, perennial and herbaceous, largely as a result of the sampling criteria.

The Sb contents are low or very low in both crops and vegetables of S1, S2 and S3 (Table 2), ranging from 0.02 to 2.02 mg/kg, and are lowest in economically important plant parts (i.e. leaves in lettuce, fruits in pepper and tomato, seeds in corn, wheat and sunflower, as well as leaves and shoots in alfalfa). Antimony levels are low in plants growing in the old field (Table 3), ranging from <0.02 to 5 mg/kg, except for *C.*

Table 2. Sb contents (mg/kg) in plant species of S1, S2 and S3 (cultivated sites)

Sampling sites and plant species	Leaves	Leaves and shoots	Fruits	Seeds	Flowers
S1					
Zea mays	0.20			<0.02	
Triticum aestivum				0.03	
S2					
Helianthus annuus	0.08			<0.02	
Medicago sativa		0.16			
S3					
Lactuca sativa	0.15				
Solanum melangena	0.37				
Cucurbita pepo	2.02				1.74
Capsicum annuum	1.81		0.32		
Lycopersicon esculentum	0.71		0.02		

Table 3. Sb contents (mg/kg) in plant species of S4 (old field)

Plant species	Leaves	Shoots	Leaves and shoots	Roots	Inflorescences	Fruits
Rubus ulmifolius		0.41	0.70			
Sonchus asper			0.02	0.50		
Medicago hispida			0.88	1.16		
Bromus hordeaceus			0.02		1.58	
Bromus madritensis			0.26		0.07	
Helichrisum italicum			1.15	0.57	0.22	
Phalaris coerulescens			0.89	2.59	<0.02	
Avena fatua			0.33	2.72	<0.02	
Achillea ageratum	5.05			1.45		
Brassica napus	1.88	<0.02		0.72		
Lupinus albus	0.90	1.18		1.45		0.14
Urospermum dalechampii	0.76			1.00	0.31	
Coleostephus myconis	0.63	0.27		26.01	0.51	
Rumex crispus	2.01	<0.02		0.39	0.39	
Anchusa italica	5.05	0.23		0.42	0.86	

myconis with an Sb concentration in the roots of 26 mg/kg. In particular, the contents are extremely low (<0.02–0.51 mg/kg) in almost all inflorescences and shoots sampled in the old field.

The data from plants on mine dumps reveal a rather varied situation (Table 4): most species have contents less than 10 mg/kg, some have contents between 10 and 100 mg/kg (generally higher in roots) and three species (*A. ageratum*, *P. lanceolata* and *Mentha aquatica*) have contents exceeding 100 mg/kg. In the last three species, Sb is mainly accumulated in the roots, reaching the highest values in *A. ageratum* (323 mg/kg) and *M. aquatica* (358 mg/kg).

Finally, very high antimony levels are found in the plant species growing at the old mineral processing tailing pond (Table 5). Six of them concentrate antimony in their epigeal parts and three in the roots. *A. ageratum*, *S. vulgaris* and *P. lanceolata* exceed the threshold of 1000 mg/kg and thus can be considered hyperaccumulators (Brooks et al., 1977). *A. ageratum* accumulates Sb mainly in the basal leaves and inflorescences and then in the shoots. *S. vulgaris* accumulates the element (1195 mg/kg) in old shoots (same-year shoots sampled at the end of the growing season have an Sb concentration of 229 mg/kg) and in leaves. *P. lanceolata* mainly accumulates Sb in the roots and then in leaves. High antimony levels are also found in shoots of *D. hirsutum* and in roots of *Phragmites australis* and *Aster squamatus.*

In general, only in a few plant species (*A. ageratum*, *A. squamatus*, *Bromus hordeaceus*, *Dactylis hispanica* and *Foeniculum vulgare*) are the structures involved in reproduction, i.e. inflorescences or seeds, significantly more enriched in Sb than the other parts of the plant.

Table 4. Sb contents (mg/kg) in plant species of S5 (mine dumps)

Plant species	Leaves	Shoots	Leaves and shoots	Roots	Inflorescences	Seeds
Achillea ageratum	108.28			323.40		
Plantago lanceolata	45.91			198.87		
Mentha aquatica	94.56	58.29		358.14		
Galactites tomentosa	45.04	7.24		84.14		
Silene vulgaris	56.41	30.86		15.49		
Coleostephus myconis	36.68	10.83		72.70	16.56	
Sylibum marianum	17.75	7.43		22.31		
Trifolium pratense	2.11	0.69		18.85		
Conyza bonariensis	15.62	4.40		12.49		
Dorycnium hirsutum	8.13	6.80		12.45		
Melilotus officinalis	8.26	2.67		10.00		
Lepidium campestre	5.98	1.19		9.12	1.10	
Dipsacus fullonum	0.41	0.05		4.51		
Reichardia picroides	9.20	0.22		2.73		
Ranunculus velutinus	3.19	1.12		2.50		
Sinapis arvensis	9.74	0.52		2.48		
Hedysarum coronarium	3.22	1.22		2.14		
Inula viscosa	5.96	0.91		1.10		
Hypericum perforatum	3.56	2.67		0.36		
Ulmus minor	4.02	0.10				
Sambucus ebulus	3.38	3.86		1.44		
Cistus salvifolius	5.72	1.06				
Trifolium incarnatum	2.03	1.74			1.82	
Rosa canina	3.12	1.31				
Dactylis hispanica	1.02	0.72			6.20	
Medicago sativa			8.13	12.07		
Lotus corniculatus			0.69	1.96		
Sanguisorba minor			1.17	0.55		
Spartium junceum		3.43, 0.93[a]				
Foeniculum vulgare		0.90		1.42		13.83
Quercus ilex	2.72					
Quercus cerris	0.49					1.00
Holoschoenus vulgaris		4.72				
Agrostis stolonifera			11.30			

[a] Young green shoots.

5.3. Relationships between the Sb contents in plants and soils

The mean Sb contents in the soils of the sampling sites (total, soluble and extractable fractions) are strongly correlated with the values in the most Sb-enriched parts of the plants, as well as with the values in the roots, shoots and leaves considered separately (Table 7). Conversely, the ability of the sampled species to concentrate antimony in their tissues, as expressed by the PCC values, is independent of the soluble Sb

Table 5. Sb contents (mg/kg) in plant species of S6 (mineral processing tailing pond)

Plant species	Leaves	Shoots	Roots	Inflorescences	Rhizomes
Achillea ageratum	1350.50	197.22	215.84	1090.37	
Silene vulgaris	829.42	1195.60	237.74		
Plantago lanceolata	507.72		1130.08		
Phragmites australis	26.71	13.96	797.06		12.86
Dorycnium hirsutum	197.76	460.71	62.76		
Aster squamatus		17.80	375.79	75.52	
Atriplex patula	238.47	211.09	44.60		
Inula viscosa	194.48	34.85	21.41		
Melilotus alba	122.78	32.94	81.94		

(absence of correlation) but positively affected by the extractable fraction. However, if we consider re-sampled species, such as *A. ageratum* and *P. lanceolata*, it can be seen that the highest accumulation rates in plants are reached in soils with intermediate extractable antimony contents (Table 8). Sensitivity to the extractable Sb form is greatest for roots and weakest for shoots (Table 7).

Table 6. Number of sampled plant species per taxonomic, morphological, life and habitus trait

Class	
Monocotyledones	10
Dicotyledones	49
Family	
Compositae	13
Leguminosae	11
Gramineae	9
Cruciferae	3
Fagaceae	3
Rosaceae	3
Solanaceae	3
Another fourteen families	14
Root morphology	
Tap rooted	41
Hair rooted	9
Intermediate rooted	9
Life span	
Annual	23
Perennial or biennial	36
Habitus	
Herb	51
Shrub	5
Tree	3

Table 7. Rank correlation coefficients between Sb content in soils and Sb content in the plants, plant concentration coefficients and biological absorption coefficients[a]

Plant aspects	Sb content in soil Total	Soluble	Extractable
[Sb]			
In plant parts with highest [Sb]	0.726***	0.778***	0.754***
In roots	0.693***	0.515***	0.703***
In shoots	0.738***	0.779***	0.783***
In leaves	0.734***	0.563***	0.777***
PCC			
For plant parts with highest [Sb]		ns	ns
For roots		ns	0.552***
For shoots		ns	0.309*
For leaves		ns	0.419**
BAC			
For plant parts with highest [Sb]	ns		
For roots	−0.450***		
For shoots	ns		
For leaves	0.780***		

[a] * $P < 0.05$; ** $P < 0.01$; *** $P < 0.001$; ns, not significant.

Correlations between the BACs and the total Sb in soil occur only for roots (negative) and leaves (positive). The different sign of the correlations can be attributed to the fact that the frequency of roots and leaves, respectively, decreases and increases with the increase in the total Sb in soils from S4 to S6 (Tables 3–5).

The situation changes somewhat when the plant-soil correlations are analyzed according to the fifteen plant-soil couples of Table 9. On the basis of these values, the BACs are independent of the Sb in soil (computation not shown here). The capacity of the plant to concentrate Sb (PCC) is negatively correlated with the soluble fraction in soil (r from -0.558 to -0.732, P from <0.05 to <0.01); only the antimony contents of the roots and leaves are correlated with the soil content (r from 0.701 to 0.875, P from <0.01 to <0.001). In re-sampled species (*A. ageratum, P. lanceolata, I. viscosa* and *S. vulgaris*), the BACs and PCCs of individual samples differ from two to more than ten times in each species; the most stable values of BAC are found in *A. ageratum* for roots, ranging from 0.021 to 0.035. However, as can be seen from Table 9, the re-samplings associated with the most Sb-enriched soils always have higher or very much higher Sb contents.

5.4. Antimony translocation in plant species

In the sampled species, the translocation of antimony from the roots to the epigeal parts appears not to be affected by the Sb levels in the roots and soils. The correlations

Table 8. Plant concentration coefficients in re-sampled plant species with couple plant–soil

Plant species	Sampling sites	Soil extractable Sb (mg/kg)	PCC		
			Roots	Shoots	Leaves
Achillea ageratum	(S5)	2.90	2.42		3.22
Achillea ageratum	(S5)	11.63	27.81		9.31
Achillea ageratum	(S6)	792.60	0.87	0.25	1.70
Inula viscosa	(S5)	0.91	1.21	1.00	6.55
Inula viscosa	(S6)	792.60	0.03	0.04	0.25
Plantago lanceolata	(S5)	3.00	3.28		8.41
Plantago lanceolata	(S5)	5.20	38.24		8.83
Plantago lanceolata	(S6)	139.00	8.13		3.65
Silene vulgaris	(S5)	1.78	8.70	17.34	31.69
Silene vulgaris	(S6)	139.00	1.71	8.60	5.97

between the antimony levels and the plant translocation coefficients are always non-significant (computation not shown here).

Plant species with a translocation coefficient to an epigeal part ≥2 are listed in Table 10. *Anchusa italica* is the species with the highest translocation to the leaves, which are 12 times richer in Sb than the roots: it is followed by *Hypericum perforatum* for translocation both to leaves and shoots. In the sampling site with the highest Sb contents in soil (S6), *I. viscosa* and *A. ageratum* exhibit a high translocation to leaves, *Atriplex patula* both to leaves and shoots, and *D. hirsutum* and *S. vulgaris* mostly to shoots. *A. ageratum* and *F. vulgare* (not listed) are particularly able to accumulate this element into the structures involved in reproduction: inflorescences in *A. ageratum* and

Table 9. Sb contents (mg/kg) in plant–soil couples

Plant species	Sb in plants (mg/kg)			Sb in soils (mg/kg)		
	Roots	Shoots	Leaves	Total	Soluble	Extractable
Achillea ageratum	7.01		9.34	202.80	0.17	2.90
Achillea ageratum	323.40		108.28	15112.90	0.97	11.63
Achillea ageratum	215.84	197.22	1350.50	9157.50	31.40	792.60
Aster squamatus	375.79	17.80		9157.50	31.40	792.60
Dactylis hispanica		0.72	1.02	288.18	0.11	1.72
Dorycnium hirsutum	62.76	460.71	197.76	3380.70	3.20	41.60
Inula viscosa	1.10	0.91	5.96	54.31	0.01	0.91
Inula viscosa	21.41	34.85	194.48	9157.50	31.40	792.60
Melilotus alba	81.94	32.94	122.78	9157.50	31.40	792.60
Plantago lanceolata	9.85		25.22	192.40	0.15	3.00
Plantago lanceolata	198.87		45.91	2795.53	1.83	5.20
Plantago lanceolata	1130.08		507.72	6529.70	13.00	139.00
Reichardia picroides	2.73	0.22	9.20	79.63	0.08	1.53
Silene vulgaris	15.49	30.86	56.41	5104.96	0.76	1.78
Silene vulgaris	237.74	1195.60	829.42	6529.70	13.00	139.00

seeds in *F. vulgare* (the latter with a translocation coefficient of 9.74). The ability of *Rumex crispus* to translocate antimony towards the leaves is also noteworthy. Altogether, 42% of the plant species have lower Sb contents in the epigeal parts than in the roots. Translocation coefficients are especially low in *Sonchus asper* (0.04), *C. myconis* (0.01–0.02), *Dipsacus fullonum* (0.01–0.09) and *P. australis* (0.02–0.03).

5.5. Effects of some plant traits on Sb content and translocation in plant species

The different aspects of each plant trait considered (root morphology, life span, habitus, and systematic rank) have no effect on the Sb contents in the plants nor on the translocation coefficients; the U values of the Mann–Whitney test are always non-significant. Nevertheless, the species enriched in Sb ([Sb] $\geqq$ 100 mg/kg) and best able to accumulate the element in their epigeal parts are all herbaceous, tap-rooted and Dicotyledones, with the exception of *P. australis*. Most of them are perennials and about a third of them are Compositae (Tables 4 and 5 for antimony content and Appendix A for species attributes).

6. Discussion and conclusions

6.1. Antimony in soil and plant species

The amount of the phytoavailable Sb fraction found in soils agrees with the moderate mobility of this element inferred by Kabata-Pendias and Pendias (1984). The differences among the various sampling sites in the correlation between the total Sb in soils and its soluble and extractable fractions can be explained by some general features of the sites. The proportion of organic matter in the soil, which is able to immobilize antimony by complexation (Crecelius et al., 1984), is likely higher at the cultivated sites and old field than in the thin soils of the other sites. Thus, a larger proportion of the mobilized Sb can be adsorbed and accumulated by the organic matter in the cultivated soils, which explains the positive correlation found only between extractable and total soil Sb. The thin soils on mine dumps, rising from stony and coarse materials, have a poor capacity to retain mobile antimony. On the other hand, they still contain enough mobilizable Sb to explain why only the soluble fraction is correlated with total Sb. At the mineral processing tailing pond, where there is a fraction of clayey particles able to adsorb antimony, both the soluble and extractable forms are correlated with the total amount of Sb.

In general, the concentration in soil of the mobile form of an element has a strong positive correlation with its concentration in plants. In our case, the antimony levels in plants appear to be strongly affected also by the total Sb. This is probably due to the high gradient of Sb contents, from cultivated fields (29–145 mg/kg) to mining sites (up to 16 389 mg/kg).

With the increase of Sb phytoavailability, the specificity of various plant species to the element's uptake by the roots becomes progressively more evident. At the mine dumps, with an average of about 3 mg/kg of extractable antimony in the soil, the Sb level in most plant species remains below 10 mg/kg, while in about one-third of them

Table 10. Antimony translocation coefficients from the roots to the epigeal parts in several plant species

Plant species	Sampling site	Direction of Sb translocation[a]			
		R → L	R → S	R → S&L	R → I
Anchusa italica	S4	12.02	0.55		2.05
Hypericum perforatum	S5	9.89	7.42		
Inula viscosa	S6	9.08	1.63		
Inula viscosa	S5	5.42	0.83		
Dorycnium hirsutum	S6	3.15	7.34		
Achillea ageratum	S6	6.26	0.91		5.05
Achillea ageratum	S4	3.48			
Atriplex patula	S6	5.35	4.73		
Silene vulgaris	S6	3.49	5.03		
Silene vulgaris	S5	3.64	1.99		
Rumex crispus	S4	5.15			1.00
Sinapis arvensis	S5	3.93	0.21		
Reichardia picroides	S5	3.37	0.08		
Brassica napus	S4	2.61	0.03		
Sambucus ebulus	S5	2.35	2.68		
Sanguisorba minor	S5			2.13	
Helichrisum italicum	S4			2.02	0.39

[a] R, roots; L, leaves; S, shoots; I, inflorescences.

it is within 10–100 mg/kg and exceeds this figure in only a few species. In the plant-soil couples from the mineral processing tailing pond, with a high concentration of extractable Sb in the soil (798 mg/kg, on average), the highest antimony content in the plants ranges from 123 to 1350 mg/kg. However, when Sb phytoavailability exceeds a certain level the plants appear to activate effective mechanisms that restrict absorption of the element, as showed by lower PCCs.

The antimony levels in cultivated species most probably constitute a minor risk both for human and animal consumption. The risk of Sb transfer to food chains also appears not to be very high, in view of the usually low Sb contents that we found in plants and the low bioaccumulation properties inferred for this element (Ainsworth et al., 1990b, 1991). However, such a risk must be considered more carefully for the species growing at the mineral processing tailing pond, which have Sb contents near or above 1000 mg/kg.

The PCCs indicate that the extractable antimony is the only form that affects the ability of plants to accumulate the element. In agreement with the general behavior of vascular species for element uptake (Fergusson, 1990), the roots are the most responsive part (stronger correlation with extractable antimony) and the shoots are the least sensitive one. Nevertheless, the site of preferential antimony accumulation in the plant shifts from the roots to the epigeal parts (mainly leaves) as the Sb content in the soil increases. For example, as the soil Sb level rises from the old field to the tailing pond, the BACs for the roots decrease (negative correlation with total antimony in the soil), while those for the leaves increase (positive correlation). According to Edwards et al.

(1998), the BACs might also predict the concentration of an element in a particular plant species growing on a particular soil when the soil concentration of the element is known. However, our results suggest that this use of the coefficient might be unreliable since the BACs of individual re-samplings of the same plant species clearly differ. This is found for *A. ageratum*, *I. viscosa*, *P. lanceolata* and *S. vulgaris*. Obviously, the risk of misleading information becomes lower as the number of species sampled in this kind of survey increases. This statement is strengthened by our study, since the subset of 15 plant-soil couples (certainly an individually precise datum) turned out to be much less informative (see Section 5.3).

From a systematic point of view, the differences between taxa are not statistically significant. However, among Monocotyledones (almost all Gramineae), the leaves and shoots of *Agrostis stolonifera* have appreciably higher Sb contents (11.3 mg/kg) than the other species, which agrees with the good ability of this species to tolerate other phytotoxic elements (Wu and Antonovics, 1975; Wu et al., 1975). Dicotyledones show both low and high antimony levels. However, all the Sb-rich species belong to this class of vascular plants. A very similar pattern was found by Kabata-Pendias et al. (1993) in an analysis of metal uptake and tolerance in some legumes and Monocotyledones. In our study, the Compositae and Leguminosae showed a varied distribution of the Sb content. The former are more numerous among the species with the highest Sb contents and the two families together include half of these species. A life span longer than one year appears to be important in this regard, as does a tap-root morphology of the root system. On the latter point, an abundant cortex parenchyma, favorable to the formation of an apoplastic and/or synplastic ion stock (Kupka, 1992), can be an important factor affecting the antimony content in plants. Therefore, in order to determine antimony levels in the phytosphere, it is especially important when sampling herbaceous plants to pay attention to the perennial and tap-rooted Compositae and Leguminosae.

Finally, our few data on shrub and tree species (0.41–5.72 mg/kg) are comparable to those found in other mineralized areas (Shacklette et al., 1978).

6.2. Antimony distribution in plant species and some related aspects

In general, the distribution of Sb through the various parts of the plant appears under biological control. Neither the phytoavailability of antimony in the soil nor its level in the roots are important (absence of correlation). However, plant species sampled at two or three sites are not completely indifferent to the increase of available Sb in the soil. They (*A. ageratum*, *I. viscosa*, *P. lanceolata* and *S. vulgaris*) show higher translocation coefficients of the element from the roots to the epigeal parts. Hence, according to the general rule, both biotic and abiotic factors appear to contribute to the antimony distribution within the plants.

It is well known that the distribution of the chemical elements in a plant is relevant to its physiology, to its defense against parasites and herbivores, to the element's entry into food chains, etc. Perennial herbs (*A. ageratum*, *I. viscosa*, *S. vulgaris*, *Reichardia picroides*, *H. perforatum*) or woody species (*Ulmus minor*, *Cistus salvifolius*, *Rosa canina*) concentrate antimony in epigeal parts which are lost at the end of the growing season. In this way, these plants rid themselves each year of part of the absorbed antimony. Along with other species that translocate antimony towards the epigeal parts, they are potentially the main sources for the entry of Sb into food chains via invertebrate and vertebrate herbivores. They also contribute to a more rapid and homogeneous diffusion of Sb into the organic compartment of the pedosphere by wind and water dispersion, as well to a change of the element's speciation in the environment and its uptake by plants (Ernst, 1990).

However, at the sites richest in Sb (S5 and S6), the wild species most palatable to livestock (*D. hispanica*, *Hedysarum coronarium*, *M. sativa*, *Sanguisorba minor*, *Trifolium incarnatum*, *Trifolium pratense*, *Spartium junceum*, green twigs; and *U. minor*, leaves) have low Sb contents in their edible parts. Only *P. lanceolata* has considerable levels in its leaves (508 mg/kg). Therefore, there does not seem to be a particular risk to sheep or other domestic animals currently grazing in the study area. Nevertheless, more detailed studies of this topic are certainly necessary.

6.3. Remarks about some plant species

Plant species, such as *P. lanceolata* and *S. vulgaris,* confirmed also for antimony their well-known tolerance for and uptake of a range of trace elements (Wu and Antonovics, 1976; Ernst, 1984; Verkleij and Prast, 1989; Paliouris and Hutchinson, 1991; Ernst et al., 1992; Schat and Vooijs, 1997).

Our data for these two species indicate a marked sensitivity to the Sb level in soils, despite their quite different physiological behavior; the former accumulates antimony in the roots and the latter in the shoots.

A new finding of our survey is the ability of *A. ageratum* to become an Sb-hyperaccumulator. Unfortunately, *P. lanceolata, S. vulgaris and A. ageratum* have not an abundant epigeal biomass and thus are not very suitable for use as phytoremediators. In this regard, plant species that yield an abundant biomass in stressed habitats, such as mine dumps and tailing ponds, and have a certain ability to accumulate antimony in epigeal parts (e.g. *I. viscosa*) could represent interesting material for attempts to induce a higher Sb uptake.

Among the most nitrophilous plant species, *Anchusa italica* and *Rumex crispus* are of certain interest on account of their ability to translocate antimony from the roots to the epigeal parts and their high above-ground biomass (especially in *R. crispus*). Among the plants collected in wet habitats – *M. aquatica*, *Holoschoenus vulgaris* and *P. australis* - the last species confirmed for antimony the high root accumulation and the low translocation to the leaves and shoots also found for other chemical elements (Ye et al., 1997). A similar behavior was also shown by *M. aquatica*, for which our research in progress has revealed a strong ability to accumulate arsenic.

Acknowledgements

This paper is a contribution to a research project: "Dispersion and transfer of metals to the biosphere in mining areas", supported by the MURST and University of Siena.

Appendix A. List of the sampled species with their taxonomic, morphological, life and habitus traits

Species	Class[a] and family	Root morphology	Life span[b] and habitus
Achillea ageratum	D – Compositae	Tap root	H – perennial
Agrostis stolonifera	M – Gramineae	Hair root	H – perennial
Anchusa italica	D – Boraginaceae	Tap root	H – annual
Atriplex patula	D – Chenopodiaceae	Tap root	H – annual
Avena fatua	M – Gramineae	Hair root	H – annual
Brassica napus	D – Cruciferae	Tap root	H – perennial
Bromus hordeaceus	M – Gramineae	Hair root	H – annual
Bromus madritensis	M – Gramineae	Hair root	H – annual
Capsicum annuum	D – Solanaceae	Tap root	H – annual
Cistus salvifolius	D – Cistaceae	Tap root	H – perennial
Coleostephus myconis	D – Compositae	Tap root	H – annual
Conyza bonariensis	D – Compositae	Tap root	H – annual
Cucurbita pepo	D – Cucurbitaceae	Tap root	H – annual
Dactylis hispanica	M – Gramineae	Hair root	H – perennial
Dipsacus fullonum	D – Dipsacaceae	Tap root	H – perennial
Dorycnium hirsutum	D – Leguminosae	Tap root	S – Perennial
Foeniculum vulgare	D – Umbrelliferae	Tap root	H – perennial
*Galactites tomentos*a	D – Compositae	Tap root	H – annual
Hedysarum coronarium	D – Leguminosae	Tap root	H – perennial
Helianthus annuus	D – Compositae	Intermediate root	H – annual
Helichrysum italicum	D – Compositae	Tap root	S – perennial
Holoschoenus vulgaris	M – Cyperaceae	Intermediate root	H – perennial
Hypericum perforatum	D – Guttiferae	Intermediate root	H – perennial
Inula viscosa	D – Compositae	Tap root	H – perennial
Lactuca sativa	D – Compositae	Tap root	H – annual
Lepidium campestre	D – Cruciferae	Intermediate root	H – annual
Lotus corniculatus	D – Leguminosae	Tap root	H – perennial
Lupinus albus	D – Leguminosae	Tap root	H – annual
Lycopersicon esculentum	D – Solanaceae	Tap root	H – annual
Medicago hispida	D – Leguminosae	Intermediate root	H – annual
Medicago sativa	D – Leguminosae	Tap root	H – perennial
Melilotus alba	D – Leguminosae	Tap root	H – annual
Melilotus officinalis	D – Leguminosae	Tap root	H – perennial
Mentha aquatica	D – Labiatae	Intermediate root	H – perennial

Appendix A (*continued*)

Species	Class[a] and family	Root morphology	Life span[b] and habitus
Phalaris coerulescens	M – Gramineae	Hair root	H – perennial
Phragmites australis	M – Gramineae	Hair root	H – perennial
Plantago lanceolata	D – Plantaginaceae	Tap root	H – perennial
Quercus cerris	D – Fagaceae	Tap root	T - perennial
Quercus ilex	D – Fagaceae	Tap root	T – perennial
Ranunculus velutinus	D – Ranunculaceae	Intermediate root	H – perennial
Reichardia picroides	D – Compositae	Tap root	H – perennial
Rosa canina	D – Rosaceae	Tap root	S – perennial
Rubus ulmifolius	D – Rosaceae	Tap root	S – perennial
Rumex crispus	D – Polygonaceae	Tap root	H – perennial
Sanbucus ebulus	D – Caprifoliaceae	Intermediate root	H – perennial
Sanguisorba minor	D – Rosaceae	Tap root	H – perennial
Silene vulgaris	D – Caryophyllaceae	Tap root	H – perennial
Sinapis arvensis	D – Cruciferae	Tap root	H – annual
Solanum melangena	D – Solanaceae	Tap root	H – annual
Sonchus asper	D – Compositae	Tap root	H – annual
Spartium junceum	D – Leguminosae	Tap root	S – perennial
Sylibum marianum	D – Compositae	Tap root	H – perennial
Trifolium incarnatum ssp. *incarnatum*	D – Leguminosae	Intermediate root	H – perennial
Trifolium pratense	D – Leguminosae	Tap root	H – perennial
Triticum aestivum	M – Gramineae	Hair root	H – annual
Ulmus minor	D – Fagaceae	Tap root	T – perennial
Urospermum dalechampii	D – Compositae	Tap root	H – perennial
Zea mays	M – Gramineae	Hair root	H – annual

[a] D, dicotyledones; M, monocotyledones.
[b] H, herb; S, shrub; T, tree.

References

Ainsworth, N., Cooke, J.A., Johnson, M.S., 1990a. Distribution of antimony in contaminated grassland: 1 – Vegetation and soils. Environ. Pollut. 65, 65–77.

Ainsworth, N., Cooke, J.A., Johnson, M.S., 1990b. Distribution of antimony in contaminated grassland: 2 – Small mammals and invertebrates. Environ. Pollut. 65, 79–87.

Ainsworth, N., Cooke, J.A., Johnson, M.S., 1991. Biological significance of antimony in contaminated grassland. Water Air Soil Pollut. 57–58, 193–199.

Barazzuoli, P., Guasparri, G., Salleolini, M., 1993. Il clima. In: Giusti, F. (Ed.). La Storia Naturale della Toscana Meridionale, Amilcare Pizzi Editore, Cinisello Balsamo, Milan, pp. 141–171.

Bowen, H.J.M., 1979. Environmental Chemistry of the Elements, Academic Press, London.

Brooks, R.R., 1972. Geobotany and Biogeochemistry in Mineral Exploration, Harper and Row, New York.

Brooks, R.R., Lee, J., Reeves, R.D., Jaffré, T., 1977. Detection of nickeliferous rocks by analysis of herbarium specimens of indicator plants. J. Geochem. Explor. 7, 49–77.

Coughtrey, P.J., Jackson, D., Thorne, M.C., 1983. Radionuclide Distribution and Transport in Terrestrial and Aquatic Ecosystems, Vol. 3, A.A. Balkema, Rotterdam.

Crecelius, E.A., Johnson, C.J., Hofer, G.C., 1984. Contamination of soils near a copper smelter by arsenic, antimony and lead. Water Air Soil Pollut. 3, 337–342.

Dehm, R.M., Klemm, D.D., Muller, C., Wagner, J., Weber-Diefenbach, K., 1983. Exploration for antimony in Southern Tuscany. Italy. Mineralium Deposita 18, 423–434.

Dessau, G., Duchi, G., Stea, B., 1972. Geologia e depositi minerari della zona Monti Romani-Monteti. Mem. Soc. Geol. It. 11 (2), 217–260.

Edwards, S.C., MacLeod, C.L., Lester, J.N., 1998. The bioavailability of copper and mercury to the common nettle (*Urtica dioica*) and the earthworm *Eisenia ferida* from contaminated dredge spoil. Water Air Soil Pollut. 102, 75–90.

Ernst, W.H.O., 1984. Metal tolerance in *Plantago lanceolata*. Nieuwsbrief, 15–19.

Ernst, W.H.O., 1990. Mine vegetation in Europe. In: Shaw, A.J. (Ed.). Heavy Metal Tolerance in Plants. Evolutionary Aspects, CRC Press, Boca Raton, FL, pp. 21–37.

Ernst, W.H.O., Verkleij, J.A.C., Schat, H., 1992. Metal tolerance in plants. Acta Bot. Neerland. 41, 229–248.

Fergusson, J.E., 1990. The Heavy Elements: Chemistry, Environmental Impact and Health Effects, Pergamon Press, Oxford.

Foy, C.D., Chaney, R.L., White, M.C., 1978. The physiology of metal toxicity in plants. Annu. Rev. Plant Physiol. 29, 511–566.

Istas, J., De Temmerman, L., Dupire, S., Hoenig, M., 1982. Luchtverontreiniging door een metallurgisch bedrijf: een systematisch onderzoek. Final report of the Belgian National Research and Development Programme on Environment – Air, Brussels, pp. 1–373.

Kabata-Pendias, A., Pendias, H., 1984. Trace Elements in Soils and Plants, CRC Press, Boca Roton, FL.

Kabata-Pendias, A., Piotrowska, M., Dudka, S., 1993. Trace metals in legumes and monocotyledons and their suitability for the assessment of soil contamination. In: Markert, B. (Ed.). Plant as Biomonitors, VCH Publishers, Weinheim, pp. 485–494.

Kupka, J., 1992. Physiology of mineral nutrition of plants. In: Sebánek, J. (Ed.). Plant Physiology, Elsevier, Amsterdam, pp. 159–198.

Paliouris, G., Hutchinson, T.C., 1991. Arsenic, cobalt and nickel tolerances in two populations of *Silene vulgaris* (Moench) Garcke from Ontario, Canada. New Phytol. 117, 449–459.

Pignatti, S., 1982. Flora d'Italia. Edagricole, Bologna.

Protano, G., Riccobono, F., 1997. Environmental levels of antimony, arsenic and mercury in the Tafone mining area (Southern Tuscany, Italy). Atti. Soc. Tosc. Sc. Nat. 104, 75–83.

Protano, G., Riccobono, F., Sabatini, G., 1998. La cartografia geochimica della Toscana meridionale. Criteri di realizzazione e rilevanza ambientale attraverso gli esempi di Hg, As, Sb, Pb e Cd. In: La Cartografia Geochimica Ambientale. Gli esempi della Calabria, Monti Peloritani, Sardegna e Toscana Meridionale. Mem. Descrittive della Carta Geologica d'Italia 55, pp. 119–140, (Abstract in English).

Ragaini, R.C., Ralston, H.R., Roberts, N., 1977. Environmental trace metal contamination in Kellog, Idaho, near a lead smelting complex. Environ. Sci. Technol. 11, 773–781.

Schat, H., Vooijs, R., 1997. Multiple tolerance and co-tolerance to heavy metals in *Silene vulgaris*: a co-segregation analysis. New Phytol. 136, 489–496.

Shacklette, H.T., Erdman, J.A., Harms, T.F., 1978. Trace elements in plant foodstuffs. In: Oehme, F.W. (Ed.). Toxicity of Heavy Metals in the Environments, Part 1, Marcel Dekker, New York, pp. 25–42.

Thornthwaite, C.W., 1948. An approach toward a rational classification of climate. Geog. Rev. 38, 55–94.

Ure, A.M., Quevauviller, P., Muntau, H., Griepink, B., 1993. Speciation of heavy metals in soils and sediments. An account of the improvement and harmonization of extraction techniques undertaken under the auspices of the BCR of the Commission of the European Communities. Int. J. Environ. Anal. Chem. 51, 135–151.

Verkleij, J.A.C., Prast, J.E., 1989. Cadmium tolerance and co-tolerance in *Silene vulgaris* (Moench) Garcke (S. cucubalus (L.) Wib.). New Phytol. 111, 637–645.

Wu, L., Antonovics, J., 1975. Zinc and copper uptake by *Agrostis stolonifera*, tolerant to both zinc and copper. New Phytol. 75, 231–237.

Wu, L., Antonovics, J., 1976. Experimental ecological genetics in *Plantago*. II. Lead tolerance in *Plantago lanceolata* and *Cynodon dactylon* from a roadside. Ecology 57, 205–208.

Wu, L., Bradshaw, A.D., Thurman, D.A., 1975. The potential for evolution of heavy metal tolerance in plants. III. The rapid evolution of copper tolerance in *Agrostis stolonifera*. Heredety 34, 165–187.

Ye, Z.H., Baker, A.J.M., Wong, M.H., Willis, A.J., 1997. Zinc, lead and cadmium tolerance, uptake and accumulation by the common reed, *Phragmites australis* (Cav.) Trin. ex Steudel. Ann. Bot. 80, 363–370.

Zanzari, A.R., Caboi, R., Cidu, R., Cristini, A., Fanfani, L., Zuddas, P., 1995. Hydrogeochemistry in the abandoned mining area of Tafone Graben (Italy): environmental implications. Proc. WRI-8, 905–908.

Trace elements – Their distribution and effects in the environment
B. Markert and K. Friese, editors

Chapter 12

Foliar heavy metals in trees from a tropical rain forest in Sarawak

G. Breulmann, K. Ogino, B. Markert

Abstract

The foliar concentrations of 11 heavy metals were analyzed in 20 tree species from a mixed dipterocarp forest in the Lambir Hills National Park near Miri, Sarawak, Malaysia. The concentrations within the species differed considerably. All elements except Cu occurred in lower concentrations in overstory species than in understory species. A comparison of the overall mean concentrations to the "reference plant" proposed by Markert (Markert, B., 1996. Instrumental Element and Multi-element Analysis of Plant Samples – Methods and Applications, 2nd ed. Wiley, New York) revealed substantial differences. Ni showed a remarkable positive deviation (311%) while the deviation of 3160% found for Co was striking. This was attributed to the accumulative characteristics for Co exhibited by four species, Santiria mollis, Koilodepas laevigatum, Trigonostemon sp. nov. aff. ionthocarpus and Allantospermum borneense. The nickel concentrations appeared to be somewhat elevated in the majority of species. This corresponds with the results of a study conducted in mountain rain forest in Sri Lanka (Jayasekera, R., 1993. Concentrations of selected heavy metals in different compartments of a mountain rain forest ecosystem in Sri Lanka. In: Markert, B. (Ed.), Plants as Biomonitors. VCH Publishers, Weinheim, pp. 613–622). The highest nickel concentration was exhibited by T. sp. nov. aff. ionthocarpus with a mean of 22.10 mg/kg dry weight. While the manganese concentrations were rather high in the majority of species, Dryobalanops lanceolata and Memecylon paniculatum exhibited low concentrations with 52.77 and 30.20 mg/kg dry wt., respectively. Cr, Fe, Pb, V and Zn occurred in rather low concentrations through all species. Particularly the examined species of Euphorbiaceae were found to accumulate rather high concentrations of certain heavy metals indicating accumulative characteristics. Due to their specific environment, trees from tropical forest might have a rather different chemistry from trees in the Northern Hemisphere.

Keywords: Heavy metals; Tropical rain forest; Multi-element analysis; Accumulation; Dipterocarpaceae; Plants

1. Introduction

With respect to the dynamics of organic and inorganic matter, tropical plant communities differ greatly form plant communities elsewhere due to their specific characteristics. In contrast to temperate regions there are no marked seasonal variations in tropical rain forests. The species diversity of tropical rain forest is extraordinarily high and the number of tree species per hectare can reach more than 200. The various tropical rain forests are the most structurally complex and diverse land ecosystems that have ever existed on earth (Whitmore, 1990). The mixed dipterocarp forest in Sarawak is one of the richest tropical rain forests in the world with quite a high species diversity of trees. Forests play a crucial role in the circulation of substances in and through the atmosphere and have a large impact on its chemical composition. Due to the large internal surface area of the leaves, forests act like a filter, extracting dust particles and aerosols including pollutants from the air. The modern life style in connection with an enormous increase in the use of motor cars and aircraft, the massive increase in power generation and the industrialization of agriculture have added dramatically to the number of pollutants in the environment. The construction of higher smokestacks to reduce local pollution leads to a dispersion of emissions in wider areas. Each year thousands of man-made chemicals are released without knowing their effects on mankind and environment. There has been growing concern about environmental contamination by heavy metals (Kabata-Pendias and Pendias 1984; Adriano, 1986; Lieth and Markert 1990). Many heavy metals are phytotoxic at high concentrations. Furthermore plants act as intermediate reservoirs of heavy metals from primary sources, e.g. soils, water or air, through which they move on to other organisms.

Though several studies on mineral elements in tropical forests (Golley et al., 1975; Ashton, 1982; Pampasit et al., 1998) have been conducted, there are very few studies (Jayasekera, 1993) concerning heavy metals. This paper describes the foliar concentrations of 11 heavy metals in 20 tree species from a mixed dipterocarp forest in Sarawak, Malaysia. The concentrations given can be considered natural as this ecosystem is largely undisturbed by anthropogenic activities.

2. Methods

2.1. Study site

The research site is located in the Lambir Hills National Park (4°2′N, 113°50′E) in the north west of Borneo, about 30 km south of Miri, the capital of the Fourth Division, Sarawak. The park consists of the central portion of the Lambir Hills with the highest peak of rugged sandstone escarpment of 458 m in altitude (Yamakura et al., 1995a).

The mean annual temperature and precipitation are 26.6°C and 2680 mm, respectively. In the park a research plot of 8 ha (200 × 400 m) was set up under auspices of the Canopy Biology Program in Sarawak (CBPS). The CBPS was initiated in the international co-operative project, "Long-term Ecological Research Project at Lambir Hills National Park, Sarawak" by Forest Department Sarawak; Harvard University, USA; Ehime University, Osaka City University, Kyoto University and several other Japanese universities in 1992 (Inoue and Hamid, 1995).

The topography of the Lambir Hills National Park is characterized by a hilly and undulating terrain with steep slopes including scars of landslides. Yamakura et al. (1995b) reports that 85% of the total park area represent slopes, while the other 15% consist of ridges and valleys, which dissect the park area and seem to result in finely fragmented local habitats for plants.

Wall (1962) reported six major soil groups in the Lambir Hills National Park. These were immature alluvial soils around riverbanks, red–yellow podzols (ultisols) on slopes and ridges, podzols around the hill summit of Bukit Lambir, and immature regosols on very steep slopes. Parent materials are sandstone and/or shale, which are considered tertiary period sedimentary rock (Hirai et al., 1995). The research plot is located on a clayey soil site near the headquarters of Lambir Hills National Park. The soil heterogeneity in the park is likely high, even on the microsite scale due to the undulating terrain, steepness of slopes or edaphic heterogeneity in combination with biotic factors such as soil fauna. Detailed information on soil properties in the research plot is not yet available.

The plot is covered by a well developed mixed dipterocarp forest at altitudes of 100–200 m above sea level, which is regarded as having one of the richest lowland rain forest floras of the Old World (Inoue et al., 1994). The emergent trees are predominated by dipterocarp trees and consist of the following families: Dipterocarpaceae (*Shorea* spp., *Dipterocarpus* spp., *Dryobalanops* spp.), Leguminosae (*Koompassia excelsa*, *Sindora* spp., *Dialium* spp.) and Sapotaceae (*Madhuca crassipes*). In the second layer more families occur, e.g. Anacardiaceae, Burseraceae, Euphorbiaceae, Moraceae, Myristicaceae, Myrtaceae, etc. (Nagamasu et al., 1994).

2.2. Samples and analyses

The sampling was performed during two visits (end of 1993; end of 1994) to the Lambir Hills National Park in the central part of the research plot ($\sim$1.38 ha). In this area all trees with diameter at breast height (DBH $\sim$ 130 cm) $\geq$10 cm were tagged and recorded by Ohkubo et al. (pers. commun.). The collected leaves were cleaned with tissue paper soaked with distilled water. All samples were pre-dried in a field drying-oven for 48 h. After drying all samples again for 48 h at 80°C at Ehime University, Matsuyama, Japan, they were homogenized with agate mortar and pestle to reduce the risk of contamination. The digestion and the instrumental measurement were carried out in different institutes in Japan and Germany. In all cases a wet digestion was performed, however, different digestion solutions and analytical instruments were utilized in accordance with the respective laboratory facilities.

The samples collected during the first visit were digested in a mixture of HNO_3 and $HClO_4$ in an aluminium block. The instrumental analysis was performed by inductively coupled plasma atomic emission spectroscopy (ICP/AES), Seiko SPS 1500 VR for Cu, Fe, Mn and Zn at the College of Agriculture, Kyoto University.

The samples collected during the second visit were digested in a mixture of H_2SO_4 and H_2O_2 (Fenton's reaction) using a sandbath. The instrumental measurement was performed by ICP/AES, Shimadzu ICPS-1000 IV, for Cu, Fe and Zn at the Faculty of Agriculture, Kochi University.

All samples were digested again in HNO_3 using high pressure Teflon vessels with quartz glass inserts. The instrumental measurement was performed by inductively coupled plasma mass spectrometry (ICP/MS), Perkin–Elmer Elan 5000 for approximately 80 chemical elements at International Graduate School (IHI) Zittau, Germany. Information on elements not discussed in this paper can be found in Breulmann et al. (1997, 1998a,b).

For several elements we performed two measurements for the same sample. These

Table 1. Certified values and values measured for Standard Reference Material NBS 1572, citrus leaves[a]

	Certified values	Obtained values at		
		Zittau	Kyoto	Kochi
As	3.10 ± 0.30	2.90	n.d.	n.d.
Cu	16.5 ± 1	13.86	13.61	12.62
Cr	0.8 ± 0.2	1.08	n.d.	n.d.
Fe	90 ± 10	90.68	82.00	72.08
Mn	23 ± 2	20.34	19.03	n.d.
Ni	0.6 ± 0.3	0.65	n.d.	n.d.
Pb	13.3 ± 2.4	11.21	n.d.	n.d.
Zn	29 ± 2	25.30	26.93	32.66

[a] The values are given in mg/kg dry weight; n.d., not determined.

values we averaged if the deviation between the values obtained was less than 10%. In case the deviation exceeded 10%, one value was selected on the basis of the Standard Reference Material for the respective measurement.

The accuracy of the digestion procedures was verified by use of Standard Reference Material 1572, citrus leaves, from the National Bureau of Standards, Washington. The certified values and the values obtained for the respective measurement are shown in Table 1. The elements not included in this table are not certified.

For the statistical evaluation of the data Student's *t*-test was applied.

3. Results

Table 2 shows the mean concentrations of all species investigated in comparison with the overall mean and the "reference plant" proposed by Markert (1996). Arsenic showed a rather high concentration in *Macaranga winkleri* with 0.71 mg/kg dry weight. Remarkably high concentrations of cobalt were found in four species, i.e. *Santiria mollis*, *Koilodepas laevigatum*, *Trigonostemon* sp. *nov. aff. ionthocarpus* and *Allantospermum borneense* with mean concentrations of 10.91, 16.27, 20.62 and 20.61 mg/kg dry weight, respectively. *T.* sp. *nov. aff. ionthocarpus* also exhibited high nickel concentrations with a mean of 22.10 mg/kg dry weight. In general the nickel concentrations appeared to be somewhat high in most species. *Mallotus penangensis* exhibited a rather high concentration of manganese with a mean of 1202.83 mg/kg dry weight. In contrast *Memecylon paniculatum* and *Dryobalanops lanceolata* showed a rather low manganese concentration with 30.20 and 52.77 mg/kg dry weight, respectively. The predominant family of Dipterocarpaceae showed a uniform pattern of comparatively low element concentrations despite of species differences.

Fig. 1 shows the fingerprint of the overall mean concentrations against the "reference plant". Fingerprint graphs are generated by standardizing the values of the "reference plant" and by graphically displaying positive and negative percentage deviations from the "normal" values of the "reference plant".

While As, Cu and Ti were in the range of the "reference plant", the elements Cr, Fe, Pb, V and Zn showed rather high negative deviations. *Pentace adenophora* showed a rather high zinc concentration (64.99 mg/kg dry weight) when compared to other species (Table 2), however, a rather normal value compared to the "reference plant". Co, Mn and Ni showed positive deviations with a striking 3160% for Co, which is clearly attributed to the high concentrations found in four species (Table 2).

Fig. 2 shows the comparison of juvenile leaves (light green color) collected from the tips of several twigs, with mature ones (deep green color) grown at the basis of the same twigs. The samples are from six species, i.e. *D. lanceolata*, *Drypetes xanthophylloides*, *Gluta laxiflora*, *K. laevigatum*, *Shorea biawak* and *T.* sp. *nov. aff. ionthocarpus*. The figures of the mean concentration for each element are shown in Table 3 with an asterisk indicating significance of the t values at 0.01 probability level. While As, Cr and Zn occurred in similar concentrations in juvenile and mature leaves, Cu and Ti were found in significantly higher concentrations in the juvenile leaves. Co, Fe, Mn, Ni, Pb and V occurred in higher concentrations in the mature leaves, with Fe, Mn, Pb and V showing significant differences.

The comparison on the basis of stratification is shown in Table 4. All elements except Cu occurred in lower concentrations in overstory species (Dipterocarpaceae and Melastomataceae) than in understory species (all other species). The differences were significant at the 0.01 probability level for all elements except Cr and Pb.

4. Discussion

This paper reports the concentrations of 11 selected heavy metals in 20 species from a tropical rain forest in Sarawak, Malaysia. The concentrations within the species differed considerably. In contrast to investigated understory species, the predominant family of Dipterocarpaceae occurring mainly in the overstory showed generally rather low element concentrations despite species differences and no accumulations. Another overstory species, *M. paniculatum* also showed rather low concentrations. The comparison of the mean concentrations in overstory and understory clearly showed that all elements, except Cu, occurred in higher concentrations in the understory. The differences were significant at the 0.01 probability level for all elements except Cr and Pb (Table 4). This was also observed for other chemical elements not discussed in this paper (Breulmann et al., 1998a,b) and corresponds with the results of studies conducted in other tropical forests in Thailand (Pampasit et al., 1998) and Panama (Golley et al., 1975). The reason for the higher concentrations in understory species is not yet known, however it is suggested that biota living in profusion on the understory trees are instrumental in concentrating mineral elements in lower strata of the forest, possibly due to factors like lower radiation or higher humidity.

The comparison of the overall mean concentrations with the "reference plant", which is based on concentrations of plants from Central Europe and North America revealed substantial differences. A striking positive deviation was found for Co with 3160%. This is clearly attributed to the accumulative characteristic for Co occurring in four species, i.e. *S. mollis*, *K. laevigatum*, *T.* sp. *nov. aff. ionthocarpus* and *A. borneense*. All other species showed cobalt concentrations in the range of the "refer-

Table 2. Mean concentrations of selected heavy metals in all species compared to overall mean and "reference plant" (mg/kg dry weight)

Family/species	As	Co	Cr	Cu	Fe	Mn	Ni	Pb	Ti	V	Zn
Anacardiaceae											
Gluta laxiflora	0.24	0.24	0.30	5.05	17.15	248.25	1.11	0.29	2.37	0.02	6.76
Burseraceae											
Santiria mollis	0.06	10.91	0.68	5.82	42.38	258.33	7.42	0.34	2.69	0.03	12.73
Dipterocarpaceae											
Dipterocarpus pachyphyllus	0.02	0.04	0.50	4.88	27.58	243.10	2.91	0.17	2.41	0.01	8.82
Dryobalanops aromatica	0.02	0.05	0.53	4.54	38.43	228.00	2.65	0.11	1.53	0.03	10.29
Dryobalanops lanceolata	0.03	0.17	0.49	7.09	40.25	52.77	4.95	0.64	1.81	0.03	14.76
Shorea beccariana	0.18	0.09	0.32	5.77	27.04	204.00	1.96	0.16	1.81	0.01	7.55
Shorea biawak	0.03	0.13	0.46	5.08	57.92	377.40	1.75	0.40	1.85	0.02	12.26
Shorea bullata	0.04	0.09	0.40	5.26	34.18	151.13	2.10	0.26	1.98	0.02	8.66
Shorea ferruginea	0.04	0.27	0.43	7.21	35.49	369.20	3.66	0.43	1.98	0.02	9.86
Shorea macroptera ssp. *macropterifolia*	0.08	0.22	0.41	6.46	43.14	375.57	2.77	0.25	2.05	0.02	12.10
Vatica micrantha	0.05	0.23	0.60	10.61	30.20	442.67	7.07	0.28	3.38	0.02	8.84

Euphorbiaceae											
Drypetes xanthophylloides	0.08	0.22	0.55	4.73	46.05	195.33	3.39	0.41	2.59	0.03	12.81
Koilodepas laevigatum	0.26	16.27	0.45	4.58	43.62	639.19	6.61	0.49	3.26	0.02	16.84
Macaranga winkleri	0.71	0.08	0.41	7.76	62.71	262.00	1.72	0.27	3.24	0.08	12.66
Mallotus penangensis	0.03	0.05	0.35	5.38	44.09	1202.83	1.34	1.08	2.63	0.04	9.51
Trigonostemon sp. *nov. aff. Ionthocarpus*	0.19	20.62	0.75	3.94	42.67	565.32	22.10	0.36	3.22	0.04	18.47
Ixonanthaceae											
Allantospermum borneense	0.33	20.61	0.46	3.24	32.70	338.00	4.31	0.30	1.79	0.02	14.07
Melastomataceae											
Memecylon paniculatum	0.08	0.02	0.51	6.69	18.84	30.20	0.57	0.20	0.84	0.01	4.31
Tiliaceae											
Pentace adenophora	0.05	0.56	0.76	8.06	43.37	864.67	8.72	0.53	2.05	0.04	64.99
Sloanea javanica	0.04	0.48	0.54	5.17	34.34	397.10	2.74	0.21	2.12	0.03	9.35
Overall mean	0.14	6.52	0.51	5.54	39.32	408.74	6.16	0.39	2.53	0.03	13.72
Reference plant	0.1	0.2	1.5	10	150	200	1.5	1	5	0.5	50

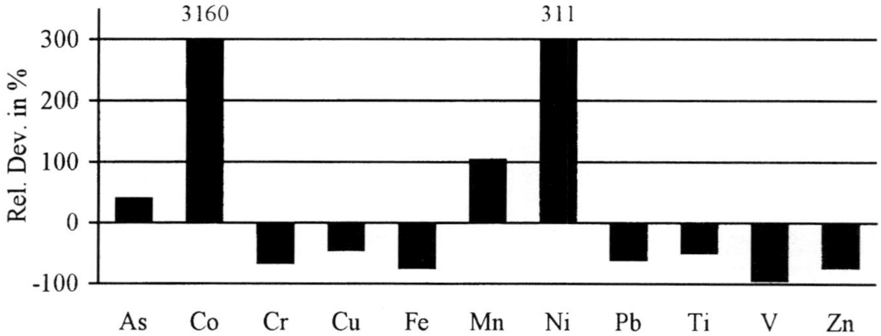

Figure 1. Fingerprint of the overall mean concentrations. The fingerprint shows relative deviation from the "reference plant" proposed by Markert (1996).

ence plant" as can be seen from Table 2. Most higher plants are reported to contain less than 1 mg/kg dry wt. of Co (Adriano, 1986; Kovacs et al., 1994). Jayasekera (1993) reported somewhat higher cobalt concentrations (1.86 mg/kg dry weight) in *Allophylus varians* (Sapindaceae), a density-dominant tree species from a mountain rain forest in Sri Lanka, however the concentrations found in the above-mentioned four species were substantially higher (Table 2). The status of cobalt as a micronutrient is not yet certain, however, it has been shown that deficiency of cobalt can lead to chlorosis and leaf-drop (Berrie et al., 1987). Co is essential for the nitrogen fixing system of *Rhizobium* bacteria living in symbiosis with Leguminosae (Epstein, 1972; Markert, 1998). Non-nodule nitrogen fixing bacteria living on and in the leaves have been reported for several non-Leguminosae species (Jones, 1982; Whitmore, 1985). No information was available concerning the above-mentioned species; however, higher Co concentrations could be beneficial for the N fixation.

Ni and Mn were also found in elevated concentrations, however, the deviations

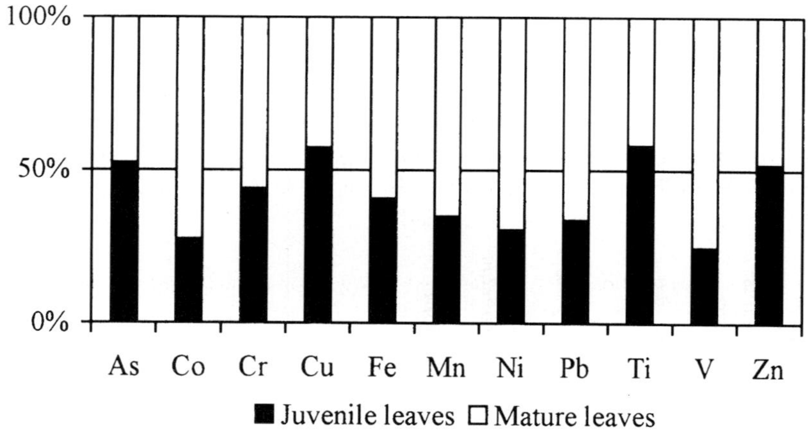

Figure 2. Comparison of the mean concentrations of selected heavy metals in juvenille and mature leaves of six species. The values are given in percentage of total concentration in both compartments.

Table 3. Mean concentrations of selected heavy metals in juvenile and mature leaves of six species (mg/kg dry weight)

	As	Co	Cr	Cu[a]	Fe[a]	Mn[a]	Ni	Pb[a]	Ti[a]	V[a]	Zn
Juvenile leaves	0.21	3.49	0.45	6.52	29.76	242.19	3.48	0.24	3.70	0.01	15.83
Mature leaves	0.19	9.14	0.57	4.85	43.39	450.17	7.89	0.47	2.67	0.03	14.65

[a] Significance of *t* values at 0.01 probability level.

were less remarkable. Jayasekera (1993) reported higher nickel concentrations in all five examined tree species from tropical forest in Sri Lanka and suggested it might be attributed to a higher uptake of nickel at lower pH values prevailing in the soil. The uptake of nickel is promoted by pH values below 6 (Streit and Stumm, 1993). The soil pH in the Lambir Hills National Park was in the acidic range from 4.2 to 5.2 (Hirai et al., 1995), while it was 4.0–6.0 in Sri Lanka. Ni probably tends to be accumulated in relatively high amounts in tropical trees. So far Ni is known to be essential for only one process in plants, the activation of the enzyme urease (Ernst, 1994).

Mature leaves showed higher concentrations of most heavy metals compared to juvenile leaves. While As, Cr and Zn occurred in only slightly higher concentrations in the mature leaves, the differences for Fe, Mn, Pb and Ti were significant at the 0.01 probability level. Cu and As were significantly higher in the juvenile leaves. The significantly higher Cu concentration in the juvenile leaves might be related to the higher growth activity of young leaves as copper containing enzymes hold key positions in proteinsynthesis, auxinmetabolism or formation of pigments (Sholnik, 1984).

Out of the 20 species examined, especially species of Euphorbiaceae showed rather high concentrations for certain heavy metals indicating accumulative characteristics, i.e. *K. laevigatum* for Co and Mn, *M. winkleri* for As, *M. penangensis* for Mn and *T.* sp. *nov. aff. ionthocarpus* for Co and Ni. Hegnauer (1966) showed that the chemical constituents of Euphorbiaceae are as heterogeneous as is their morphology.

Forest chemistry appears to be related to climate (mainly amounts of precipitation), geographical location (e.g. maritime influences) and the availability of elements in the soil (Jayasekera, 1993). Furthermore organisms are quite diverse and highly adaptive in element requirements and particularly in their tolerances. Due to the specific environment in an ecosystem without marked seasonal variations, predominantly poor soils and factors like high levels of precipitation, humidity or temperature, trees from tropical rain forests might have a rather different chemistry from trees of the Northern Hemisphere. In order to understand the chemistry of heavy metals particularly in tropical forests more information is needed.

Table 4. Mean concentrations of selected heavy metals in understory and overstory (mg/kg dry weight)

	As[a]	Co[a]	Cr	Cu[a]	Fe[a]	Mn[a]	Ni[a]	Pb	Ti[a]	V[a]	Zn[a]
Understory	0.20	11.21	0.54	4.90	41.97	539.42	8.31	0.43	2.87	0.03	16.15
Overstory	0.05	0.13	0.47	6.42	35.73	230.71	3.23	0.33	2.06	0.02	10.41

[a] Significance of *t* values at 0.01 probability level.

References

Adriano, D.C., 1986. Trace Elements in the Terrestrial Environment. Springer-Verlag, New York.

Ashton, P.S., 1982. Dipterocarpaceae. In: Van Steenis, C.G.G.J. (Ed.), Flora Malesiana, Ser. 1, Vol. 9. pp. 237–552.

Berrie, G.K., Berrie, A., Eze, J.M.O., 1987. Tropical Plant Science. Longman, Hong Kong, pp. 163–166.

Breulmann, G., Markert, B., Leffler, U.S., Weckert, V., Herpin, U., Lieth, H., Ogino, K., Ashton, P.S., La Frankie, J.V., Lee, H.S., Ninomiya, I., 1997. Multielement analysis in different compartments of Euphorbiaceae species from a tropical rain forest ecosystem in Sarawak, Malaysia. Trop. Ecol. 38 (2), 181–191.

Breulmann, G., Ogino, K., Ninomiya, I., Ashton, P.S., La Frankie, J.V., Leffler, U.S., Weckert, V., Lieth, H., Konschak, R., Markert, B., 1998a. Chemical characterisation of Dipterocarpaceae by use of chemical fingerprinting – a multielement approach at Sarawak. Malaysia. Sci. Total Environ. 215, 85–100.

Breulmann, G., Ogino, K., Markert, B., Leffler, U.S., Herpin, U., Weckert, V., Konschak, R., Kikugawa, Y., Ohkubo, T., 1998b. Comparison of chemical elements in Dipterocarpaceae and Euphorbiaceae from a tropical rain forest in Sarawak. Malaysia. Sci. Total Environ. 225, 231–240.

Epstein, E., 1972. Mineral Nutrients of Plants: Principles and Perspectives. Wiley, New York.

Ernst, W.H.O., 1994. Sampling of plants for environmental trace analysis in terrestrial, semiterrestrial and aquatic environments. In: Markert, B. (Ed.), Environmental Sampling for Trace Analysis. VCH Publishers, Weinheim, pp. 381–394.

Golley, F.B., McGinnis, J.T., Clements, R.G, Child, G., Duever, M.J., 1975. Mineral cycling in a tropical moist forest ecosystem. University of Georgia Press, Athens, GA.

Hegnauer, R., 1966. Chemotaxonomie der Pflanzen (Chemo-taxonomy of Plants), vol. 4. Birkhauser, Basel.

Hirai, H., Matsumura, H., Hirotani, H., Sakurai, K., 1995. Soils and the distribution of Dryobalanops aromatica and D. lanceolata in mixed dipterocarp forest – a case study at Lambir Hills National Park, Sarawak, Malaysia. In: Lee, H.S., Ashton, P.S., Ogino, K. (Eds.), Long Term Ecological Research of Tropical Rain Forest in Sarawak II-3. pp. 92–99.

Inoue, T., Hamid, A.A., 1995. Introduction to the canopy biology program in Sarawak. In: Lee, H.S., Ashton, P.S., Ogino, K. (Eds.), Long Term Ecological Research of Tropical Rain Forest in Sarawak II-3. pp. 120–131.

Inoue, T., Yumoto, T., Hamid, A.A., Ogino, K., Lee, H.S., 1994. Construction of a canopy observation system in the tropical rainforest in Sarawak. In: Inoue, T., Hamid, A.A. (Eds.), Long-term Study of Dipterocarp Forests in Sarawak 1. pp. 4–18.

Jayasekera, R., 1993. Concentrations of selected heavy metals in different compartments of a mountain rain forest ecosystem in Sri Lanka. In: Markert, B. (Ed.), Plants as Biomonitors. VCH Publishers, Weinheim, pp. 613–622.

Jones, K., 1982. Nitrogen fixation in the canopy of temperate forest trees: a re-examination. Ann. Bot. 34, 239–244.

Kabata-Pendias, A., Pendias, H., 1984. Trace Elements in Soils and Plants. CRC Press, Boca Raton, FL.

Kovacs, M., Penksza K., Turscanyi, G., Kaszab, L., Toth, S., Szöke, P., 1994. Comparative investigation of the distribution of chemical elements in an Aceri tatarico Quercetum plant community and in stands of cultivated plants. In: Markert, B. (Ed.), Environmental Sampling for Trace Analysis. VCH Publishers, Weinheim, pp. 435–442.

Lieth, H., Markert, B., (Eds.), 1990. Element Concentration Cadasters in Ecosystems, Methods of Assessment and Evaluation. VCH Publishers, Weinheim.

Markert, B., 1996. Instrumental Element and Multi-element Analysis of Plant Samples – Methods and Applications, 2nd revised ed. Wiley, New York.

Markert, B., 1998. Distribution and biogeochemistry of inorganic chemicals in the environment. In: Schürmann, G., Markert, B. (Eds.), Ecotoxicology. Wiley, New York, and Spektrum Akademischer Verlag, pp. 165–222.

Nagamasu, H., Momose, K., Nagamitsu, T., 1994. Flora of the canopy biology plot and surrounding areas in Lambir Hills National Park, Sarawak. In: Inoue, T., Hamid, A.A. (Eds.), Long-term Study of Dipterocarp Forests in Sarawak 1. pp. 47–56.

Pampasit, S., Khamyong, S., Breulmann, G., Ninomiya, I., Ogino, K., 1998. Mineral element accumulation in soils and trees in Tropical Hill Evergreen Forest, Northern Thailand. Tropics 8, in press.

Sholnik, M.Y.A., 1984. Trace Elements in Plants. Elsevier, Amsterdam.

Streit, B., Stumm, W., 1993. Chemical properties of metals and the process of bioaccumulation in terrestrial plants. In: Markert, B. (Ed.), Plants as Biomonitors. VCH Publishers, Weinheim, pp. 31–62.

Wall, J.R.D., 1962. Report on the Reconnaissance Soil Survey of Bekenu-Niah-Suai Area, Fourth Division, Department of Agriculture Report 35, Agriculture Department, Sarawak, Kuching.

Whitmore, T.C., 1985. Tropical Rain Forests of the Far East. Oxford University Press, Oxford, pp. 123–136.

Whitmore, T.C., 1990. An Introduction to Tropical Rain Forests. Oxford University Press, Oxford, pp. 9–36.

Yamakura T., Kanzaki, M., Itoh, A., Ohkubo, T., Ogino, K., Chai, E., Lee, H.S., Ashton, P.S., 1995a. Forest architecture of Lambir Rain Forest revealed by a large-scale research plot. I. Topography of the plot. In: Lee, H.S., Ashton, P.S., Ogino, K. (Eds.), Long Term Ecological Research of Tropical Rain Forest in Sarawak II-3. pp. 2–20.

Yamakura, T., Kanzaki, M., Itoh, A., Ohkubo, T., Ogino, K., Chai, E., Lee, H.S., Ashton, P.S., 1995b. Topography of a large-scale research plot established within a tropical rain forest at Lambir, Sarawak. Tropics 5, Japan Society of Tropical Ecology, Kagoshima, Japan, pp. 41–56.

Trace elements – Their distribution and effects in the environment
B. Markert and K. Friese, editors

Chapter 13

Trace metals in tree rings: what do they tell us?

Jürgen Hagemeyer

Abstract

Dendroanalysis (dendrochemistry) is the retrospective biomonitoring of trace metal pollution using the distributions of metals in tree rings. The value of the method is questioned, since trace metal distributions in wood are unstable. The distributions are affected by physiological processes in the life wood tissue. The transition from sapwood to heartwood affects the element distributions. Furthermore, the element patterns depend on the cation exchange capacity of the wood tissue. Currently it seems not possible to decipher the information which may be stored in the distribution patterns of trace metals in stems of trees. Experimental approaches are described that may help to extract information from trace metals in tree rings.

Keywords: cation exchange, Cd, dendroanalysis, dendrochemistry, heartwood, Pb, retrospective biomonitoring, sapwood, tree rings, wood, Zn

1. Introduction

This chapter deals with a method of retrospective biomonitoring of trace metal pollution using chemical analyses of tree rings, known as dendroanalysis. The usefulness of dendroanalytical results is disputed (Hagemeyer, 1993). Experimental tests of the method are presented and its applicability is critically examined. Furthermore, prospects for future studies of trace metals in trees are outlined.

2. Dendroanalysis – retrospective biomonitoring with tree rings

2.1. Origin and aims of dendroanalysis

The continuously increasing pollution of the environment with potentially toxic trace elements, among them heavy metals, is a problem of particular concern (Nriagu, 1992, 1996). However, one may ask, whether part of the reported increase results from the application of more and more sensitive analytical techniques in recent years. Such methods enable us to find most of the elements of the periodic table in all kinds of analyzed materials (Markert, 1992, 1993). Therefore, the question arises: is the environment today actually more polluted with heavy metals than it was in the past? To find an answer, we need methods to determine pollution levels of the past as a comparison basis for recent levels. Technical approaches, like direct recordings of emissions and of atmospheric dry or wet deposition do not help to solve the problem, since sophisticated pollution monitoring devices are only available for relatively short periods of time.

A useful method of monitoring variations in pollution levels over decades or centuries can be *biomonitoring* (Martin and Coughtrey, 1982; Wittig, 1993; Markert et al., 1997a,b). This includes the use of plants or animals as living accumulators of toxic substances. In particular, retrospective biomonitoring can provide information about environmental conditions in the past (Markert, 1994). The wood of trees seems to be a suitable material in this respect, since trees growing in temperate, seasonal climates produce conspicuous annual growth rings. These rings can be precisely dated and the year of growth can be assigned to each ring (Schweingruber, 1988, 1996). This essential discovery was first put down in writing by the famous Florentine Leonardo da Vinci, some 500 years ago (translation in Ludwig, 1970). The series of wood rings of old trees extend far back into the past. Such trees can thus provide a stratified, dated material for long-term retrospective studies. Another advantage for the purpose of biomonitoring is, that wood is easy to sample and to analyze with routine methods. Accordingly, since the beginning of the 1960s (Hagemeyer, 1993), a field of research developed, which became known as *dendroanalysis* (Gilboy et al., 1976) or *dendrochemistry* (Amato, 1988). Dendroanalysis is the retrospective biomonitoring of the environmental trace metal pollution with chemical analyses of annual growth rings of trees.

2.2. An example of a dendroanalytical study

How is a dendroanalytical study carried out? As a first step a suitable sample tree is selected at the site of the investigation. Wood samples can be obtained in different ways: with an increment borer, which is frequently used in silvicultural growth studies, a radial wood core can be extracted from the living stem. The core represents a horizontal narrow radius of the tree stem showing the growth rings. Usually the core diameter is 5 mm. An alternative way is the felling of the tree and the cutting of a stem cross-section. The wood rings are dated, i.e. to each ring the year of its formation is assigned. The material is divided into single annual rings or into groups of rings, depending on their size and the intended resolution of the element patterns. Trace element concentrations in the samples are analyzed by means of any of the numerous available techniques. The obtained concentration data are presented graphically over a time axis showing the growth years of the wood rings. The resulting curves are radial distribution patterns of elements in tree stems. Such patterns were repeatedly used as chronologies of variations in environmental pollution levels (see Section 2.3).

An example of a distribution pattern obtained in the described way is the radial distribution of Cd in the stem of an oak tree growing near Bielefeld, northern Germany (Fig. 1). Particularly high Cd concentrations occur in annual rings grown in the middle of the 1970s. Concentrations in the outer rings as well as in inner stem parts are much lower. Peak Cd concentrations appear close to the border between sapwood and heartwood. Sapwood is the outer portion of the stem which contains living parenchymatic cells: its main purpose is the conduction of xylem sap from roots to leaves and the storage of reserve material (Kaennel and Schweingruber, 1995). Heartwood is the

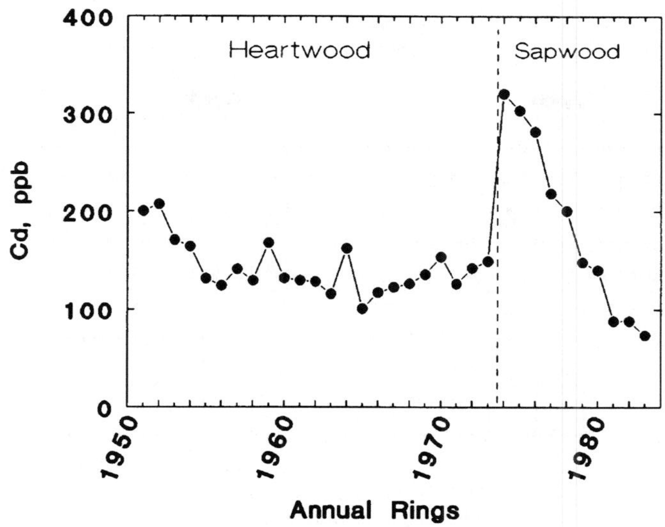

Figure 1. Radial distribution of Cd in stem wood of an oak tree (*Quercus robur*) from Bielefeld, Germany. The sample was taken at breast height with an increment borer. The broken vertical line indicates the border between sapwood and heartwood. After Hagemeyer and Breckle (1986).

dead central portion of the stem; in oak trees it is readily distinguished from sapwood by the darker color. When a mature tree grows, the number of rings in the sapwood remains more or less constant. Thus, while the stem is growing in diameter from year to year, the heartwood border also progresses outwards. What this means for the element distribution is described in Section 3.1.

With respect to dendroanalytical biomonitoring, the question is: what can such an element distribution as shown in Fig. 1 tell us? Was there an outburst of Cd pollution in Bielefeld in the 1970s. We have to ask what information may be stored in such distribution patterns. Basically the question is: Is dendroanalysis a useful method?

2.3. Dendroanalysis in the literature

In the scientific literature more than 160 publications are available, which present radial distributions of elements in stems of all kinds of tree species from sites all over the world – references in Hagemeyer (1993, 1997). More recent examples are the reports of Garbe-Schönberg et al. (1997), Jonsson et al. (1997), Oliveira et al. (1997), Schaumlöffel et al. (1998) and Watmough and Hutchinson (1999). Reviews of this field of research were compiled by Cutter and Guyette (1993), Hagemeyer (1993) and Forget and Zayed (1995). Most of the well known trace metals as well as more exotic elements have been examined already (Fig. 2). Also radial distributions of isotopic ratios of H, C, N and O in tree stems were studied repeatedly (Leuenberger et al., 1999). Such data are used to infer climatic or other past environmental conditions (Feng and Epstein, 1995; Loader and Switsur, 1995; Poulson et al., 1995; Sheu et al., 1996; Hemming et al., 1998). These investigations are usually not considered as dendroanalytical studies. Elements like H, C, N and O are part of the chemical

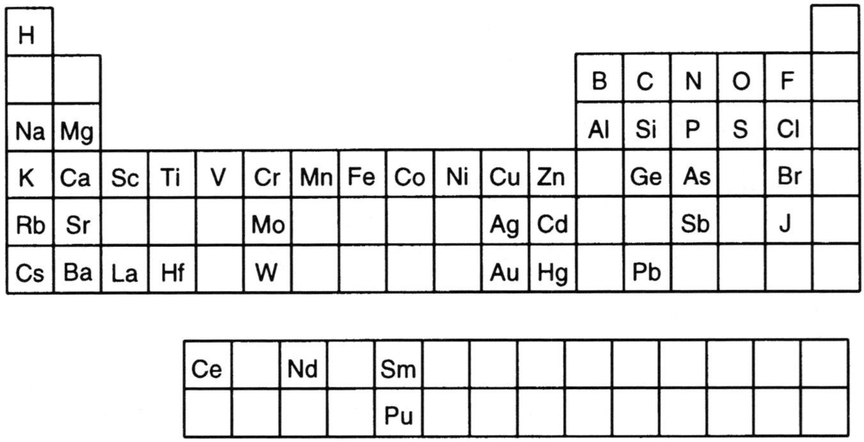

Figure 2. Periodic table showing elements for which radial distribution patterns in tree stems were published, independent of tree species and site. References in Hagemeyer (1993, 1997).

compounds forming the wood tissue, e.g. lignin, cellulose or other polysaccharides (Fengel and Wegener, 1989). They are, thus, firmly incorporated into the wood structure and problems of mobility as described below (Section 3.2) are not encountered.

Only in very few of the above mentioned studies the validity and usefulness of the dendroanalytical method of biomonitoring trace metal pollution was seriously checked.

3. A test of the dendroanalytical method

In the presented study, a test of the validity of the method was attempted. A direct, experimental test is hardly feasible. It would require, for instance, regular collection of soil samples over a period of many years. Afterwards, the soil concentrations must be compared with the concentrations of the studied elements in tree rings of the corresponding years. In this way a possible correlation between element availability in the environment and concentrations in the stem wood might be uncovered. Such a long-term investigation would require excessive time. Only a thorough exploration of the physiology and internal functioning of the trees offers a solution of the problem.

A first step is the formulation and examination of essential biological preconditions or premises of the method (Hagemeyer, 1993, 1997). For a successful application of radial element distributions for retrospective biomonitoring, the following preconditions should be valid (Hagemeyer, 1993, 1997):

1. Radial element distributions in tree stems should not change during long periods of time. The distribution patterns must be stable during the time interval between element accumulation in the growth rings and the time of sampling perhaps decades or centuries later.
2. After accumulation of the elements in the annual rings no remobilization or retranslocation should occur.

These preconditions were tested experimentally in order to find out, if and to what extend they are valid.

3.1. Test of the long-term stability of element distributions

The question was: can the radial distributions change during long periods of time? Therefore, the same oak trees were sampled and analyzed twice with an interval of 10 years. Of particular interest was the influence of the heartwood border on element concentrations. An example of the results is shown in Fig. 3. In the decade between the two samplings the heartwood border shifted about 10 growth rings outwards, as the tree grew in diameter (see Section 2.2). In both samples, the peaks of Cd concentrations were close to the heartwood border, wherever it was located at the time of sampling. The same was found in other oak trees from the same site (Hagemeyer, 1995). Thus, it is concluded that the Cd peak had moved during the studied decade. The Cd distributions in oak trees are not stable for long time periods. Obviously the distributions are affected by internal, physiological processes, like the conversion of sapwood to heartwood. Such distributions do not store information about the trends in environmental Cd pollution.

From these static pictures of Cd distributions it is not clear whether the Cd peaks were actually pushed outwards before the advancing heartwood border. An alternative possibility is, that such peaks might develop rapidly anew wherever the heartwood border is located at a certain time. This is a question of turnover rates of Cd in the wood – so far, dynamic evaluations of the distributions are not possible.

Nevertheless, the outlined results are against the dendroanalytical method. The precondition of long-term stability of the distributions is not valid. The findings

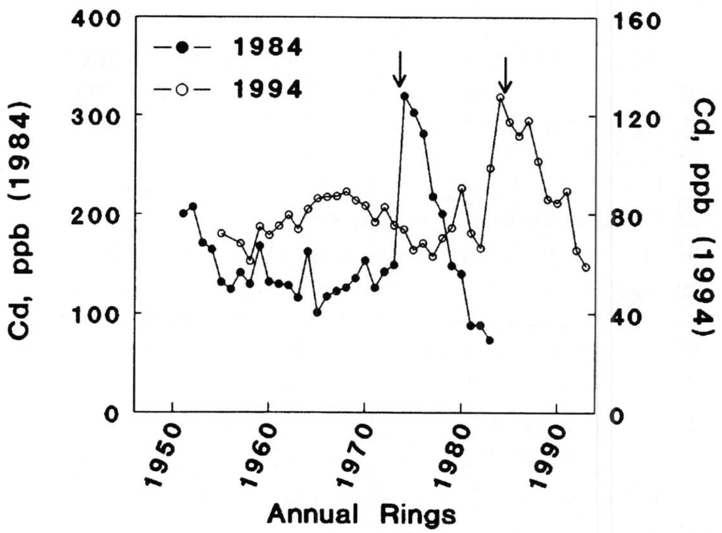

Figure 3. Radial distributions of Cd in stem wood of an oak tree (*Q. robur*) from Bielefeld, Germany. The tree was sampled twice with an interval of 10 years. Wood samples were taken at breast height with an increment borer. Arrows indicate the position of the border between sapwood and heartwood at the sampling dates. After Hagemeyer (1995).

indicate a strong mobility of metals in wood. Accordingly, this was investigated in the next part of the study.

3.2. The mobility of trace metals in wood

In order to assess the binding strength and potential mobility of cations in wood, stem wood samples of pine trees (*Pinus sylvestris*) and of oak trees (*Quercus petraea*) were extracted with various agents. Among these were H_2O and 0.01 M solutions of acetic acid, malic acid, citric acid, EDTA, $CaCl_2$ or $MgCl_2$ (Balk and Hagemeyer, 1994; Hagemeyer and Shin, 1995). Such extractants were selected to cover a wide range of binding strength to metal cations. The molecules of the organic acids contain from one (acetic acid) to four (EDTA) carboxylic groups. In the same order the stability constants of complexes formed by these molecules with metal cations increase. The salt solutions were used to study the cation exchange of trace metals against Ca^{2+} and Mg^{2+} ions. The aim was to determine the binding strength and potential mobility of Cd and Pb in the wood of two different tree species.

Wood samples were taken from stem sections cut near the stem base. The samples were obtained at different positions on the stem cross-section along radii from the stem center to the outermost rings adjacent to the bark. In this way, radial distribution patterns were determined. Extractable portions of Cd and Pb were expressed as percentages of the total element concentrations in the wood.

In pine and oak wood up to 10% of Cd were extractable with H_2O (Figs. 4 and 5). With the other agents larger portions of the elements were extracted. Their efficiency depended on the stability constants of complexes formed by the extractant molecules

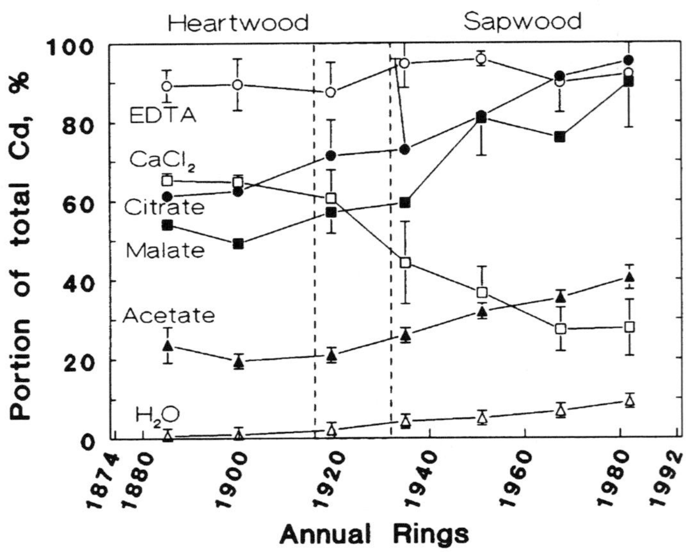

Figure 4. Radial distributions of percentages of Cd extracted from wood of pine trees (*P. sylvestris*) with various solutions. The broken vertical lines outline the transition between sapwood and heartwood. Means ± SD of three trees. After Hagemeyer and Shin (1995).

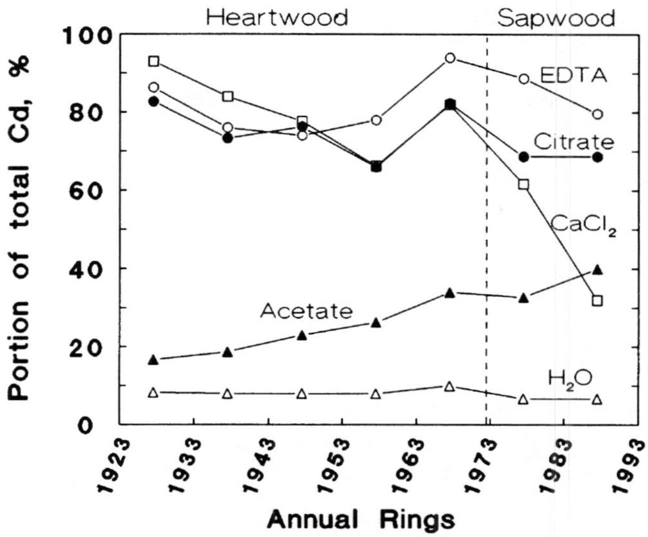

Figure 5. Radial distributions of percentages of Cd extracted from wood of oak trees (*Q. petraea*) with various solutions. The broken vertical line indicates the position of the border between sapwood and heartwood. Means of three trees. After Balk and Hagemeyer (1994).

with Cd or Pb. Differences in the extracted portions between sapwood and heartwood were obvious. The element Pb was stronger bound in the wood than Cd. From pine wood, Pb was only extractable with $CaCl_2$ and EDTA and in smaller portions than Cd (Fig. 6). In the case of oak wood, Pb was extractable with citric acid, $CaCl_2$ and EDTA, again in smaller portions than Cd (Fig. 7).

These results suggest, that different portions of Cd and Pb occur in the wood of pine and oak trees (Balk and Hagemeyer, 1994; Hagemeyer and Shin, 1995):

1. A *mobile, water-soluble portion*; it includes cations which are mobile in the xylem sap. Up to 10% of the total Cd belong to this portion, whereas no Pb was detected.
2. An *exchangeably bound portion*, which is extractable with complexing agents, like organic acids, or with salt solutions. This portion of the elements is exchangeable against H^+ or cations. It is assumed to be bound to fixed negative charges in walls of xylem vessels or tracheids. Cadmium in this portion reaches up to 90% and Pb up to 85%, depending on the radial position in the stems.
3. An *immobile portion*, which is not extractable with the applied agents. It is conceivable in various forms: precipitated as insoluble salts, like Pb-phosphate, strongly bound to constituents of the wood tissue or in some way physically not accessible for the extractants. This portion contains up to 25% of the Cd and up to 60% of the Pb. Again, the stronger binding of Pb in the wood is obvious.

From these findings we can conclude, that the studied elements are not completely immobile in wood. Remobilization and retranslocation after the deposition in parti-

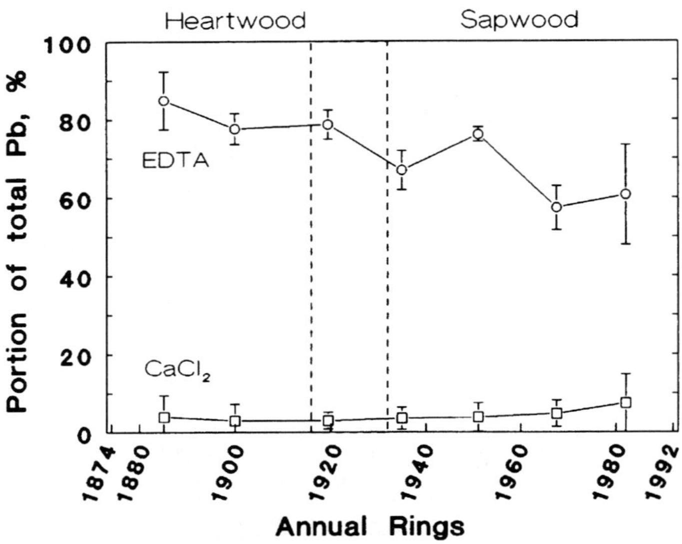

Figure 6. Radial distributions of percentages of Pb extracted from wood of pine trees (*P. sylvestris*) with various solutions. The broken vertical lines outline the transition between sapwood and heartwood. Means ± SD of three trees. After Hagemeyer and Shin (1995).

cular growth rings are possible (Hagemeyer, 1997). Seasonal variations in element distributions, which were observed in other studies (Hagemeyer et al., 1992, 1994; Hagemeyer and Schäfer, 1995) are, thus, explicable.

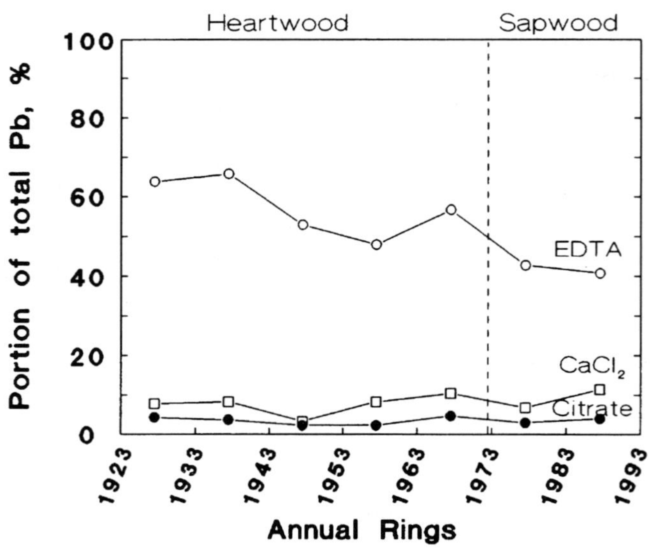

Figure 7. Radial distributions of percentages of Pb extracted from wood of oak trees (*Q. petraea*) with various solutions. The broken vertical line indicates the position of the border between sapwood and heartwood. Means of three to four trees. After Balk and Hagemeyer (1994).

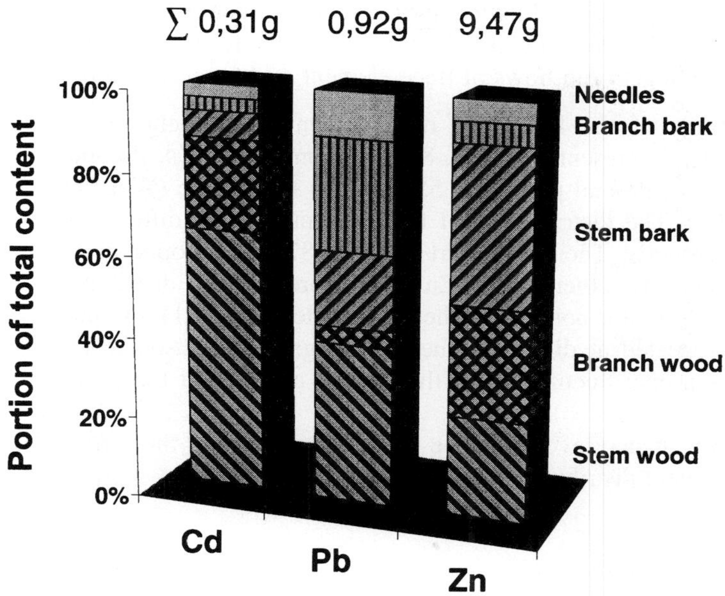

Figure 8. Distribution of Cd, Pb and Zn in above-ground parts of a 55-year-old spruce tree (*Picea abies*) from an uncontaminated site in Bielefeld, Germany. The stem height was 24.5 m and diameter at breast height was 25 cm. After Walkenhorst and Hagemeyer (1997).

4. General conclusions

What do we learn from the above outlined studies about the dendroanalytical method? Radial element distributions can change during long periods of time. Mobile portions of elements in wood indicate the possibility of remobilization and of retranslocation. Therefore, the use of element distributions for retrospective biomonitoring of trace element pollution appears questionable. At our current state of knowledge we do not understand the information which may be stored in the element distributions. We should not try to use them for biomonitoring, if we do not have additional supporting evidence from other, independent methods to verify the results (Brackhage et al., 1996).

Nevertheless, in combination with additional studies (Hagemeyer, 1997) the presented results provided much information about transport and storage of elements in trees. The most interesting result is the observation of the strong mobility of metal cations in trees.

5. Outlook

Many questions are still open, which should be addressed in further studies:

1. The quantification of the mobility of different cations in various types of wood, i.e. ring- and diffuse-porous as well as conifer wood.

2. The characterisation of binding partners/sites for cations in the xylem sap (liquid phase) as well as in walls of xylem vessels or tracheids (solid phase).
3. Turnover budgets and flows of trace element *quantities* in complete trees.

An example of the latter, the measurement of element *quantities* instead of concentrations is presented in Fig. 8. Total contents of Cd, Pb and Zn were determined in above-ground parts of a 55-year-old spruce tree (Walkenhorst and Hagemeyer, 1997). The distributions of the elements among different parts of the tree varied significantly. The largest part of Cd (>87%) was found in wood of stem and branches. On the other hand, Zn was more evenly distributed. The needles contained only minor portions of the studied elements (<11%). Such measurements should be repeated at different times of the growing season in order to elucidate flows and seasonal fluctuations in the storage capacity of the different parts of the trees.

In this way it may eventually be possible to decipher the information, which is stored in element distributions in trees.

References

Amato, I., 1988. Tapping tree-rings for the environmental tales they tell. Anal. Chem. 60, 1103–1107.

Balk, G., Hagemeyer, J., 1994. Extraction of Cd and Pb from stem wood of oak trees (*Quercus petraea* [MATT.]LIEBL.) with various solutions. Angew. Bot. 68, 191–194.

Brackhage, C., Hagemeyer, J., Breckle, S.W., Greszta, J., 1996. Radial distribution patterns of Cd and Zn in stems of Scots pine (*Pinus sylvestris* L.) trees analyzed 12 years after a contamination event. Water Air Soil Pollut. 90, 417–428.

Cutter, B.E., Guyette, R.P., 1993. Anatomical, chemical, and ecological factors affecting tree species choice in dendrochemistry studies. J. Environ. Qual. 22, 611–619.

Feng, X., Epstein, S., 1995. Carbon isotopes of trees from arid environments and implications for reconstructing atmospheric CO_2 concentration. Geochim. Cosmochim. Acta 59, 2599–2608.

Fengel, D., Wegener, G., 1989, Wood. Chemistry, Ultrastructure, Reactions. W. de Gruyter, Berlin, 613 pp.

Forget, E., Zayed, J., 1995. Tree-ring analysis for monitoring pollution by metals. In: Lewis, T.E.(Ed.), Tree Rings as Indicators of Ecosystem Health. CRC Press, Boca Raton, FL, pp. 157–176.

Garbe-Schönberg, C.D., Reimann, C., Pavlov, V.A., 1997. Laser ablation ICP-MS analyses of tree-ring profiles in pine and birch from N Norway and NW Russia – a reliable record of the pollution history of the area? Environ. Geol. 32, 9–16.

Gilboy, W.B., Tout R.E., Spyrou, N.M., 1976. Dendroanalysis: the study of trace elements in tree rings. Proceedings of the ERDA X- and Gamma-Ray Symposium, Ann Arbor, MI. pp. 164–165.

Hagemeyer, J., 1993. Monitoring trace metal pollution with tree rings: a critical reassessment. In: Markert, B. (Ed.), Plants as Biomonitors. Verlag Chemie, Weinheim, pp. 541–563.

Hagemeyer, J., 1995. Radial distributions of Cd in stems of oak trees (*Quercus robur* L.) re-analyzed after 10 years. Trees 9, 200–203.

Hagemeyer, J., 1997. Die Variabilität der Elementverteilungen im Stammholz von Bäumen. Untersuchungen zur dendroanalytischen Bioindikation. Diss. Bot. 279, 1–84.

Hagemeyer, J., Breckle, S.W., 1986. Cadmium in den Jahrringen von Eichen: Untersuchungen zur Aufstellung einer Chronologie der Immissionen. Angew. Bot. 60, 161–174.

Hagemeyer, J., Schäfer, H., 1995. Seasonal variations in concentrations and radial distribution patterns of Cd, Pb and Zn in stem wood of beech trees (*Fagus sylvatica* L.). Sci. Total Environ. 166, 77–87.

Hagemeyer, J., Shin, K., 1995. Extraction of Cd and Pb from stem wood of pine (*Pinus sylvestris* L.) trees with various solutions. Angew. Bot. 69, 55–59.

Hagemeyer, J., Lülfsmann, A., Perk, M., Breckle, S.W., 1992. Are there seasonal variations of trace element concentrations (Cd, Pb, Zn) in wood of *Fagus* trees in Germany? Vegetatio 101, 55–63.

Hagemeyer, J., Schäfer, H., Breckle, S.W., 1994. Seasonal variations of nickel concentrations in annual xylem rings of beech trees (*Fagus sylvatica* L.). Sci. Total Environ. 145, 111–118.

Hemming, D.L., Switsur, V.R., Waterhouse, J.S., Heaton, T.H.E., Carter, A.H.C., 1998. Climate variation and the stable carbon isotope composition of tree ring cellulose: an intercomparison of *Quercus robur*, *Fagus sylvatica*, and *Pinus sylvestris*. Tellus 50B, 25–33.

Jonsson, A., Eklund, M., Håkansson, K., 1997. Heavy metals of the 20th century recorded in oak tree rings. J. Environ. Qual. 26, 1638–1643.

Kaennel M., Schweingruber F.H., 1995. Multilingual Glossary of Dendrochronology. Paul Haupt Publishers, Berne, 467 pp.

Leuenberger, M., Borella, S., Stocker, T., Saurer, M., Siegwolf, R., Schweingruber, F., Matyssek, R., 1999. Stable Isotopes in Tree Rings as Climate and Stress Indicators. VDF Hochschulverlag AG, Zürich, 202 pp.

Loader, N.J., Switsur, V.R., 1995. Reconstructing past environmental change using stable isotopes in tree-rings. Bot. J. Scot. 48, 65–78.

Ludwig, H., 1970. Lionardo da Vinci. Das Buch von der Malerei. Otto Zeller Verlag, Osnabrück, pp. 402–403.

Markert, B., 1992. Presence and significance of naturally occurring chemical elements of the periodic system in the plant organism and consequences for future investigations on inorganic environmental chemistry in ecosystems. Vegetatio 103, 1–30.

Markert, B., 1993. Instrumental analysis of plants. In: Markert, B., (Ed.) Plants as Biomonitors. Verlag Chemie, Weinheim, pp. 65–103.

Markert, B., 1994. Biomonitoring – quo vadis. UWSF Z. Umweltchem. Ökotox. 6, 145–149.

Markert, B., Oehlmann, J., Roth, M., 1997a. Biomonitoring von Schwermetallen - eine kritische Bestandsaufnahme. Z. Ökol. Naturschutz 6, 1–8.

Markert, B., Oehlmann, J., Roth, M., 1997b. General aspects of heavy metal monitoring by plants and animals. In: Subramanian, K.S., Iyengar, G.V. (Eds.), Environmental Biomonitoring – Exposure Assessment and Specimen Banking. ACS Symp Ser 654, American Chemical Society, Washington, DC, pp. 18–29.

Martin, M.H., Coughtrey, P.J., 1982. Biological Monitoring of Heavy Metal Pollution. Land and Air. Applied Science Publishers, London, 475 pp.

Nriagu, J.O., 1992. Toxic metal pollution in Africa. Sci. Total Environ. 121, 1–37.

Nriagu, J.O., 1996. A history of global metal pollution. Science 272, 223–224.

Oliveira, H., Fernandes, E.A.N., Ferraz, E.S.B., 1997. Determination of trace elements in tree rings of *Pinus* by neutron activation analysis. J. Radioanal. Nucl. Chem. 217, 125–129.

Poulson, S.R., Chamberlain, C.P., Friedland, A.J., 1995. Nitrogen isotope variation of tree rings as a potential indicator of environmental change. Chem. Geol. (Isotope Geosci. Sect.) 125, 307–315.

Schaumlöffel, J.C., Filby, R.H., Moore, B.C., 1998. Ponderosa pine tree rings as historical monitors of zinc and cadmium pollution. J. Environ. Qual. 27, 851–859.

Schweingruber, F.H., 1988. Tree Rings: Basics and Applications of Dendrochronology. D. Reidel Publishing Company, Dordrecht, 276 pp.

Schweingruber, F.H., 1996. Tree Rings and Environment. Dendroecology. Paul Haupt Publishers, Berne, 609 pp.

Sheu, D.D., Kou, P., Chiu, C.H., Chen, M.J., 1996. Variability of tree-ring $\delta^{13}C$ in Taiwan fir: growth effect and response to May–October temperatures. Geochim. Cosmochim. Acta 60, 171–177.

Walkenhorst, A., Hagemeyer, J., 1997. Distribution of Cd, Pb, and Zn in above ground parts of a 55-year-old spruce tree (*Picea abies* [L.] KARST.). Angew. Bot. 71, 164–167.

Watmough, S.A., Hutchinson, T.C., 1999. Change in the dendrochemistry of sacred fir close to Mexico City over the past 100 years. Environ. Pollut. 104, 79–88.

Wittig, R., 1993. General aspects of biomonitoring heavy metals by plants. In: Markert, B. (Ed.), Plants as Biomonitors. Verlag Chemie, Weinheim, pp. 3–27.

IV
Zoological samples

Trace elements – Their distribution and effects in the environment
B. Markert and K. Friese, editors

Chapter 14

Mercury induced alterations in the energetics of hepatopancreas of two freshwater molluscs, *Pila globosa* and *Lamellidens marginalis*

B. Sivaramakrishna, K. Radhakrishnaiah

Abstract

Exposure of two freshwater molluscs, Pila globosa (snail) and Lamellidens marginalis (mussel), to an acute (4.0 mg/l) and a subacute (0.8 mg/l) concentrations of mercury resulted in distinct changes in energy patterns of hepatopancreas at days 1, 2, and 3 in acute and 5, 15, and 30 in subacute concentrations. The changes were as follows: (a) Mercury accumulated progressively over time of exposure to both acute and subacute concentrations, more in the former, but the rate of accumulation increased in acute and decreased in the subacute concentration. In either concentrations the mussels accumulated more metal than the snails due to their morphological differences. (b) The levels of sodium, potassium and calcium ions decreased progressively over time of exposure to acute concentration with a corresponding decrease in Na^+–K^+, Mg^{2+} and Ca^{2+} ATPase activities in both snails and mussels. The levels also decreased on day 5 in subacute exposure, however the animals tried to rise them on further exposure by increasing their energy out put through compensatory mechanisms; but the snails could not succeed as noticed by the drastic decrease in the levels on day 30. (c) The level of glucose progressively decreased with a corresponding decrease in the activity of glycogen phosphorylase and increase in the level of glycogen in snails under acute stress. The level of lactate, activities of lactate dehydrogenase (LDH) and succinate dehydrogenase (SDH) and the rate of oxygen consumption also gradually decreased with the accumulation of pyruvate. These results indicated uncoordinated stimulation of glycogenesis and progressive suppression of anaerobic glycolysis and oxidative metabolism. In contrast to it, glycogenolysis was noticed in mussels exposed to acute stress; and the oxidative metabolism though suppressed an elevation was observed in anaerobic glycolysis. (d) In subacute concentration though the glycogenolysis was initially stimulated in snails but led to glycogenesis on prolonged exposure. In mussels also glycogenolysis was initially elevated but reached near to normal on day 30; with a gain of energy by elevating both anaerobic glycolysis and oxidative metabolisms. (e) The results thus indicated that the mussels though accumulated high metal content they could resist it, and even could adapt to subacute toxic stress by elevating the levels of energetics. Such a type of adaptive response was not observed in snails.

Keywords: Mercury; Acute toxicity; Subacute toxicity; *Pila globosa*; *Lamellidens marginalis*; Accumulation; Ions; ATPases; Glycolysis; Oxidative metabolism

1. Introduction

Mercury, a group IIB metal, is one of the most toxic pollutants among many heavy metals. The total production of it greatly increased during this century with the increase in industrialization (Krishna Kumar et al., 1990). It has wide spread use in the synthesis of pesticides and pharmaceuticals and is consumed in the production of more than 3000 different alloys, organic and inorganic chemicals, petrochemicals, fertilizers, welding products, motor vehicles, paints and paper. It is now mainly used in the electrolytic production of chlorine and caustic soda (Shaw and Panigrahi, 1990).

Mercury is the most potential biocide to aquatic fauna even at fairly low concentrations. Its concentration in aquatic ecosystem, particularly freshwaters, is rapidly increasing world-wide; almost doubling every decade with the discharge of the effluents from the industries involved in the production of above said products. It is reported that mercury concentration is higher in freshwaters, ranging from 0.02 to 0.1 mg/l, than in marine waters and from 0.01 to 0.03 mg/1, and its effect on organisms of coastal waters is also less compared to the effect on freshwater forms (Moore and Ramamoorthy, 1984). Mercury has high affinity for sulfhydryl groups and also combines with phosphoryl, carboryl, amide and amine groups. It is able to bind to multiple high and low affinity sites of DNA molecule and produce confirmational changes in it (Lavie and Eviatar, 1987). The degree of interference of it, however, differs among different animals and also among different organs of the same animal (Sivaramakrishna et al., 1991).

The level of accumulation of any metal into an animal/organ is one of the deciding factors of its interaction with the active sites. Molluscs have the ability to accumulate heavy metals in their tissues along with the ingestion of mucus laden (Phillips, 1977) to higher levels than the concentration in their ambient waters (Nammalwar, 1983). The uptake in gastropods and bivalves is of first order kinetics wherein the rate at which a metal enters the organism is proportional to its level in ambient water (Metayer et al., 1990). A few reports on accumulation of mercury and other heavy metals in freshwater gastropods and bivalves (V-Balough et al., 1988; Sreedevi et al., 1992; Sivaramakrishna and Radhakrishnaiah, 1995) explained that their productivity has been heavily disturbed leading to an imbalance in freshwater food web. Further, as these molluscs are consumed by higher tropic levels, including human beings, the metals accumulated in them are getting transported with biomagnification (Micallef and Tyler, 1990).

Experiments concerning the effects of heavy metals on osmo- and iono-regulation are of recent interest (Suresh et al., 1995). Mercury can produce structural changes in cell membrane by affecting its permeability and its entry inside the cell alters a number of enzymatic reactions and induces an imbalance in the levels of vital inorganic cations. As freshwater animals are hyperosmotic, they gain water osmotically and tend to loose solutes by diffusion. In the regulation of hyperosmotic phenomenon, sodium, potassium and calcium ions play a significant role (Balavenkatasubbaiah et al., 1984). ATPases are one of the enzyme complexes serve for the transport of ions by altering the membrane permeable properties. Na^+–K^+ ATPase is the enzyme which also serves as an index of metabolic alertness of tissues. Mg^{2+} ATPase is involved in energy synthesis, maintenance of mitochondrial integrity and oxidative phosphorylation (Tucker and Matte, 1980). The hydrolysis of ATP and calcium transport system are associated with Ca^{2+} stimulated ATPase, which is a single polypeptide and transports two calcium ions per molecule of ATP (Moore and Dedman, 1982). Freshwater molluscs could concentrate high calcium to maintain a state of saturation and partially use it in shell formation (Hemelraad et al., 1990). Any change in the activities of ATPase complex could cause differential status of tissue electrolytes, and bring about the loss of functional efficiency of cells (Degani and Warburg, 1984; Harper, 1986).

For any cellular activity energy is a vital force which occupies a key position in the metabolic machinery. Hence, study on the level of energetics has gained importance in the physiology of an animal. Carbohydrates release the electromagnetic energy locked up in the monosaccharides, which transforms itself into the "phosphate bond" energy (ATP) before its utilization in any work. ATP synthesis in biological systems mainly depends on the phosphorylation of ADP associated with glycolysis and biological oxidation, involving the citric acid cycle and electron transport system. Any disturbances in these cycles lead to alterations in the energy budget of the cell (Lehniger, 1984).

Studies on molluscs in the past touched on the different aspects of energetics in relation to the influence of natural environmental factors (Guruprasada Rao and Nandakumar, 1982; Kulkarni and Utkar, 1983; Kalarani et al., 1984; Kulkarni and Bhatlawande, 1987). These studies provided information that glucose is the principal sugar serving the tissues as a major metabolic fuel. Glucose available to molluscs through the active absorption from the digested food stuffs, and is also formed from glycogen, amino acids and glycerol though glycogenolysis and gluconeogenesis under the conditions of aestivation (Asha Devi, 1983; Hanumantha Raju et al., 1989). Glycogen, commonly called animal starch, is the main storage polysaccharide and a great source of haemolymph glucose. Alterations in the glycogen under situations of stress have also been reported in snails and mussels, and the significant depletion in tissue glycogen is said to reflect a state of strenuous activity on the part of these animals (Kulkarni and Utkar, 1983; Kalarani et al., 1984; Hemelraad et al., 1990). The phosphorylase is the enzyme needed for the metabolic breakdown of glycogen; hence estimation of glycogen phosphorylase activity assumes considerable importance in the studies involving liver and muscle glycogen levels.

On glycolysis the breakdown of glucose results in the formation of pyruvate. However, under anaerobic conditions, it converts into lactate: the rate of conversion can be considered as an index of physiological stress on the animal. A few reports on marine bivalves though indicated that the succinate and alanine, not lactate, are the major end products of anaerobic metabolism (Bayne, 1976), the available studies on freshwater gastropods and bivalves reported that lactate is the major end product (Asha Devi, 1983; Krishna Mohan et al., 1987; Hemelraad et al., 1990). Lactate dehydrogenase (LDH) is the key enzyme responsible for the conversion of pyruvate to lactate, which is located at the vital point between glycolysis and tricarboxylic acid cycle (TCA) cycle. Because of its strategic location and its relation to Cori cycle, it is likely that any fluctuation in the cellular environment alters the activity of this enzyme. Oxidation of pyruvate through the TCA cycle produces more energy; if the tissue consumes enough oxygen it enters the TCA cycle after decarboxylation to acetyl Co-A. Succinate dehydrogenase (SDH), a flavin-linked enzyme of the TCA cycle, catalyses the reversible oxidation of succinate to fumarate and serves as a link between electron transport and oxidative phosphorylation (Singer et al., 1973). Since the activity of SDH in mitochondria is far greater than the other enzymes of TCA cycle (Madanmohan Das and Venkatachari, 1984), an insight into the alterations of this enzyme activity may be taken to reflect the function of the TCA cycle in different organs of the freshwater molluscs. Presence of pyruvate, lactate, LDH and SDH in a number of tissues of *Pila globosa* and *Lamellidens marginalis* and their role on exposure of these animals to pesticides are well reported (Rajeshwara Rao et al., 1983; Kalarani et al., 1984).

The literature available on the effects of heavy metals on different freshwater animals indicated noticeable changes in the levels of ions, ATPases and some substrate and enzymes of glycolytic and oxidative metabolisms (Sharma, 1984; Radhakrishnaiah and Busappa, 1986; Hilmy et al., 1987; Hemelraad et al., 1990; Radhakrishnaiah et al., 1992; Suresh et al., 1993, 1995). Most of these reports, however, are confined to freshwater fishes; the studies of this nature on freshwater molluscs are rather limited (Guruprasada Rao and Nandakumar, 1982; Ravinder Reddy et al., 1986; Hemelraad et al., 1990; Sivaramakrishna et al., 1991). It shows that though molluscs are used as biomarkers in metal pollution, no detailed studies have been made in them comparing the level of accumulation of any metal with the energetics of them exposed to acute and subacute concentrations, particularly to highly toxic metals like mercury. Such studies are required in order to understand whether these animals possess inherent abilities to resist the imposed toxic stress or not. Hence, to bridge this gap, present investigation is taken up by estimating the levels of sodium, potassium and calcium ions, and glucose, glycogen, pyruvate, and lactate and the activities of Na^{+}–K^{+}, Mg^{2+} and Ca^{2+} ATPases, glycogen phosphorylase, SDH and LDH, in addition to the measurement of the rate of oxygen consumption, in the hepatopancreas, an important organ for all metabolic activities, of *P. globosa* and *L. marginalis* on exposure to an acute and a subacute concentration of mercury. The changes in the said parameters are correlated with the amount of accumulation of mercury to evaluate their levels of energetics to toxic stress. As the period of exposure has considerable influence on the metabolism of an animal (Sivaramakrishna et al., 1992), the present study is carried out on days 1, 2, and 3 in acute and 5, 15, and 30 in subacute concentrations to observe the short-term and long-term response patterns.

2. Materials and methods

Freshwater snails and mussels weighing 15 ± 1 g were collected from local ponds and wells and were maintained in the laboratory in $3' \times 2' \times 1'$ glass tanks, ten in each. Water from local well was used for their maintenance. It had a pH of 7.6 ± 0.2, a total hardness of 100 ± 5 mg/l $CaCO_3$, a temperature of 280 ± 10°C and a dissolved oxygen content of 5.79 ± 0.4 mg/l. Water was changed once a day. Snails and mussels were fed with hydrilla and freshwater plankton, respectively, ad libitum. The animals were allowed to adapt themselves to laboratory conditions for 10 days. Later, groups of 30 each were exposed to different concentrations of mercury, prepared from pure mercuric chloride, ranging from 2 to 8 mg/l, and the mortality rate was observed at 120 h of exposure. The LC_{50}s for 120 h were derived from percent and probit mortality versus log concentration curves (Finney, 1971) and were subsequently verified by Dragstedt and Behren's method (Carpenter, 1975). Animals in freshwater containing 1 ml of HCl per 10 l, to nullify the chloride effect, served as controls. The $LC_{50}s/120$ h obtained were 4.5 mg/l for snails and 3.5 mg/l for mussels; which indicated species-specificity to mercury toxicity. Hence, the present study was made in the snails and mussels exposed to one acute (4.0 mg/l), an average of the two $LC_{50}s$, and a subacute (0.8 mg/l), one fifth of the acute, concentrations in order to assess the species specific differential responses to toxic stress. The studies were made at the end of each exposure period 1, 2, and 3 days in acute and 5, 15, and 30 days in subacute concentrations by isolating the hepatopancreas of snails and mussels from the animals survived 9, 7, and 5 at the respective periods in acute and 10, 9, and 8 in subacute concentrations. Similarly, the hepatopancreas from snails and mussels maintained in normal freshwater were served as control. In the hepatopancreas of both control and metal treated animals the following parameters were measured/estimated using standard experimental procedures.

2.1. Measurement of accumulation and its rate

The organ was weighed, kept in a hot air oven at 850°C for 24 h and digested with concentrated perchloric acid, nitric acid and sulphuric acid mixture in 1:2:1 ratio (Humpries, 1956). The acid digest was neutralized with 20% sodium hydroxide, diluted with distilled water and analysed for mercury by atomic absorption spectrophotometer (Varian Techtron Model 1000) using a mercury cathode lamp at the wavelength of 253.7 nm. The amount of mercury accumulated is expressed as mgHg/g wet weight of the organ. The values obtained were converted as percentages over the respective controls. The rate of accumulation was calculated by using the formula $\Delta c/t$, where Δc is the amount of change in concentration over control and t is the time of exposure.

2.2. Estimation of sodium, potassium and calcium ions and Na^+–K^+, Mg^{2+}, Ca^{2+} ATPases

The organ isolated was subjected to wet digestion in 1:1 concentrated perchloric acid and nitric acid and evaporated to 100°C (Dall, 1967). The residues were dissolved in distilled water and the levels of sodium, potassium and calcium were determined with a flame photometer (Elico Pvt. Ltd., Model CL-22A). The ion levels are expressed as mg/g wet weight of the organ.

The activities of Na^+–K^+, Mg^{2+} and Ca^{2+} ATPases were estimated in the organ by using the method of Watson and Beamish (1981). A 1% tissue homogenate was prepared in 0.25 M ice-cold sucrose solution and centrifuged for 10 min. The supernatent was taken and three sets of incubation mixtures were prepared. In a total volume of 2 ml for all the three sets, 100 mM disodium ATP, 20 mM KCl, 100 mM NaCl, 3 mM $MgCl_2$, and 0.3 ml enzyme extract were added. Three millimolar $MgCl_2$ (for Na^+–K^+ ATPase) to the first set, 1 Mm ouabain (for Mg^{2+} ATPase) to the second set and 5 mM $CaCl_2$ (for Ca^{2+} ATPase) to the third set were added and incubated at 37°C for 15 min. The reaction was stopped by adding 2 ml of cold 10% trichloracetic acid. The inorganic phosphates liberated were estimated by the method of Fiske and Subba Row (1925). For this, to an aliquot of the above reaction mixture 4.5 ml of 0.44% ammonium molybdate and 0.2 ml of 1-amino-2-naphthol-4-sulphonoic acid were added. The mixture was kept in boiling water for 10 min, cooled under tap water, and made to 10 ml with distilled water. The colour developed was read at 660 nm. The first set gives the total ATPase activities of Na^+–K^+ and Mg^{2+}, the second set gives only the Mg^{2+} ATPase activity and the third set gives Ca^{2+} ATPase. These enzyme activities are expressed as μM Pi liberated/mg protein per h. Protein content was estimated by using the folinphenol reagent method (Lowry et al., 1951).

2.3. Measurement of tissue oxygen consumption

The rate of oxygen uptake was measured in a Gilson 5/6 oxygraph. An oxygen electrode was prepared by adding a drop of potassium chloride to the probe and was inserted into the water jacketed cell and calibrated with air saturated water taking oxygen concentration at 25°C as 0.265 mm/1.0 ml of air saturated water. The water was removed from reaction vessel and 2.0 ml of 0.05 M phosphate buffer (pH 5.5) was added. After 5 min, 50 mg of tissue slice was added to reaction vessel and the rate of oxygen depletion from the reaction medium was followed. The rate of oxygen consumption is expressed as μM O_2 consumed/g^{-1} wet weight per min^{-5}.

2.4. Biochemical assays of carbohydrate metabolism

The glucose content in the organ was determined by colorimetric method (Nelson and Somogyi, 1952). 4.5 ml of 10% trichloracetic acid was added to 0.5 ml of 5% tissue homogenate prepared in pure distilled water and the mixture was centrifused for 10 min. One milliliter of alkaline copper reagent was added to 1 ml of supernatent, shaken and heated in boiling waterbath for 20 min. On cooling, 1 ml of arsenomolybdate colour reagent was added. The colour developed was read at 540 nm. Glucose standards were run simultaneously. The glucose content is expressed as mg/g wet weight.

The glycogen content was estimated by using the anthrone reagent method (Carroll et al., 1956). The organ was digested with 3 ml of hot 30% potassium hydroxide (Hassid and Abraham, 1957). After cooling 3.75 ml of absolute ethanol was added to the digestate. The mixture was kept overnight in refrigerator. Later it was centrifuged for 15 min. Ten milliliters of warm distilled water was added to the residue to dissolve the precipitated glycogen. 1.8 ml of distilled water and 0.5 ml of 2% anthrone reagent (dissolved in 72% concentrated sulphuric acid) were added to 0.2 ml of this, and kept in boiling water for 10 min, cooled under tap water, made to 10 ml with distilled water and the optical density of the colour developed was measured at 620 nm. The glycogen content is expressed as mg/g wet weight.

The activity of glycogen phosphorylase (1-4-glucan: orthophosphate glucosyl transferase, EC 2.4.1.1) was estimated by using the method of Sutherland (1955). Tissue homogenate (5%) prepared in 0.17 sodium fluoride (pH 6.5) was centrifuged at 1500 rev./min for 15 min, thus extracting the enzyme into the supernatent. It was diluted four times with the cold sodium fluoride solution. The incubation mixture consisted of 0.2 ml of 2% glycogen, 0.4 ml diluted enzyme and 0.2 ml of 0.016 M glucose-1-phosphate and 0.004 M adenosine-5-monophosphate (1:1 ratio). The mixture was incubated at 37°C for 15 min and the reaction was stopped by adding 5 ml of 10% trichloracetic acid. The inorganic phosphates liberated were estimated by using the method of Fiske and Subba Row (1925). the optical density of the colour developed was measured at 660 nm. The activity is expressed as μM Pi liberated/mg protein per h.

The pyruvate content in the organ was estimated by using a method of Friedman and Hangen (1942). A 5% tissue homogenate was prepared in 10% trichloracetic acid and centrifuged (3500 rev./min, 15 min). One milliliter of 0.001 M 2,4-dinitro-phenyl-hydrazine and 3 ml of 4 N sodium hydroxide were added to 1 ml of supernatent, and the colour developed was read at 540 nm. The pyruvate content is expressed as mg protein/g wet weight.

The lactate content in the organ was estimated by using the modified method of Huckabee (1961). A 5% tissue homogenate was prepared in 10% trichloracertic acid and centrifuged (3000 rev./min, 15 min) One milliliter of 20% copper sulphate was added to 1 ml of supernatent, and made up to 10 ml with distilled water. Then 1 g of powdered calcium hydroxide was added, kept at room temperature for an hour with intermittent shaking and centrifuged (3000 rev./min, 10 min). 0.5 ml of 4% copper sulphate and 6.0 ml of concentrated sulphuric acid were added to 1 ml of supernatent, kept in a boiling water for 6.5 min, cooled and 0.1 ml of 1.5% *p*-hydrophenyl (prepared in 0.5% of sodium hydroxide) was added. After 30 min the precipitate formed was placed in boiling water for 90 s, cooled and the colour was read at 560 nm. The lactate content is expressed as mg lactate/g wet weight.

The succinate dehydrogenase (succinate: acceptor Oxido-reductase, EC 1.3.99.1) activity estimated by using the method of Nachlas et al. (1960). Tissue homogenate (5%) was prepared in 0.25 M sucrose was centrifuged at 3000 rev./min for 10 min, and the supernatent was taken as source of enzyme. The incubation mixture consisted of 2 ml of 0.4 M phosphate buffer (pH 7.7), 0.2 ml of 0.2 M sodium succinate, 1 ml of 0.004 M 2-(*p*-indophenyl)-3-*p*-nitrophenyl-5-phenyl tetrazoliumchloride (INT), 0.1 ml of 0.005 M phenazine methosulphate and 0.5 ml of enzyme preparation. The mixture was incubated at 37°C for 30 min, and the reaction was stopped by adding 6 ml of glacial acetic acid. The formozan formed was extracted into 6 ml of toluene overnight at 0°C and the colour was read at 495 nm. The enzyme activity is expressed as μM formozan/mg protein per h.

The lactate dehydragenase (L-lactate NAD–Oxido-reductase, EC 1.1.2.7) activity was estimated by using modified method of Govindappa and Swami (1965). Tissue homogenate (5%) prepared in 0.25 M sucrose was centrifused at 2500 rev./min for 15 min, and the supernatent was taken as the source of enzyme. The incubation mixture consisted of 1 ml of 0.0001 M nicotinamide adenine dinucleotide, 1 ml of 0.004 M INT and 0.5 ml of enzyme preparation. The mixture was incubated at 370C for 30 min, and the reaction was stoped by adding 6 ml of glacial acetic acid. The formozan formed was extracted into 6 ml of toluene overnight at 0°C and the colour was read at 495 nm. The enzyme activity is expressed as μM formozan/mg protein per h.

2.5. Statistical analysis

For each estimation/measurement, the mean of the values obtained from six control and/or experimental animals was taken into consideration. Standard deviations were calculated and Wilcoxon's *t*-test was employed to derive the level of significance between the means of controls and experimentals. The level at 5% or below was considered significant.

3. Results

The data presented in Table 1 show a significant increase in the amount of accumulation of mercury in the hepatopancreas of snails and mussels on all the days of exposure to the acute and subacute concentrations of mercury: relatively, with a higher accumulation in former concentration than in the latter. In acute concentration the amount of metal accumulation and the rate of it increased over time of exposure in both the animal groups in the order: day 1 < 2 < 3. In subacute concentration though the amount of accumulation of mercury increased over time of exposure in both snails and mussels (day 5 < 15 < 30), the rate of accumulation progressively decreased over time of exposure (day 5 > 15 > 30). On any corresponding day of exposure, either in acute or in subacute concentrations, the mussels accumulated significantly more concentration of mercury than the snails, with a higher rate of it.

Table 1. Accumulation of mercury (mg/g wet weight) in hepatopancreas of snails and mussels on exposure to the acute and subacute concentrations of mercury[a,b]

Animal	Control	Period of exposure (days)					
		Acute			Subacute		
		1	2	3	5	15	30
Snails	0.182	0.541	1.725	3.452	0.554	0.446	0.613
SD	0.025	0.062	0.190	0.390	0.042	0.059	0.075
%[c]		200.4	858.3	1817.7	94.4	145.1	238.9
Rate		0.360	0.772	1.09	0.034	0.018	0.014
Mussels	0.164	0.754	2.322	4.390	0.474	0.625	0.958
SD	0.019	0.084	0.312	0.440	0.032	0.048	0.089
%[c]		370.5	1355.2	2645.8	189.0	281.1	484.2
Rate		0.592	1.079	1.409	0.062	0.031	0.026

[a] Each value is based on six measurements.
[b] The differences between control and experimentals are statistically significant ($P < 0.05$).
[c] Percentage increase over control.

The data summarized in the Tables 2 and 3 revealed a significant decrease in the levels of sodium, potassium and calcium ions in the hepatopancreas of snails and mussels on all the days of exposure to acute concentration of mercury. Corresponding to it, the activities of Na^{+}–K^{+}, Mg^{2+} and Ca^{2+} ATPases also significantly decreased in the hepatopancreas of both snails and mussels. The percentage of decrease in all these parameters progressed over time of exposure in the order day $1 < 2 < 3$; but the magnitude of it on any day of exposure was significantly greater in the hepatopancreas of snails than the mussels. In subacute concentration also the ion levels and activities of ATPases decreased in snails and mussels, with an exception of a significant increase on day 30 in mussels. The degree of decrease, however, was significantly less than in acute concentration. Further, the percentage of decrease was in the order: day $5 > 15 <$

Table 2. Levels of sodium (Na^{+}), potassium (K^{+}) and calcium (Ca^{2+}) ions (mg/g wet weight) in the hepatopancreas of snails and mussels on exposure to the acute and subacute concentrations of mercury[a,b]

Parameter/animal	Control	Period of exposure (days)					
		Acute			Subacute		
		1	2	3	5	15	30
Na^{+}							
Snails	5.46	2.54	1.89	1.63	3.95	4.16	2.71
SD	0.62	0.35	0.28	0.21	0.44	0.45	0.38
%[c]		−53.4	−65.3	−70.2	−27.6	−23.9	−50.4
Mussels	5.35	3.18	2.51	1.95	3.61	4.58	5.91[d]
SD	0.62	0.48	0.38	0.21	0.39	0.52	0.58
%[c]		−40.5	−53.2	−63.2	−35.4	−14.5	+10.3
K^{+}							
Snails	5.65	2.71	2.27	1.72	3.86	4.23	3.06
SD	0.56	0.31	0.29	0.18	0.31	0.45	0.41
%[c]		−52.2	−59.9	−69.9	−31.8	−25.2	−45.8
Mussels	5.43	3.23	2.79	2.21	3.29	4.41	6.09[d]
SD	0.64	0.42	0.32	0.31	0.38	0.46	0.61
%[c]		−40.5	−48.7	−59.4	−39.4	−18.9	+12.2
Ca^{2+}							
Snails	8.18	4.08	3.29	2.68	5.49	6.08	4.28
SD	0.92	0.45	0.36	0.30	0.59	0.62	0.43
%[c]		−50.1	−59.9	−67.3	−32.8	−25.7	−47.7
Mussels	7.95	5.09	4.35	3.82	4.81	6.35	8.99
SD	0.82	0.52	0.44	0.39	0.49	0.66	0.92
%[c]		−35.9	−45.3	−51.9	−39.5	−20.1	+13.1

[a] Each value is based on six estimations.
[b] The differences between control and experimentals are statistically significant ($P < 0.05$).
[c] Percent change over control.
[d] Not significantly different from control ($P > 0.05$).

Table 3. Activities of Na^+–K^+, Mg^{2+} and Ca^{2+} ATPases (μM Pi formed/mg protein per h) in the hepatopancreas of snails and mussels on exposure to the acute and subacute concentrations of mercury[a,b]

Parameter/animal	Control	Period of exposure (days)					
		Acute			Subacute		
		1	2	3	5	15	30
Na^+–K^+ATPase							
Snails	18.14	7.67	4.68	2.16	12.17	13.48	10.48
SD	2.15	0.78	0.49	0.25	1.33	1.48	1.15
%[c]		−57.7	−74.2	−88.1	−32.9	−25.7	−42.2
Mussels	18.08	10.09	7.11	4.48	10.99	14.39	20.03[d]
SD	2.15	1.11	0.85	0.49	1.21	1.58	2.21
%[c]		−44.2	−60.8	−75.2	−39.2	−20.4	+10.8
Mg^{2+} ATPase							
Snails	8.19	3.27	1.86	1.23	5.31	6.09	4.30
SD	0.92	0.35	0.22	0.16	0.58	0.63	0.45
%[c]		−60.1	−77.3	−85.0	−35.2	−25.6	−47.5
Mussels	8.18	4.24	2.81	1.87	4.52	6.52	9.17
SD	0.91	0.48	0.31	0.21	0.49	0.68	0.94
%[c]		−48.2	−65.7	−77.2	−44.8	−20.3	+12.1
Ca^{2+} ATPase							
Snails	7.14	2.96	1.77	1.34	4.94	5.44	3.68
SD	0.75	0.31	0.21	0.16	0.59	0.62	0.39
%[c]		−58.5	−75.2	−81.3	−30.8	−23.9	−48.5
Mussels	6.88	3.75	2.59	2.01	4.18	5.79	7.69[d]
SD	0.62	0.39	0.28	0.21	0.42	0.59	0.78
%[c]		−45.5	−62.3	−70.8	−39.2	−15.8	+11.8

[a] Each value is based on six estimations.
[b] The differences between controls and experimentals are statistically significant ($P < 0.05$).
[c] Percentage change over control.
[d] Not significantly different from control ($P < 0.05$).

30 in snails; whereas, it was in the order: day 5 > 15 with an increase on day 30 in mussels.

From the data presented in Table 4 it is observed that the level of glucose and the activity of glycogen phosphorylase decreased significantly with an increase in the level of glycogen in the hepatopancreas of snails on exposure to acute concentration of mercury. Either the decrease or increase was progressive over time of exposure (day 1 < 2 < 3). Interestingly, an opposite trend was observed in the hepatopancreas of mussels with a significant increase in the level of glucose and the activity of glycogen phosphorylase and a decrease in the level of glycogen. The changes were progressive over time of exposure (day 1 < 2 < 3). In subacute concentration the level of glucose and the activity of glycogen phosphorylase increased on day 5 with a decrease in the

Table 4. Levels of glucose and glycogen (mg/g wet wt) and the activity of glycogen phosphorylase (μM Pi liberated/mg protein per h) in the hepatopancreas of snails and mussels on exposure to the acute and subacute concentrations of mercury[a,b]

Parameter/animal	Control	Period of exposure in days					
		Acute			Subacute		
		1	2	3	5	15	30
Glucose							
Snails	11.06	6.56	4.88	3.52	15.33	11.91[c]	9.25
SD	1.28	0.85	0.67	0.45	1.68	1.53	1.02
%[d]		−40.6	−55.8	−68.1	+38.6	+7.7	−16.4
Mussels	14.28	20.30	22.07[c]	23.77	18.62	17.26	15.32[c]
SD	1.67	2.35	2.48	2.56	1.94	1.63	1.67
%[d]		+42.2	+54.6	+66.5	+30.4	+20.9	+7.3
Glycogen							
Snails	71.67	99.69	109.3	118.3	60.35	92.22	99.41
SD	6.41	12.4	12.9	12.5	7.24	11.1	11.9
%[d]		+39.1	+52.6	+65.6	−15.8	+28.6	+38.7
Mussels	54.28	34.68	28.01	19.16	38.86	45.33	50.59[c]
SD	7.95	4.28	4.05	2.65	4.66	5.44	6.07
%[d]		−36.1	−48.4	−64.7	−28.4	−16.5	−6.8
Glycogen phosphorylase							
Snails	2.78	1.941	1.596	1.277	3.36	1.92	1.66
SD	0.37	0.22	0.23	0.14	0.40	0.21	0.19
%[d]		−30.2	−42.6	−54.1	+20.9	−31.1	−40.2
Mussels	2.16	3.149	3.577	3.802	2.920	2.709	2.329[c]
SD	0.25	0.33	0.42	0.35	0.35	0.33	0.41
%[d]		+45.8	+65.6	+76.1	+35.5	+25.5	+7.8

[a] Each value is based on six estimations.
[b] The differences between controls and experimentals are statistically significant ($P < 0.05$).
[c] Not significantly different from control ($P > 0.05$).
[d] Percentage change over control.

level of glycogen in snails. But, on further exposure, the glycogen level progressively increased on days 15 and 30 with a decrease of phosphorylase activity and glucose level. Even in mussels the glucose level and glycogen phosphorylase activity increased on day 5 with a decrease in glycogen level. Similar trend was also noticed even on days 15 and 30, but with a lower degree, in the order: day $5 > 15 > 30$; thereby all the levels reached near to normal on day 30.

The level of pyruvate decreased on day 1 but progressively increased on days 2 and 3 in the hepatopancreas of snails exposed to acute concentration. The level of lactate and the activity of LDH significantly increased on days 1 and 2 (day $1 > 2$) with a decrease on day 3. In mussels also the pyruvate level decreased on day 1, which furthered high on day 2, but increased significantly on day 3. The level of lactate and the activity of LDH increased on all days of exposure with the degree in the order: day $1 < 2 > 3$. Thus on day 3 the increase in pyruvate level was high in the tissue of snails than of mussels, with a decrease in lactate level and LDH activity. Whereas, in mussels the LDH activity showed a high elevation along with the increase in lactate level. In subacute concentration both the levels of pyruvate and lactate and the activity of LDH increased on all days of exposure; but on any day the increase in pyruvate level was high in mussels than in snails vice versa in the percentage increase of lactate level and LDH activity. Further, the increase in the levels of pyruvate and lactate and the activity of LDH was progressive over time of exposure in snails (day $5 < 15 < 30$) whereas the order was day $5 < 15 > 30$ in mussels (Table 5).

The rate of oxygen consumption and the activity of SDH in the hepatopancreas of snails and mussels significantly decreased relative to their respective control on all the days of exposure to acute concentration of mercury. The degree of decrease was progressive in the order $1 < 2 < 3$; but on any corresponding day of exposure it was high in mussels than in snails. Even in subacute concentration of mercury a decrease was observed in the rate of oxygen consumption and SDH activity on day 5 in both the groups of animals (higher in mussels than in snails). on further exposure of them the magnitude of suppression decreased on day 15; but on day 30 again a higher degree of suppression was noticed in snails whereas the levels increased in mussels (Table 6).

4. Discussion

Molluscs, among invertebrates, accumulate heavy metals in their tissues to several orders of magnitude above the ambient medium concentrations. (Coimbra and Caraca, 1990). Presence of mucus sheet over the body and high lipid content could be responsible for the sequester of both particulate and soluble forms of heavy metals from habitat (Satyanarayana Rao and Balaji, 1990). Hepatopancreas is considered as the central metabolic organ (Moore and Ramamoorthy, 1984) and is the counter part of the vertebrate liver (Ramana Rao et al., 1990). It plays a dual role as a secretory site and storage organ, and also primary site for detoxification of any toxicant (Hutterer et al., 1969). More accumulation of mercury in it indicates the active role of this organ in metal storage and detoxification processes.

Accumulation of mercury in greater amounts in hepatopancreas of both snails and

Table 5. Levels of pyruvate and lactate (mg/g wet weight) and the activity of LDH (μM formozan/mg protein per h) in the hepatopancreas of snails and mussels on exposure to acute and subacute concentrations of mercury[a,b]

Parameter/animal	Control	Period of exposure (days)					
		Acute			Subacute		
		1	2	3	5	15	30
Pyruvate							
Snails	2.02	1.75	2.52	3.01	2.53	2.80	2.47
SD	0.22	0.19	0.28	0.31	0.32	0.31	0.35
%[c]		−13.4	+24.8	+49.0	+25.4	+38.7	+22.3
Mussels	1.86	1.43	1.14	2.36	2.52	2.65	2.79
SD	0.16	0.15	0.13	0.31	0.28	0.30	0.29
%[c]		−23.4	−38.9	+26.9	+35.2	+42.4	+49.8
Lactate							
Snails	1.08	1.59	1.43	1.20[d]	1.42	1.49	1.57
SD	0.21	0.18	0.16	0.09	0.17	0.18	0.17
%[c]		+47.8	+32.6	+11.1	+30.9	+38.8	+45.6
Mussels	1.03	1.40	1.49	1.45	1.25	1.35	1.30
SD	0.12	0.16	0.19	0.17	0.14	0.16	0.14
%[c]		+36.3	+44.5	+40.9	+20.5	+30.6	+26.4
LDH							
Snails	0.024	0.034	0.030	0.018	0.033	0.040	0.045
SD	0.003	0.004	0.003	0.002	0.004	0.004	0.005
%[c]		+40.1	+25.0	−25.6	+37.5	+66.7	+87.5
Mussels	0.020	0.030	0.034	0.027	0.025	0.030	0.027
SD	0.003	0.002	0.003	0.003	0.003	0.003	0.003
%[c]		+50.0	+70.0	+35.0	+28.4	+50.0	+35.0

[a] Each value is based on six estimations.
[b] The differences between controls and experimentals are statistically significant ($P < 0.05$).
[c] Percentage change over control.
[d] Not significantly different from control ($P > 0.05$).

Table 6. Rate of oxygen consumption (μM O_2 consumed/g wet weight per min) and the activity of SDH (μM formozan/mg protein per h) in the hepatopancreas of snails and mussels on exposure to the acute and subacute concentrations of mercury[a,b]

Animal/parameter	Control	Period of exposure (days)					
		Acute			Subacute		
		1	2	3	5	15	30
Oxygen consumption							
Snails	1.400	0.619	0.473	0.345	0.970	1.113	0.715
SD	0.180	0.078	0.052	0.036	0.106	0.122	0.078
%[c]		−55.8	−66.2	−75.4	−30.7	−20.5	−48.9
Mussels	1.350	0.478	0.290	0.174	0.692	1.006	1.496[d]
SD	0.173	0.057	0.032	0.018	0.176	0.12	0.178
%[c]		−64.6	−78.5	−87.1	−48.7	−25.5	+10.8
SDH							
Snails	0.168	1.07	0.083	0.062	0.117	0.154	0.092
SD	0.018	0.012	0.009	0.007	0.013	0.017	0.009
%[c]		−36.5	−50.9	−62.8	−30.6	−20.7	−45.6
Mussels	0.135	0.071	0.052	0.033	0.077	0.114	0.53[d]
SD	0.016	0.009	0.007	0.004	0.008	0.013	0.17
%[c]		−47.4	−62.2	−75.9	−42.9	−15.8	+13.2

[a] Each value is based on six estimations.
[b] The differences between controls and experimentals are statistically significant ($P < 0.05$).
[c] The percentage change over control.
[d] Not significantly different from control ($P > 0.05$).

mussels at acute concentration could be due to more bioavailability of metal ions, speedy absorption onto the ctenidium and further diffusion of them into the inner tissues. Hence, the amount of accumulation of mercury progressively increased over time of exposure. The increase in the rate of accumulation could be due to its domination over elimination. A more than finite limit of accumulation of mercury in hepatopancreas could lead to the total poisoning of these animals on acute toxic stress. Absorption of metal ions by the mucus sheets could be one of the major routes for the increase in the amount of mercury in hepatopancreas (Krishna Kumar et al., 1990). Cunningham and Tripp (1975) stated that 57% of all the body mercury in *Crassostrea virginica* was in hepatopancreas. Relative to the level of accumulation of mercury at acute concentration, the less amount of accumulation of it in the hepatopancreas of both the snails and mussels exposed to subacute concentration could be due to low bioavailability of metal ions in their ambient medium. Although the total amount of mercury increased with the increase in exposure period, decrease in the rate of accumulation indicated the activation of metal elimination processes on prolonged exposure. Cunningham and Tripp (1975) have also stated that subacute levels of mercury could stimulate the regulatory mechanisms in bivalves.

The ambient medium concentration of mercury though same either in acute or in subacute concentration, high amount of mercury deposition in hepatopancreas of mussels than of snails on any corresponding day of exposure could be due to greater flux of polluted water onto the soft parts of mussels. As the mussels are filter feeders, they invariably have to allow more quantum of water to flow onto the mantle and ctenidium. The ctenidium being bipectinate in nature, the surface area of it is relatively greater thereby more diffusion of metal ions is possible. In snails the circumference of univalve shell mouth is less and the ctenidium being monopectinate, the intake of water and flow of it onto this soft tissue is relatively less. In addition, the sensing of water quality by osphradium could also control the entry of excess of polluted water. Thus, these morphological variations between these two groups of animals could be the possible reasons for the differential mercury accumulation in hepatopancreas.

Osmo- and ion-regulation play a pivotal role in the cellular metabolism of an animal and its imbalance leads to changes in various physiological and biochemical functions (Moorthy et al., 1984). For maintenance of osmo-concentration of body fluids, sodium, potassium and calcium ions are essential and their role in this aspect has been well demonstrated in molluscs (Schoffeniels and Gilles, 1972). The ATPase enzyme complex system helps in the uptake of such vital ions from the external medium into the body fluids.

In the present study the changes in the ion levels and associated ATPase activities in hepatopancreas of snails and mussels exposed to mercury could be correlated to the differential permeability properties of hepatic cell membranes on toxic stress. The gradual net loss of these ions in acute concentration could be due to progressive tissue damage. Higher concentrations of mercury might have destabilized the membrane bound enzymes and related energy yielding processes (Yadwad et al., 1990). As the active transport of ions is an energy-dependent process, suppression of ATPase activities could reveal the decreased production of ATP; hence a progressively lowered uptake of ions is noticed under acute toxic stress. Sodium, potassium and calcium being involved in the maintenance of mitochondrial integrity (Bygrave, 1978), the decrease in these ion, levels and ATPase enzyme activities in hepatopancreas is known to cause a direct adverse effect on mitochondria; the metal affects the metabolism of hepatic cells and disrupt the functions of hepatopancreas (Dhavale et al., 1988). Greater suppression over time of exposure in acute concentration indicated that the binding of metal to active sites of membrane bound enzymes is not only dependent on the concentration of the metal but also on the duration of exposure (Sivaramakrishna et al., 1991; Suresh et al., 1995).

The decrease in ion levels and suppression of ATPase activities in snails and mussels exposed to subacute concentration on day 5 also indicated the inhibition of ionic exchange under mercury stress. But, as the stress is subacute, these animals made an effort to recover from the impact on further exposure; thereby the degree of suppression of ATPase activities decreased on day 15 with a little rise in the levels of ions relative to day 5. The activation of detoxification mechanisms and metal elimination processes might have facilitated for such metabolic compensation. Interestingly, the ability of compensation improved in mussels on further exposure of them to 30 days as revealed by the attainment of normal levels of ions and ATPases in their hepatopancreas; whereas the snails could not tolerate the prolonged impact of stress, hence the level of ions and ATPases in them exhibited a significant decrease.

A decrease in glucose level and the activity of phosphorylase along with an increase in the glycogen level in the hepatopancreas of snails exposed to acute concentration of mercury indicated the stimulation of glycogenesis in this organ. During acute mercury stress snails failed to utilize glucose for energy releasing purposes, but, converted it to glycogen, may be due to uncoordinated stimulation of glycogenesis. Such a strategic response appears specific to gastropods, those generally undergo aestivation or hibernation during unfavourable conditions (Asha Devi, 1983). As against to snails, significant decrease in glycogen content with the corresponding increase in glucose level and the activity of phosphorylase in the hepatopancreas of mussels exposed to acute concentration of mercury indicated the stimulation of glycogenolysis, may be by the activation of a hyperglycemic principle specific to bivalves (Ahmad and Chaplin, 1984; Krishna Mohan et al., 1987). Hence, the mussels responded to acute mercury stress differently from snails. Both these shifts could be considered metabolic imbalances, which might have occurred under the influence of catecholamines as well as glucocorticoides (Hemelraad et al., 1990). Involvement of such endocrine principle in glycogen metabolism was further supported by the works conducted in *Lymnea stagnalis* (Hemminga et al., 1984) and *P. globosa* (Naidu et al., 1985). The manifestation of both glycogenesis and glycogenolysis have been observed in tissues of a few gastropods and bivalves exposed to pesticides, and it has been reported that these manifestations are species specific (Ramana Rao and Ramamurthi, 1980; Kalarani et al., 1984).

In subacute concentrations of mercury the snails initially exhibited glycogenolysis and this may be to derive energy from glucose to meet the toxic stress. But on an exposure of 15 days glycogenesis is triggered and progresses up to day 30, which ascertains their specific response pattern to mercury as noticed in acute concentration. Probably they could not tolerate even the subacute mercury stress for a longer period. The mussels also initially showed the breakdown of glycogen to obtain necessary energy from glucose, but on prolonged exposure the attainment of normal levels in glucose, glycogen and glycogen phosphorylase activity suggested that they are able to adapt to the subacute toxic stress of mercury, which is not seen in acute concentration where the stimulation of glycogenolysis is progressive over the period of exposure.

Progressive suppression of oxygen consumption, and SDH activity, a key enzyme of the TCA cycle, in the hepatopancreas of snails and mussels exposed to acute concentration of mercury indicated the inhibition of oxidative metabolism by the action of toxic metal ions. Decreased oxygen consumption could be due to the alterations in the membrane permeability and damage to hepatocytes; and the suppression of SDH activity could be due to the binding of metal ions to sulfhydryl groups of this enzyme resulting in the loss of its affinity to the substrate. Such observations were also reported in the freshwater fish *Notopterus notopterus* treated with mercury (Verma and Ramesh Chand, 1986). A decrease in SDH activity uncouples the oxidative phosphorylation and causes an impairment in energy releasing processes. These changes are further supported by the suppression of ATPase enzyme complex in both the animal groups under acute mercury stress. In snails, as the glucose is converted to glycogen the entry of it into glycolysis is suppressed on exposure of them to acute mercury stress. In mussels though glycogen is mobilized to meet the energy demands: the animals could not derive sufficient energy due to the decreased oxidative metabolism, hence glucose levels increased considerably. An initial shift to anaerobic glycolysis with the rise of LDH activity, however, is observed in both the animal groups, probably to derive energy to meet the acute toxic stress. Even the elevation of less efficient anaerobic glycolysis is further suppressed in snails indicating their less ability of adaptation. The gradual accumulation of pyruvate and the decrease in the level of lactate and LDH activity in snails indicated the impairment of both oxidative metabolic cycle and anaerobic glycolysis. This type of impairment is also reported in the *P. globosa* under pesticide stress (Rajeshwara Rao et al., 1987). Probably, under such situations the hepatopancreas of snails might have restored to the glycogenesis. In mussels though the oxidative metabolism is suppressed, they could operate the anaerobic glycolysis considerably for a longer duration though the elevation of LDH activity and lactate level decreased with an increase in pyruvate level on day 3 of exposure. Increase in pyruvate and lactate levels could also compensate the loss of ions and may help in osmoregulatory phenomenon (Balavenkatasubbaiah et al., 1984).

A decrease in the rate of oxygen consumption and SDH activity in the hepatopancreas of snails and mussels on day 5 of exposure to subacute concentration also indicated the suppression of oxidative metabolic enzymes under mercury stress. As the interaction is dependent on the amount of metal ions available, it is in less intensity compared to acute toxic stress. A decrease in oxidative metabolism induced an elevation in anaerobic glycolysis in order to derive the required energy to resist the imposed toxic stress. On further exposure to 15 days the suppression of oxidative metabolism decreased with greater elevation of anaerobic glycolysis in snails (a compensatory step to adapt to subacute toxic stress). However, after a further exposure of 30 days they failed to recover from the state of metabolic unstability as evident by a steep decrease in the rate of oxygen consumption and SDH activity. However, a progressive elevation in anaerobic glycolysis indicated their dependency more on it to meet the required energy demands. Hence, a progressive increase is observed in LDH activity and lactate and pyruvate levels in them. Compared with snails, the hepatopancreas of mussels exhibit a progressive recovery from the suppression of its rate of oxygen consumption and SDH activity became almost normal after a prolonged exposure of 30 days. The activation of detoxification mechanisms and the synthesis of metal binding proteins might have prevented the toxic ions to interact with the active sites of oxidative enzymes (Harrison et al., 1987; Anderson et al., 1989). Corresponding to the regain in the oxidative metabolism, a regression is observed in the elevation of anaerobic glycolysis indicating the attainment of metabolic homeostasis. Elevations in lactate levels could lead to its oxidation and more lipid biosynthesis. This type of metabolic compensation in subacute concentration is a positive response of these animals to combat the induced chronic toxic stress.

The overall study revealed a few interesting differences in the energetics of snails and mussels on exposure to mercury. One difference is a high accumulation of mercury in the hepatopancreas of mussels than of snails under similar ambient medium concentration. The second is the stimulation of glycogensis in snails but glycogenolysis in mussels. The third is a greater ability of compensation in countering the toxic stress, either acute or subacute, in mussels than in snails, though the former group could accumulate high amount of metal. This is clear in subacute toxic stress, wherein the shift in the levels of ions, glucose and glycogen, activities of ATPase enzyme complex and SDH and the rate of oxygen consumption reached near to control level in the hepatopancreas of mussels on prolonged exposure. The snails though accumulated relatively less amount of mercury in its hepatopancreas, but failed to activate the compensatory mechanisms even under subacute toxic stress. Thus a species specificity is observed in the adaptive abilities of snails and mussels to mercury; snails showed higher sensitivity than mussels even though the former group could survive in relatively greater concentration due to its morphological specializations.

References

Ahmad, T.A., Chaplin, A.E., 1984. The effect of experimental conditions on the activity of pyruvate kinase from the aductor muscle of the mussel *Mytilus edulis* (L). J. An. Morphol. Physiol. 31, 107–118.

Anderson, R.A., Eriksen, K.D.H., Bakke, T., 1989. Evidence of presence of low molecular weight non-metallothionein like metal binding proteins in marine gastropod *Nassarius reticulatus (L)*. Comp. Biochem. Physiol. B. Comp. Biochem. 94, 285–292.

Asha Devi, S., 1983. A comparative study of lactate dehydrogenase in heart and foot muscle of aestivating *Pila globosa*. Comp. Physiol. Ecol. 8, 89–91.

Balavenkatasubbaiah, M., Raghupathi, M., Purushotham, K.R., Ramamurthi, R., 1984. Neuroendocrine principle regulating water and ionic balance in the Indian apple snail *Pila globosa*. Ind. J. Exp. Biol. 22, 460–462.

Bayne, B.L., 1976. In: Oxidative phosphorylation and respiratory metabolism. Bayne, B.L. (Ed.), Marine Mussels – Their Ecology and Physiology. Cambridge University Press, Cambridge, pp. 230.

Bygrave, F.L., 1978. Mitochondria and the control of intracellular calcium. Biol. Rev. 53, 43–46.

Carpenter, P.L., 1975. In: Immunology and Serology, 3rd ed. W.B.S. Saunder, Philadelphia, PA, p. 254.

Carroll, N.V., Longley, R.W., Roe, J.H., 1956. Glycogen determination in liver and muscle by use of anthrone reagent. J. Biol. Chem. 220, 583–593.

Coimbra, J., Caraca, S., 1990. Accumulation of Fe, Zn, Cu and Cd during the different stages of reproductive cycle in *Mytilus edulis*. Comp. Biochem. Physiol. C. Comp. Pharmacol. Toxicol. 95 (2), 265–270.

Cunningham, P.A., Tripp, M.R., 1975. Factors affecting accumulation and removal of mercury from tissues of the American oyster. *Crassostrea virginica*. Mar. Biol. 31, 321–324.

Dall, W., 1967. Hypo-osmoregulation in crustacea. Biochem. Physiol. 21, 653–678.

Degani, G., Warburg, M.R., 1984. Changes in concentrations of ions and urea in both plasma and muscle in a dehydrated hydid anuran. Comp. Biochem. Physiol. 77, 357–360.

Dhavale, D.M., Masurekar, V.B., Giridhar, B.A., 1988. Cadmium induced inhibitio of Na^{+}-K^{+} ATPase activity in tissues of crab *Scylla serrata*. Bull. Environ. Contam. Toxicol. 40, 759–763.

Finney, D.J., 1971. In: Probit Analysis, 3rd ed. Cambridge University Press, London, p. 333.

Fiske, C.H., Subba Row, Y., 1925. The colorimetric determination of phosphorous. J. Biol. Chem. 66, 375–400.

Friedman, T.E., Hangen, G.E., 1942. Collection of blood for the determination of pyruvic and lactic acids. J. Biol. Chem. 144, 67–77.

Govindappa, S., Swami, K.S., 1965. Electrophoretic characteristics of subcellur components and their relation to enzyme activities in amphibian muscle fibres. Ind. J. Exp. Biol. 3, 209–212.

Guruprasada Rao, M., Nandakumar, N.V., 1982. The tannery industrial effluent effect on succinate dehydrogenase activity pattern in a freshwater snail *Pila globosa*. Proc. Ind. Acad. Sci. 91, 427–431.

Hanumantha Raju, P., Ramesh Reddy, G., Veera Babu, G.R., Peddanna, P., Chetty, C.S., 1989. Glucose-6-phosphate dehydrogenase activity in aestivating freshwater snail *Pila globosa* (Swainson). Possible neroendocrine involvement. Proc. Ind. Natl. Sci. Acad. 55, 27–30.

Harper, H.A., 1986. In: Biologic oxidation. Martin, D.W., Mayes, P.A., Rodwell, V.W. (Ed.), Harper's Review of Biochemistry, Vol. 20. Lange Medical Publications, Singapore, pp. 255–279.

Harrison, F.L., Watness, K., Nelson, D.A., Miller, J.E., Calabrese, A., 1987. Mercury metal binding-proteins in slipper limpet. *Crepidula fornicata* exposed to increased soluble mercury. Estuaries 10, 78–83.

Hassid, W.Z., Abraham, S. 1957. In: Phosphorylase activity. Colowick, S.P., Kalpan, N.O. (Ed.), Methods in Enzymology, Vol. 3. Academic Press, New York, pp. 34.

Hemelraad, J., Holwerda, D.A., Herwig, H.J., Zandee, D.I., 1990. Effects of cadmium in freshwater clams: interaction with energy metabolism in *Anodonta cygnea*. Arch. Enviro. Contam. Toxicol. 19, 660–703.

Hemminga, M.A., Maaskant, J.J., Vanderplas, J., Gabbotta, P.A., 1984. The hyperglycemic factor of the CNS of the freshwater snail *Lymnea stagnalis*: interaction with glucose stimulation of glycogen synthesis and evidence for its release during anaerobiosis. Gen. Comp. Endocrinol. 59, 301–307.

Hilmy, A.M., Eldomiaty, N.A., Daabees, A.Y., Moussa, F.I., 1987. Short-term effect of mercury on survival, behaviour, bioaccumulation and ionic patterns in the catfish (*Clarias lazera*). Comp. Biochem. Physiol. C. Comp. Pharmacol. Toxicol. 87, 303–308.

Huckabee, W.E., 1961. Relationship of pyruvate and lactate during anaerobic metabolism. Am. J. Physiol. 200, 1169–1179.

Humpries, E.C., 1956. Mineral contents and ash analysis. In: Peach, K., Franchy, M.V. (Ed.), Modern Methods of Plant Analysis, Vol. 1. Springer-Verlag, Berlin, pp. 468–502.

Hutterer, F., Klion, F.M., Wengraf, A., Schaffner, F., Poper, H., 1969. Hepatocellular adaptation and injury-structural and biochemical changes following dieldrin and methyl butter yellow. Lab. Invest. 20, 455–464.

Kalarani, V., Chandrasekhara Reddy, D., Padmanabha Naidu, B., 1984. Impact of endosulfon toxicity on certain aspects of carbohydrate metabolism of the freshwater snail *Pila globosa* (Sainson). Geobios. 11, 197–200.

Krishna Kumar, P.K., Damodaran, R.D., Nambisan, P.N.K., 1990. Accumulation, distribution and depuration of mercury in the green mussel *Perna viridis* (Linnaeus). Proc. Ind. Acad. Sci. 99, 345–352.

Krishna Mohan, P., Venkatarami Reddy, L., Ravishankara, C., Indira, K., 1987. Metabolic consequences of methylparathion exposure in the bivalve *Lamellidens marginalis* (Lamarck). Bull. Environ. Contam. Toxicol. 38, 509–514.

Kulkarni, A.B., Bhatlawande, S.B., 1987. Carbohydrate variations induced by optic tentacle in a terrestrial snail *Cerastus moussonianus*. Ind. J. Comp. An. Physiol. 5, 1–5.

Kulkarni, A.B., Utkar, V.N., 1983. Effect of molluscicide on the biochemical composition of the tissues in the freshwater prosobranch snail *Vivipaprus bengalensis* (Lamarck). J. Environ. Biol. 4, 215–220.

Lavie, B., Eviatar, N., 1987. Differential fitness of allelic isoenzymes in the marine gastropods *Littorina punctata* and *L. neritoides* exposed to the environmental stress of the combined effects of cadmium and mercury pollution. Environ. Manag. 11, 345–350.

Lehniger, A.L., 1984. Biochemistry. 2nd, 340.

Lowry, O.H., Rosenbrough, N.J., Farr, A.L., Randall, R.J., 1951. Protein measurement with Folin phenol reagent. J. Biol. Chem. 193, 265–275.

Madanmohan Das, V., Venkatachari, S.A.T., 1984. Influence of varying oxygen tension on the oxygen consumption of the freshwater mussel *Lamellidens marginalis* (Lamarck) and its relation to body size. Veliger 26, 305–310.

Metayer, C., Amiartiquet, C., Baud, J.P., 1990. Species related variations of Silver bioaccumulation and toxicity to three marine bivalves. Water Res. 14 (8), 995–1002.

Micallef, S., Tyler, P.A., 1990. Effect of mercury and selinium on the gill function of *Mytilus edulis*. Mar. Pollut. Bull. 21 (6), 288–292.

Moore, P.B., Dedman, J.R., 1982. Calcium binding proteins and cellular regulation. Life Sci. 31, 2937–2946.

Moore, J.W., Ramamoorthy, S., 1984. In: Desato, R.S. (Ed.), Heavy Metals in Natural Waters – Applied Monitoring and Impact Assessment. Springer-Verlag, New York, pp. 77–99.

Moorthy, K.S., Kasi Reddy, B., Swami, K.S., Chetty, C.S., 1984. Changes in respiration and ionic content in tissues of freshwater mussel exposed to methyl parathion toxicity. Toxicol. Lett. 21, 287–292.

Nachlas, M.M., Margulies, S.I., Seligman, A.M., 1960. A colorimetric method for the estimation of succinate dehydrogenase activity. J. Biol. Chem. 235, 499.

Naidu, R.C., Reddy, B.K., Swami, K.S., Chetty, C.S., 1985. Neuroendocrine regulation of phosphorylase activity in the Indian apple snail *Pila globosa* (Swainson). Natl. Acad. Sci. Lett. 8, 8–13.

Nammalwar, P., 1983. Heavy metal pollution in the marine environment. Sci. Rep. 31, 158–160.

Nelson, N., Somogyi, M., 1952. A photometric adoptation of Somogyi method for the determination of glucose. J. Biol. Chem. 151, 375–380.

Phillips, D.J.H., 1977. The use of biological indicator organisms to monitor trace metal pollution in marine and estuarine environments – a review. Environ. Pollut. 13, 281–318.

Radhakrishnaiah, K., Busappa, B., 1986. Effect of cadmium on the carbohydrate metabolism of the freshwater field crab *Oziotelphusa senex senex* (Fabricius). Environ. Biol. 7, 17–21.

Radhakrishnaiah, K., Venkataramana, P., Suresh, A., Sivaramakrishna, B., 1992. Effects of lethal and sublethal concentrations of copper on glycolysis in liver and muscle of the freshwater teleost *Labeo rohita* (Hamilton). J. Environ. Biol. 13, 63–68.

Rajeshwara Rao, M., Sivaprasad Rao, K., Srinivasulu Reddy, M., Sambasiva Rao, K.R.S., Mal Reddy, N., 1983. Toxicity of phenthoate and changes in the organic constituents of the snail *Pila globosa* (Swainson) under sublethal and lethal impacts. Geobios. 10, 250–253.

Rajeshwara Rao, M., Vijaya Kumar, K., Ramesh Babu, K., Chandrasekhara Reddy, A., 1987. Elsan induced alterations in selected tissue dehydrogenase activity levels of the freshwater snail *Pila globosa*. Environ. Ecol. 5, 83–86.

Ramana Rao, M.V., Ramamurthi, R., 1980. Effect of sublethal concentration of sumithion on some biochemical constituents of the freshwater snail *Pila globosa* (Swainson). Geobios. 7, 247–250.

Ramana Rao, K.V., Surendranath, P., Kodavanti, P.R.S., 1990. Levels of transaminases in tissues of the penaeid prawn, *Metapenaeus monoceros* (Fabricius) following sublethal kelthane exposure. Bull. Environ. Contam. Toxicol. 44, 114–120.

Ravinder Reddy, T., Vijaya Kumar, N., Chari, N., 1986. Effect of mercuric chloride on carbohydrate metabolism of mantle, foot and gill of freshwater mussel *Parreysia reyosa*. Environ. Biol. 7, 12–14.

Satyanarayana Rao, K., Balaji, M., 1990. Accumulation of copper by *Mytilopsis sallei* an important fouling bivalve at Visakhapatnam harbour. Ind. J. Comp. An. Physiol. 8 (2), 1–3.

Schoffeniels, E., Gilles, R., 1972. Ion regulation and osmoregulation in mollusca. In: Florikin, M., Scheer, B.T. (Ed.), Chemical Zoology, Vol. 1. Academic Press, New York, pp. 393–418.

Sharma, K.C., 1984. Effect of mercury pollution on the general biology and carbohydrate metabolism of a freshwater murrel *Channa punctatus* (Bloch). Geobios. 11, 122–127.

Shaw, B.P., Panigrahi, A.K., 1990. Geographical distribution of mercury around a chloralkali factory. J. Environ. Biol. 8, 277–281.

Singer, T.P., Kearney, E.B., Ackrell, B.A.C., 1973. Newer knowledge on the regulatory properties of succinate dehydrogenase. Mech. Bionerg. Proc. Invt. Conf., 22, 485–498.

Sivaramakrishna, B., Radhakrishnaiah, K., 1995. Mercury accumulation in lepatopancreas and foot of *Pila globosa* (snail) and *Lamellidens marginalis* (mussel) at lethal and sublethal concentrations. Proc. Ind. Natl. Sci. Acad. 61, 437–440.

Sivaramakrishna, B., Radhakrishnaiah, K., Suresh, A., 1991. Assessment of mercury toxicity by the changes in oxygen consumption and ion levels in the freshwater snail *Pila globosa* and the mussel *Lamellides marginalis*. Bull. Environ. Contam. Toxicol. 46, 913–920.

Sivaramakrishna, B., Suresh, A., Radhakrishnaiah, K., 1992. Comparative study of the acute toxicity of mercury, in the freshwater snail *Pila globosa* (Swainson) and mussel *Lamellidens marginalis* (Lamarck). Acta Ciencia Indica 18, 83–86.

Sreedevi, P., Suresh, A., Sivaramakrishna, B., Prabhavathi, B., Radhakrishnaiah, K., 1992. Bioaccumulation of nickel in the organs of the freshwater fish, *Cyprinus carpio*, and the freshwater mussel, *Lamellidens marginalis*, under lethal and sublethal nickel stress. Chemosphere 24, 29–36.

Suresh, A., Sivaramakrishna, B., Radhakrishnaiah, K., 1993. Effect of lethal and sublethal concentrtions of cadmium on the eneretics in the gills of fry and fingerlings of *Cyprinus carpio*. Bull. Environ. Contam. Toxicol. 51, 920–926.

Suresh, A., Sivaramakrishna, B., Radhakrishnaiah, K., 1995. Cadmium induced changes in ion levels and ATPase activities in the muscle of the fry and fingerlings of the freshwter fish, *Cyplinus carpio*. Chemosphere 30 (2), 367–375.

Sutherland, E.W., 1955. Polysaccharide phosphorylase, liver. In: Clowick, S.P., Kaplan, N.O. (Eds.), Methods in Enzymology, Vol. I. Academic Press, New York, pp. 215.

Tucker, R.K., Matte, A., 1980. Invitro effects of cadmium and lead on ATPase in gill of the rock crab *Cancer irroratus*. Bull. Environ. Contam. Toxicol. 24, 847–852.

V-Balough, K., Fernandez, D.S., Salanki, J., 1988. Heavy metal concentrations in *Lymnea stagnalis* (L) in the environments of lake Balaton (Hungary). Water Res. 226, 1205–1210.

Verma, S.R., Ramesh Chand, P., 1986. Toxicity effects of mercuric chloride on a few enzymes of carbohydrate metabolism of *Notopterus notopterus*. Ind. J. Environ. Health 28, 1–7.

Watson, T.A., Beamish, F.W.H., 1981. The effects of Zn on branchial adenosine triphosphatase enzymes in vitro from rainbow trout *Salmo gairdneri*. Comp. Biochem. Physiol. 68, 167–173.

Yadwad, V.B., Kallapur, V.L., Basalingappa, S., 1990. Inhibition of gill Na^+-K^+ ATPase activity in dragon fly larva *Pantala flavesens* by endosultan. Bull. Environ. Contam. Toxicol. 44, 585–589.

Trace elements – Their distribution and effects in the environment
B. Markert and K. Friese, editors

Chapter 15

Effects of mercury on glutathione and glutathione-dependent enzymes in catfish (*Ictalurus melas* R.)

A.-C. Elia, A.-J.-M. Dörr, L. Mantilacci, M.-I. Taticchi, R. Galarini

Abstract

Concentrations of mercury that show acute and subacute toxicity in freshwater teleost (Ictalurus melas R.) have been determined during static experiments in aquaria. The LC_{50} at 96 h was found to be 0.57 mg/l Hg^{2+}. Thus, catfish were exposed to selected subacute concentrations of Hg^{2+} (0.1, 0.2 and 0.4 mg/l) for 96 h and tissues were analyzed for metal accumulation. Gills, kidneys, liver and muscle were the tissues in which the mercury was stored in a dose-dependent manner. Biochemical responses of glutathione and several glutathione-dependent enzymes were evaluated in the liver of catfish after laboratory exposure to subacute concentrations of mercury during the static short-term experiment. Glutathione levels increased, while the hepatic specific activities of glyoxalase I, glyoxalase II, glutathione peroxidases (either selenium-dependent or selenium-independent enzymes) and glutatione S-transferases decreased. Glutathione reductase did not show significant variation in activity. The selected biochemical parameters are sensitive to assess the toxic effects of mercury in catfish exposed to subacute mercury concentrations.

Keywords: Glutathione; Glutathione-dependent enzymes; Glyoxalases; Mercury toxicity; *Ictalurus melas*

1. Introduction

The catfish (*Ictalurus melas* R.), a bottom-dwelling fish, is widely distributed in natural or artificial lakes and often lives and thrives under extreme conditions of water pollution (Scott and Crosman, 1973).

Mercury is a highly toxic, non-essential and persistent heavy metal which contaminates the aquatic environment, and pollution may be caused in part by the atmospheric transport of anthropogenically mobilized Hg species, or through natural weathering processes (Watras and Huckabee, 1994). Mercury exerts different toxicity on aquatic organisms related to its chemical species. It has been reported that organic mercury is more toxic than inorganic mercury in fish; this higher toxicity is assigned to the high rate of absorption and lipid solubility in the former (Moore and Ramamoorthy, 1983). Uptake of mercury by fish can occur through the gills or the food chain (Cope et al., 1990).

Several studies have been undertaken to report the levels of inorganic mercury having lethal or sublethal effects on fish. The lethal concentrations ranged from about 0.5 to 1 mg/l Hg^{2+} for certain fish. Concentrations are generally in function of the species of fish and of the experimental conditions of exposure. Sublethal effects may occur at levels of exposure less than 0.05 mg/l Hg during a long term exposure (Menezes and Qasim, 1984; Kirubagaran and Joy, 1988; Prasad, 1994).

Many pollutants in aquatic environments have been shown to exert toxic effects related to oxidative stress in fish, when the antioxidative and detoxifying enzymes are not able to protect them efficiently. Glutathione and glutathione-dependent enzymes constitute an important pathway of the antioxidant and detoxification defenses in fish as well as in other organisms.

Mercury induces lipid peroxidation and oxidative stress and involves changes in the glutathione level (Fukino et al., 1984; Rana et al., 1995).

Glutathione provides the cell with reducing power, protects protein sulfhydryl groups and interacts with heavy metal ions, xenobiotics, and other toxic substances (Fuhr and Rabenstein, 1973). Moreover, glutathione acts as a substrate or cofactor for several enzymes which are involved in the detoxification of toxic compounds. Among these enzymes, the selenium-dependent glutathione peroxidase (EC 1.11.1.9) reduces either hydrogen peroxide or organic peroxides (Mannervik, 1985b), while the selenium-independent glutathione peroxidase, an isoenzyme of glutathione *S*-transferases (EC 2.5.1.18), reduces organic hydroperoxides but not hydrogen peroxide (Mannervik, 1985a).

The ubiquitous glutathione *S*-transferases are responsible for the conjugation of xenobiotics containing reactive electrophilic centers to the thiol group of reduced glutathione. The conjugation to glutathione decreases their reactivity, making these molecules more water soluble, and more suited for excretion (Boyland and Chasseaud, 1969). The glutathione *S*-transferases are a multigene family encoding for many isoenzymes with variable specifities for many substrates differing in their electrophilic centers and molecular structures.

There are other glutathione-dependent enzymes involved in detoxification, such as glutathione reductase (EC 1.6.4.2) and the glyoxalase system.

This system is composed of two enzymes, glyoxalase I (EC 4.4.1.5) and glyoxalase II (EC 3.1.2.6). Glyoxalase I acts upon the equilibrium adduct of methylglyoxal (or other toxic 2-ketoaldehydes) and reduced glutathione, catalyzing the formation of the thioester *S*-D-lactoylglutathione. Glyoxalase II catalyses the hydrolysis of the thioester to regenerate reduced glutathione and liberates D-lactate. There have been many ideas regarding the functions of the glyoxalase system. It has been shown that one of the functions is certainly the protection against 2-ketoaldehydes, cytotoxic and reactive compounds formed in cellular oxidative processes (Schauenstein et al., 1977).

Although certain glutathione-dependent enzymes have been studied in fish (Chatterjee and Bhattacharya, 1984; Gallagher and Di Giulio, 1992; Hasspielier et al., 1994), only a few reports have appeared concerning these enzymes in *I. melas* related to mercury (Heisinger and Scott, 1985), and no reports exist regarding glyoxalases in fish exposed to mercury.

The aim of the research described herein is the determination of acute and subacute doses of mercury in static experiments with catfish exposed in aquaria at different concentrations. The accumulation of mercury in different organs and tissues has been determined as well as the disappearance of mercury because of abiotic trapping in the used aquaria.

Catfish were exposed to subacute concentrations of mercury for 96 h and the resulting levels of glutathione and of some glutathione-dependent enzymes were evaluated in liver. Glyoxalase I, glyoxalase II, selenium-dependent and -independent glutathione peroxidases and glutathione reductase were measured. The present research also reports the kinetic study with several substrates of hepatic cytosolic glutathione *S*-transferases in order to find a possible variation in substrate specificities of transferase pool related to mercury exposure.

2. Methods

2.1. Animals and experimental design

Catfish of mixed sex (mean total length 21.6 ± 1.0 cm; mean weight 120.08 ± 16.26 g), were purchased from a local fish farm (Trasimeno Lake, Umbria, Central Italy), and kept for 15 days acclimation time in a 500 l aerated flow-through holding tank provided with dechlorinated, carbon filtered city tap water (pH 7.8, hardness 21.16°F, dissolved oxygen 8.55 mg/l, alkalinity 200 mg/l as $CaCO_3$, conductivity 294 μS/cm, temperature 16°C). Catfish were fed daily with commercial pellets during this period until 24 h prior to the experiments.

To assess the acute toxicity of mercury on catfish, an experiment was performed under static test conditions to determine the 96 h LC_{50} and its 95% confidence limits. Eight groups of six fish were exposed to a geometric series of mercury concentrations of 0, 0.3, 0.36, 0.44, 0,53, 0.64, 0,78 and 0.94 mg/l Hg^{2+}. An additional bioassay was performed using eight groups of six fish exposed to linear series of concentrations (0.0, 0.05, 0.1, 0.2, 0.3, 0.4, 0.5 and 0.6 mg/l Hg^{2+}). The 96 h LC_{50} value was determined according to the method of Lichfield and Wilcoxon (1949). On the basis of this test, mercury concentrations below lethal concentrations were chosen for the following experiments.

Before the start of static experiments with mercury on fish, a preliminary study was conducted on the mercury solution in a glass aquaria. It was provided with 60 l of acclimation water in order to verify the effective permanence in solution of Hg^{2+} during the 96 h of static test by sampling the water at each metal concentration (0.1, 0.2 and 0.4, mg/l) every 24 h. The water samples were treated with a solution of 0.25 M H_2SO_4 and 0.3 M NaCl and stored at 4°C until analysis.

The subacute toxicity of mercury in catfish was investigated with an experiment of short-term exposure. Catfish were divided into four experimental groups of six specimens each and exposed to specific metal concentrations (0.0, 0.1, 0.2, and 0.4 mg/l Hg^{2+}) for 96 h, in glass aquaria with acclimation water. The aquaria were maintained at ambient room temperature during the experimental period and the fish were not fed during the treatment. Samples of water of each treatment aquaria were collected every 24 h for analysis of mercury.

After the exposure period at the specific metal concentrations, the fish from the experimental and control groups were sacrificed by cervical dislocation, weighed, measured, and gills, liver, kidneys and muscle were removed and washed with sodium chloride cold solution (9 g/l). The livers were subdivided into aliquots for chemical and biochemical analyses, while the other tissues were utilized only for the chemical analysis. The tissues for the biochemical and chemical analyses were immediately stored at −80 and −20°C, respectively, until used.

2.2. Mercury analysis

Tissues and water samples were analyzed by cold vapor atomic absorption spectrometry (CV-AAS) using a Perkin–Elmer 5100 PC instrument.

The gills, liver, kidneys and muscle were wet digested with 10 ml of a concentrated sulfuric and nitric acid mixture (1:1 v/v) at 50°C for 1.5 h in suitable flasks. Digests were then cooled and 50 ml of 4% potassium permanganate was carefully added. Flasks were heated for a further 0.5 h at 50°C and cooled. The excess permanganate was removed with 10 ml of 30% of hydroxylammonium chloride and 30% of sodium chloride. Mercury was reduced with a $SnCl_2$ solution (5%) and purged with N_2 in a closed system (MHS-10). The values obtained were compared with the standard solutions. Blanks and controls were carried out at the same time.

2.3. Biochemical analyses

Total glutathione (GSH + 2GSSG) was determined on the hepatic deproteinized supernatant. Liver was weighed and homogenized with an Ultra-Turrax in four volumes of 1 M $HClO_4$, 2 mM EDTA. Samples were maintained for 30 min on ice and centrifuged at 30 000 × *g* for 20 min. Total glutathione content was determined by the enzymatic method according to Akerboom and Sies (1981). The pellets were dissolved in 1 M NaOH and the total protein concentrations were determined by the method of Lowry et al. (1951), employing bovine serum albumin as a protein standard. The glutathione levels are reported as μmol/g of tissue (wet weight).

The study of enzymatic activities was carried out on the hepatic cytosolic fractions. Weighed tissues were homogenized with a Ultra-Turrax in ten volumes of 20 mM Tris buffer (pH 7.8) containing 100 μM phenylmethylsulphonyl fluoride (PMSF). Centrifugation was carried out at 4°C at 100 000 × *g* for 60 min and the supernatant was utilized for enzymatic activities determination.

Glyoxalase I activity was determined according to Oray and Norton (1977) by following at 240 nm ($\varepsilon = 3.37\ mM^{-1}\ cm^{-1}$) the formation of *S*-D-lactoylglutathione from the hemimercaptal adduct of methylglyoxal (MG) and reduced glutathione (GSH). The mixture assay of 1 ml contained 100 mM sodium phosphate buffer (pH 6.8), 1 mM GSH and 1 mM MG and sample.

Glyoxalase II activity was determined according to Principato et al. (1987) at 412 nm ($\varepsilon = 13.6\ mM^{-1}\ cm^{-1}$) by monitoring GSH formation in the presence of 5,5′-dithio-bis-2-nitrobenzoic acid (DTNB) and *S*-D-lactoylglutathione (LSG). The assay of 1 ml contained 100 mM MOPS buffer (pH 7.0), 0.2 mM DTNB, 0.5 mM LSG and sample.

Glutathione *S*-transferase activities with the substrates 1-chloro-2,4-dinitrobenzene (CDNB), bromosulfophthalein (BSP), 1,2-epoxy-3-(*p*-nitrophenoxy)-propane

(ENPP), *p*-nitrophenyl acetate (NPA), *p*-nitrobenzyl chloride (NBC), were measured according to published procedures (Habig et al., 1974; Habig and Jakoby, 1981). The assay of glutathione *S*-transferase activity with CDNB measures the formation of the conjugate with GSH at 340 nm ($\varepsilon = 9.6\ mM^{-1}cm^{-1}$). The assay of 1 ml contained 100 mM sodium phosphate buffer (pH 6.5), 1 mM GSH, 1 mM of CDNB and the sample. With the other substrates, samples were incubated and increases in absorbance measured constantly. The incubation medium contained in a final volume of 1 ml was: (a) 100 mM sodium phosphate buffer (pH 6.5), 5 mM GSH, 0.3 mM ENPP and sample ($\lambda = 360$ nm, $\varepsilon = 0.5\ mM^{-1}\ cm^{-1}$); (b) 100 mM sodium phosphate buffer (pH 7.5), 5 mM GSH, 0.03 mM BSP and sample ($\lambda = 330$ nm, $\varepsilon = 4.5\ mM^{-1}\ cm^{-1}$); (c) 100 mM sodium phosphate buffer (pH 6.5), 5 mM GSH, 0.25 mM NBC and sample ($\lambda = 310$ nm, $\varepsilon = 1.9\ mM^{-1}\ cm^{-1}$); (d) 100 mM sodium phosphate buffer (pH 7.0), 0.5 mM GSH, 0.2 mM NPA and sample ($\lambda = 400$ nm, $\varepsilon = 8.79\ mM^{-1}cm^{-1}$).

Glutathione reductase activity was determined according to Chung et al. (1991), by following the decrease in absorbance at 340 nm ($\varepsilon = -6.22\ mM^{-1}cm^{-1}$) due to the oxidation of NADPH. The assay of 1 ml contained 100 mM sodium phosphate buffer (pH 7.0), 1 mM GSSG, 60 μM NADPH and sample.

Glutathione peroxidases (Se-dependent and -independent enzymes) activities toward cumene hydroperoxide and H_2O_2 as substrates were determined by the coupled enzyme method of Lawrence and Burk (1976) with minor modifications. The standard assay contained 100 mM sodium phosphate buffer (pH 7.5), 1 mM EDTA, 0.12 mM NADPH, 1 mM NaN_3 (for H_2O_2 assay), 2 mM GSH, 0.8 mM cumene hydroperoxide or 0.4 mM H_2O_2, 1 unit glutathione reductase and sample. The oxidation of NADPH was followed at 340 nm ($\varepsilon = -6.22\ mM^{-1}\ cm^{-1}$).

Enzyme assays were performed at a constant room temperature of 20°C. All assays were conducted in duplicate or triplicate and blanks were subtracted from the sample absorbance. The values reported are the results of averages of the activities obtained.

Protein concentration supernatant fractions were determined by the method of Lowry et al. (1951). The specific activities are reported as nmol/min per mg of protein.

Statistical analysis was made using a computer-assisted single-factor ANOVA and Newman–Keuls' test multiple comparison.

3. Results and discussion

Fish are considered as sentinel organisms in the aquatic ecosystem to assess the environmental contamination of water by several compounds. Fish can be chronically exposed to different pollutants and can bioaccumulate by direct exposure or through the food chain.

Mercury (Hg^{2+}) is a toxic ion that can be found in water environments. However, the mercury content that exists in insoluble form is not biologically relevant, so only the soluble ion concentrations are challenging living organisms such as fish.

Because only the soluble mercury species are effective, the percentage of mercury remaining soluble in water with respect to the starting metal levels in treatment aquaria with and without fish were monitored every 24 h during the static test (results

Table 1. Kinetics of the percentage concentration of mercury in water, with respect to the starting levels, with and without fish[a]

Time (h)	Hg^{2+} (mg/l)		
	0.1	0.2	0.4
0	100 (100)	100 (100)	100 (100)
24	93 (78)	93 (71)	87 (76)
48	86 (65)	90 (65)	85 (61)
72	81 (50)	87 (54)	81 (57)
96	78 (16)	76 (13)	75 (13)

[a] Numbers in parentheses are values with fish.

are given in Table 1). The numbers without parentheses reflect the actual level of mercury dissolved in our experimental glass aquaria. As shown, the levels of mercury measured in each treatment aquarium without fish were slightly decreased by about 10% after 24 h and by about 25% after 96 h at all mercury concentrations used. This might be due to adsorption, precipitation or some other factor. On the other hand the concentrations in water during the catfish exposure decreased from about 25% after 24 h to about 85% after 96 h. Thus, the decline of mercury level in the treatment aquarium is due mainly to the accumulation of mercury in catfish tissues.

The concentrations of total mercury in different tissues of treated and untreated catfish after 96 h Hg^{2+} exposure are reported in Table 2. As evidenced, the accumulation of mercury by *I. melas* results in a significant increase of the total mercury content in all tissues with respect to control tissues. At 0.1 mg/l Hg^{2+}, no significant differences were obtained between kidneys, liver and gills: at 0.2 mg/l of metal, there was a significant difference in mercury concentration between kidneys and liver ($P < 0.01$) and between kidneys and gills ($P < 0.05$); at 0.4 mg/l there was a significant difference in mercury concentration between kidneys and liver ($P < 0.001$) and between kidneys and gills ($P < 0.001$). These results show that kidneys accumulate more of the heavy metal with respect to the other organs at the two highest mercury concentrations. Fish, treated with mercuric chloride under different experimental conditions, exhibited the highest metal level in kidneys, followed by gills and liver (Walczak et al., 1986; Jorgensen and Heisinger, 1987).

Table 2. Mercury concentrations (mg/kg wet weight) in tissues of *Ictalurus melas* exposed to mercury for 96 h[a]

	Hg^{2+} (mg/l)			
	0	0.1	0.2	0.4
Gills	0.14 ± 0.05	4.19 ± 0.45**	4.71 ± 0.36**	8.03 ± 0.98**
Liver	0.12 ± 0.04	2.48 ± 0.72*	2.81 ± 0.25*	6.48 ± 0.99**
Kidneys	0.12 ± 0.03	4.43 ± 0.51**	7.14 ± 0.74**	12.92 ± 1.75**
Muscle	0.05 ± 0.01	0.72 ± 0.14**	0.66 ± 0.05**	0.93 ± 0.13**

[a] Values are mean ± SE ($n = 6$). Symbols indicate significant differences between treatments and control. * $P < 0.05$; ** $P < 0.01$.

The 96 h LC_{50} value for catfish exposed to Hg^{2+} under static test was determined to be 0.57 mg/l, with upper and lower 95% confidence limits of 0.64 and 0.51 mg/l, respectively. In our experimental conditions signs of mortality were evident at 0.5 mg/l Hg^{2+} and 100% mortality occurred after exposure at 0.8 mg/l Hg^{2+}.

These results indicate that mercury concentration from 0.5 mg/l to higher levels are able to induce acute toxicity in catfish after 96 h exposure. Lower doses of mercury (0.1, 0.2 and 0.4 mg/l) resulted in accumulation but were subacute after 96 h exposure. These subacute doses of mercury were used for the determination of molecular parameters in catfish liver. Liver has been chosen because of its larger dimension; in fact, it was possible to use a single liver for both mercury content measurements and glutathione and enzymes determinations.

Glutathione level determination in liver from 96 h mercury-exposed fish and controls fish is summarized in Table 3. The total glutathione level (GSH + 2GSSG) is reported as μmol/g wet weight and the results presented in the table demonstrate that in our experimental conditions, mercury caused significant changes in glutathione levels in the liver of metal-treated catfish, with respect to the controls. In our study, in exposed catfish, glutathione levels in liver increased significantly by about 20% at 0.2 mg/l and 25% 0.4 mg/l Hg^{2+} concentrations ($P < 0.05$, $P < 0.01$, respectively), when compared to the controls. Increased glutathione levels have always been found in several fish exposed to mercury at different concentrations and for different times. *Mugil cephalus* treated with 0.2 mg/l of mercury for 10 days exhibits an increased hepatic glutathione level (Thomas and Wofford, 1984). Chatterjee and Bhattacharya (1984) also found that short-term subacute mercury levels induce an increase of hepatic glutathione in *Anabas testiduneus* exposed to 0.166 mg/l Hg^{2+} for 48 h. The only decrease of glutathione during long term mercury treatments has been reported by Rana et al. (1995) in *Channa punctatus*. Also, increased hepatic levels of glutathione have been reported for *Ictalurus punctatus* exposed to the fungicide chlorothalonil for 6 days (Gallagher et al., 1992) or to the

Table 3. Levels of glutathione (expressed as GSH equivalents, GSH + 2GSSG) and of glutathione-dependent enzymes in the liver of *Ictalurus melas* exposed to mercury for 96 h[a]

	Hg^{2+} (mg/l)			
	0	0.1	0.2	0.4
Glutathione	1.14 ± 0.04	1.18 ± 0.11	1.45 ± 0.06*	1.54 ± 0.09**
Glyoxalase I	221.89 ± 4.04	245.12 ± 12.19	195.78 ± 17.92	160.31 ± 2.42**
Glyoxalase II	140.39 ± 8.31	120.02 ± 13.06	101.16 ± 9.33*	97.66 ± 7.81*
Glutathione reductase	22.76 ± 4.65	40.39 ± 4.69	34.05 ± 6.33	35.43 ± 8.81
Glutathione peroxidase (H_2O_2)	109.26 ± 16.94	103.41 ± 17.98	40.79 ± 6.12**	45.41 ± 2.81**
Glutathione peroxidase (CuOOH)	84.81 ± 7.27	61.58 ± 9.84	43.92 ± 5.85**	46.12 ± 8.86**

[a] Values given as mean ± SE ($n = 6$). Glutathione is reported in μmol/g of wet weight. Enzymatic activities are reported in nmol/min per mg of protein. Symbols indicate significant differences between treatments and control. * $P < 0.05$; ** $P < 0.01$.

sediments contaminated with various aromatic hydrocarbons for 28 days (Di Giulio et al., 1993) and for three species of flatfish sampled from polluted sites in Puget Sound (Stein et al., 1992). However, this increase could be due to the enhanced hepatic uptake of amino acid substrates and the activity of biosynthetic enzymes as suggested by Thomas and Wofford (1984) and Gallagher et al. (1992).

The glutathione-dependent enzymes, glyoxalase I and glyoxalase II, glutathione reductase and glutathione peroxidase, were studied in 96 h mercury-treated fish and control fish. The results are shown in Table 3. The present study contains the first results of the level of glyoxalase enzymes in *I. melas*. The specific activity of glyoxalase I was about 1.5-fold greater than the specific activity of glyoxalase II. These enzymes have been found markedly influenced by subacute mercury concentrations. In fact, glyoxalase I decreased by about 25% ($P < 0.01$) at the highest mercury concentration, while glyoxalase II decreased by an average of 30% at 0.2 and 0.4 mg/l Hg^{2+} ($P < 0.05$). Since metals are known to increase peroxidative processes in fish, they probably also cause an indirect enhancement of 2-ketoaldehydes formation that is known to be formed during lipid peroxidation (Schauenstein et al., 1977). In this respect, the decrease of glyoxalases system could represent an inhibition of these metabolic detoxification pathways and increase the toxicity of 2-ketoaldehydes.

The activity of glutathione reductase in the liver of *I. melas* seems to be higher compared to the activity of the enzyme in the related species, *I. punctatus* and *Ameriurus nebulosus* (Gallagher and Di Giulio, 1992; Hasspielier et al., 1994). The comparative analysis suggests that *I. melas* has a greater capacity to regenerate reduced glutathione from its oxidized form and then ensure higher reduced glutathione levels and counteract the oxidative stress better than channel catfish and brown bullhead. However, as evidenced in Table 3, even if the means of the enzymatic activities are increased, no significant changes in activity were observed during the experiment. Thus, this enzyme seems to not behave as a good marker for mercury damage.

Glutathione acts as a cellular antioxidant through the glutathione peroxidase-catalyzed enzymatic reduction of hydroperoxides to corresponding alcohols and this activity is due to a selenium-dependent enzyme whose presence has been demonstrated in many organisms, including fish (Lawrence and Burk, 1976; Bell et al., 1987). Moreover, in several organisms a further enzyme, called selenium-independent glutathione peroxidase, is also present (Heisinger and Dawson, 1983; Gallagher and Di Giulio, 1992). It is catalyzed by different members of the multigenic family of glutathione *S*-transferases. Their presence greatly potentiates the ability of the cell/organism to detoxify itself of organic hydroperoxides. Glutathione peroxidases activities towards cumene hydroperoxide and hydrogen peroxide in the liver of exposed-catfish and control fish are reported in Table 3. Mercury caused a decrease of about 60% of selenium-dependent glutathione peroxidase activity with hydrogen peroxide as substrate at the two highest mercury concentrations ($P < 0.01$), thus showing a reduced antioxidant defense at cellular level. The selenium-independent glutathione peroxidase specific activity with cumene hydroperoxide as substrate was also significantly decreased after mercury treatment at 0.2 and 0.4 mg/l by an average of 45% ($P < 0.01$) in liver with respect to the control. However, other authors, in different experimental conditions, reported that the coelomic injections of mercury doses did not modify either selenium-dependent or selenium-independent glutathione peroxi-

Table 4. Substrates studies of glutathione S-transferase enzyme in the liver of *Ictalurus melas* exposed to mercury[a]

	Hg^{2+} mg/l			
	0	0.1	0.2	0.4
1-Chloro-2,4-dinitro-benzene	324.31 ± 35.69	176.36 ± 14.08*	157.01 ± 11.27*	167.72 ± 4.06*
1,2-Epoxy-3-(*p*-nitro-phenoxy)-propane	11.92 ± 0.35	10.71 ± 1.19	7.22 ± 0.73*	6.36 ± 0.85*
p-Nitrobenzyl chloride	6.79 ± 0.61	8.88 ± 0.29	7.36 ± 1.21	7.14 ± 1.18
p-Nitrophenyl acetate	3.28 ± 0.36	3.38 ± 0.51	2.75 ± 0.72	2.17 ± 0.36
Bromosulfophthalein	n.d.	n.d.	n.d.	n.d.

[a] Values given as mean ± SE ($n = 6$). Enzymatic activities are reported in nmol/min per mg of protein. n.d., not detectable. * Significant differences between treatments and control ($P < 0.01$).

dase activities in *I. melas* (Heisinger and Scott, 1985). The inhibition of glutathione peroxidases activities could play a significant role in the development of mercury toxicity in catfish in our experimental conditions.

In Table 4 the activities of glutathione *S*-transferases determined in hepatic cytosol of catfish with the substrates CDNB, BSP, ENPP, NBC and NPA are reported. Glutathione *S*-transferases showed activity towards almost all the substrates assayed. CDNB was the best substrate, followed by ENPP, NBC and NPA. In contrast, the enzyme did not exhibit activity towards BSP. In the present study the glutathione *S*-trasferases activities with CDNB as substrate were markedly decreased at all mercury concentrations used by an average of about 50% ($P < 0.01$) in the liver of exposed catfish when compared to the control. Moreover, the results evidenced lower enzyme activities with the increase of mercury concentrations using ENPP of about 40% at 0.2 and 0.4 mg/l Hg^{2+} ($P < 0.01$), while no change was found with the NBC and NPA as substrates. This result can be important because the epoxides are formed during oxidative stress conditions. The important role of glutathione *S*-transferases in detoxification is linked to the broad spectrum of possible electrophilic substrates, including alkyl and aryl halides, organophosphorus and organochlorines, epoxides and organic hydroperoxides (Chasseaud, 1979). Multiple hepatic isoenzymes have been already reported for fish (Ramage et al., 1986; Elia et al., 1999). Even if the glutathione *S*-transferases multiple forms in catfish are not resolved in our experimental conditions, we cannot, however, completely exclude variations in the pattern of isoenzymes and this variation can render less efficient the detoxification action of glutathione S-transferases during oxidative stress conditions. The observed decline in activity could be related either to the effects of inhibition of enzyme synthesis or an inhibition of enzymatic activity because of the formation of protein disulfide bridges, with consequent diminution of catalytic capability. Although, there are few data in the literature on glutathione *S*-trasferases in fish exposed to mercury, the results obtained previously in field investigations (Jedamski-Grymlas et al., 1995) confirmed our observed data.

Our results indicate that the decrease in the activities of catfish enzymes involved in the inactivation of reactive molecules formed during oxidative stress (i.e. 2-ketoaldehydes and epoxides) can worsen the effect of the oxidative damage induced by mercury. At this stage of the research we do not know whether there is a direct enzyme inactivation by mercury or a more subtle effect at transcriptional or translational level.

Finally, these results showed that selected indexes – glutathione, glutathione *S*-transferases, glutathione peroxidases, and glyoxalase I and II – are very sensitive to harmful levels of mercury and emphasize the utility of biochemical responses as tools for assessing contaminant exposures and mercury toxicity in fish.

Acknowledgements

This work was supported by Ministero della Sanità, Italy (Ricerca corrente no. 9/1996). The authors thank Professor G. Principato for his critical review of the manuscript.

References

Akerboom, T.P.M., Sies, H., 1981. Assay of glutathione disulfide, and glutathione mixed disulfide in biological samples. Meth. Enzymol. 71, 373–382.

Bell, J.G., Cowey, C.B., Andron, J.W., Pirie, B.J.S., 1987. Some effects of selenium deficiency on enzyme activities and indices of tissue peroxidation in Atlantic salmon parr (*Salmo salar*). Aquaculture 65, 43–54.

Boyland, E., Chasseaud, L.F., 1969. The role of glutathione and glutathione S-transferases in mercapturic acid biosynthesis. Adv. Enzymol. 32, 173–179.

Chasseaud, L.F., 1979. The role of glutathione and glutathione S-transferase in the metabolism of chemical carcinogens and other electrophilic agents. Adv. Cancer Res. 29, 175–274.

Chatterjee, S., Bhattacharya, S., 1984. Detoxification of industrial pollutants by the glutathione-S-transferase system in the liver of *Anabas testudineus* (Bloch). Toxicol. Lett. 22, 187–198.

Chung, P.M., Cappel, R.E., Gilbert, H.F., 1991. Inhibition of glutathione disulfide reductase by glutathione. Arch. Biochem. Biophys. 288, 48–53.

Cope, W.G., Wiener, J.G., Rada, R.G., 1990. Mercury accumulation in yellow perch in Wisconsin seepage lakes: relation to lake characteristics. Environ. Toxicol. Chem. 9, 931–940.

Di Giulio, R.T., Habig, C., Gallagher, E.P., 1993. Effect of Black rock Harbor sediments on indices of biotrasformation, oxidative stress, and DNA integrity in channel catfish. Aquat. Toxicol. 26, 1–22.

Elia, A.C., Mantilacci, L., Taticchi, M.I., Principato, G., 1999. Characterization of the major glutathione transferases in carp liver. Exp. J. Zool. 284, 130–136.

Fukino, H., Hirai, M., Hsueh, Y.M., Yamane, Y., 1984. Effect of zinc pretreatment on mercuric chloride-induced lipid peroxidation in the rat kidney. Toxicol. Appl. Pharmacol. 3, 395–401.

Fuhr, B.J., Rabenstein, D.L., 1973. Nuclear magnetic resonance studies of the solution chemistry of metal complexes. IX. The binding of cadmium, zinc, lead and mercury by glutathione. J. Am. Chem. Soc. 95, 6944–6950.

Gallagher, E.P., Di Giulio, R.T., 1992. A comparison of glutathione-dependent enzymes in liver, gills and posterior kidney of channel catfish. Comp. Biochem. Physiol. 102C, 543–547.

Gallagher, E.P., Canada, A.T., Di Giulio, R.T., 1992. The protective role of glutathione in chlorothalonil-induced toxicity to channel catfish. Aquat. Toxicol. 23, 155–168.

Habig, W.H., Pabst, M.J., Jakoby, W.B., 1974. Glutathione S-trasferases. The first enzymatic step in mercapturic acid formation. J. Biol. Chem. 249, 7130–7139.

Habig, W.H., Jakoby, W.B., 1981. Assay for differentiation of glutathione S-trasferases. Meth. Enzymol. 77, 398–405.

Hasspielier, B.M., Behar, J.V., Di Giulio, R.T., 1994. Glutathione-dependent defence in channel catfish (*Ictalurus punctatus*) and brown bullhead (*Amerius nebulosus*). Ecotoxicol. Environ. Safety 28, 82–90.

Heisinger, J.F., Dawson, S.M., 1983. Effect of selenium deficiency on liver and blood glutathione peroxidase activity in the black bullhead. J. Exp. Zool. 225, 325–327.

Heisinger, J.F., Scott, L., 1985. Selenium prevents mercuric chloride induced acute osmoregulatory failure without glutathione peroxidase involvement in the black bullhead (Ictalurus melas). Comp. Biochem. Physiol. 80C, 295–297.

Jedamski-Grymlas, J., Kammann, U., Tempelmann, A., Karbe, L., Siebers, D., 1995. Biochemical responses and environmental contaminants in breams (Abramis brama L.) caught in the river Elbe. Ecotoxicol. Environ. Safety 31, 49–56.

Jorgensen, D., Heisinger, J.F., 1987. The effects of selenium on the distribution of mercury in the organs of the black bullhead (*Ictalurus melas*). Comp. Biochem. Physiol. 87C, 181–186.

Kirubagaran, R., Joy, K.P., 1988. Toxic effects of three mercurial compounds on survival, and histology of the kidney on the catfish Clarias batrachus (L.). Ecotoxicol. Environ. Safety 15, 171–179.

Lawrence, R.A., Burk, R.F., 1976. Glutathione peroxidase activity in selenium-deficient rat liver. Biochem. Biophys. Res. Commun. 71, 592–598.

Lichfield, J.T., Wilcoxon, F., 1949. A simplified method of evaluating effects experiments. J. Pharmacol. Exp. Ther. 96, 99–113.

Lowry, O.H., Rosenbrough, N.J., Farr, A.L., Randall, R.J., 1951. Protein measurement with the folin phenol reagent. J. Biol. Chem. 193, 266–275.

Mannervik, B., 1985a. The isoenzymes of glutathione transferase. Adv. Enzymol. Relat. Areas Mol. Biol. 57, 357–417.

Mannervik, B., 1985b. Glutathione peroxidase. Method Enzymol. 113, 490–495.

Menezes, M.R., Qasim, S.Z., 1984. Effects of mercury accumulation on the electrophoretic patterns of the serum, haemoglobin and eye lens proteins of *Tilapia mossambica* (Peters). Water Res. 18, 153–161.

Moore, J.W., Ramamoorthy, S., 1983. Heavy Metals in Natural Waters. Springer-Verlag, New York, pp. 125–160.

Oray, B., Norton, S.J., 1977. Two step purification of mouse liver glyoxalase I and evidence of its dimeric constitution. Biochim. Biophys. Acta 483, 203–208.

Prasad, M.S., 1994. Effect of short-term exposure to mercuric chloride on the air-breathing catfish, *Heteropneustes fossilis*. I. Light microscopic study of the gill. Biomed. Environ. Sci. 7, 327–336.

Principato, G.B., Rosi, G., Talesa, V., Giovannini, E., Uotila, L., 1987. Purification and characterization of two forms of GII from the liver and brain of Wistar rats. Biochim. Biophys. Acta 911, 349–355.

Ramage, P.I.N., Rae, G.H., Nimmo, I.A., 1986. Purification and properties of the hepatic glutathione S-trasferases of the Atlantic salmon (*Salmo solar*). Comp. Biochem. Physiol. 83B, 23–29.

Rana, S.V.S., Singh, R., Verma, S., 1995. Mercury-induced lipid peroxidation in the liver, kidney, brain and gills of a fresh water fish, *Channa punctatus*. Jpn. J. Ichthyol. 42, 255–259.

Scott, W.B., Crosman, E.J., 1973. Fisheries Research Board of Canada, Ottawa XVIII. pp. 966.

Schauenstein, E., Esterbauer, H., Zollner, H., 1977. Aldehydes in Biological Systems. Their Natural Occurrence and Biological Activities. Pion., London, pp. 121–126.

Stein, J.E., Collier, T.K., Reichert, W.L., Casillas, E., Hom, T., Varanasi, U., 1992. Biondicators of contaminant exposure and sublethal effects: studies with benthic fish in Puget Sound. Washington. Environ. Toxicol. Chem. 11, 701–714.

Thomas, P., Wofford, H.W., 1984. Effects of metal and organic compounds on hepatic glutathione, cysteine and acid soluble thiol levels in mullet (*Mugil cephalus* L.). Toxicol. Appl. Pharmacol. 76, 172–182.

Walczak, B.Z., Hammer, U.T., Huang, P.M., 1986. Ecophysiology and mercury accumulation of rainbow trout (*Salmo gairdneri*) when exposed to mercury in various concentrations of chloride. Can. J. Fish. Aquat. Sci. 43, 710–714.

Watras, C.J., Huckabee, J.W., 1994. Mercury Pollution: Integration and Synthesis. Lewis, Boca Raton, FL.

Trace elements – Their distribution and effects in the environment
B. Markert and K. Friese, editors

Chapter 16

Body burden with metals and detoxifying abilities of the grasshopper – *Chorthippus brunneus* (Thunberg) from industrially polluted areas

Maria Augustyniak, Pawet Migula

Abstract

Relations between detoxifying enzymes – glutathione peroxidase (GPOX), glutathione S-transferase (GST), catalase (CAT), carboxylesterases (CarE) and exposure to cadmium of Chorthippus brunneus living in areas under different pollution impact were studied. Activity patterns of GST and CAT were cadmium dependent. Insects from a population living in a weakly polluted site exposed to cadmium showed lower GST and CAT activity in the midgut and gonads but GPOX and CarE was higher and positively correlated with the level of cadmium in consumed grass and in their tissues. Effects of cadmium were more pronounced in females. Inhibitory effects of cadmium on GST and CAT activity in grasshoppers from a less polluted site was compensated by the GPOX with a high H_2O_2 affinity. Insects from heavily polluted site expressed higher involvement of catalase, suggesting their better detoxifying abilities. CarE showed higher activity rates with increased pollution gradient in insects from five different populations and with increased exposure to cadmium in laboratory experiments.

Keywords: Grasshopper; Cadmium; Detoxifying enzymes; Metal tolerance

1. Introduction

Many industrial contaminants are to some extent biodegradable but some of them, such as heavy metals, are highly persistent in the environment. A reduction of their harmful effects is possible only through their inactivation (Chang, 1996; Migula, 1996a). A biological significance for some of these heavy metals is still doubtful and they are highly toxic (Hopkin, 1989; Migula, 1996b; Simkiss and Taylor, 1996). The relations between many biochemical and physiological processes and concentration of such metals as cadmium or lead in tissues of animals or in their diet have been identified. Also mechanisms of these relationships are well recognized (Chang, 1996; Migula and Głowacka, 1996; Simkiss and Taylor, 1996; Migula et al., 1997). Many animals living in highly degraded environments are able to cope with increased pollution, demonstrating various compensatory mechanisms and some of them even genetic adaptations to high levels of such pollutants as heavy metals or pesticides (Posthuma and Van Straalen, 1993; Van Straalen, 1994; Migula et al., 1996).

The selective pressure from excessive amounts of certain chemicals in the environment, such as heavy metals, may induce similar adaptive mechanisms as these supporting life under extreme natural environmental factors (Posthuma and Van Straalen, 1993; Van Straalen, 1994; Sibly, 1996; Simkiss and Taylor, 1996; Migula et al., 1997). Animals developed various strategies for living in environments polluted with organic compounds or heavy metals (Hopkin, 1989, 1994; Nuorteva, 1990; Migula et al., 1999). Many of them are functioning at the subcellular level and depend on the efficiency of detoxifying mechanisms.

Insects may use various strategies, that may differ in relation to their position in a food chain, life span, the *r* or *k* strategy (Heliövaara and Väisänen, 1993; Walker et al., 1996). It was proved that some herbivorous insects, such as grasshoppers (*Chorthippus* sp.) are able to select qualitatively their diet, by e.g. reduced consumption of leaves with higher metal content (Migula et al., 1990; Migula and Bińkowska, 1993). Such findings were made for insects pre-exposed in natural conditions to excessive amounts of such metals as cadmium or lead. Insects are not equipped with chaemosensilla sensitive to heavy metals and selection of the quality of eaten material is probably due to recognition of other changes in taste of leaves.

Most of metal ions are not absorbed and can easily pass through the alimentary tract. Also during moulting and metamorphosis a part of unwanted metals could be eliminated from the organism (Lindqvist, 1994). *Chorthippus* sp. may use two strategies: physiological – based on a quick elimination of excessive amounts of ingested metals; and behavioural – based on the avoidance mechanism (Migula, 1996b). The first one dominates in populations living in unpolluted areas. Insects from industrially polluted regions may reveal both mentioned strategies as they tolerate higher concentrations of Pb and Cd in consumed leaves (Migula and Bińkowska, 1993).

Existence of a given population in the environment contaminated with heavy metals suggests also a presence of other efficient detoxifying mechanisms. Insects may store metals in spherocrystals formed in the intestinal epithelium or in Malpighian tubules, bind metals to metallothioneins or glycoproteins (Jeantet et al., 1977; Martoja et al., 1983; Hopkin, 1989). They need also efficient enzymatic detoxifying systems (Chraścina et al., 1996; Migula, 1996b, 1997). This was the main reason of undertaken comparative studies focused on detoxifying abilities in various groups of arthropods (Chraścina et al., 1996; Kędziorski et al., 1996; Migula and Głowacka, 1996, 1998; Wilczek and Migula, 1996; Migula et al., 1999). Enzymatic systems involved in the reduction of metal toxicity are generally known, but the role for many of them remain unclear. Moreover, still open problem for the discussion among ecotoxicologists is the extrapolation of laboratory results to the natural conditions (Newman and Jagoe, 1996; Walker et al., 1996). Studies, that help in maintaining the relation between the exposure to xenobiotics and the detoxifying capabilities of natural populations may help in predicting their fate in specific environmental conditions (Migula et al., 1999).

Promising results of our previous studies on orthopterans stimulated us to study what are their detoxifying capabilities in relation to varying environmental load with pollutants. The grasshopper, *Chorthippus brunneus* used in our earlier studies, was selected as a test organism as its biology and ecology is relatively well known. Insects are easy to maintain in laboratory conditions, are abundant in the field and sensitive to metals (Moriarty, 1969; Fisher, 1981; Monk, 1985; Cherrill and Begon, 1989; Migula et al., 1990; Migula and Bińkowska, 1993).

Our preliminary experiments showed that glutathione *S*-transferase (GST) in *C. brunneus* was negatively correlated with metal contamination in their habitats and in the eaten plants (Augustyniak and Migula, 1996). We suggested that excess of metals was responsible for the inhibition of GST, in a similar way as this was demonstrated for mice exposed to cadmium and nickel by İşcan et al. (1995). This enzyme was also inhibited in house crickets exposed to cadmium (Kędziorski, 1993) but other studies showed rather its stimulation - in the red wood ants, *Formica polyctena* (Migula and Głowacka, 1996) or in spiders (Wilczek and Migula, 1996), all from the sites under the pressure of industrial pollutants.

Metals may affect the structure of cellular membrane, stimulating lipoperoxidative processes induced by reactive forms of oxygen (Viarengo, 1989). Superoxide dismutase (SOD), glutathione peroxidase (GPOX), glutathione reductase (GR) or catalase (CAT) are well known scavengers of free radicals (Palace and Kleverkamp, 1993; Felton and Summers, 1995; Viarengo et al., 1995). Data on the activity of GPOX in insects are also controversial. Ahmad et al. (1989) showed very low activity of Se-GPOX in some lepidopteran and dipteran species and a high activity of Se-insensitive GPOX. In this case insects had high activity of CAT which may compensate lack of Se-dependent GPOX. How far these enzymes are altered in grasshoppers from metal polluted sites has not been recognized yet.

The gut epithelium is the main barrier for metals ingested with food. Efficient detoxifying systems in these cells are of a special importance as they indirectly decide on what kind of genetic material will be successfully transferred to the next generation (Konno and Shishido, 1992). As metals are not detoxified enzymatically they can influence directly or indirectly the functional abilities in some of these systems. Relations between the level of environmental contamination and activity patterns of such enzymes as glutathione *S*-transferase, carboxylesterases or enzymes acting as free radical scavengers remain unclear.

The main aims for this study were as follows:

1. To characterize relations between body burdens of *C. brunneus* with metals and the level of metal contamination in their natural biota.
2. To check whether intensity of selected enzymatic processes responsible for biotransformation of xenobiotics differ in relation to the level of industrial pollution.
3. To determine whether differences of activity patterns of selected enzymes between grasshoppers are gender and organ dependent.

Grasshoppers from the populations living in weakly and highly polluted sites were also used to study the effects of short and long-term exposure to cadmium. We compared whether metal loads in the intestinal epithelium and gonads differ between these two groups of insects exposure to different doses of cadmium in their diet for a short time and for the full time of larval development. We checked also the relations of time-dependent cadmium load and changes in the detoxifying enzyme activity patterns.

2. The area, materials and methods

2.1. The area

Field studies were carried out at five plots, selected on the basis of data on environmental pollution collected from national monitoring stations. The criteria of site selection were the level of fallen and suspended dust particles, their heavy metal contents (lead, cadmium, zinc, nickel, copper), distance from larger industrial plants, geographic localization of the biotopes, abundance of *C. brunneus* in selected plots, similarity of plant communities and presence of similar grass species – potential hosts of grasshoppers. Two study areas were selected in the heart of Upper Silesian industrial region – the most polluted part in southern Poland, Katowice-Szopienice (further the shorter name Szopienice will be used) and Sosnowiec. Two other sites were selected on the edges of the Upper Silesia: Tychy and Kokotek, both moderately polluted with heavy metals. The reference site was in unpolluted Roztoki Górne (further named Roztoki) in Bieszczady Massif (Beskidy Mountains) in south-eastern Poland, close to the border with Ukraine. For comparison of the environmental quality and air pollution intensity a cumulative I_{15} index based on the data for 15 various pollutants with known permissible levels, established in Polish environmental legislative regulation: the annual fallen dust particles, lead and cadmium fall out, average concentration of SO_2, phenols, formaldehyde, ammonia, Pb, Cd, Mn in suspended dust particles, and benzo[*a*]pyrene in the air. The following equation was used for calculation of the index:

$$I_{15} = C_1/C_{p1} + C_2/C_{p2} + \cdots + C_n/C_{pn}$$

where: I_{15} is the index of pollution, C_1, C_2, C_n are the mean concentrations of particular contaminants in a given area; C_{p1}, C_{p2}, C_{pn} are the maximal permissible levels of contaminants (SANEPID, 1998; Godzik et al., 1999).

Data for the reference site are lacking, due to its localization outside the monitoring area. The I_{15} index, calculated for experimental areas, is equal for Katowice-Szopienice (22.7), Sosnowiec (12.8), Kokotek (9.8) and Tychy (8.2) (SANEPID, 1998). For laboratory feeding experiments with cadmium insects were collected in Szopienice and Tychy.

2.2. Insects

The grasshopper *C. (Glyptobothrus) brunneus* (Orthoptera, Acrididae) is a cosmopolite species living in open grasslands (Moriarty, 1969; Cherrill and Begon, 1989). Eggs are obligatorily diapausing. With increased density higher mortality and cannibalism is induced. The most sensitive to environmental pollutants are the second instar nymphs (Wall and Begon, 1986). They are also present in industrially altered environments, used as bioindicator species for metal contamination (Hunter et al., 1987a,b; Migula et al., 1990; Migula and Bińkowska, 1993).

Adult grasshoppers were collected at each of the selected areas during June and July and kept in large plastic containers, 10–15 individuals in each. They were fed on

Dactylis glomerata leaves collected at their sites before they had been used for enzymatic assays or metal content analysis.

Short-term effects of cadmium were studied in adult insects caught at sites in Tychy and Szopienice. After 7 days of acclimation to laboratory conditions insects were selected into four groups. Ten to fifteen individuals were kept in each container, and received leaves of *D. glomerata* pre-soaked for 24 h in water solution containing 0, 1, 2 and 10 μmol $CdCl_2$. Concentration of cadmium determined in offered leaves was as follows: control, 0.36, C1, 2.85, C2, 7.26 and C3, 36.05 μg Cd/g dry weight. The photoperiod was 16L:8D with artificial illumination added to the natural light. Temperature was 25.3 ± 1.5°C and 69 ± 5.3% relative humidity.

For the assessment of long-term effects of cadmium nymphs of the first stage were collected in May from experimental plots in Tychy and Szopienice and kept in laboratory as long as they reached the adult stage. Insects were divided into similar four groups (separately for each site) and fed in the same way as described above.

2.3. Enzyme activity assays

Before preparation insects were anaesthetized on ice and their midgut and gonads were isolated in a cold medium of 1.15% KCl (Benke and Wilkinson, 1971). Washed intestines, ovaries and testes isolated from about ten individuals were used as one sample. The isolates were weighed and homogenized in 5 ml 0.05 M phosphate buffer (pH 7.4) using Teflon-glass homogenizer, at 0–4°C. Homogenates were filtered and centrifuged at 15 000 × *g* for 10 min (Kotze and Rose, 1987). The supernatant was used as a crude enzyme solution. The assays were generally done in seven replications for each combination with site, experimental group, sex and organs.

Catalase (CAT) [EC 1.11.1.6] was measured by determining the reduction of UV absorbancy in the presence of H_2O_2 at 230 nm using, UV spectrophotometer (Chance, 1954; Aebi, 1984). The specific activity unit was defined as the enzyme equivalent reducing 1 mM H_2O_2/min per mg protein.

Glutathione peroxidase (GPOX) [EC 1.11.1.9] was determined indirectly, according to Paglia and Valentine (1967), coupled with the activity of glutathione reductase. Reduction of glutathione in the presence of NADPH changes the absorbance, measured spectrophotometrically at 340 nm. Two substrates were used: H_2O_2 and cumene. The specific activity was expressed in nmol NADPH/min per mg protein.

Glutathione reductase (GR) [EC 1.6.4.2] was measured spectrophotometrically at 340 nm, as in Racker (1955). The activity unit was expressed as a change of 0.001 unit of absorbance/min per mg protein.

Glutathione *S*-transferase (GST) [EC 2.5.1.18] was assayed measuring the conjugation of CDNB (1-chloro-2,4-dinitrobenzene) or DCNB (1,2-dichloro-4-nitrobenzene) as the substrates (Yu, 1982) with glutathione. Changes in absorbance at 345 nm for DCNB and 340 nm for CDNB were recorded. The extinction coefficients were used for conversion of readings to mmol of substrate/min per mg protein (Habig et al., 1974).

Carboxylesterases (CarE) [EC 3.1.1.1] were measured with α-naphtyl as a substrate at 600 nm, according to the method of van Asperen (1962).

The protein concentration was determined by the method of Bradford (1976) using bovine albumin as a standard.

2.4. Metal content analysis

Four metals, cadmium, lead, zinc and copper, were measured in tissues of grasshoppers collected from five sites and in homogenized leaves of 14 species of grasses selected as potential host plants of *C. brunneus* (*Agropyron repens*, *Agrostis vulgaris*, *Anthoxanthum odoratum*, *Calamagrostis epigeios*, *Carex hirta*, *Cynosurus cristatus*, *D. glomerata*, *Deschampsia caespitosa*, *Festuca ovina*, *Festuca rubra*, *Juncus macer*, *Lolium perenne*, *Poa annua*, *P. pratensis*).

Content of cadmium was determined in samples from a host plant (*D. glomerata*) used in feeding laboratory experiments. Metal content was also measured in homogenates from the intestine and gonads using the same material as was used for the enzyme assays.

All samples were dried at 70°C, weighed and mineralized in 200 μl of suprapure grade nitric acid and perchloric acid mixed as 4:1. Digests were analyzed for their metal content using the atomic absorption spectrophotometer Phillips SP 9 Pye Unicam and Solaar Unicam 939 with a graphite furnace PU-93 090X as described elsewhere (Wilczek and Migula, 1996).

The data were expressed as arithmetic means ± standard error. Significance of analyzed variables were analyzed using the *t*-test. For the analysis of significance between sites and experimental groups the ANOVA and least square difference tests were applied using STATISTICA 4.0 and Statgraphics 5.1 packages for PC. Coefficients of correlation and regression equations were calculated for enzyme activity and metal loads in insects and eaten grass.

3. Results

3.1. Concentration of metals and enzyme activity in C. brunneus from variously polluted sites

3.1.1. Metal contents

Cadmium content was the lowest in the midgut and gonads of grasshoppers from the reference site (Roztoki). Testes of insects from Sosnowiec accumulated the highest level of this metal. Insects from Szopienice had most of cadmium accumulated in the intestine (Table 1).

The accumulation of cadmium was gender dependent. Content of lead was more variable and was the highest always in insects from Sosnowiec, with higher levels in the midgut of males (Table 2). The level of lead in ovaries and testes was also extremely high.

The highest level of zinc was always found in samples of the midgut and gonads from Szopienice, with contrasting low levels found in insects collected from Sosnowiec (Table 3). The level of zinc in samples from the reference site was comparable to

Table 1. Concentration of cadmium (mean ± SE in μg/g dry weight) in the midgut cells and gonads of *C. brunneus* from five sites[a]

Site	Midgut/male	Midgut/female	Testes	Ovaries
Roztoki	4.81 ± 0.27	5.35 ± 0.15	4.21 ± 0.19*	2.52 ± 0.22*
Tychy	7.31 ± 0.42**	4.67 ± 0.12**	5.36 ± 0.15***	3.29 ± 0.16***
Kokotek	7.35 ± 0.36†	5.52 ± 0.19†	5.37 ± 0.11††	4.02 ± 0.36††
Sosnowiec	7.49 ± 0.37	7.06 ± 0.44	7.98 ± 0.17	6.67 ± 0.66
Szopienice	8.56 ± 0.23	9.54 ± 0.62	5.39 ± 0.58†††	3.69 ± 0.28†††
Homogeneity	$\mathrm{R} < \underline{\mathrm{T} < \mathrm{K} < \mathrm{S}} < \mathrm{Sz}$	$\underline{\mathrm{T} < \mathrm{R} < \mathrm{K}} < \mathrm{S} < \mathrm{Sz}$	$\mathrm{R} < \underline{\mathrm{T} < \mathrm{K} < \mathrm{Sz}} < \mathrm{S}$	$\underline{\mathrm{R} < \mathbf{T}} < \mathbf{Sz} < \mathbf{K} < \mathrm{S}$

[a] R, Roztoki; T, Tychy; K, Kokotek; S, Sosnowiec; Sz, Szopienice. Homogenous groups placed in the increased order of the means are underlined and bold ($P > 0.05$; $N = 7$). The same symbols in rows denote significant differences between sexes for a given organ.

Table 2. Concentration of lead (mean ± SE in μg/g dry weight) in the midgut cells and gonads of *C. brunneus* from five sites[a]

Site	Midgut/male	Midgut/female	Testes	Ovaries
Roztoki	30.16 ± 3.19	28.94 ± 2.11	30.07 ± 2.32*	21.40 ± 1.58*
Tychy	43.20 ± 1.79**	31.73 ± 3.47**	41.52 ± 4.61	35.20 ± 4.39
Kokotek	31.94 ± 4.55***	18.00 ± 0.64***	25.16 ± 3.45	21.33 ± 2.97
Sosnowiec	52.51 ± 11.02	41.55 ± 6.51	45.92 ± 3.98	52.51 ± 11.43
Szopienice	51.72 ± 3.81†	29.46 ± 2.23†	33.40 ± 3.11	23.92 ± 3.80
Homogeneity	R < K < **T** < **Sz** < **S**	K < R < Sz < T < S	K < R < **Sz** < **T** < S	K < R < **Sz** < **T** < S

[a] R, Roztoki; T, Tychy; K, Kokotek; S, Sosnowiec; Sz, Szopienice. Homogenous groups placed in the increased order of the means are underlined and bold ($P > 0.05$; $N = 7$). The same symbols in rows denote significant differences between sexes for a given organ.

Table 3. Concentration of zinc (mean ± SE in μg/g dry weight) in the midgut cells and gonads of *C. brunneus* from five sites[a]

Site	Midgut/male	Midgut/female	Testes	Ovaries
Roztoki	77.46 ± 6.65*	102.35 ± 6.73*	76.14 ± 3.14**	97.29 ± 3.68**
Tychy	74.19 ± 4.63***	110.52 ± 3.98***	88.37 ± 3.30	72.32 ± 12.93
Kokotek	88.45 ± 4.95†	128.53 ± 7.12†	97.33 ± 4.81	104.17 ± 1.85
Sosnowiec	74.11 ± 12.78††	114.42 ± 8.07††	63.44 ± 6.31†††	92.54 ± 5.05†††
Szopienice	163.34 ± 4.81#	237.97 ± 10.63#	116.68 ± 5.14	133.32 ± 9.36
Homogeneity	S < T < R < K < Sz	R < **T** < **S** < **K** < Sz	S < **R** < **T** < *K* < Sz	T < **S** < **R** < **K** < Sz

[a] R, Roztoki; T, Tychy; K, Kokotek; S, Sosnowiec; Sz, Szopienice. Homogenous groups placed in the increased order of the means are underlined, bold and italic ($P > 0.05$; $N = 7$). The same symbols in rows denote significant differences between sexes for a given organ.

these observed in insects collected from moderately polluted sites. Copper was accumulated in a similar way in the intestine as in gonads. The highest was in samples from the most polluted site in Szopienice, but the lowest was often in insects from Tychy, not as it was expected from the reference site (Table 4).

In samples of grasses from Szopienice the level of cadmium was ten times higher than in a comparable material from the reference site (Table 5). Also other metals from the reference site were at the lowest concentration. Grasses collected at moderately polluted sites were often homogenous in respect to their metal content, but were higher than stated in plants from Roztoki.

The level of zinc in grasses was positively correlated with the amounts of zinc in the midgut cells of both sexes ($r = 0.98$; $P < 0.05$) and in ovaries of insects ($r = 0.89$; $P < 0.05$).

Bioconcentration factor calculated for grasshoppers from various sites shows clearly differences between sites indicating the highest differences between the reference and the most polluted site (Table 6).

3.1.2. Enzyme activity

GST activity (against CDNB) was always higher in the midgut of males, except the samples from Szopienice. Insects from the reference site had higher GST activity than these from polluted sites, with differences exceeding 60% (Table 7). GST activity was higher in testes than in the intestine. In ovaries of insects from Sosnowiec was nearly 6-fold lower than that showed by grasshoppers from Szopienice ($P < 0.001$). Differences in GST activity between insects from Sosnowiec and Roztoki were insignificant.

CAT activity was much lower in males, independent of the analyzed organs and site. Despite the females from Sosnowiec, ovaries showed higher CAT activity than the midgut (Table 8). The midgut cells of males from Szopienice had the highest CAT activity, three to six times higher than in other groups ($P < 0.001$). Much higher activity of CAT were also in gonads of insects from less polluted or unpolluted sites.

Activity of CarE was higher in the midgut of males. The highest level was measured in males from Kokotek, while the lowest in males from Sosnowiec (Table 9). Testes showed higher CarE activity than ovaries. Generally, gonads demonstrated lower CarE activity than the midgut cells. Reversibly, gonads of grasshoppers from Sosnowiec had the highest activity level of CarE among the analyzed groups.

The activity of CarE correlate positively with cadmium level across gonads, with $r = 0.95$ ($P < 0.01$) for ovaries and $r = 0.91$ ($P < 0.05$) for testes.

3.2. *Effects of short-term exposure to cadmium on detoxifying enzymes and the accumulation of metal*

Cadmium load in the midgut cells of grasshoppers from less polluted site (Tychy) raised proportionally to its concentration in the eaten grass, without significant differences between males and females (Tables 10 and 11).

The correlation coefficient for the accumulation of Cd in the intestine of grasshoppers from Szopienice against cadmium content in their diet was lower when compared with this for Tychy (Table 12). Maximal differences did not exceed

Table 4. Concentration of copper (mean ± SE in μg/g dry weight) in the midgut cells and gonads of *C. brunneus* from five sites[a]

Site	Midgut/male	Midgut/female	Testes	Ovaries
Roztoki	8.48 ± 1.02	14.05 ± 2.12	35.30 ± 2.03*	27.13 ± 0.86*
Tychy	10.00 ± 1.09	11.05 ± 0.89	29.52 ± 3.13**	5.42 ± 0.60**
Kokotek	11.56 ± 0.95***	23.30 ± 2.90***	33.78 ± 0.87†	11.31 ± 1.11†
Sosnowiec	24.36 ± 2.34	18.69 ± 1.36	33.01 ± 4.19††	24.69 ± 3.61††
Szopienice	27.37 ± 1.36	31.88 ± 2.75	51.44 ± 2.90†††	18.74 ± 1.69†††
Homogeneity	R < T < K < S < Sz	T < **R** < **S** < K < Sz	T < S < K < R < Sz	T < K < Sz < S < R

[a] R, Roztoki; T, Tychy; K, Kokotek; S, Sosnowiec; Sz, Szopienice. Homogenous groups placed in the increased order of the means are underlined and bold ($P > 0.05$; $N = 7$). The same symbols in rows denote significant differences between sexes for a given organ.

Table 5. Content of four metals (mean ± SE in μg/g dry weight) in mixture of grasses, a potential food of *C. brunneus* from five sites[a]

Metal	Roztoki	Tychy	Kokotek	Szopienice	Sosnowiec	Homogeneity
Pb	0.40 ± 0.10	0.81 ± 0.16	0.90 ± 0.16	1.21 ± 0.13	1.06 ± 0.13	RG < **T** < **K** < **S** < **Sz**
Cu	2.61 ± 0.30	4.96 ± 0.26	3.51 ± 0.16	5.39 ± 0.28	3.54 ± 0.36	RG < K < S < T < Sz
Cd	0.33 ± 0.07	1.81 ± 0.32	1.24 ± 0.24	3.98 ± 0.20	1.89 ± 0.23	RG < K < T < S < Sz
Zn	151.4 ± 13.1	130.3 ± 7.4	162.6 ± 19.5	440.2 ± 46.4	175.4 ± 9.8	T < RG < K < S < Sz

[a] R, Roztoki; T, Tychy; K, Kokotek; S, Sosnowiec; Sz, Szopienice. Homogenous groups placed in the increased order of the means are underlined and bold ($P > 0.05$; $N = 7$).

Table 6. Bioconcentration factor for selected metals in the midgut cells and gonads of *C. brunneus* from five sites in relation to the level of metal in the consumed food

Site	Midgut/male	Testes	Midgut/ female	Ovaries
Cadmium				
Roztoki	14.57	12.75	16.21	7.63
Tychy	4.07	2.94	2.58	1.81
Kokotek	5.92	4.33	4.45	3.24
Sosnowiec	3.96	4.22	3.73	3.52
Szopienice	2.15	1.35	2.39	0.92
Lead				
Roztoki	75.40	75.17	72.35	53.50
Tychy	53.30	51.25	39.17	43.40
Kokotek	35.48	27.95	20.00	23.70
Sosnowiec	49.53	43.32	39.20	49.50
Szopienice	42.74	27.60	24.30	19.70
Zinc				
Roztoki	0.51	0.50	0.67	0.64
Tychy	0.56	0.67	0.84	0.55
Kokotek	0.54	0.60	0.79	0.64
Sosnowiec	0.42	0.36	0.65	0.53
Szopienice	0.37	0.26	0.53	0.30
Copper				
Roztoki	0.32	13.53	5.38	10.39
Tychy	2.01	5.94	2.23	1.09
Kokotek	3.29	9.62	6.63	3.22
Sosnowiec	6.88	9.32	5.28	6.97
Szopienice	5.08	6.54	5.91	3.47

200% in females and were less than a 100% in males. The accumulation of cadmium in gonads of grasshoppers from both sites remained on a very similar level. Bioconcentration coefficient for the midgut of insects from both sites was similar in groups exposed to the highest dose of cadmium, but higher in other groups for Szopienice. Calculations made for testes indicated higher indices for males from Szopienice and similar values for ovaries.

GST activity in the midgut of both sexes from Tychy was negatively correlated with the cadmium load and was more pronounced in females from the C_3 group (Tables 11 and 12). There was lack of significance for this relationship in males from Szopienice, and significantly higher levels were only in females representing the C_3 group.

Exposure to cadmium caused a significant reduction of GST activity in ovaries of *C. brunneus* from Tychy. For testes this was observed in insects kept on leaves with the highest concentration of cadmium. The response of insects from Szopienice was different, since a slight increase of GST activity has been indicated for testes with lack of significant changes in GST activity in ovaries (Fig. 1).

Table 7. Activity of glutathione *S*-transferase (mean ± SE in μmol CDNB/min per mg protein) in the midgut cells and gonads of *C. brunneus* from sites that differ in the level of industrial pollution[a]

Site	Midgut/male	Midgut/female	Testes	Ovaries
Roztoki	34.1 ± 2.1	28.3 ± 2.4	62.2 ± 1.7*	16.8 ± 1.9*
Tychy	30.0 ± 3.0**	21.7 ± 1.7**	68.6 ± 2.0***	52.8 ± 2.5****
Kokotek	20.7 ± 1.5#	15.8 ± 1.3#	50.5 ± 2.2##	34.3 ± 2.7##
Sosnowiec	12.8 ± 0.8###	8.1 ± 1.0###	71.7 ± 5.2†	9.7 ± 0.6†
Szopienice	20.5 ± 1.4	22.3 ± 1.7	46.3 ± 3.9	57.8 ± 6.4
Homogeneity	S < Sz < K < T < R	S < K < T < Sz < R	Sz < K < R < T < S	S < R < K < T < Sz

[a] R, Roztoki; T, Tychy; K, Kokotek; S, Sosnowiec; Sz, Szopienice. Homogenous groups placed in the increased order of the means are underlined ($P > 0.05$; $N = 7$) The same symbols in rows denote significant differences between sexes for a given organ.

Table 8. Activity of catalase (mean ± SE in mmol H_2O_2/min per mg protein) in the midgut cells and gonads of *C. brunneus* from sites which differ in the level of industrial pollution[a]

Site	Midgut/male	Midgut/female	Testes	Ovaries
Roztoki	20.2 ± 2.0*	804.4 ± 36.5*	20.4 ± 0.5**	1560.9 ± 115.0**
Tychy	11.6 ± 1.7***	824.9 ± 38.2***	72.4 ± 5.9#	1409.7 ± 85.9#
Kokotek	20.1 ± 2.1##	584.8 ± 47.0##	52.5 ± 5.9###	785.6 ± 65.8###
Sosnowiec	22.8 ± 1.2†	305.5 ± 23.7†	2.3 ± 0.4††	200.7 ± 27.2††
Szopienice	64.9 ± 2.9†††	729.4 ± 20.9†††	48.0 ± 7.2****	1241.9 ± 85.2****
Homogeneity	T < K < R < S < Sz	S < K < Sz < R < T	S < R < Sz < K < T	S < K < Sz < **T** < **R**

[a] R, Roztoki; T, Tychy; K, Kokotek; S, Sosnowiec; Sz, Szopienice. Homogenous groups placed in the increased order of the means are underlined and bold ($P > 0.05$; $N = 7$) The same symbols in rows denote significant differences between sexes for a given organ.

Table 9. Activity of carboxylesterases (mean ± SE in μmol α-naphtyl/min per mg protein) in the midgut cells and gonads of *C. brunneus* from five sites[a]

Site	Midgut/male	Midgut/female	Testes	Ovaries
Roztoki	1.27 ± 0.04*	0.98 ± 0.09*	0.19 ± 0.02**	0.17 ± 0.01**
Tychy	1.46 ± 0.23***	0.87 ± 0.05***	0.23 ± 0.01#	0.15 ± 0.01#
Kokotek	2.06 ± 0.09##	1.30 ± 0.08##	0.24 ± 0.01###	0.17 ± 0.01###
Sosnowiec	0.58 ± 0.14†	1.20 ± 0.08†	0.30 ± 0.02	0.28 ± 0.02
Szopienice	1.36 ± 0.05††	0.89 ± 0.07††	0.21 ± 0.01	0.19 ± 0.01
Homogeneity	S < R < Sz < T < K	T < Sz < **R** < **S** < K	R < **Sz** < **T** < **K** < S	T < **K** < **R** < **Sz** < S

[a] R, Roztoki; T, Tychy; K, Kokotek; S, Sosnowiec; Sz, Szopienice. Homogenous groups placed in the increased order of the means are underlined and bold ($P > 0.05$; $N = 7$) The same symbols in rows denote significant differences between sexes for a given organ.

Table 10. Bioconcentration index for the midgut cells and gonads of grasshoppers from Tychy and Szopienice kept for 7 days or during the whole larval development (l.-term) on a diet containing different concentrations of cadmium (C_1–C_3)

Group	Midgut/male		Testes		Midgut/female		Ovaries	
	7 days	l.-term	7 days	l.-term	7 days	l.-term	7 days	l.-term
Tychy								
Control	16.8	4.1	10.4	3.1	25.4	7.4	9.3	4.0
C_1	2.6	1.2	1.3	0.4	3.8	1.5	1.2	0.6
C_2	1.2	2.6	0.5	0.4	1.7	2.8	0.5	0.3
C_3	0.4	1.0	0.1	0.1	1.2	0.9	0.1	0.1
Szopienice								
Control	20.9	8.6	13.4	2.3	25.4	9.0	9.9	3.4
C_1	3.1	1.1	1.6	1.7	3.6	1.5	1.2	0.9
C_2	1.9	3.9	0.7	0.5	1.9	4.7	0.5	0.4
C_3	0.4	1.1	0.1	0.1	0.7	1.6	0.1	0.2

Cadmium treated insects from Tychy had catalase activity 35–45% lower than the reference group. For groups from Szopienice the opposite pattern was observed – the exposure to cadmium stimulated CAT activity (Fig. 2). In gonads of insects from less polluted sites CAT activity was negatively correlated with the Cd load, but positively correlated in case of ovaries from Szopienice (Table 12). A correlation between CAT activity and Cd exposure in testes was not found.

GPOX activity measured against H_2O_2 in the midgut of both sexes from Tychy strongly increased, parallel with the cadmium load of the epithelial cells (Table 12).

For the same organ from Szopienice a similar pattern of GPOX activity was found, with slightly indicated positive correlation for males and generally lower activity level of this enzyme (Fig. 3; Table 12). Induction of GPOX in ovaries was stated only in group C3 from both sites. Increased GPOX activity was more pronounced in testes of males from Szopienice. The use of cumene as another substrate showed similar functional patterns of GPOX affected by cadmium and quantitative differences indicated only effects of the another substrate used (Fig. 4). GR showed a reduced activity in the midgut of females from both sites. In males the effect of cadmium was less visible, showing both the inhibition (Tychy) and a strong stimulation in C_2 group from Szopienice (Fig. 5). In testes of insects from Tychy population a binomial character of Cd induced changes in the enzyme activity was observed. In a comparable material from Szopienice significantly lower activity predominated, with no effects caused by cadmium exposure. Reduced GR activity after cadmium treatment was shown in ovaries of females from Tychy.

Activity of carboxylesterases in the midgut of males from both sites was strongly stimulated. The same was observed in testes (Fig. 6). A strong positive correlation between CarE activity and cadmium load was indicated for the midgut of females from Szopienice, without significant changes of this parameter in material from Tychy (Table 12). A weak but positive correlation was also indicated for gonads and

Table 11. Concentration of cadmium in the midgut cells and gonads of adult *C. brunneus* from two sites kept for 7 days on *D. glomerata* with different concentrations of cadmium[a]

Group/site	Midgut/male	Midgut/female	Testes	Ovaries
Tychy				
K	6.03 ± 0.96*	9.09 ± 0.35*	3.73 ± 0.88	3.34 ± 0.62
C1	7.35 ± 0.44**	10.79 ± 0.30**	3.72 ± 0.36	3.34 ± 0.48
C2	8.64 ± 1.12	12.14 ± 1.41	3.65 ± 0.48	3.71 ± 0.43
C3	14.32 ± 0.57***	41.72 ± 7.09***	3.70 ± 0.39	3.96 ± 0.06
Homogeneity	K < C1 < C2 < C3	K < C1 < C2 < C3	C2 < C3 < C1 < K	K = C1 < C2 < C3
Szopienice				
K	7.51 ± 0.35	9.10 ± 0.61	4.80 ± 0.36*	3.55 ± 0.30*
C1	8.69 ± 0.48	10.16 ± 0.47	4.62 ± 0.35	3.50 ± 0.29
C2	9.34 ± 0.38**	14.35 ± 1.05**	4.97 ± 0.53	3.66 ± 0.33
C3	14.14 ± 1.33***	26.94 ± 1.88***	4.51 ± 0.34	3.85 ± 0.17
Homogeneity	K < C1 < C2 < C3	K < C1 < C2 < C3	C3 < C1 < K < C2	C1 < K < C2 < C3

[a] Values are mean ± SE in μg/g dry weight. $N = 4$–7 determinations for each organ. K, control group; C1–C3, experimental groups kept on grass with different concentration of cadmium (C1, 2.85, C2, 7.26 and C3, 36.05 μg Cd/g dry weight, respectively). Homogenous groups within a given organ and sex are underlined ($P > 0.05$). The same symbol in the rows denotes significant differences between sexes for a given organ ($P < 0.05$).

Table 12. Correlation coefficients (*r*) and angular coefficients (*b*) of regressions of cadmium in a grasshopper organs, their food (μg/g dry weight) and enzyme activity in grasshoppers from two sites exposed for 7 days to cadmium[a,b]

Site	Organ/gender	*Y*	*X*	*b*	*r*	*P*
T	Midg/M.	Cadmium	Cd – grass	2.19	0.99	<0.01
T	Midg/F	Cadmium	Cd – grass	9.38	0.99	<0.01
Sz	Midg/M.	Cadmium	Cd – grass	0.17	0.99	<0.01
Sz	Midg/F	Cadmium	Cd – grass	0.49	0.99	<0.01
T	Midg/F	GPOX (H_2O_2)	Cd – midgut	22.61	0.98	<0.05
T	Midg/F	GPOX (H_2O_2)	GPOX (cumin)	1.21	0.99	<0.01
T	Midg/F	GPOX (H_2O_2)	GST	−3.11	−0.97	<0.01
T	Midg/M	GPOX (H_2O_2)	Cd – midgut	18.21	0.90	<0.05
T	Midg/M	GPOX (H_2O_2)	GPOX (cumin)	0.59	0.96	<0.01
Sz	Midg/F	GPOX (H_2O_2)	Cd – midgut	0.66	0.99	<0.01
Sz	Midg/F	GPOX (H_2O_2)	GPOX (cumin)	0.79	0.92	<0.05
Sz	Midg/M	GPOX (H_2O_2)	CarE	10.32	0.98	<0.01
T	Ovar/F	GPOX (H_2O_2)	GPOX (cumin)	1.99	0.99	<0.01
Sz	Tes/M	GPOX (H_2O_2)	GPOX (cumin)	1.45	0.96	<0.01
T	Midg/F	GPOX (cumin)	Cd – midgut	18.34	0.98	<0.01
T	Midg/M	GPOX (cumin)	Cd – midgut	31.86	0.98	<0.01
T	Midg/F	GPOX (cumin)	GST	−2.51	−0.95	<0.01

[a] Various units as described in Section 2.

[b] Regression follows the model $Y = a \pm bX$; *P*, level of significance for difference of *r* from zero. Abbreviations: T, Tychy; Sz, Szopienice; M, male; F, Female; Midg, midgut cells; Ovar, ovaries; Tes, testes.

cadmium in both sexes. This was more pronounced in samples from Szopienice (Table 12).

3.3. Effects of long-term exposure to cadmium on detoxifying enzymes and accumulation of metal

Male grasshoppers from Tychy kept during the larval developmental on a diet containing excess of cadmium accumulated in the midgut 2.5, 13 and 24.5 times more cadmium in groups C1–C3 respectively than did the control insects (Table 13). A similar relation between metal content in eaten grass and in the intestine was documented for the females. Insects from Szopienice had more cadmium in the midgut than insects from Tychy. Metal load in gonads of both sexes was significantly higher in males from groups C2–C3, but in ovaries only in insects exposed to the highest dose of cadmium added to consumed grass. Transfer of metal to gonads in grasshoppers from Szopienice was slightly higher when compared with a similar material from Tychy (Table 13).

Enzymatic assays did not confirm differences in the activity of GST measured against DCNB in the intestine of both sexes (Fig. 7). Insects from Szopienice showed generally higher activity of this enzyme, and positive correlation with the metal load ($P < 0.05$). There was lack of significant changes in GST activity measured in

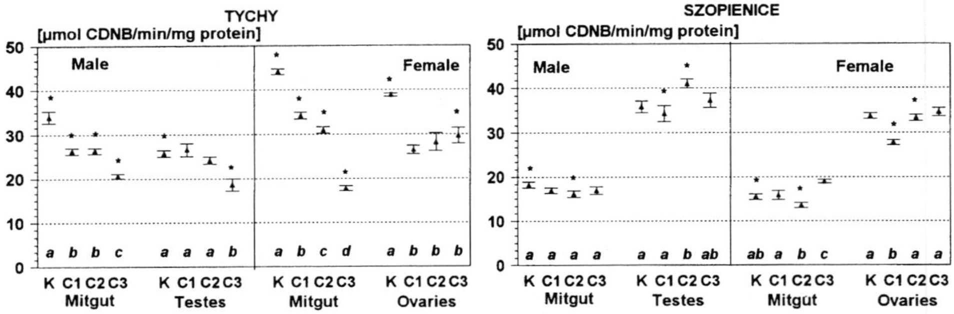

Figure 1. Activity of GST in the midgut cells and gonads of adult *C. brunneus* from two sites kept for 7 days on *D. glomerata* with different concentrations of cadmium. Means ± SE. Different letters denote significant differences between experimental groups within the same organ and gender. Stairs indicate significant differences between gender within the same organ and experimental group. K, control group; C1–C3, experimental groups kept on grass with different concentration of cadmium (C1, 2.85, C2, 7.26 and C3, 36.05 μg Cd/g dry weight).

testes of males from both sites and in ovaries of females from Szopienice. Induction of this enzyme by cadmium was found only in ovaries of females from Tychy exposed to the lowest of the used metal dose.

In case of CarE there was lack of significant differences between the mean activity in the midgut of males from both sites. Reversibly, females from both sites showed a stimulation of CarE activity in this organ (Fig. 8). Ovaries of females from Szopienice had carboxylesterases unaffected by cadmium, with higher activity level than ovaries of females from Tychy. Induction of this enzyme by cadmium was observed in testes, with more pronounced effects in males from Szopienice (Fig. 8).

4. Discussion

4.1. To what extent the accumulation of metals in grasshoppers may reflect the environmental contamination?

Our studies demonstrated that burden with metals was the highest in grasshoppers from Szopienice. Only higher amounts of lead were accumulated in insects from Sosnowiec. The reason of this was in a close neighbourhood a double carriage way, with a high traffic intensity of many cars and trucks using unleaded fuel.

It has been shown that insects differ in abilities to accumulate metals (Hunter et al., 1987a,b; Hopkin, 1989; Migula et al., 1990; Wilczek and Migula, 1996; Głowacka et al., 1997). Van Capelleveen (1987) clearly demonstrated that higher tolerance of the terrestrial isopod *Porcellio scaber* from polluted sites reflected its ability to accumulate cadmium and zinc. This was also true for strongly accumulating metals orthopterans, *Gryllus* sp. and *Melanoplus* sp., living in ash heaps from electric power plants (Brieger et al., 1992). These species are postulated as good bioindicator species for polluted areas (Schmidt et al., 1992).

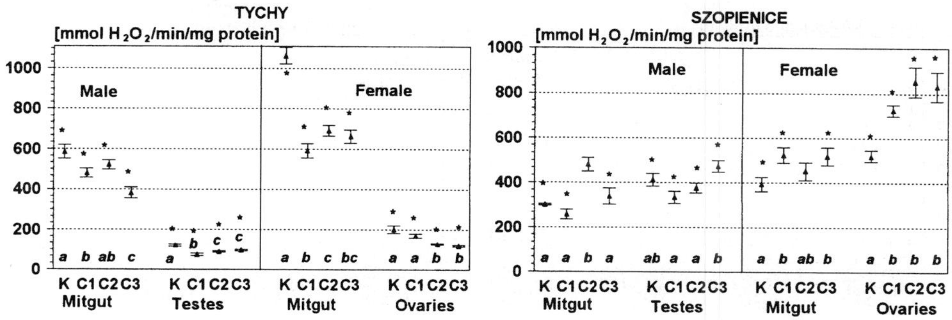

Figure 2. Activity of catalase in the midgut cells and gonads of adult *C. brunneus* from two sites kept for 7 days on *D. glomerata* with different concentrations of cadmium. Explanations as in Fig. 1.

Comparisons of body zinc in *C. brunneus* with the metal content in their habitats showed a close relationship between these variables, especially for the areas exposed to contaminants from non-ferrous metal plants from their vicinity (Szopienice). Grasshoppers collected from Kokotek showed also high zinc loads in gonads resulted from direct influence of the zinc plant in the vicinity (Miasteczko Śląskie).

Metal load in grasses collected in selected experimental plots showed the highest levels in Szopienice (Table 5). A comparison of bioconcentration factors shows that these metals are partly transferred along the food chains (Dallinger, 1993). Differences between sites may reflect the fact that grasshoppers can selectively eat only some parts of the leaves, with lower amounts of metals, as this was demonstrated in earlier studies (Migula and Bińkowska, 1993). Also caterpillars of *Deilephila elpenor* kept on leaves with cadmium left uneaten parts where concentration of this metal was twice as high as the average (Nuorteva et al., 1992).

Heavy metals are unevenly distributed within tissues. High amounts are typical for the intestine, as these epithelial cells are highly exposed to metals (Migula et al., 1990; Lindqvist, 1993). Through the elimination of degenerated midgut cells loaded with

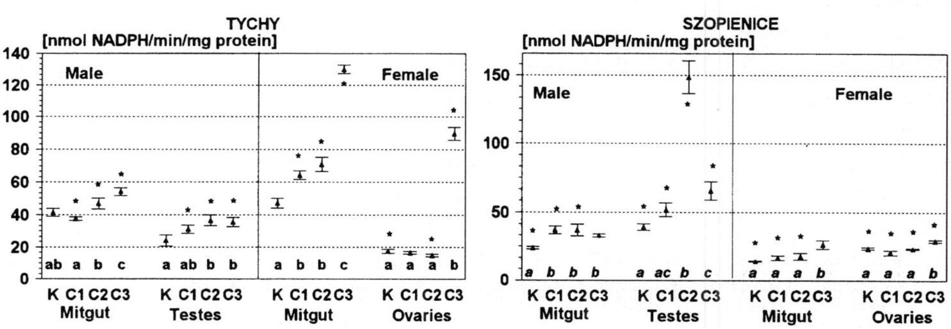

Figure 3. Activity of glutathione peroxidase measured against H_2O_2 in the midgut cells and gonads of adult *C. brunneus* from two sites kept for 7 days on *D. glomerata* with different concentrations of cadmium. Explanations as in Fig. 1.

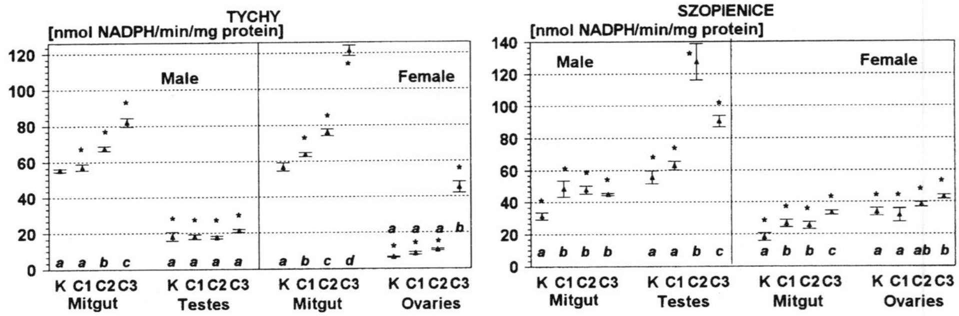

Figure 4. Activity of glutathione peroxidase measured against cumene hydroperoxide in the midgut cells and gonads of adult *C. brunneus* from two sites kept for 7 days on *D. glomerata* with different concentrations of cadmium. Explanations as in Fig. 1.

metals insects are able to reduce their excess, protecting other parts of the body (Hopkin, 1990; Rabitsch, 1997). Such mechanism may be used only to a certain extent. At higher concentrations metals may reach other organs, also gonads, transported with external body fluids (Hunter et al., 1987a; Hopkin, 1989; Dallinger, 1993). In *C. brunneus* such increased accumulation of metals was observed in gonads, with the exception of insects from Szopienice. For these insects the bioconcentration index for cadmium was two to ten times lower as compared with insects from other less polluted sites. A similar trend was also observed for other metals, but with less visible differences. This means that also at the level of a population grasshoppers have different abilities to accumulate metals, with higher tolerance to metals of insects from a heavily polluted site. Occasional contacts with the metal contaminated food may not be dangerous for grasshoppers. A short-time feeding with excessive amounts of cadmium showed that transfer of this metals to gonads was insignificant.

Excess of cadmium in the diet of *Porcellio scaber* from three sites that varied in metal pollution level was accumulated in a similar rate (Van Capelleveen, 1987).

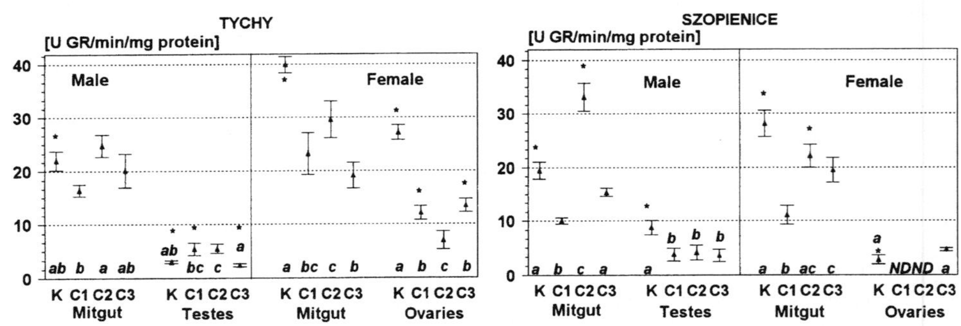

Figure 5. Activity of glutathione reductase in the midgut cells and gonads of adult *C. brunneus* from two sites kept for 7 days on *D. glomerata* with different concentrations of cadmium. Explanations as in Fig. 1.

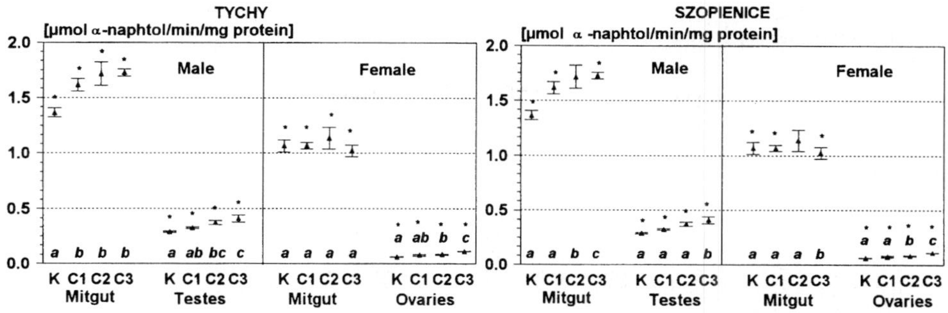

Figure 6. Activity of carboxylesterases measured in the midgut cells and gonads of adult *C. brunneus* from two sites kept for 7 days on *D. glomerata* with different concentrations of cadmium. Explanations as in Fig. 1.

Insects from unpolluted site consumed less food and were grown slower. Also in grasshoppers fed on cadmium the content of this metal in the midgut cells was similar in individuals originated from both polluted and less polluted sites, reflecting its concentration in leaves, and the bioconcentration index was always higher for the intestine of insects from Szopienice, but not for the gonads. Insects from Tychy accumulated less cadmium but reached the maturity later than insects from Szopienice. This latter population showed better adaptation to heavy metal loads.

4.2. Detoxifying abilities of C. brunneus and metal pollution

The enzymes selected for bioassays in our studies are known as very sensitive for various environmental pollutants, including metals (Yu, 1988; Ishikawa and Kubota, 1991; Migula et al., 1993; Kędziorski et al., 1996). Comparative studies demonstrated a close relationship between the environmental conditions and the activity of GST, with distinct differences even among closely related species (Stenersen et al., 1987; Isman et al., 1996). Our results indicated that differences between sexes should be also taken into account, as in most cases males had higher level of GST activity. Activity patterns of GST in insects from industrially polluted areas may result from a phenotypic response of this inducible system, with increased activity when exposed to a xenobiotic (Yu, 1982, 1988) or, reversibly, decreased activity with mechanisms engaged in reducing the cellular level of glutathione are intensified. Díaz-Cruz et al. (1997) showed that cadmium may form different complexes with glutathione that are pH dependent ($Cd[GSH]_2$ or $Cd_2[GSH]_2$). Such complexes were formed under cadmium and/or nickel exposure (İşcan et al., 1995) with concomitant decrease of the GSH level and fall in the level of GST (against DCNB) in liver of mice exposed to cadmium and nickel, but increased GST activity when exposed to the surplus of both metals. Cadmium reduced GST activity in aquatic invertebrates (Almar et al., 1987) or insects (Kędziorski, 1993; Migula and Głowacka, 1996) where cadmium had stronger inhibitory effects than lead. Metals are presumed as direct inhibitors of active centre of the enzyme (Kędziorski, 1993). Kirby and Ottea (1995) showed that many inducers may control expression of GST that are species dependent. In case of *C. brunneus*

Table 13. Concentration of cadmium in the midgut cells and gonads of adult *C. brunneus* from two sites fed during a whole larval development with *D. glomerata* containing different concentrations of cadmium[a]

Group/site	Midgut/male	Midgut/female	Testes	Ovaries
Tychy				
K	1.46 ± 0.22*	2.64 ± 0.39*	1.13 ± 0,28	1.42 ± 0.33
C1	3.55 ± 0.48	4.35 ± 0.45	1.18 ± 0,22	1.71 ± 0.18
C2	19.19 ± 2.33	20.62 ± 3.48	2.72 ± 0,75	2.46 ± 0.34
C3	35.87 ± 4.75	34.44 ± 4.63	2.65 ± 0,27	3.37 ± 0.50
Homogeneity	K < C1 < C2 < C3	K < C1 < C2 < C3	K < **C1** < **C3** < **C2**	K < C1 < **C2** < **C3**
Szopienice				
K	3.02 ± 0.22	3.23 ± 0.22	0.84 ± 0.08	1.21 ± 0.29
C1	3.28 ± 0.48	4.17 ± 0.32	3.05 ± 0.30	2.68 ± 0.32
C2	28.99 ± 2.13	34.22 ± 1.94	3.95 ± 0.34**	3.02 ± 0.15**
C3	41.19 ± 4.79***	58.84 ± 4.75***	4.61 ± 0.49	5.72 ± 0.49
Homogeneity	K < C1 < C2 < C3	K < C1 < C2 < C3	K < C1 < **C2** < **C3**	K < C1 < C2 < C3

[a] Values are mean ± SE in μg/g dry weight. $N = 4$–7 determinations for each organ. K, control group; C1–C3, experimental groups kept on grass with different concentration of cadmium (C1, 2.85, C2, 7.26 and C3, 36.05 μg Cd/g dry weight, respectively). Homogenous groups within a given organ and sex are underlined and bold ($P > 0.05$). The same symbol in the rows denotes significant differences between sexes for a given organ ($P < 0.05$).

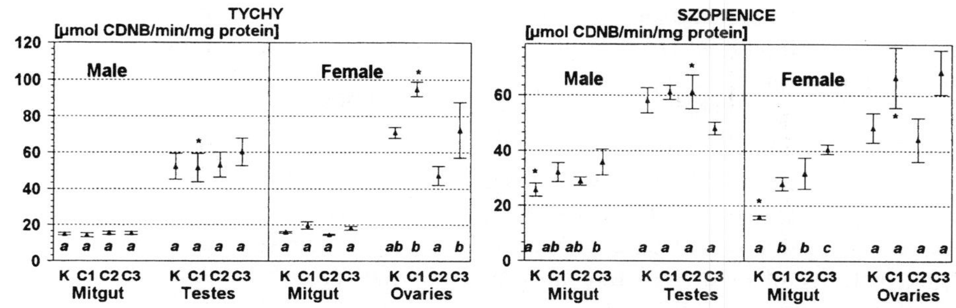

Figure 7. Activity of GST in the midgut cells and gonads of adult *C. brunneus* from two sites fed during a whole larval development with *D. glomerata* containing different concentrations of cadmium. Explanations as in Fig. 1.

from polluted sites this was also a result of metal pollution. Conjugation processes activated by GST were more efficient in gonads of studied grasshoppers. Also Konno and Shishido (1992) found high activity of this enzyme in Malpighian tubules and testes of *Periplaneta americana*. Intensified reactions protecting produced genetic material during the breeding period against cancerogenic and mutagenic agents were also demonstrated or other enzymes, such as catalase and esterases (Lee and Berenbaum, 1992) and explaining differences of GST level between localities.

Carboxylesterases in the midgut of grasshoppers did not correlate well with the degree of pollution. This enzyme can be more altered by factors other than metals, such as organic compounds or the composition of diet. Kędziorski et al. (1996) demonstrated that in insects fed with excess of cadmium CarE in the midgut of *Acheta domestica* was inhibited in all developmental stages but in older larvae of *Gromphadorhina portentosa* was stimulated. Marczyk et al. (1993) and Wilczek and Migula (1996) showed lower CarE activity in spiders from polluted sites. Opposite to this was increased enzyme activity found in the red wood ants *Formica polyctena* from metal-polluted areas and a high positive correlation between Cd, Pb, Zn and Cu content in ants with their CarE or GST activity (Migula and Głowacka, 1996). Similar stimulation of CarE activity by cadmium was obtained

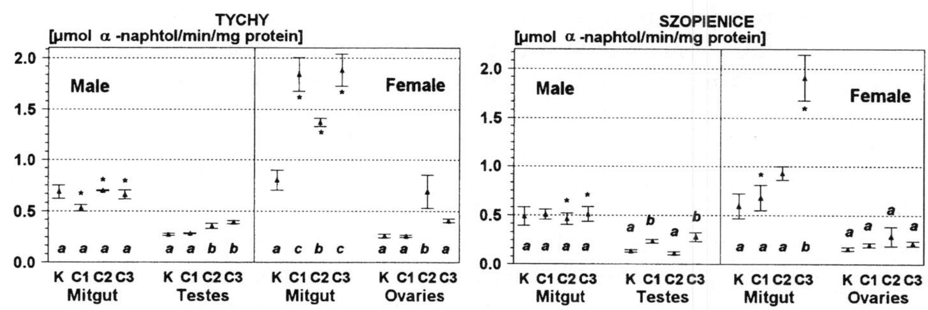

Figure 8. Activity of carboxylesterases in the midgut cells and gonads of adult *C. brunneus* from two sites fed during a whole larval development with *D. glomerata* containing different concentrations of cadmium. Explanations as in Fig. 1.

in *C. brunneus* after exposure to cadmium in laboratory experiments in both short and long-term treatment (Figs. 6 and 8). This kind of response is more comparable with reactions of such herbivores as *A. domestica* than with hunting spiders or omnivorous insects, excluding ants.

Metals (Cd, Cu, Cr, Fe, Pb, Hg, Ni or V) through catalytic activity in the Fenton reaction may act as prooxidants (Ahmad, 1995). We expected that catalase may reveal higher activity in grasshoppers from more contaminated sites, but this was not proved (Table 8). In case of males a low CAT activity was compensated by higher activity of other free radical scavengers and probably higher production of glutathione (Viarengo, 1989; Viarengo et al., 1997). CAT is inhibited in vitro by cadmium (Shukla et al., 1989). In the liver of freshwater fish cadmium was also responsible for the inhibition of CAT but stimulated SOD activity (Palace and Kleverkamp, 1993). Mechanism of cadmium toxicity for CAT is made by their binding with –SH groups. CAT has six such groups. The inhibition is indirect as these groups are not in the its active centre. Low activity levels of CAT in *C. brunneus* from polluted sites might result from higher levels of metals and increased pool of glutathione or other antioxidants such as ascorbate were probably used in compensatory mechanisms reducing high oxidative stress, as laboratory experiments indicated a stimulatory effects of cadmium on glutathione peroxidase activity (Figs. 3 and 4).

4.3. Sex and detoxifying capabilities of C. brunneus

Most of the enzymatic assays in insects were done without taking care on sexual differentiation of the material. Studying GST and CarE in five species of *Melanoplus* grasshoppers, Isman et al. (1996) did not differentiate samples. In orthopterans females are often larger than males and consume more food, cumulating higher amounts of toxins that need to be detoxified. Also reproductive processes differ between sexes. CarE were at a similar level in salivary glands of the dipteran *Aedes aegypti* but 2–3 days after eclosion increased in females. In males it remained at the same level (Argentine and James, 1995). Konno and Shishido (1992) found twice as high activity of GST in testes than in ovaries of *P. americana*. In studied grasshoppers CarE and GST was generally higher in males (Tables 7 and 9). Sexual behaviour of these insects suggest higher energetic output in aggressively active males (Cade, 1985) compensated with larger amounts of eaten contaminated leaves that partially explain gender related differences in activity of CarE and GST. This is in agreement with the results Frohlich (1990) who studied bees *Megachile rotundata* where production of sperm needed higher activity of many enzymes, including esterases. In *C. brunneus* exposed to cadmium CarE or GST were higher in females, opposite to the situation realized in field sampling. Probably a similar mechanism is involved as in *M. rotundata*, where activity of these enzymes decreased with the age of insects. This was not observed in the field as the insects were collected in the beginning of the reproductive phase in June–July and laboratory experiments were carried out later when insects in natural conditions laid cocoons with eggs.

4.4. Short and long term exposure to cadmium and detoxifying capabilities of C. brunneus

Cadmium was selected for laboratory experiments as its toxic effects on insects were well proved (Martoja et al., 1983; Hopkin, 1989; Heliövaara and Väisänen, 1993). Its effect on production of free radicals has been discussed above. By elimination of zinc from metallothioneins cadmium may indirectly stimulate reactive oxygen particles (Viarengo, 1989). Direct inhibitory effects on detoxifying enzymes have been well documented (Byczkowski and Sorenson, 1984; Palace and Kleverkamp, 1993). Enzyme activity pattern under cadmium stress shows a cumulative effects of many interacting factors.

In *C. brunneus* cadmium generated free radicals causing an increase of GPOX activity. For many years the presence of GPOX against H_2O_2 has been neglected (Ahmad et al., 1988). More recent studies showed its activity of about 20 nmol NADPH/min per mg protein in Papilio polyxenes, and much higher (47 nmol/min per mg protein) in last stage caterpillars of *Spodoptera eridaria* (Ahmad et al., 1989). Mean activity level of this enzyme measured in *C. brunneus* was in control groups in the range of 14.07–47.06 nmol/min per mg protein, placing this species among the insects with the higher activity of this enzyme. Activity of GPOX was similar when cumene as another substrate was used, proving directly that in both reactions the same, Se-dependent enzyme, was involved.

A similar pattern of GPOX, GST and CAT for *C. brunneus* from Tychy population exposed shortly to cadmium was observed as in the mollusc *Mytilus galloprovincialis* from waters polluted with metals (Regoli and Principato, 1995). In this species placed in water highly polluted with metals the activity of GPOX significantly increased. Reversibly, CAT and GST of control bivalves transferred to metal polluted water significantly decreased after 1, 3 and 6 days. In *C. brunneus* from a polluted site GST activity remained at the same level in males or increased in females exposed to cadmium. Insects from the reference site showed a decrease of this enzyme activity (Fig. 1). This was probably due to a higher glutathione level in grasshoppers from Szopienice that helped them in binding and eliminating cadmium outside the cells (Viarengo, 1989; Díaz-Cruz et al., 1997). Viarengo et al. (1997) demonstrated that Cu or Hg at 100 nM and 1 μM concentration significantly reduced microsomal detoxifying enzymes and addition of glutathione eliminated this inhibitory effect. Probably in grasshoppers from Tychy reduced GST activity resulted from their deficiency in GSH, while insect from Szopienice pre-adapted to metals were better equipped with this tripeptide.

GR co-operates with GPOX and GST in regeneration of reduced glutathione (Pardini, 1995). Reduced GR activity might be expected when glutathione partly bound with cadmium is eliminated from the cell. Our experiments partly confirm this assumption. Intestinal GR had lower activity in insects exposed to cadmium. Similar effects were shown in studies of Regoli and Principato (1995) with reduced glutathione level and fall in GR activity in hepatopancreas of *M. galloprovincialis*, after its short exposure to metals.

Comparing short-term effects of cadmium on insects from both populations we may assume that activity patterns for GST and CAT are different, but similar are for GPOX,

GR and CarE (Figs. 1–6), demonstrating higher metal tolerance in grasshoppers from highly polluted site (Szopienice).

A long-term effect of metal may more easily develop compensatory reactions increasing their tolerance within a given genetic range (Donker and Bogert, 1992; Posthuma and Van Straalen, 1993). Higher tolerance limits were indicated in soil dwelling invertebrates, such as *Porcellio scaber* from the areas contaminated with metals exposed to zinc and cadmium (Van Capelleveen, 1987). The same isopod species originated from the vicinity of non ferrous plant, lead mining and a reference site exposed in the laboratory to cadmium indicated a similar pattern of metal accumulation in all groups (Donker and Bogert, 1992). Lack of developmental changes between these groups was explained by more efficient detoxification in isopods from more contaminated areas. Grasshoppers from Szopienice accumulated more cadmium than did insects from Tychy when exposed to the comparable amounts of this metal in the eaten grass (Fig. 8). In such conditions activity of CarE increased in both populations, but the activity of GST only in the midgut cells, without significant alterations in gonads of grasshoppers from Szopienice. Insects from a reference group showed a reduction of GST after 7 days of exposure and remained unchanged when cadmium was in excess throughout the larval development. These results indicate existing compensatory mechanisms in both populations studied, more stronger pronounced in a population exposed to higher metal stress in natural conditions.

Our results confirmed earlier statements that orthopterans are very resistant to man-made environmental stressing factors. Brieger et al. (1992) showed that this herbivore group is the most abundant in artificial ash heaps from electric power station. This is probably a result of their pre-adaptation from natural toxins produced by their host plant species (Lee, 1991; Lindqvist, 1992; Dallinger, 1993) which appear effective also in detoxification of anthropogenic pollutants. Our results are not sufficient enough to prove that differences in detoxifying abilities between analyzed populations are only through compensatory mechanisms or developing adaptations. Genetic adaptations to metals in terrestrial animals was proved only for some soil invertebrates discussed elsewhere (Posthuma and Van Straalen, 1993; Van Straalen, 1994). In herbivores such changes are slower than in more vulnerable carnivorous or saprophagic species (Migula, 1996a). Nevertheless results of this study showed a larger range of cadmium tolerance in grasshoppers living in highly polluted site. Assessment of tolerance abilities in consecutive generations in further studies should clarify whether alterations induced by industrial contaminants may act in a comparable way to other natural factors as genetic inducers involving genetic adaptations.

Acknowledgements

This study was financially supported by Polish National Scientific Committee (grant KBN No. P 04F 037 13).

References

Aebi, H., 1984. Catalase *in vitro*. Method Enzymol. 105, 121–126.

Ahmad, S., 1995. Oxidative stress from environmental pollutants. Arch. Insect Biochem. Physiol. 29, 135–157.

Ahmad, S., Pritsos, C.A., Bowen, S.M., Heisler, C.R., Blomquist, G.J., Pardini, R.S., 1988. Subcellular distribution and activities of superoxide dismutase, catalase, glutathione peroxidase, and glutathione reductase in the southern armyworm, *Spodoptera eridaria*. Arch. Insect Biochem. Physiol. 7, 173–186.

Ahmad, S., Beilstein, M.A., Pardini, R.S., 1989. Glutathione peroxidase activity in insects: a reassessment. Arch. Insect Biochem. Physiol. 12, 31–49.

Almar, M.A., Diaz-Mayans, J., Romero, F.J., 1987. Glutathione content and GSH S-transferase activity in midgut gland of *Procambarus clarkii*. Sex differences, theeffect of fasting, and their implications in cadmium toxicity. Comp. Biochem. Physiol. 87C (2), 433–435.

Argentine, J.A., James, A.A., 1995. Characterization of a salivary gland-specific esterase in the vector mosquito, *Aedes aegypti*. Insect Biochem. Mol. Biol. 25, 621–630.

Augustyniak, M., Migula, P., 1996. Patterns of glutathione S-transferase activity as a biomarker of exposure to industrial pollution in the grasshopper *Chorthippus brunneus* (Thunberg). Studia Soc. Sci. Torunensis 4, 9–15.

Benke, G.M., Wilkinson, C.F., 1971. Microsomal oxidation in the house cricket, *Acheta domesticus* (L.). Pesticide Biochem. Physiol. 1, 19–31.

Bradford, M.M., 1976. A rapid and sensitive method for the quantitation of microgram quantities of protein utilizing the principle of protein dye-binding. Anal. Biochem. 72, 248–254.

Brieger, G., Wells, J.R., Hunter, R.D., 1992. Plant and animal species composition and heavy metal content in fly ash ecosystems. Water Air Soil Pollut. 63, 87–103.

Byczkowski, J.Z., Sorenson, J.R.J., 1984. Effects of metal compounds on mitochondrial function: a review. Sci. Total Environ. 37, 33–62.

Cade, W.H., 1985. Insects mating and courtship behaviour. In: Kerkut, G.A., Gilbert, L. A. (Eds.), Comprehensive Insect Physiology Biochemistry and Pharmacology. Pergamon Press, Oxford, pp. 591–661.

Chance, B., 1954. The Mechanisms of Enzyme Action. Johns Hopkins Press, Baltimore, MA.

Chang, L.W., 1996. Toxicology of Metals. CRC Lewis Publishers, Boca Raton, FL.

Cherrill, A.J., Begon, M., 1989. Timing of life cycles in a seasonal environment: the temperature-dependence of embryogenesis and diapause in a grasshopper (*Chorthippus brunneus* Thunberg). Oecologia (Berlin) 78, 237–241.

Chraścina, M., Kafel, A., Migula, P., 1996. Patterns of detoxifying enzymes in larval stage of *Smerinthus ocellatus* L. exposed to cadmium, tocopherol or quercitin. Studia Soc. Sci. Torunensis 4, 31-37.

Dallinger, R., 1993. Strategies of metal detoxification in terrestrial invertebrates. In: Dallinger, R., Rainbow, P.S. (Eds.), Ecotoxicology of Metals in Invertebrates. Lewis, New York, pp. 245–289.

Díaz-Cruz, M.S., Mendieta, J., Tauler, R., Esteban, M., 1997. Cadmium-binding properties of glutathione: a chemometrical analysis of voltametric data. J. Inorg. Biochem. 66, 1997.

Donker, M.H., Bogert, C.G., 1992. Adaptation to cadmium in three populations of the isopod *Porcellio scaber*. In: Donker, M.H. (Ed.), Physiology of Metal Adaptation in the Isopod *Porcellio scaber*. Vrije Universiteit, Enschede, pp. 75–84.

Felton, G.W., Summers, C.B., 1995. Antioxidant systems in insects. Arch. Insect Biochem. Physiol. 29, 187–197.

Fisher, Z., 1981. Energy budget of grasshopper *Chorthippus* sp. in industrial environment. Pol. Ecol. Stud. 7, 5–14.

Frohlich, D.R., 1990. Substrate specificity of esterases in a solitary bee, *Megachile rotundata* (Hymenoptera: Megachilidae): variability in sex, age and life stage. Biochem. Syst. Ecol. 18, 539–547.

Głowacka, E., Migula, P., Nuorteva, S.-L., Nuorteva, P., Tulisalo, E., 1997. Psyllids as potential source of heavy metals for predators. Arch. Environ. Contam. Toxicol. 32, 376–382.

Godzik, S., Kubiesa, P., Staszewski, T., Szdzuj, J., 1999. Ecological problems of the Katowice administrative district. In: Peakall, D.B., Walker, C.H., Migula, P. (Eds.), Biomarkers: a Pragmatic Basis for Remediation of Severe Pollution in Eastern Europe, Vol. 4. Kluwer, Dodrecht, pp. 49–73.

Habig, W.H., Pabst, M.J., Jacoby, W.B., 1974. Glutathione S-transferases: the first enzymatic step in mercapturic acid formation. J. Biol. Chem. 249, 7130–7139.

Heliövaara, K., Väisänen, R., 1993. Insects and Pollution. CRC Press, Ann Arbor, MI.

Hopkin, S.P., 1989. Ecotoxicology of Metals in Terrestrial Invertebrates. Elsevier, New York.

Hopkin, S.P., 1990. Critical concentrations, pathways of detoxification and cellular ecotoxicology of metals in terrestrial arthropods. Funct. Ecol. 4, 321–327.

Hopkin, S.P., 1994. In situ biological monitoring of pollution in terrestrial and aquatic ecosystems. In: Calow, P. (Ed.), Handbook of Ecotoxicology. Blackwell, Oxford, pp. 397–427.

Hunter, B.A., Hunter, L.M., Johnson, M.S., Thompson, D.J., 1987. Dynamics of metal accumulation in the grasshopper *Chorthippus brunneus* in contaminated grasslands. Arch. Environ. Contam. Toxicol. 16, 711–716.

Hunter, B.A., Johnson, M.S., Thompson, D.J., 1987. Ecotoxicology of copper and cadmium in a contaminated grassland ecosystem. J. Appl. Ecol. 24, 573–586.

Íşcan, M., Çoban, T., Eke, B.C., 1995. Differential responses of hepatic monooxygenases and glutathione S-transferases of mice to a combination of cadmium and nickel. Comp. Biochem. Physiol. 111C, 61-68.

Ishikawa, Y., Kubota, T., 1991. Effect of host plant and dietary quercetin on antioxidant enzymes in onion and seedcorn maggots, *Delia antigua* and *D. platura* (Diptera: Anthomyiidae). Appl. Entomol. Zool. 26 (2), 245-253.

Isman, M.B., Feng, R., Johnson, D.L., 1996. Detoxicative enzyme activites in five species of field-collected melanopline grasshoppers (Orthoptera: Acrididae). Can. Entomol. 128, 353–354.

Jeantet, A.J., Ballan-Dufrancaise, C., Martoja, R., 1977. Insect resistance to mineral pollution. Rev. Ecol. Biol. Sol 14, 563–582.

Kędziorski, A., 1993. Oddziaływanie kadmu i ołowiu na aktywność wybranych enzymów detoksykacyjnych w końcowych etapach rozwoju świerszcza domowego, *Acheta domesticus* L. Ph.D. Thesis. University of Silesia, Katowice.

Kędziorski, A., Nakonieczny, M., Migula, P., Grzesiak, M., 1996. Age-dependent activity patterns of carboxylesterases in insects exposed to cadmium. Studia Soc. Sci. Torunensis 4, 71-76.

Kirby, M.L., Ottea, J.A., 1995. Multiple mechanisms for enhancement of glutathione S-transferase activities in *Spodoptera frugiperda* (Lepidoptera: Noctuidae). Insect Biochem. Mol. Biol. 25, 347–353.

Konno, Y., Shishido, T., 1992. Distribution of glutathione S-transferase activity in insect tissues. Appl. Entomol. Zool. 27, 391–397.

Kotze, A.C., Rose, H.A., 1987. Purification and properties of glutathione S transferases from the larvae of the Australian sheep blowfly, *Lucilia cuprina* (Wiedemann). Insect Biochem. 19, 703–713.

Lee, K., 1991. Glutathione S-transferase activities in phytophagous insects: induction and inhibition by plant phototoxins and phenols. Insect Biochem. 21, 353–361.

Lee, K., Berenbaum, M.R., 1992. Ecological aspects of antioxidant enzymes and glutathione-S-transferases in three *Papilio* species. Biochem. Syst. Ecol. 20, 197–207.

Lindqvist, L., 1992. Accumulation of cadmium, copper and zinc in five species of phytophagous insects. Environ. Entomol. 21, 160–163.

Lindqvist, L., 1993. Individual and interspecific variation in metal concentration in five species of bumblebees (Hymenoptera: Apidae). Environ. Entomol. 22, 1355–1357.

Lindqvist, L., 1994. Metal uptake and accumulation during growth of *Aglais urticae* (Lepidoptera: Nymphalidae) larvae. Environ. Entomol. 23, 975–978.

Marczyk, G., Migula, P., Trzcionka, E., 1993. Physiological responses of spiders to environmental pollution in the Silesian region (Southern Poland). Sci. Total Environ. (Suppl.), 1315–1322.

Martoja, R., Bouquegneau, J.M., Verthe, C., 1983. Toxicological effects and storage of cadmium and mercury in an insect Locustra migratoria (Orthoptera). J. Invert. Pathol. 42, 17–32.

Migula, P., 1996a. Constraints in the use of bioindicators and biomarkers in ecotoxicology. In: Van Straalen, N., Krivolutsky, D.A. (Eds.), Bioindicator Systems for Soil Pollution, Vol. 3. Kluwer, Dodrecht, pp. 17–31.

Migula, P., 1996b. Trace metals in animals: an overview on essentiality transport, accumulation and adaptive mechanisms. Biol. Bull. Poznań 33, 9–13.

Migula, P., 1997. Molecular and physiological biomarkers in insects as the tools in environmental hazard assessment. Acta Phytopatol. Entomol. Hungarica 32, 231–243.

Migula, P., Bińkowska, K., 1993. Feeding strategies of grasshoppers (*Chorthippus* sp.) on heavy metal contaminated plants. Sci. Total Environ. (Suppl.), 1071–1083.

Migula, P., Głowacka, E., 1996. Heavy metals as stressing factors in the red wood ants (*Formica polyctena*) from industrially polluted forests. Fresenius J. Anal. Chem. 354, 653–659.

Migula, P., Głowacka, E., 1998. Cadmium stress and biomerkers of exposure in two populations of the red wood ants (*Formica aquilonia* and *F. polyctena*). In: Migula, P., Doleżych, B., Nakonieczny, M., (Eds.), Proceedings of 2nd International Conference. Trace Elements: Effects on Organisms and Environment, EcoEdycja, Katowice, pp. 167–173.

Migula, P., Doleżych, B., Kielan, Z., Łaszczyca, P., Howaniec, M., 1990. Reakcje na stresy organizmów zwierzęcych na terenach zanieczyszczonych działalnoşcią przemysłu. In: Godzik, S. (Eds.), Środowisko Przyrodnicze Regionów Szczególnego Zagrożenia. SGGW, Warszawa, pp. 108–129.

Migula, P., Nuorteva, P., Nuorteva, S.-L., Głowacka, E., Oja, A., 1993. Physiological distribution in ants (*Formica aquilonia*) from excess of cadmium and mercury in Finnish forest. Sci. Total Environ. (Suppl.), 1305–1314.

Migula, P., Kafel, A., Bednarska, K., 1996. Environmental pollution and allozyme Variation in the red wood ant (*F. polyctena*). Studia Soc. Sci. Torunensis 4, 77–82.

Migula, P., Głowacka, E., Nuorteva, S.-L., Nuorteva, P., Tulisalo, E., 1997. Time related effects of intoxication with cadmium and mercury in the red wood ants (*Formica aquilonia*). Ecotoxicology 6, 307–320.

Migula, P., Augustyniak, M., Łaszczyca, P., Wilczek, G., 1999. Validation of selected biomarkers in invertebrates from the polluted Silesian Region. In: Peakall, D.B., Walker, C.H., Migula, P. (Eds.), Biomarkers: a Pragmatic Basis for Remediation of Severe Pollution in Eastern Europe, Vol. 5. Kluwer, Dodrecht, pp. 75–90.

Monk, K.A., 1985. Effect of habitat on the life history strategies of some British grasshoppers. J. Anim. Ecol. 54, 163–177.

Moriarty, F., 1969. The Laboratory breeding and embryonic development of *Chorthippus brunneus* Thunberg (Orthoptera: Acrididae). Proc. R. Soc. London (A) 44, 25–34.

Newman, M.C., Jagoe, C.H., 1996. Ecotoxicology: a Hierarchical Treatment. Lewis, New York.

Nuorteva, P., 1990. Metal distribution patterns and forest decline. Seeking Achilles' heels for metals in finish forest biocenoses. Publications of Department Environmental Conservation 11. University of Helsinki, Helsinki.

Nuorteva, P., Nuorteva, S.-L., Mathur, B., 1992. Excretory inhibition of bioaccumulation in Lepidoptera. In: Bohac, J.; Ruzicka, V. (Eds.), Bioindicatores Deteriorisationis Regionis. Ceské Budejovice, pp. 362–367.

Paglia, D.E., Valentine, W.N., 1967. Studies on the quantitative and qualitative characterization of erythrocyte glutathione peroxidase. J. Lab. Clin. Med. 70, 158–169.

Palace, V.P., Kleverkamp, J.F., 1993. Variation of hepatic enzymes in three species of freshwater fish from precambrian shield lakes and the effect of cadmium exposure. Comp. Biochem. Physiol. 104C, 147–154.

Pardini, R.S., 1995. Toxicity of oxygen from naturally occurring redox-active pro-oxidants. Arch. Insect Biochem. Physiol. 29, 101–118.

Posthuma, L., Van Straalen, N.M., 1993. Heavy metal adaptation in terrestrial invertebrates: a review of occurrence, genetics, physiology and ecological consequences. Comp. Biochem. Physiol. 106C, 11–38.

Rabitsch, W.B., 1997. Tissue-specific accumulation patterns of Pb, Cd, Cu, Zn, Fe, and Mn in workers of three ant species (Formicidae, Hymenoptera) from a metal-polluted site. Arch. Environ. Contam. Toxicol. 32, 172–177.

Racker, E., 1955. Glutathione reductase. Method Enzymol. 2, 722.

Regoli, F., Principato, G., 1995. Glutathione, glutathione-dependent and antioxidant enzymes in mussel, *Mytilus galloprovincialis*, exposed to metals under field and laboratory conditions: implications for the use of biochemical biomarkers. Aquat. Toxicol. 31, 143–164.

SANEPID, 1998. Zanieczyszczenie atmosfery w województwie katowickim w latach 1996–1997. Wojewódzka Stacja Sanitarno-Epidemiologiczna, Katowice.

Schmidt, G.H., Ibrahim, N.M.M., Abdallah, M.D., 1992. Long-term effects of heavy metals in food on developmental stages of *Aiolopus thalassinus* (Saltatoria: Acrididae). Arch. Environ. Contam. Toxicol. 23, 375–382.

Shukla, G.S., Hussain, T., Srivastava, F.S., Chandra, S.V., 1989. Glutathione peroxidase and catalase in liver, kidney, testis and brain regions of rat following cadmium exposure and subsequent withdrawal. Indust. Health 27, 59–69.

Sibly, R.M., 1996. Effects of pollutants on individual life histories. In: Newman, M.C., Jagoe, C.H. (Eds.), Ecotoxicology: a Hierarchical Treatment. Lewis, New York, pp. 197–224.

Simkiss, K., Taylor, M.G., 1996. Transport of metals across membranes. In: Tessier, A., Turner, D.R., (Eds.), Metal Speciation and Bioavailability in Aquatic Systems. Wiley, New York. pp. 1–44.

Stenersen, J., Kobro, M., Bjerke, M., Arend, U., 1987. Glutathione transferases in aquatic and terrestrial animals from nine phyla. Comp. Biochem. Physiol. 86C, 73–82.

Van Asperen, K., 1962. A study of housefly esterases by means of a sensitive colorimetric method. J. Insect Physiol. 1, 19–31.

Van Capelleveen, H.E., 1987. The ecotoxicity of zinc and cadmium for the woodlouse *Porcellio scaber* Latr. 1: consumption, growth and accumulation. In: Van Capelleveen, H.E. (Ed.), Ecotoxicity of Heavy Metals for Terrestrial Isopods. Free University Press, Amsterdam, pp. 21–41.

Van Straalen, N.M., 1994. Heavy metal adaptation in terrestrial arthropods - pysiological and genetic aspects. Proc. Exp. Appl. Entomol. 5, 3–17.

Viarengo, A., 1989. Heavy metals in marine invertebrates: mechanisms of regulation and toxicity at the cellular level. Aquat. Sci. 1, 295–317.

Viarengo, A., Bettella, E., Fabbri, R., Burlando, B., Lafaurie, M., 1997. Heavy metal inhibition of EROD activity in liver microsomes from the bass *Dicentrarchus labrax* exposed to organic xenobiotics: role of GSH in the reduction of heavy metal effects. Mar. Environ. Res. 44, 1–11.

Viarengo, A., Canesi, L., Martinez, P.G., Peters, L.D., Livingstone, D.R., 1995. Pro-oxidant processes and antioxidant defence systems in the tissues of the Antarctic scallop (*Adamussium colbecki*) compared with the Mediterranean scallop (*Pecten jacobaeus*). Comp. Biochem. Physiol. 111B, 119–126.

Walker, C.H., Hopkin, S.P., Sibly, R.M., Peakall, D.B., 1996. Principles of Ecotoxicology. Taylor & Francis, London.

Wall, R., Begon, M., 1986. Population density, phenotype and mortality in the grasshopper *Chorthippus brunneus*. Ecol. Entomol. 11, 445–456.

Wilczek, G., Migula, P., 1996. Metal body burdens and detoxifying enzymes in spiders from industrially polluted areas. Fresenius J. Anal. Chem. 354, 643–647.

Yu, S.J., 1982. Host plant induction of glutathione S-transferase in the fall armyworm. Pesticide Biochem. Physiol. 18, 101–106.

Yu, S.J., 1988. Enzyme Induction as Detoxication Mechanism in Phytophagous insect. Symposium on Insect Biochemistry and Physiology, pp. 49–68.

Trace elements – Their distribution and effects in the environment
B. Markert and K. Friese, editors

Chapter 17

Environmental monitoring of heavy metals with magpie (*Pica pica*) feathers – an example of Polish polluted and control areas

K. Dmowski

Abstract

The magpie (Pica pica) fulfils most of the criteria for a bioindicator species. The bird has widespread occurrence, relatively small home range and sedentary behaviour. During the period 1989–1993 samples of external tail feathers from 165 individuals living in 20 Polish regions were collected. The areas differed in type of industrialization and urbanization, e.g. the vicinities of various smelters (Zn, Pb, Cr, Cu, Fe), an oil refinery, lignite-fired power plants, a Hg lamp factory, dumping grounds and agricultural areas. Twenty-one elements were determined in feather vanes (Pb, Ba, Cd, Zn, Mn, Fe, Hg, Tl, Sb, Sn, Mo, Sr, Rb, Se, As, Ge, Cu, Ni, Co, Cr, V) by inductively coupled plasma mass spectrometry (ICP-MS). The results were generally in accordance with the lists of elements emitted in the course of technological processes applied by particular works and they show that the surroundings of zinc smelters are much more polluted, especially with Pb, Cd, Zn, Sn, As in comparison to the regions so far considered as seriously threatened. The described method enables to visualize such differences and to control the changes in emissions of many elements from year to year. For example the analyses of magpie feathers provided evidence for very dangerous levels of thallium in several Polish regions and have stimulated thorough studies there. This study examined also the multi-elemental composition of the samples from each area. Mean concentrations of the elements from one region arranged in sun ray diagrams show characteristic patterns ("elemental profiles") typical for the region. Low concentrations noted for agricultural areas gave a pattern focused in the centre of the plot with a small stripe for mercury. For the most polluted region (Szopienice – the district of Katowice city close to a zinc smelter) maximum concentrations of Pb, Cd, Zn, Cu, Ge, Sn, Sb, Se were found and the elemental profile covered a large area of the sun ray diagram. Transposition of results of feather analyses to internal tissues is also possible to a certain extent, at least with respect to the most toxic elements as lead, cadmium or thallium. All these data can be obtained without any harm to animals.

Keywords: Biomonitoring; Birds; Feathers; Heavy metals; Inductively coupled plasma mass spectrometry; Pollution; Magpie

1. Introduction

The magpie (*Pica pica*) fulfils most of the criteria for accumulative bioindicators. The bird:

- has widespread occurrence (it appears in whole Europe, northern Africa, large part of Asia and western part of North America)
- has relatively small home range and sedentary behaviour (individuals stay for many years within the same area estimated to cover 10–50 ha. and visit only winter roosting places not far from the nesting territory thus they do not cumulate any pollutants during spring or autumn migrations)
- is common and easy to catch with a standardized method (magpie population increases in Poland permanently, so much so that it is one of the few unprotected birds)
- has well-known biology
- is easy to keep in laboratory (certain aspects have to be tested in artificial conditions)
- tends to accumulate certain xenobiotics

Magpie feathers have been tested for around a dozen years and finally tail feathers were recommended as analytical material for the assessment of heavy metal pollution (Gast, 1984; Müller et al., 1984; Hahn et al., 1985, 1989, 1993; Kooiker, 1986, 1987; Kühnast and Ellenberg, 1990; Dmowski and Golimowski, 1993; Golimowski and Dmowski, 1995, 1997). Studies of feathers are particularly important nowadays, in the face of increasing protests against the killing of animals (even for obtaining internal tissues for scientific purposes). It may be assumed that magpie feathers can be used instead of internal tissues as alternative, convenient samples to obtain information on pollution without any harm to animals.

However, studies of magpie feathers have, until now, concerned mostly a few elements such as lead, cadmium and mercury. The objective of the presented research was to evaluate how suitable the method is for a broader spectrum of elements found in many Polish areas which differ in type of industrialization and urbanization. Also the studies aimed at testing the severity of contamination in certain areas, both multi-elemental and regarding individual elements

2. Methods

2.1. Study areas

In total, magpies were captured in 32 regions of Poland with air or soil pollution of various degree caused by heavy metals, organic pesticides, PCBs, or radionuclides. The presented data regards 20 areas influenced by heavy metal emissions and compared to unpolluted regions. Birds were collected in areas (Fig. 1) surrounding:

- Heavy metal smelters
  (1) the zinc smelter "Szopienice" in Katowice *(Zn1)*
  (2) the largest Polish zinc smelter "Miasteczko Slaskie" (*Zn2*)
  (3) the zinc smelter "Boleslaw" near Olkusz (*Zn3*)
  (4) a lead smelter (closed before 1989) in Olawa (*Pb*)
  (5) a copper smelter in Glogow (*Cu*)
  (6) the chromium smelter "Siechnice" (closed in 1990) near Wroclaw (*Cr*)

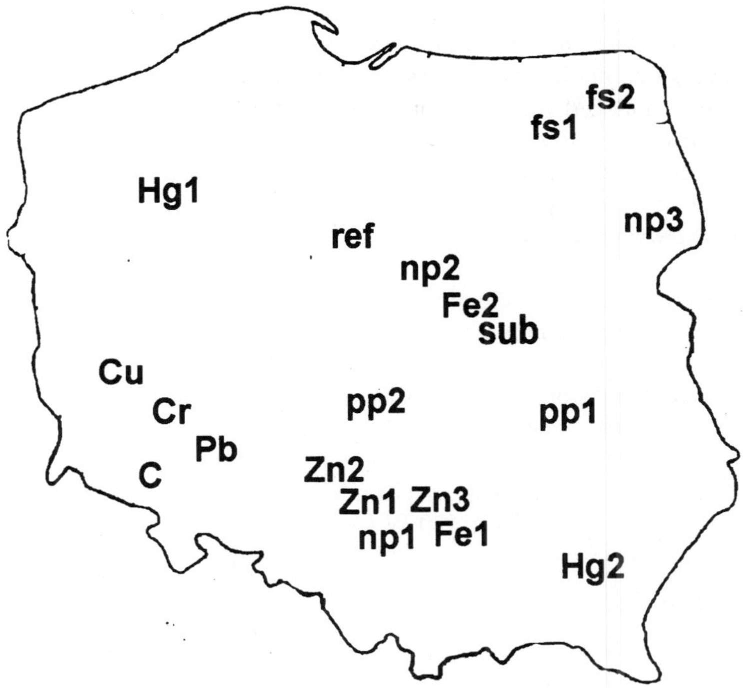

Figure 1. Localities of bird captures.

(7) the largest Polish steelworks in Nowa Huta near Cracow (*Fe1*)
(8) the steelworks in Warsaw (*Fe2*)

- Other industrialized areas
  (9) coking plants in Walbrzych (*C*)
  (10) the largest Polish oil refinery in Plock (*ref*)
  (11) a lignite-fired power plant in Kozienice (*pp1*)
  (12) the largest Polish lignite-fired power plant in Belchatow (*pp2*)
  (13) the suburban small city Miedzeszyn near Warsaw (*sub*)
  (14) mercury lamp factory waste dump in Pila (*Hg1*)
  (15) a mercury lamp factory (closed in 1989) in Rzeszow (*Hg2*)
- Potentially unpolluted regions
  (16) villages near the Ojcow National Park (*np1*)
  (17) villages near the Kampinos National Park (*np2*)
  (18) villages near the Białowieza National Park (*np3*)
  (19) field station of Warsaw University, lake area near Mikolajki (*fs1*)
  (20) field station of Warsaw University, lake area near Elk (*fs2*)

Within the *Zn1* and *Zn2* areas, magpies were captured in 2 years, 1989 and 1993. Within the *Cr* area feathers were collected in 1990, the last year of operation of the factory, and in 1991 and 1992, after the factory had been closed. In other areas the samples were taken only in 1 year – all at the beginning of the 1990s. Magpies were always collected in the same season (May–June, shortly before feather moulting) with a modified "bal-chatri" method (Bub, 1986). In the vicinity of a chosen magpie nest a cage containing another live magpie individual was set up. The local birds defending their territories attacked the outsiders and were captured in angling thread loops installed in grass around the cage. Tail feathers were torn out and stored in polyethylene bags. In the vicinity of the mentioned areas 165 individuals were trapped in total.

2.2. Analytical procedure

The tail feathers (exclusively the most external rectrices R_6) were placed in polyethylene vessels containing detergent, Triton X-100 suprapure, and put into an ultra-sonic bath for 15 min. Then the samples were rinsed with NANOpure water for 2 min and dried 24 h at 65°C in a drying oven. Vanes were separated from their shafts with a titanium knife and used for further analysis. The whole vanes were mineralized in Teflon vessels with HNO_3 (Merck, Suprapure) in the Milestone 1200 Mega microwave oven. An eight-step digestion program (23 min in total: 250 W for 2 min; 0 W for 30 s; 250 W for 10 min; 0 W for 30 s; 450 W for 5 min; 0 W for 30 s; 600 W for 3 min; ventilation for 2 min) was used resulting in clear solutions. The samples were diluted to the mark in 50 ml volumetric flasks.

Concentrations of 21 elements (Pb, Ba, Cd, Zn, Mn, Fe, Hg, Tl, Sb, Sn, Mo, Sr, Rb, Se, As, Ge, Cu, Ni, Co, Cr, V) were determined with a PlasmaQuad 2 Turbo Plus inductively coupled plasma mass spectrometer (VG Elemental, Winsfort, Cheshire, UK) in the Institute of Analytical Chemistry of the Karl-Franzens University (Graz, Austria). A Meinhard concentric glass nebulizer, type Tr-30-A3, and a double pass Scott-type spray chamber were used. Nickel sampling cones with an orifice of 1.00 mm diameter and nickel skimmer cones with an orifice diameter of 0.75 mm were used for ion sampling. The uptake time of 60 s was used, each sample being measured for 60 s, twice. Quantification was performed by external calibration. Concentration of 50 ng/ml of indium was added to each solution (standards and samples) to correct for instrumental instabilities. To correct for chloride interferences at 51 V and 75 As the interference correction provided from the PQVisons software was used. The accuracy of the determinations was examined by parallel analysis of reference material (Human Hair, GBW 09101, Shanghai Nuclear Centre, China).

All results are expressed in μg/g on a dry weight basis. Tukey test was used to compare differences between the studied areas at the level $P = 0.05$.

3. Results and discussion

3.1. Pollution of the areas

Very enhanced concentrations of some elements, especially Pb, Cd, Zn, Sn, Sb, As were found in the feather samples from the vicinities of all the studied zinc smelter areas (*Zn1*, *Zn2*, *Zn3*) (Figs. 2 and 3). Some of them were at least a dozen and sometimes scores of times higher than, e.g. those in the samples from the surroundings of the *Fe1* or *Fe2* steelworks, and lignite-fired power plants *pp1* or *pp2*, so far considered

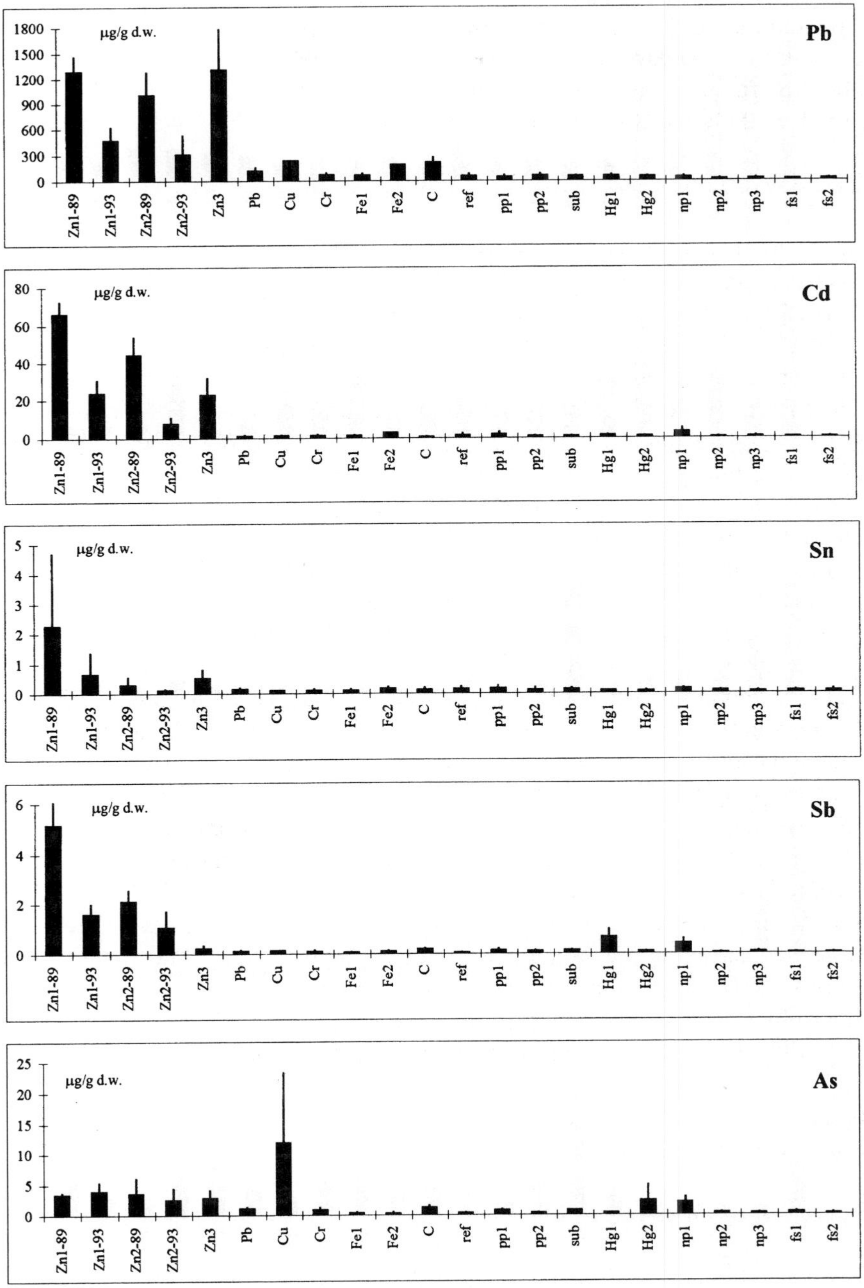

Figure 2. Pb, Cd, Sn, Sb, and As concentrations (mean ± SD) in magpie feathers from the studied areas.

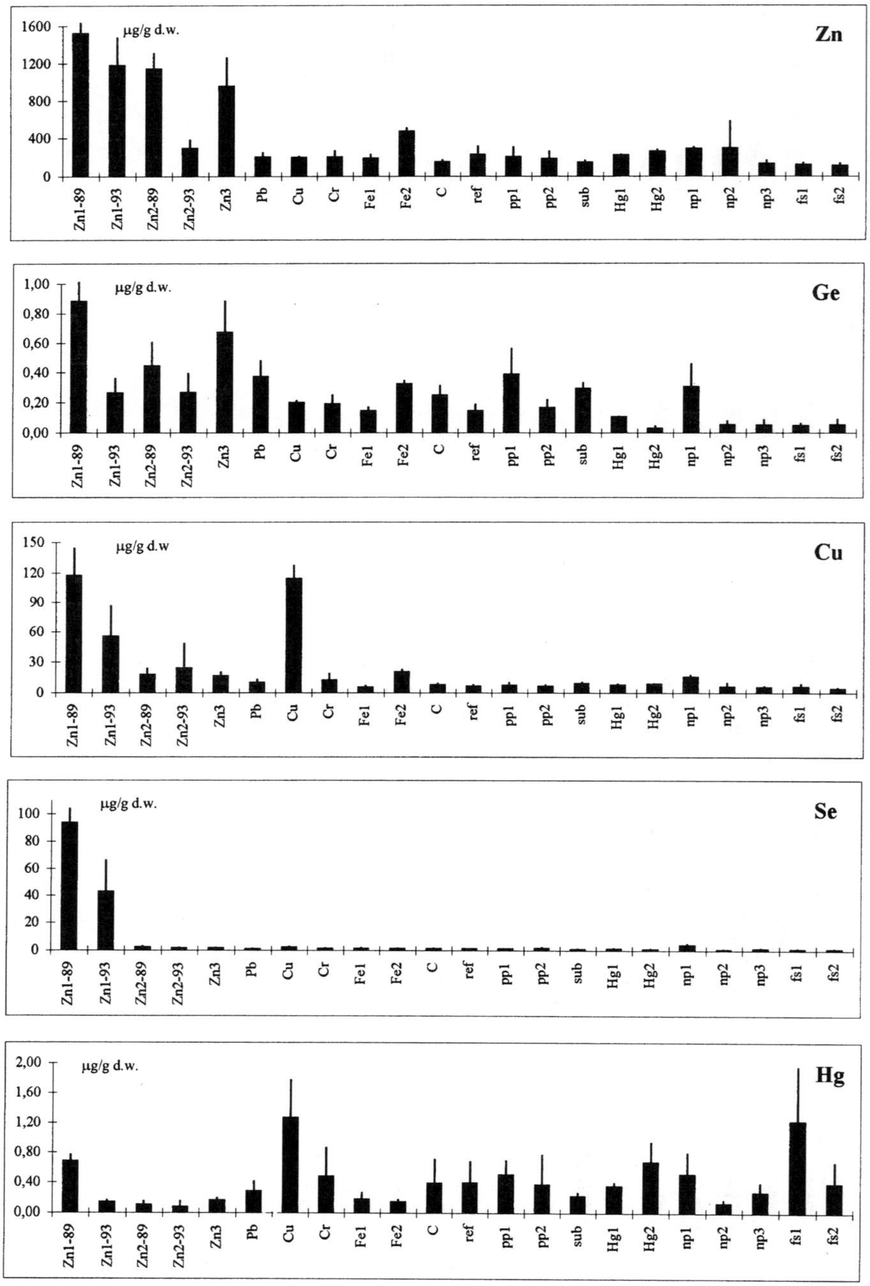

Figure 3. Zn, Ge, Cu, Se, and Hg concentrations (mean ± SD) in magpie feathers from the studied areas.

as seriously polluted with heavy metals. Particularly high were the levels of lead (up to 1000–1400 μg/g dry weight on average; maximum value ~2090 μg/g in the *Zn3* area) and cadmium (up to 25–65 μg/g on average; maximum value ~75 μg/g in the *Zn1*-89 area). Copper and selenium concentrations were very elevated in the *Zn1* area in both periods of sample collections.

With respect to some elements particularly high concentrations were noted as unique to certain areas, for example, thallium in the *Zn3* area, barium in the *Zn2* area (but only in 1989), vanadium in the *ref* area, and chromium in the *Cr* area (Fig. 4).

The data were generally in accordance with the lists of elements emitted in the course of technological processes applied by particular works. The main zinc ores used in Poland such as sphalerite, ZnS, and galman, $ZnCO_3 + ZnSiO_4 \cdot H_2O$ are almost always accompanied by lead and cadmium compounds, and very often by copper, arsenic and antimony (Jedrzejowski, 1987). Germanium and tin compounds occur as additional components of sulphidic zinc and lead ores.

Unexpectedly enhanced barium concentrations present in 1989 in the *Zn2* area were probably caused by emissions from Ba processing in a factory located 3 km from the zinc smelter in the direction of the prevailing wind. The factory was closed at the beginning of the 1990s and this was clearly reflected in the substantial decrease of barium in feathers in 1993 to the background level. Permanent very high levels of copper and selenium in a vicinity of the zinc smelter *Zn1* resulted probably from copper ores processing in the region (production of rolled sheets); a share of Se in a mass of refining sludges can reach up to 5% (Jedrzejowski, 1987).

Feathers exposed to the emissions within the *Cu* area were found to have very elevated concentrations of copper, arsenic and mercury. Arsenic is known to be a constituent of copper ores and can be easily converted to the volatile arsenic trioxide at smelting temperatures. The presence of large amounts of vanadium in crude oils that was reflected in the samples from the *ref* area is also well known. The concentrations of V in feathers were correlated with the distance of magpie territories from a certain section of the oil refinery (Dmowski and Golimowski, 1998).

As expected, the highest concentrations of chromium were noted in the samples from the *Cr* area (Fig. 4). The large standard deviation of results is due to a comprising of three years data. Feather analyses from 1990, 1991, and 1992 reflected that chromium concentrations in the environment have decreased gradually.

Extremely enhanced levels of thallium in the feather samples from the *Zn3* area was a quite startling. This highly toxic element is even used in rat poison mixtures. Thallium can be commercially obtained mostly as a by-product of zinc and lead ore processing (Kelner, 1994) and certain amounts of Tl in the environment ought to be expected also close to other zinc smelters, but only in the *Zn3* area floatation failings are kept in an elevated place – inside a 70-m-high mine dump. Drying dust is easily spread by the wind over the whole of the surrounding area. The presence of extremely large amounts of thallium in this region demonstrated by the magpie biomonitoring studies was further confirmed by analyses of the failings (149 μg/g; Institute of Geology of the Polish Academy of Sciences, unpublished data) and analyses of thallium levels in the bodies of small mammals (Dmowski et al., 1998). Rodents with apparent hair loss were found which is known as one of the basic signs of thallium poisoning.

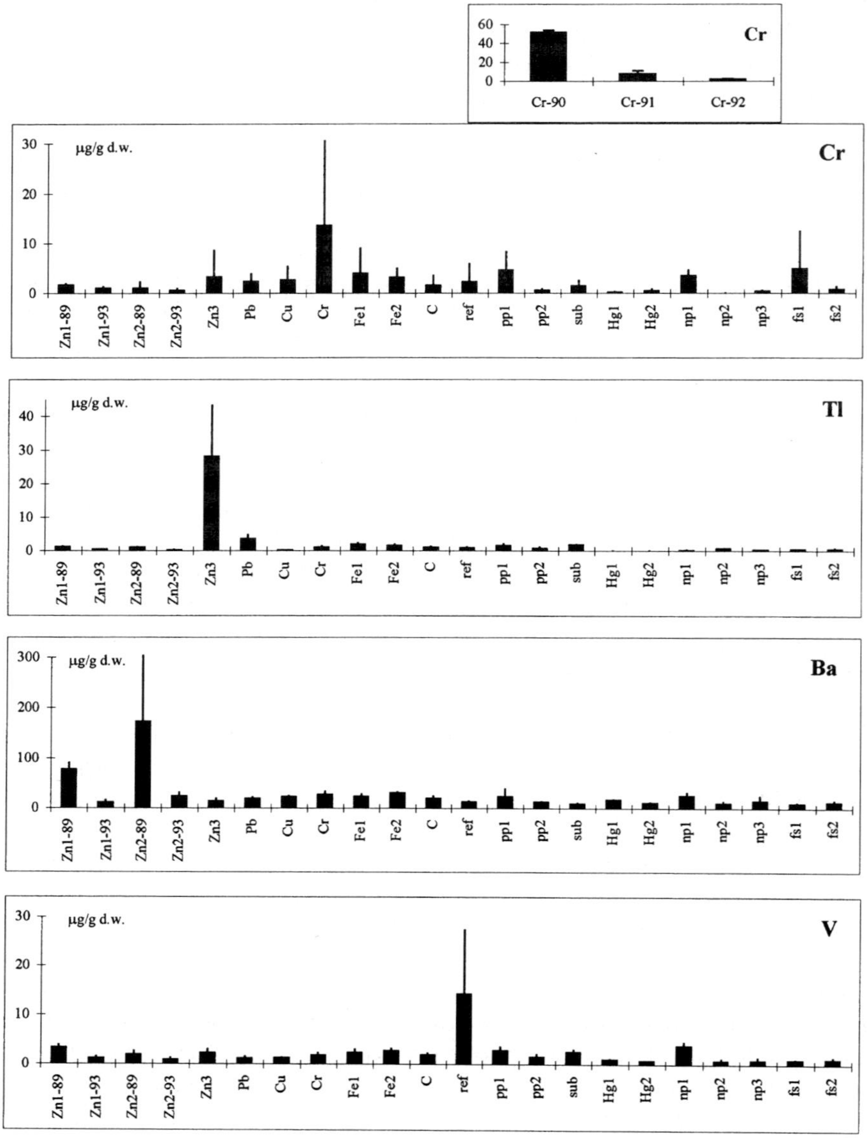

Figure 4. Cr, Tl, Ba, and V concentrations (mean ± SD) in magpie feathers from the studied areas.

The only element accumulated in raised amounts in the samples from the potentially unpolluted agricultural Polish regions was mercury (Fig. 3). The enhanced Hg levels in magpie feathers may reflect the further use of fungicides containing mercury compounds for seed dressing, as this was described elsewhere (Golimowski and Dmowski, 1995). There is also a possibility that the levels of mercury in magpie feathers from the *np* and *fs* areas represent the most "natural" amounts of these elements entering the feathers from the circulatory system during their growth.

Perhaps in more contaminated regions this element does not enter feathers to such a high degree because the chemical bonds it prefers to form are already occupied by other heavy metals, which are absent, or present only in trace amounts in the unpolluted areas.

The studies carried out in 1992 of feathers of magpies from the *Hg1* and *Hg2* areas did not confirm the presence of distinctly high amounts of mercury in the environment, probably due to its evaporation (Tukey test; the *Hg2* area = *fs1* area = *Cu* area > *Hg1* area). Large amounts of broken glass from mercury lamp factories were stored here at the end of 1980s. The factory in the *Hg2* area was completely closed in 1989.

The presented bioindication method is distinctly not suitable for the elements as follows: Co, Fe, Mn, Mo, Ni, Rb, Sr, The mean concentrations of most of them did not show great variability between the investigated areas (Fig. 5). Levels found in samples from the industrialized areas were similar to the unpolluted ones. In addition nickel and rubidium exhibited high differences of results within the samples from one area (expressed by large standard deviations)

The analysis of samples from Polish national parks has shown that concentrations of many elements were distinctly higher in the *np1* area in comparison to other selected parks (see also Dmowski, 1997). With respect to most of them (Ba, Cd, Cu, Ge, Ni, Pb and Sn) the concentrations were even equal to those found in the surroundings of the *Fe1* steelworks. Thus the high level of contamination in the area is probably caused by emissions from the *Fe1* steelworks and Cracow agglomeration located 20–25 km SE of the park, as well as from the *Zn3* area. The Ojców National Park was regarded to be the most highly polluted park in Poland at the end of the 1970s and the beginning of the 1980s on the basis of heavy metal analyses in selected species of mosses (Grodzińska, 1978; Grodzińska, 1985). Magpie feather biomonitoring studies confirm this.

3.2. Origin of measured elements

Many investigators in the past assumed that their measurements of heavy metal contents in feathers represented mostly the so-called endogenous input, which is a reflection of the physiological and environmental conditions to which the birds were exposed during the period of feather growth. At present it is known that heavy metals measured in the feathers have double origin (endogenous and exogenous), with the exogenous one prevailing with respect to many elements (Bortolotti and Barlow, 1985; Goede and de Voogt, 1985; Weyers et al., 1988; Hahn et al., 1989, 1993; Hartner et al., 1992).

The endogenous process consists of the entry of elements into the organism through the alimentary tract. They are transported in the blood and then taken up into the structure of the feathers during their growth. The exogenous way is based on the deposition of the elements on the feather surface. There is such a dense reticulum of tiny filaments that complete removal of the microparticles is impossible even by treatment with ether, acetone, detergents, EDTA, ultrasound, etc. The cleaning procedure is much easier with respect to shafts, therefore, levels of many heavy metals measured in keratin of shafts are distinctly lower as compared to the vanes.

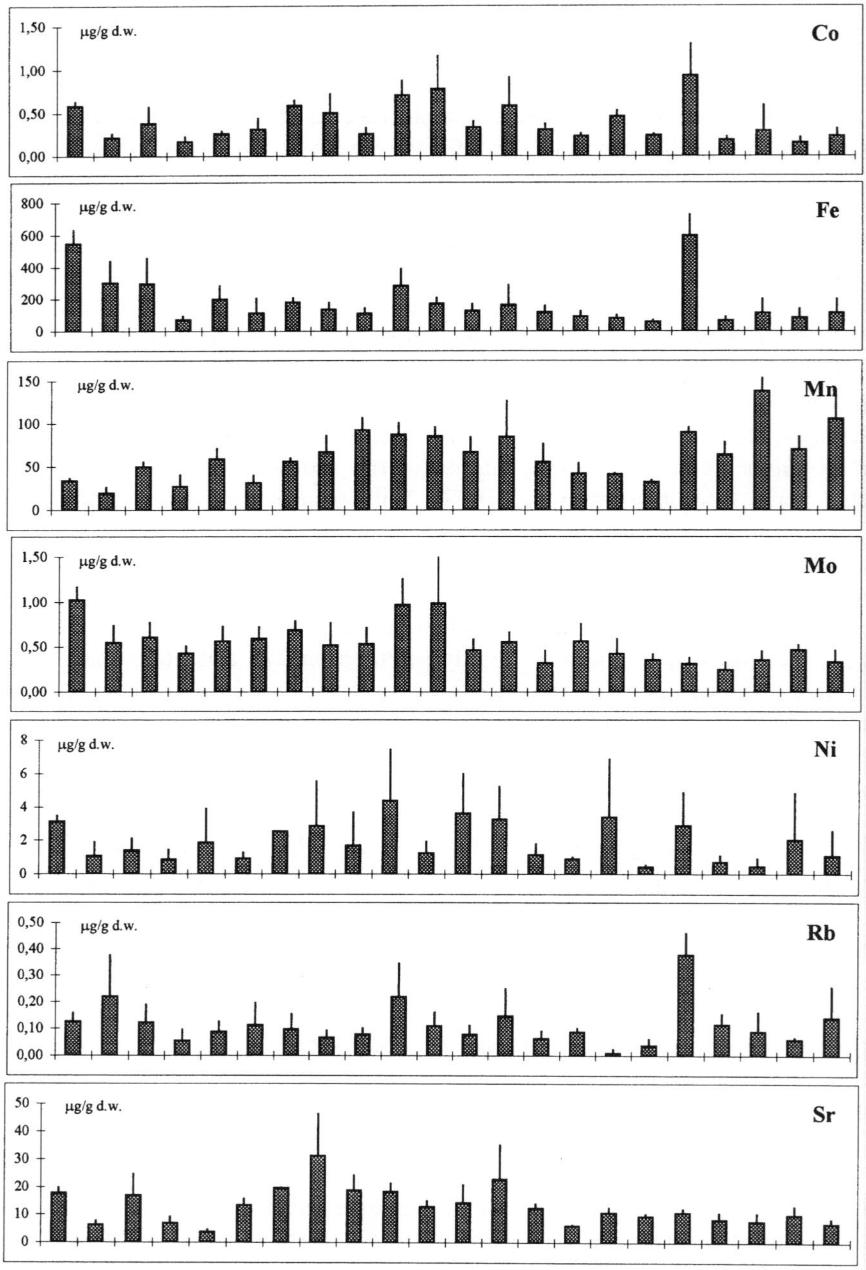

Figure 5. Co, Fe, Mn, Mo, Ni, Rb, and Sr concentrations (mean ± SD) in magpie feathers from the studied areas.

None of the current methods of cleaning feathers are capable of removing all external contaminants (Weyers et al., 1988; Hahn et al., 1989, 1993; Hennig, 1992), and the experiments on cleaning of plumage of Polish magpies with ether, acetone, detergents, and bidistilled water were also not satisfactory (author's own data). Concentrations of many elements measured in "cleaned" bird feathers are the combination of both endogenous and mostly exogenous contributions. The domination of the exogenous way leads to the effect where the levels of metals measured in feathers correspond much more with data on deposition (emission) or content in the air than with resources of elements in the environment accessible to organisms by food. This was also proved by very high correlation between lead and cadmium content in magpie tail feathers from Silesian towns and size of emission monitored very precisely in the region (Fig. 6) (Dmowski and Golimowski, 1993).

The domination of the exogenous way with regard to most of the elements causes also that feathers of migratory or nomadic birds can not be applied for this kind of bioindication studies. Analysis of feathers ought to be limited to settled bird species or young individuals of other species but still remaining within the breeding territory of their parents.

The only exception among the elements important from the ecotoxicological point of view and with prevailing endogenous way of accumulation, is mercury. The element expresses a tendency to enter feathers during their growth, and its external deposition on the surface of feather can be neglected because of significant volatility caused by permanent transformations of the element.

Nevertheless if standardized conditions of collection and cleaning are applied (identical periods of capture, the same types and parts of feathers taken for analysis) a comparison of the results is possible regarding other elements with considerable share of exogenous deposition of pollutants on the surface of feathers. The most convenient period to trap magpies extends from May through June, which is their breeding season. Only during this time will the birds aggressively confront other magpies entering their territory. The feathers collected during this season just before the moult represent an exposition to air-borne emissions for almost 1 year. External

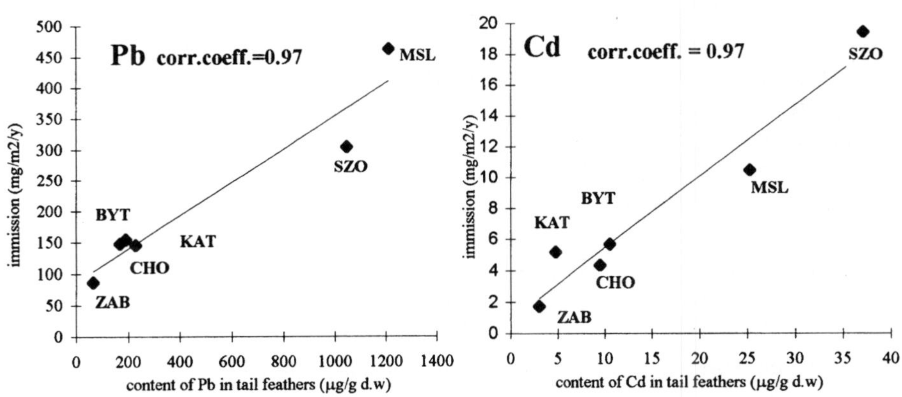

Figure 6. Correlation between Pb and Cd in tail feathers and size of emission. Data from Silesian towns (ZAB-SZO). After Dmowski and Golimowski (1993).

tail feathers (rectrices), especially R_6 seem to be best suited as bioindicative material of all contour feathers. The rectrices closer to the centre of the tail are more often damaged than R_6 and sometimes their moult starts during the trapping season.

With reference to many elements no distinct differences in results for individuals inhabiting a single area were found, which was reflected by small error bars representing the standard deviation. A sample consisting of five to seven individuals appears to be representative for a defined region. The age of individuals captured in May–June ought to be greater than 2 years (easy to distinguish due to different size of stripes on wing feathers; feathers of magpies with immature plumage (1 year) are exposed to contaminants several months longer than older ones).

3.3. Monitoring of environmental changes

A reduction in the level of many heavy metals was noted in samples from the *Zn2* area between 1989 and 1993 (Figs. 2–4). It was probably connected with the installation of a new air filter system, a distinct decrease in production and the application of new economic motivating factors for the recycling of wastes (Managers of the Zinc Smelter, pers. commun.). The concentrations of Pb, Zn, Cd and Sn in the samples of magpie feathers from 1993 had decreased to 31, 25, 17 and 35%, respectively, in comparison to the feathers collected in 1989. Presumably similar reasons can explain decrease of concentrations of certain heavy metals in the samples from the *Zn1* area between 1989 and 1993. The above-mentioned results, as well as examples of barium decrease in the *Zn2* area or gradual chromium decrease in the *Cr* area, provide evidence that the studies of magpie feathers enable the possibility of control of the changes in emissions of many elements from year to year. Since the method is harmless to the birds samples can be repeatedly obtained for many years from the same individuals. We may treat their rectrices like renewable samplers of 1-year exposition to certain pollutants, e.g. as important as lead or cadmium.

Application of the inductively coupled plasma mass spectrometry (ICP-MS) equipment allows us to study not only contamination of an area with individual element but also to estimate the risk of multi-elemental exposure. This is quite well illustrated by presentation of so-called "elemental profiles" (Fig. 7). The data regarding one single element are recalculated in such way that the maximum concentration in feathers from all studied areas is treated as 100% and the other ones take on respective, proportional values. After recalculations of all elements values regarding the area we want to study are selected and arranged in sun ray diagrams (Excel 5.0). The picture consists of many axes – each for one of all considered elements and creates a characteristic pattern called "elemental profile", typical for the region. Elements presenting a similar tendency for accumulation in feathers from unpolluted and polluted regions (Co, Fe, Mn, Mo, Ni, Rb, Sr) have been excluded from the calculations.

Low concentrations of many elements typical for agricultural lands give a pattern focused in the centre of the sun ray plots, with a small stripe for mercury. Regions affected by the same type of industry have similar elemental profiles. If the areas surrounding the zinc smelters can be regarded as the most polluted with heavy metals in Poland, the primacy of pollution degree among them belongs to the *Zn1* area (Szopienice – the district of the Katowice city, located in the centre of the Upper

Silesian region). The concentrations of many elements were the highest there. Maximum concentrations of Pb, Cd, Zn, Cu, Ge, Sn, Sb, Bi were found in 1989 and the elemental profile covered a large area of the sun ray diagram. In comparison to the *Zn1* area, samples from the neighbourhoods of some steelworks and lignite-fired power plants, considered until now as highly contaminated, reveal elemental profiles of much more moderate dimension. The method of multi-elemental profiles helps to visualize such differences, in some cases more clearly than cluster analysis which was performed in parallel.

Significant long-term changes of emission can also be reflected in the elemental profiles. For example, an installation of more effective filters and 70% decrease in production in the *Zn2* area in the early 1990s caused notable reduction in the size of the profile. There were probably similar reasons for differences between *Zn1*-89 and *Zn1*-93, nevertheless the data from this town is still alarming (Fig. 7). Nowadays, the district of Szopienice requires special treatment and routine multi-elemental checking, particularly taking into account the longevity and severity this pollution represents and the vicinity of human agglomeration

3.4. Relations: feathers/internal tissues

Transposition of results of feather analysis to internal tissues is one of fundamental topics with regard to an application of magpie feathers as bioindicators of heavy metal pollution. This appears to be very limited because of different origin of incorporated heavy metals: a dominance of exogenous deposition in feathers and endogenous way with respect to the internal tissues. Fig. 8 compares merits and demerits of the analysis of discussed tissues – a scatter of results regarding individuals of different age from a single study area (arrows 1), a reflection of pollution variabilities between areas (arrows 2) and a transposition of feather data to internal tissues (arrow 3).

Magpie feathers collected in May–June represent, thanks to a regular moulting rhythm, 1-year exposition independently from an age of captured adult "$>$2 years" birds (n, $n + 1$, $n + 2$, etc.). The problem of a scatter of results theoretically does not exist. Much larger one is observed in internal tissues due to various amounts of pollutants cumulated in older and younger tissues. The accumulation rate can be additionally not uniform from year to year because of a gradual filling out (comp. the arrows 1). The described above studies of magpie rectrices have proved that feather analysis can reflect differences of heavy metal pollution between areas equally well as internal tissues (the arrows 2).

Comparison of concentrations found in feathers and internal tissues of magpie individuals has been possible because a part of the birds captured in Poland (mostly males) was sacrificed for studies of internal tissues as kidneys, livers, femur bones, breast muscles, and brains. The results have shown that transposition of results of feather analyses to internal tissues is possible, at least with respect to the most environmentally harmful elements such as lead, cadmium or thallium. The proportionalities of pollution of the studied areas mirrored by tail feather samples were similar to the proportionalities expressed by some internal tissues. Complete data regarding all

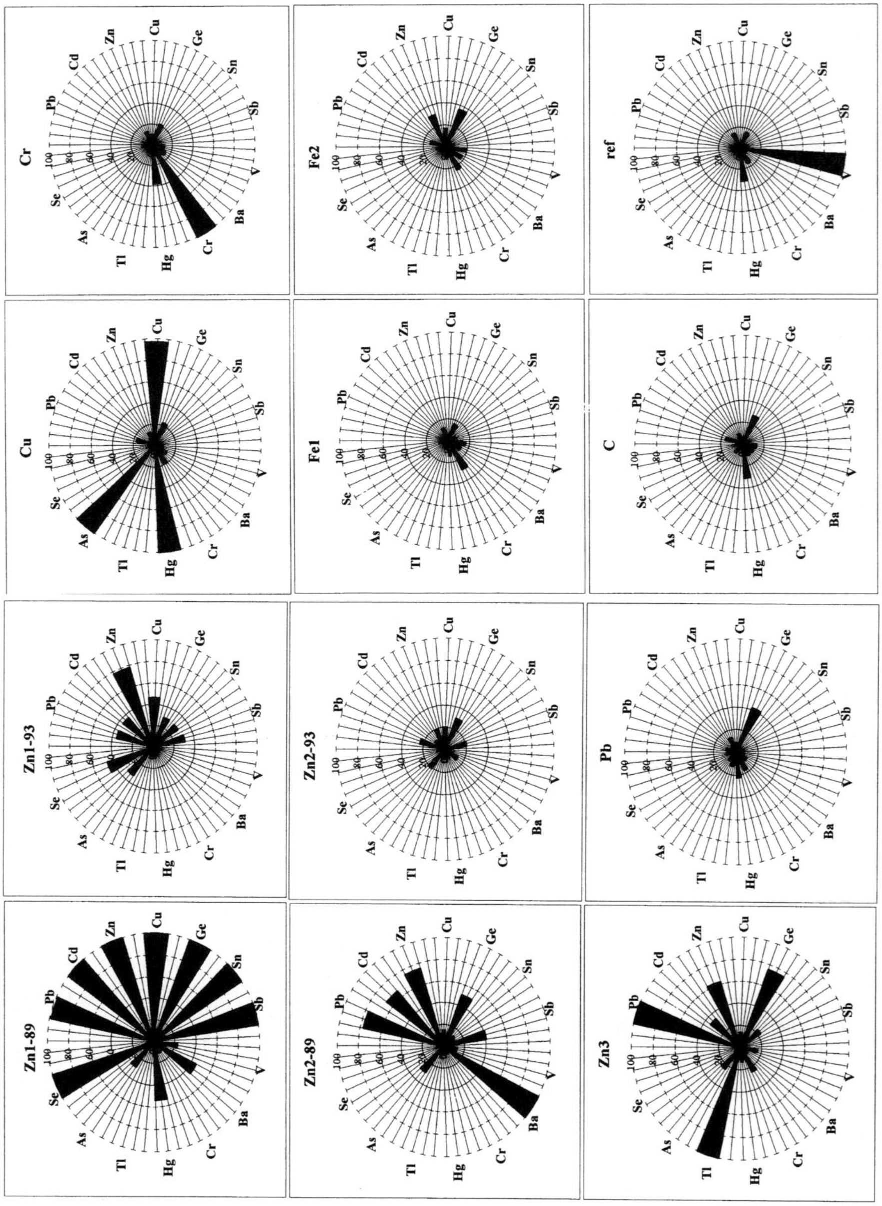

Figure 7. Elemental profiles of the studied areas.

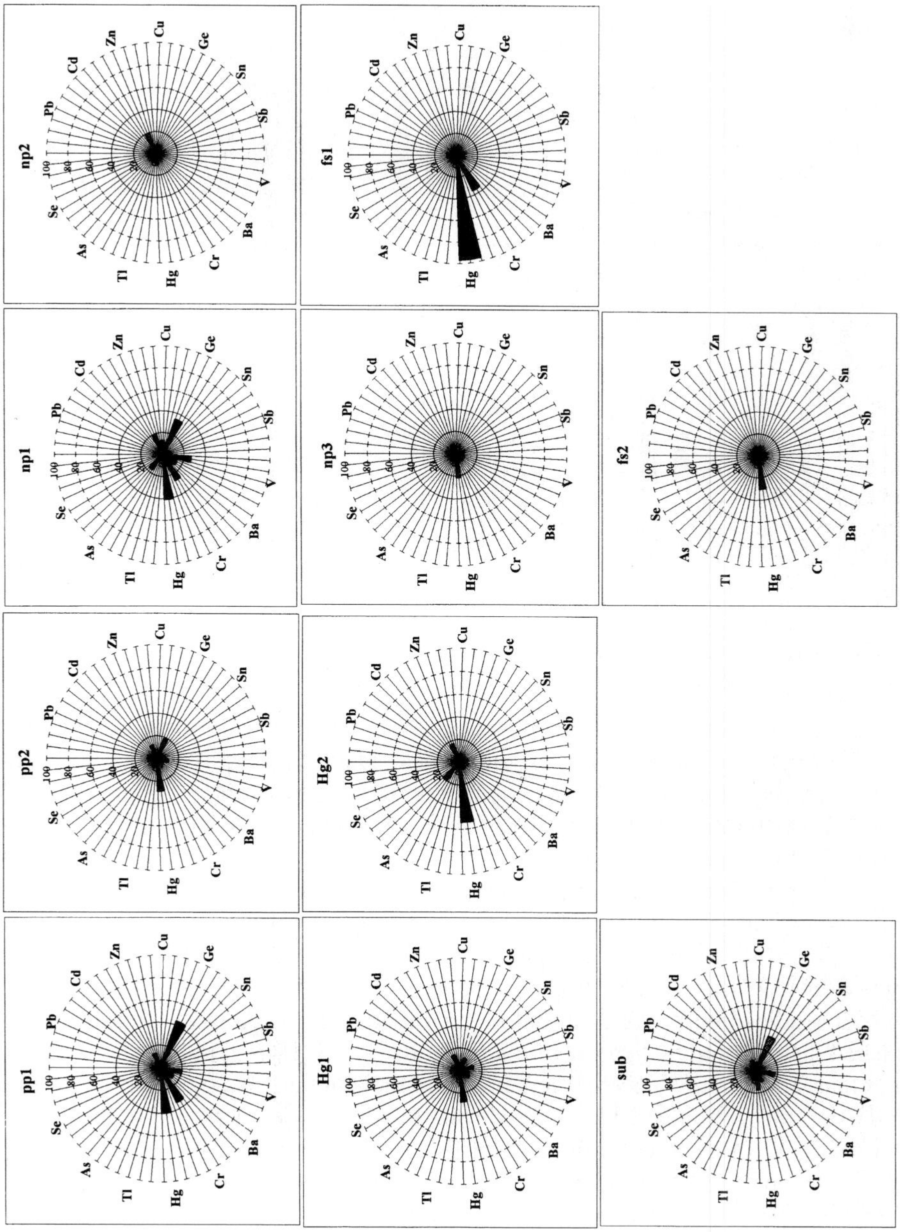

Figure 7. (continued).

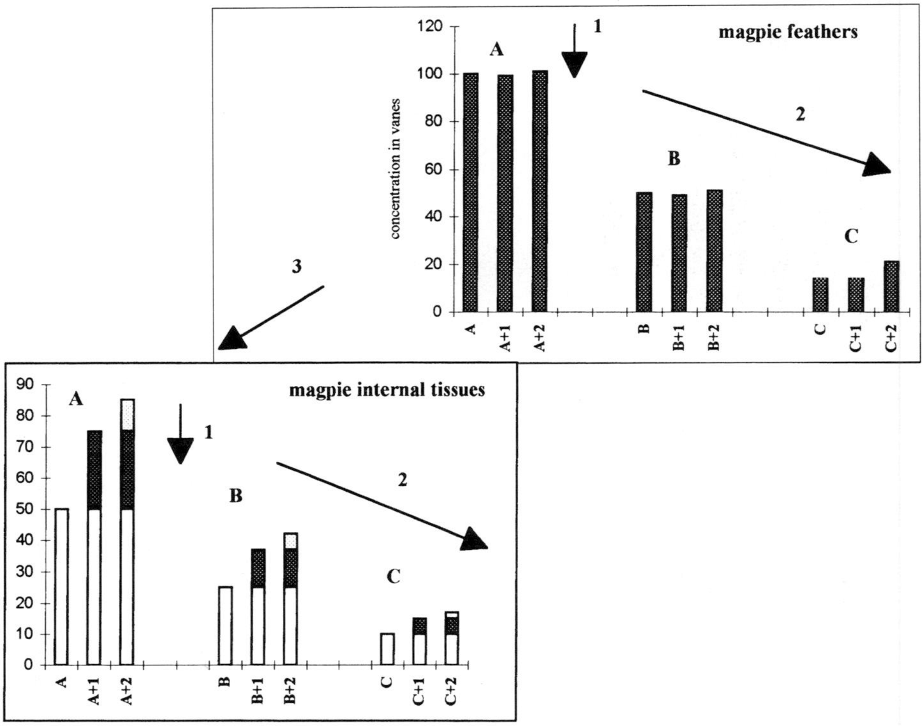

Figure 8. Comparison of feathers and internal tissues as a material for analysis.

tissues and many elements will be published soon. Examples are provided here of comparisons between magpie feather and liver contents of lead (correlation coefficient, 0.86) and thallium (correlation coefficient, 0.90). These two examples (Fig. 9) have been also chosen for presentation of some advantages as well certain inconveniences of feather analyses.

As we know feathers reflect better levels of emission, while internal tissues reflect a pool of elements accessible for animals in the environment through consumption and drinking. This difference, which explains the lack of a decrease of lead content in magpie livers from the *Zn1* and *Zn2* areas between 1989 and 1993, is well-mirrored in the feather studies and proved by routine air monitoring. Although pollution levels diminished distinctly in this period, amounts of heavy metals in soil (the main abiotic element that stores many harmful elements) and the amounts made further accessible by food to animals, remained at the same level or even slightly increased so the studies of liver (and also other internal tissues) have registered no reduction of pollution.

Deposition on feather surfaces reflects also better changes of emission of some bioelements, such as, e.g. Cu in comparison to internal tissues. The accumulation of these elements in internal tissues is very often equal independently from a pollution degree.

Nevertheless in certain cases feather studies in contradistinction to studies of some internal tissues can not reveal a presence of important elements in the environment.

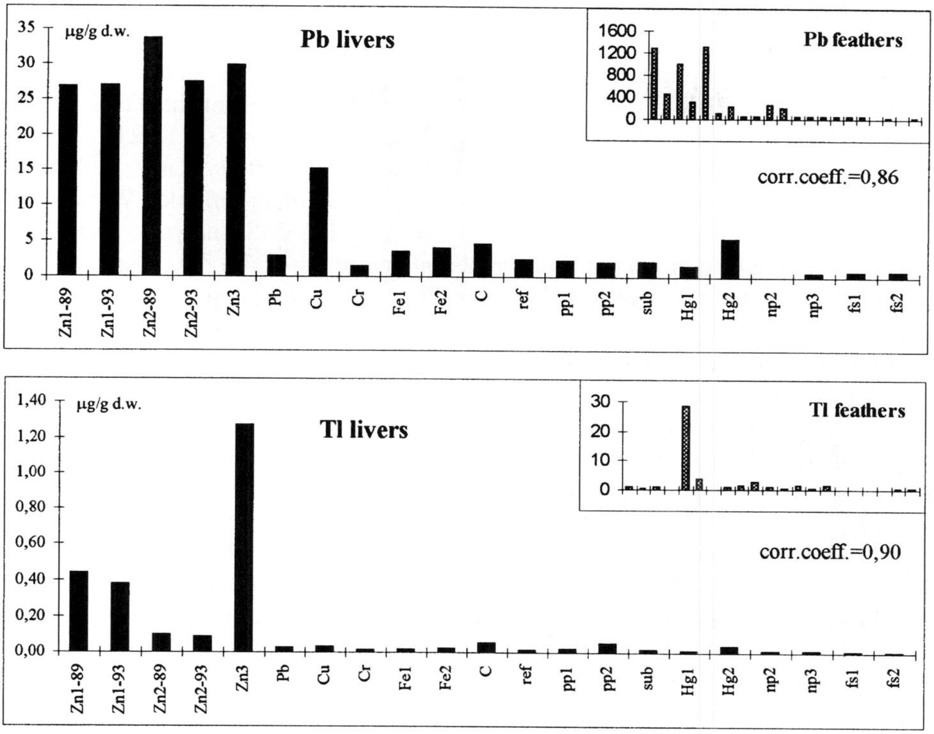

Figure 9. Comparison of Pb and Tl concentrations (mean values) in tail feathers and livers of magpies from the study areas.

For example significant thallium concentrations were found only in magpies feathers from the *Zn3* area while livers (Fig. 9) as well other four internal tissues registered equally large amounts in the *Zn1* area. This was probably caused by the different ways in which thallium spreads in the both areas. In the *Zn1* area floatation failings are not kept in elevated dump hills but in a pit below a ground level. Tl could spread out from this not as a dust but, for example, through insects or other animals eaten further by magpies. The dust not taken into the air could not be deposited on feather surfaces. Magpies living a certain distance from the source of thallium pollution could accumulate the element in internal tissues and simultaneously not absorb it from the air onto the plumage.

The presented comparisons provide evidence that feather analyses can supply important information on the pollution of the environment with heavy metals (mostly an emission rate) but can not be treated as fully alternative method to analyses of internal tissues. The method should be applied as a preliminary, non-destructive stage in the control of an environment. Enhanced concentration of harmful elements in feathers are a warning signal and should stimulate further studies including also analyses of internal tissues.

Acknowledgements

This study was supported in part through a grant of the Polish Committee of Scientific Studies KBN 6 6361 91 02 and a mobility grant from the TEMPUS JEP 3572/93. I am very grateful to the following persons: Professor Kurt Kalcher from the Institute of Analytical Chemistry of the Karl-Franzens University (Graz, Austria) for facilities during my stays in Austria and valuable, benevolent remarks; Walter Gössler, M.Sc. from the same Institute for facilities of ICP-MS analyses; eng. Justyn Sołtysik from the Department of Environmental Protection of the Zinc Smelter "Miasteczko Slaskie" for his inappreciable help in organization of studies in the Upper Silesia region; Professor Jerzy Golimowski from the Faculty of Chemistry of the Warsaw University for a manifold help; Dr Kolanowska from the Ministry of Health, Dr Polkowski from the Ministry of Environmental Protection, Dr Swiatczak from the Institute of Occupational Medicine in Łódź, and Mr Kogut from the mercury lamp factory in Rzeszow for data helpful during a selection of study areas; M. Barkowska, M.Sc. from the Department of Ecology of the Warsaw University for her help of elaboration of statistical data; M. Golachowski, M.Sc. from the Department of Ecology of the Warsaw University for his help in the translation of the text; the staff of the Department of Breeding at the Zoological Garden in Łódź and the Zoological Garden of Polish Fauna in Bydgoszcz for providing living magpies as a bait.

Appendix A. Concentrations (mean values ± SD) of the elements in magpie tail feathers from the studied areas (*Zn1-89–fs2*)

Area	Vanadium	Chromium	Cobalt	Nickel	Copper	Germanium
Zn1-89	3.24 ± 0.59	1.60 ± 0.26	0.59 ± 0.04	3.10 ± 0.36	117.46 ± 26.14	0.88 ± 0.13
Zn1-93	1.13 ± 0.40	0.97 ± 0.27	0.21 ± 0.05	1.06 ± 0.80	54.93 ± 31.17	0.26 ± 0.10
Zn2-89	1.86 ± 0.77	1.04 ± 1.25	0.38 ± 0.19	1.38 ± 0.71	17.53 ± 5.68	0.44 ± 0.16
Zn2-93	0.84 ± 0.36	0.53 ± 0.34	0.17 ± 0.05	0.83 ± 0.58	23.89 ± 24.00	0.27 ± 0.13
Zn3	2.28 ± 0.69	3.28 ± 5.27	0.26 ± 0.03	1.83 ± 2.07	16.04 ± 3.45	0.67 ± 0.21
Pb	1.11 ± 0.40	2.36 ± 1.53	0.31 ± 0.13	0.89 ± 0.35	9.79 ± 2.54	0.37 ± 0.10
Cu	1.21 ± 0.13	2.61 ± 2.75	0.59 ± 0.06	2.52 ± 0.00	114.27 ± 12.96	0.20 ± 0.01
Cr	1.78 ± 0.51	13.61 ± 20.07	0.50 ± 0.22	2.84 ± 2.69	11.89 ± 6.66	0.19 ± 0.06
Fe1	2.36 ± 0.62	3.96 ± 5.11	0.25 ± 0.08	1.67 ± 2.01	5.57 ± 1.26	0.14 ± 0.03
Fe2	2.72 ± 0.44	3.20 ± 1.77	0.70 ± 0.17	4.38 ± 3.03	20.06 ± 2.24	0.32 ± 0.02
C	1.86 ± 0.36	1.56 ± 1.93	0.78 ± 0.38	1.22 ± 0.68	7.45 ± 1.38	0.25 ± 0.06
ref	14.23 ± 13.16	2.23 ± 3.66	0.33 ± 0.07	3.62 ± 2.34	6.31 ± 1.19	0.14 ± 0.04
pp1	2.84 ± 0.77	4.61 ± 3.68	0.59 ± 0.32	3.25 ± 1.98	7.31 ± 2.43	0.39 ± 0.17
pp2	1.51 ± 0.59	0.54 ± 0.26	0.30 ± 0.07	1.11 ± 0.66	6.20 ± 1.29	0.17 ± 0.05
sub	2.50 ± 0.48	1.48 ± 1.07	0.23 ± 0.03	0.88 ± 0.10	8.99 ± 1.34	0.29 ± 0.04
Hg1	1.06 ± 0.10	0.21 ± 0.03	0.46 ± 0.06	3.39 ± 3.44	7.60 ± 0.95	0.10 ± 0.00
Hg2	0.71 ± 0.07	0.42 ± 0.43	0.24 ± 0.02	0.39 ± 0.11	8.50 ± 0.32	0.03 ± 0.01
np1	3.80 ± 0.74	3.54 ± 1.04	0.93 ± 0.36	2.87 ± 1.98	15.20 ± 1.89	0.30 ± 0.15
np2	0.78 ± 0.35		0.18 ± 0.03	0.67 ± 0.39	5.93 ± 3.28	0.05 ± 0.03
np3	0.81 ± 0.65	0.39 ± 0.18	0.29 ± 0.30	0.43 ± 0.45	5.53 ± 0.82	0.05 ± 0.04
fs1	0.95 ± 0.08	4.85 ± 7.64	0.15 ± 0.06	2.00 ± 2.84	5.96 ± 2.68	0.04 ± 0.02
fs2	1.00 ± 0.40	0.72 ± 0.57	0.23 ± 0.09	1.07 ± 1.47	4.00 ± 1.08	0.05 ± 0.03

Area	Arsenic	Selenium	Rubidium	Strontium	Molybdenum	Tin
Zn1-89	3.39 ± 0.28	93.34 ± 10.25	0.12 ± 0.03	17.62 ± 2.00	1.02 ± 0.15	2.26 ± 2.42
Zn1-93	3.87 ± 1.43	42.79 ± 23.19	0.22 ± 0.16	6.13 ± 1.52	0.55 ± 0.19	0.65 ± 0.70
Zn2-89	3.47 ± 2.46	2.03 ± 0.49	0.12 ± 0.07	16.79 ± 7.76	0.61 ± 0.16	0.29 ± 0.25
Zn2-93	2.44 ± 1.89	1.28 ± 0.38	0.05 ± 0.04	6.70 ± 2.34	0.43 ± 0.08	0.10 ± 0.04
Zn3	2.78 ± 1.22	1.23 ± 0.44	0.09 ± 0.04	3.66 ± 0.83	0.56 ± 0.16	0.51 ± 0.29
Pb	1.07 ± 0.18	0.68 ± 0.21	0.11 ± 0.08	13.37 ± 2.32	0.59 ± 0.13	0.13 ± 0.06
Cu	11.90 ± 11.28	2.06 ± 0.34	0.10 ± 0.06	19.67 ± 0.16	0.68 ± 0.10	0.09 ± 0.00
Cr	0.81 ± 0.42	0.93 ± 0.25	0.06 ± 0.03	31.26 ± 15.02	0.52 ± 0.25	0.09 ± 0.04
Fe1	0.33 ± 0.12	1.16 ± 0.35	0.08 ± 0.02	18.89 ± 5.32	0.53 ± 0.18	0.09 ± 0.05
Fe2	0.24 ± 0.20	0.95 ± 0.23	0.22 ± 0.13	18.32 ± 3.01	0.97 ± 0.28	0.15 ± 0.06
C	1.16 ± 0.32	1.05 ± 0.19	0.11 ± 0.05	12.83 ± 2.25	0.98 ± 0.51	0.10 ± 0.08
ref	0.32 ± 0.09	0.85 ± 0.30	0.08 ± 0.03	14.39 ± 6.42	0.46 ± 0.12	0.13 ± 0.09
pp1	0.71 ± 0.18	0.92 ± 0.20	0.15 ± 0.10	22.94 ± 12.28	0.55 ± 0.11	0.16 ± 0.09
pp2	0.27 ± 0.07	1.35 ± 0.50	0.06 ± 0.03	12.38 ± 1.59	0.32 ± 0.14	0.09 ± 0.08
sub	0.74 ± 0.04	0.82 ± 0.25	0.09 ± 0.01	5.98 ± 0.35	0.56 ± 0.19	0.13 ± 0.04
Hg1	0.30 ± 0.04	1.01 ± 0.25	0.01 ± 0.01	10.65 ± 1.87	0.42 ± 0.16	0.08 ± 0.01
Hg2	2.34 ± 2.51	0.74 ± 0.12	0.04 ± 0.02	9.34 ± 0.83	0.35 ± 0.06	0.06 ± 0.04
np1	2.01 ± 0.85	3.52 ± 1.20	0.38 ± 0.08	10.67 ± 1.39	0.31 ± 0.07	0.13 ± 0.03
np2	0.30 ± 0.09	0.50 ± 0.23	0.11 ± 0.04	8.17 ± 2.36	0.25 ± 0.08	0.08 ± 0.01
np3	0.20 ± 0.09	0.88 ± 0.36	0.09 ± 0.07	7.63 ± 2.68	0.35 ± 0.10	0.04 ± 0.04
fs1	0.35 ± 0.16	0.77 ± 0.32	0.06 ± 0.01	9.90 ± 3.19	0.47 ± 0.05	0.06 ± 0.02
fs2	0.18 ± 0.11	0.58 ± 0.30	0.14 ± 0.12	6.69 ± 1.68	0.33 ± 0.13	0.06 ± 0.06

Area	Antimony	Thallium	Mercury	Iron	Manganese	Zinc
Zn1-89	5.15 ± 0.89	1.02 ± 0.17	0.68 ± 0.08	547.3 ± 83.1	33.7 ± 2.5	1516.7 ± 138.4
Zn1-93	1.60 ± 0.39	0.37 ± 0.08	0.14 ± 0.02	301.3 ± 137.1	19.5 ± 6.2	1176.5 ± 293.8
Zn2-89	2.10 ± 0.44	0.87 ± 0.14	0.10 ± 0.05	295.4 ± 161.7	49.4 ± 6.0	1136.3 ± 168.8
Zn2-93	1.06 ± 0.65	0.17 ± 0.08	0.07 ± 0.07	70.7 ± 20.0	26.8 ± 13.4	289.1 ± 90.3
Zn3	0.22 ± 0.10	28.13 ± 15.19	0.16 ± 0.03	198.7 ± 82.2	58.3 ± 12.2	948.1 ± 307.6
Pb	0.11 ± 0.03	3.48 ± 1.20	0.28 ± 0.13	110.9 ± 91.4	30.8 ± 8.8	199.2 ± 45.7
Cu	0.12 ± 0.01	0.09 ± 0.03	1.27 ± 0.50	179.8 ± 26.2	55.9 ± 3.4	197.2 ± 9.7
Cr	0.09 ± 0.05	0.97 ± 0.40	0.48 ± 0.38	134.1 ± 41.3	66.3 ± 19.2	200.9 ± 60.6
Fe1	0.05 ± 0.02	1.84 ± 0.46	0.17 ± 0.09	108.1 ± 33.3	92.2 ± 13.8	184.2 ± 39.8
Fe2	0.10 ± 0.03	1.53 ± 0.31	0.14 ± 0.02	282.3 ± 106.2	86.9 ± 13.8	467.3 ± 32.0
C	0.19 ± 0.04	0.93 ± 0.25	0.39 ± 0.32	169.9 ± 37.6	84.6 ± 10.6	146.2 ± 19.6
ref	0.04 ± 0.02	0.78 ± 0.24	0.39 ± 0.28	125.4 ± 41.9	66.4 ± 17.5	223.6 ± 86.1
pp1	0.12 ± 0.08	1.47 ± 0.54	0.50 ± 0.19	162.0 ± 123.3	84.1 ± 42.0	201.9 ± 99.2
pp2	0.09 ± 0.04	0.68 ± 0.35	0.37 ± 0.40	117.7 ± 39.4	55.1 ± 20.8	183.5 ± 76.4
sub	0.12 ± 0.02	1.78 ± 0.11	0.22 ± 0.04	90.5 ± 31.3	41.5 ± 12.1	144.6 ± 19.0
Hg1	0.65 ± 0.31	0.04 ± 0.01	0.35 ± 0.05	78.4 ± 18.2	41.2 ± 0.9	218.1 ± 5.9
Hg2	0.08 ± 0.01	0.06 ± 0.01	0.68 ± 0.26	54.7 ± 11.1	32.1 ± 2.4	258.5 ± 19.1
np1	0.39 ± 0.19	0.13 ± 0.08	0.51 ± 0.29	592.8 ± 129.3	89.4 ± 5.6	285.1 ± 17.6
np2	0.02 ± 0.01	0.64 ± 0.08	0.12 ± 0.04	63.4 ± 24.3	63.4 ± 14.5	292.8 ± 281.0
np3	0.06 ± 0.04	0.23 ± 0.05	0.27 ± 0.12	111.6 ± 86.2	138.1 ± 113.0	128.1 ± 33.3
fs1	0.02 ± 0.02	0.41 ± 0.01	1.22 ± 0.72	80.3 ± 55.0	69.3 ± 15.0	117.7 ± 19.7
fs2	0.02 ± 0.01	0.47 ± 0.23	0.38 ± 0.29	110.9 ± 86.4	105.1 ± 33.7	107.8 ± 27.7

Area	Cadmium	Barium	Lead
Zn1-89	65.95 ± 6.45	77.0 ± 13.6	1282.8 ± 177.1
Zn1-93	23.47 ± 6.95	10.8 ± 4.2	463.4 ± 155.5
Zn2-89	44.02 ± 9.39	171.7 ± 130.7	1002.0 ± 271.3
Zn2-93	7.36 ± 3.17	23.4 ± 6.8	305.4 ± 217.8
Zn3	22.54 ± 8.97	13.6 ± 5.2	1301.9 ± 474.1
Pb	1.07 ± 0.41	18.7 ± 3.1	116.6 ± 37.3
Cu	1.27 ± 0.10	22.0 ± 1.8	229.3 ± 3.4
Cr	1.24 ± 0.42	26.6 ± 6.5	67.5 ± 23.0
Fe1	1.22 ± 0.18	23.1 ± 4.5	61.3 ± 21.8
Fe2	2.65 ± 0.05	30.0 ± 2.3	181.9 ± 2.9
C	0.50 ± 0.16	19.6 ± 4.7	209.3 ± 58.8
ref	1.30 ± 0.86	12.2 ± 2.2	51.5 ± 28.8
pp1	1.78 ± 1.13	23.4 ± 15.3	43.3 ± 18.5
pp2	0.81 ± 0.23	12.1 ± 1.3	55.1 ± 22.3
sub	0.69 ± 0.17	9.3 ± 1.1	51.0 ± 7.7
Hg1	1.10 ± 0.28	16.8 ± 1.1	50.1 ± 19.5
Hg2	0.76 ± 0.13	10.9 ± 0.8	46.9 ± 4.2
np1	2.96 ± 1.75	24.8 ± 6.5	8.3 ± 16.0
np2	0.30 ± 0.06	10.0 ± 3.1	15.7 ± 8.4
np3	0.45 ± 0.40	14.5 ± 10.1	20.1 ± 13.8
fs1	0.26 ± 0.03	8.5 ± 2.5	13.4 ± 1.4
fs2	0.33 ± 0.25	11.8 ± 3.8	18.8 ± 9.0

References

Bortolotti, G., Barlow, J., 1985. Neutron activation analysis of bald eagle feathers: analytical precision and sources of sampling variation. Can. J. Zool. 63, 2707–2718.

Bub, H., 1986. Vogelfang und Vogelberingung (Bird trapping and ringing). Part III, Ziemsen Verlag, Wittenberg-Luthenstadt.

Dmowski, K., 1997. Biomonitoring with the use of magpie Pica pica feathers: heavy metal pollution in the vicinity of zinc smelters and national parks in Poland. Acta Ornithol. 32, 15–23.

Dmowski, K., Golimowski, J., 1998. Badania biomonitoringowe w rejonie Plocka (Biomonitoring studies in the Plock region). In: Makles Z. (Ed.) Analytical aspects of waste monitoring from a fuel and electricity industry. Commission of Waste Analysis of the Committee of Analytical Chemistry of the Polish Academy of Science, Warsaw, pp. 73–80.

Dmowski, K., Golimowski, J., 1993. Feathers of magpie (Pica pica) as a bioindicative material for heavy metal pollution assessment. Sci. Total Environ. 139/140, 251–258.

Dmowski, K., Kozakiewicz, A., Kozakiewicz, M., 1998. Small mammal populations and community in conditions of extremely high thallium contamination in the environment. Ecotoxicol. Environ. Safety 41, 2–7.

Gast, F., 1984. Die Elster (Pica pica) als Bioindikator für die Belastung von Nahrungsnetzen durch Umweltchemikalien (Magpie as bioindicator of contamination of food webs with environmental chemicals). Ph.D. Thesis. Saarbrücken University, Germany.

Goede, A., de Voogt, P., 1985. Lead and cadmium in waders from the Dutch Wadden Sea. Environ. Pollut. A 37, 311–322.

Golimowski, J., Dmowski, K., 1995. Voltammetric determination of mercury in bird feathers for biomonitoring studies. Anal. Chem. (Warsaw) 40, 13–19.

Grodzińska, K., 1978. Mosses as bioindicators of heavy metal pollution in Polish national parks. Water Air Soil Pollut. 9, 83–97.

Grodzińska, K., 1985. Heavy metal pollution of Polish national parks. In: Grodzińska, K., Olaczek, R. (Eds.). Threat to Polish National Park, Panstwowe Wydawnictwo Naukowe, Warsaw, pp. 23–35.

Hahn, E., Ostapczuk, P., Ellenberg, H., Stoeppler, M., 1985. Environmental monitoring of heavy metals with birds as pollution integrating biomonitors. II. Cadmium, lead and copper in Magpie (Pica pica) feathers from a heavily polluted and a control area. In: Proceedings of the International Conference on Heavy Metals in the Environment. Athens, Greece. pp. 724–726.

Hahn, E., Ostapczuk, P., Stoeppler, M., Ellenberg, H., 1989. Schwermetalle in Elsterfedern-Zur Frage nach den Anteilen von exogener und endogener Einlagerung in die Federn bei Zink. Cadmium, Blei, Kupfer, Nickel und Kobalt (Heavy metals in feathers of magpie – Question of exogenous and endogenous share of zinc, cadmium, lead, copper, nickel, and cobalt in feathers). Ökologie der Vögel 11, 265–281.

Hahn, E., Hahn, K., Stoeppler, M., 1993. Bird feathers as bioindicators in areas of the German Environmental Specimen Bank – bioaccumulation of mercury in food chains and exogenous deposition of atmospheric pollution with lead and cadmium. Sci. Total Environ. 139/140, 259–270.

Hartner, L., Hueber, N., Schreiber, H., 1992. Uber die Eignung der Vogelfedern als Bioindikator (Application of bird feathers as bioindicators). Hohenheimer Umwelttagung 24, 75–91.

Hennig, V., 1992. Elementanalysen von Kolkraben (Corvus corax) und Wanderfalkenfedern (Falco p. peregrinus): Moglichkeiten und Grenzen der Bioindikation (Analysis of elements in feathers of raven and peregrine falcon: possibilities and limits of bioindication). Hohenheimer Umwelttagung 24, 93–110.

Jedrzejowski, J., 1987. Procesy przemyslowe a zanieczyszczenie srodowiska: przemysl hutniczy i cementowy (Industrial processes and environmental pollution; metallurgy and cement industry), Panstwowe Wydawnictwo Naukowe, Warsaw.

Kelner, M., 1994. Thallium. In: Seiler, H., Sigel, A., Sigel, H. (Eds.). Handbook on Metals in Clinical and Analytical Chemistry, Marcel Dekker, New York, pp. 601–610.

Kooiker, G., 1986. Schwermetalle in Federn von Elstern (Pica pica) (Heavy metals in magpie feathers). J. Ornithol. 127, 25–36.

Kooiker, G., 1987. Schwermetalle in Eiern und Federn von Elstern (Pica pica): Ein Beitrag zur Bioindikation (Heavy metals in magpie eggs and feathers; A contribution to bioindication). Osnabrücker naturwissenschaftliche Mitteilungen 13, 57–64.

Kühnast, O., Ellenberg, H., 1990. Schwermetalluntersuchungen (Cadmium, Kupfer, Blei) in Federn von Elster (*Pica pica*) und Habicht (*Accipiter gentilis*) als flächenhaft integrierendes Biomonitoring für Luft-Schadstofeinträge im südöstlichen Schleswig-Holstein (Studies of heavy metals in magpie and goshawk feathers as area integrating biomonitoring of air pollution in southeastern Schleswig-Holstein) Corax 13, 309–325.

Müller, P., Dmowski, K., Gast, F., Hahn, E., Wagner, G., 1984. Zum Problem der Bioindikation von standortspezifischen Schwermetallbelastungen mit Vogelfedern (Bioindication of local heavy metal pollution). Wissenschaft und Umwelt 3, 139–144.

Weyers, B., Gluck, E., Stoeppler, M., 1988. Investigation of the significance of heavy metal contents of blackbird feathers. Sci. Total Environ. 77, 61–67.

Trace elements – Their distribution and effects in the environment
B. Markert and K. Friese, editors

Chapter 18

Lead, zinc and cadmium in biological tissues of sheep bred in a polluted area

G. Chessa, G. Calaresu, G. Ledda, M.C. Testa, A. Orrù

Abstract

Lead, cadmium and zinc levels in biological tissues of sheep bred in south-west Sardinia were reported. The monitored area shows a natural and unnatural pollution: the environmental impact of Pb and Zn smelter refinery and carbon mining activities on heavy metals contamination has been reported for long time. Since 1990 the area was recognized as a ''risk area'' and studies to develop an adequate pollution-control system will take place, using recently-allocated government funds. In this work, nine clinically normal sheep from the polluted area and seven from an agricultural area with minimum pollution were tested. Lead, cadmium and zinc concentrations were determined in muscle meat, liver, kidney, bone, milk, blood and wool. The analyses were performed with electrotermal atomic spectrometry with Zeeman correction on acid-mineralized solution obtained from a microwave system. Quality control procedure were assured by use of validated procedure and, where possible, certified reference materials. The obtained results show the distribution of heavy metals in target accumulation organs and biological tissues.

Keywords: Lead; Cadmium; Zinc; Environmental pollution; Liver; Kidney; Bone; Wool; Blood

1. Introduction

Lead, cadmium and zinc levels in biological tissues of sheep bred in contaminated environment were determined. The studied area, situated in south-west Sardinia shows natural and unnatural pollution due to geological properties and anthropogenic activities, carbon-field and ores have been processed here for lead and zinc since the Roman age. An important industrial estate with one of the most important Al, Pb, Zn and Cd smelting refineries in Europe is located in the area.

Gas and liquid emissions and mine tailing are the principal pollution sources: sulfur dioxide release, particulate, diffused dust, cooling wastewater, ash production, fluorine, lead, cadmium and zinc emissions.

Since 1990 the area was recognized as a ''risk area'' and studies to develop an adequate pollution-control system will take place, using recently-allocated government funds.

The average of lead and cadmium concentrations, obtained over 4-year surveillance programs (1994–1998), frequently exceed the permissible tolerances of 0.5 mg/kg for lead and 1 mg/kg for cadmium of liver from sheep slaughtered in this area. A monitoring study performed in the spring of 1997, on lead and cadmium traces on sheep milk showed a mean concentration of 349 μg/l Pb and of 4.8 μg/l Cd, that were several times higher than in the case of sheep milk from an agricultural region (control).

To improve knowledge about the distribution of heavy metals in target accumulation organs and biological tissues, nine clinically normal sheep from this polluted area and seven from an agricultural area (control) with minimum pollution, were tested. Lead, cadmium and zinc concentrations were determined in muscle meat, liver, kidney, bone, milk, blood and wool.

2. Methods

All glassware and PTFE digestion vessels used were soaked in nitric acid and rinsed with distilled water.

All reagents used for sample digestion and for standards preparation were Baker Instra Analyzed grade.

Electrotermal and flame atomic adsorption spectrophotometry used were Perkin–Elmer Instruments. The microwave oven for sample digestion was CEM MDS 2000 with temperature and pressure control.

The study group comprised 16 clinically normal sheep, nine from polluted area and seven from an agricultural area with minimum pollution (control). The target samples were collected at the time of slaughtering.

Blood was collected in heparinized glass tubes and milk was collected into acid-washed poly-propylene test tubes. Organs and tissues collected were stored in polythene bags. All samples were stored at −20°C until analysis.

Weighed tissue samples (kidney, liver, meat, bone and wool) of approximately 1 and 0.5 g CRM 185 bovine liver were placed in PTFE vessels, 5 ml of concentrated HNO_3 and 5 ml of distilled H_2O were added. A microwave oven program with temperature and pressure settings of 150°C and 130 p.s.i., respectively, was performed for all analyzed matrices. The digestion solutions were diluted to a final volume of 50 ml: all dilution, excluding blood samples, were performed with 1% HNO_3. Zinc analysis was performed on a Flame AAS 2100 (Perkin–Elmer), with an air–acetylene flame. Lead and cadmium analysis was performed on a GF-AAS 5100 ZL (Perkin–Elmer). Analysis on GF-AAS were carried out under stabilized temperature platform furnace (STPF) conditions.

Before GFAAS analysis milk samples were diluted 10-fold with distilled water and whole blood 10-fold with 0.01% Triton X-100 as well as reconstituted CRM 195 lyophilized bovine blood material.

Quality control measurements were performed using a validated inter-analytical methods. Accuracy for blood and liver matrices was determined by analysis of international reference materials, Bovine Blood CRM 195 and Bovine Liver CRM 185

The obtained quality control performances are reported in Table 1.

Table 1. Concentration of lead and cadmium and zinc obtained from analysis of blood and liver reference materials[a]

	Lead		Cadmium		Zinc	
	Certified value	Actual value	Certified value	Actual value	Certified value	Actual value
Bovine blood (ng/ml) CRM 195	416 ± 9	400 ± 10	5.37 ± 0.24	6.00 ± 0.18	Not certified	–
Bovine liver (mg/g) CRM 185	0.501 ± 0.027	0.57 ± 0.11	0.298 ± 0.025	0.319 ± 0.042	142 ± 3	138 ± 2

[a] Results are the mean of triplicate analysis ± 1 SD.

3. Results and discussion

The obtained results are reported in the following tables: medians and non-parametric statistical tests were used because data did not approach a Gaussian distribution.

Tables 2–4 show the Pb, Cd, and Zn content of sheep biological tissue from animals of the risk and control areas.

The non-parametric Mann–Whitney *U*-test results, reported as ''yes'' or ''no'', give statistically significant differences between the two groups studied. Target matrices, for heavy metal accumulation, present higher values for animals from the risk area than those from the region control, confirming that grazing animals readily pick up contaminants from the environment on which they live and feed.

Elevated Pb and Cd levels have been recorded in soft tissues from exposed sheep. The median of Pb and Cd kidney values were 0.75 and 9.945 mg/kg, respectively; the Cd liver value, in the range of 1.184–5.270 mg/kg, is significantly high and exceeds the permissible tolerances fixed at 1 mg/kg in national surveillance programs.

As reported by many authors (Carson et al., 1973; Falandysz, 1991; Leita et al.,

Table 2. Lead content of sheep biological tissues

	Risk area			Control area			
	X_{min}	X_{max}	Median	X_{min}	X_{max}	Median	U^a
Milk (ng/ml)	<1	23	13	<1	24	<1	Yes
Blood (ng/ml)	9	46	16	10	29	15	No
Wool (mg/kg)	6.58	54.97	18.64	6.53	11.29	8.82	Yes
Liver mg/kg)	0.27	0.85	0.49	0.15	0.26	0.21	Yes
Kidney (mg/kg)	0.42	1.06	0.75	0.24	0.76	0.36	Yes
Bone (mg/kg)	1.78	22.14	16.22	2.25	3.88	2.95	Yes
Muscle (mg/kg)	0.17	0.63	0.32	0.23	0.62	0.38	No

[a] Mann–Whitney 95%.

Table 3. Cadmium content of sheep biological tissues

	Risk area			Control Area			
	X_{min}	X_{max}	Median	X_{min}	X_{max}	Median	U^a
Milk (ng/ml)	0.1	1.2	0.5	0.6	1.3	1.0	No
Blood (ng/ml)	0.9	2.0	1.2	0.8	2.8	1.3	No
Wool (mg/kg)	0.264	2.263	1.367	0.177	0.512	0.364	Yes
Liver (mg/kg)	1.184	5.270	1.884	0.341	2.580	0.595	Yes
Kidney (mg/kg)	3.870	48.530	9.945	1.990	15.720	3.590	Yes
Bone (mg/kg)	0.007	0.015	0.012	0.009	0.016	0.013	No
Muscle (mg/kg)	0.015	0.036	0.019	0.013	0.098	0.025	No

[a] Mann–Whitney 95%.

1991) a large portion of adsorbed lead, retained in the soft tissue initially, is later accumulated in the skeleton. High values of this chronic cumulative toxicant were recorded in analyzed bone. A high Pb median of 16.22 mg/kg was observed in risk area with maximum level reaching 22.14 mg/kg.

No statistically significant differences were observed for the Pb, Cd and Zn content of muscle from the areas studied, confirming that muscle is not an accumulating tissue.

While heavy metal levels in target organs are a result of a long-term, chronic exposure, the elemental content of unwashed sheep wool reflects an exposure to dust, particulate or general recent fallout pollution. The results for animals from the risk area are similar to reported data regarding pollution due to motor vehicle emission as alongside a busy motorway (Ward and Savage, 1994).

In conclusion, this study confirmed that organs involved in detoxifying metabolism are also accumulation sites for toxic metals. Monitoring programs using these biological tissue may provide information to heavy metal body burden for long-term exposure; wool analysis, which is a non-destructive matrix representative of recent pollution exposure could be used to perform a rapid survey. Finally, reported values can be used to develop power tools such as related databases or expert systems, to help agencies or scientists who perform environmental studies.

Table 4. Zinc content of sheep biological tissues

	Risk area			Control area			
	X_{min}	X_{max}	Median	X_{min}	X_{max}	Median	U^a
Wool (mg/kg)	95.69	238.77	145.25	81.60	133.48	97.61	Yes
Liver mg/kg)	33.97	44.73	36.98	28.63	47.21	38.16	No
Kidney (mg/kg)	31.27	52.08	35.14	24.42	36.84	30.24	Yes
Bone (mg/kg)	84.25	113.94	100.9	88.25	131.03	95.73	No
Muscle (mg/kg)	36.17	54.29	44.13	25.22	82.54	52.6	No

[a] Mann–Whitney 95%.

References

Carson, T.L., Van Gelde, G.A., Buck, W.B., Hoffman, L.J., 1973. Effects of low level Pb ingestion in sheep. Clin. Toxicol. 6, 389–397.

Falandysz, J., 1991. Manganese, copper, zinc, iron, cadmium, mercury and lead in muscle meat, liver and kidneys o poultry, rabbit and sheep slaughtered in the northern part of Poland. Food Addit. Contam. 8, 71–83.

Leita, L., Enne, G., DeNobili, M., Baldini, M., Sequi, P., 1991. Heavy metal bioaccumulation in lamb and sheep bred in smelting and mining areas of S.W. Sardinia (Italy). Bull. Environ. Contam. Toxicol. 46, 887–893.

Ward, N.I., Savage, J.M., 1994. Elemental status of grazing animals located adjacent to the London Orbital (M25) motorway. Sci. Total Environ. 146/147, 185–189.

Trace elements – Their distribution and effects in the environment
B. Markert and K. Friese, editors

Chapter 19

Essential and toxic elements in roe deer blood (Siena County, Italy)

F. Baroni, G. Protano, F. Riccobono, G. Sabatini

Abstract

The levels of zinc, copper, arsenic, cadmium, lead, mercury and antimony were determined in the whole blood of 730 roe deer harvested during the 1997 hunting season in Siena County. For the essential elements copper and zinc, the median values are 0.71 and 2.85 μg/ml, respectively. The geochemistry of the lithological background and the age and sex of the animals seem to influence the concentrations of these elements in the blood. Lead contents were detected in most of the samples and the median value is 96 ng/ml. Arsenic, mercury and cadmium levels are mostly below the instrumental detection limit; the estimated median values suggest metal levels comparable to (Hg) or lower than (As and Cd) those reported for normal human blood. Antimony was detected in half of the blood samples, which raises the question of the element's mobility in oxygenated superficial systems. The scattered high values of mercury and antimony appear to be related to the still active hydrothermal circulation in the region and thus with a natural source of heavy metal contamination.

Keywords: Wildlife; Roe deer; Blood; Baseline values; Heavy elements; Essential elements; Toxic elements; Environmental pollution

1. Introduction

For a long time, numerous living organisms have been successfully employed to test or monitor the quality of the environment with respect to various noxious substances, including the so-called ''xenobiotic'' heavy elements. In some cases, the use of wildlife species, particularly mammals, has been decisive in revealing environmental damage and in tracing the paths of specific contaminants (Frøslie et al., 1984; Bobek et al., 1991). Although the assessment of exposure for wildlife and humans differs in several significant ways, the information collected from mammals can provide important insight into the risks to human health. This is especially significant in heavily populated areas and avoids the necessity for the invasive sampling of humans.

On account of their territorial behavior and dietary habits, some mammalian species are particularly suitable for an appraisal of the distribution of heavy elements within a territory and thus of the related risk of exposure. The roe deer (*Capreolus capreolus L.*), the European member of the Odocoileini (the white-tailed deer, *Odocoileus virginianus,* is a North American member), possesses these characteristics and, particularly where important populations of it are present, appears to be a good biogeochemical tool for the assessment of environmental quality. Indeed, roe deer tissue, e.g. antler, hair, liver and kidney, has often been used by researchers to indicate the extent of environmental damage induced by a number of human activities, such as the pollution from smokestack emissions around industrial plants.

In this paper, we describe one approach to the problem, which takes advantage of a fortunate situation present in southern Tuscany. This part of central Italy hosts a numerous and rapidly-growing roe deer population. A density of 37 roes/km^2 has been reported by San José et al. (1997), who studied the grouping behavior of these ungulates over an area of 1320 ha. However, several zones with an estimated density up to 70 roes/km^2 have been pointed out by the personnel of the Wildlife Office (Mazzoni della Stella, pers. commun.). To maintain the roe deer population within appropriate density limits, the Wildlife Office, which strictly regulates all game harvesting, promoted selective hunting of this species in the early 1990s.

As a result of the hunting regulations, it is possible to know the exact topographic point where each animal was shot. Moreover, each hunter is obliged to hand in the head of the slaughtered animal for the control of sex and age by the Wildlife Office staff.

In southern Tuscany, the possibility to collect such data is coupled with a good geochemical knowledge of the region. Indeed, high-resolution geochemical maps (based on stream sediments) and copious information on rocks, soils, water geochemistry and heavy elements contents in wild plant species are available for this territory (Baroni et al., 1994; Protano and Riccobono, 1997a,b; Protano et al., 1998).

This especially favorable situation encouraged us to undertake a research project aimed at assessing the biological responses in mammals to the impact with a relatively well known geochemical environment. The specific objectives of the project were: (1) to obtain reliable baseline values of essential and toxic heavy elements in roe deer tissues, and (2) to appraise the exposure of wildlife to several pronounced geochemical anomalies for xenobiotic elements in southern Tuscany.

The geochemical anomalies in this area are mostly, albeit not exclusively, linked to the historical mining and smelting activities carried out from the Etruscan Age until the latter part of the twentieth century. The mining district practically coincides with the western half of southern Tuscany (Fig. 1), while the eastern half (Siena County) represents a decidedly cleaner zone. Thus, we chose Siena County as the site of research to achieve the first objective of our project, namely to obtain baseline values for some of the essential and toxic heavy elements in roe deer blood.

In this paper, we report our findings from analyses of the whole blood of 730 roes shot in Siena County during the 1997 hunting season (in the period 1 August–30 September). The elemental analyses were extended to 18 chemical species. However, here we limit the discussion to the essential elements Zn and Cu and the highly toxic elements As, Cd, Hg, Pb and Sb.

Although we found high metal levels in some samples, we will avoid discussion of them here. Indeed, further and more specific research is required to shed light on this topic, which will be dealt with in a future work.

2. Sampling and analytical methods

This work could not have been undertaken without the help of many hundreds of hunters in Siena County, who generously gave us their full collaboration. Since we were dealing with shot animals, the hunters were instructed, shortly before the hunting season began, how to collect blood and other tissues from the roe deer so as to avoid any sample contamination arising from the impact of the bullet. Each hunter was

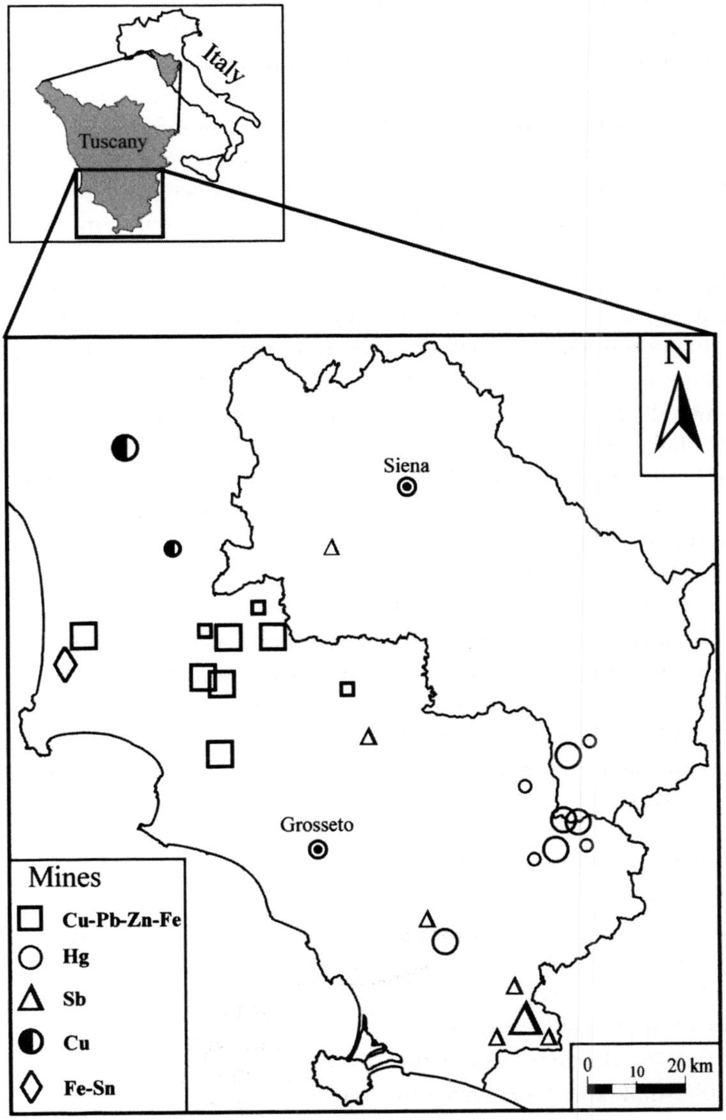

Figure 1. Map showing the main mining centers in southern Tuscany.

furnished with HDPE tubes (10 ml) and was told to fill a tube with blood drawn from the untouched heart or jugular vein and then to store it in a freezer.

However, since the hunters were non-specialized personnel, we also faced the real possibility of some incorrect samplings. In fact, some analytical values for Pb were clearly misleading, being incompatible with life (probably a result of blood collection from the wound or from the thoracic cavity during the butchering of the animal). These samples were obviously discarded but they did highlight the possible existence of other, better camouflaged, erroneous data. Thus we were faced with a clear decision: either renounce the Pb data (and possibly that for Sb since a 3% antimony alloy is often used in bullets) or try to rid the sample of possible biased data. We opted for the second solution because of the importance of lead from an environmental point of view. This choice was made easier by the well known high concentrations of lead in mammalian soft tissues, such as liver and kidney (Ma, 1996). Thus we checked livers of roes with lead contents in blood (PbB) significantly higher than baseline values of herbivores like cattle and horse, and samples showing incongruous values were discarded. When following this procedure, one must be conscious that any non-steady-state situation is most probably overlooked. However, in similar cases, it seems to be the best possible compromise.

Whole blood samples were subjected to hot digestion for 1 h with a $HNO_3 + H_2O_2$ mixture in Pyrex flasks connected to a Friederich's refrigeration apparatus.

Ultrapure reagents (BDH) were used for the attacks, while bidistilled water was prepared from deionized water obtained from a Seradast USF system.

The elemental analyses were carried out at various laboratories with different analytical techniques. Copper and zinc were analyzed by ICP with a Varian Liberty 220 instrument at the Enel S.p.A. laboratories of Larderello (PI). Antimony, mercury and arsenic were determined in our laboratories with a Perkin–Elmer 5000 AAS equipped with FIAS using the hydride method. Lead and cadmium were analyzed at the Enel S.p.A. laboratories of Piacenza with a Perkin–Elmer Elan 5000 ICP-MS equipped with two mass flow controllers and a Perkin–Elmer Gem-Tip crossflow nebulizer. To enhance the sensitivity for cadmium, we also employed a CETAC U5000AT ultrasonic nebulizer (Cetac Technologies Inc., Omaha, NE) as an alternative introduction system to the pneumatic nebulizer.

Blood standards for AAS consisted of international reference material from the CBR of Brussels and Nycomed (Norway).

3. Outline of the study area

The County of Siena coincides with the north-eastern part of southern Tuscany (Fig. 1), which in turn is defined as the part of Tuscany south of the Cecina river. This region features a varied landscape, which mirrors the complex geological history, although hilly morphological shapes prevail.

A large part of the area (about 50%) is covered by thick coppices, with oak, holm-oak, chestnut and pine being the most abundant tree species (Fig. 2). The remaining territory is cultivated or used as pasture. On the plain, intensive cultivation of maize, wheat and sunflower are the main agricultural activities, while the cultivation on the hills of grapes and olive-trees produces world-famous wines and olive oil.

Industrial activities in Siena County are not prominent and restricted to a few urban areas. Known sources of heavy element emissions are rather limited and mostly situated at the county borders.

The geology of the area is quite complex from both the stratigraphic and structural point of view. Hence we will avoid a detailed illustration, since it would be beyond the scope of this work, and will limit our discussion to the main lithological features of the area. Fig. 3 is a simplified lithological map of Siena County obtained from a geological survey at 1:25 000 scale.

The central part of the area and the south-western end are mostly occupied by Lower Pliocene clayey terrains. They represent a graben structure filled by the deepest sediments deposited during the last marine transgression that followed the postorogenic collapse of the Apennine mountain chain. Arenaceous terrains mostly crop out along the eastern border of the area, Mesozoic limestones lie in the north-western side, and lithologically very heterogeneous flysches mainly occur in the western part. Small scattered outcrops of magmatic rocks complete the general picture.

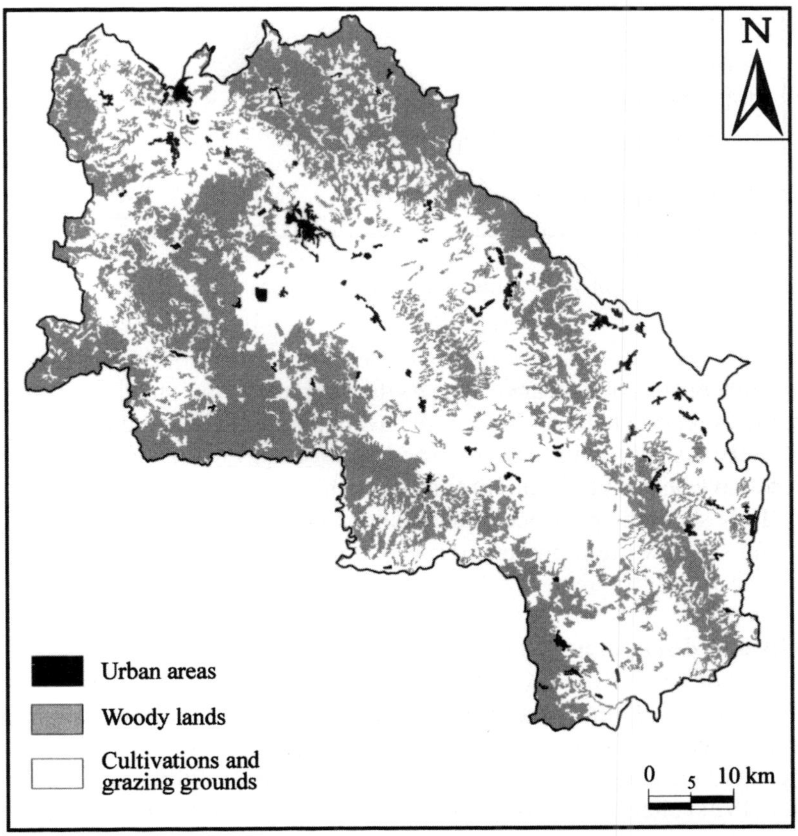

Figure 2. Map showing the land use in Siena County.

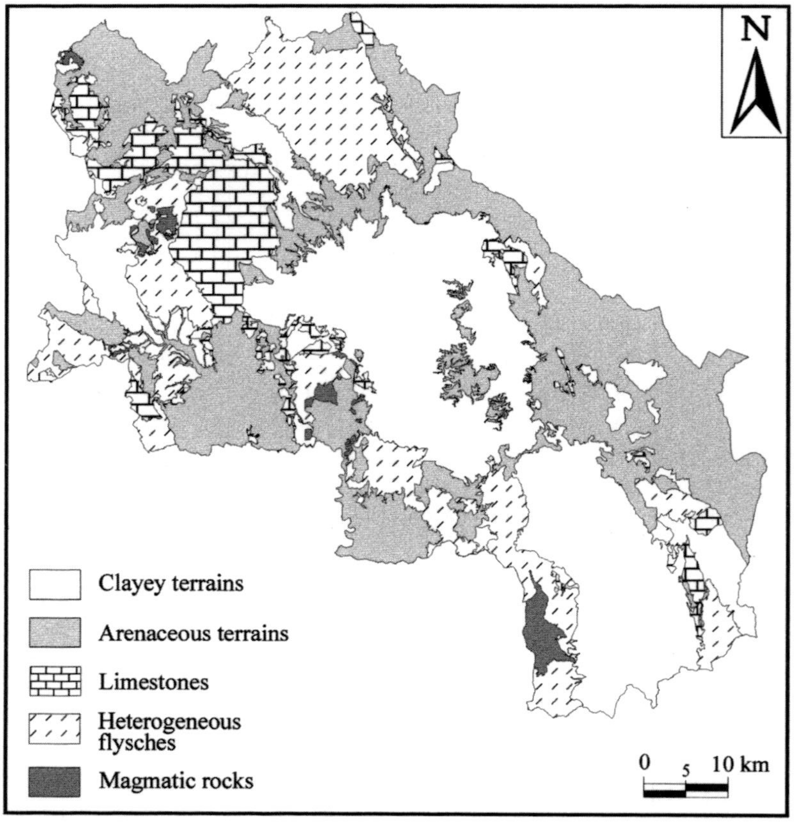

Figure 3. Lithological map of Siena County.

The geochemistry of the area appears to be rather consistent with the general characteristics reported above. Indeed, the available geochemical data reveal a rather clean territory, usually with background values for heavy elements. However, some exceptions have been singled out by our geochemical maps, based on stream sediment surveys. They show the presence of anomalies for Sb and Hg (Fig. 4) near the north-western and south-western borders, respectively.

4. Results and discussion

The terms ''essential'' and ''toxic'' might seem rather self-explanatory with reference to the role played by a chemical element in the biosphere. However, if we try to assign a more precise meaning to the words, the matter becomes a little more complicated.

Kieffer (1991) cites 37 chemical species as usually present in the elemental composition of the human body. Among them there are elements abundant in nature, such as calcium, sodium and potassium, as well as rather rare ones, such as iodine and cobalt,

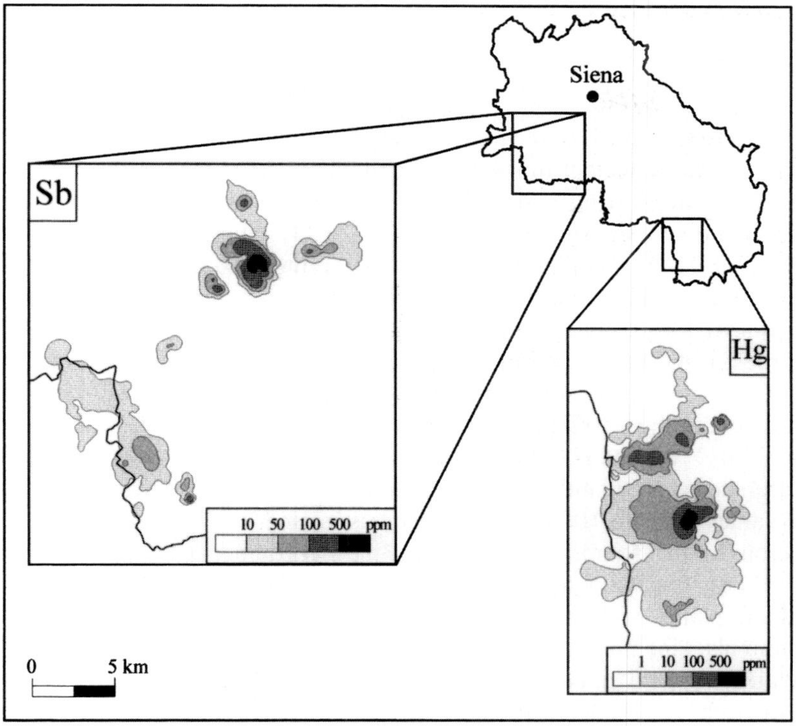

Figure 4. Examples of the few geochemical anomalies (Sb and Hg) present in Siena County, as revealed by stream sediment surveys.

which have obvious organic or essential metabolic functions. Other trace elements, such as selenium, are necessary for living organisms within a quite narrow range of concentrations, while an excessive or deficient intake could be harmful. Some trace elements, such as mercury, are particularly harmful to mammals and seem to have no known metabolic functions. We refer to these elements as ''xenobiotic''.

For a number of other rare chemical species, however, it is very difficult to give any indication about their relationships with the complex metabolic mechanisms of living organisms. The lack of knowledge is in part due to the fact that we have only recently been able to analyze ultratraces in tissues. This coupled with our still largely incomplete understanding of many biochemical processes undoubtedly limits our possibility to predict both the fate and functions of numerous chemical elements in the biochemistry of living organisms. On the other hand, life evolved within a quite complex chemical system where all elements were certainly present and we will probably be able to revise our biogeochemical tables in the near future.

Here we discuss the contents found in the whole blood of roes of two biochemically important heavy metals, zinc and copper, and try to establish their baseline values. We also comment on the contents in the same blood samples of five elements (As, Cd, Pb, Hg and Sb) whose noxious effects are now well-known.

4.1. Essential elements

4.1.1. Zinc

Zinc belongs to Group IIB of the periodic table and is a bluish-white metal used (albeit unknowingly) since ancient times for the production of brass.

The main mineral from which zinc is extracted is the sulfide ZnS (sphalerite), although in some deposits the carbonate $ZnCO_3$ (smithsonite) and the silicate $Zn_4(OH)_2Si_2O_7{\cdot}H_2O$ (calamine or hemimorphite) may also be abundant.

The average crustal abundance of zinc is estimated at about 70 ppm (Greenwood and Earnshaw, 1984), but there are significantly different concentrations in different types of rock. Limestones and sandstones have the lowest contents of this metal, which usually range from 20 to 40 ppm. Decidedly higher values can be found in clays, frequently exceeding 100 ppm. Among the magmatic rocks, since the mantle-''compatibility'' of zinc is comparable to that of iron (Ringwood, 1979), the highest values are shown by mantle differentiates, such as gabbros and basalts, for which average values close to 100 ppm have been estimated.

Numerous human sources of zinc input to the environment can be listed (Nriagu and Pacyna, 1988), the main ones being smelting plants, the production of protective coatings and paints, and organometals used in agriculture.

In mammals, zinc intake is largely due to food ingestion, while drinking water and inspired air are considered quite unimportant sources. The gastro-intestinal absorption rate seems rather variable, with suggested values ranging from 20 to 80%. Zinc is recognized to be essential to animal life and the metal is found in all mammalian tissues and body fluids. In humans, zinc is essential for growth, development and reproduction (Ohnesorge and Wilhelm, 1991).

The homeostatic regulation mechanism usually prevents zinc intoxication (a rather rare event), while a deficiency of the metal may be a more frequent source of disorders in animals (Miller, 1970).

Zinc contained in the blood is a very small part of the total metal content in the body (<0.5%), both in humans (National Research Council, 1978) and in animals, such as sheep (Grace, 1983). Most of the zinc in the blood (>75%) is associated with the red cells (Hambridge et al., 1986).

Table 1 reports the zinc contents in blood of 730 roe deer harvested during the same hunting season in Siena County. The statistical parameters are shown for the entire sample as well as for subsamples based on the age and sex of the animals. On account of the limited home-range of this species, we could also relate the blood zinc values (as well as the Cu and Pb ones) of many animals to the lithology of the environment in which they lived.

As expected, the frequency distribution of the data has a lognormal, positive skewed, form. The range of the zinc values (0.99–5.69 μg/ml), but mainly the median (2.85 μg/ml), are lower than those of humans. According to Minoia et al. (1990), the zinc concentration in human blood ranges from 3.5 to 8.8 μg/ml, with an average of 6.34 μg/ml.

The median value of zinc differs with the age of the animals; the difference is small but significant (on account of the high number of analyses). Young animals have

Table 1. Zinc concentrations (μg/ml) in roe deer blood

	No. of samples	Median	Range
Entire sample	730	2.85	0.99–5.69
Age			
Fawn	63	2.92	1.42–4.88
Yearling	299	2.92	1.08–5.39
Adult	368	2.79	0.99–5.69
Sex			
Male	376	2.92	1.03–5.69
Female	354	2.75	0.99–5.39
Background lithology			
Clayey terrains	162	3.04	1.18–5.69
Arenaceous terrains	204	2.77	1.13–5.39
Limestones	62	2.67	1.08–5.24
Heterogeneous flysches	71	2.67	1.03–5.23
Mafic magmatic rocks	3	–	2.99–4.03

higher median values than adults. Indeed, the fawn sample has the highest minimum value. This agrees with the data of Lindeman et al. (1971) and Keen and Hurley (1980) suggesting a slight decline in blood zinc values with age in both humans and animals.

There is a greater difference in the median of zinc concentrations between the sexes, with value higher in males (2.92 μg/ml). We cannot provide a clear explanation for this and pass the problem to the physiologists. Nevertheless, it could be related to the fact that most of the zinc in the blood is linked to erythrocytes and that some mammal species (e.g. humans) show a sex difference in the number of red cells.

Before discussion of the other data reported in Table 1, it would be useful to examine Fig. 5, which illustrates the median zinc values found in whole blood samples divided according to the fifteen hunting areas of Siena County. As shown by the rod length, there is a great variability of the median values of the samples collected in the various hunting areas.

Nevertheless, if we consider the data on the basis of the lithological characteristics of these areas, we can acquire further important informations. The full gray rods in the graph of Fig. 5 represent three selected hunting areas, which can be considered good examples of three different lithologies. The highest median value of 3.24 μg/ml is associated with an eminently clayey area (Crete Ovest), while the values of 2.56 μg/ml (Val di Chiana Nord) and 2.64 μg/ml (Val di Merse Ovest) refer to areas dominated by arenaceous rocks and limestones, respectively.

This trend appears to be confirmed by the analysis of the samples reliably associated with the different lithologies (Table 1). Indeed, roes harvested on clayey terrains have the highest median value, while those shot on limestones and heterogeneous flysches have the lowest ones. High values, even though referring to only three samples, come from roe deer which lived on mafic magmatic rocks.

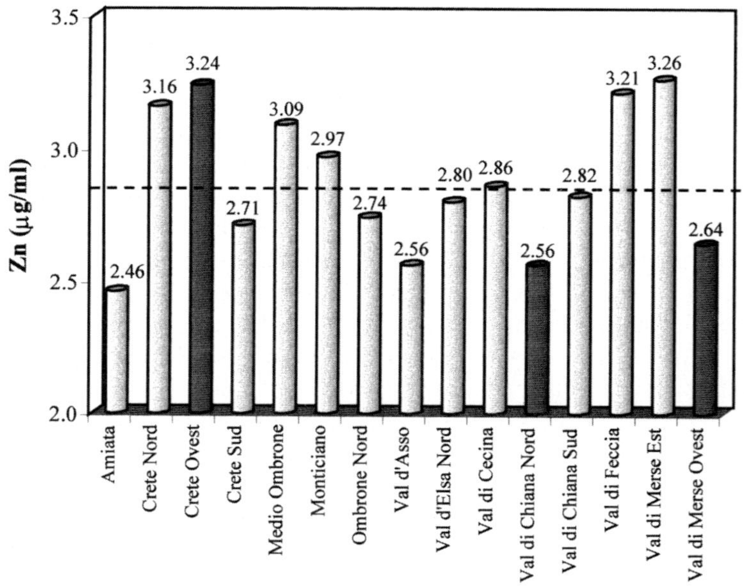

Figure 5. Median values of zinc found in roe deer blood in the fifteen hunting areas of Siena County.

It is well known that several key factors (e.g. mineralogical soil composition, pH, redox condition and bioaccumulation in plants) can strongly influence the intake of zinc (as well as of other trace elements) by herbivores, and thus its concentration in the biological fluids and tissues of these animals. Therefore, the clear influence of the background lithology on the metal levels in roe deer blood may in some way seem surprising. We will avoid engaging in a long discussion of this subject, but instead wish to underline that the entire sample confirms a very clean area with respect to this heavy element.

4.1.2. Copper

Together with silver and gold, copper is known as "coinage metal" because of its common occurrence in the elemental, or "native", form. Copper was used by ancient peoples as early as 5000 years BC and by about 3500 BC the reduction of its ores with charcoal was carried out.

Its relative abundance in the earth's crust is about 68 ppm, mainly as sulfide, oxide and carbonate. The major ores are $CuFeS_2$ (chalcopyrite), Cu_2S (chalcocite), Cu_2O (cuprite) and $Cu_2CO_3(OH)_2$ (malachite).

Copper distribution among the most common rock types (Faure, 1991) is similar to that of zinc: basic igneous rocks and clays have the highest values (around 100 and 50 ppm, respectively), while concentrations in limestones are lower (around 5 ppm). Sandstones are, of course, more variable, but generally show intermediate values.

Copper is an essential element for both plants and animals, even though it may be toxic in some cases. Concentrations of Cu in plants usually range from 5 to 20 mg/kg and the element functions as part of the prosthetic groups of enzyme systems (Baker and Senft, 1995).

In animals, copper plays some specific physiological roles, forming part of many proteins, e.g. the enzymes: ceruloplasmin, monoamine oxidase, superoxide dismutase (SOD) and cytochrome *c* oxidase. Copper deficiency is a more common metabolic alteration than copper toxicosis and anemia is the common expression in most of the studied species. Ruminants appear to be particularly susceptible to copper toxicity and are the only animals in which significant, even lethal, effects can occur (Davis and Mertz, 1986). According to Davis and Mertz (1986), the normal range of copper in the blood of healthy animals is 0.5–1.5 μg/ml. However, Minoia et al. (1990) report a slightly different range in humans, from 0.535 to 1.94 μg/ml. In mammalian blood, 50% of the copper is usually contained in the erythrocytes, where it appears to be more or less firmly bound (Davis and Mertz, 1986).

Table 2 reports the statistics of copper concentrations in the whole blood of roe deer. The median value of the entire sample is 0.71 μg/ml, which is in the middle of the normal range for healthy animals cited above. However, about 3.7% of the values are below the lower limit of the reference range, i.e. less than 0.5 μg/ml Cu.

Interactions with other elements can affect copper absorption; in particular, a strong antagonism between copper and zinc has been clearly established, according to Davis and Mertz (1986), in numerous animal experiments. Therefore, we checked the zinc contents in the low Cu blood samples, but found no relationship between low Cu levels and Zn contents. The Cu/Zn plot in Fig. 6 (for the entire sample) does not indicate any inverse correlation of the two elements.

Young animals have little higher median values than the adults. However, the fawn sample also has the highest minimum value. This trend seems to agree with the data of

Table 2. Copper concentrations (μg/ml) in roe deer blood

	No. of samples	Median	Range
Entire sample	730	0.71	0.13–1.68
Age			
Fawn	63	0.72	0.48–1.48
Yearling	299	0.72	0.13–1.65
Adult	368	0.70	0.22–1.68
Sex			
Male	376	0.73	0.34–1.66
Female	354	0.69	0.13–1.68
Background lithology			
Clayey terrains	162	0.76	0.34–1.68
Arenaceous terrains	204	0.70	0.13–1.65
Limestones	62	0.68	0.25–1.66
Heterogeneous flysches	71	0.70	0.37–1.61
Mafic magmatic rocks	3	–	0.65–1.30

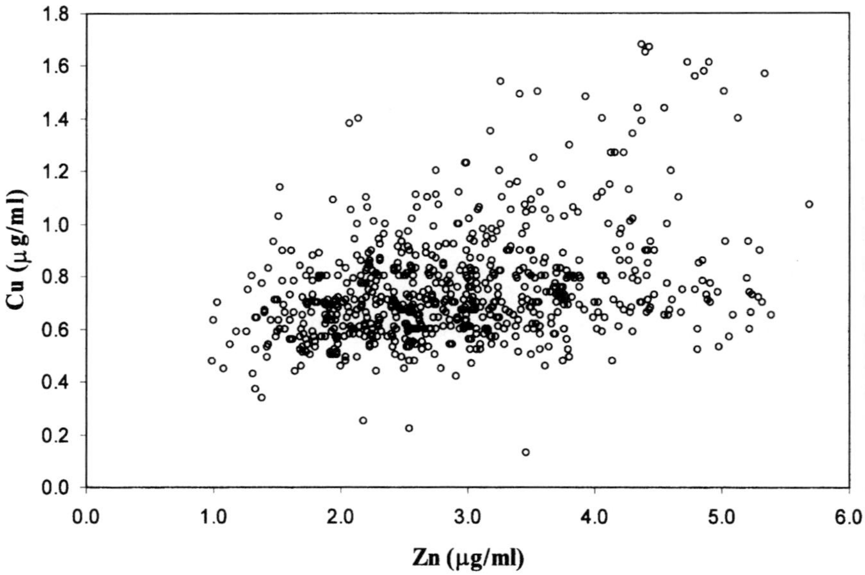

Figure 6. Copper vs. zinc contents in roe deer blood (entire sample).

Suttle (1973) which showed that the young suckling lambs absorb copper at a higher rate than adult sheep.

The median and (in part) range values indicate a slightly higher copper concentration in the blood of males. Although the sex difference is rather small, it seems to mirror the same trend found for zinc.

As was seen for zinc, the background lithology seems to influence the copper levels in roe deer blood, albeit to a much smaller extent. The highest median value (0.76 μg/ml) was found in roes which lived on clayey terrains, in which copper is normally the most enriched among the chalcophile elements (Faure, 1991). Likewise, the values for the (unfortunately very few) roes that lived on mafic magmatic rocks seem to be relatively high.

4.2. Toxic elements

For all the toxic elements, except lead, the values of most of the samples were below the detection limits (even though the limits are rather low). Therefore, it was impossible to treat statistically these data as made for Zn and Cu. Nevertheless, it was possible to infer, with a reasonable approximation, the baseline values of these elements in the roe deer blood. For lead, only about 5% of the samples had values below the detection limit; hence the statistical analysis was not significantly affected.

4.2.1. Arsenic

The content of arsenic in the earth's crust is about 2 ppm. However, in sedimentary rocks such as clays, it can reach much higher values (13 ppm).

The toxicity of arsenic is certainly better known among the general public than that of other heavy elements on account of its ancient use as a poison.

''Black foot'' disease and skin and lung cancer are serious pathologies linked to As exposure. One of the mechanisms of the harmful activity of arsenic is the competition of arsenate with phosphate in the synthesis of adenosine-5-triphosphate or ATP (Andreae, 1986).

In mammals, arsenic intake derives from food and drinking water, inspired air and absorption through the skin. The chemical form of arsenic and its solubility affect the absorption of the element in mammals and its transfer into the bloodstream. According to Leonard (1991), the normal arsenic content in human blood is about 4 ng/ml. However, an average arsenic value of 7.9 ng/ml and a range of 0.4–70.5 ng/ml were reported by Minoia et al. (1990), who analyzed blood samples of 470 healthy Italians.

The arsenic contents in the blood of roe deer of Siena County are usually extremely low. In fact, in most samples (92%), they fall below the detection limit (2 ng/ml). In most of the other samples, the As values are in the range of 2–10 ng/ml. A very small number of samples have higher values, even up to 100 ng/ml.

4.2.2. Cadmium

The average content of cadmium in the earth's crust is around 0.16 ppm and its distribution in nature is linked to Zn minerals.

The environmental concern about this element has arisen quite recently, since over half the cadmium ever produced has been refined in the last few decades (Nriagu, 1979). The main sources of Cd contamination in soils are metalliferous mining and smelting, heavy applications of sewage sludge, P-fertilizer and input from atmospheric deposition.

The metal is readily translocated to the tops of plants after absorption through the roots (Alloway, 1995).

In animals, cadmium can be introduced to the body from the air, food and water and is toxic to all mammalian systems whether ingested, injected or inhaled (Kostiel, 1986). Although this ''technological'' pollutant can be extremely noxious to animals in general, it is particularly so in mammals since the metal tends to accumulate in critical soft tissues like liver and kidney. Acute Cd toxicosis may cause serious diseases, e.g. the so-called ''itai-itai'', or even lead to death (Nogawa, 1981).

Numerous analyses of cadmium levels in human blood have shown a range of 0.1–5.5 ng/ml (Minoia et al., 1990), although a wider range of 0.06–15.9 ng/ml was given by Forstner (1980).

The cadmium concentrations in the roe deer blood samples are mostly very low, with about 70% of them below the detection limit (2 ng/ml). More than 50% of the detectable levels are within a narrow range of 2–4 ng/ml. Most of the other samples have values in the range of 10–50 ng/ml, with only four of them slightly exceeding this range.

4.2.3. Lead

Along with zinc and copper, lead constitutes the triad of ''base metals''. This element has also been used since very early times (7000–5000 BC); in particular, the Romans used it extensively for water pipes and plumbing.

The most important Pb ore is the heavy black mineral galena (PbS), while other ore minerals are anglesite ($PbSO_4$), cerussite ($PbCO_3$) and pyromorphite ($Pb_5(PO_4)_3Cl$).

The estimated average content in the earth's crust is about 13 ppm. In igneous rocks, the average values range from 0.5 ppm in ultrabasic rocks to 19 ppm in granitic rocks. This distribution is explained by the diadochic substitution of potassium by lead, owing to the similar ionic radius of the two elements. In sedimentary rocks, clays have the highest contents (average, 20 ppm), while the values in both sandstones and limestones are normally below 10 ppm.

Today, lead is one of the most noxious and widespread environmental pollutants among the heavy elements. Both the metal and its inorganic compounds can be considered harmful to the environment. Particularly dangerous are the lead organo-compounds, such as tetramethyllead (TML) and tetraethyllead (TEL), which have been extensively used as anti-knock agents in gasoline.

Although lead pollution effectively began early in the history of mankind, the human input of this metal to the environment increased sharply in the twentieth century (Fergusson, 1990). For example, the burning of lead-supplemented fuels has caused the release into the atmosphere of huge quantities of lead-containing particles and aerosols (and their subsequent wet and dry deposition in waters and soils).

Lead is neither an essential nor useful element for plants and animals.

The uptake of Pb by plant roots and its translocation to the shoots varies seasonally. In broad terms, there is a positive relationship between the concentration in the soil and that in the plant (Davies, 1995). A review of the evolutionary and physiological responses of plants to heavy elements, such as lead, was given by Turner (1994).

Lead is toxic to animals. Nevertheless, the lead tolerance threshold appears to vary greatly from one species to another; indeed, pigs, goats and rabbits appear to be rather resistant to lead (Quaterman, 1986). Link and Pensinger (1966) reported no effects in pigs even at Pb levels of 1.3 μg/ml in blood and 210 μg/g in liver and kidney. It must be said, however, that lead toxicity in mammals is difficult to diagnose (Ma, 1996). However, it can be revealed by biochemical markers, such as the ALAD (δ-aminolevulinic acid dehydrase) concentration in erythrocytes which sharply decreases very early in lead toxicosis (WHO, 1977).

Normal PbB baseline values in mature mammals, such as cattle, are about 20–60 ng/ml (Pripc-Majic et al., 1990). According to Piomelli et al. (1980), PbB levels in humans from remote areas range from 30 to 50 ng/ml. A mean value of 157 ng/ml and a range of 30–390 ng/ml were reported by Minoia et al. (1990) who analyzed the whole blood of 959 selected (i.e. not professionally exposed) healthy persons in northern Italy.

The lowest reported PbB levels associated with clinical lead toxicosis in mammals (observed in both cattle and horses) are about 350 ng/ml, while the highest level of

PbB in cattle at which clinical signs of lead toxicosis have not been noted is 290 ng/ml (Ma, 1996).

Lead intake in mammals is caused by both food ingestion and inhaled air containing lead-bearing particles and aerosols. The latter pathway is a two-stage process: the deposition of very small lead-bearing materials in the lungs and then the absorption of the metal which ultimately enters the bloodstream.

Food ingestion can also certainly play a prominent role. Grazing herbivores, such as cattle or horse, may ingest contaminated soil particles along with the grass. However, since the roe deer is essentially a browser, this form of lead intake is probably unimportant.

As expected, there is only a small percentage (Table 3) of blood samples with values below the instrumental detection limit for the element (2 ng/ml).

Neither age nor sex seems to influence the median values, which are not much different from the entire sample value of 96 ng/ml. Conversely, animals living on different background lithologies once again show some differences in Pb concentration in the blood. Roes harvested on clayey terrains have the highest median value while those shot on limestones have the lowest one.

As can be seen from ranges, several blood samples have rather high Pb contents.

4.2.4. Mercury

Mercury is another metal that was well-known even in antiquity, being used in amalgams for the extraction of metals and directly as the mineral cinnabar, widely used as a pigment (vermilion).

Like the previously discussed elements, Hg is geochemically chalcophile and it is

Table 3. Lead concentrations (ng/ml) in roe deer blood

	No. of samples	Range	% of data < d.l.	Median
Entire sample	730	<2–394	5.0	96
Age				
Fawn	63	<2–383	12.5	90
Yearling	299	<2–393	3.1	94
Adult	368	<2–394	5.0	98
Sex				
Male	376	<2–394	6.1	96
Female	354	<2–393	3.8	97
Background lithology				
Clayey terrains	162	<2–392	6.0	113
Arenaceous terrains	204	<2–393	3.7	96
Limestones	62	3–393	0.0	68
Heterogeneous flysches	71	<2–357	10.2	75
Mafic magmatic rocks	3	9–85	–	–

mainly concentrated in sulfides. However, its abnormally low melting point, coupled with the strong tendency to be stabilized as native metal, leads mercury to be easily enriched in the vapor phase and thus often to be concentrated in geothermal fluids.

The average crustal value of mercury is around 0.08 ppm (it is a very dispersed element) and, once again, clays are the richest rocks (around 0.4 ppm) among the different lithotypes, igneous and sedimentary.

The element enters the environment through natural and human emissions, the latter being the predominant source (Nriagu, 1989).

Mercury is a biologically non-essential heavy metal. It exists in a range of inorganic and organic forms with varying degrees of toxicity.

Mercury has been involved in very serious outbreaks of widespread heavy metal poisoning. In fact, it has been responsible for health problems and thousands of deaths in Japan, Iraq, Pakistan, Canada, etc.

The most insidious mercury toxin is methylmercury, which accumulates in various organs and is able to cross both the placental and blood-brain barriers to harm the fetus (Craig, 1986; Fergusson, 1990).

The inhalation of gaseous elemental mercury Hg° is also prejudicial to health, since the metal vapor crosses the blood-brain barrier and once oxidized in the brain remains in this organ to produce neurological damage.

Not much is known about mercury uptake and metabolism in wild plant species. However, accumulation of the metal has been observed in mushrooms and aquatic plants (Kabata-Pendias and Pendias, 1984). Most studies of agricultural crops, with Hg loads far greater than those encountered in normal conditions, demonstrate the tendency for mercury to accumulate in the roots (Steinnes, 1995).

Little is known about the mercury contents in whole blood of mammals, such as wild herbivores. However, much information is available regarding marine mammals and humans.

In humans, mercury intake from terrestrial food and drinking water is normally negligible and thus low blood values are expected. A range of 0.5–17.3 ng/ml of Hg in the blood of healthy persons was reported by Minoia et al. (1990). Higher values (1–42 ng/ml) were given by Clarkson (1986) on the basis of analyses of blood samples of 679 residents in Saskatchewan (Canada).

In over 90% of the roe deer blood samples, the mercury levels are below the detection limit (10 ng/ml) of our analytical method. The detected levels range from 10 to 188 ng/ml, and some tens of samples have relatively high Hg values, i.e. between 25 and 100 ng/ml.

In an attempt to explain these objectively high values, we checked all the information for the area in our database and we have singled out a likely reason for these anomalous samples. The sketch map of Fig. 7 indicates both the fault system and the thermal springs present in the north-western part of Siena County, as well as the sites of the highest blood levels for Hg (and also Sb). There is an obvious close spatial correspondence among all these elements.

The thermal fluids that presently rise from the deep through this fault system are consanguineous with the fluids that created, a few hundred thousand years ago, very important Hg–Sb ore deposits in southern Tuscany. Moreover, our still unpublished data on the mercury contents in the air in the vicinity of various thermal springs of the

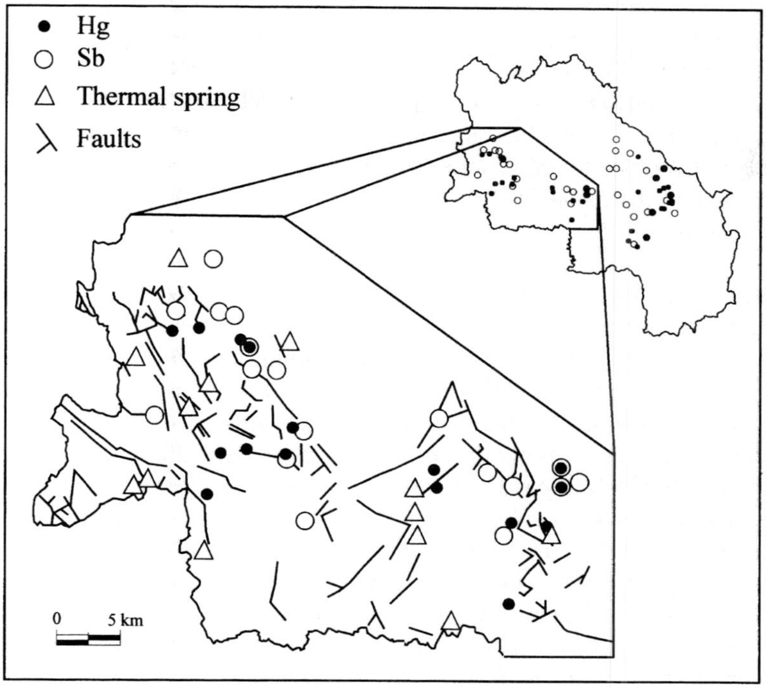

Figure 7. Map showing the correlation of tectonics with the highest values of Hg and Sb in roe deer blood.

region show absolutely high metal values. Since the blood Hg concentration is influenced by the levels of the metal in food and inspired air, we believe that the natural hydrothermal mercury input to the environment of the area satisfactorily explains the presence of the anomalous samples.

4.2.5. Antimony

From a geochemical point of view, the behavior of antimony is in many respects parallel to (even if distinct from) that of arsenic. This situation, which is mostly due to the chemical characteristics of the two elements, is reflected in a large number of similar compounds.

Stibnite (Sb_2S_3) is by far the most important antimony ore, while valentinite (Sb_2O_3) and cervantite (Sb_2O_4) are the most common oxide minerals formed by weathering.

The average content in the earth's crust is around 0.2 ppm and once again the richest rocks are clays (1.5 ppm).

Volcanoes are the main natural sources of antimony, while smelters and coal-fired power plants are the principal human sources. In addition, wind-borne dust particles and water drainage from mine-waste dumps may be important sources of the spread of the element in Sb-mining zones (Protano and Riccobono, 1997b).

Antimony is not an essential element for plants, animals and humans. The toxic properties of antimony are similar to those of arsenic. The trivalent oxidation state is more harmful than the pentavalent one, since it interferes with the bonding of the HS-groups of enzymes (Luckey and Venugopal, 1978).

Although antimony is not essential to plants, its soluble forms in soil can be readily taken up by the roots (Edwards et al., 1995). According to Fowler and Goering (1991), antimony concentrations in plant and animal tissues are typically low in areas without human sources of the element.

Reliable antimony contents in mammalian blood are available only for humans. Hamilton et al. (1973) reported an average value of Sb in human blood of about 5 ng/ml. More recently, Minoia et al. (1990) found an average value of about 2 ng/ml and a range of 0.03–5 ng/ml.

Rather surprisingly in view of the common natural association of Sb with As and Hg, only half of the Sb values in the roe deer blood fall below the detection limit (2 ng/ml). About 50% of the detectable values are in the range of 2–20 ng/ml. Most of the others range from 20 to 100 ng/ml, but several are even higher (up to 500 ng/ml).

Thus the environmental impact of antimony seems to be rather important, certainly greater than that of its chemically similar companion, arsenic. Some brief observations might clarify the situation. Fig. 7 indicates the sites of the highest blood levels of Sb (as well as Hg). It can be seen that their locations agree well with both the fault system and the related points of rising hydrothermal fluids.

In this regard, we should emphasize that, in addition to the well-known geochemical tendency of the two elements to travel together in hydrothermal fluids, there is also ample evidence in southern Tuscany of the contemporaneous deposition of mercury and antimony ores. Moreover, Protano et al. (1998) showed that very recent and still active hydrothermal circulation is responsible for some weak but spatially diffuse geochemical anomalies for Sb in this region.

The Sb values in the roe deer blood indicate a generalized availability of this element to be metabolized by organisms. This would appear to be in contrast to the presumed immobile nature of Sb in oxidizing environments (Edwards et al., 1995) and to the apparently poor availability of the much more abundant arsenic.

However, Vink (1996) recently stressed the broad field occupied by the ionic species SbO_3^-(aq) in the Eh-pH space under oxidizing conditions, be they alkaline or acidic. This substantially rectifies earlier Eh-pH diagrams which suggested that antimony is practically immobile under the same conditions.

5. Conclusions

First, we must underline once again the preliminary nature of this study. We would also emphasize that our objective was to establish the baseline values in roe blood of several heavy elements, taking into account various local factors but mainly the geochemical and geological background of the area.

A general conclusion arising from the entire sample is the confirmation of the substantially clean nature of the study area (Siena County). This is particularly true for Zn and Cu. Hence the influence of background lithology on roe deer blood levels is still perceptible in the small (but significant, in view of the large sample size) differences among subsamples representing different terrains. This also appears to apply to Pb despite the actual masking caused by the well-known environmental diffusion of this pollutant worldwide.

With regard to the baseline values, we must stress the great difficulty in establishing reliable figures, even if based on large sample sizes as in our case. This is especially difficult for toxic elements, for a number of reasons. First, we cannot exclude (in fact, it is almost certain) that a part of the overall data (albeit a small part) is an expression of an anomalous environmental situation. Only after we have identified and removed these data we will have at our disposal a set of data really suitable for the purpose. For this reason, but mainly because of the lognormal distribution of the element contents in our population (as usually occurs), we have made and will make reference only to median values.

A second difficulty comes from the limits of analytical detection. We have seen that for Hg, As, Cd and also Sb, the roe deer blood levels are below these limits in most of the samples. However, since these limits are rather low, it is possible to draw some relevant conclusions, even though they might be revised during the future development of the research program.

Zinc and copper require only a brief comment. The baseline values around the respective medians of 2.85 and 0.71 μg/ml appear to be reliable in view of the quasi-normal distribution of the data and their small limits of variations. For Zn the datum is well under those computed for human blood.

The median value of lead (96 ng/ml) is somewhat lower than the average concentration in human blood, although it is higher than those of mammals in remote regions (40 ng/ml). It is probably not by chance that the value of the mode in our Pb frequency distribution is around 30–40 ng/ml.

The median value of arsenic is certainly lower than 2 ng/ml since the levels in over 90% of the samples fall below this value. It appears certain then that the concentration of this element in our roe deer blood samples is very low when compared with the values for human blood.

Also for cadmium the median value is below 2 ng/ml. Thus, a slightly lower content than that of human bloods can be inferred.

The same can be said for mercury, even though the higher detection limit (10 ng/ml) renders the estimated median value of the population even less definite. However, the analytical sensitivity is sufficient to place our data within the range of variation of normal human blood, probably in the lower part.

For antimony, somewhat more than 50% of the roe deer blood levels are below 2 ng/ml; hence its median value must be near this figure, which is comparable to the estimated content in normal human blood. However, the data for this element suggest a situation that is in some way peculiar, especially if compared with that of arsenic (see the discussion in the relevant section above).

As mentioned previously, we have intentionally avoided discussion of the samples with high values of the toxic elements. We will deal with this topic in future papers, which will also benefit from data for other roe deer tissues. We have made only one exception regarding some high values of Hg and Sb found in various part of the study area. The reason for this is the evident correlation of the high values of the two metals with late tectonics and particularly with the still active hydrothermal circulation.

Acknowledgements

We are indebted to many people who helped us to perform this research. First, many thanks are due to the hundreds of hunters in Siena County who collected the roe deer tissue samples. We gratefully acknowledge the staff of the Wildlife Office and the ATC 17-18-19 of Siena County for their invaluable help, as well as Enel S.p.A. for the supply of analytical facilities. We warmly thank Dr Roberto Mazzoni della Stella (Wildlife Office), Ing. Aldo Baldacci, Dr Giorgio Culivicchi and Dr Claudio Donati (Enel S.p.A. of Pisa and Larderello), Ing. Edmondo Joannilli and Dr Maurizio Bettinelli (Enel S.p.A. of Piacenza). Finally, special thanks are due to Paola Aliciati, Dr. Luigi di Lella, Alessandro Terrosi and Maurizio Pinassi of the Institute staff.

References

Alloway, B.J., 1995. Cadmium. In: Alloway, B.J. (Ed.), Heavy Metals in Soils, 2nd ed. pp. 122–151.

Andreae, M.O., 1986. Organoarsenic compounds in the environment. In: Craig, P.J. (Ed.), Organometallic Compounds in the Environment. Longman, New York, pp. 198–228.

Baker, D.E., Senft, J.P., 1995. Copper. In: Alloway, B.J. (Ed.), Heavy Metals in Soils, 2nd ed. pp. 179–205.

Baroni, F., Protano, G., Riccobono, F., 1994. Mercury content of the rocks of Tuscany. A geochemical contribution to geology of Hg-ores. Atti Acad. Sci. Siena XV (XIII), 59–67.

Bobek, B., Perzanowski, K., Regelin, W.L., 1991. Global Trends in Wildlife Management. Transactions Vol. 1. 18th IUGB Congress. Swiat Press, Krakow-Warszawa.

Clarkson, T.W., 1986. Mercury. In: Mertz, W. (Ed.), Trace Elements in Human and Animal Nutrition, 5th ed. pp. 417–428.

Craig, P.J., 1986. Organomercury compounds in the environment. In: Craig, P.J. (Ed.), Organometallic Compounds in the Environment. Longman, New York, pp. 65–110.

Davies, B.E., 1995. Lead. In: Alloway, B.J. (Ed.), Heavy Metals in Soils, 2nd ed. pp. 206–223.

Davis, G.K., Mertz, W., 1986. Copper. In: Mertz, W. (Ed.), Trace Elements in Human and Animal Nutrition, 5th ed. pp. 301–363.

Edwards, R., Lepp, N.W., Jones, K.C., 1995. Other less abundant elements of potential environmental significance. In: Alloway, B.J. (Ed.), Heavy Metals in Soils, 2nd ed. pp. 306–310.

Faure, G., 1991. Principles and Applications of Inorganic Geochemistry. Macmillan, New York.

Fergusson, J.E., 1990. The heavy Elements: Chemistry, Environmental Impact and Health Effects. Pergamon Press, Oxford.

Forstner, U., 1980. Cadmium. In: Huntzinger, O. (Ed.), Handbook of Environmental Chemistry, 3A. Springer-Verlag, Berlin, pp. 59–107.

Fowler, B.A., Goering, P.L., 1991. Antimony. In: Merian, E. (Ed.), Metals and Their Compounds in the Environment. VCH Publishers, Weinheim, pp. 743–749.

Frøslie, A., Norheim, G., Rambaek, J.P., Steinnes, E., 1984. Levels of trace elements in liver from norvegian mouse, reindeer and red deer in relation to atmospheric deposition. Acta Vet. Scand. 25, 333–345.

Grace, N.D., 1983. Amount and distribution of mineral elements associated with fleece free empty body weight gains in the grazing sheep. NZ J. Agric. Res. 26, 59–70.

Greenwood, N.N., Earnshaw, A., 1984. Chemistry of the Elements. Pergamon Press, Oxford.

Hambridge, K.M., Casey, C.E., Krebs, N.F., 1986. Zinc. In: Mertz, W. (Ed.), Trace Elements in Human and Animal Nutrition, 5th ed. pp. 1–109.

Hamilton, E.I., Minski, M.J., Cleary, J.J., 1973. The concentration and distribution of some stable elements in healthy human tissues from the United Kingdom. An environmental study. Sci. Total Environ. 1, 341–374.

Kabata-Pendias, A., Pendias, H., 1984. Trace Elements in Soils and Plants. CRC Press, Boca Raton, FL.

Keen, C.L., Hurley, L.S., 1980. Developmental changes in concentrations of iron, copper and zinc in mouse tissues. Mech. Age. Dev. 13 (2), 161–167.

Kieffer, F., 1991. Metals as essential trace elements for plants, animals and humans. In: Merian, E. (Ed.), Metals and Their Compounds in the Environment. VCH Publishers, Weinheim, pp. 481–489.

Kostiel, K., 1986. Cadmium. In: Mertz, W. (Ed.), Trace Elements in Human and Animal Nutrition, 5th ed. pp. 319–345.

Leonard, A., 1991. Arsenic. In: Merian, E. (Ed.), Metals and Their Compounds in the Environment. VCH Publishers, Weinheim, pp. 751–773.

Lindeman, R.D., Clark, M.L., Colmore, J.P., 1971. Influence of age and sex on plasma and red-cell zinc concentrations. J. Gerontol. 26, 358–363.

Link, R.P., Pensinger, R.R., 1966. Lead toxicosis in swine. Am. J. Vet. Res. 27 (118), 759–763.

Luckey, T.D., Venugopal, B., 1978. Metal Toxicity in Mammals, Vol. 2. Plenum Press, New York.

Ma, W.C., 1996. Lead in mammals. In: Heinz, G.H., Redmon-Norwood, W. (Eds.), Environmental Contaminants in Wildlife, SETAC Special Publications Series. CRC-Lewis, Boca Raton, FL, pp. 281–296.

Miller, W.J., 1970. Zinc nutrition of cattle: a review. J. Dairy Sci. 53 (8), 1123–1135.

Minoia, C., Sabbioni, E., Apostoli, P., Pietra, R., Pozzoli, L., Gallorini, M., Nicolau, G., Alessio, L., Capodaglio, E., 1990. Trace elements reference values in tissues from inhabitants of the European community. I. A study of 46 elements in urine, blood and serum of Italian subjects. Sci. Total Environ. 95, 89–105.

National Research Council, 1978. Zinc. Committee of Medical and Biological Effects of Environmental Pollutants. National Academy of Sciences, Washington DC.

Nogawa, K., 1981. Itai-Itai disease and follow up studies. In: Nriagu, J.O. (Ed.), Cadmium in the Environment, Part II – Health Effects. Wiley, New York, pp. 1–37.

Nriagu, J.O., 1979. Global inventory of natural and anthropogenic emission of trace metals to the atmosphere. Nature 279, 409–411.

Nriagu, J.O., 1989. A global assessment of natural sources of atmospheric trace metals. Nature 338, 47–49.

Nriagu, J.O., Pacyna, J.M., 1988. Quantitative assessment of worldwide contamination of air, water and soils by trace metals. Nature 333, 134–139.

Ohnesorge, F.K., Wilhelm, M., 1991. Zinc. In: Merian, E. (Ed.), Metals and Their Compounds in the Environment. VCH Publishers, Weinheim, pp. 1308–1341.

Piomelli, S., Corash, L., Corash, M.B., Seaman, C., Mushak, P., Glover, B., Padgett, R., 1980. Blood lead concentrations in a remote Himalayan population. Science 210, 1135–1137.

Pripc-Majic, D., Karacic, V., Skender, L., 1990. A follow-up study of lead adsorption on cows as an indicator of environmental lead pollution. Bull. Environ. Contam. Toxicol. 45, 19–24.

Protano, G., Riccobono, F., 1997a. Antimony and arsenic contents in fresh waters and stream sediments of the Monti Romani area (Southern Tuscany, Italy). Miner. Petrogr. Acta 40, 213–226.

Protano, G., Riccobono, F., 1997b. Environmental levels of antimony, arsenic and mercury in the Tafone mining area (Southern Tuscany, Italy). Atti Soc. Tosc. Sc. Nat. 104, 75–83.

Protano, G., Riccobono, F., Sabatini, G., 1998. La cartografia geochimica della Toscana meridionale. Criteri di realizzazione e rilevanza ambientale attraverso gli esempi di Hg, As, Sb, Pb e Cd. In: La Cartografia Geochimica Ambientale. Gli esempi della Calabria, Monti Peloritani, Sardegna e Toscana meridionale. Mem. Des. Carta Geologica d'Italia 55, 119–140, (abstract in English).

Quaterman, J., 1986. Lead. In: Mertz, W. (Ed.), Trace Elements in Human and Animal Nutrition, 5th ed. pp. 281–317.

Ringwood, A.E., 1979. Origin of the Earth and Moon. Springer-Verlag, New York.

San José, C., Lovari, S., Ferrari, N., 1997. Grouping in roe deer: an effect of habitat openness or cover distribution? Acta Theriolog. 42 (2), 235–239.

Steinnes, E., 1995. Mercury. In: Alloway, B.J. (Ed.), Heavy Metals in Soils, 2nd ed. pp. 245–259.

Suttle, N.F., 1973. Effects of the age and weaning on the apparent availability of dietary copper to young lambs. Proc. Nutr. Soc. 32 (1), 24A–25A.

Turner, A.P., 1994. The responses of plants to heavy metals. In: Ross, S.M. (Ed.), Toxic Metals in Soil-Plant Systems. Wiley, Chichester, pp. 153–187.

Vink, B.W., 1996. Stability relations of antimony and arsenic compounds in the light of revised and extended Eh-pH diagrams. Chem. Geol. 130, 21–30.

WHO, 1977. Environmental Health Criteria. 3. Lead. World Health Organization, Geneve.

Trace elements – Their distribution and effects in the environment
B. Markert and K. Friese, editors

Chapter 20

Cadmium residue in placenta tissue and umbilical cord blood

Vidovic M. Milka, Dusan I. Bogunovic, Divna M. Vidovic

Abstract

Cadmium concentrations were determined in aerosol samples collected in the period 1994–1997 in the areas of Kikinda and Novi Sad towns. In Kikinda, a town with developed metal and building industries located in a residental zone, they ranged from 0.00 to 1010 μg/m^2 per day. In Novi Sad, an urban environment with developed industry located in a residental zone, cadmium concentrations ranged in this period from 0.00 to 13.3 μg/m^2 per day. With the goal of estimating the effect of elevated cadmium concentrations in air on the fetus during the period of intrauterine growth, samples were collected and Cd concentrations were determined in the placenta and umbilical cord blood of newborn children from these areas. Cadmium concentrations in placentas of puerperae residing in the Kikinda area ranged from 0.006 to 0.124 μg/g, while the levels for Novi Sad ranged from 0.035 to 0.119 μg/g. It was noticed that cadmium concentrations in placentas of puerperae from the Novi Sad area were statistically significantly higher than in placentas of puerperae from the Kikinda area ($t = 2.76$, $P = 0.0075$). Cadmium concentrations in umbilical cord blood of newborn children from the Kikinda and Novi Sad areas ranged from 0.0010 to 0.0079 μg/ml and from 0.0006 to 0.0021 μg/ml, respectively, whereby it was found that the mean cadmium concentration in umbilical cord blood of newborn children from the Kikinda area was higher than cadmium concentration in umbilical cord blood of newborn children from the Novi Sad area.

Keywords: Cadmium; Placenta; Umbilical cord blood; Aerosediment

1. Introduction

The main routes for cadmium to enter an organism are through the respiratory and digestive tracts. Cadmium is a cumulative poison; 5–6% of the total cadmium intake through the digestive tract is absorbed while the rest is eliminated in faeces. About 90% of resorbed cadmium is bound to particular proteins (metallothionein) and deposited in liver, kidneys, bones and spleen (Lehman and Poisner, 1984). Excretion of cadmium is a very slow process and amounts to 0.01% of daily body deposit, and its biological half-life is 10–30 years.

Acute effects of oral intake of cadmium are excess salivation, nausea, vomiting, abdominal pains, diarrhea and vertigo. Intake of cadmium by inhalation is absorbed in an organism at a level of 20–40% and therefore cadmium concentration in the air is very important. This absorption depends on toxicity modifiers such as smoking, alcoholism and nutrition. Acute effects of inhalation are bronchitis, pneumonitis and toxemia of the liver.

Chronic effects of exposure to elevated cadmium concentrations lead to lung emphysema, renal damage and softening of bones. The atomic diameter of cadmium is very similar to that of calcium which leads to the substitution of calcium by cadmium and its building into the bones which results in osteomyelitis and osteoporosis. Experiments carried out on animals showed cancerous effects of cadmium and its influence on zinc metabolism (Lehman and Poisner, 1984).

In a pregnant woman, cadmium is retained in the placenta causing specific damage to the placenta itself and its function (Copius-Peereboom et al., 1979; Lehman and Poisner, 1984). Cell mechanisms and factors affecting transplacental transport of Cd are not sufficiently known. It is suggested in the literature that maternal hypertension and reduction of fetus weight at birth are associated with elevated Cd levels in the neonatal period. Cadmium permeability through the placenta depends also on zinc and copper concentrations in the mother. Elevated levels of cadmium in the fetus cause teratogenic and fetotoxic changes (Thieme et al., 1977; Friberg and Vather, 1983; Lehman and Poisner, 1984; Fergusson, 1990; WHO, 1992).

2. Methods

The occurrence of elevated cadmium concentrations in air determined by analysis of aerosediments (settlings from air) in the Kikinda and Novi Sad areas, and literature data on toxicological properties of cadmium have prompted the authors to monitor cadmium concentration in umbilical cord blood of newborn children and in placentas of puerperae residing in the Kikinda and Novi Sad areas before and during their pregnancy.

Aerosediment samples were continuously collected in the Kikinda and Novi Sad areas, at different sites, for the period from 1994 to 1997. In this period, 194 samples were collected in the Kikinda area and 703 samples were collected in the Novi Sad area. Each sample was collected for 30 days following a standard sediment sampling method. Cadmium concentrations were determined using atomic absorption spectrophotometry (AAS) on a graphite furnace and expressed as mean monthly values in $\mu g/m^2$ per day (WHO, 1987).

Cadmium in umbilical cord blood and placental tissue was determined for 31 samples collected at the Gynecology Maternity Ward in Kikinda Hospital, and 30 samples collected at the Gynecology Maternity Clinic in Novi Sad, in the period 1996–1997. The samples of hepatizated blood and placental tissue were broken down by nitrous acid using a wet process, the fats were extracted using diethyl ether, and the cadmium concentration was determined by atomic absorption spectrophotometry on a graphite furnace (Vather, 1982; Stoeppler, 1983; Beauty, 1988).

Statistical processing used the methods of central tendency (arithmetic mean $\bar{x}$ and standard deviation, SD) and determination of statistical significance level (Student's *t*-test).

3. Results and discussion

The cadium concentration in air in the Kikinda area for the period 1994–1997, shown in Table 1, are considerably above the maximum threshold levels (MTI). The mean annual cadmium concentration in air in the Kikinda area in 1994 was 48.9 μg/m^2 per day, the percentage of the mean monthly concentration above threshold level being 67%. In 1995 the mean annual level was 35.8 μg/m^2 per day, the percentage of the mean monthly concentrations above threshold level being 44%. In 1996 the mean annual cadmium concentration in air in the Kikinda area was 16.08 μg/m^2 per day, the percentage of the mean monthly concentrations above threshold level being 38.4%. In 1997 the mean annual cadmium concentration in air was 2.54 μg/m^2 per day, the percentage of mean monthly concentrations above threshold limit being 16.6%. The great difference between the minimum and the maximum of mean monthly levels indicates discontinuity in emissions of this heavy metal.

Cadmium concentrations in air in the Novi Sad area were considerably lower, and in 1994 the percentage of measurements above threshold levels was zero. In 1995 the mean annual cadmium concentration in air was below the threshold level, and the percentage of mean monthly concentrations above threshold level was 4%. This was the case in 1996 also, with the exception that the percentage of mean monthly concentrations above threshold level was 6.58%. In 1997 the mean annual cadmium concentration was 4.1 μg/m^2 per day, the percentage of mean monthly concentrations above threshold limit being 1.04%.

Fig. 1 compares the cadmium concentration in placentas of puerperae from the Kikinda area in the period 1996–1997 with the mean cadmium concentration given by the World Health Organization (Environmental Health Criteria 134, Cadmium) of 0.0125 μg/g (WHO, 1992). Cadmium concentration in placentas of puerperae from the Kikinda Maternity Hospital ranged from 0.006 to 0.124 μg/g, the mean level being 0.061 μg/g. In one sample, the cadmium concentration was below the mean cadmium levels according to WHO.

The cadmium concentration in placentas of puerperae at Clinic in Novi Sad in the period 1996–1997 was compared with the mean cadmium concentration according to WHO (0.0125 μg/g) (WHO, 1992). The cadmium concentration in placentas of these puerperae ranged from 0.035 to 0.119 μg/g, and the mean concentration was 0.0749

Table 1. Cadmium level as a fraction of settlings in air in the Kikinda and Novi Sad areas in the period 1994–1997

	Cd μg/m^2 per day (Kikinda)				Cd μg/m^2 per day (Novi Sad)			
	1994	1995	1996	1997	1994	1995	1996	1997
Minimum	0.9	0	0.02	0.006	<5	<5	0	4
Maximum	313	1010	133.74	15.35	<5	<5	13.2	11
$\bar{x}$/year	49	35.8	16.08	2.54	<5	<5	3.7	4.1
% over MTI	67	44	38.4	16.6	0	4	6.58	1.04
MTI	5	5	5	5	5	5	5	5

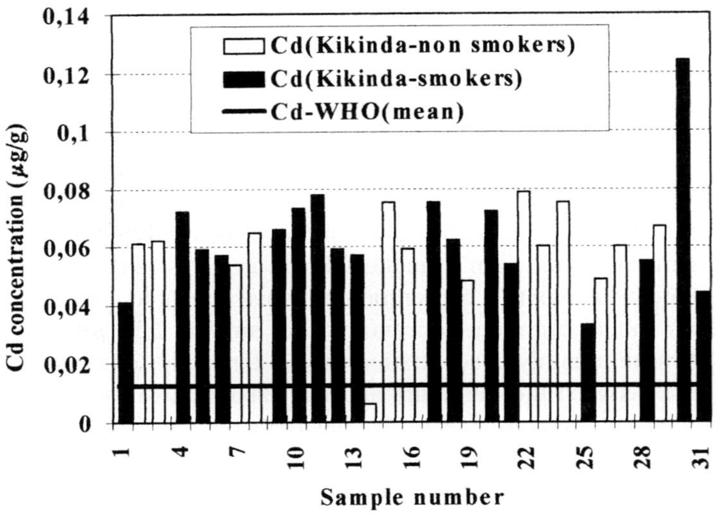

Figure 1. Cadmium concentration in placenta samples from the Kikinda area.

μg/g. In all samples, the cadmium concentration was higher than the mean cadmium concentration according to (WHO, 1992).

Based on the mean cadmium concentration in placentas of puerperae, a statistically significantly higher cadmium concentration in placentas of puerperae from Novi Sad in comparison with those from Kikinda can be seen ($t = 2.76$, $P = 0.0075$), which is illustrated in Fig. 3.

Figs. 1 and 2 compare the cadmium concentration in placentas of puerperae who smoked before and during their pregnancy and those who did not smoke, with mean levels given in EHC-134 according to WHO. The mean cadmium concentration in

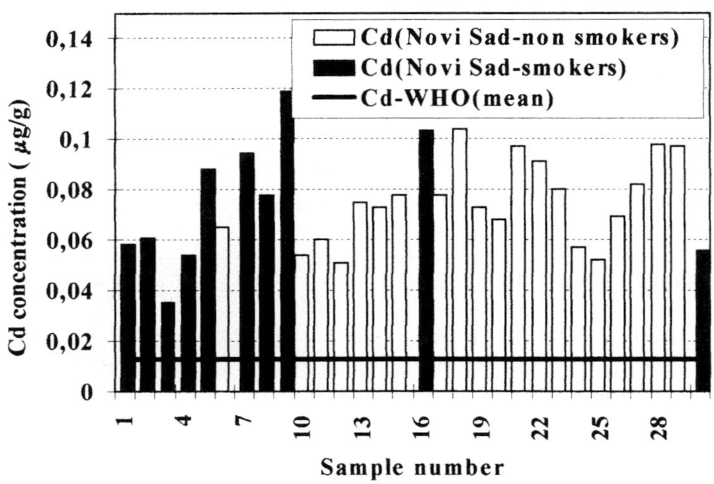

Figure 2. Cadmium concentration in placenta samples from the Novi Sad area.

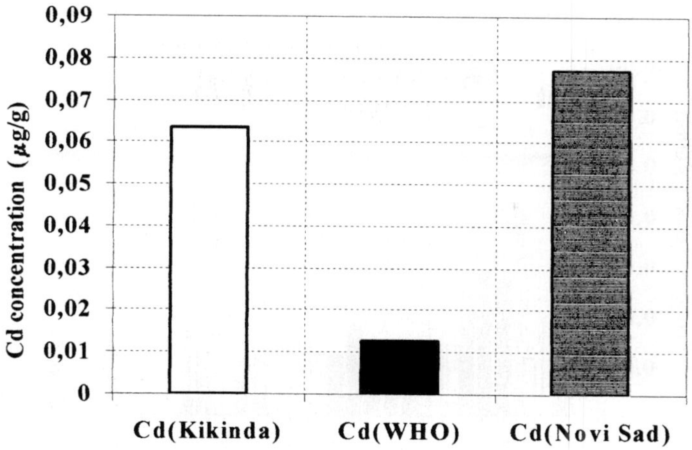

Figure 3. Mean cadmium concentration in placenta.

placentas of puerperae who smoked before and during their pregnancy was 0.0635 μg/g in the Kikinda area and 0.0772 μg/g in the Novi Sad area. In puerperae who did not smoke during their pregnancy, the mean cadmium concentration in placentas was 0.0585 μg/g in the Kikinda area and 0.0751 μg/g in the Novi Sad area. It can be seen that the mean cadmium concentration in placentas of puerperae who smoked before and during their pregnancy in Kikinda area is 5 times higher than mean world level (0.0125 μg/g), and 6.2 times in the Novi Sad area. In purperae who did not smoke, the mean cadmium concentration in the Kikinda area is 4.6 times higher than the mean world level (0.0125 μg/g), and 6.0 times in the Novi Sad area.

The results of cadmium concentration in umbilical cord blood of newborn children in the Maternity Hospital in Kikinda and the Novi Sad Clinic in the period 1996–1997 are shown in Figs. 4 and 5. The levels obtained were compared with the mean levels of cadmium in blood of newborn children (0.0052 μg/ml) and the mean levels of cadmium in blood of pregnant women (0.0051 μg/ml) obtained in the tests conducted by Roels and co-workers (Lauwerys et al., 1978; Roels et al., 1978; WHO, 1992).

The cadmium concentration in umbilical cord blood of newborn children in the Maternity Hospital in Kikinda ranged from 0.0010 to 0.0079 μg/ml, with a mean value $\bar{x} = 0.00183$ μg/ml. The number of samples analyzed was 31. In one case the concentration of cadmium in umbilical cord blood of newborn children was higher than the mean tabular values of cadmium concentration in blood of newborn children and blood of pregnant women (Lauwerys et al., 1978; Roels et al., 1978; WHO, 1992). Records show that this pregnant woman smoked a lot of cigarettes during her pregnancy.

The cadmium concentration in umbilical cord blood of newborn children at Novi Sad Clinic ranged from 0.0000 to 0.0021 μg/ml, with a mean value $\bar{x} = 0.0015$ μg/ml. The number of samples analyzed was 30. Cadmium concentration in umbilical cord blood of newborn children was lower than the mean tabular values of cadmium concentration according to Roels and co-workers (Lauwerys et al., 1978; Roels et al., 1978; WHO, 1992).

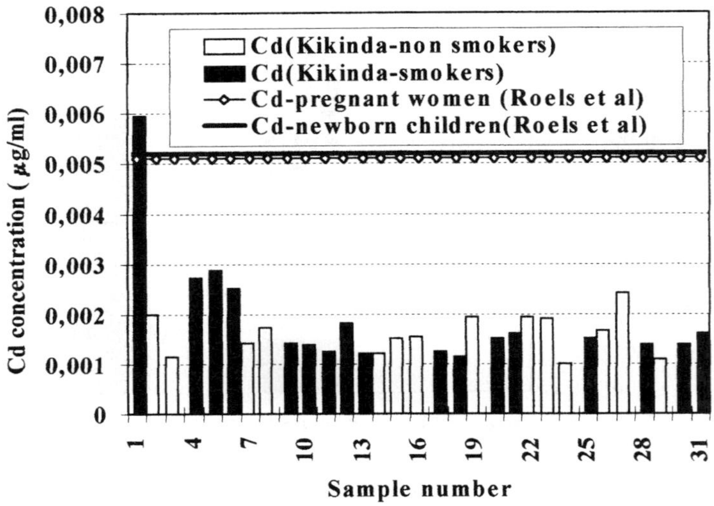

Figure 4. Cadmium concentration in umbilical cord blood of newborn children (Kikinda).

It can be seen that the mean cadmium concentration in umbilical cord blood of newborn children in the Kikinda area is higher than the concentration in blood of newborn children from Novi Sad but the difference is not statistically significant ($t = 1.34$, $P = 0.183$).

Comparing the results of cadmium concentration in placentas and blood of newborn children from the Kikinda area with those from the Novi Sad area, the reverse relationship is noticed (Figs. 3 and 6). In order to find an explanation for this, and based on knowledge from the literature that there is metabolic antagonism between cadmium and zinc at tissue level, and that zinc has an antitoxic affect in relation to cadmium,

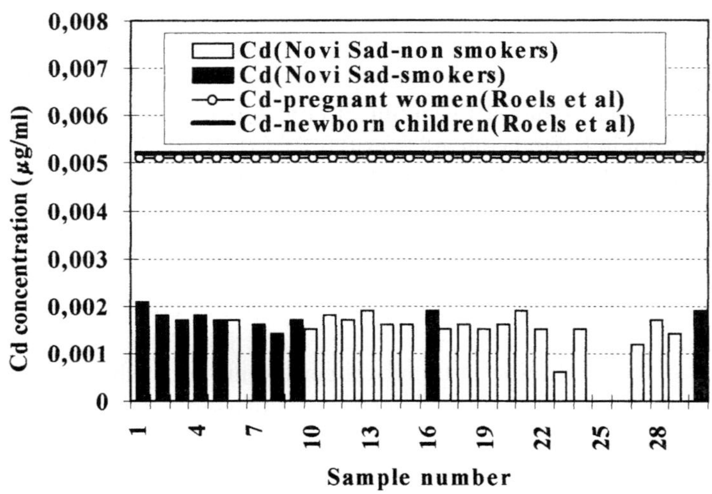

Figure 5. Cadmium concentration in umbilical cord blood of newborn children (Novi Sad).

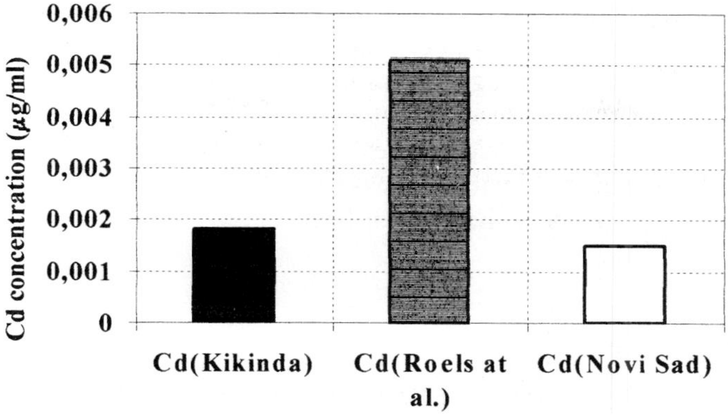

Figure 6. Mean cadmium concentration in umbilical cord.

zinc concentration in aerosediments of the tested areas, and the relation between the concentration in placenta and umbilical cord blood in tested samples were analyzed simultaneously.

The mean annual concentration of zinc in aerosediments in the Kikinda area ranged from 408 μg/m^2 per day in 1994, falling to 69.72 μg/m^2 per day in 1997. In Novi Sad, zinc concentration in aerosediments ranged from 60 μg/m^2 per day in 1994, increasing to 119.9 μg/m^2 per day in 1997.

The mean zinc concentration in placental tissue in the samples from the Kikinda area was 3.199 μg/g, while the mean zinc concentration in samples from the Novi Sad area was lower, i.e. 2.478 μg/g which is statistically significant ($t = 2.37$, $P = 0.0206$).

In umbilical cord blood, the mean zinc concentration in the samples from the Kikinda area was 1.406 μg/ml, while in the samples from the Novi Sad area it was higher, i.e. 1.600 μg/ml, which is not statistically significant.

The results show that the mean annual concentrations of cadmium and zinc in air in the Kikinda area were considerably higher in the period than those in the Novi Sad area. The elevated levels of cadmium and zinc in air in the Kikinda area were reflected in cadmium and zinc concentrations in placenta and umbilical cord blood of newborn children from Kikinda; this is shown in Table 2, which compares the cadmium and zinc concentrations in placenta and umbilical cord blood of newborn children from

Table 2. Mean cadmium and zinc concentrations in placenta and umbilical cord blood

	Kikinda	Novi Sad
Placenta		
Cd (μg/g)	0.061	0.074
Zn (μg/g)	3.199	2.472
Umbilical cord blood		
Cd (μg/ml)	0.0018	0.0015
Zn (μg/ml)	1.406	1.600

Novi Sad. Such a relationship of cadmium concentration in placenta and umbilical cord blood of newborn children can be explained by the influence of zinc on the permeability of cadmium through the placenta (Forstner, 1980; Kägi and Hapke, 1984; Fergusson, 1990; Sandstead, 1997).

Since cadmium has teratogenic and fetotoxic effects, the results described above suggest the need to perform detailed assessment of the health risks of the population being exposed and to increase the level of health.

References

Beauty, D.R., 1988. Concepts, Instrumentation and Techniques in Atomic Absorption Spectrophotometry. Perkin Elmer, Norwalk, CT.

Copius-Peereboom, J.W., de Voogt, P., van Hattum, B., van der Velde, W., Copius-Peereboom-Stegman, J.H.J., 1979. The use of the human placenta as a biological indicator for cadmium exposure. In: International Conference on Management and Control of Heavy Metals in the Environment, London. CEP Consultants, Edinburgh, pp. 8–10.

Fergusson, J.E., 1990. The Heavy Elements: Chemistry, Environmental Impact and Health Effects. Pergamon Press, Oxford, pp. 461–551.

Forstner, U., 1980. Cadmium. In: Huntzinger, O. (Ed.), Handbook of Environmental Chemistry, vol. 3A. Springer-Verlag, Berlin, pp. 59–107.

Friberg, L., Vather, M., 1983. Assessment of exposure to lead and cadmium through biological monitoring results of a UNEP/WHO global study. Environ. Res. 30, 95–128.

Kägi, J.H.R., Hapke, H.J., 1984. Biochemical interactions of mercury, cadmium and lead. In: Nriagu, J.O. (Ed.), Changing Metal Cycles and Human Health. Springer-Verlag, Dahlem Kanferenzen, pp. 237–250.

Lauwerys, R., Buchet, J.-P., Roels, H., Hubermont, G., 1978. Placental transfer of lead, mercury, cadmium and carbon monoxide in women. Environ. Res. 15, 278–289.

Lehman, L.D., Poisner, A.M., 1984. Induction of metallothionein synthesis in cultured human trophoblasts by cadmium and zinc. J. Toxicol. Environ. Health 14, 419–432.

Roels, H.A., Buchet, J.-P., Bernard, A., Hubermont, G., Lauwerys, R.R., Masson, P., 1978. Investigation of factors influencing exposure and response to lead, mercury and cadmium in man and animals. Environ. Health Perspect. 25, 91–96.

Sandstead, H.H., 1997. Nutrient interactions with toxic elements. In: Goyer, R.A., Mehlman, M.A. (Eds.), Toxicology of Trace Elements. Wiley, New York, pp. 241–256.

Stoeppler. M., 1983. General analytical aspects of the determination of lead, cadmium, and nickel in biological fluids. In: Facchetti, S. (Ed.), Analytical Techniques for Heavy Metals in Biological Fluids. Elsevier, Amsterdam.

Thieme, R., Schramel, P., Kurz, E., 1977. Trace element concentration in the human placenta in a heavily contaminated environment. Geburtsh. Frauenheilkd. 37, 756–761.

Vather, M. (Ed.), 1982. Assessment of Human Exposure to Lead and Cadmium through Biological Monitoring. Report prepared for the United Nations Environment Programme and the World Health Organization. National Swedish Institute of Environmental Medicine and Department of Environmental Hygiene, Karolinska Institute, Stockholm, 136 pp.

WHO, 1987. Air Quality Guidelines for Europe, vol. 23. World Health Organization, Regional Office for Europe, Copenhagen, pp. 200–209.

WHO, 1992. International Programme on Chemical Safety: Environmental Health Criteria 134, Cadmium. World Health Organization, Geneva, pp. 67–174.

V
Soils and sediments

Trace elements – Their distribution and effects in the environment
B. Markert and K. Friese, editors

Chapter 21

Geochemical behaviour of trace metals in freshwater sediments

Gerard A. van den Berg, J.P. Gustav Loch, Lambertus M. van der Heijdt, John J.G. Zwolsman

Abstract

Pore-water concentration profiles provide insight into diagenetic processes taking place in aquatic sediments. Due to the high input of reactive organic matter and the relatively low availability of external oxidators, redox zones generally occur on a millimetre to centimetre scale in organic-rich freshwater sediments. Distinct peaks in dissolved trace metal concentrations measured immediately below the sediment–water interface, can be mainly attributed to degradation of organic matter. The resulting concentration gradients between the pore water and the water column indicate that upward fluxes of trace metals from the sediment may contribute to concentrations in the surface water. An excess of acid-volatile sulphide (AVS) over reactive metals (SEM) in the anoxic sediment layer is a strong indication that trace metals may be effectively scavenged by sulphides, thus reducing their pore-water concentrations. However, supersaturation of the pore water with respect to metal sulphides indicates intensive complexation of trace metals with dissolved organic compounds. Given the large vertical variation in redox potential with depth, mixed homogenized sediment samples (0–10 cm) are generally not suited for the assessment of potential toxicity of sediments with respect to trace metals. Temporal variation in pore-water chemistry reflects environmental and intrinsic changes. These may be related to fluctuation in temperature, the supply of degradable organic matter, microbiological activity in the sediment, diffusion of solutes, mixing and reaction rates. As a result of this variation in the intensities of biogeochemical processes, (bio)availability of trace metals may fluctuate considerably. Therefore, in dynamic depositional environments it is necessary to account for non-steady state to determine diffusive exchange between sediments and surface water, as well as potential toxicity of sediments with respect to trace metals.

Keywords: Acid-volatile sulphide; Bioavailability; Pore water; Redox; Sediment; Trace metals

1. Introduction

Contaminated sediments are considered a large ecotoxicological risk to the aquatic environment (Burton, 1992). They may act as important sources or sinks of contaminants, e.g. trace metals, in aquatic systems (e.g. Gaillard et al., 1986; Calmano et al., 1993). As particle–water interactions influence the mobilization of trace metals in sediments (Hamilton-Taylor and Davison, 1995; Hong et al., 1995), (bio)availability of trace metals is influenced by the extent of their sorption and (co)precipitation with the solid phase. Pore-water concentrations are generally considered to be the most sensitive parameters of environmental change (Adams, 1994), serving as useful indicators of the solid phases controlling the fate of trace metals in sediments. If sampling takes place at high resolution, pore-water characteristics provide insight into (bio)geochemical processes in sediments and interactions between sediments and the overlying water column (Elderfield et al., 1981; Gaillard et al., 1989; Santschi et al., 1990). Pore-water composition is also considered a useful indicator for potential toxicity of sediments (Swartz et al., 1985), although the exact pathway of uptake of contaminants by aquatic organisms is not yet well known and may vary considerably.

Recent sedimentation basins in delta areas are characterized by a relatively large input of reactive organic matter (OM), in combination with highly dynamic conditions. Degradation of OM results in redox gradients in sediments. As a result, a gradual vertical change in solid phases controlling the solubility of trace metals may be observed. Secondary processes, directly or indirectly influenced by the primary diagenetic redox reactions driven by degradation of organic matter, may result in postdepositional mobilization of trace metals (Williams, 1992). Secondary processes in sediments include complexation of metals with dissolved organic and inorganic ligands, modification of surface properties of adsorptive particles, and formation and dissolution of metal precipitates (Kersten and Förstner, 1995). Because the behaviour of trace metals in sediments is to a large extent controlled by the redox cycles of the major biogeochemical elements (Shaw et al., 1990), the latter need to be quantified in order to provide a rational basis for managing contaminated water systems (Förstner et al., 1994).

The purpose of this study was to describe the major (bio)geochemical cycles in organic-rich freshwater sediments, and to relate the effects of these early diagenetic processes on the mobilization and (bio)availability of trace metals in the surface sediments. Results may be used to predict potential toxicity of sediments with respect to trace metals. This study is based on sediment cores taken on permanently submerged locations in former tidal channels in the delta of the rivers Rhine and Meuse in The Netherlands (Van den Berg, 1998). Solid-phase and pore-water characteristics were determined to identify the main geochemical processes in the sediments studied. Due to low current velocities, these channels are actively silting up with fine-grained sediments. Sediments in the study area are heavily contaminated by trace metals, and may contain up to approximately 1200 mg kg^{-1} Zn, 13 mg kg^{-1} Cd, 250 mg kg^{-1} Pb, 60 mg kg^{-1} Ni and 110 mg kg^{-1} Cu, reflecting the serious metal pollution history of the rivers Meuse and Rhine (Van Eck et al., 1997).

2. Materials and methods

2.1. Sample collection and handling

Sediment cores were collected in Plexiglas tubes with an inner diameter of 15 cm and a length of 60 cm. Sampling took place in November 1995 and June 1996, reflecting the winter and summer situation. Bottom water remained on the sediment cores in the tubes to prevent oxidation of the sediment. Pore-water pH was measured on board with a MI-410/MI-415 pH micro-electrode (Microelectrode Inc.) equipped with a micromanipulator. To avoid disturbance of the sediment, pH measurements were conducted in a core different from the one used for pore-water and solid-phase extraction. The collected cores were stored at 4°C. Bottom water was sampled at the same locations 0.5 m above the sediment–water interface in June 1996. Collected bottom water was filtered through a 0.45 μm membrane filter (Sartorius) and analyzed on board for temperature, oxygen, pH and salinity. Suspended matter was collected by continuous flow-through centrifugation.

Within 48 h after collection, the top 7–10 cm of the sediment cores were sectioned into 0.5- or 1.0-cm slices. Laboratory subsampling took place under strictly anaerobic conditions in a nitrogen-filled glove box to prevent oxidation of reduced compounds (Förstner and Salomons, 1980). Pore water was sampled by pressure filtration, according to the method described by De Lange (1992), and collected in high density polyethylene bottles. Part of the pore water was acidified immediately with concentrated HNO_3 suprapur (1 ml HNO_3 per 50 ml pore water). Sediment and non-acidified pore-water samples were kept at 4°C under a nitrogen atmosphere until extraction and analysis. All materials used (bottles, filters and filtration blocks) were acid cleaned prior to use.

2.2. Solid phase analysis

Freeze-dried sediment samples were digested in a mixture of concentrated $HClO_4$, HNO_3 and HF, and finally dissolved in 1 M HCl. Elemental concentrations in the acid solution were determined by inductively coupled plasma atomic emission spectrometry (Perkin Elmer Optima 3000). The accuracy of the solid-phase extractions and analyses was validated by simultaneously extracting and analyzing international standard sediments. Moisture content of the sediment samples, necessary to calculate concentrations on a dry weight basis, was determined by measuring the weight loss after freeze drying. Measurement of organic C (OrgC) content in sediment and suspended matter was made with a CNS analyzer (Carlo Erba Strumentazione nitrogen analyzer 1500). Inorganic carbon was removed before analysis by shaking the sediment sample in 1 M HCl.

Extraction of AVS and SEM took place within 24 h after subsampling. Sectioned sediment samples were analyzed for acid-volatile sulphide (AVS) and simultaneously extracted metals (SEM), according to the method described by Henneke et al. (1991) with slight modification for this purpose. AVS and SEM are both operationally defined parameters. AVS is defined as the amount of sulphides released during cold acid extraction and SEM is defined as the sum of (reactive) trace metals (usually Zn, Cd, Pb, Cu and Ni) liberated during AVS extraction. About 1 g of wet sediment was weighed exactly and used for extraction of AVS. As proposed by Cornwell and Morse (1987), sediment samples were acidified during 1 h with a 10-ml 6 M HCl solution to form H_2S which was subsequently collected in a 1 M NaOH solution. The sulphide concentration in the NaOH solution was measured with an EG and G Princeton Applied Research model 384 B polarograph with a static mercury drop (EG and G Parc model 303A) using square wave voltammetry. The acidified sediment suspension in the reaction flask was filtered through 0.45-μm cellulose acetate filters (Sartorius) for measurement of SEM. Concentrations of Zn, Cd, Cu, Pb and Ni in the filtrates were analyzed with an inductively coupled plasma atomic emission spectrometer (Perkin Elmer Optima 3000). Duplicate extractions and analyses of AVS and SEM showed that, using the method described, the results are well reproducible.

2.3. Pore water analysis

Collected pore water was analyzed for concentrations of major elements by ICP-AES. Zn was measured by flame atomic adsorption spectrometry (Perkin Elmer 4000) using deuterium background correction. Dissolved Cd, Pb, Ni and Cu concentrations were measured with hydride graphite furnace AAS (Perkin Elmer Z3030/HGA600/AS60) using Zeeman background correction. Matrix modifiers (ammonium phosphate/magnesium nitrate for Cd and Pb; magnesium nitrate for Ni) were used to correct for matrix effects. Ammonium was determined by autoanalyzer (Technicon TRAACS 800) in November 1995 and spectrophotometrically according to the Berthelot method in June 1996. Sulphate and nitrate were measured in non-acidified pore water by ion-chromatography (Dionex QIC analyzer). Alkalinity was determined with the Gran plot method after titration with 0.01 M HCl. Dissolved organic carbon (DOC) was measured by the high-temperature catalytic oxidation method (Shimadzu Total Organic Carbon Analyzer TOC-500) after pre-treatment with phosphoric acid to remove inorganic carbon.

3. General pore-water chemistry

3.1. Introduction

Degradation of OM is the main biogeochemical process taking place in recently deposited sediments (Berner, 1980). Depth distributions of microbially mediated electron transfer reactions within sediments are determined by their corresponding redox potentials (Stumm and Morgan, 1996). A distinction between an oxic layer (oxic respiration), a suboxic layer (denitrification, Mn(III,IV) and Fe(III) reduction) and an anoxic layer (sulphate reduction and methanogenesis) is usually made (Froelich et al., 1979).

Pore-water concentration profiles of redox-sensitive ions (nitrate, Mn^{2+}, Fe^{2+}, sulphate and sulphide) and nutrients (ammonium and phosphate) demonstrate the effects of degradation of OM. In freshwater sediments, the redox zones generally occur on a millimetre to centimetre scale due to the high input of reactive OM and the relatively low availability of external oxidators, especially nitrate and sulphate, compared to marine systems. A typical feature for organic-rich freshwater sediments deposited in aerobic surface waters, is the presence of anaerobic conditions close to the sediment–water interface (SWI). This is indicated by the absence of dissolved oxygen and the presence of reduced solutes (e.g. Mn^{2+}, Fe^{2+} and sulphides) in the pore water. Secondary redox reactions, like oxidation of reduced pore-water and solid-phase constituents, and other postdepositional processes, like precipitation-dissolution and adsorption-desorption, may also significantly influence these changes in pore-water concentrations with depth.

3.2. Early diagenetic reactions

Pore-water characteristics are shown in Fig. 1. In organic-rich freshwater sediments, aerobic conditions are generally limited to the upper millimetres, as may also be inferred from the absence of a well developed reddish-brown surface layer. Oxygen consumption in sediments is not only due to oxic respiration, but oxidation of reduced solid-phase constituents (e.g. FeS) and pore-water species (e.g. Fe^{2+}, Mn^{2+}, ammonium, sulphide and dissolved methane) also consumes oxygen. As a result of these secondary redox reactions, oxic respiration may only account for a relatively small amount of total oxygen reduction in sediments (Wang and Van Cappellen, 1996). The presence of dissolved oxygen in the top few millimetres of the sediment may significantly decrease the diffusive flux of reduced pore-water constituents to the bottom water.

Comparable to oxic respiration, denitrification is limited to the top layer of the sediment. As shown in Fig. 1, nitrate is completely reduced within 1 cm below the

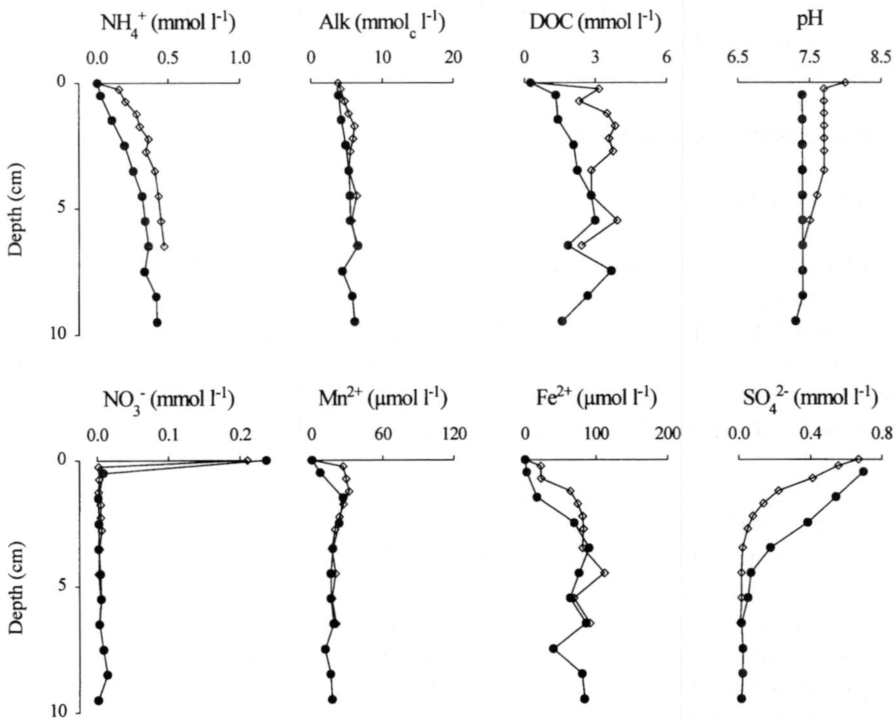

Figure 1. Typical pore-water characteristics of organic-rich freshwater sediments. The solid dots represent concentrations during winter; open diamonds represent concentrations during summer. Dissolved concentrations in the water column are plotted at zero depth.

SWI due to the low nitrate flux into the sediment and the high degradation rate of OM in freshwater sediments.

A shift of Mn(III,IV) and Fe(III) solid phases (mainly Mn- and Fe-(hydr)oxides) to Mn(II) and Fe(II) solid phases (e.g. $MnCO_3$, $FeCO_3$ and FeS) may be expected with depth as a result of redox processes taking place in the sediment. Below the SWI, a maximum in the pore-water concentration of Mn^{2+} is observed. Deeper in the sediment, pore-water Mn^{2+} concentrations decrease to an asymptotic value. Calculations of the degree of saturation of the pore water with respect to different minerals indicate that rhodochrosite ($MnCO_3$) may to a large extent control this pore-water concentration. Vertical separation between pore-water build-up of Mn^{2+} and Fe^{2+} may be caused primarily by oxidation of dissolved Fe^{2+} by MnO_2 (Postma, 1985; Wang and Van Cappellen, 1996).

Equilibria involving Mn minerals are generally reached faster than those involving Fe minerals (Davison, 1993). Consequently, a gradual increase in pore-water Fe^{2+} with depth, even below the zone of sulphate reduction, may be explained by heterogeneity of the sediment and differences in dissolution behaviour of Fe(III)-(hydr)oxide phases (Wersin et al., 1991; Postma, 1993). Supersaturation of the anaerobic pore water with respect to siderite ($FeCO_3$) and vivianite ($Fe_3(PO_4)_2$) also favours a non-equilibrium behaviour of Fe. Due to bioturbation and physical mixing of the sediment, pore-water Fe^{2+} and Mn^{2+} and Mn(II)- and Fe(II)-containing solid phases may be oxidized and re-precipitated as (coatings of) amorphous Mn- and Fe-(hydr)oxides. Due to the highly adsorptive surface properties of secondary (hydr)oxides, these may act as effective scavengers of trace metals. When the aerobic surface layer has minimal thickness or is absent, diffusion of pore-water Mn^{2+} and Fe^{2+} may take place to the water column. Cycling of Mn and Fe between the sediment and the bottom water may result in enrichment of the suspended matter (SPM) with respect to Mn and Fe (Dehairs et al., 1989).

Respiratory sulphate reduction ideally takes place when all other electron acceptors are exhausted, but significant overlap may occur between the zones of microbial Fe(III) reduction and sulphate reduction due to kinetic constraints, as discussed before. Sulphate concentrations typically decrease to zero within the upper sediment layer (Fig. 1). In freshwater sediments, reduced S formed mainly by reduction of pore-water sulphate, is predominantly present as inorganic S in the form of AVS. Although pyrite is the most stable sulphide mineral, its formation in permanently submerged freshwater sediments is subject to controversy (Rickard et al., 1995). Because, contrary to marine sediments (S-dominated), there is an excess of Fe^{2+} liberation over HS^- production in freshwater sediments (Fe-dominated), $FeCO_3$ as well as FeS may control pore-water Fe^{2+} concentrations in the anoxic sediment layer.

The presence of ammonium in the pore water is predominantly related to degradation of N-bearing organic substrates and may, therefore, be used as a measure for OM degradation. Pore-water ammonium profiles show a typical increase with depth. As shown in Fig. 1, the slope of the curve decreases with depth indicating a decrease in OM degradation rate due to preferential degradation of reactive OM constituents near the SWI. The presence of methanogenesis is indicated by the continuous production of NH_4^+ with depth, even below the zone of sulphate reduction. Because in freshwater sediments the supply of external oxidizing agents is limited, methanogenesis is often the dominant process of OM degradation beyond a few centimetres below the SWI (Capone and Kiene, 1988). Methane is effectively oxidized in the aerobic surface layer. In addition, methane diffusing upwards may become a source of carbon for sulphate-reducing bacteria at the sulphate–methane interface (Devol, 1983; Iversen and Jörgensen, 1985), although this process may be of limited importance in freshwater sediments. In organic-rich sediments, dissolved methane concentrations may often exceed the solubility limit, resulting in methane bubble formation (Martens and Klump, 1980).

Because $CaCO_3$ dissolution buffers the pH in carbonate-bearing sediments, pore-water pH typically remains at near-neutral conditions in the surface sediments (Boudreau, 1987). The extent of $CaCO_3$ dissolution increases with integrated rates of O_2 reduction and methanogenesis, which are the main producers of CO_2 driving $CaCO_3$ dissolution.

Fluctuations in surface water concentrations are attributed to changes in the degree of oxygenation due to periodic stratification of the water column (Balistrieri et al., 1992; Hines et al., 1997), river discharge (Van der Weijden and Middelburg, 1989) and primary production (Zwolsman, 1994). In comparison, temporal, often seasonal, changes in concentrations of pore-water solutes can be linked predominantly to a difference in temperature of the bottom water and upper sediment layer, which may be reflected by changes in OM degradation rates, metabolic activity in the sediment, diffusivity of pore-water solutes, and reaction and bioturbation rates (e.g. Berner, 1980). Within the surface sediments, steeper gradients of redox-sensitive ions are generally measured during summer than during winter. In Fig. 1, a steeper Fe^{2+} concentration gradient is measured at the SWI during summer. This agrees well with an upward shift of the peak in the Mn^{2+} concentration profiles. Additionally, during summer, sulphate concentrations in the pore water show a significantly larger negative gradient at the SWI. The effects of changes in temperature on the intensities of (bio)geochemical processes in sediments have also been described, for example near-shore sedimentary coastal areas (Klump and Martens, 1989), salt-marshes (Kostka and Luther, 1995) and lakes (Holdren and Armstrong, 1986). However, temperature effects in periodically saturated environments, like salt marshes and wetlands, may be outweighed by short-term effects like episodic inundations (Van den Berg et al., 1998b) and plant growth (Oenema, 1988; Hines et al., 1989). In dynamic environments, variation in the intensity of (bio)geochemical processes may significantly affect the behaviour and bioavailability of, for example nutrients and trace metals in sediments, and their interactions between sediments and surface- and groundwater (Carignan and Lean, 1991; Sherman et al., 1994). Consequently, in dynamic depositional environments fluxes of solutes between the sediment and the surface water are not calculated properly when assuming steady state.

The effects of the individual primary and secondary (bio)geochemical processes can be approximated with transport models describing early diagenetic processes in recent sediments (e.g. Rabouille and Gaillard, 1991; Dhakar and Burdige, 1996; Soetaert et al., 1996; Van Cappellen and Wang, 1996). These models have been developed to identify environmental and biogeochemical factors controlling the chemical dynamics of sediments and to predict the behaviour and speciation of redox-sensitive ions. One-dimensional modelling may predict considerable vertical overlap between the degradation pathways. This reflects high mixing intensities of sediments (Boudreau, 1998) in combination with limiting values of the respiratory electron acceptors (Wang and Van Cappellen, 1996). Additionally, considerable overlap may result from lateral heterogeneity of the sediment, as proposed by Brandes and Devol (1995); Wang and Van Cappellen (1996) and the presence of a wide range of mineral reactivities (Postma and Jakobsen, 1996).

The multicomponent reaction/transport model STEADYSED1 (Van Cappellen and Wang, 1996) explicitly accounts for the individual OM degradation pathways and secondary (redox) reactions. The model is capable of predicting the significance of the respiratory pathways, as well as the secondary reactions. The results of a modelling exercise, described by Van den Berg (1998), show that the measured pore-water profiles of redox-sensitive species can be reproduced reasonably well (Fig. 2). The model predictions indicate that methanogenesis is the major pathway of OM degradation below 3 cm from the SWI. The other oxidative pathways are limited to the top 3 cm of the sediment. Differences in profiles of redox-sensitive ions among sediment cores are explained by depositional conditions, particularly sediment accumulation rate, supply of OM and macrofaunal activity.

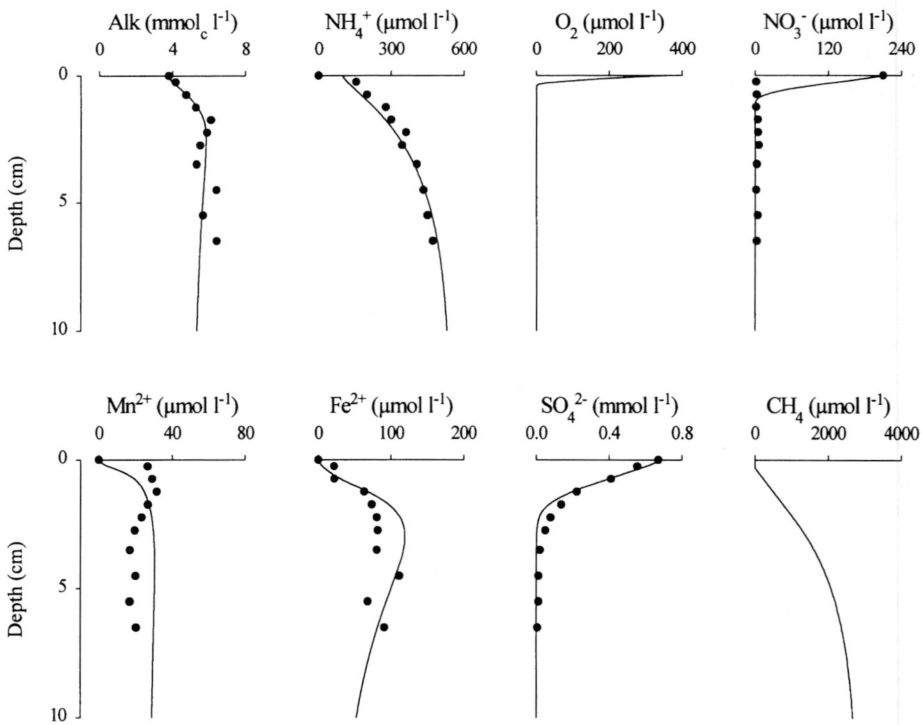

Figure 2. Measured concentrations in the pore water (solid dots) and modelled concentration profiles (straight line).

4. Redox-sensitive behaviour of trace metals

4.1. Introduction

Because fast reliable measurements of pore-water concentrations are not possible without serious laboratory precautions, equilibrium partition coefficients between solid-phase contents and dissolved concentrations in the pore water are regularly used to estimate pore-water concentrations. Although great variability may be expected in natural environments, both in time and space, constant partition coefficients are generally assumed. To classify sediments and SPM according to sediment quality criteria in The Netherlands, Van der Kooij et al. (1991) included a factor to account for differences in binding capacity of SPM and sediments. SPM is generally more enriched in trace metals compared to the underlying sediment due to preferential deposition of more coarsed-grained material. It should be noted that solid-phase characteristics often do not provide any information about the effects of diagenetic processes, because trace metals are present primarily in the solid-phase (less than 0.1% is generally present in the pore water). Fig. 3 illustrates that spatial and vertical variation in metal contents can be attributed mainly to a variation in grain size. Although a good correlation is shown between metals and OrgC in Fig. 3, it should be recognized that this does not imply that organic matter is the active adsorptive phase for trace metals, given the linear correlation of OrgC with Al, Fe and Mn. The latter may be due to the presence of OM and oxides in the finest fraction as well as formation of organic and oxide coatings on clay minerals (Horowitz and Elrick, 1987; Tessier et al., 1996).

4.2. Distribution of trace metals in the pore water

It is well known that mobilization of trace metals in sediments is strongly dependent on redox gradients (e.g. Hamilton-Taylor and Davison, 1995). Trace metal concentrations in pore water commonly decrease with depth, suggesting a source near the SWI and a removal process deeper in the sediment (Gaillard et al., 1986; Walton-Day et al., 1990). In the oxic and suboxic sediment layers, adsorption of trace metals to OM and oxyhydroxide surfaces is the main process controlling pore-water concentrations of trace metals (Tessier et al., 1993). The SWI is characterized by the presence of degradable OM and reactive (hydr)oxides, which can be rapidly broken down or transformed into new phases. As a result of degradation of organic (partly biogenic) compounds, which have concentrated metals during previous biological uptake and by binding to particle surface sites (Sigg, 1985), trace metals may be released into solution. This may result in a concentration peak of dissolved trace metals directly below the SWI, as shown in Fig. 4. To improve resolution of this surface peak, sampling with high-resolution techniques (e.g. Zhang et al., 1995) may give more detail. Remobilization of trace metals at the SWI and a diffusive flux between sediments and the overlying water column (benthic flux) may potentially influence concentrations in the surface water (Van den Berg et al., 1998c). Trace metals remobilized by decomposition of OM near the SWI, may initially be scavenged by Mn- and Fe-(hydr)oxides (McKee et al., 1989). Due to the presence of (hydr)oxides, the (sub)oxic zone of the

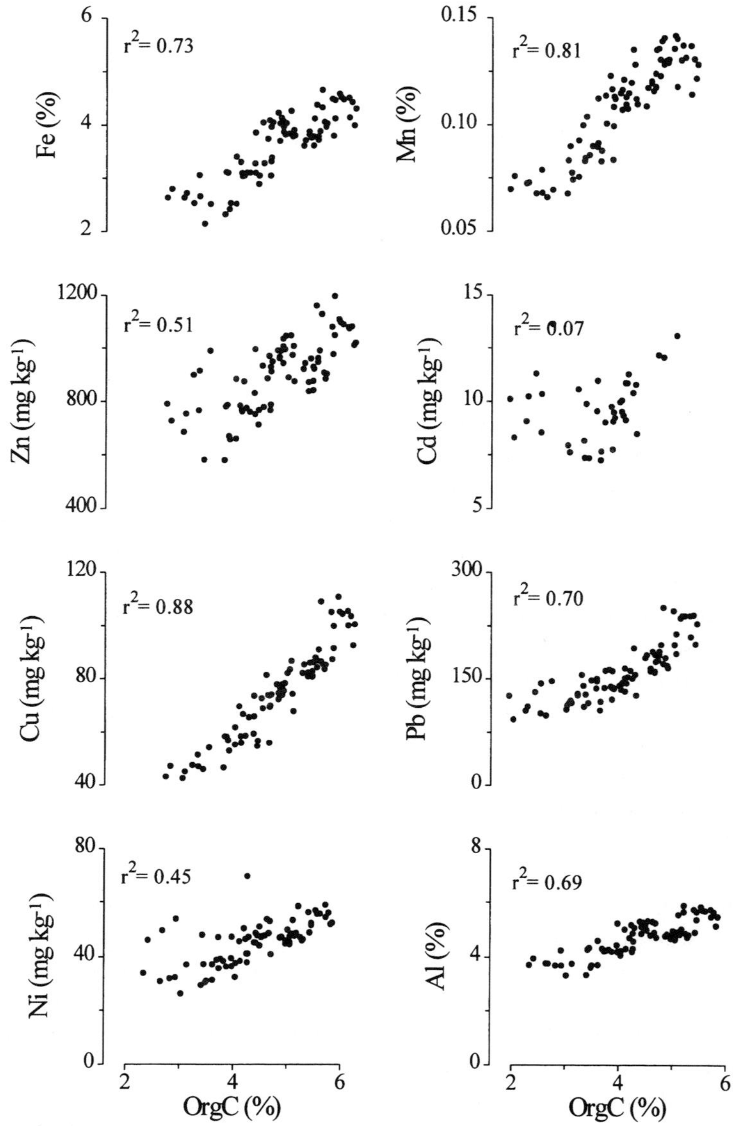

Figure 3. Scatter plots of measured solid-phase metal contents versus OrgC content.

sediment can also function as a (temporary) adsorbing layer for trace metals that are transported upwards. Deeper in the profile, increased dissolved metal concentrations may be caused by reduction of Mn- and Fe-(hydr)oxides or oxidation of metal sulphides. Seasonal fluctuation in temperature, deposition rates and primary production of biomass may result in variation in pore-water concentrations of trace metals (Song and Müller, 1995).

Thermodynamic equilibrium calculations are often employed to investigate solubi-

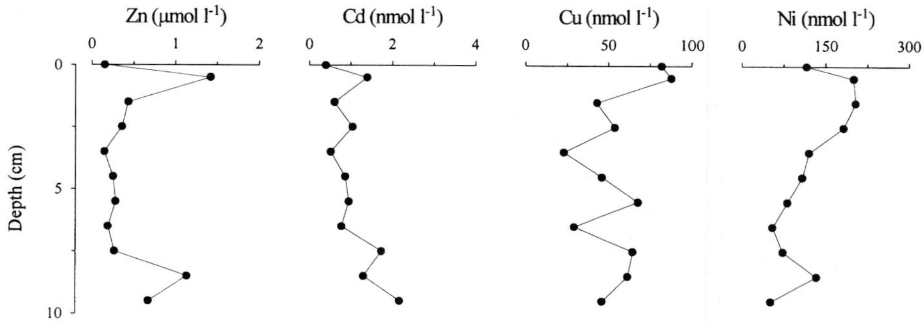

Figure 4. Profiles of dissolved trace metal concentrations in pore water.

lity and speciation of trace metals in sulphide-bearing sediments. Relatively low concentrations of dissolved sulphide (on the order of 0.1–1 μmol l^{-1}) in anoxic pore water may be explained by relatively high concentrations of Fe^{2+} in (Fe-dominated) freshwater sediments (Carignan and Nriagu, 1985). Equilibrium calculations, based solely on inorganic speciation, indicate supersaturation of the anoxic pore water with respect to crystalline metal sulphides, whereby saturation indices roughly decrease in the order CuS $>$ PbS $>$ ZnS $\approx$ CdS $>$ NiS. Thus, pore-water concentrations of the respective trace metals are significantly higher than would be predicted from their sulphide solubility. However, calculated saturation indices provide no direct evidence for the formation of sulphides, because, for example kinetic factors are not included. It should also be taken into account that the solubility of sulphides may vary significantly between crystalline and amorphous forms (Davison, 1991). Supersaturation of the anoxic pore water with respect to sulphide precipitates of trace metals suggests that the dissolved concentrations of trace metals are controlled by formation of their respective discrete sulphide minerals (e.g. Lee and Kittrick, 1984; Van den Berg et al., 1998b), rather than by co-precipitation or adsorption to iron-sulphides (Davison et al., 1992; Morse and Arakaki, 1993; Cooper and Morse, 1998).

DOC concentrations may be significantly higher (up to a factor of 10) in the pore water than in the overlying water column (Fig. 1). Complexation of trace metals with dissolved organic matter may, therefore, increase dissolved concentrations of trace metals (Santschi et al., 1997; Skrabal et al., 1998). Discrepancies between pore-water concentrations of trace metals and equilibrium concentrations may, therefore, be an indication for intense complexation with, for example humic substances, organic thiols, amino acids and polysulphides in the pore water (Boulègue et al., 1982; Giblin et al., 1986; Shea and MacCrehan, 1988). The stability of organic-metal complexes increases according to the Irving–Williams series (Tipping and Hurley, 1992).

4.3. Association of trace metals with AVS and potential toxicity of sediments

The most efficient fixation of trace metals in sediments takes place by association with sulphide. Although AVS is a metastable reduced sulphur form, it may be the controlling sediment property for metal (bio)availability in (anoxic) sediments. AVS includes amorphous iron monosulphide (FeS), crystalline mackinawite and pyrrhotite (FeS), greigite (Fe_3S_4), as well as ZnS, CdS, etc. In sulphidic environments, trace metals can compete with or displace Fe(II) from previously formed iron(mono)sulphides to form trace metal sulphide precipitates (Di Toro et al., 1992). Based on laboratory toxicity experiments, Di Toro et al. (1992) introduced the ratio of SEM to AVS as a useful tool for predicting metal bioavailability and, consequently, potential sediment toxicity. When the ratio of the molar concentrations of SEM and AVS is smaller than one, there is no acute toxicity for aquatic organisms. The sediment is considered potentially toxic when the molar ratio SEM to AVS exceeds one.

Vertical profiles of AVS and SEM, as well as dissolved sulphate concentrations in freshwater sediments are illustrated in Fig. 5. Although SEM is relatively constant with depth, indicating input of SPM having constant trace metal content, AVS increases dramatically in the zone, where sulphate reduction takes place. Increasing AVS with depth is explained by in situ production (precipitation) during reduction of pore-water sulphate. The vertical variability of AVS in the sediments is very likely to influence the (bio)availability of trace metals (Van den Berg et al., 1998a). In the sediment core presented, SEM/AVS decreases from values greater than one in the surface sediments to values lower than one in the anoxic sediment layer. Ratios lower than one suggest that trace metals are associated with AVS and dissolved concentrations in the pore water should be controlled by their respective sulphide precipitates, as has been discussed before.

The concurrent release of sulphide (AVS) and reactive metals (SEM) during extraction with HCl cannot be considered as evidence that those particular metals were bound by sulphide in the sediment (Allen et al., 1993). Various studies show that

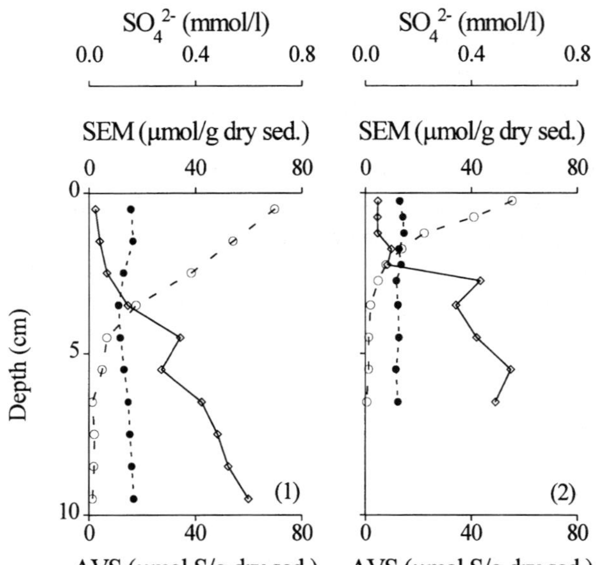

Figure 5. Depth profiles of SEM (solid dots), AVS (open diamonds) and dissolved pore-water sulphate (open dots) during winter (1) and summer (2).

AVS is not the sole partitioning phase for predicting the acute toxicity, especially for Cu, which may be strongly associated with sediment organic matter (e.g. Ankley et al., 1993). Although Huerta-Diaz et al. (1993) have presented experiments to measure the quantity of trace metals associated with AVS and pyrite in sediments, selective extraction techniques for prediction of solid phases controlling pore-water concentrations of trace metals in (partly) anaerobic sediments solid phases are still absent (Wallmann et al., 1993).

Similar to the major redox-sensitive ions, discussed before, the quantity and distribution of AVS in sediments depend upon, for example OM input, sediment burial, sulphate diffusion, sediment mixing and sediment temperature (Morse et al., 1987; Matisoff and Holdren, 1995). Leonard et al. (1993) have shown that variations in AVS concentrations in lake sediments are directly correlated with temperature of the overlying lake water. Fig. 5 clearly shows that a higher sulphate reduction rate is reflected in AVS present closer to the SWI. Large fluctuations in AVS have also been measured in sediments deposited in lakes with a seasonally anoxic water column (Howard and Evans, 1993).

5. Conclusions

Internal biogeochemical processes may influence the behaviour and mobilization of trace metals in sediments, determine the availability of trace metals to aquatic organisms and control the exchange of trace metals between the sediment and the water column. A gradual change in solid phases controlling the solubility of trace metals is generally observed in sediments. Due to OM degradation and release of associated trace metals, concentration peaks of dissolved trace metals are measured at the SWI. Because (acid-volatile) sulphide is an effective scavenger of trace metals in the anoxic sediment layer only, concentration gradients are measured in the pore water, potentially affecting (bio)availability of trace metals in sediments. Consequently, mixed homogenized sediment samples (e.g. 0–10 cm) are generally not suited to assess potential toxicity of sediments with respect to trace metals. Additionally, temporal fluctuation in, for example sediment temperature and OM input may influence the behaviour of trace metals in sediments. Dissolved trace metal concentrations are significantly increased by complexation with organic ligands in the pore water. The effects of organic complexation on bioavailability of trace metals in sediments need further investigation.

References

Adams, D.D., 1994. Sediment pore water sampling. In: Mudroch, A., MacKnight, S.D. (Eds.), Handbook of Techniques for Aquatic Sediments Sampling, 2nd ed. Lewis, Boca Raton, FL, pp. 171–202

Allen, H.E., Fu, G., Deng, B., 1993. Analysis of acid-volatile sulfide (AVS) and simultaneously extracted metals (SEM) for the estimation of potential toxicity in aquatic sediments. Environ. Toxicol. Chem. 12, 1441–1453.

Ankley, G.T., Mattson, V.R., Leonard, E.N., West, C.W., Bennett, J.L., 1993. Predicting the acute toxicity of copper in freshwater sediments: evaluation of the role of acid-volatile sulfide. Environ. Toxicol. Chem. 12, 315–320.

Balistrieri, L.S., Murray, J.W., Paul, B., 1992. The cycling of iron and manganese in the water column of Lake Sammamish, Washington. Limnol. Oceanogr. 37, 510–528.

Berner, R.A., 1980. Early Diagenesis: A Theoretical Approach. Princeton University Press, Princeton, NJ.

Boudreau, B.P., 1987. A steady-state diagenetic model for dissolved carbonate species and pH in the porewaters of oxic and suboxic sediments. Geochim. Cosmochim. Acta 51, 1985–1996.

Boudreau, B.P., 1998. Mean mixed depth of sediments: the wherefore and the why. Limnol. Oceanogr. 43, 524–526.

Boulègue, J., Lord III, C.J., Church, T.M., 1982. Sulfur speciation and associated trace metals (FeCu) in the pore waters of Great Marsh, Delaware. Geochim. Cosmochim. Acta 46, 453–464.

Brandes, J.A., Devol, A.H., 1995. Simultaneous nitrate and oxygen respiration in coastal sediments: evidence for discrete diagenesis. J. Mar. Res. 53, 771–797.

Burton Jr., G.A., 1992. Assessing contaminated aquatic sediments. Environ. Sci. Technol. 26, 1862–1863.

Calmano, W., Hong, J., Förstner, U., 1993. Binding and mobilization of heavy metals in contaminated sediments affected by pH and redox potential. Water Sci. Technol. 28, 223–235.

Capone, D.G., Kiene, R.P., 1988. Comparison of microbial dynamics in marine and freshwater sediments: contrasts in anaerobic carbon catabolism. Limnol. Oceanogr. 33, 725–749.

Carignan, R., Lean, D.R.S., 1991. Regeneration of dissolved substances in a seasonally anoxic lake: the relative importance of processes occurring in the water column and in the sediments. Limnol. Oceanogr. 36, 683–707.

Carignan, R., Nriagu, J.O., 1985. Trace metal deposition and mobility in the sediments of two lakes near Sudbury, Ontario. Geochim. Cosmochim. Acta 49, 1753–1764.

Cooper, D.C., Morse, J.W., 1998. Extractability of metal sulfide minerals in acidic solutions: application to environmental studies of trace metal contamination within anoxic sediments. Environ. Sci. Technol. 32, 1076–1078.

Cornwell, J.C., Morse, J.W., 1987. The characterization of iron sulfide minerals in anoxic marine sediments. Mar. Chem. 22, 193–206.

Davison, W., 1991. The solubility of iron sulphides in synthetic and natural waters at ambient temperature. Aquat. Sci. 53, 309–329.

Davison, W., 1993. Iron and manganese in lakes. Earth-Sci. Rev. 34, 119–163.

Davison, W., Grime, G.W., Woof, C., 1992. Characterization of lacustrine iron sulfide particles with proton-induced X-ray emission. Limnol. Oceanogr. 37, 1770–1777.

Dehairs, F., Baeyens, W., Van Gansbeke, D., 1989. Tight coupling between enrichment of iron and manganese in North Sea suspended matter and sedimentary redox processes: evidence for seasonal variability. Estuar. Coast. Shelf Sci. 29, 457–471.

De Lange, G.J., 1992. Shipboard routine and pressure filtration system for pore water extraction from suboxic sediments. Mar. Geol. 109, 77–81.

Devol, A.H., 1983. Methane oxidation rates in the anaerobic sediments of Saanich Inlet. Limnol. Oceanogr. 28, 738–742.

Dhakar, S.P., Burdige, D.J., 1996. A coupled, non-linear, steady-state model for early diagenetic processes in pelagic sediments. Am. J. Sci. 296, 244–265.

Di Toro, D.M., Mahony, J.D., Hansen, D.J., Scott, K.J., Carlson, A.R., Ankley, G.T., 1992. Acid volatile sulfide predicts the acute toxicity of cadmium and nickel in sediments. Environ. Sci. Technol. 26, 96–101.

Elderfield, H., McCaffrey, R.J., Luedtke, N., Bender, M., Truesdale, V.W., 1981. Chemical diagenesis in Narragansett Bay sediments. Am. J. Sci. 281, 1021–1055.

Förstner, U., Salomons, W., 1980. Trace metal analysis on polluted sediments: Assessment of sources and intensities. Environ. Tech. Lett. 1, 494–505.

Förstner, U., Calmano, W., Hong, J., Kersten, M., 1994. Effects of redox variations on metal speciation: implications on sediment quality criteria assessment, Special Publication 154. The Royal Society of Chemistry, London, pp. 83–102.

Froelich, P.N., Klinkhammer, G.P., Bender, M.L., Luedtke, N.A., Heath, G.R., Cullen, D., Dauphin, P., Hammond, D., Hartman, B., Maynard, V., 1979. Early oxidation of organic matter in pelagic sediments of the eastern equatorial Atlantic: suboxic diagenesis. Geochim. Cosmochim. Acta 43, 1075–1090.

Gaillard, J.F., Jeandel, O., Michard, G., Nicolas, E., Renard, D., 1986. Interstitial water chemistry of Villefranche Bay sediments: trace metal diagenesis. Mar. Chem. 18, 233–247.

Gaillard, J.F., Pauwels, H., Michard, G., 1989. Chemical diagenesis in coastal marine sediments. Oceanol. Acta 12, 175–187.

Giblin, A.E., LutherIII, G.W., Valiela, I., 1986. Trace metal solubility in salt marsh sediments contaminated with sewage sludge. Estuar. Coast. Shelf Sci. 23, 477–498.

Hamilton-Taylor, J., Davison, W., 1995. Redox-driven cycling of trace elements in lakes. In: Lerman, A., Imboden, D.M., Gat, J.R. (Eds.), Physics and Chemistry of Lakes. Springer, Berlin, pp. 217–263.

Henneke, E., Luther III, G.W., De Lange, G.J., 1991. Determination of inorganic sulphur speciation with polarographic techniques: some preliminary results for recent hypersaline anoxic sediments. Mar. Geol. 100, 115–123.

Hines, M.E., Faganeli, J., Planinc, R., 1997. Sedimentary anaerobic microbial biogeochemistry in the Gulf of Trieste, northern Adriatic Sea: influences of bottom water oxygen depletion. Biogeochemistry 39, 65–86.

Hines, M.E., Knollmeyer, S.L., Tugel, J.B., 1989. Sulfate reduction and other sedimentary biogeochemistry in a northern New England salt marsh. Limnol. Oceanogr. 34, 578–590.

Holdren, G.C., Armstrong, D.E., 1986. Interstitial ion concentrations as an indicator of phosphorus release and mineral formation in lake sediments. In: Sly, P.G. (Ed.), Sediments and Water Interactions. Springer, New York, pp. 133–147.

Hong, J., Calmano, W., Förstner, U., 1995. Interstitial waters. In: Salbu, B., Steinnes, E. (Eds.), Trace Elements in Natural Waters. CRC Press, Boca Raton, FL, pp. 117–150.

Horowitz, A.J., Elrick, K.A., 1987. The relation of stream sediment surface area, grain size and composition to trace element chemistry. Appl. Geochem. 2, 437–451.

Howard, D.E., Evans, R.D., 1993. Acid-volatile sulfide (AVS) in a seasonally anoxic mesotrophic lake: seasonal and spatial changes in sediment AVS. Environ. Toxicol. Chem. 12, 1051–1057.

Huerta-Diaz, M.A., Carignan, R., Tessier, A., 1993. Measurement of trace metals associated with acid volatile sulfides and pyrite in organic freshwater sediments. Environ. Sci. Technol. 27, 2367–2372.

Iversen, N., Jörgensen, B.B., 1985. Anaerobic methane oxidation rates at the sulfate-methane transition in marine sediments from Kattegat and Skagerrak (Denmark). Limnol. Oceanogr. 30, 944–955.

Kersten, M., Förstner, U., 1995. Speciation of trace metals in sediments and combustion waste. In: Ure, A.M., Davidson, C.M. (Eds.), Chemical Speciation in the Environment. Blackie, London, pp. 234–275.

Klump, J.V., Martens, C.S., 1989. The seasonality of nutrient regeneration in an organic-rich coastal sediment: kinetic modelling of changing pore-water nutrient and sulfate distributions. Limnol. Oceanogr. 34, 559–577.

Kostka, J.E., Luther III, G.W., 1995. Seasonal cycling of Fe in saltmarsh sediments. Biogeochemistry 29, 159–181.

Lee, F.Y., Kittrick, J.A., 1984. Electron microprobe analysis of elements associated with zinc and copper in an oxidising and an anaerobic soil environment. Soil Sci. Soc. Am. J. 48, 548–554.

Leonard, E.N., Mattson, V.R., Benoit, D.A., Hoke, R.A., Ankley, G.T., 1993. Seasonal variation of acid volatile sulfide concentration in sediment cores from three northeastern Minnesota lakes. Hydrobiologia 271, 87–95.

Matisoff, G., Holdren Jr., G.R., 1995. A model for sulfur accumulation in soft water lake sediments. Water Resources Res. 31, 1751–1760.

Martens, C.S., Klump, J.V., 1980. Biogeochemical cycling in an organic-rich coastal marine basin – 1. Methane sediment-water exchange processes. Geochim. Cosmochim. Acta 44, 471–490.

McKee, J.D., Wilson, T.P., Long, D.T., Owen, R.M., 1989. Pore water profiles and early diagenesis of Mn, Cu, and Pb in sediments from large lakes. J. Great Lakes Res. 15, 68–83.

Morse, J.W., Arakaki, T., 1993. Adsorption and coprecipitation of divalent metals with mackinawite (FeS). Geochim. Cosmochim. Acta 57, 3635–3640.

Morse, J.W., Millero, F.J., Cornwell, J.C., Rickard, D., 1987. The chemistry of hydrogen sulfide and iron sulfide systems in natural waters. Earth-Sci. Rev. 24, 1–42.

Oenema, O., 1988. Diagenesis in subrecent marine sediments in the eastern Scheldt, southwest Netherlands. Neth. J. Sea. Res. 22, 253–265.

Postma, D., 1985. Concentration of Mn and separation from Fe in sediments – 1. Kinetics and stoichiometry of the reaction between birnessite and dissolved Fe(II) at 10°C. Geochim. Cosmochim. Acta 45, 687–698.

Postma, D., 1993. The reactivity of iron oxides in sediments: a kinetic approach. Geochim. Cosmochim. Acta 57, 5027–5034.

Postma, D., Jakobsen, R., 1996. Redox zonation: equilibrium constraints on the Fe(III)/SO_4-reduction interface. Geochim. Cosmochim. Acta 60, 3169–3175.

Rabouille, C., Gaillard, J.F., 1991. Toward the EDGE: early diagenetic global explanation: a model depicting the early diagenesis of organic matter, O_2, NO_3, Mn, and PO_4. Geochim. Cosmochim Acta 55, 2511–2525.

Rickard, D., Schoonen, M.A.A., LutherIII, G.W., 1995. Chemistry of iron sulfides in sedimentary environments. In: Vairavamurthy, M.A., Schoonen, M.A.A. (Eds.), Geochemical Transformations of Sedimentary Sulfur, ACS Symposium Series, vol. 612. American Chemical Society, Washington, DC, pp. 168–193.

Santschi, P., Höhener, P., Benoit, G., Buchholtz-ten Brink, M., 1990. Chemical processes at the sediment-water interface. Mar. Chem. 30, 269–315.

Santschi, P.H., Lenhart, J.J., Honeyman, B.D., 1997. Heterogeneous processes affecting trace contaminant distribution in estuaries: the role of natural organic matter. Mar. Chem. 58, 99–125.

Shaw, T.J., Gieskes, J.M., Jahnke, R.A., 1990. Early diagenesis in differing depositional environments: the response of transition metals in pore water. Geochim. Cosmochim. Acta 54, 1233–1246.

Shea, D., MacCrehan, W.A., 1988. Role of biogenic thiols in the solubility of sulfide minerals. Sci. Total Environ. 73, 135–141.

Sherman, L.A., Baker, L.A., Weir, E.P., Brezonik, P.L., 1994. Sediment pore-water dynamics of Little Rock Lake, Wisconsin: geochemical processes and seasonal and spatial variability. Limnol. Oceanogr. 39, 1155–1171.

Sigg, L., 1985. Metal transfer mechanisms in lakes: the role of settling particles. In: Stumm, W. (Ed.), Chemical Processes in Lakes. Wiley, New York, pp. 283–310.

Skrabal, S.A., Donat, J.R., Burdige, D.J., 1998. Fluxes of copper-complexing ligands from estuarine sediments. Limnol. Oceanogr. 42, 992–996.

Soetaert, K., Herman, P.M.J., Middelburg, J.J., 1996. A model for early diagenetic processes from the shelf to abyssal depths. Geochim. Cosmochim. Acta 60, 1019–1040.

Song, Y., Müller, G., 1995. Biogeochemical cycling of nutrients and trace metals in anoxic freshwater sediments of the Neckar river, Germany. Mar. Freshwater Res. 46, 237–243.

Stumm, W., Morgan, J.J., 1996. Aquatic Chemistry, Chemical Equilibria and Rates in Natural Waters. Wiley, New York.

Swartz, R.C., Ditsworth, G.R., Schults, D.W., Lamberson, J.O., 1985. Sediment toxicity to a marine infaunal amphipod: cadmium and its interaction with sewage sludge. Mar. Environ. Res. 18, 133–153.

Tessier, A., Couillard, Y., Campbell, P.G.C., Auclair, J.C., 1993. Modeling Cd partitioning in oxic lake sediments and Cd concentrations in the freshwater bivalve *Anodonta grandis*. Limnol. Oceanogr. 38, 1–17.

Tessier, A., Fortin, D., Belzille, N., De Vitre, R.R., Leppard, G.G., 1996. Metal sorption to diagenetic iron and manganese oxyhydroxides and associated organic matter: narrowing the gap between field and laboratory measurements. Geochim. Cosmochim. Acta 60, 387–404.

Tipping, E., Hurley, M.A., 1992. A unifying model of cation binding by humic substances. Geochim. Cosmochim. Acta 56, 3627–3641.

Van Cappellen, P., Wang, Y., 1996. Cycling of iron and manganese in surface sediments: a general theory for the coupled transport and reaction of carbon, oxygen, nitrogen, sulfur, iron and manganese. Am. J. Sci. 296, 197–243.

Van den Berg, G.A., 1998. Geochemical behaviour of heavy metals in a sedimentation area of the rivers Rhine and Meuse, Ph.D. thesis, Utrecht University, The Netherlands.

Van den Berg, G.A., Loch, J.P.G., Van der Heijdt, L.M., Zwolsman, J.J.G., 1998a. Vertical distribution of acid-volatile sulfide and simultaneously extracted metals in a recent sedimentation area of the river Meuse in The Netherlands. Environ. Toxicol. Chem. 17, 758–763.

Van den Berg, G.A., Loch, J.P.G., Winkels, H.J., 1998b. Effect of fluctuating hydrological conditions on the mobility of heavy metals in soils of a freshwater estuary in The Netherlands. Water Air Soil Pollut. 102, 377–388.

Van den Berg, G.A., Loch, J.P.G., Zwolsman, J.J.G., Van der Heijdt, L.M., 1998c. Non-steady state behaviour of heavy metals in contaminated freshwater sediments. Water Sci. Technol. 37, 39–46.

Van der Kooij, L.A., Van de Meent, D., Van Leeuwen, C.J., Bruggeman, W.A., 1991. Deriving quality criteria for water and sediment from the results of aquatic toxicity tests and product standards: application of the equilibrium partitioning method. Water Res. 25, 697–705.

Van der Weijden, C.H., Middelburg, J.J., 1989. Hydrogeochemistry of the river Rhine: long term and seasonal variability, elemental budgets, base levels and pollution. Water Res. 23, 1247–1266.

Van Eck, B.T.M., Zwolsman, J.J.G., Saeijs, H.L.F., 1997. Influence of compartmentalization on the water quality of reservoirs: lessons learned from the enclosure of the Haringvliet estuary, The Netherlands. In: Dix-neuvième Congrès des Grands Barrages, vol. III. Commission Internationale des Grands Barrages, Florence, France, pp. 647–669.

Wallmann, K., Kersten, M., Gruber, J., Förstner, U., 1993. Artifacts in the determination of trace metal binding forms in anoxic sediments by sequential extraction. Int. J. Environ. Anal. Chem. 51, 187–200.

Walton-Day, K., Filipek, L.H., Papp, C.S.E., 1990. Mechanisms controlling Cu, Fe, Mn, and Co profiles in peat of the Filson Creek Fen, northeastern Minnesota. Geochim. Cosmochim. Acta 54, 2933–2946.

Wang, Y., Van Cappellen, P., 1996. A multicomponent reactive transport model of early diagenesis: application to redox cycling in coastal marine sediments. Geochim. Cosmochim. Acta 60, 2993–3014.

Wersin, P., Höhener, P., Giovanoli, R., Stumm, W., 1991. Early diagenetic influences on iron transformations in a freshwater lake sediment. Chem. Geol. 90, 233–252.

Williams, T.M., 1992. Diagenetic metal profiles in recent sediments of a Scottish freshwater loch. Environ. Geol. Water Sci. 20, 117–123.

Zhang, H., Davison, W., Miller, S., Tych, W., 1995. In situ high-resolution measurements of fluxes of Ni, Cu, Fe and Mn and concentrations of Zn and Cd in porewaters by DGT. Geochim. Cosmochim. Acta 59, 4181–4192.

Zwolsman, J.J.G., 1994. Seasonal variability and biogeochemistry of phosphorus in the Scheldt estuary, south-west The Netherlands. Estuar. Coast. Shelf Sci. 39, 227–248.

VI
Radionuclides

Trace elements – Their distribution and effects in the environment
B. Markert and K. Friese, editors

Chapter 22

Spatial variability of ^{137}Cs and ^{40}K activity concentrations in soils and plants of alpine pastures: effects of micro- and mesotopography

B.P. Albers, R. Rackwitz, S. Kleinschroth, K. Bunzl

Abstract

The radiocesium activity in soils and plants of undisturbed environments often shows an exceptionally high variability in soils and in the above-ground vegetation. To examine the underlying reasons, we selected a 1 ha alpine pasture area on a slope with 100 randomly selected sampling plots and assessed the ^{137}Cs and ^{40}K activity concentrations in soil and plants. The variability of the radiocesium values was high (coefficient of variance 43% for soil and 82% for plants) and the minimum number of samples taken was 57 for soil and 82 for plants, if an error in the median of 20% at the 95% confidence level is tolerable. Geostatistical evaluation of the data revealed that the activity concentrations of radiocesium in soils is randomly distributed, at least for a distance between 5 and 90 m, but the natural occurring ^{40}K radionuclide showed a spatial dependence with a range of 75 m. In contrast, the radiocesium and also the potassium concentrations in mixed vegetation had a clear spatial dependency with a range of about 20 m (^{137}Cs) or 25 m (K) and are not statistically significant associated with the ^{137}Cs or K content in soil. Therefore changing plant communities are more responsible for the radiocesium concentration in vegetation than the soil concentration itself. On a mesoscale (100 m) level, no significant accumulation of radiocesium in relation to different slope conditions was found. Small scale (m) hollows, however, contained a higher amount of Chernobyl-derived ^{137}Cs than small scale collines, but there was no statistically significant differences found for radiocesium from global fallout. It is concluded that the inhomogenous distribution of radiocesium is caused by a run-off phenomenon during the deposition of Chernobyl-derived ^{137}Cs in May 1986 rather than by sediment redistribution in the last 35 years. The consequences of these results for soil-to-plant radiocesium transfer models based on soil/plant radiocesium concentrations or transfer factors are discussed.

Keywords: Spatial variability; Cesium-137; Potassium-40; Alpine soil and plants; Topography

1. Introduction

A considerable fraction of the atmospheric radiocesium release (about 85×10^{15} Bq) from the Chernobyl accident in April 1986 was deposited in Europe (NEA Committee on Radiation Protection and Public Health, 1995). During the first days after the Chernobyl disaster, the meteorological situation in Central Europe with local showers after a period of dryness caused an inhomogenous deposition of radionuclides on a *macroscale* (landscape) level (SSK, 1987). Schimmack et al. (1997)

found for ten different grasslands around Munich, Germany, a total activity of Chernobyl-derived ^{137}Cs deposited in the soil varying between 7 and 50 kBq m^{-2}, even though some of the sites are located only a few kilometers apart. De Roo (1991) also reported a high variability for ^{134}Cs within a catchment area of 225 ha in the Southern Netherlands. But not only different water intensities of the local rainstorms created macroscale variability, alternating morphological landscape properties could also lead to different radiocesium deposition. Bunzl and Kracke (1988) found ^{137}Cs deposition for forest sites 1.65 times higher than at undisturbed grassland sites in the neighborhood.

The effects of topography on the spatial distribution of soil radiocesium distribution on a *mesoscale* level (field scale, 100 m to 1 km) was studied in several investigations, because ^{137}Cs redistribution has been used to retrospectively quantify erosion rates for periods up to 35 years (e.g. Sutherland, 1996). Most of these studies were carried out in cultivated soils (e.g. Montgomery et al., 1997), but this approach is not suitable when a high spatial variability in the Chernobyl associated deposition is present (Higgitt et al., 1992). However, the spatial distribution of ^{137}Cs in soils on a field scale level in undisturbed grassland environments is not well documented. Some more detailed studies were carried out by Lance et al. (1986), who found that the ^{137}Cs distribution in soil in a basin was random and Sutherland (1991, 1994), reported that the pattern of bomb ^{137}Cs in undisturbed fields was also random and that sediment redistribution can be negligible over the past 35 year in native fields in Saskatchewan, Canada. Systematic investigations about the spatial distribution of radiocesium in plants and their dependence on the soil radiocesium activity in seminatural environments are rarely documented in the literature. McGee (1992, 1996) investigated the relationship between topography and ^{137}Cs fallout in soils and plants of Irish montane peatlands and reported that the deposition values were not correlated with plant concentrations.

The effect of the topography on a *microscale* level (m) to the radiocesium distribution in a cultivated field was investigated by Kachanoski et al. (1985). They reported that the spatial distribution was significantly related to the past surface curvature. Nyhan et al. (1983) described a decreasing trend in the spatial variance component with increasing field sampling volume for Trinity soils (25 ml with a coefficient of variance (CV) of 64% to 1500 ml with a CV of 17%, $N = 10$), which indicates a small scale spatial dependence of ^{137}Cs.

The purpose of this study was to systematically investigate the spatial distribution of radiocesium in soil and plants in a seminatural environment like an alpine pasture. Recent transect studies showed a high variability in soil and plant radiocesium distribution (Albers et al., 1998), but the reasons for these results remained unknown. Therefore we decided to select a 1 ha area on a slope, where 100 randomly selected sampling plots soil and plant samples were collected. Geostatistical techniques were used to quantify the data (for a review see Goovaerts, 1998). Applied techniques in geostatistics like variograms are widely used to characterize the spatial variability of natural soil parameters (e.g. Marriott et al., 1997) or fallout pollutants (e.g. lead, Arrouays et al., 1996). Potassium (1 g stable K contains always 31.3 Bq ^{40}K) in many cases shows similarities in its environmental behavior, therefore we compare the radiocesium activity concentrations in soil and plants with the potassium distribution. Another main focus of the present study centered on the effect of topography on the spatial distribution of radiocesium. For this purpose, the geomorphology of the slope was carefully defined using differential GPS and stereo aerial photographs. These techniques allowed us to relate the measured radionuclide variables not only to the mesorelief (100 m) of the slope, but additionally give an answer to the question of the distribution on a microscale (m) level.

2. Materials and methods

2.1. Site

The sampling site is located about 180 km south-east of Munich in the Berchtesgaden National Park between the mountains Jenner (1800 m), Watzmann (2700 m) and Hagengebirge (2200 m). The area belongs to the Kalkalpen (Jurassic limestone) and the slopes are built up by micaceous and marl limestones. The mean annual temperature at an altitude of about 1550 m is 2.2°C with precipitation of 1757 mm in 1997, where 50% falls as snow.

The meadow investigated at Priesbergalm is situated at 47°33′ 14″N; 13°00′47″ E between the lake Koenigssee and the Jenner mountain. The soil is classified as a Dystric Cambisol according to the FAO system with a pH of 4.5 and an organic matter content of about 19%. The clay content is about 33% and the CEC is 116 mmol kg^{-1}: The vegetation can be described according to Oberdorfer (1993) as a Alchemillo cynosuretum, which is one of the most typical plant associations for alpine pastures in the Alps. In places where cattle destroyed the vegetation, a Rumicetum alpini was found. A detailed description of the soil and vegetation is given by Albers et al. (1998).

2.2. Sampling design

Within a 100 × 100 m area, 100 sampling points were generated with a computer using a stratified random sampling method. The grid (10 × 10 m) was measured in the field using differential GPS and from the edge of this grid each sampling point was located without any consideration of the site situation. The site morphology was measured with stereo aerial photographs using a SD2000 (Zeiss, Germany) within 3 × 3 m grid. Three repeated measurements of a total of 1200 data points resulted in a precision to estimate the total altitude at each point of the grid with a standard deviation of ± 10 cm.

In June 1997, the soil samples were taken from the upper 10 cm at each point using a plastic corer with a diameter of 8 cm. The herbage of 1 m^2 was cut using electric hand shears 2 cm above ground to avoid soil contamination.

2.3. Quantitative analysis

The 10 cm soil column was divided into three parts (0–3.3; 3.3–6.7; 6.7–10 cm), weighed, dried at 105°C, reweighed to obtain the dry bulk density and passed through a 2 mm sieve. The gravel content was quantitatively separated from the fine fraction.

The soil analyses were carried out (pH, water content, loss on ignition) using standard methods (Rump and Krist, 1992). The plant material was weighed, dried at 70°C, reweighed and milled using a special plant mill (Retsch SM1; Retsch, Haan, Germany).

^{137}Cs, ^{134}Cs and ^{40}K in soil and plants were determined by direct gamma-spectrometry using high purity Ge detectors and a multichannel analyzer. Detectors were calibrated with standards made up to the same geometry as the samples. Counting times were adjusted up to 5000 min to provide 1 s errors below 2%. All measured activities were corrected for radioactive decay to the reference date of 1 June 1996. The isotopic ratio $^{137}Cs/^{134}Cs$ at the time of the Chernobyl deposition (30 April 1986) was taken as 1.75 (Hoetzl et al., 1987). With this information, it is possible to calculate the contribution of ^{137}Cs from the earlier global fallout of weapon testing to the total amount of ^{137}Cs in the soil. All data were expressed as Bq kg^{-1} on a dry weight basis or as kBq m^{-2}. The aggregated soil-to-plant transfer factor (T_{ag}) was calculated using the ^{137}Cs concentration in plants (Bq kg^{-1}) as related to a mean activity per unit area of ^{137}Cs (Bq m^{-2}) in the top layer (0–10 cm) of the soil.

2.4. Statistical analysis

The Mann–Whitney *U*-test was used to test for differences between two samples that might represent different populations. The frequency distribution of the variables was examined by applying the Kolmogorov–Smirnov test. The 95% confidence interval of the median was calculated according to the equation

$$h = \frac{n - 1.96\sqrt{n-1}}{2} \tag{1}$$

The number of independent observations required for estimating the true median for a log-normal distribution was calculated according to Hale (1972) (also given in Gilbert, 1987, p. 174).

$$n = \frac{Z^2_{1-\alpha/2} s^s_y}{[\ln(d+1)]^2 + Z^2_{1-\alpha/2} s^s_y / N} \tag{2}$$

where d is the prespecified relative error in the estimated median that can be tolerated, $100(-\alpha)\%$ is the percent confidence required so that this error is not exceeded, and s^s_y is the variance of the ln transformed values.

The central tool for the geostastical procedure is the semivariogram, which measures the degree of spatial dependence or autocorrelation in data arising from the underlying spatial structure in the variation (Oliver and Webster, 1987). The semivariance is defined as

$$\gamma(h) = 0.5N(h)\sum(z_i - z_{i+h})^2 \tag{3}$$

where $\gamma(h)$ is the semivariance for interval distance class h, z_i is the measured sample value at point i, z_{i+h} is the measured sample value at point $i + h$ and $N(h)$ is the total number of sample couples for the lag interval h.

In accordance with our sampling design, the lag class distance interval was chosen as 10 m. The active lag distance was 90 m which results in $N_{min} = 92$ data pairs in the first class and $N_{max} = 697$ for the isotropic analysis. Spherical and exponential models were used to fit the data points, which can be described based on three parameters:

nugget variance: the *y*-intercept of the model,
sill: the model asymptote,
range: where the distance over spatial dependence is apparent.

The formula for the spherical and the exponential models are given in the manual of the program GS+ (Gamma Design Software, 1998). The range in the exponential model is defined as the point at which the model includes 95% of the sill. If no model can be adapted to the data points ($r^2 < 0.2$), the spatial distribution of the results within the range investigated was defined as random.

3. Results and discussion

3.1. Variability of radiocesium and radiopotassium in soil

The mean ^{137}Cs activity concentration within the 1 ha area was 29.7 kBq m^{-2}, although it should be noted that 14% of the total contamination originated from the weapons fallout (Table 1). Due to the high annual precipitation (about 2000 mm/year) it was expected that the radiocesium would be rather evenly distributed within the 0–10 cm depth layer of the soil (Schimmack et al., 1997). But 10 years after the Chernobyl accident, about 60% of the total radiocesium is still concentrated in the first 3.3 cm of the soil. The high amount of micaceous clay in combination with a relatively high organic matter content are able to fix radiocesium very strongly (Livens and Loveland, 1988), so that only 8% of the radiocesium migrated to deeper layers below 6.7 cm. Within the 100 soil samples, the total ^{137}Cs activity in the first 10 cm layer varied more than one order of magnitude, resulting in a CV of 43%. The distribution of

Table 1. Statistical parameter of radionuclide variables in soil and plants[a]

Variables	Mean	SD	Median	Confid.	Range	Distr.
Soil (0–10 cm)						
$^{137}Cs_{Tot}$ (kBq m^{-2})	29.7	12.7	26.0	24.4–31.7	1.6–70.0	Log-normal
$^{137}Cs_{Ch}$ (kBq m^{-2})	25.6	11.8	22.5	19.6–25.7	1.6–63.0	Normal
$^{137}Cs_{G}$ (kBq m^{-2})	4.0	2.0	3.8	3.4–4.3	0–10.3	Normal
^{40}K (kBq m^{-2})	23.2	5.2	22.6	21.0–23.7	13.5–44.8	Normal
Plants						
^{137}Cs (Bq kg^{-2})	49.5	40.5	35.4	29.8–45.5	4.5–205	Log-normal
^{40}K (Bq kg^{-2})	662	147.8	634	612–685	320–1115	Normal
T_{ag} (m^2 $kg^{-2} \times 10^{-3}$)	2.9	10.4	1.3	1.1–1.7	0.16–105	Log-normal

[a] Mean, standard deviation (SD), median, 95% confidence limit of the median (Confid.), range and distribution (Distr.) of radionuclide variables ($N = 100$). Indices: Tot, total; Ch, Chernobyl; G, global fallout; T_{ag}, aggregated transfer factor.

the natural radionuclide ^{40}K was much more homogenous (CV 23%). If one allows for the establishment of baselines of radiocesium in the soil, an error in the *median* activity concentration of 20% (10%) at the 95% confidence level, the minimum number of soil samples to be taken can be estimated as 57 (208) for the 100 × 100 m area. Sutherland (1994) investigated an undisturbed forest field, 50 × 50 m, with 36 samples and estimated that for the radiocesium concentration (CV 18%) a minimum number of 13 soil samples is required and for a cultivated field within a 500 × 500 m square 32 soil samples is required (allowable 10% error in the mean; 95% confidence interval). This difference shows clearly the comparatively high number of samples to be taken for such alpine pastures in order to obtain a representative estimation of the median, or, if the values are normally distributed, for the mean.

3.2. Effects of micro- and mesorelief for soil radiocesium distribution

The distribution of the total radiocesium activity concentrations in soil at each sampling point in comparison to the morphology of the slope investigated is given in Fig. 1. As suggested from the figure, the radiocesium distribution reveals no clear patchiness. But in contrast to these, the topography of the slope shows at least two features, which possibly may produce localized areas of ^{137}Cs depletion and areas of ^{137}Cs enrichment within the field by water flow in small rills.

At a mesoscale level (100 m), the inclination is steeper in the upper part and lower part (about 28%) than between 1530 to 1540 m altitude (about 20%). At the topographical shoulder in the middle of the slope, rainwater accumulates or runs slower than at the steeper parts and the resulting deposition of cesium is expected to be higher.

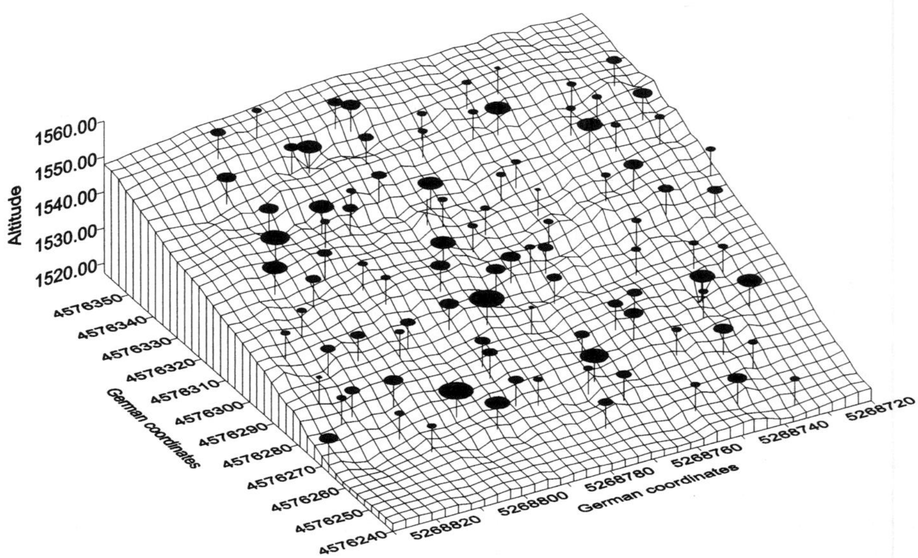

Figure 1. Arrangement of the randomly selected sampling plots and the geomorphology of the slope using a 3 × 3 m grid. The diameter of the circles are proportional to the values for the total radiocesium concentration in the soil between <10 and 70 kBq m^{-2}.

But results of the Mann–Whitney *U*-tests indicated that there were no significant differences in radiocesium activity between steeper and flatter parts of the slope on a 95% level (not shown). This supports our results in previous transect studies, where a significant accumulation of ^{137}Cs in hollows was also not detectable, even at a much steeper slope (Albers et al., 1998). However, Kachanoski (1987) found a 26% increase in ^{137}Cs downslope in a permanent grassed site and De Roo (1991) grouped ^{137}Cs activity from the global fallout in six geomorphological groups and also found significant differences between the maximum slope angle locations (lower activity) and the waning slopes (higher activities), but other differences within the classes were not significant on a 95% level.

At a microscale level (m) each sampling plot is situated in a small scale hollow, more or less at the same level as its neighborhood or at a significantly higher elevation than four other points within a 3 m radius. In that way we grouped the sampling plots into three categories (Fig. 2). The radiocesium activity concentration derived from the Chernobyl fallout was the highest in small scale hollows with significant differences (*U*-test, $P < 0.05$) to the sampling plots situated at small collines (Fig. 2). The same trend was found for the bomb fallout, but the differences were significant only at $P = 0.0734$. The water flow due to the severe rainfall event after the Chernobyl accident transported the radiocesium to the small scale depressions, where it was fixed by clay minerals or organic matter. Later sediment movement can be neglected because (i) no significant accumulation of radiocesium from the global fallout is detectable in the hollows and (ii) the soil and vegetation properties cause a high aggregate stability which protects the soil against erosion (Forster and Zech, 1992).

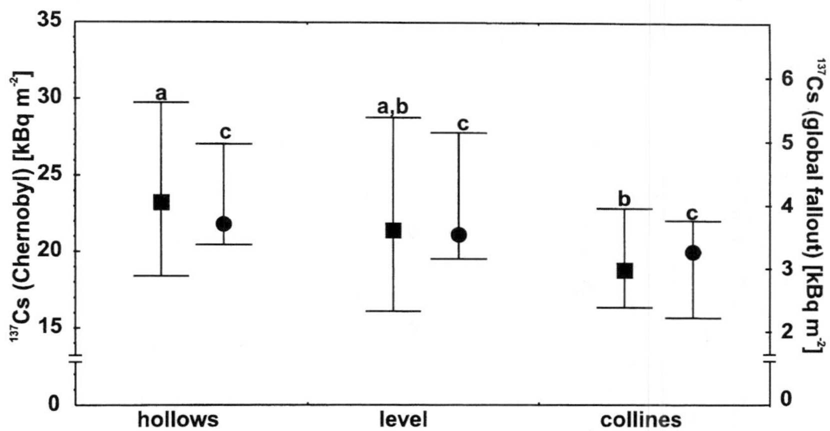

Figure 2. Median and 95% confidence interval for radiocesium concentration in soil derived from the Chernobyl accident (squares, left *y*-axes) and weapons test (circles, right *y*-axes). Groups followed by the same characters are not significant different (*U*-test, $P > 0.05$). Hollows (1) are defined where at least three from definite measured four points within a 3 m radius have a significantly higher altitude (>10 cm) than the sampling plot, collines (3) are defined where at least three points have a significant lower altitude (>10 cm) than the sampling plot. Plots which are not defined in that way (more or less on the same level) are grouped in class 2. $N = 30$ for each class, 10 points are excluded due to missing measurements within a 3 m radius.

3.3. Variability of radiocesium and radiopotassium in plants

According to the soil data, the ^{137}Cs in vegetation (mean 49.5 Bq kg^{-1}) also showed a wide variability of two orders of magnitude resulting in a CV of about 82% (Table 1); whereas the CV of the ^{40}K distribution in plants (mean 662 Bq kg^{-1}) was found to be much smaller with 22%. To estimate the median value of radiocesium in plants within 1 ha with a tolerated error of 20%, a minimum of 82 samples (m^2) is required. The highest variability was found within the calculated aggregated transfer factor (T_{ag}) resulting in a CV of about 350%, but no significant relationship of the radiocesium concentrations between soils and plants was found ($P > 0.05$; Spearman Rank test). A missing correlation between soil and plant radiocesium content will cause serious problems with the use of T_{ag} values for the evaluation of radionuclide transfer (McGee et al., 1996). However, a significant, but negative correlation of potassium and radiocesium in plants (Spearman $R = -0.5387$; $P < 0.001$) reveals a physiological interrelationship between both elements. The missing correlation between soil and plant radiocesium content is possibly rather caused by the individual root depth in connection with the inhomogenous distribution and availability of ^{137}Cs with a peak in the first centimeter of the soil.

3.4. Geostatistical evaluation of the data

To test whether there are anisotropic effects on the data, directional variograms are calculated in four directions to identify anisotropy. Even though Webster and Oliver (1992) recommended 400 data points to quantify anisotropic effects, we applied this technique to our dataset of 100 samples to examine at least whether any strong anisotropic effects are apparent. For this purpose, we created some anisotropic variograms from the variables (N_{min} 27; N_{max} 183) and did not find any evidence for

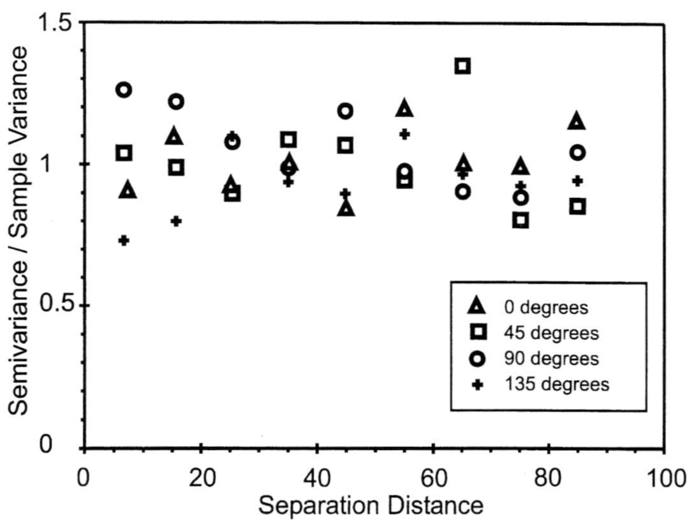

Figure 3. Normalized anisotropic semivariogram for the total radiocesium concentration in soil.

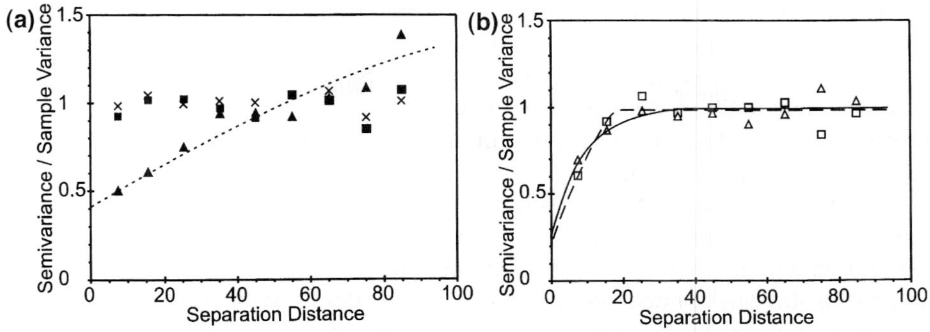

Figure 4. (a) Normalized isotropic variograms for soil variables. Black squares, ^{137}Cs from the Chernobyl fallout; crosses, ^{137}Cs derived from the global fallout; black triangles, ^{40}K. The values for ^{40}K are fitted best with an exponential model (dashed line). (b) Normalized isotropic variograms and best fitted model for plant variables. Open squares, ^{137}Cs; open triangles, ^{40}K. The values for radiocesium are fitted with a spherical model (dashed line), the one for ^{40}K with an exponential model (solid line).

anisotropy (Fig. 3), although the area investigated is situated on a slope. Additionally, recent transect studies within this area revealed no significant long-term transport of radionuclides along the slopes as previously thought (Albers et al., 1998), and also Sutherland (1994) does not find any significant differences down-slope and across-slope transects in an undisturbed field. Because of this evidence, isotropic properties were assumed for the variables.

The experimental isotropic variograms for each variable are shown in Fig. 4a (soil) and Fig. 4b (plants). To compare the semivariograms of each variable, the data are normalized (division of the semivariance by the sample variance). The main properties of each variogram are given in Table 2. Note that in contrast to Fig. 4 the model parameters in Table 2 originated from the original values.

The variograms of the soil variables ^{137}Cs derived from the Chernobyl accident and from the global fallout showed no spatial dependence within our range limit between 5 and 90 m (Fig. 4a). It was not possible to fit a model (spherical, exponential, linear) to the values with an $r^2 > 0.2$ indicating a spatial dependence below 5 m or a random

Table 2. According to the Kolmogorov–Smirnov test (0.05 level). Best fit model variograms and fitted parameters for different soil and plant variables

Parameter	Model	Nugget variance C_0	Sill $C_0 + C$	Range (m)	r^2 of model
$^{137}Cs_{Tot}$ (soil)	Random	164	164	–	<0.200
$^{137}Cs_{Ch}$ (soil)	Random	142	142	–	<0.200
$^{137}Cs_{G}$ (soil)	Random	4	4	–	<0.200
^{40}K (soil)	Exponential	11	45	75	0.895
^{137}Cs (plant)	Spherical	320	1613	20	0.810
^{40}K (plant)	Exponential	5380	21720	25	0.754
T_{ag}	Random	113	113	–	<0.200

distribution of radiocesium in the soil. No spatial dependence for radiocesium was reported by Sutherland (1994) for uncultivated soil and Bachhuber et al. (1987) for global fallout radiocesium in cultivated soil. De Roo (1991), investigating a 225 ha catchment area in the Netherlands, observed that the Chernobyl ^{134}Cs derived activity showed a large spatial variability, but did find a spatial dependency also for weapon ^{137}Cs with a range of 275 m.

Naturally occurring soil variables like pH, carbon or nitrogen in most cases showed a spatial dependency below 100 m, as reported in several studies (e.g. Marriott et al., 1997 for upland grassland). As expected, the semivariogram for the potassium distribution (Fig. 4a) can be fitted with an exponential model ($r^2 = 0.895$) with a calculated range of spatial dependence of 75 m (Table 2), also other measured soil variables (not shown) in that area showed a finite limit to the range of spatial dependence (e.g. soil organic matter with a range of 21 m).

In strong contrast to the radiocesium concentration in soil, the activity concentrations of ^{137}Cs and also ^{40}K in plants showed a clear spatial behavior (Fig. 4b). Both variables could be fitted with a spherical (^{137}Cs) or exponential (^{40}K) model and reached a plateau at 20 m, respectively, at 25 m (Table 2). Visual inspection of the size of individual plant communities, especially alternating Alchemillo cynosuretum and Rumectum alpinii associations, also showed a patchiness of this order. Additionally, the proportion of the nugget value to the sill (20% for ^{137}Cs and 24% for ^{40}K) were in the same range. These results clearly revealed that the spatial radiocesium and, with some restrictions, the potassium concentrations in mixed vegetation is dependent on the patchiness of the plant communities and not from the concentration of radiocesium or potassium in the soil.

4. Conclusions

The variability of radiocesium concentration in undisturbed environments such as alpine pastures is very high and a high minimum of soil and plant samples would be required to estimate the radiocesium distribution median with a 95% confidence interval and a tolerable error of at least 20%. Low cost sampling with only a few samples would result in a biased estimate of the real radionuclide contamination. There is some evidence that the high variability is caused more by microscale runoff phenomenon directly after the Chernobyl accident rather than by long-term erosion processes. The high spatial variability of the ^{137}Cs, the random distribution and the concentration in the first centimeter of the soil possibly causes the missing correlation of the radiocesium activity concentrations in soils and plants and therefore does not allow a soil-to-plant transfer factor to be calculated.

Although there was no detectable patchiness of the radiocesium distribution in soil within the area, the geostatistical evaluation of the plant data clearly revealed a spatial dependency with a range of about 20 m, which was nearly the same as for the potassium concentration (25 m). This supported the general conclusions that in undisturbed environments with a spatial and vertical irregular distribution of radiocesium in soil, the physiological and ecological properties of special plant communities or individual plants are responsible for the uptake of radiocesium in plants and, if such an area is used for beef and milk production, consequently to the human food chain. A model based on deposition data or the soil-to-plant transfer concept does not necessarily provide reliable estimation of the plant radionuclide concentrations. Caution is advisable to apply such models if (i) the topography of the area generates a high variability in soil radiocesium distribution, (ii) the vertical distribution in the root zone is inhomogenous and (iii) the plant community or individual plants alternate in a relatively short distance (e.g. 20 m).

Acknowledgements

The authors would like to thank Dr. Zierl and the staff from the Nationalpark Berchtesgaden for permission to take samples and for providing helpful information, Thomas Hahn (University of the Federal Armed Forces, Munich) for exact measurements of the sampling plots and Dr. Foerster (LM University of Munich) for his help with the evaluation of the aerial photographs. We also thank Wolfgang Schultz for help with the fieldwork and Folker Dietl for his assistance with the determination of the radionuclide activities. This work was carried out as a part of the EC-project ''Long term dynamics of radionuclides in semi-natural environments: derivations of parameters and modelling'' in the EC framework ''Nuclear fission safety''.

References

Albers, B.P., Rackwitz, R., Schimmack, W., Bunzl, K., 1998. Transect survey of radiocesium in soils and plants of two alpine pastures. Sci. Total Environ. 216, 159–172.

Arrouays, D., Mench, M., Amans, V., Gomez, A., 1996. Short-range variability of fallout Pb in a contaminated soil. Can. J. Soil Sci. 76, 73–81.

Bachhuber, H., Bunzl, K., Schimmack, W., 1987. Spatial variability of fallout-${}^{137}Cs$ in the soil of a cultivated field. Environ. Mon. Assess. 8, 93–101.

Bunzl, K., Kracke, W., 1988. Cumulative deposition of ${}^{137}Cs$${}^{238}Pu$, ${}^{239-240}Pu$ and ${}^{241}Am$ from global fallout in soils from forest, grassland and arable land in Bavaria (FRG). J. Environ. Radioact. 8, 1–14.

De Roo, A.P.J., 1991. The use of ${}^{137}Cs$ as a tracer in an erosion study in South Limburg (The Netherlands) and the influence of Chernobyl fallout. Hydrol. Processes 5, 215–227.

Forster, J.C., Zech, W., 1992. Organische Substanz und Aggregierung in Böden unter Almweide und Bergwald im Nationalpark Berchtesgaden. Mitt. Dtsch. Bodenkd. Ges. 68, 161–164.

Gamma Design Software, 1998. Manual to the GS+ Geostastistics for the Environmental Sciences, version 3.11.3, Plainwell, MI.

Gilbert, R.O., 1987. Statistical Methods for Environmental Pollution Monitoring. Van Nostrand Reinhold, New York.

Goovaerts, P., 1998. Geostatistical tools for characterizing the spatial variability of microbiological and physico-chemical soil properties. Biol. Fertil. Soils 27, 315–334.

Hale, W.E., 1972. Sample size determination for the log-normal distribution. Atmos. Environ. 6, 419–422.

Higgitt, D.L., Froehlich, W., Walling, D.E., 1992. Applications and limitations of Chernobyl radiocaesium measurements in a Carpathian erosion investigation. Poland. Land Degrad. Rehabil. 3, 15–26.

Hoetzl, H., Rosner, G., Winkler, R., 1987. Ground depositions and air concentrations of Chernobyl fallout radionuclides at Munich-Neuherberg. Radiochim. Acta 41, 181–190.

Kachanoski, R.G., 1987. Comparison of measured soil 137-caesium losses and erosion rates. Can. J. Soil Sci. 67, 199–203.

Kachanoski, R.G., Rolston, D.E., DeJong, E., 1985. Spatial variability of a cultivated soil as affected by past and present microtopography. Soil Sci. Soc. Am. J. 49, 1082–1087.

Lance, J.C., McIntrye, S.C., Nane, J.W., Rousseva, S.S., 1986. Measuring sediment movement at low erosion rates using caesium-137. Soil Sci. Soc. Am. J. 50, 1303–1309.

Livens, F.R., Loveland, P.J., 1988. The influence of soil properties on the environmental mobility of caesium in Cumbria. Soil Use Mgmt. 4, 69–75.

Marriott, C.A., Hudson, G., Hamilton, D., Neilson, R., Boag, B., Handley, L.L., Wishart, J., Scrimgeour, C.M., Robinson, D., 1997. Spatial variability of soil total C and N and their stable isotopes in an upland Scottish grassland. Plant Soil 196, 151–162.

McGee, E.J., Colgan, P.A., Dawson, D.E., Rafferty, B., O'Keefe, C., 1992. Effects of topography on caesium-137 in montane peat soils and vegetation. Analyst 117, 461–464.

McGee, E.J., Johanson, K.J., Keatinge, M.J., Synnott, H.J., Colgan, P.A., 1996. An evaluation of ratio systems in radioecological studies. Health Phys. 70, 215–221.

Montgomery, J.A., Busacca, A.J., Frazier, B.E., McCool, D.K., 1997. Evaluating soil movement using cesium-137 and the revised universal soil loss equation. Soil Sci. Soc. Am. J. 61, 571–579.

NEA Committee on Radiation Protection and Public Health, 1995. Chernobyl, Ten Years On Radiological and Health Impact. OECD/NEA, Paris.

Nyhan, J.W., White, G.C., Schofield, T.G., Trujillo, G., 1983. An evaluation of soil sampling for ^{137}Cs using various field-sampling volumes. Health Phys. 44, 541–552.

Oberdorfer, E., 1993. Süddeutsche Pflanzengesellschaften. Fischer Verlag, Stuttgart.

Oliver, M.A., Webster, R., 1987. The elucidation of soil pattern in the Wyre Forest of the West Midlands, England. II. Spatial distribution. J. Soil Sci. 38, 293–307.

Rump, H.H., Krist, H., 1992. Laborhandbuch für die Untersuchung von Wasser, Abwasser und Boden. VCH Verlagsgesellschaft, Weinheim.

Schimmack, W., Flessa, H., Bunzl, K., 1997. Vertical migration of Chernobyl-derived radiocesium in Bavarian grassland soils. Naturwissenschaften 84, 204–207.

SSK, 1987. Auswirkungen des Reaktorunfalls in Tschernobyl auf die Bundesrepublik Deutschland. Zusammenfassender Bericht der Strahlenschutzkommission (SSK). Gustav Fischer Verlag, Stuttgart.

Sutherland, R.A., 1991. Examination of caesium-137 areal activities in control (uneroded) locations. Soil Technol. 4, 33–50.

Sutherland, R.A., 1994. Spatial variability of ^{137}Cs and the influence of sampling on estimates of sediment redistribution. Catena 21, 57–71.

Sutherland, R.A., 1996. Caesium-137 soil sampling and inventory variability in reference locations: a literature survey. Hydrol. Processes 10, 43–53.

Webster, R., Oliver, M.A., 1992. Sample adequately to estimate variograms of soil properties. J. Soil Sci. 43, 177–192.

Trace elements – Their distribution and effects in the environment
B. Markert and K. Friese, editors

Chapter 23

The fate of several radionuclides derived from atmospheric fallout in a river watershed

T. Matsunaga

Abstract

The redistribution of atmospherically derived radionuclides by two Japanese rivers has been carried out. This work concerns their potential health impact related to the redistribution and the geochemical cycle of radionuclides in river watersheds, including connected water bodies. The chapter focuses on cesium-137, a radionuclide originating mainly from nuclear weapon tests fallout and partly from the Chernobyl accident. Lead-210 and beryllium-7 of natural origins are also considered in some analyses. A load of fluvial discharge of these radionuclides in particulate form exhibited a log–log relationship with the river water flow rate. A budget analysis for ^{137}Cs in the Kuji River watershed showed that approx. 0.05% of the accumulated ^{137}Cs in the watershed was discharged annually during the period 1987–1988. In the Kuji River watershed, the particulate form was dominant (90% or more) in the annual load of fluvial discharge of ^{137}Cs, ^{210}Pb, and ^{7}Be. This finding, along with results from previous studies by others in other watersheds, suggests that this dominance is coupled with the concentration of suspended solids, which is derived from the geohydrological features of a watershed. A budget analysis for ^{137}Cs in the Hi-i river watershed of Shimane in Japan, where the contribution of the Chernobyl fallout is minor, revealed a temporal decrease in the rate of fluvial discharge of ^{137}Cs from ~1% in the 1950s and 1960s, when the fallout of ^{137}Cs was greatest, to ~0.15% in the 1990s, when the fallout has become much less. This decrease may be due to progressive infiltration and fixation of ^{137}Cs in the soil layers within the watershed.

Keywords: Fallout; Fluvial discharge; Budget analysis; ^{137}Cs; ^{210}Pb; ^{7}Be

1. Introduction

There are several pathways in the secondary migration of atmospherically derived radionuclides on the ground: infiltration to deeper soil layers and groundwater; resuspension to the atmosphere; turnover including vegetation; and redistribution to surface water bodies. The significance of each pathway depends on the chemistry of each radionuclide, environmental conditions, and the time scale of interest. Among the secondary migration pathways, redistribution from the ground soil to the downstream region of a river is of special importance. It has an influence on man via contamination of water bodies and bottom sediments, especially in the downstream region. These regions can include lakes and even coastal areas, where the population is sometimes larger than upstream, as is the case in Japan, provoking increased concern with public health. For example, following the Chernobyl accident, radioactive contamination in the lower reaches of the River Severn in Wales (UK) occurred due to contamination of suspended sediment that was transported from the upper channels, where there was much more fallout (Walling et al., 1989). Berkovski et al. (1996) reported that the health impact through aquatic pathways due to the release of ^{90}Sr and ^{137}Cs from the Chernobyl accident accounts for about 30% of the collective Chernobyl-related radiation exposures in the population along the Dniepr River cascade in Ukraine.

The significance of the fluvial redistribution process can also be seen in studies of erosion and sedimentation processes. The global fallout of ^{137}Cs can be used to trace eroded sediment because of its strong adsorption to soil constituents, adequate half-life ($T_{1/2} = 30.17$ years), ease of detection, ubiquity, and regionally uniform ground distribution (Brown et al., 1981; Spomer et al., 1985; Bonnet, 1990; Walling and Bradley, 1990).

The present study reports measurement methodology and features of the migration behavior of atmospherically derived radionuclides (^{137}Cs, ^{210}Pb, and ^{7}Be) and includes data from previous studies (Matsunaga et al., 1991, 1995, 1999) of two Japanese watersheds. In addition to the good traceability of ^{137}Cs, the choice of this radionuclide was rational for us because it is a major radionuclide and important in a nuclear accident because of its abundance in nuclear reactors and its volatility. Though, the present report is for radionuclide elements, the present findings could provide suggestions for the fate of other trace elements.

2. Origin of the radionuclides studied

2.1. *Cesium-137*

In Japanese watersheds, ^{137}Cs is mainly derived from the radioactive fallout of nuclear fission during weapon tests in the atmosphere (Wise, 1980). The majority of the tests were conducted during the period 1950–1962. The total injection of ^{137}Cs to the atmosphere from these tests is estimated to be 960 PBq (10^{15} Bq) (UNSCEAR, 1982). The Chernobyl accident in 1986 resulted in several times greater fallout of ^{137}Cs in European countries compared to the weapon fallout (Nuclear Energy Agency, 1995). However, it increased only about 4% of the integrated deposition density of ^{137}Cs in Japan at the time of accident (1986).

2.2. *Lead-210 and beryllium-7*

Lead-210 formed in the atmosphere via ^{222}Rn which diffuses from the earth's surface is adsorbed to aerosols or dust particles, and gradually deposits on to the earth's surface. It is not in radioequilibrium with ^{226}Ra, and is referred to as "unsupported ^{210}Pb". Thus, after being deposited, atmospherically derived ^{210}Pb can be distinguished as an excess of associated ^{226}Ra, and can be used as a radioactive tracer. An extensive discussion of the origin and the scientific applications of radioactive lead has been provided by Wise (1980). Beryllium-7 is formed by cosmic ray spallation of

^{14}N and ^{16}O in the atmosphere and has also been used as a convenient radiotracer (Matsunaga et al., 1995).

3. Method of collection of ^{137}Cs from river water

3.1. General strategy

In order to investigate the migration behavior of aquatic radionuclides, it is necessary to distinguish those that occur associated with suspended solids (''suspended form'' or ''particulate form'') from those ''dissolved'' in water. While the definition of each form is only an operational one, they are still useful for characterizing the behavior of radionuclides in river water. Fig. 1 presents data for radionuclides in river water affected by the Chernobyl accident (Matsunaga et al., 1998). The radionuclides can be categorized according to their apparent ''distribution ratios'' (concentration ratios) between ''suspended'' (collected on a 1.0-μm pore size filter) and ''dissolved'' (passed through a 0.20-μm filter) forms. The particulate activity collected on 0.45- and 0.20-μm filters was not incorporated in the ratio in Fig. 1, since its contribution was found to be small in this case.

One should also note the large volume of water that is necessary for detection of radionuclides. Recent freshwater bodies normally contain only about 0.1–1 mBq/l of global-fallout ^{137}Cs. Therefore, volumes of 200 l or greater must be used in order to

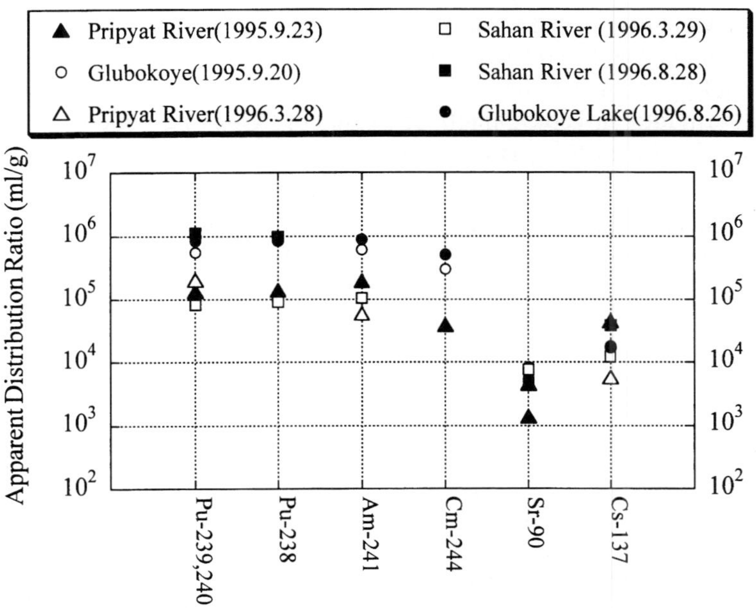

Figure 1. Apparent distribution ratio between concentrations of particulate and dissolved radionuclides in rivers and a lake in the vicinity of the Chernobyl nuclear power plant. Reprinted from Matsunaga et al. (1998).

avoid serious counting errors in radioactivity measurement. In order to complete this large volume sampling, the following method has been used by many researchers. Filters are used for on-site collection to collect suspended solids, and an adsorption unit is used to collect dissolved ^{137}Cs. Detailed discussions on the sampling methodology of radionuclides in aquatic environments are found in Von Gunten and Beneš (1995).

3.2. Collection of particulate ^{137}Cs

Filtration is the most common method for collection of particulate radionuclides, although care must be taken to avoid clogging of a filter. For this reason, a depth-type filter in a cartridge model is convenient, which provides a large effective filtration area. In our case, a depth-type filter in a cotton yarn, wound cartridge was used. The cotton is easily incinerated and gives a small size of sample. This results in a high counting efficiency in the measurement of ^{137}Cs or any other gamma-ray emitters with a germanium detector. Digestion of organic material by incineration is also necessary for the radiochemical analysis of ^{90}Sr, Pu, and Am isotopes. One disadvantage of using cotton is the uncertainty of its cut-off pore size. Although material with nominal pore sizes of 1–100 μm is commercially available, the pore sizes are only approximate measures. Table 1 shows data on ^{137}Cs collected in the Kuji River watershed. The lower values under high flow conditions are probably due to an increase in the concentration of fine particles during flooding (Matsunaga et al., 1991). Table 2 shows the distribution of particulate radionuclides around the Chernobyl Power Plant in Ukraine. The data suggest that more than 90% of the particulate ^{137}Cs was found in a 1-μm filter, so that a 1-μm cartridge filter is adequate in this case. However, it should be noted that the collection efficiency can vary markedly due to differences in sampling and hydrological conditions. For instance, the concentrated distribution of ^{137}Cs in a 1-μm filter in Table 2 might be a result of "clogging" of a filter which leads to filtration of a part of finer particles.

Table 1. Experienced collection efficiencies using a depth-type filter[a]

Sample	Collection efficiency	Evaluation method
Kuji river water (low-flow condition)	83% (by ^{137}Cs)	Comparison with flocculation by poly-aluminum chloride ($AlCl_3$)
Kuji river water (high-flow condition)	65% (by ^{137}Cs)	Comparison with sedimentation in a tank followed by decantation
Latex balls[b]	2–5 μm, 82% 5–10 μm, 93% 10–20 μm, 97% 20–30 μm, 98%	Particle counting by electron microscopy

[a] A cartridge-type filter made of cotton yarn having a nominal pore size of 1 μm (TCW-1CSS, TOYO Roshi. Ltd.).

[b] Data provided by the manufacturer (Toyo Roshi. Co. Ltd.).

Table 2. Percent distribution of particulate ^{137}Cs over three successive cartridge-type filters of different sizes for surface waters around the Chernobyl Nuclear Power Plant[a]

Location	Collected percent distribution (samples collected in 1995)[b]		
	1 μm filter	0.45 μm filter	0.20 μm filter
River Pripyat	97	3	<0.2
River Uzh	98	1.7	0.3
Lake Glubokoye	92	4	4

[a] Reprinted and modified from Matsunaga et al. (1997) with special permission from the publisher. Materials of filters: 1 μm: cotton yarn; 0.45 μm and 0.20 μm: acetate membrane.

[b] Relative values, taking the sum of the radioactivity detected on the three filters was 100%.

Cartridge-type membrane filters made of membranes are available which have clearly defined pore sizes and large filtration areas, but their incineration is problematic due to the polypropylene/polyethylene used as supporting materials for the membranes. A continuous flow centrifuge can also be used to collect suspended solid from a large volume of water (Lammers, 1968; Walling and Bradley, 1990; Joshi et al., 1991). In this method, water samples are delivered to the centrifuge via a peristaltic pump and the process of centrifugation (18000 × *g*, in the present study) separates the solid from the liquid. Solid samples are recovered from the inside of the centrifuge rotor. The advantages of this method are: (1) to eliminate the problem of filter clogging; and (2) to be able to collect intact (or less altered) solid samples. The second point makes physicochemical analyses possible, such as mineral identification and observation with an electron microscopy, which are impossible for incinerated samples. This method resulted in a collection efficiency of ~90%, in terms of weight, for solids larger than 0.45 μm in our analysis of Kuji River water.

The disadvantages are the necessity for time-consuming centrifugation in the laboratory. In our experience, a flow rate of 10 l/h was maintained to supply water to the centrifuge, although it is not necessarily the maximum rate. Given a suspended solid concentration of 10 mg/l, it would take 10 h to obtain 1 g of suspended solid. Additionally, it requires transportation of large volumes of water to the laboratory. There are several manufacturers of the type of centrifuge described here, although not all of them provide for temperature control, which is sometimes important since the temperature of the samples in the centrifuge tends to increase during centrifugation. For example, we observed warming of the sample to 28°C at the outlet of the centrifuge, while it was supplied at 6°C. In specific objectives, this problem of warming of the sample might need to be considered.

3.3. Collection of dissolved ^{137}Cs

Adsorbents have been widely used to collect dissolved ^{137}Cs in both fresh and seawater. Ferrocyanide compounds are known as specific adsorbents for cesium (Krylov et al., 1973; Folsom et al., 1975; Mann and Casso, 1984) and are used in two forms, granular or impregnated on fiber. Milham and Kantelo (1984) used a granular form of potassium cobalt ferrocyanide (KCFC) to collect ^{137}Cs in the Savannah river with a sensitivity of 0.001 pCi/l (0.037 mBq/l) although their flow rate was only 30–60 ml/min. To obtain a higher flow rate, Higuchi and Nonaka (1985) developed a novel system which employed an acrylic fiber impregnated with KCFC (US Patent No. 4,720,422) and allowed a flow rate of 8 l/min with a collection efficiency of 94% for ^{137}Cs in seawater. In the study describe below, we used this KCFC impregnated fiber. The fiber was incinerated in an electric furnace (450°C, 8 h) to prepare a sample for gamma spectrometry. The evaporation loss of volatile ^{137}Cs during incineration was less than 1% and the detection limit for dissolved ^{137}Cs was 0.05 mBq/l. Other methods, such as those using manganese dioxide-impregnated filters are known to be effective for creating pre-concentrations of dissolved plutonium and americium (Von Gunten and Benes, 1995). Zirconium ferrocyanide (Egorova et al., 1977; A.V. Stepanov, pers. commun.) is sometimes preferred to KCFC because it has less potassium, which can lower the detection limit of ^{137}Cs because of Compton scattering of the gamma ray of naturally contained ^{40}K in the adsorbent. Marinin (1999) has recently developed a sorbent that concentrates radiocesium and radiostrontium from seawater.

4. Features of fluvial discharge of ^{137}Cs, ^{210}Pb, and ^{7}Be

4.1. Study area

The Kuji River is located in the northern part of the Kanto Plains, Japan (Fig. 2). It originates from the Yamizo mountain area in the center of Japan, which is an area of 1490 km^2 (hills, 80.6%; plains, 16.9%; streams, 2.5%) characterized by afforestation and numerous agricultural fields. The mean annual precipitation of the watershed is 1360 mm. There are no nuclear facilities along the river except at its mouth.

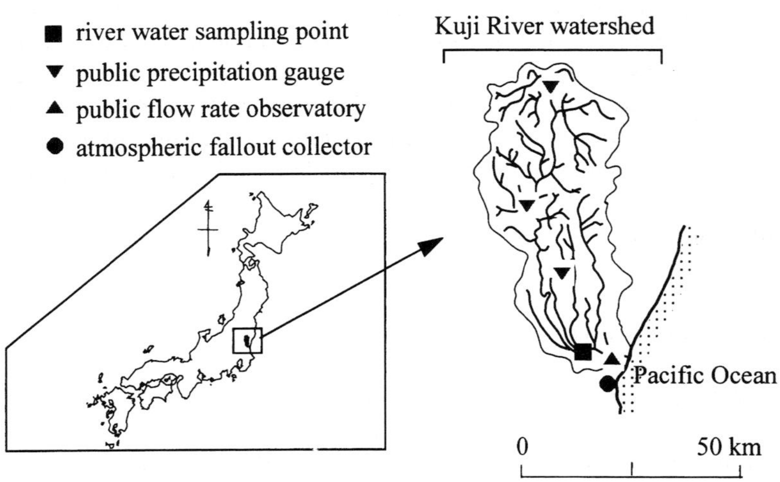

Figure 2. Location of the Kuji River watershed and related observation points.

4.2. Methodology

Water samples of 500–2000 l were pumped at 2–5 l/min. using an in situ sampling system for particulate forms of ^{137}Cs, ^{210}Pb, and ^{7}Be; and for the dissolved form of ^{137}Cs, as described above. We did not attempt to sample dissolved ^{210}Pb or ^{7}Be. All sampling was performed at the downstream end of the river, 10 km from the sea. There was no tidal influence at the sampling location. Field experiments were done under both low (10–20 m^3/s) and high (up to 80 m^3/s) water flow conditions. Representative water samples over the cross-section of the flow at the sampling point were obtained for measurements of total particulate concentration, ignition loss, particle size distribution, and dissolved inorganic ions at 10 points on a transverse of the river. Flow velocity and dimensions of the flow cross-section were also measured. Atmospheric deposition samples were collected monthly in 1987 and 1988 in Tokai village, 7 km from the sampling site on the Kuji River (Fig. 2) (Matsunaga et al., 1995; Ueno et al., unpublished data).

4.3. Analysis of radioactivity

The radioactivity of ^{137}Cs in the ash of the cartridge-type filter and a KCFC-impregnated fiber was determined by gamma-ray spectrometry with a Ge(Li) detector, and the detection levels were corrected to take into account the collection efficiency, which was estimated by the activity of ^{137}Cs in a few samples collected by flocculation or by sedimentation which were thought to exhibit complete recovery.

The detection limits were 0.05 mBq/l and 0.02 mBq/l for dissolved and particulate ^{137}Cs, respectively. The activity of ^{7}Be was determined by gamma-ray spectrometry simultaneously with ^{137}Cs. The activity of unsupported ^{210}Pb was then determined from the difference between the total activity of ^{210}Pb and the activity of ^{226}Ra, as deduced from the measurement of ^{214}Pb. The samples were sealed in a gas-tight plastic vessel with an o-ring, and allowed to stand for at least 2 weeks before the measurement to attain radioequilibrium between ^{226}Ra and ^{214}Pb via gaseous ^{222}Rn. The total activity of ^{210}Pb in the sample was determined by gamma-ray spectrometry using a hyperpure germanium detector with a beryllium window, which was calibrated with certified U and Th ore standards (Yanase and Sekine, 1995). Hereafter, unsupported ^{210}Pb will be written simply as ^{210}Pb in this chapter, unless specified. The detection limits were approximately 0.1 and 0.01 Bq/g for total ^{210}Pb and ^{7}Be, respectively.

4.4. Evaluation of the load of fluvial discharge

4.4.1. Empirical relationship for the daily load

A dataset of measured concentrations of the radionuclides and the flow rate of river water was used to derive a relationship between the load of fluvial discharge of the radionuclides and the water flow rate. The load was obtained from the concentration of radionuclides in each form (dissolved or particulate) multiplied by the flow rate. Results showed that the load conforms empirically to the following relationship:

$$L = aQ^b \tag{1}$$

where L is the load of fluvial discharge of the dissolved form of ^{137}Cs and the particulate form of ^{137}Cs, ^{210}Pb, and ^{7}Be (Bq/km^2 per day); Q is the flow rate (m^3/s) at the sampling site; and a and b are the constants determined for each of the radioactive species in the watershed. It was found that the load of suspended solids also follows the same form as Eq. (1). These relationships are shown in Fig. 3. This type of relationship is common in the fluvial discharge of dissolved materials (Walling and Webb, 1983) and of nutrients (Ebise, 1991). The similarity in the graphs (variable b) in Fig. 3 for suspended solids and particulate radioactivities is noteworthy, and is due to the fact that the concentrations of the particulate radionuclides per unit mass of suspended solids was relatively independent of the flow rate, while the concentrations of suspended solids were more dependent on changes in the flow rate. The similarity in particulate radioactivities may have occurred by chance, because the characteristics of suspended solids were different under high- and low-flow conditions. The median particle sizes in terms of volumetric distribution of suspended solids were 10–19 μm and 9–13 μm under normal- and high-flow conditions, respectively. The organic matter content in these solids was between 16 and 45% (mean 26%) as the loss on ignition under normal-flow conditions, and between 15 and 22% (mean 19%) under high-flow conditions.

4.4.2. Evaluation of annual load and related analyses

4.4.2.1. ^{137}Cs We applied Eq. (1) to estimate the load of annual discharge radionu-

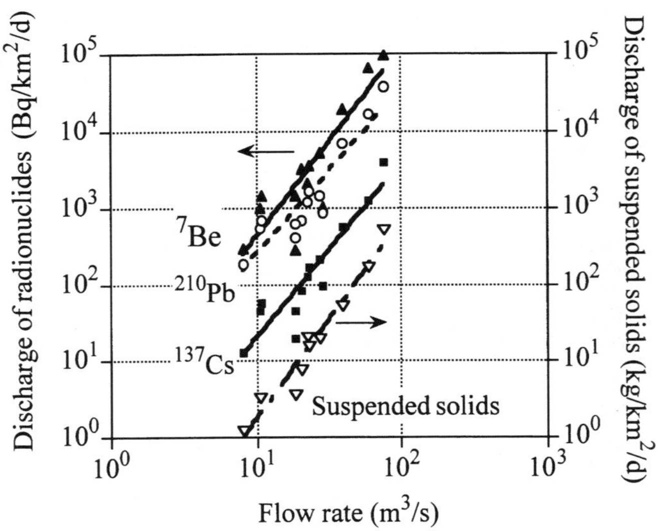

Figure 3. Dependency of the load of fluvial discharge of radionuclides on that of suspended solids (Kuji River, 1987–1988). The coefficients a and b (see text) are: ^{7}Be, $a = 1.5$, $b = 2.5$; ^{210}Pb, a = 1.7, $b = 2.2$; ^{137}Cs, $a = 0.1$, $b = 2.3$. Reproduced and modified from Matsunaga et al. (1995).

clides in the watershed. Records of daily averaged flow rates at the nearest flow gauge operated by the Ministry of Construction, which is 2 km downstream from the sampling location (Fig. 1), were used to determine Q, with appropriate correction for our sampling site. Estimated mean loads in 1979–1988 were 1.6 and 0.14 MBq/km^2 year for particulate and dissolved forms of ^{137}Cs, respectively. In the Kuji River watershed, the contribution of the particulate ^{137}Cs was estimated to be ~90% of the total annual discharge of ^{137}Cs. It should be noted that this result was substantially derived from a large discharge of particulate ^{137}Cs under high water flow conditions following a precipitation event.

The accumulated fallout of ^{137}Cs in the surface soil of the Kuji River watershed was estimated to be 3.7×10^3 MBq/km^2 at the end of 1988, based on the reported fallout of ^{137}Cs (Kasai et al., 1986; Hirose et al., 1987; Yanase et al., unpublished data). This estimate is comparable to that reported for the accumulated ^{137}Cs deposition in the northern hemisphere, e.g. 2.68×10^3–3.58×10^3 MBq/km^2 in Germany at the end of 1984 (Bunzl and Kracke, 1988). The contribution from the Chernobyl accident to the accumulated ^{137}Cs in the Kuji River watershed is as low as 4% (Yanase et al., unpublished data). Any decrease in soil radioactivity due to infiltration to deep geologic media and resuspension to the atmosphere was not calculated due to the strong adsorption property of cesium and the well-vegetated nature of the watershed. Fig. 4 shows a schematic representation for determining a budget of ^{137}Cs. The estimated annual load of fluvial discharge through the river is 1.7 MBq/km^2, corresponding to a fraction of about 0.05% of the accumulated ^{137}Cs in the whole watershed at the time. Although the fraction itself is not large, the relative importance of this fluvial discharge among the secondary migration pathways of this nuclide is significant.

4.4.2.2. ^{210}Pb and ^{7}Be The annual budget of ^{210}Pb and ^{7}Be in the Kuji river watershed was determined using methods similar to those for ^{137}Cs. The annual load of discharge was estimated empirically from the river water flow rate and the input flux of atmospheric fallout of the radionuclides was determined by experimental

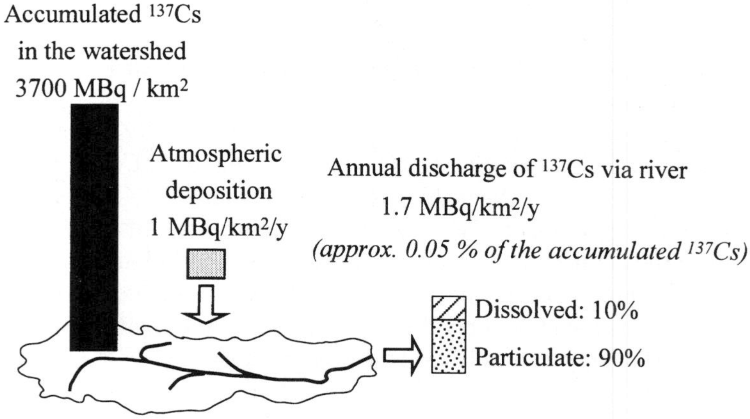

Figure 4. Scheme for an annual budget of atmospherically derived ^{137}Cs in the Kuji River watershed (1987–1988).

observations. The proportions of annual fluvial discharge to the atmospheric flux were estimated to be approximately 6.4% and 3.2% for ^{210}Pb and ^{7}Be, respectively. The loads of dissolved ^{210}Pb and ^{7}Be were estimated by using previously reported distribution coefficients from the literature (Lewis, 1977; Santschi et al., 1983; Lupien and Grondin, 1984; Brunskill and Wilkinson, 1987; Dominik et al., 1987; Joshi et al., 1991). The data suggest that approximately 90% of ^{210}Pb and ^{7}Be are discharged in particulate form in the Kuji River watershed. Dominik et al. (1987) reported that about 85% of ^{210}Pb was discharged in particulate form in the Rhone River, which is reported to contain a high concentration of suspended particles. On the other hand, Joshi et al. (1991) found that the dissolved form of predominant (about 77% of total annual load) in the Ottawa River, where the concentration of suspended particles is as low as 1.0–4.4 mg/l throughout the year. The combined data for the Kuji River, Rhone River, and Ottawa River suggest that the importance of suspended solids in the fluvial discharge of atmospherically derived radionuclides river watersheds. These radionuclides reach the watershed as atmospheric particles, and they have high affinities to particles in the atmosphere.

5. Temporal change in the rate of fluvial discharge of atmospherically derived ^{137}Cs

5.1. Study area

The preceding section describes methods for determining the current rate of fluvial discharge of ^{137}Cs in the Kuji River watershed. In the present section, we describe a method for determining historical discharge rates for the Hi-i River watershed in Japan (Matsunaga et al., 1999). The watershed is located in the eastern part of the Shimane Prefecture (Fig. 5); it has an area of 2070 km^2 and includes Lake Shinji (surface area, 79.7 km^2; maximum depth, 5.5 m) at the lower part of the watershed, which is fed by a subwatershed of approximately 915.3 km^2.

5.2. Methodology

A sediment core was sampled from the middle of Lake Shinji in 1994 and the radioactivity concentrations of ^{137}Cs and ^{210}Pb were analyzed. Details of the sampling and radioactivity measurement are described by Ueno et al., 1997; an interpretative method is described below (see Fig. 6). The depth profile of ^{137}Cs in the lake bottom sediment was dated using a ^{210}Pb profile. This dated profile of ^{137}Cs was then assigned to each year by interpolation of the measured concentrations of ^{137}Cs in a sediment core to estimate the annual sedimentation flux of ^{137}Cs. The fallout records of ^{137}Cs in the studied area (Shimane Prefecture) were obtained from the literature (Kasai et al., 1986; MRI, 1996; Yanase et al., unpublished data). A similar method of interpretation was adopted by McHenry et al. (1973) and was explained by Wise (1980).

The sedimentation flux was considered to be affected by two factors: direct deposition on the lake and erosion from the ground in the lake catchment. The fallout of ^{137}Cs was found to be negligible compared with the sedimentation flux of ^{137}Cs in lake Shinji during the 1990s. Therefore, the present sedimentation can be attributed to the

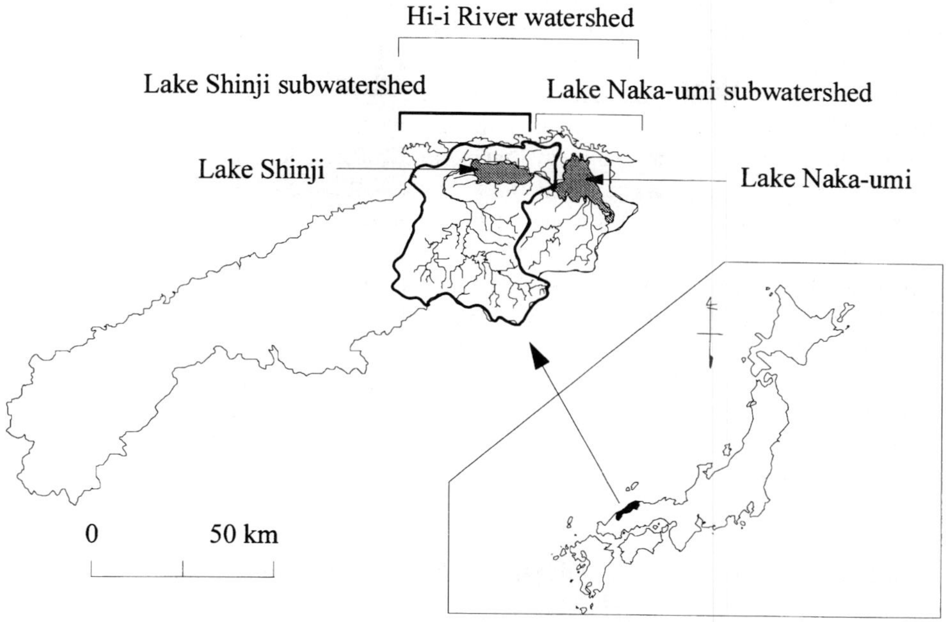

Figure 5. Location of the Hi-i River watershed and Lake Shinji.

erosional contribution. This contribution was found to correspond to 0.15% per year of the accumulated ^{137}Cs in the ground of the subwatershed of Lake Shinji in the period 1990–1993.

Assuming this percentage was consistent in past years, the erosional contribution was estimated from the accumulated ^{137}Cs in the ground which was calculated as an integrated fallout until the year of interest, with a correction for radioactive decay. Since Lake Shinji has a channel of outflow, some portion of the fallout on the lake might have escaped. Two cases were considered to account for the effect of escape on the direct fallout contribution. The first is an extreme case, in which there is no escape (100% accumulation, Case 1). The second is based on an analysis of the Chernobyl accident, although the Chernobyl contribution was minimal in this area. The atmospheric fallout of ^{137}Cs and the accumulation in lake bottom sediment due to the accident was evaluated from the records of atmospheric fallout and the dated sedimentation profile of ^{137}Cs. Comparison of these values showed that about 40% of the Chernobyl fallout on the lake remained in the bottom sediment (40% accumulation, Case 2). It was considered that Case 2 would be more probable than Case 1.

5.3. Examination of the historical fluvial discharge of ^{137}Cs

A sum of the estimated erosion and fallout contributions was compared with the experimentally obtained sedimentation flux of ^{137}Cs. The fluxes were grouped as follows:

- 1954–1969: period of most extensive atmospheric flux;

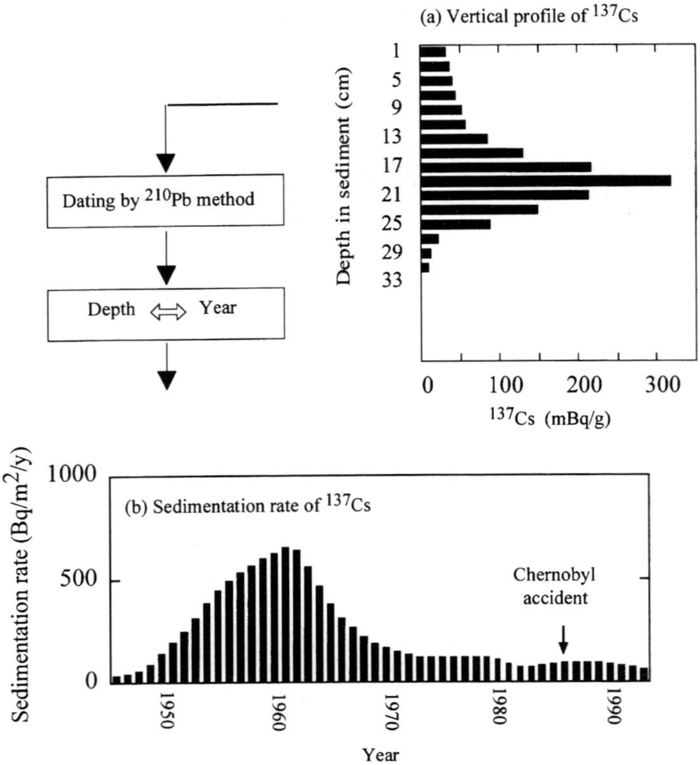

Figure 6. Estimated sedimentation flux of ^{137}Cs in the bottom sediment of Lake Shinji.

- 1970–1979: period of steadily declining atmospheric flux, with a small increase in the later years due to several new weapon tests;
- 1980–1989: period of further decline of atmospheric flux, except for an abrupt increase due to the Chernobyl accident in 1986;
- 1990–1993: period of nearly constant levels, with the least flux.

The results indicate that the estimated sum flux can match the experimental flux fairly well, except for the period 1950–1969 in both cases (Fig. 7, Case 1 not shown). The estimated sum in Case 2 was somewhat smaller than the experimental flux due, we concluded, to an underestimate of the erosional contribution. When the erosional contribution was increased to 1% of the accumulation in the ground, the estimation reproduces the experimental value quite well (Fig. 7). The fluvial discharge must have been greater in the 1950–1969 period than in the later periods. This effect may be due to the following reasons. The top layer of the ground, which was exposed to erosion, must have been most contaminated in the 1950–1960s when the fallout was greatest, and fixation of the ^{137}Cs to soil constituents, such as clay minerals, likely progressed with time such that the ^{137}Cs became fixed and the rate of its fluvial discharge thereby decreased. Bunzl et al., 1994 described a temporal decrease over a few decades in the mobility of global fallout ^{137}Cs in soil layers, resulting from a gradual fixation. Fig. 8 is

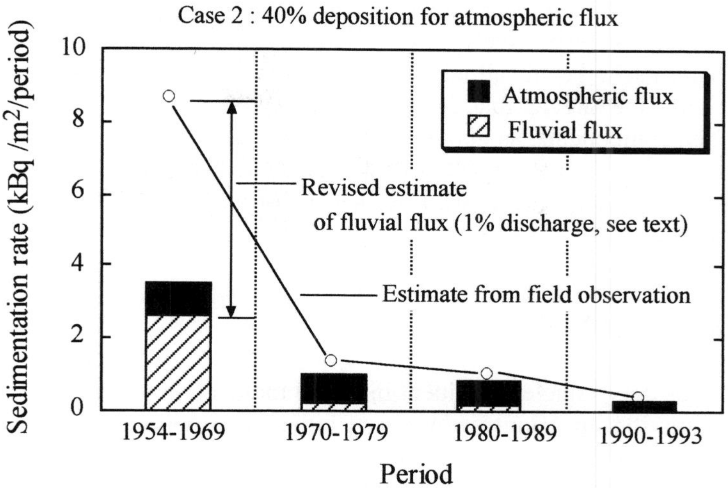

Figure 7. Estimated contributions of erosion and atmospheric fallout to the sedimentation of ^{137}Cs in the bottom sediment of Lake Shinji compared with observed sedimentation rate (solid line) based on the measured amount of ^{137}Cs in a sediment core collected in 1994.

a schematic illustration of the fluvial discharge of ^{137}Cs in the Hi-i river and Lake Shinji system.

Yamagata et al. (1963) formulated a fluvial discharge of fallout ^{137}Cs consisting of two components derived from the run-off of (a) fresh precipitation and (b) cumulative deposition, whereas those components are combined as a single component in the present analysis. As common values for Japanese rivers in the period of their study,

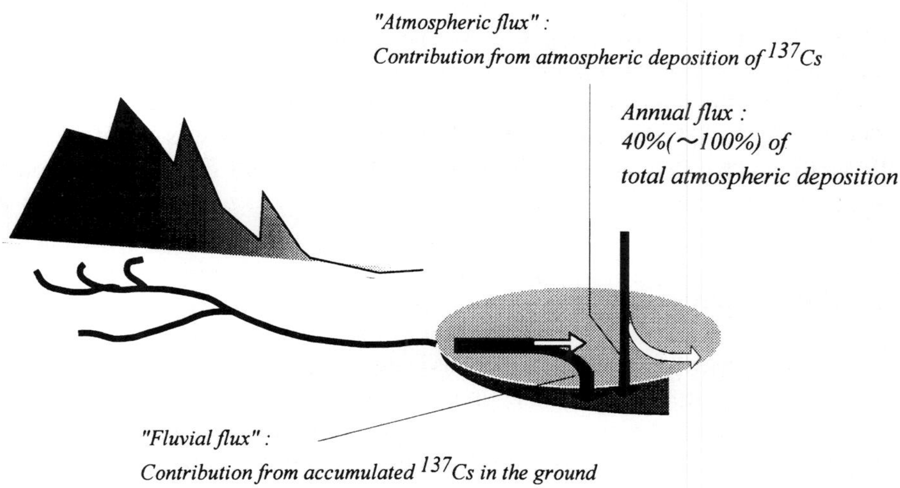

Figure 8. Schematic illustration for the transport of atmospherically derived ^{137}Cs in the Hi-i river watershed, Lake Shinji system. Annual flux: 1980s, 0.15% of the accumulated ^{137}Cs in the ground; 1950–1960s, ~1%.

1960–1962, they assigned 1.3% and 0.06% as the annual rate of fluvial discharge of the two components, respectively. Our analysis is similar to that of Yamagata et al., i.e. the fluvial discharge of fallout ^{137}Cs could contain a fresh component with an annual discharge rate greater than 1% and an accumulated component of less than 1%. The combined components, as in our case, would have a temporal dependency on the fallout intensity. A similar model for the erosional transport of fallout ^{137}Cs was established by Carlsson (1978), who studied the turnover of fallout ^{137}Cs in a small lake in Sweden.

6. Concluding remarks

When one considers a nuclear accident in which radioactive materials are released to the atmosphere, the important pathways of radiation exposure to the public change over time elapsed after the release. In the first stage, during and after the release, the dominant exposure will be from direct exposure to the radioactive plume and from the radioactive materials deposited on the ground. This stage may last from several days to several weeks. Thereafter, concerning the long-lived radionuclides with strong adsorption to the ground soil, the ground will act as a secondary source of radioactive contamination.

The instances described here indicate the important role of fluvial discharge in the secondary transport of atmospherically derived, highly adsorptive radionuclides in a river watershed. A particulate form of radionuclides associated with suspended solids in river water is especially important. It was shown that the form can account for 90% of the annual discharge of ^{137}Cs (mainly from the fallout from nuclear weapon testing), as was observed in the Kuji River watershed. The importance of this mechanism has also been stressed by other researchers (e.g. Walling et al., 1989).

The present study described an alternative approach to the analysis of the fluvial discharge of ^{137}Cs in a river watershed. This method is useful to trace temporal changes in the fluvial discharge in a lake system, although it requires a few assumptions relating to the atmospheric fallout flux. The present findings suggest that the contribution of fluvial discharge to the sedimentation of ^{137}Cs was about 1% in the 1950–1960s, and has decreased to 0.15% in the 1990s, possibly due to the fixation and covering of accumulated ^{137}Cs in ground soil.

Acknowledgements

The following work cited in this paper has been conducted in a cooperative research program between the Japan Atomic Energy Research Institute and the Chernobyl Scientific and Technical Centre for International Research, the State Enterprise for Region Monitoring Environment and Dosimetric Monitoring of Ukraine: Matsunaga et al. (1997, 1998). The author is grateful to Dr H. Amano for his helpful comments on the manuscript. The author is also indebted to Dr N. Yanase, Mr T. Ueno, Dr Y. Kobayashi, Dr K. Sekine, Mr A. Kasai, Mr R.L.R. Chandradjith, Dr M. Okumura and Dr H. Hashitani for their collaboration throughout the relevant studies cited in this chapter. Use of KCFC fiber was kindly allowed by Drs H. Higuchi and N. Nonaka. Critical reviewing of the manuscript by Prof. Dr H.R. Von Gunten is greatly appreciated. Prof. Dr N. Senesi kindly provided an occasion for a poster presentation of the study in the Hi-i river watershed.

References

Berkovski, V., Ratia, G., Nasvit, O., 1996. Internal doses to Ukrainian populations using Dnieper river water. Health Phys. 71, 37–44.

Bonnet, P.J.P., 1990. A review of the erosional behaviour of radionuclides in selected drainage basins. J. Environ. Radioact. 11, 251–266.

Brown, R.B., Kling, G.F., Cutshall, N.H., 1981. Agricultural erosion indicated by ^{137}Cs redistribution: II. Estimates of erosion rates. Soil Sci. Soc. Am. J. 45, 1191–1197.

Brunskill, G.J., Wilkinson, P., 1987. Annual supply of ^{238}U, ^{234}U, ^{230}Th, ^{226}Ra, ^{210}Pb, ^{210}Po, and ^{232}Th to Lake 239 (Experimental Lakes Area, Ontario) from terrestrial and atmospheric sources. Can. J. Fish. Aquat. Sci. 44, 215–230.

Bunzl, K., Kracke, W., 1988. Cumulative deposition of ^{137}Cs, ^{238}Pu, $^{239+240}$Pu and ^{241}Am from global fallout in soils from forest, grassland and arable land in Bavaria (FRG). J. Environ. Radioact. 8, 1–14.

Bunzl, K., Förster, H., Kracke, W., Schimmack, W., 1994. Residence times of fallout $^{239+240}$Pu^{238}Pu, ^{241}Am and ^{137}Cs in upper horizons of an undisturbed grassland soil. J. Environ. Radioact. 22, 11–27.

Carlsson, S., 1978. A model for the movement and loss of ^{137}Cs in a small watershed. Health Phys. 34, 33–37.

Dominik, J., Burrus, D., Vernet, J.-P., 1987. Transport of the environmental radionuclides in an alpine watershed. Earth Planet. Sci. Lett. 84, 165–180.

Ebise, S., 1991. Storm run-off loading of nutrients and pollutants in upper rivers. Jpn. J. Limnol. 52, 241–253.

Egorova, N.V., Krylov, V.N., Pitalev, V.G., Stepanov, A.V., 1977. Inorganic sorbents in the radiochemical analysis of seawater, IV. Influence of the conditions of synthesis of zirconium ferrocyanide on the sorption of cesium-137. Radiokhimiya 20, 737–741.

Folsom, T.R., Hansen, N., Tatum, T.J., Hodge, V.F., 1975. Recent improvements in methods for concentrating and analyzing radiocesium in seawater. J. Radiat. Res. 16, 19–27.

Higuchi, H., Nonaka, N., 1985. In-situ sampling method of radionuclides in seawater. Jpn. Chem. Anal. Centre Publ. 10, 65–71.

Hirose, K., Aoyama, M., Katsuragi, Y., Sugimura, Y., 1987. Annual deposition of Sr-90, Cs-137 and Pu-239, -240 from the 1961–1980 nuclear explosions: a simple model. J. Meteor. Soc. Jpn. 65, 259–277.

Joshi, S.R., Mcrea, R.C., Shukla, B.S., Roy, J.-C., 1991. Partitioning and transport of lead-210 in the Ottawa river watershed. Water Air Soil Pollut. 59, 311–320.

Kasai, A., Sekine, K., Imai, T., Amano, H., Yanase, N., Matsunaga, T., 1986. Survey Data of Radionuclides in Environmental Materials (I), JAERI-M 86-047. Japan Atomic Energy Res. Inst., Tokai, Ibaraki.

Krylov, V.N., Gedeonov, L.I., Rakov, N.A., Trofimov, A.M., 1973. Inorganic sorbents in the radiochemical analysis of seawater, 1. Determination of cesium-137 and strontium-90. Radiokhimiya 15 (5), 654–658.

Lammers, W.T., 1968. The distribution of cobalt-60, ruthenium-106, and cesium-137 among suspended and dissolved particles in White Oak Lake K-1758. Union Carbide Corporation, Oak Ridge, TN.

Lewis, D.L., 1977. The use of ^{210}Pb as a heavy metal tracer in the Susquehanna River system. Geochim. Cosmochim. Acta 41, 1557–1564.

Lupien, M.D., Grondin, D., 1984. Natural radionuclides in the St. Lawrence River. Water Pollut. Res. J. Can. 19, 120–128.

Mann, R.D., Casso, S.A., 1984. In situ chemisorption of radiocesium from seawater. Marine Chem. 14, 307–318.

Marinin, D., 1999. New sorbent for analytical concentrating of radionuclides from seawater, Int. Symp. on Radiation Safety Management, International Atomic Energy Agency, Taejon, Korea (submitted).

Matsunaga, T., Amano, H., Yanase, N., 1991. Discharge of dissolved and particulate ^{137}Cs in the Kuji River, Japan. Appl. Geochem. 6, 159–167.

Matsunaga, T., Amano, H., Ueno, T., Yanase, N., Kobayashi, Y., 1995. The role of suspended particles in the discharge of ^{210}Pb and ^{7}Be within the Kuji River watershed, Japan. J. Environ. Radioact. 26, 3–17.

Matsunaga, T., Ueno, T., Nagao, S., Onuma, Y., Amano, H., Watanabe, M., Kovalyov, A.V., Tkachenko, Yu.V., Sukhoruchkin, A. K.. Kazakov, S.V., 1997. Migration behavior of the released radionuclides in the river system in the exclusion zone of the Chernobyl nuclear power plant. In: One Decade after Chernobyl: Summing up the Consequences of the Accident, IAEA-TECDOC-964, vol. 2. International Atomic Energy Agency, Vienna, pp. 172–177.

Matsunaga, T., Ueno, T., Amano, H., Tkachenko, Yu., Kovalyov, A.V., Watanabe, M., Onuma, Y., 1998. Characteristics of Chernobyl-derived radionuclides in particulate form in surface waters in the exclusion zone around the Chernobyl Nuclear Power Plants. J. Contam. Hydrol. 35, 101–113.

Matsunaga, T., Ueno, T., Chandradjith, R.L.R., Amano, H., Okumura, M., Hashitani, H., 1999. A study on ^{137}Cs and mercury contamination in lake sediments. Chemosphere 39, 269–283.

McHenry, J.R., Ritchie, J.C., Gill, A.C., 1973. Accumulation of fallout cesium-137 in soil and sediments in selected watersheds. Water Resource Res. 9, 676–686.

Milham, R.C., Kantelo, M.V., 1984. Measurement of Low Levels of Cesium-137 in Water, DP-MS-84-48, Savannah River Laboratory, Aiken, CA.

Nuclear Energy Agency, 1995. Chernobyl Ten Years On – Radiological and Health Impact. Nuclear Energy Agency, Organization for Economic Co-operation and Development, Paris.

Santschi, P.H., Li, Y.-H., Adler, D.M., Amdurer, M., Bell, J., Nyffeller, U.P., 1983. The relative mobility of natural (Th, Pb and Po) and fallout (Pu, Am, Cs) radionuclides in the coastal marine environment: results from model ecosystems (MERL) and Narragansett Bay. Geochim. Cosmochim. Acta 47, 201–210.

Spomer, R.G., McHenry, J.R., Piest, R.F., 1985. Sediment movement and deposition using cesium-137 tracer. Trans. ASAE 28, 767–772.

Ueno, T., Amano, H., Chandrajith, R.L.R., Okumura, M., 1996. Vertical distribution of radionuclides and mercury in Lake sediments. In: Proc. Int. Workshop on the Fate of Mercury in Gold Mining and Measures to Control the Environmental Pollution in Various Countries, NIMD 001-1997. National Institute for Minamata Disease.

UNSCEAR, 1982. Ionizing radiation: sources and biological effects, United Nations, Annex E, Section J. United Nations Scientific Committee on the Effects of Atomic Radiation, pp. 220–222.

Von Gunten, H.R., Beneš, P., 1995. Speciation of radionuclides in the environment. Radiochim. Acta 69, 1–29.

Walling, D.E., Bradley, S.B., 1990. Some applications of cesium-137 measurements in the study of erosion, transport and deposition. In: Erosion, Transport and Deposition Processes. IAHS Publ. no. 189, pp. 179–203.

Walling, D.E., Webb, B.W., 1983. The dissolved loads of rivers: a global overview. In: Dissolved Loads of Rivers and Surface Water Quantity/Quality Relationships. IAHS Publ. no. 141, pp. 3–20.

Walling, D.E., Rowan, J.S., Bradley, S.B., 1989. Sediment-associated transport and redistribution of Chernobyl fallout radionuclides. In: Hadley, R.F., Ongley, E.D. (Eds.), Sediment and the Environment. IAHS Publ. no. 184, pp. 37–45.

Wise, S.M., 1980. Caesium-137 and lead-210: a review of the techniques and some applications in geomorphology. In: Cullingford, R.A., Davidson, D.A., Lewin, J. (Eds.), Timescales in Geomorphology. Wiley, Chichester, Chapter 9.

Yamagata, N., Matsuda, S., Kodaira, K., 1963. Run-off of cesium-137 and strontium-90 from rivers. Nature 200, 668–669.

Yanase, N., Sekine, K., 1995. Measurement of uranium series radionuclides in rock and groundwater at the Koongarra ore deposit, Australia by gamma spectrometry. Mat. Res. Soc. Symp. 353, 1235–1242.

Author Index

Subject index